AF412265

Springer

Berlin
Heidelberg
New York
Barcelona
Hong Kong
London
Milan
Paris
Singapore
Tokyo

60 PROGRESS IN BOTANY

Genetics
Cell Biology and Physiology
Systematics and
Comparative Morphology
Ecology and
Vegetation Science

Edited by

K. Esser, Bochum
J. W. Kadereit, Mainz
U. Lüttge, Darmstadt
M. Runge, Göttingen

 Springer

With 45 Figures

ISSN 0340-4773
ISBN 3-540-64689-2 Springer-Verlag Berlin Heidelberg New York

The Library of Congress Card Number 33-15850

Cover design: Design & Production, Heidelberg
Typesetting: M. Masson-Scheurer, Homburg, Saar
SPIN 10646109   31/3137 - 5 4 3 2 1 0 - Printed on acid-free paper

Contents

Review

**Chloroplast Movement: from Phenomenology
to Molecular Biology**
By Wolfgang Haupt (With 3 Figures)

Genetics

**Recombination:
Organelle DNA of Plants and Fungi: Inheritance
and Recombination**
By Heike Röhr, Ursula Kües, and Ulf Stahl (With 3 Figures)

Mutation:
Nuclear and Plastomic Transformation of Higher Plants
Using Microprojectile Bombardment 88
By Christer Jansson and Pirkko Mäenää (With 2 Figures)

Extranuclear Inheritance:
Genetics and Biogenesis of Mitochondria 99
By Thomas Lisowsky, Karlheinz Esser, Torsten Stein, Elke Pratje,
and Georg Michaelis (With 4 Figures)

Genetics of Phytopathology: Phytopathogenic Bacteria 119
By Holger Jahr, Rainer Bahro, and Rudolf Eichenlaub

Plant Breeding: Male Sterility in Higher Plants – Fundamentals and Applications 139
By Frank Kempken and Daryl Pring

Plant Breeding:
Genetic Mapping in Woody Crops 167
By Eva Zyprian

Cell Biology and Physiology

Plant Water Relations 193
By Rainer Lösch

Secondary Plant Substances: Sesquiterpenes 341
By Horst-Robert Schütte (With 10 Figures)

Systematics and Comparative Morphology

**Systematics and Evolution of the Algae: Phylogenetic Relationships
of Taxa Within the Different Groups of Algae** 369
By Hans R. Preisig

Systematics of the Pteridophytes 413
By Stefan Schneckenburger

Lichenized and Lichenicolous Fungi 1995–96 438
By Harrie J. M. Sipman (With 2 Figures)

Ecology and Vegetation Science

Mycorrhizae: Ectotrophic and Ectendotrophic Mycorrhizae 471
By Reinhard Agerer

List of Editors

Professor Dr. Dr. h. c. mult. K. Esser, Lehrstuhl für Allgemeine Botanik,
Ruhr Universität, Postfach 10 21 48, D-44780 Bochum, Germany
Phone: +49-234-7002211; Fax: +49-234-7094211
e-mail: karl.esser@ruhr-uni-bochum.de

Professor Dr. J. W. Kadereit, Institut für Spezielle Botanik
und Botanischer Garten, Universität Mainz,
Saarstraße 21, D-55099 Mainz, Germany
Phone: +49-6131-392533; Fax: +49-6131-393524
e-mail: kadereit@goofy.zdv.uni-mainz.de

Professor Dr. U. Lüttge, TU Darmstadt, Institut für Botanik,
FB Biologie (10), Schnittspahnstraße 3-5, D-64287 Darmstadt, Germany
Phone: +49-6151-163200; Fax: +49-6151-164808
e-mail: luettge@bio.tu-darmstadt.de

Professor Dr. M. Runge, Lehrstuhl für Geobotanik,
Systematisch-Geobotanisches Institut der Universität,
Untere Karspüle 2, D-37073 Göttingen, Germany
Phone: +49-551-395721; Fax: +49-551-395749
e-mail: mrunge@gwdg.de

Wolfgang Haupt was born on 24 January 1921 in Bonn, Germany. During his military service, he started studying biology, chemistry and physics in a prisoner-of-war camp in France (1946/47), continuing thereafter at the universities of Erlangen and Tübingen. In 1952 he received his doctoral degree (Dr. rer. nat.) in Botany, supervised by E. Bünning (Tübingen), and in 1957 he habilitated at Universität Tübingen, where he became Privatdozent. In 1962 he was appointed o. Professor der Botanik, Universität Erlangen-Nürnberg. In 1966 and 1970 he declined professorships offered at Heidelberg and Tübingen, respectively; in 1988 he retired.

His research activities in plant physiology included: initiation of flowering (1951–1965), induction of polarity by light (1957–1962), chloroplast orientation to light (1958–1984), phytochrome research (1960–1992), light control of fern-spore germination (1983–1994). He also worked with scientists in or from Denmark, France, Greece, Italy, Japan, Pakistan, Poland, Romania, Sweden, and the USA.

W. Haupt published two books: *Bewegungen der Pflanzen* (1977) and Encyclopedia of Plant Physiology, vol VII, *Movements* (coeditor, 1979). In addition, he contributed chapters to textbooks, proceedings and monographs; among others, he wrote the chapter Bewegungen/Physiology of Movements in 21 volumes of Fortschritte der Botanik/Progress in Botany (1958–1984). A substantial part of his research papers appeared in *Planta* and *Zeitschrift für Pflanzenphysiologie/Journal of Plant Physiology*. He served on the Editorial Boards of several international journals, and he was a member in the organisation and program committees for the XIV International Botanical Congress Berlin, 1987.

W. Haupt was active in local, regional and national science committees, in particular the Deutsche Forschungsgemeinschaft and Alexander-von-Humboldt Stiftung. Moreover, in 1975, he was a founding member of the European Communities Biologists Association (ECBA). W. Haupt was president of the Deutsche Botanische Gesellschaft (1979–1985) and of the Verband Deutscher Biologen (1969–1976); later, he became an honorary member in both societies.

He was elected to the Deutsche Akademie der Naturforscher Leopoldina (Halle) in 1975, and in 1988 to the Royal Physiographic Society, Lund (Sweden). In 1984, he was awarded the Finsen medal for contributions to photobiology.

His continuous activities over nearly half a century were possible thanks to his good fortune in marrying, in 1950, Gerda (née Rohde). As a counterpoise to his scientific specialisation, she and their four children enriched his life with a broad cultural background and thus promoted his creativity.

Chloroplast Movement: from Phenomenology to Molecular Biology

By Wolfgang Haupt

1 Introduction

In the second half of the 19th century, botanists discovered and became interested in the displacement and reorientation of chloroplasts in the cell, particularly with respect to light. As usual, in a first period of research, as much information as possible was collected about the phenomenon in its huge diversity throughout the plant kingdom. This period peaked in the monograph by Senn (1908), which became the fundament for subsequent additional and supplementary work in the **descriptive** sense, and which is still a valuable source of detailed information.

Although the early authors, and particularly Senn, already posed basic questions on the **causality** of oriented chloroplast movement, comprehensive physiological research began in the 1950s. At the beginning of this second period, main research centred on the first step in the light-controlled responses, viz. **perception of light**, with emphasis on the photoreceptor pigments. In these topics, substantial knowledge has accumulated, and on this sound basis, molecular and genetic approaches have recently been started.

On the other hand, research on the **mechanism of movement** and its control have become effective only in the recent decade or two, when "classical" approaches could be combined with molecular ones. This combined research now leads to steadily increasing knowledge about ultrastructure as well as the biophysics and biochemistry of the motility system (the motor apparatus) and of the controlling factors.

Many detailed and comprehensive, as well as summarising, reviews have appeared in the past decades, which are cited, e.g., in Haupt and Scheuerlein (1990), Nagai (1993), Wada et al. (1993), Wagner (1995) and

---

In 1957, when the author gave a lecture for his higher qualification, he selected the topic CHLOROPLAST MOVEMENT; he presented the state of the art and pointed out problems to be analysed. Afterwards, one of the professors commented: "You have done a fairly good job; but it is a pity that you have selected such an old-fashioned topic, which has no perspective for modern research".

The reader of the following chapter may decide whether in retrospect the critical professor was right.

Yatsuhashi (1996). The present review tries to inlcude the historical aspect, starting with the descriptive level, continuing to the physiological approach, which finally extends to molecular genetics.

2 The Period of Phenomenological Research

Böhm (1856) was probably the first to report on changes in the intracellular chloroplast distribution as depending on the light conditions. As a general rule, the pattern of intracellular distribution in high-intensity light (which corresponds to direct sunlight) is different from that in lower intensities (as typical for an overcast sky), and these patterns have a relation to both **light direction** and **neighbouring cells** (e.g. Frank 1871; Stahl 1880). In detail, however, these patterns exhibit a huge diversity, depending on taxonomy and cell morphology.

In his comprehensive monograph, Senn (1908) pointed out common principles on the one hand, and classified the manifold phenomena on the other (supplemented by Senn 1919). Accordingly, all the various patterns can be described by a **logical and comprehensive terminology** (e.g. *epistrophe, parastrophe*, etc.). This sophisticated terminology is still useful for phenomenological research. For physiological investigations, however, which aim at analysing single cases or at finding common rules, a more **general and unifying terminology** is preferable (cf. Haupt 1959a) and will be used in this review, viz. *low-* and *high-intensity arrangement (movement, response)*, besides the *dark arrangement* if occurring, although *fluence rate* is physically more correct than *intensity* (cf. Wada et al. 1993).

In the early time, there was already another unifying terminology, which, however, appears to be outdated and should now be avoided: superficially, the rearrangement of chloroplasts recalls the orientation of microorganisms with respect to light direction and was therefore called **phototaxis**; but this term tacitly implies that the chloroplasts themselves perceive the light direction and respond by an oriented movement individually and actively in a "resting" cytoplasmic environment; it neglects the respective alternatives, although these questions had already been discussed very early.

This concern about the "phototaxis" concept becomes evident if one follows the rearrangement in a "standard" object, e.g. *Lemna* or *Funaria*. In high-intensity light, the chloroplasts assemble at those walls that are parallel to the light direction, i.e. at the **anticlinals**. When in low-intensity light they rearrange to the **periclinal walls**, some of them migrate to the proximal wall, i.e. towards the light source, but others to the distal wall, i.e. away from the light. This would be hard to understand if the chloroplasts were perceiving the light direction by themselves and orienting accordingly, part of them with positive and others with negative "phototaxis", and vice versa in high-intensity movement. Indeed, as early as 1880, Stahl speculated about "attraction sites" in the cytoplasm, generated by unequal light distribution, to which the chloroplasts respond *chemotactically* rather than *phototactically*. This would mean, of course, sensing of light outside the chloroplasts.

Continuing along this line, Senn's research became already physiological. He demonstrated that it is not really the **direction of light**, but its **distribution within the cell** that determines the pattern of chloroplast arrangement, so that in low-intensity light chloroplasts assemble at the sites of highest illumination, whereas in high intensities they escape from these sites. Senn therefore asked how the vectorial information of the light beam is transformed into an **intracellular gradient of light absorption**. He pointed out that the most important factors determining this transformation are **refraction** and **reflection** of light at the optical boundaries (environment/wall/cytoplasm/vacuole), due to differences in the index of refraction. For various cell types and dimensions, he constructed light paths and found excellent consistency between **predicted intensity distribution** and **arrangement of chloroplasts**. Thus, the chloroplasts respond to optical gradients in the cell, and species-specific differences are mainly due to differences in the geometrical and optical properties of the cells.

For most of the cells, the role of reflection can be neglected, and the results of **refraction** can be reduced to two types (Senn 1908). (1) As plant cells usually act as **collecting lenses**, at least in air, part of the distant region (as related to the light source) receives more light than the proximal region. Accordingly, in some particular cases the chloroplast(s) prefer this **"focus" region** in low-intensity light (e.g., in the green algae *Ulva* and *Hormidium* and in the epidermis of *Selaginella*). (2) As a rule, however, another effect is by far more important: due to the refraction, the regions at the flanks are **bypassed** by direct light, the gradient distal/proximal vs. flanks directs the chloroplasts to the usual low-intensity **arrangement at the distal and proximal wall** (see above); thus, no discrimination is made between light beams propagating in two opposite directions.

As a complication, these patterns can be modified by effects from **neighbouring cells** (Senn 1908). Such effects can be separated from light effects by analysing the **dark arrangement**, which is frequently found. The factors responsible for these patterns are still unknown; they may be, according to Senn, substances that are locally accumulated or depleted by the metabolism of neighbouring cells. Our present knowledge has not substantially exceeded that general view; only few recent more detailed investigations are available (e.g. Rüffer et al. 1981). Lack of general interest may be due to the fact that most of the modern model systems either have no specific dark positioning (e.g. *Mougeotia*), or have a random distribution in darkness (e.g., *Vaucheria, Adiantum*), thus obviously not being under the influence of asymmetric factors other than light. Moreover, experimentally, it is much easier to vary the light conditions precisely in space, time and intensity than to vary "internal conditions" i.e. in the surrounding tissue. Finally, the movement to the dark arrangement ususally requires more time than the response to the light.

The above-mentioned question whether chloroplasts move **actively** in the cytoplasm, or are transported **passively** by masses of cytoplasm, was already discussed by the early investigators. Although occasionally filamentous structures have been observed and thought to exert pulling forces on the chloroplasts (e.g., Knoll 1908), there were also doubts about this function (e.g., Boresch 1914, Voerkel 1934). However, Senn had already pointed out that this "active-passive" question need not be an "either-or" one. Instead, he put forward the "peristromium hypothesis", according to which a cytoplasmic "envelope" around the chloroplast is responsible for the movement; thus chloroplasts are passively moved by a cytoplasmic structure, but still their main cytoplasmic "environment" is resting. Indeed, Strugger (1956) claimed to have found the peristromium; but later ultrastructural investigations with improved techniques failed to confirm it, and it can be considered as outdated. Recent "as-well-as" views will be presented in Section 3.b.

As to the **photobiology** of chloroplast movement, the early authors were already interested in the effective spectral ranges, as far as theoretical and methodical backgrounds allowed it. According to Senn, in most cases the response is restricted to the short-wavelength part of visible light. For *Funaria*, this was confirmed and slightly improved by Voerkel (1934), who used broad-band gelatine filters. He found **blue and blue-green light** the most effective spectral ranges, and this reminded him of phototaxis and phototropism. On the other hand, Senn had also pointed out the remarkable exception that the alga *Mougeotia* is sensitive to **red light** for its low-intensity response, whereas the high-intensity response fits in the general blue-light rule.

In conclusion, until the middle of our century, an amazing amount of descriptive information was accumulated, and in many of the publications physiological questions were touched upon and respective investigations initiated. Subsequently, this developed to successful analytical research, on which the following sections will be centred.

3 Physiological and Molecular Approach

At the start of the new era, two main questions were posed:

a) **Perception**: how is the light signal perceived (Sect. 3.a)?
b) **Response**: what mechanism underlies the movement (Sect. 3.b)?

From progress in these two topics, viz., input and output, a third question is becoming accessible:

c) **Transduction**: what steps can be found between input and output, forming the sequence of events in the "black box" (Sect. 3.c)?

As a basis of recent research, quantification of chloroplast arrangement and respective responses has been substantially improved by **methods of photometric recording** (e.g., Pfau et al. 1974; Zurzycki et al. 1983), which can be processed by computer programmes. Besides, most of the approaches to particular questions are based on new or at least improved experimental methods, as will become evident in the respective sections.

a) Perception

Basic questions are the **nature of the photoreceptor pigment** and its **localisation** or compartmentation in the cell (α). Unexpectedly, in several cases there are **multiple photoreceptor substances**, working alternatively or in series (β). Most important for an oriented response is the **perception of light direction** (γ). Finally, first promising results of **genetic approaches** at the level of perception will be presented (δ).

α) The Photoreceptor Pigment

As mentioned in Section 2, in most cases the response is restricted to **short wavelengths.** The first precise **action spectrum** was established by Zurzycki (1962b) for *Lemna trisulca*, showing peaks near 450 and 370 nm. Thereafter, similar results were obtained in a huge diversity of taxonomic groups and ecotypes such as algae, mosses and pteridophytes, mono- and dicots, water plants and land plants, for low- as well as for high-intensity movement (e.g., Fischer-Arnold 1963; Zurzycki 1967a; Lechowski 1972; Inoue and Shibata 1973). They all strongly recall those for other blue-light responses in plants and fungi, suggesting a common universal photoreceptor pigment, usually called **cryptochrome.** Although the action spectra and other evidence (see below) strongly point to a flavin compound as responsible for perception, other blue-light and UV-absorbing pigments cannot be excluded yet. For recent progress, see Section 3.a.δ.

The exceptional dependence of the low-intensity response on **red light** in the alga *Mougeotia* (Sect. 2) was a challenge to test the hypothesis that here photoperception is located in the chloroplast proper, with chlorophyll as the photoreceptor pigment. This hypothesis, however, was disproved by Haupt (1959b), whose experiments by chance fell into the period of fast-growing phytochrome research. The action spectrum, as well as a typical reversible red/far-red antagonism, indicated **phytochrome** as photoreceptor pigment.

The cell morphology of several species, particularly of *Mougeotia*, allowed for partial irradiations of the cell with **microbeams,** with or without touching the chloroplast. As a result, the photoreceptor pigment was **located in the cytoplasm** rather than in the moving chloroplasts, for phytochrome in *Mougeotia* (Bock and Haupt 1961) as well as for cryptochrome in *Vaucheria sessilis* and *Selaginealla martensii* (Fischer-Arnold 1963; Mayer 1964). Incidentally, the former result was the first approach to intracellular location or compartmentation of functional phytochrome.

Using polarised light, a strong **action dichroism** was found in *Mougeotia*, with maximal response if the electrical vector vibrates perpendicularly to the axis of the cylindrical cell (Haupt 1960). Supplemented by **polarised microbeams,** Haupt and Bock (1962) concluded that the transition moments of phytochrome are oriented parallel to the cell surface and to single-handed helical lines around the cell.

However, this **surface-parallel orientation** is restricted to the red-absorbing form of phytochrome, Pr. From otherwise unexplained results and based on conclusions of Etzold (1965) for polartropism in fern protonemata, Haupt (1968) postulated that in *Mougeotia* phytochrome in its far-red absorbing form (Pfr) is oriented with its **transition moments normal to the surface,** and this has been proven with polarised microbeams (Haupt 1970a). Later, with preparations of immobilised oat phytochrome, such a **change in molecular orientation** upon photoconversion was demonstrated directly in vitro, although only by 32° rather than by 90° (Sundqvist and Björn 1983). Nevertheless, all theoretical conclusions based on the 90° change can be fitted also with the 32° change (Björn 1984) or even less, if some boundary conditions are met (Nakasako et al. 1990).

After the initial experiments of Mayer (1964) in *Selaginella martensii*, an **action dichroism** was found also in nearly all **blue-light-responding species** investigated so far, which always indicated dichroic orientation parallel to the cell surface for cryptochrome as well (Zurzycki and Lełatko 1969). The most spectacular result was obtained in the lower epidermis of *Sambucus nigra*, with only few chloroplasts, which exhibit a **dichroic "micropattern"** along the strongly folded walls (Fig. 1).

In *Funaria* the action dichroism disappears below 400 nm (Zurzycki 1967b). This was taken as strong support for cryptochrome being a **flavin,** because in a flavin molecule the transition moments for the two absorption peaks in UV (260 and 370 nm) are tilted by 30° to 50° against that for 450 nm (Kurtin and Song 1968). Accordingly, Zurzycki (1972) suggested that the transition moments for the UV are inclined by about 45° to the surface, thus excluding any directional preference of absorption.

As an additional argument in favour of flavin, Mayer (1966) reported in *Selaginella martensii* an inhibition by **iodide,** which is known as a **quencher** of the triplet excited state of flavins, and this inhibition has been found also in other species (see below). Unfortunately, however, the inhibition is restricted neither to blue light, nor to the perception proper. Rather, the response to red light in *Mougeotia* is affected as well, even

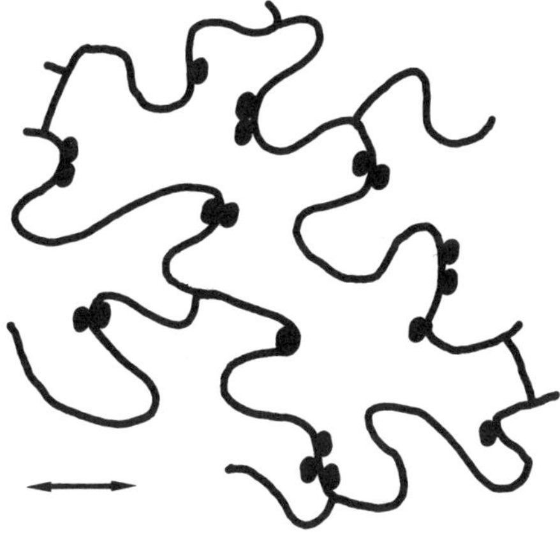

Fig. 1. High-intensity chloroplast arrangement in the lower epidermis of *Sambucus nigra* in polarised light. Notice the location at those parts only where the electric vector (*double arrow*) is perpendicular to the cell wall (least absorbing sites). (After Zurzycki and Lełatko 1969)

when iodide is applied in darkness *after* the light signal (Schönbohm 1967). Yet, although it does not support the flavin hypothesis, inhibition by iodide can be a useful handle in particular cases (cf. Sect. 3.a.β).

Coming back to the exceptional red-light sensitivity for the low-intensity movement of *Mougeotia*, the closely related unicellular *Mesotaenium* shares the main features with *Mougeotia* (Haupt and Thiele 1961). The same is true for fern protonemata, as shown by Yatsuhashi et al. (1985) for *Adiantum capillus-veneris*, although here the surface-parallel transition moments of Pr are not restricted to a single-handed helix (Yatsuhashi et al. 1987a, b). As ferns can contain more than one phytochrome species (e.g., Maucher et al. 1992; see Sect. 3.a.δ), and because of multiple phytochrome responses of protonemata in development (e.g. Wada et al. 1983; Hayami et al. 1992), *Adiantum* might become a model system to relate particular phytochrome species to particular responses as a promising step towards analysing the way in which phytochrome controls the response.

Phytochrome-controlled chloroplast movement has recently been discovered also in epidermal cells of *Vallisneria gigantea* (Izutani et al. 1990). This calls for reinvestigating scattered earlier reports on marginal "chlorophyll effects" (e.g. Schweickerdt 1928; Seitz 1967a).

β) Multiple Photoreceptor Pigments

The results as reported in the preceding section suggested that there are two photoreceptor pigments which control chloroplast movement alternatively, depending on the species, viz. cryptochrome and phytochrome. However, in particular species, two or more photoreceptor pigments can effectively perceive the light signal and trigger the response independently of each other. Still more complicated, in some particular cases two photoreceptor pigments or their products have to interact in order to bring about the response.

Two or More Photoreceptor Pigments, Acting Independently of Each Other. Phytochrome and Cryptochrome. Besides the phytochrome-typical main peak in red light, the action spectrum for the low-intensity movement in *Mougeotia* has also a smaller peak in the blue region (Haupt 1959b). After some discussions as to whether or not this effect can be mediated by the short-wavelength absorption band of phytochrome (Weisenseel 1968; Haupt 1971; Hartmann and Cohnen Unser 1973), a **separate blue-light receptor** was convincingly demonstrated (Gabryś et al. 1984): on a background irradiation with strong far-red, the response can no longer be induced by red light, whereas the blue-light effect is not impaired. This blue-light effect shows a typical **cryptochrome action spectrum** (Walczak et al. 1984).

Corresponding conclusions were derived for the protonema of *Adiantum capillus veneris* (Yatsuhashi 1996). The two-photoreceptor model in *Adiantum* is further supported by the situation in the protonema of *Pteris vittata*, which orients its chloroplasts exclusively to blue light, although it makes use of phytochrome for other responses. This may indicate differences in phytochrome species or in their localisation between the two ferns *Adiantum* and *Pteris* (Kadota et al. 1989).

In conclusion, the peculiarity of the red-light-sensitive systems is not "phytochrome *instead* of cryptochrome" but "phytochrome *in addition* to cryptochrome".

The evidence for two separate photoreceptor pigments may not exclude some relations between them. For *Mougeotia*, a highly speculative interpretation had been proposed by Haupt (1971): could a protein exist with two chromophoric groups, viz. phytochromobilin and flavin? No idea was presented as to how to follow up this strange line, and no further discussion has taken place since. Surprisingly, however, this view appears to be reviving, as will be shown in Section 3.a.δ.

Cryptochrome and Photosynthetic Pigments. In *Vallisneria spiralis*, the induction or enhancement of **rotational streaming** has been used as a **model for oriented chloroplast movement,** as in this species both responses share the same photoreceptor systems (Seitz 1967a). In the action spectrum, inhibition by iodide is restricted to the main peak, whereas other regions are hardly affected, as well as the marginal effect of red light (Seitz 1967b). Although the inhibition by iodide is **not proof** for a **flavin** photoreceptor (see above), the wavelength dependence of the inhibition indicates two different photoreceptor pigments in *Vallisneria*, and this is supported by additional wavelength-dependent observations. Because of its additional red-light sensitivity, the second pigment is assumed to be **chlorophyll,** and indeed, the inhibitor of photosynthesis,

DCMU (dichlorophenyl dimethylurea) inhibits the response as induced by the respective wavelengths.

Depending on the intensity range, *Hormidium flaccidum* makes use of at least two photo-receptor pigments for its low-intensity response, viz. cryptochrome in the cytoplasm and photosynthetic pigments in the chloroplast. This has been concluded from different principles of perception of light direction (Sect. 3.a.γ), from different dichroic orientation, and from different sensitivity to iodide and to inhibitors of photosynthesis (Scholz 1976b).

Interaction of Photoreceptor Systems. In the blue-light-responding plants, low-intensity and high-intensity movement make use of the same photoreceptor pigment. A quite different situation is found in *Mougeotia*, as had already been reported by Senn (1908): whereas for **low-intensity movement** red light is the most effective spectral range, the **high-intensity movement** is restricted to blue light. In the latter response, however, there is, in fact, a rather complicated interaction of photoreceptor-pigment systems, which will be discussed next.

In a first step Schönbohm (1963) found what appeared to be a strong potentiation of the blue-light effect by red light via phytochrome, either given simultaneously, or in a series of alternating short pulses. In fact, however, phytochrome mediates the **orienting light signal**, whereas a blue-light photoreceptor acts as an **intensity-dependent switch**, which acts independently of light direction and determines that the chloroplast exposes its profile rather than its face to the red light (Schönbohm 1966). Accordingly, in monochromatic blue light the **directional signal** is mediated via the blue-light absorption band of phytochrome and the **switch effect** again via the separate blue-light receptor (Schönbohm 1980). As to the latter, the action spectrum as well as the wavelength-dependent action dichroism recall **cryptochrome** and suggest a **flavin-like substance** in the cytoplasm (Schönbohm 1968, 1971).

The high-intensity movement of *Mesotaenium caldariorum* depends as well on an interaction of phytochrome and cryptochrome (Gärtner 1970). The low-intensity movement, however, differs from that in *Mougeotia*, where a single red pulse can induce a full response. In *Mesotaenium*, instead, red light has to be given continuously or as repeated pulses, until the new orientation has been reached (Haupt and Thiele 1961). Obviously, in *Mesotaenium*, the gradient of the otherwise long-living Pfr loses its activity within minutes, and this has been interpreted by a poorly understood **"aging of Pfr"** (Haupt and Reif 1979) or by a rapid **diffusion of Pfr** in this rather small cell (Herrmann and Kraml 1997). Interestingly, an almost ineffective single red pulse can be made fully effective if followed by a blue pulse, which by itself is completely ineffective (Kraml et al. 1988). In addition, this latter also accelerates the movement during its irradiation time. As in the blue-red interaction of *Mougeotia*, the direction of the blue pulse has no bearing on the response.

It is tempting to recall the dual-chromophore hypothesis (Haupt 1971; see above) for the interactions in *Mougeotia* and *Mesotaenium*. However, these interactions do not involve any coupling at the level of photo-receptor pigments. Rather, at least one of the respective interacting factors is an early product of photoperception, as will be shown in Section 3.c.

γ) Perception of Light Direction

Given the fact that usually the light signal is perceived in the cytoplasm, perception of light direction requires a **cytoplasmic gradient of light absorption**. According to Senn (1908), this gradient is brought about mainly by light refracction, sometimes modified by reflection (Sect. 2). The questions remained whether this theoretical conclusion can be confirmed experimentally, and whether light refraction (and reflection) is sufficient to explain perception of light direction. An additional question concerns the absorption gradient in continuous saturating light, i.e. under natural conditions.

Refraction. **The role of refraction** was confirmed by Scholz (1976a) for *Hormidium flaccidum*, using the **inversion experiment** of Buder (1918). If surrounded by air, the cylindrical cell of this alga acts as a collecting lens, the single large chloroplast slides along the wall to the "focus region", i.e. distal with respect to the light source. If, however, the cell is **surrounded by oils**, the chloroplast prefers the proximal side, and the percentage of this preference is a linear function of the **refractive index** of the oils. Accordingly, in water with its refraction index close to that of cytoplasm, there is little preference for either side.

However, in intensity ranges where the "second photoreceptor" (Sect. 3.a.β) controls the response, the chloroplast always goes to the proximal side, irrespective of the refraction index of the surrounding medium. This latter is to be expected if the photoreceptor pigment is located in the chloroplast, with the absorption gradient resulting from **attenuation** in the photosynthetic pigments. This assumption is consistent with arguments presented in Section 3.a.β.

For **cells in water**, i.e., under conditions of only small refraction effects, it can be predicted that not only the **focusing effect** becomes ineffective, but that also the **bypassing effect** cannot be a good basis for reliable perception of light direction (cf. already Senn 1908). Accordingly, an **additional principle** for generating an absorption gradient has to be expected at least for cells in water.

Dichroic Orientation. Jaffe (1958) had pointed out the **role of dichroic orientation** of photoreceptor molecules for gradients of light absorption in polarised light. After the discovery of dichroic orientation of phytochrome molecules in *Mougeotia* (Haupt 1960; Sect. 3.a.α), it was demonstrated that also in unpolarised light such orientation leads to a gradient of light absorption, which is related to the light direction (Haupt 1965; cf. also Kraml 1994). With **surface-parallel orientation of the transition moments** (e.g. Pr in *Mougeotia*), the resulting **tetrapolar absorption gradient** is similar to that due to the bypassing effect, and as with the latter, no discrimination is made between light beams coming from two opposite sides. Accordingly, both mechanisms can coact and support each other, as has been shown in a computation for a model cell by Gabrys-Mizera (1976). The **dichroic effect** is thought to be the **main mechanism for cells in water** to detect the light direction, as it does not depend on the refractive index of the cell environment (cf. Zurzycki and Lełatko 1969).

As a general conclusion, dichroic orientation of photoreceptor molecules is an important factor for perception of light direction, and this, in turn, requires an underlying **cytoplasmic structure**, which in *Mougeotia* was supported experimentally by Haupt and Wirth (1967). Originally, for these structures the cell membrane was assumed to be a good candidate; but today the **cytoskeleton** is preferred as more appropriate for anchoring the photoreceptor pigment (cf. Sect. 3.a.δ).

Saturation. Whatever the gradient results from, there is the problem of how the **absorption gradient** is maintained **in saturating light**. This problem is particularly serious in *Mougeotia*. Photoconversion of phytochrome (Pr $\rightarrow$ Pfr) saturates in a wavelength-dependent steady state, which can be obtained by light pulses of only seconds (Haupt 1959b; Kraml and Schäfer 1983) and which therefore will be reached very soon even in the less-absorbing spectral regions. Moreover, the **life time of Pfr** is much longer than the whole response of the chloroplast (e.g. Wagner and Klein 1981; Kraml et al. 1987), and thus there appears to be no chance for a persistent gradient that can control the response. Yet, in nature, chloroplasts respond to continuous saturating light. The solution to this paradox has been found in the **different orientation of Pr and Pfr** (see above, Sect. 3.a.α), with the former parallel, the latter normal to the cell surface. As a result, the **photoequilibrium** between Pr and Pfr becomes **dependent on** the pigment's differential **dichroic orientation** and hence is different at the flanks as compared to the proximal and distal surfaces; the gradient can never be levelled, either in polarised, or in unpolarised light (e.g., Haupt 1970b).

Although this conclusion is theoretically convincing and fits all observations, it has not yet been possible to make differential measurements of Pfr in different regions of a single cell.

δ) Genetic Approach to the Photoreceptor Problem

In recent years, genetic approaches such as, e.g. the use of mutants and transgenic plants, cloning, sequencing and in-vitro expression of genes, have opened new insights into photoreceptor research in general and also into the field of chloroplast movement.

Phytochrome. In higher plants, a **diversity of phytochromes** exists, and even in a single species multiple phytochromes have been found. In *Arabidopsis thaliana*, for example, at least five phytochromes can be distinguished, which probably differ in their functions, mediating different photomorphogenetic responses (cf. Whitelam and Devlin 1997). This raises the question as to the number of phytochromes in those plants that make use of red light for their chloroplast orientation.

In *Mougeotia*, only **one phytochrome gene** has been found so far. The amino-acid sequence reveals no hydrophobic domain in the apoprotein that could make it a transmembrane protein (Winands et al. 1992). This **excludes** a primary action of Pfr in *Mougeotia* as a **membrane effector** (e.g. as a calcium channel), as had occasionally been suggested in the past (e.g. Haupt and Weisenseel 1976, based on Hendricks and Borthwick 1967). On the other hand, Winands and Wagner (1996) found sequences that correspond to **microtubule-binding sites**; this is important in view of the localisation (membrane vs. cytoskeleton; Sect. 3.a.γ). Moreover, the phytochrome gene in *Mougeotia* is **autoregulated by light**, with its expression substantially slowed down by Pfr, the active form of phytochrome.

In *Mesotaenium*, lack of hydrophobic transmembrane-binding domains had been reported alreay by Kidd and Lagarias (1990), and gene expression is autoregulated by light as in *Mougeotia* (Lagarias et al. 1995). The **six phytochrome genes** discovered so far are so closely related to each other that virtually **one functional phytochrome species** can be assumed, as in *Mougeotia* (Wu and Lagarias 1997).

In fern protonemata, more than one phytochrome species has been found. After conclusions from partial sequences in *Anemia phyllitidis* (Maucher et al. 1992), **three phytochrome genes** in *Adiantum* were sequenced to full length (Wada et al. 1997). According to experiments with mutants (Kadota, cited in Wada et al. 1997), chloroplast orientation is coupled with photo/polarotropism, whereas spore germination is mediated by a different phytochrome. As the latter response lacks the action dichroism, a **different compartmentation** can be assumed. These **multiple phytochromes** are particularly interesting in comparison with *Pteris*, where chloroplast orientation is restricted to blue light (see Sect. 3.a.β).

It may be added that in a phylogenetic tree the phytochromes of *Mougeotia*, *Mesotaenium* and *Adiantum* are more closely related to each other than to either of the phytochromes of higher plants (Winands and Wagner 1996).

Cryptochrome. True progress in identifying **cryptochromes** did not start before genetic approaches were applied, most successfully again in *Arabidopsis*. Two cryptochrome genes were demonstrated there which code for **two distinctly different pigments** (Ahmed and Cashmore 1993).

The proteins are related to the photolyases; one of them has **two chromophoric groups**, viz. a flavin and probably a deazaflavin or a pterin (Cashmore 1997). Thus, the old controversy flavin vs. pterin as the chromophore of cryptochrome appears in a new light. Briggs and Liscum (1997) extended their respective investigations to the phototropism problem.

These studies in *Arabidopsis* are a promising basis for **cryptochrome research in chloroplast movement**. In *Mougeotia*, a sample from a cDNA library shows homologies with a fragment of a cryptochrome gene in *Arabidopsis* (Brunner and Wagner 1996). Thus, in the near future, cryptochrome gene(s) of *Mougeotia* may become isolated and expressed; but as cryptochrome lacks the handle of photochromicity, it is a difficult task to prove the photoreceptor function of a particular cryptochrome as long as no mutants are available.

In *Adiantum*, **five different cryptochromes** were reported (Wada et al. 1996). As now mutants can be made available in ferns (Wada et al. 1997), it should become possible to assign chloroplast movement to one of the cryptochromes.

Most interestingly, using genetic approaches in cooperation with Wada, W. R. Briggs (pers. comm.) found evidence for a **hybrid photoreceptor protein** in *Adiantum*, a "superchrome". This "anomalous phytochrome" shows high **phytochrome homology** through about 560 amino acids; but on the other hand, it reveals a high homology also with a newly analysed kinase, including a repeated domain that is supposed to be a **flavin-binding site**. After arguments have been presented for proteins with two different blue-light-absorbing chromophores (see above), it is no longer "nonsense" to suggest also a hybrid photoreceptor that simultaneously functions as phytochrome and as cryptochrome, as had been speculated by Haupt (1971; cf. Sect. 3.a.β).

b) Response: Mechanics of Movement

The early question "active or passive" (Sect. 2) was probably too simple, imprecise, and not completely logical. In fact, there are several aspects to be separated from each other. "Active or passive" concerns already the very first step, i.e. perception of the light signal. As was shown above (Sect. 3.1.α), this perception is usually located in the cytoplasm, hence to be considered "passive" in terms of the early investigators. For the response, there are two aspects, which overlap partially.

– The more descriptive question (α): do the chloroplasts migrate relative to a stationary endoplasm, or are they transported by moving

portions of endoplasm? The terms *independent* or *dependent* may be more adequate for this alternative.
- The more analytical question (β): where is the motor apparatus localised, what is its nature, and what kind of force is generated?

α) Dependent or Independent Movement?

As a model system, Jarosch (1956) **isolated cytoplasmic drops** from characean algae. Whereas in intact cells chloroplasts are firmly anchored in the ectoplasm and hence do not participate in the rotational streaming of the cytoplasm, they move and rotate in the drops, and the opposite displacement of neighbouring particles (microsomes) shows clearly that these chloroplasts move relative to the surrounding cytoplasm. Thus, in principle, chloroplasts can move *independently* (actively in the former terminology). However, this cannot be extrapolated uncritically.

For the high-intensity movement in *Lemna*, Zurzycka and Zurzycki (1957) could show, with **cinematographic methods**, that groups of chloroplasts always move together for a while; they obviously **depend** on the motion of surrounding cytoplasm (passive). In these examples, the chloroplasts moved more slowly than "microsomes", as their frictional resistance in the thin cytoplasmic layer is larger than that of smaller particles. On the other hand, individual chloroplasts frequently escape from their "environment" and join a neighbouring group; this particular movement might be **independent** of the surrounding endoplasm.

As a third example, two responses of *Vaucheria sessilis* will be discussed. In the cylindrical coenocytic "cell", cytoplasm streams forth and back along **longitudinal tracks**, together with its organelles, including chloroplasts. If a small region is irradiated with a **microbeam** of blue light, this acts as a **"light trap"**: streaming stops here, but outside this region it continues and brings further masses of cytoplasm into the field. As a result, many chloroplasts that enter the field by chance are "trapped" (Fischer-Arnold 1963). Thus, this kind of orientation pattern is fully **dependent** on cytoplasmic streaming.

The **normal orientation movement** in *Vaucheria*, however, proceeds in **azimuthal** rather than in longitudinal direction, and no respective cytoplasmic streaming has been reported yet. It can be expected, therefore, that in *Vaucheria* **dependent as well as independent** chloroplast rearrangements are possible, and this has been analysed in some detail in the protonema of *Adiantum capillus-veneris*, which exhibits responses similar to *Vaucheria* (Kadota and Wada 1992a).

In this protonema, an obvious difference has been found between chloroplasts participating in the **longitudinal** cytoplasmic streaming and those orienting in **azimuthal** direction. In the latter case, no particles or organelles are seen to move with the chloroplasts. Moreover, in their azimuthal orientation movement, chloroplasts migrate in a relatively straightforward way, although more slowly than the longitudinally streaming cytoplasm;

this excludes a trial-and-error mechanism of orientation and is in favour of an **independent movement** of the chloroplasts.

Thus, dependent and independent movement appear to contribute to the rearrangement of chloroplasts in the cell to various degrees, and this has to be considered when asking about the motor apparatus.

β) The Motor Apparatus

Zurzycki (1962a) critically summarised six hypotheses as to the mechanics of chloroplast movement, which had been put forward during the first half of our century. Most of them suffered from lack of sound knowledge about respective cell ultrastructures. Only the old view of fibrous structures has survived as a basis for modern approach.

Cinematographic analysis allowed the movement of each chloroplast to be followed in detail (Zurzycka and Zurzycki 1957). In some cases (e.g. *Elodea* and *Lemna*) the chloroplasts approached their high-intensity arrangement only **by biasing their random movement**. Remarkably, however, the return movement to the low-intensity arrangement proceeded **in rather direct ways,** as if an elastic force brought the chloroplasts back. This initiated experiments in *Mougeotia*, in which by proper irradiation protocols the ribbon-shaped chloroplast was caused to rotate up to five full revolutions; but afterwards there was neither a relaxing "backward" movement, nor a preference in direction for a further orientation movement (Haupt and Heymann 1967). Thus, **no elastic force can be found** in *Mougeotia*, and no additional evidence in favor of it has been presented in *Lemna*.

As a structural appraoch, Jarosch (1958) observed, in his isolated drops of cytoplasm from *Chara* (Sect. 3.b.α), **fibrous structures** that were attached to the moving chloroplasts. Besides, also **isolated fibres** were found, along which "microsomes" were displaced in a direction opposite to the fibre's movement. Admittedly, this concerns a system in which normally chloroplasts do not migrate. However, observations of Schönbohm (1970) with **interference contrast** in intact cells of *Mougeotia* and *Funaria* appear well related to orientation movement. In *Mougeotia*, these fibres extend during movement from the edge of the large chloroplast into the cytoplasm, particularly **in the advancing direction** of the chloroplast. This suggests that they would pull the latter (cf. also Schönbohm 1972, 1974). With the electron microscope, respective structures have been demonstrated in more detail in the cytoplasm close to the chloroplast's edge (Wagner and Klein 1978).

As to the nature of these fibrous structures, **actin microfilaments** appeared to be a promising candidate, after actin-myosin interaction had been established as underlying cytoplasmic streaming in characean algae (cf. Kamiya 1959). Indeed, the rather specific inhibitor of actin activity,

cytochalasin B, was found to **block the movements** in *Mougeotia* (Wagner et al. 1972; Schönbohm 1973a), *Funaria* (Schönbohm 1973a), *Vaucheria* (Blatt and Briggs 1980) and *Adiantum* (Kadota and Wada 1992a). In *Mougeotia*, **reversibility of this inhibition** was demonstrated; the "memory" to the inducing light signal was stored in the gradient of active phytochrome, Pfr, and hence response could still take place after the inhibitor was washed out (Wagner et al. 1972). In addition, with the **arrowhead decoration** by heavy meromyosin or S1 fragments of myosin, actin microfilaments could be demonstrated directly in cell homogenates and in protoplasts of *Mougeotia* (Marchant 1976; Klein et al. 1980) as well as in intact cells of *Vaucheria* (Blatt et al. 1980) and *Adiantum* (Kadota and Wada 1989). Finally, the "pulling fibres" of Schönbohm, mentioned above, were characterised as actin by **fluorescein-conjugated phalloidin** (Mineyuki et al. 1995). Thus, actin as underlying the mechanism of chloroplast movement is now a widely accepted hypothesis, but, in detail, questions remain open.

The role of **myosin** in chloroplast movement is not as strongly established. Since **sulph-hydryl groups** are involved in the ATPase function of myosin, SH-group inhibitors have been tested and found to impair chloroplast orientation in *Mougeotia* (Schönbohm 1969), *Funaria* (Schönbohm 1972) and *Vallisneria* (Seitz 1970). This rather unspecific effect has been confirmed, e.g. in *Mougeotia* and *Adiantum*, with the more specific myosin inhibitor N-ethyl maleimide (cf. Wagner et al. 1978; Kadota and Wada 1992b). Although even this latter substance is not strictly specific (Wada et al. 1993), the hypothesis of actin-myosin interaction appears reasonable.

In consequence, the next question asked how **actin-myosin interaction** can bring about chloroplast displacement. For cytoplasmic streaming in *Nitella*, it was already known that actin is anchored to the stationary ectoplasm and/or to the stationary chloroplasts in it, and that myosin ATPase resides in the moving endoplasm, probably bound to small organelles or particles (summarised, e.g. by Seitz 1979). Sliding of actin and myosin along each other generates the **shearing force**. Whether the resulting movement is cytoplasmic streaming without or with chloroplasts or movement of chloroplasts only, depends on the localisation of chloroplasts in ecto or endoplasm and on the localisation of actin and myosin at respective structures or organelles (cf. Seitz 1979).

On this basis, the "light-trap" effect in *Vaucheria sessillis* (Sect. 3.b.α) is explained. In darkness, **cable-like structures** can be visualised, obviously anchored in the ectoplasm and oriented in the direction of cytoplasmic streaming. These fibres appear to be identical with **bundles of actin**, demonstrated by Blatt and Briggs (1980). The actin filaments are supposed to interact with **myosin** at the surface of organelles, thus generating the **shearing force** for cytoplasmic streaming. Upon local blue-light irradiation, the **fibres disintegrate** into a faint network (reticulation), which obviously is no longer capable of sustaining a uniform mass streaming. After light-off, soon the "cables" reappear, and streaming of endoplasm and chloroplasts is resumed. With **cytochalasin B**, the reticulation is obtained also in the dark regions, and all movement stops; but the antagonist of this inhibitor, **phalloidin**, keeps the fibres intact and inhibits reticulation even in light, thus not allowing cytoplasm and chloroplasts to stop in the light field.

In addition to a structure that transforms the motive force into movement, it is important that the motive force overcomes a certain resistance, which may be due to **anchoring** of the chloroplasts to the stationary phase of cytoplasm (ectoplasm). This anchoring can be controlled by light, as shown for *Lemna trisulca* (Zurzycki 1960) and *Vallisneria gigantea* (S. Takagi, pers. comm.), where chloroplasts can be centrifuged much more easily during their orientation movement than in their stationary arrangement before or thereafter. In contrast, however, in *Mougeotia* the chloroplast is anchored to the ectoplasm more strongly during its movement. This is a typical far-red reversible **phytochrome effect** with short-term kinetics similar to that of the low-intensity movement (Schönbohm 1970, 1973b); it is abolished by cytochalasin B and hence due to actin (Schönbohm and Meyer-Wegener 1989). Thus, the light effect and the function of anchoring in movement cannot be described in a unifying hypothesis, and this justifies an additional or alternative approach.

If actin-myosin interaction were the primary basis of movement in *Mougeotia*, the **microfilaments** would have to be oriented **parallel to the surface** and in the hoop direction (i.e. azimuthal). Although this appears to hold for the early findings of Schönbohm (1974) and again for the observations of Mineyuki et al. (1995), both reported above, the ultrastructural investigations of Wagner and Klein (1978) revealed evidence for a **radial component in their microorientation**. Moreover, Grolig and Wagner (1988) pointed out that the **speed of movement** is 2 to 3 orders of magnitude slower than in examples of cytoplasmic streaming, where actin and myosin slide along each other. The authors therefore propose another mechanism in *Mougeotia*. According to this, microfilaments connecting the chloroplast with the ectoplasm are in a steady state of fastening and loosening at their **ectoplasmic anchoring sites**. If the abundance of the latter is locally reduced by Pfr, as supposed by Wagner and Klein (1981), the steady state would be asymmetrically biased, and eventually this would have the same result as the shearing force. However, it is an open question how this view can be compatible with the above-mentioned fastening by Pfr. This controversy calls for comparative experiments under strictly identical conditions with identical strains of *Mougeotia*.

At this stage, it might be helpful to consider the function of **microtubules**, which are abundant in *Mougeotia* (Foos 1970). Their depolymerisation by, e.g., **colchicine**, does not impair the movements in *Mougeotia* (Foos 1971; Schönbohm 1973a) and *Adiantum*

(Kadota and Wada 1992a). Rather, **these inhibitors** can even **accelerate** the movement in *Mougeotia*, suggesting that microtubules act to strengthen the resistance against movement (Serlin and Ferrell 1989). Accordingly, inactivation of microtubules by strong blue light appears to be causally related to the speeding up of the high-intensity movement as compared to the low-intensity movement (Al-Rawass et al. 1997). Besides this indirect function in movement, there might be some exceptional cases where microtubules serve the function of the motor apparatus, as reported for the coenocytic alga *Dichotomosiphon* (Maekawa et al. 1986).

As a general conclusion, there are several **prospective points of attack** for controlling signals, and this has to be kept in mind when trying to disentangle the transduction chain (Sect. 3.c).

c) Signal Transduction

Perception of the light signal and activity of the motor apparatus are linked together by a **sequence of events.** Although in no case has such a **transduction chain** been completely analysed, there are a few examples where **early steps** after signal perception are known or have been postulated theoretically from respective experiments (**forward approach** into the black box, Sect. 3.c.α and β). On the other hand, in some examples, factors have been found that can control the motor apparatus and thus could be candidates for the **terminal steps** in the transduction chain (**backward approach**, Sect. 3.c.γ). Only preliminary speculations are available about an **integrated model** or about steps in the centre of the black box (Sect. 3.c.δ).

α) Forward Approach: Early Reactions After Photoperception

There are reactions to the respective light signals that share dependence on fluence (rate) and wavelength with the chloroplast rearrangement. Two of them are likely to be related to the latter.

Vaucheria sessilis. Upon local irradiation with blue light, an **outward current** starts within 5 s at that site (Blatt et al. 1981). This is almost certainly due to a **proton extrusion**, but, in addition, a controlling function of calcium appears to be involved. The dependence on fluence and wavelength coincides with that for the **reticulation of actin bundles** (see above Sect. 3.b.β) and subsequent **chloroplast aggregation**, respectively, and the current clearly precedes the latter. It might be speculated that protons and/or calcium are controlling factors in the transduction chain, acting on the integration/disintegration of the actin bundles, thus affecting the motive force. However, the temporal relation between current and reticulation has not yet been established.

Mougeotia. Facilitated **calcium fluxes** have been found as a far-red reversible effect of red light, thus mediated by phytochrome. This suggests a close relation to the Pfr-controlled chloroplast movement. The observations, however, do not yet allow a unifying model of Pfr action. On the one hand, Dreyer and Weisenseel (1979) found a strongly **enhanced uptake** of radioactive calcium. On the other hand, Roux (1984) reported a **release from the cell**. Finally, Jacobshagen et al. (1986) demonstrated **intracellular release** of calcium from osmiophilic organelles close to the chloroplast's edge, resulting in **increase of cytoplasmic calcium**. The calcium effects appear to precede the terminal response, as postulated for a step in the transduction chain. Additional arguments (pro and con) on calcium as an internal signal will be discussed in the backward approach (Sect. 3.c.γ).

In both these cases, it cannot be decided yet whether the early reactions are indeed steps in the transduction chain, or side effects only. For the examples presented next, there is little doubt that the theoretically concluded internal signals are steps in the transduction chain; however, their nature is completely unknown.

β) Forward Approach: Analysis of Aftereffects

Although most of the blue-light-dependent systems require the controlling light to be present during almost the whole response time, **aftereffects of light pulses in subsequent darkness** have occassionally been reported (e.g. Seitz 1967a). With sensitive photometric registrations, aftereffects have been observed in *Lemna* even after pulses in the range of seconds (Zurzycki et al. 1983). Their kinetics allowed **early internal signals** (S) resulting from photoperception to be postulated, which can sustain the response during their lifetime of a few minutes (Fig. 2A). For a complete response, requiring up to 1 h, continuous or repeated production of S is necessary.

In low-intensity movement of *Mesotaenium*, a single red pulse becomes effective only if followed by blue (cf. Sect. 3.a.γ). From time-resolved irradiation protocols, a sequence of two internal signals, S1 and S2, has been hypothesised (Fig. 2B; Kraml et al. 1988; Herrmann and Kraml 1997). It should be stressed that the **directionality** of S2 is exclusively **determined** by S1 and hence **by the red-light pulse**, whereas **blue light acts in a scalar way**. The model (Fig. 2B) does not yet include the full response in continuous red light without blue.

The most complicated situation has been reported for *Mougeotia*. In contrast to the above-mentioned systems, in *Mougeotia* the responses can proceed to the final stage in darkness after a **short light pulse** (Moore 1888; Lewis 1898). For the red-light-induced low-intensity response, pulses even shorter than milliseconds are effective (e.g.,

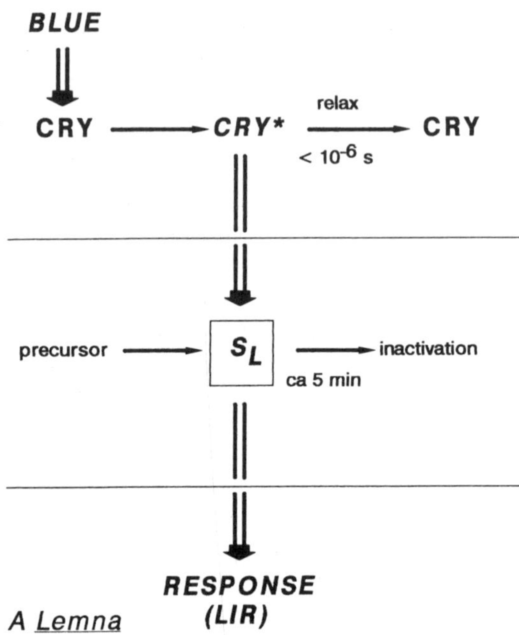

BLUE
CRY
CRY*
relax
< 10^-6 s
CRY
precursor
S_L
inactivation
ca 5 min
RESPONSE
(LIR)
A Lemna

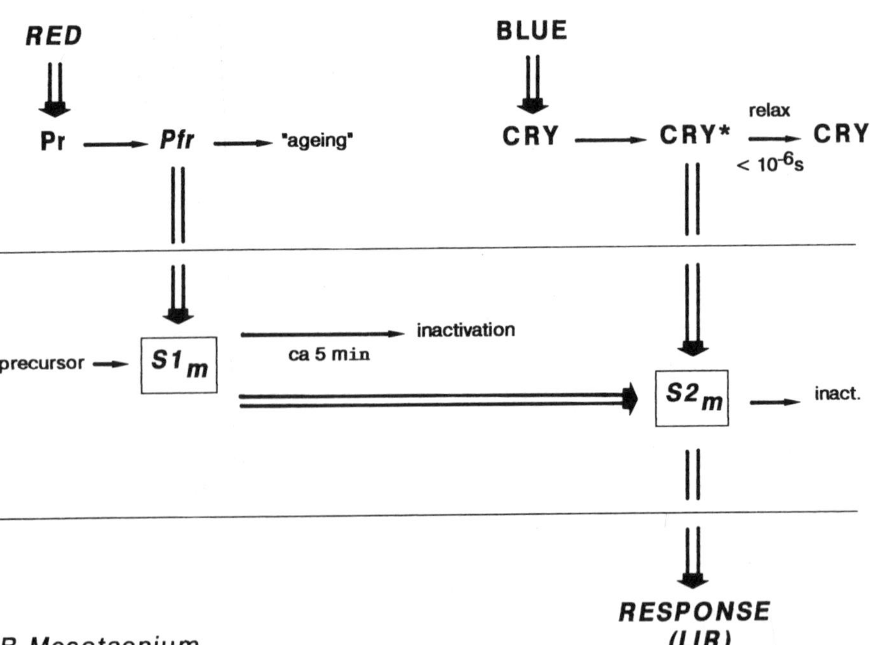

RED
Pr
Pfr
"ageing"
BLUE
CRY
CRY*
relax
< 10^-6 s
CRY
precursor
S1_m
inactivation
ca 5 min
S2_m
inact.
RESPONSE
(LIR)
B Mesotaenium

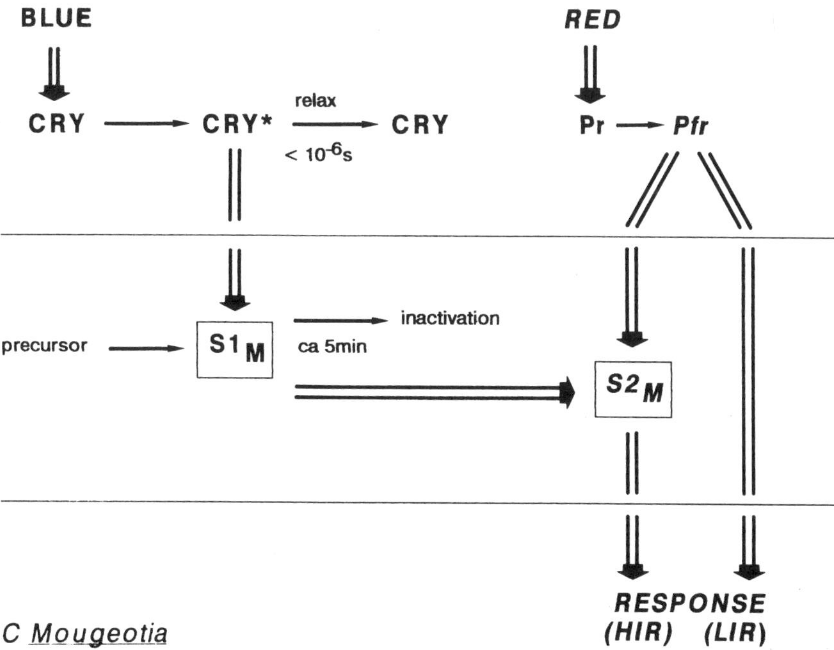

Fig. 2A–C. Early steps in the transduction chain. *CRY* Cryptochrome (CRY* = excited state); *Pr, Pfr* phytochrome in the red-absorbing and far-red-absorbing form; *S, S1* and *S2* hypothetical internal signals as postulated from the kinetics of aftereffects, with the subscripts *L, m* and *M* for *Lemna* (**A**), *Mesotaenium*, (**B**) and *Mougeotia* (**C**), respectively. *HIR* and *LIR* high-intensity and low-intensity movement; *single-line arrows* transformation of respective substances; *double-lined arrows* control of transformations or of reactions; *italics* substance or signal acts in a vectorial way (i.e. dependent on the light direction); otherwise scalar. *Horizontal lines* separate perception, transduction, and response from each other. (After data and conclusions of Zurzycki et al. 1983; Herrmann and Kraml 1997; Gabryś et al. 1985)

Scheuerlein and Braslavsky 1987). In contrast, for the blue-light-induced high-intensity movement, **full response** to a light pulse has been found only under particular conditions, and it requires a pulse in the range of minutes (Schönbohm 1965, 1987). This latter does not allow early short-term transduction steps to be reliably separated from perception. However, **transient responses** can be photometrically recorded after a single sequence of a blue and a red pulse in the range of seconds. Here, too, the kinetics allow for concluding on a **sequence of early events** with at least one **scalar** and one **vectorial internal signal** (Fig. 2C; Gabryś et al. 1985).

In the model, the internal signal S2 tries to act in opposite direction to Pfr, i.e. directing the chloroplast to the profile position. In this competition, usually the effect of Pfr prevails; only if Pfr is abolished by far-red within 40 min after the formation of S2, the latter can diret the chloroplast to the high-intensity position.

For future research, it is a challenge to **identify the internal signals** S, S1, S2 and to compare them in the various systems. Particularly interesting is the similar lifetime of S in *Lemna* and S1 in the two algae, although it is scalar in *Mougeotia*, but otherwise vectorial.

γ) Backward Approach

If we take it for granted that the motive force is mainly based on actin microfilaments, whatever the detailed mechanism (Sect. 3.b.β), the final step of transduction has to be the **control of actin activity** in a broad sense – or more precisely: a differential control of it in space. Theoretically, such a control can concern several levels: **polymerisation** of G-actin to F-actin and vice versa; **integration** of microfilaments into "active" bundles and disintegration (reticulation); **anchoring** of actin to and loosening from respective sites or activation/inactivation of these anchoring sites; **activation/inactivation** of myosin; **actin-myosin interaction.**

One promising candidate for the control of most of these steps is **calcium**, particularly as **calmodulin** and other calcium-regulated proteins have been found in *Mougeotia* (cf. Wagner et al. 1984; Roberts 1989). Indeed, several observations suggest a role of calcium and calmodulin in control of movement in *Mougeotia*, although in detail there is much diversity, as also mentioned in the forward approach (Sect. 3.c.α). On the one hand, **deprivation of calcium** from the culture medium and in consequence also from the cells has strongly reduced phytochrome-mediated low-intensity response in *Mougeotia* (Wagner and Klein 1978). Moreover, under highly specific conditions, Serlin and Roux (1984) could elicit chloroplast rotation in *Mougeotia* by local application of a **calcium ionophore** in a calcium-containing medium. On the other hand, **no adverse effect on low-intensity movement** could be found by Schönbohm et al. (1990), if calcium was withdrawn from the medium or if calcium uptake was blocked by various inhibitors. In contrast, the **high-intensity movement** appears to **depend strongly on calcium influx** (E. Schönbohm, pers. comm.). In an advanced view, these discrepancies need not exclude a controlling role of calcium in general terms, if its release from internal stores is additionally taken into account as an important factor (Russ et al. 1991, confirming earlier assumptions of Wagner and Klein 1981; cf. also Sect. 3.c.α).

Similar conclusions have been derived for the role of calcium in *Lemna* (Tłatka and Gabryś 1993). However, to make the situation still more complicated, Schönbohm et al. (1990) found strong **inhibition of photosynthesis** by calcium deficiency. Thus, dependence of movement on calcium could be an **unspecific effect** on the energy-providing system.

δ) Hypothetical Transduction Chains

Since calcium has been found as well in the forward as in the backward approach, it was tempting to propose, for *Mougeotia* a working hypothesis based on calcium (Haupt 1980), in spite of the above-mentioned uncertainties. Its main steps were:

Pfr gradient → local change of calcium fluxes → calcium redistribution → local calcium-calmodulin interaction → local control of actin and/or myosin activity → movement.

Unfortunately, this model has subsequently been included in several publications, although already in 1981 various shortcomings and new results led Wagner and Klein to a more complex, **two-branched model** (Fig. 3). According to this, Pfr acts in a twofold way, viz. **scalar via calcium**, thus enabling the motor apparatus to become effective at all (kinetic branch), and **vectorial via an unknown messenger**, controlling the motor apparatus locally so as to produce an **oriented response** (orienting branch), e.g. via controlling the anchoring sites for actin (cf. Sect. 3.b.β). In this model, the most interesting question is still open: what are the transduction steps in the vectorial branch that link the oriented signal perception with the local effects on the motor apparatus?

Similarly, two separate phytochrome effects have recently been proposed by Yatsuhashi and Kobayashi (1993) for *Dryopteris sparsa*. Here, too, **two transduction chains** are thought to coact or interact as to control movement and orientation of the chloroplasts in the protonema; but in detail, the respective internal signals are still unknown.

Considering the partly contradicting statements about calcium (Sect. 3.c.α and γ), it may be useful to pay attention also to the "**phenolic compounds**" in *Mougeotia*. Their release from and uptake into cytoplasmic vesicles can be under the control of light and calcium, and they can strongly affect chloroplast movement (Schönbohm and Schönbohm 1984). Thus, for the two-branched model a dual function of calcium might be considered.

An additional modification of the model might involve the **microtubules** (cf. Sect. 3.b.β). As their activity can be regulated by calcium as well as by light, the resulting change in resistance against actin-based movement could be an important checkpoint in signal transduction, and this might even represent the "kinetic branch" in Fig. 3 (cf. Wagner et al. 1992). Interestingly, the genetic approach has recently revealed a **domain** in *Mougeotia* **phytochrome** that may have **microtubule-binding properties** (Winands and Wagner 1996; Sect. 3.a.δ). Moreover, as microtubules can be controlled also by **magnesium**, this until now neglected ion should be included in future research (cf. Tłaka and Gabryś 1993).

Besides the two-branched model in *Mougeotia* with its possible modifications, there are approaches in blue-light-absorbing systems, which, however, have not yet evolved to complete models. According to Seitz (1972; summarised 1979), **ATP** may be a controlling factor in *Vallisneria spiralis*. In a gradient of light absorption, local ATP concentrations

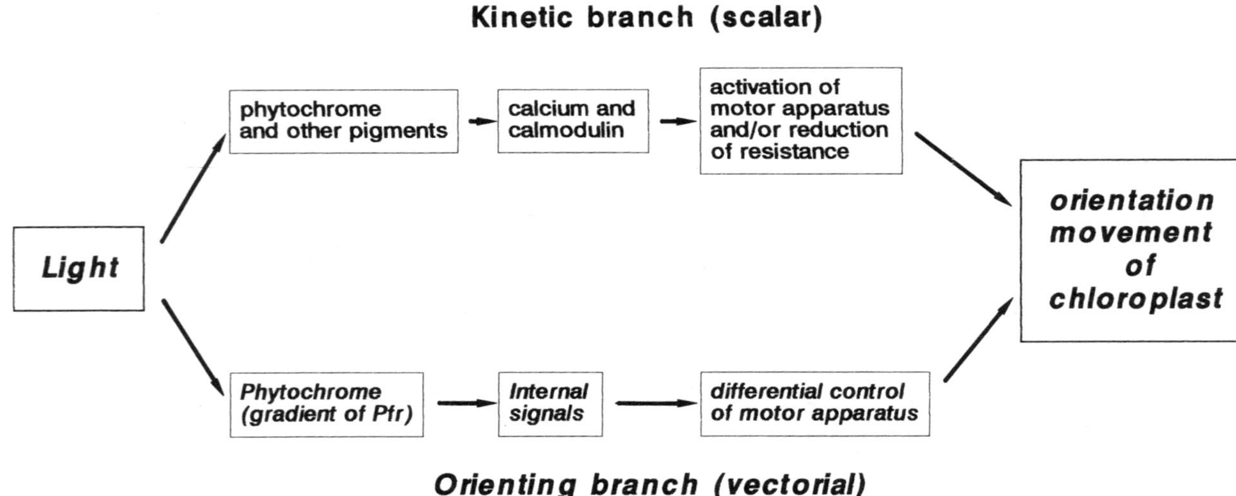

Fig. 3. The two-branched model of signal transduction in *Mougeotia*. Vectorial steps are indicated by *italics*. (After Wagner and Klein 1981; Grolig and Wagner 1988; Wagner et al. 1992)

can be modified via photosynthesis and photorespiration, depending on fluence rate and spectral range. This is thought to result in **differential energising of the motor apparatus** and/or **differential fastening of actin microfilaments** to the ectoplasm, and the chloroplasts orient in response to this **ATP gradient**. However, full consistency with parameters of photoreception has not yet been reached.

Another metabolic approach was proposed by Zurzycki (1972) in *Lemna*. **Glycine oxidase** can be activated by light absorption in its flavin component, and as a result the steady-state conditions in the **glycine-glyoxylate cycle** are changed. One of the products of this cycle might be a controlling factor.

In none of the models discussed above could it be demonstrated that the gradient of the substance in question does not level faster than it takes to convey its information to the next step of the transduction chain. This, however, is an essential requirement for a realistic model.

4 Concluding Remarks

This chapter has reported on substantial progress in the physiology of chloroplast movement. Nevertheless, important questions are still unsolved. The whole **sequence of events** is far from being understood, although models have evolved which are continuously improving. One most important result is the doubt as to a **simple and uniform signal-transduction chain**, which might have to be replaced by a **complex signal-transducing network.**

Complexity is already found in the first step, the **perception of the light signal. Two photoreceptor pigments** can operate independently of each other, even in a single species, indicating **two separate input sites.** Thus, "either-or" has to be replaced by "as-well-as". Moreover, in some cases, **coaction** or **interaction** of two photoreceptors or of their products is important. Next, sensing of **light direction** is not restricted to one principle only.

Similarly, the **mechanism of movement** may be rather complex, with **actin-myosin interaction, anchoring of microfilaments**, and **activity of microtubules** as basic factors, which makes it puzzling to find the crucial point of attack for controlling signals.

First insight into the **transduction processes** suggests the coexistence of at least two little understood sequences of events that in some way have to **converge** to **control the response**. It might be that still more complexity will be found.

Two more **general views** can be added:

- It is unlikely that in each system the **relative contribution of the various components** is identical. Thus, even if in one system a substantial part of the whole physiology has been elucidated, this must not be generalised uncritically to other systems.

- The complex perception-response system of chloroplast orientation may become **a model for other responses,** e.g. phototropism or gravitropism, suggesting that here, too, a linear, uniform signal-transduction chain with only one input site and one output mechanism might be a too simple approach. Considering higher levels of complexity may facilitate, then, the access to open problems.

A few particular **problems** in light-controlled chloroplast orientation are evident:

1. A series of questions concern **perception.** In spite of the successful genetic approach, the **cryptochrome problem** is still widely unsolved. What is the particular cryptochrome? Do all "cryptochrome plants" use the same pigment?

   As to **phytochrome** as photoreceptor pigment: is its contribution to chloroplast movement indeed an exception of very few species, although nearly all others also contain phytochrome? May such differences be due to different intracellular location of phytochrome? What is the basis of occassionally observed fast **loss of activity of the Pfr gradient,** although Pfr is considered as a long-living and long-acting signal and although its association with the cytoskeleton appears to exclude levelling of the gradient by diffusion? Might there be a steady-state exchange between differently compartmented phytochromes?

   What are the **primary reactions** initiated by the excited cryptochrome or by the active phytochrome Pfr? What is the **molecular basis of interaction** between the sequences started by cryptochrome and phytochrome in cases where such interaction is necessary or has specific effects?

   One fundamental question concerns the **opposite direction of response** in low- vs. high-intensity light, a similar problem as, e.g. in phototropism. A promising handle for a first approach might be the exceptional case where an **additional light signal** (independent of its direction) determines the **sense of response** with respect to the directional signal.

2. The **transduction** chain is still an almost "black" box – so black that good questions are not yet easy to ask. Assuming that the **two-branched model** in *Mougeotia* is a realistic approach, the question arises about transduction of the **directionality of the light signal.** Moreover, the possible **role of calcium** in this complex model is still open, and it can even be discussed whether calcium is only a **permissive condition** rather than a **messenger** as part of the transduction chain.

   Besides, the existence of **multiple input sites** (e.g. cryptochrome and phytochrome) calls for detecting the **point of fusion** in the transduction chain(s). On the other hand, multiple responses as output (e.g. chloroplast orientation, phototropism, photomorphogenesis) indicate **branching of the transduction chain**. These questions appear to be a promising "classical" field for future genetic approach.

3. For the **response**, the answer to the above-mentioned fundamental questions (p. 27) might depend not only on the **species** (e.g. *Vaucheria* vs. *Mougeotia*), but also on the **particular response** in question (e.g. low- vs. high-intensity response), and even the **input site** (see above) might be important.

4. Finally, interesting questions concern the **ecological views**, which have not been treated in this review. Intuitively, orientation movement of chloroplasts is thought to **optimise photosynthesis** by regulating the light-harvesting area in the cell. There is, however, only little experimental evidence (e.g. Zurzycki 1955). In contrast, Nultsch could unambiguously show, in marine green and brown algae, that photosynthesis is not influenced *directly* by the arrangements of chloroplasts or phaeoplasts, respectively. Instead, the high-intensity arrangement appears to **protect the photosynthesis apparatus** against damaging irradiation and thus to have a *long-term effect* on photosynthesis (Nultsch et al. 1981; Hanelt and Nultsch 1991). This raises two questions: can this result be generalised, even to **higher plants**? And what is the aim of the **low-intensity movement**, if photosynthesis proceeds as well in the high-intensity arrangement? This calls for including neighbouring fields of research.

Thus, many questions are still open and are a challenge for future investigations with both physiological and genetic approaches. Cooperation with other fields in photobiology (e.g. phototropism, photomorphogenesis) will certainly continue to be of mutual benefit.

References

Ahmed M, Cashmore AR (1993) HY4 gene of *A. thaliana* encodes a protein with characteristics of a blue-light photoreceptor. Nature 336:162–166

Al-Rawass B, Grolig F, Wagner G (1997) High irradiance blue light affects cortical microtubules in the green Alga *Mougeotia scalaris*. Plant Cell Physiol 38:882–886

Björn LO (1984) Light-induced linear dichroism in photoreversibly photochromic sensor pigments. V Reinterpretation of the experiments on in vivo action dichroism of phytochrome. Physiol Plant 60:369–372

Blatt MR, Briggs WR (1980) Blue-light-induced cortical fiber reticulation concomitant with chloroplast aggregation in the Alga *Vaucheria sessilis*. Planta 147:355–362

Blatt MR, Wessels NK, Briggs WR (1980) Actin and cortical fiber reticulation in the siphonaceous alga *Vaucheria sessilis*. Planta 147:363–375

Blatt MR, Weisenseel MH, Haupt W (1981) A light-dependent current associated with chloroplast aggregation in the Alga *Vaucheria sessilis*. Planta 152:513–526

Bock G, Haupt W (1961) Die Chloroplastendrehung bei *Mougeotia*. III. Die Frage der Lokalisierung des Hellrot-Dunkelrot-Pigmentsystems in der Zelle. Planta 57:518–530

Böhm JA (1856) Beiträge zur näheren Kenntnis des Chlorophylls. Sitzungsber Akad Wiss Wien Math Nat Kl 22:479

Boresch K (1914) Über fadenförmige Gebilde in den Zellen von Moosblättern und Chloroplastenverlagerung bei *Funaria*. Z Bot 6:97–156

Briggs WR, Liscum E (1997) The role of mutants in the search for the photoreceptor for phototropism in higher plants. Plant Cell Environ 20:768–772

Brunner K, Wagner G (1996) Cryptochrome of the green alga *Mougeotia scalaris*: a member of the HY4/photolyase-family. In: Senger H (ed) The blue light syndrome III. Phillips-Universität, Marburg, Book of Abstracts, p 43

Buder J (1918) Die Inversion des Phototropismus bei *Phycomyces*. Ber Dtsch Bot Ges 36:104–105

Cashmore AR (1997) The cryptochrome family of photoreceptors. Plant Cell Environ 20:764–767

Dreyer EM, Weisenseel MH (1979) Phytochrome-mediated uptake of calcium in *Mougeotia* cells. Planta 146:31–39

Etzold H (1965) Der Polarotropismus und Phototropismus der Chloronemen von *Dryopteris filix-mas*. (L.) Schott. Planta 64:254–280

Fischer-Arnold G (1963) Untersuchungen über die Chloroplastenbewegung bei *Vaucheria sessilis*. Protoplasma 56:495–520

Foos K (1970) Mikrotubuli bei *Mougeotia* spec. Z Pflanzenphysiol 62:201–203

Foos K (1971) Untersuchungen zur Feinstruktur von *Mougeotia* spec. und zum Bewegungsmechanismus des Chloroplasten. Z Pflanzenphysiol 64:369–386

Frank B (1871) Über lichtwärts sich bewegende Chlorphyllkörner. Bot Z 29:209–225

Gabryś H, Walczak T, Haupt W (1984) Blue-light-induced chloroplast orientation in *Mougeotia*. Evidence for a separate sensor pigment besides phytochrome. Planta 160:21–24

Gabryś H, Walczak T, Haupt W (1985) Interaction between phytochrome and the blue light photoreceptor system in *Mougeotia*. Photochem Photobiol 42:731–734

Gabryś-Mizerza H (1976) Model considerations of the light conditions in noncylindrical plant cells. Photochem Photobiol 24:453–461

Gärtner R (1970) Die Bewegung des *Mesotaenium*-Chloroplasten im Starklichtbereich. II. Aktionsdichroismus und Wechselwirkungen des Photoreceptors mit Phytochrom. Z Pflanzenphysiol 63:428–443

Grolig F, Wagner G (1988) Light-dependent chloroplast reorientations in *Mougeotia* and *Mesotaenium*: biased by pigment-related plasmalemma anchorage sites to actin filaments? Bot Acta 101:2–6

Hanelt D, Nultsch W (1991) The role of chromatophore arrangement in protecting the chromatophores of the brown alga *Dictyota dichotoma* against photodamage. J Plant Physiol 138:470–475

Hartmann KM, Cohnen Unser I (1973) Carotenoids and flavins versus phytochrome as the controlling pigment for blue-UV-mediated photoresponses. Z Pflanzenphysiol 69:109–124

Haupt W (1959a) Chloroplastenbewegung. In: Bünning E (ed) Handbuch der Pflanzenphysiologie, vol. 17/1. Bewegungen. Springer, Berlin Göttingen Heidelberg, pp 278–317

Haupt W (1959b) Die Chloroplastendrehung bei *Mougeotia* I. Über den quantitativen und qualitativen Lichtbedarf der Schwachlichtbewegung. Planta 53:484–501

Haupt W (1960) Die Chloroplastendrehung bei *Mougeotia*. II. Die Induktion der Schwachlichtbewegung durch linear polarisiertes Licht. Planta 55: 465–479

Haupt W (1965) Perception of environmental stimuli orienting growth and movement in lower plants. Annu Rev Plant Physiol 16:267–290

Haupt W (1968) Die Orientierung der Phytochrommoleküle in der *Mougeotia*zelle: Ein neues Modell zur Deutung der experimentellen Befunde. Z Pflanzenphysiol 58:331–346

Haupt W (1970a) Über den Dichroismus von Phytochrom$_{660}$ und Phytochrom$_{730}$ bei *Mougeotia*. Z Pflanzenphysiol 62:287–298

Haupt W (1970b) Localization of phytochrome in the cell. Physiol Vég 8:551–563

Haupt W (1971) Schwachlichtbewegung des *Mougeotia*-Chloroplasten im Blaulicht. Z Pflanzenphysiol 65:248–265

Haupt W (1980) Sensory transduction and photobehaviors: final considerations and emerging themes. (Panel Discussion). In: Lenci F, Colombetti G (eds) Photoperception and sensory transduction in aneural organisms. Plenum Press, New York, pp 397–404

Haupt W, Bock G (1962) Die Chloroplastendrehung bei *Mougeotia*. IV. Die Orientierung der Phytochrom-Moleküle im Cytoplasma. Planta 59:38–48

Haupt W, Heymann N (1967) Versuche zur Mechanik der Chloroplastenbewegung von *Mougeotia*. Z Pflanzenphysiol 57:68–71

Haupt W, Reif G (1979) "Ageing" of Phytochrome P$_{fr}$ in *Mesotaenium*. Z Pflanzenphysiol 92:153–161

Haupt W, Scheuerlein R (1990) Chloroplast movement. Plant Cell Environ 13:595–614

Haupt W, Thiele R (1961) Chloroplastenbewegung bei *Mesotaenium*. Planta 56:388–401

Haupt W, Weisenseel M (1976) Physiological evidence and some thoughts on localised responses, intracellular localisation and action of phytochrome. In: Smith H (ed) Light and plant development. Butterworth, London, pp 63–74

Haupt W, Wirth H (1967) Nachweis einer Schraubenstruktur in der *Mougeotia*-Zelle. Plant Cell Physiol 8:541–543

Hayami J, Kadota A, Wada M (1992) Intracellular dichroic orientation of the blue light-absorbing pigment and the blue-absorbing band of red-absorbing form of phytochrome responsible for phototropism of the fern *Adiantum* protonemata. Photochem Photobiol 56:661–666

Hendricks SB, Borthwick HA (1967) The function of phytochrome in regulation of plant growth. Proc Natl Acad Sci USA 58:2125–2130

Herrmann H, Kraml M (1997) Time-dependent formation of Pfr-mediated signals for the interaction with blue light in *Mesotaenium* chloroplast orientation. J Photochem Photobiol 37:60–65

Inoue Y, Shibata K (1973) Light-induced chloroplast rearrangements and their action spectra as measured by absorption spectrophotometry. Planta 114:342–358

Izutani Y, Takagi S, Nagai R (1990) Orientation movements of chloroplasts in *Vallisneria* epidermal cells: different effects of light at low and high fluence rate. Photochem Photobiol 51:105–111

Jacobshagen S, Altmüller D, Grolig F, Wagner G (1986) Calcium pools, calmodulin and light-regulated chloroplast movements in *Mougeotia* and *Mesotaenium*. In: Trewavas AJ (ed) Molecular and cellular aspects of calcium in development. Plenum Press, New York, pp 201–209

Jaffe L (1958) Tropistic response of zygotes of the Fucaceae to polarized light. Exp Cell Res 15:282–299

Jarosch R (1956) Plasmaströmung und Chloroplastenrotation bei Characeen. Phyton (Argent) 6:87–107

Jarosch R (1958) Die Protoplasmafibrillen der Characeen. Protoplasma 50:93–108

Kadota A, Wada M (1989) Photoinduction of circular F-actin in a fern protonemal cell. Protoplasma 151:171–174

Kadota A, Wada M (1992a) Photo-orientation of chloroplasts in the fern *Adiantum* protonemal cell as analyzed by use of a video-tracking system. Bot Mag Tokyo 105:265–279

Kadota A, Wada M (1992b) Photoinduction of formation of circular structures by microfilaments on chloroplasts during intracellular orientation in protonemal cells of the fern *Adiantum capillus-veneris*. Protoplasma 167:97–107

Kadota A, Kohyama I, Wada M (1989) Polarotropism and photomovement of chloroplasts in the protonemata of the ferns *Pteris* and *Adiantum*: evidence of the possible lack of dichroic phytochrome in *Pteris*. Plant Cell Physiol 30:523–531

Kamiya N (1959) Protoplasmic streaming. Protoplasmatologia, vol VIII3a. Springer, Berlin Göttingen Heidelberg

Kidd DG, Lagarias JC (1990) Phytochrome from the green alga *Mesotaenium caldariorum*. J Biol Chem 265:7029–7035

Klein K, Wagner G, Blatt MR (1980) Heavy-meromyosin decoration of microfilaments from *Mougeotia* protoplasts. Planta 150:354–356

Knoll F (1908) Über netzartige Protoplasmadifferenzierungen und Chloroplastenbewegungen. Sitzungsber Akad Wien 117:1224–1241

Kraml M (1994) Light direction and polarization. In: Kendrick R, Kronenberg GHM (eds) Photomorphogenesis in plants, 2nd edn. Kluwer, Dordrecht, pp 417–445

Kraml M, Schäfer E (1983) Photoconversion of phytochrome in vivo studied by double-flash irradiation in *Mougeotia* and *Avena*. Photochem Photobiol 38:461–468

Kraml M, Leopold K, Winkler B (1987) Long-lasting activity of P_{fr} and P_{fr}-gradients in *Mougeotia*-chloroplast movement? Acta Physiol Plant 9:189–198

Kraml M, Büttner G, Haupt W, Herrmann H (1988) Chloroplast orientation in *Mesotaenium*: the phytochrome effect is strongly potentiated by interaction with blue light. Protoplasma Suppl 1:172–179

Kurtin WE, Song PS (1968) Photochemistry of the model phototropic system involving flavins and indoles. I. Fluorescence polarization and MO calculation of the direction of the electronic transition moments in flavins. Photochem Photobiol 7:263–273

Lagarias DM, Wu SH, Lagarias JC (1995) Atypical phytochrome gene structure in the green alga *Mesotaenium caldariorum*. Plant Mol Biol 29:1127–1142

Lechowski Z (1972) Action spectrum of chloroplast displacements in the leaves of land plants. Acta Protozool 11:202–209

Lewis IJ (1898) The action of light on *Mesocarpus*. Ann Bot 12:418–421

Maekawa T, Tsutsui I, Nagai R (1986) Light-regulated translocation of cytoplasm in green alga *Dichotomosiphon*. Plant Cell Physiol 27:837–851

Marchant HJ (1976) Actin in the green algae *Coleochaete* and *Mougeotia*. Planta 131:119–120

Maucher PH, Scheuerlein R, Schraudolf H (1992) Detection and partial sequence of phytochrome genes in the ferns *Anemia phyllitidis* (L.) Sw (Schizaeaceae) and *Dryopteris filix-mas* L. (Polypodiaceae) by using polymerase-chain reaction technology. Photochem Photobiol 56:759–763

Mayer F (1964) Lichtorientierte Chloroplasten-Verlagerung bei *Selaginella martensii*. Z Bot 52:346–381

Mayer F (1966) Lichtinduzierte Chloroplasten-Verlagerung bei *Selaginella martensii*. Untersuchungen zur Identifizierung des Photoreceptors durch Anwendung von Quencher-Substanzen. Z Pflanzenphysiol 55:65–70

Mineyuki Y, Kataoka H, Masuda Y, Nagai R (1995) Dynamic changes in the actin cytoskeleton during the high-fluence rate response of the *Mougeotia* chloroplast. Protoplasma 185:222–229

Moore SM (1888) The influence of light upon protoplasmic movement. J Linn Soc Bot 24:200–251, 351–389

Nagai R (1993) Regulation of intracellular movements in plant cells by environmental stimuli. Int Rev Cytol 145:251–310

Nakasako M, Wada M, Tokutomi S, Yamamoto KT, Sakai J, Kataoka M (1990) Quaternary structure of pea phytochrome I dimer studied with small-angle X-ray scattering and rotary shadowing electron microscopy. Photochem Photobiol 52:3–12

Nultsch W, Pfau J, Rüffer U (1981) Do correlations exist between chromatophore arrangement and photosynthetic activity in seaweeds? Mar Biol 62:111–117

Pfau J, Throm G, Nultsch W (1974) Recording microphotometer for determination of light induced chromatophore movements in brown algae. Z Pflanzenphysiol 71:242–260

Roberts D (1989) Detection of a calcium-activated protein kinase in *Mougeotia* by using synthetic peptide substrates. Plant Physiol 91:1613–1619

Roux SJ (1984) Ca^{2+} and phytochrome action in plants. Bioscience 34:25–29

Rüffer U, Pfau J, Nultsch W (1981) Movements and arrangements of *Dictyota* phaeoplasts in response to light and darkness. Z Pflanzenphysiol 101:283–293

Russ U, Grolig F, Wagner G (1991) Changes of cytoplasmic free calcium in the green alga, *Mougeotia scalaris*: monitored with indo-1, and effect on the velocity of the chloroplast movement. Planta 184:105–112

Scheuerlein R, Braslavsky SE (1987) Induction of chloroplast movement in the alga *Mougeotia* by polarized nanosecond dye-laser pulses. Photochem Photobiol 46:525–530

Scholz A (1976a) Lichtorientierte Chloroplastenbewegung bei *Hormidium flaccidum*: Perception der Lichtrichtung mittels Sammellinseneffekt. Z Pflanzenphysiol 77:406–421

Scholz A (1976b) Lichtorientierte Chloroplastenbewegung bei *Hormidium flaccidum*: Verschiedene Methoden der Lichtrichtungsperception und die wirksamen Pigmente. Z Pflanzenphysiol 77:422–436

Schönbohm E (1963) Untersuchungen über die Starklichtbewegung des *Mougeotia*-Chloroplasten. Z. Bot 51:233–276

Schönbohm E (1965) Die Beeinflussung der negativen Phototaxis des *Mougeotia*-Chloroplasten durch linear polarisierte langwellige Strahlung. Z Pflanzenphysiol 53:344–355

Schönbohm E (1966) Der Einfluß von Rotlicht auf die negative Phototaxis des *Mougeotia*-Chloroplasten: Die Bedeutung eines Gradienten von P_{730} für die Orientierung. Z Pflanzenphysiol 55:278–286

Schönbohm E (1967) Die Hemmung der positiven und negativen Phototaxis des *Mougeotia*-Chloroplasten durch Jodid-Ionen. Z Pflanzenphysiol 56:366–374

Schönbohm E (1968) Aktionsdichroismus bei der Starklichtbewegung des Chloroplasten von *Mougeotia* spec. Ber Dtsch Bot Ges 81:203–209

Schönbohm E (1969) Die Hemmung der lichtinduzierten Bewegung des *Mougeotia*-Chloroplasten durch p-Chlormercuribenzoat (Versuche zur Mechanik der Chloroplastenbewegung). Z Pflanzenphysiol 61:250–260

Schönbohm E (1970) Sekundärreaktionen bei der Chloroplastenbewegung. Ber Dtsch Bot Ges 83:629–632

Schönbohm E (1971) Untersuchungen zum Photoreceptorproblem beim tonischen Blaulicht-Effekt der Starklichtbewegung des *Mougeotia*-Chloroplasten. Z Pflanzenphysiol 66:20–33

Schönbohm E (1972) Experiments on the mechanism of chloroplast movement in light-oriented chloroplast arrangement. Acta Protozool 11:211–236

Schönbohm E (1973a) Kontraktile Fibrillen als aktive Elemente bei der Mechanik der Chloroplastenverlagerung. Ber Dtsch Bot Ges 86:407–422

Schönbohm E (1973b) Die lichtinduzierte Verankerung der Plastiden im cytoplasmatischen Wandbelag: Eine phytochromgesteuerte Kurzzeitreaktion. Ber Dtsch Bot Ges 86:423–430

Schönbohm E (1974) Untersuchungen zur Mechanik der lichtorientierten Chloroplastenbewegung unter besonderer Berücksichtigung der Plasmastrukturen. Leitz Mitt Wiss Techn 6:98–109

Schönbohm E (1980) Phytochrome and non-phytochrome dependent blue light effects on intracellular movements in freshwater algae. In: Senger H (ed) The blue light syndrome. Springer, Berlin Heidelberg New York, pp 69–96

Schönbohm E (1987) Movement of *Mougeotia* chloroplasts under continuous weak and strong light. Acta Physiol Plant 9:109–135

Schönbohm E, Meyer-Wegener J (1989) Actin polymerization as an essential process in light- and dark-controlled chloroplast anchorage. Biochem Physiol Pflanz 185:337–342

Schönbohm E, Schönbohm E (1984) Biophenole: Steuernde Faktoren bei der lichtorientierten Chloroplastenbewegung? Biochem Physiol Pflanz 179:489–505

Schönbohm E, Schönbohm E, Meyer-Wegener J (1990) On the signal-transduction chains of two Pfr-mediated short-term processes: increase of anchorage and movement of *Mougeotia* chloroplasts. Photochem Photobiol 52:203–209

Schweickerdt H (1928) Untersuchungen über Photodinese bei *Vallisneria spiralis*. Jahrb Wiss Bot 68:79–134

Seitz K (1967a) Wirkungsspektren für die Starklichtbewegung der Chloroplasten, die Photodinese und die lichtabhängige Viskositätsänderung bei *Vallisneria spiralis* ssp. *torta*. Z Pflanzenphysiol 56:246–261

Seitz K (1967b) Eine Analyse der für die lichtabhängigen Bewegungen der Chloroplasten verantwortlichen Photorezeptorsysteme bei *Vallisneria spiralis* ssp. *torta*. Z Pflanzenphysiol 57:96–104

Seitz K (1970) Die Starklichtbewegung der Chloroplasten von *Vallisneria* in Abhängigkeit von Hemmstoffen der oxydativen Phosphorylierung. Z Pflanzenphysiol 63:401–407

Seitz K (1972) Primary processes controlling the light-induced movement of chloroplasts. Acta Protozool 11:226–235

Seitz K (1979) Cytoplasmic streaming and cyclosis of chloroplasts. In: Haupt W, Feinleib ME (eds) Encyclopedia of plant physiology, new series vol 7. Physiology of movements. Springer, Berlin Heidelberg New York, pp 150–169

Senn G (1908) Die Gestalts- und Lageveränderung der Pflanzen-Chromatophoren. Engelmann, Leipzig

Senn G (1919) Weitere Untersuchungen über Gestalts- und Lageveränderung der Chromatophoren. IV und V. Z Bot 11:81–141

Serlin BS, Ferrell S (1989) The involvement of microtubules in chloroplast rotation in the alga *Mougeotia*. Plant Sci 60:1–8

Serlin BS, Roux SJ (1984) Modulation of chloroplast movement in the green alga *Mougeotia* by the Ca^{2+} ionophere, A23187, and by calmodulin antagonists. Proc Natl Acad Sci USA 81:6368–6372

Stahl E (1880) Über den Einfluß von Richtung und Stärke der Beleuchtung auf einige Bewegungserscheinungen im Pflanzenreich. Bot Z 38:297, 321, 345, 361, 377, 393, 409

Strugger S (1956) Elektronenmikroskopische Beobachtungen an den Chloroplasten von *Chlorophytum comosum*. Ber Dtsch Bot Ges 69:177–178

Sundqvist C, Björn LO (1983) Light-induced linear dichroism in photoreversibly photochromic sensor pigments. II. Chromophore rotation in immobilized phytochrome. Photochem Photobiol 37:69–75

Tłaka M, Gabryś H (1993) Influence of calcium on blue-light-induced chloroplast movement in *Lemna trisulca* L. Planta 189:491–498

Voerkel SH (1934) Untersuchungen über die Phototaxis der Chloroplasten. Planta 21:156–205

Wada M, Kadota A, Furuya M (1983) Intracellular localization and dichroic orientation of phytochrome in plasma membrane and/or ectoplasm of a centrifuged protonema of fern *Adiantum*. Plant Cell Physiol 24:1441–1447

Wada M, Grolig F, Haupt W (1993) Light-oriented chloroplast positioning. Contribution to progress in photobiology. J Photochem Photobiol. B 17:3–25

Wada M, Kanegae T, Nozue K, Fukada S (1996) New prospects of Photobiology. Conference Proceedings, Okazaki, NIBB (Japan)

Wada M, Kanegae T, Nozue K, Fukada S (1997) Cryptogam phytochromes. Plant Cell Environ 20:685–690

Wagner G (1995) Intracellular movement. Prog Bot 57:68–80

Wagner G, Klein K (1978) Differential effect of calcium on chloroplast movement in *Mougeotia*. Photochem Photobiol 27:137–140

Wagner G, Klein K (1981) Mechanism of chloroplast movement in *Mougeotia*. Protoplasma 109:169–185

Wagner G, Haupt W, Laux A (1972) Reversible inhibition of chloroplast movement by cytochalasin B in the green alga *Mougeotia*. Science 176:808–809

Wagner G, Klein K, Rossbacher R (1978) Strukturelle und physiologische Grundlagen der lichtorientierten Chloroplastenbewegung bei der Grünalge *Mougeotia* sp. Cytobiologie 18:198

Wagner G, Valentin P, Dieter P, Marmé D (1984) Identification of calmodulin in the green alga *Mougeotia* and its possible function in chloroplast reorientational movement. Planta 162:62–67

Wagner G, Russ U, Quader H (1992) Calcium, a regulator of cytoskeletal activity and cellular competence. In: Menzel D (ed) The cytoskeleton of the Algae. CRC Press, Boca Raton, pp 411–424

Walczak T, Gabryś H, Haupt W (1984) Flavin-mediated weak-light chloroplast movement in *Mougeotia*. In: Senger H (ed) Blue light effects in biological systems. Springer, Berlin Heidelberg New York, pp 454–459

Weisenseel M (1968) Vergleichende Untersuchungen zum Einfluß der Temperatur auf lichtinduzierte Chloroplastenverlagerungen. I. Die Wirkung verschiedener Lichtintensitäten auf die Chloroplastenanordnung und ihre Abhängigkeit von der Temperatur. Z Pflanzenphysiol 59:56–69

Whitelam GC, Devlin PF (1997) Roles of different phytochromes in *Arabidopsis* photomorphogenesis. Plant Cell Environ 20:752–758

Winands A, Wagner G (1996) Phytochrome of the green alga *Mougeotia*: cDNA sequence, autoregulation and phylogenetic position. Plant Mol Biol 32:589–597

Winands A, Wagner G, Marx S, Schneider-Poetsch HAW (1992) Partial nucleotide sequence of phytochrome from the zygnematophycean green alga *Mougeotia*. Photochem Photobiol 56:765–770

Wu SH, Lagarias JC (1997) The phytochrome photoreceptor in the green alga *Mesotaenium caldariorum*: implication for a conserved mechanism of phytochrome action. Plant Cell Environ 20:691–699

Yatsuhashi H (1996) Photoregulation systems for light-oriented chloroplast movement. J Plant Res 109:139–146

Yatsuhashi H, Hashimoto T, Wada M (1987a) Dichroic orientation of photoreceptors for chloroplast movement in *Adiantum* protonemata. Non-helical orientation. Plant Sci 51:165–170

Yatsuhashi H, Wada M, Hashimoto T (1987b) Dichroic orientation of phytochrome and blue-light photoreceptor in *Adiantum* protonemata as determined by chloroplast movement. Acta Physiol Plant 9:163–173

Yatsuhashi H, Kadota A, Wada M (1985) Blue- and red-light action in photo-orientation of chloroplasts in *Adiantum* protonemata. Planta 165:43–50

Yatsuhashi H, Kobayashi H (1993) Dual involvement of phytochrome in light-oriented chloroplast movement in *Dryopteris sparsa* protonemata. J Photochem Photobiol B 19:25–31

Zurzycka A, Zurzycki J (1957) Cinematographic studies on phototactic movements of chloroplasts. Acta Soc Bot Pol 26:177–206

Zurzycki J (1955) Chloroplast arrangement as a factor in photosynthesis. Acta Soc Bot Pol 24:27–63

Zurzycki J (1960) Studies on the centrifugation of chloroplasts in *Lemna trisulca*. Acta Soc Bot Pol 29:385–393

Zurzycki J (1962a) The mechanism of the movements of plastids: In Bünning E (ed) Handbuch der Pflanzenphysiologie, vol 17/2. Bewegungen. Springer, Berlin Göttingen Heidelberg, pp 940–978

Zurzycki J (1962b) The action spectrum for the light-dependent movements of chloroplasts in *Lemna trisulca*. Acta Soc Bot Pol 31:489–538

Zurzycki J (1967a) Properties and localization of the photoreceptor active in displacements of chloroplasts in *Funaria hygrometrica*. I. Action spectrum. Acta Soc Bot Pol 36:133–142

Zurzycki J (1967b) Properties and localization of the photoreceptor active in displacements of chloroplasts in *Funaria hygrometrica*. II. Studies with polarized light. Acta Soc Bot Pol 36:143–152

Zurzycki J (1972) Primary reactions in the chloroplast rearrangements. Acta Protozool 11:189–200

Zurzycki J, Lełatko Z (1969) Action dichroism in the chloroplasts rearrangements in various plant species. Acta Soc Bot Pol 38:493–506

Zurzycki J, Walczak T, Gabryś H, Kajfosz J (1983) Chloroplast translocations in *Lemna trisulca* L. induced by continuous irradiation and by light pulses. Kinetic analysis. Planta 157:502–510

Prof. Dr. Wolfgang Haupt
Erlenstraße 28
D-91341 Röttenbach, Germany

Edited by
K. Esser

Genetics

Recombination:
Organelle DNA of Plants and Fungi: Inheritance and Recombination

By Heike Röhr, Ursula Kües, and Ulf Stahl

1 Introduction

Most of the genetic information in eukaryotic cells is stored by chromosomal DNA in the nucleus. Yet a much smaller but nevertheless essential part of the genetic information is kept by some extrachromosomal DNA molecules situated in organelles present in the cytoplasm of the cell. Two main groups of DNA containing organelles are distinguished: *Mitochondria* are an almost universal feature of the eukaryotic cells while *chloroplasts* are found in algae and plant cells capable of photosynthesis but not in the nonphototropic fungal and animal cells. Organelle DNA is able to undergo intra- and intergenomic recombination. Here we review the present knowledge on inheritance and recombination of organelle DNA in plants and fungi as well as transmission and recombination of organelle-associated plasmids.

2 Organelle Inheritance

Mendel's laws of inheritance apply to nuclear genes. They describe the segregation of pairs of alleles and the independent assortment of unlinked genes during gametogenesis. Moreover, Mendelian genetics implies two fundamental principles: first, segregation of alleles never takes places during mitotic cell divisions, and second, genes are generally inherited biparentally. Strikingly, inheritance of mitochondrial and chloroplast genes does not obey these principles. Organelle inheritance is *non-Mendelian*, also referred to as *extranuclear* or *cytoplasmic* inheritance. The non-Mendelian character of organelle inheritance is manifested in (1) *vegetative (mitotic) segregation*, (2) *uniparental inheritance*, and (3) *intracellular selection*. Vegetative segregation, i.e., separation of alleles of organelle genes during mitosis as well as meiosis, has been observed in all organisms investigated. It is thought to be a consequence of random replication and partitioning of organelle DNA and organelles during the cell cycle. Uniparental inheritance describes the for many organisms typical observation that sexual progeny inherits organelle

Progress in Botany, Vol. 60
© Springer-Verlag Berlin Heidelberg 1999

genes solely or predominantly from one parent. The third element, intracellular selection, has classically not been considered as a principle of non-Mendelian genetics. However, increasing experimental evidence is available for intracellular selection in organelle inheritance. Depending on their genotypes, certain organelle chromosomes may replicate more or degrade less often than others, giving rise to inequalities inside a cell. Consequences of such an intracellular selection are the occasional loss of alleles and establishment of mutant alleles (further discussion in Birky 1994, 1995).

a) Mode of Inheritance

Cytoplasmic inheritance can principally follow three different modes, each referring to the origin of the organelles found within the sexual offspring: biparental, uniparental paternal, and uniparental maternal inheritance (Fig. 1). Current textbooks consider uniparental maternal inheritance as the rule of organelle inheritance. Biparental inheritance is

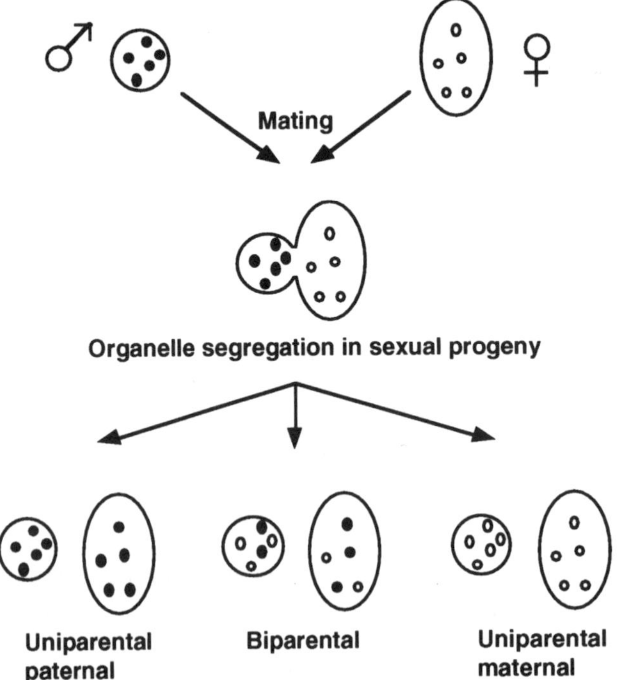

Fig. 1. Modes of organelle inheritance. An example of anisogamy which is characterized by paternal and maternal mates distinguished in their cellular size (*small circle* and *large oval*). Anisogamy in the illustrated example is not perfect, excluding presence of organelles in the smaller paternal gamete. The origins of organelles are indicated by *filled* (paternal) and *open* spots (maternal)

considered as the exception and no indication of paternal inheritance is given (Alberts et al. 1994). This view comes mainly from the occurrence of maternal mitochondrial inheritance in animals, and from the historical order in which taxa and types of organelles were studied. However, the vast, growing list regarding studies of organelle inheritance in fungi and plants brings about a novel, more subtle impression (Table 1): uniparental paternal inheritance is not as uncommon as might be thought and can even predominate in certain groups of organisms and types of organelles. It has been observed in chloroplast inheritance in most gymnosperms and in some angiosperms, but also in mitochondrial inheritance in fungi, algae, and angiosperms. Uniparental maternal inheritance appears to be the primary mode of mitochondrial and chloroplast transmission in angiosperms, but biparental inheritance has also been described a number of times. In sharp contrast to chloroplast inheritance, mitochondrial transmission in gymnosperms follows in most cases the pattern of uniparental maternal inheritance. Mitochondria in fungi are transmitted primarily uniparentally with a preference for maternal inheritance in cases of anisogamy. Organelles from algae have a mixed preference for maternal or paternal uniparental inheritance.

α) Biparental Inheritance

Biparental inheritance of organelles is a mixed (Fig. 1) but in many cases not a fixed inheritance. Biparental zygotes that transfer both maternal and parental organelles to daughter cells will be observed next to a varying distribution of maternal and paternal zygotes that transfer only one type of organelles. Sometimes, biparental inheritance is biased more or less toward one or the other parent (Tilney-Bassett 1994). The most prominent organisms for extended biparental inheritance among the fungi are the isogamous yeasts *Sacchharomyces cerevisiae* and *Schizosaccharomyces pombe* (Table 1). Of these, *S. cerevisiae* has been most thouroughly studied (Birky 1994, 1995). Genetic analysis indicates that mitochondrial transmission is not random and depends in part on the zygote bud position. Mixing of mtDNA in the zygote is not extensive and the zone of mixing is limited mainly to that portion of the cell which gives rise to medial buds. These buds normally receive heteroplasmic mtDNA populations (Zinn et al. 1987; Fig. 2). In contrast, end-buds contain mtDNA predominantly from the parent which formed that half of the zygote from which the bud arose (Azpiroz and Butow 1993; Fig. 2). Recent cytological observations provide a remarkable explanation for these patterns: parental mitochondria fuse during mating and form one continuous dynamic network in which mitochondrial proteins rapidly redistribute and colocalize throughout the whole zygote. In marked contrast, mtDNA remains in position, suggesting the presence of anchoring

Table 1. Patterns of organelle inheritance in selected fungi and plants

Species	Inheritance of mtDNA[a]	Inheritance of cpDNA[a]	References
Protista (slime molds)		Not applicable	
Didymium iridis	B		Silliker and Collins (1988)
Physarum polycephalum	U		Kawano et al. (1995)
Polysphondylium pallidium	U		Mirfakhrai et al. (1990)
Oomycota		Not applicable	
Phytophthora infestans (potato blight fungus)	U		Whittaker et al. (1994)
Phytophtora parasitica	U		Förster and Coffey (1990)
Pythium sylvaticum	M		Martin (1989)
Chytridiomycota		Not applicable	
Allomyces macrogynus x *A. arbusculus*	P		Borkhardt and Olsen (1983)
Ascomycota		Not applicable	
Aspergillus nidulans	M		Coenen et al. (1996)
Atkinsonella hypoxylon (grass pathogen)	M		van Horn and Clay (1995)
Cryphonectria parasitica (chestnut blight fungus)	M		Milgroom and Lipari (1993)
Epichloë typhina (grass endophyte)	M		Chung et al. (1996)
Neurospora crassa (red bread mold)	M		Mannella et al. (1979)
Neurospora tetrasperma	M		Lee and Taylor (1993)
Ophiostoma ulmi (Dutch elm disease)	M		Brasier and Kirk (1986)

Podospora anserina (dung fungus)	M		Belcour and Begel (1977)
Saccharomyces cerevisiae (bakers' yeast)	B		Thomas and Wilkie (1968)
Schizosaccharomyces pombe (fission yeast)	B		Thrailkill et al. (1980)
Basidiomycota		Not applicable	
Agaricus bisporus (button mushroom)	U		Jin and Horgen (1994)
Agaricus bitorquis (field mushroom)	U		Hintz et al. (1988)
Armillaria bulbosa (root-rot fungus)	U		Smith et al. (1990)
Coprinus cinereus (inky cap fungus)	U		May and Taylor (1988)
Lentinula edodes (Shiitake)	U		Fukuda et al. (1995)
Microbotryum violaceum (anther smut fungus)	U		Wilch et al. (1992)
Pleurotus ostreatus (oyster fungus)	U		Matsumoto and Fukumasa-Nakai (1996)
Schizophyllum commune (split gill)	U		Specht et al. (1992)
Stereum hirsuitum x *S. complicatum* (false turkey tail)	U		Ainsworth et al. (1992)
Ustilago maydis (corn smut fungus)	U		M. Bölker (pers. comm.)
Mitosporic fungi		Not applicable	
Torulopsis glabrata (asporogenous yeast)	U		Sriprakash and Batum (1981)
Chlorophyta (green algae)			
Chlamydomonas eugametos x *C. moewusii*	U, (mt$^+$, "M")	U, (mt$^+$, "M")	Lee et al. (1990)
Chlamydomonas reinhardtii, C. reinhardtii x *C. smithii*	U, (mt$^+$, "P")	U, (mt$^+$, "M")	Boynton et al. (1987); Beckers et al. (1991)
Monostroma latissimum		U ("M")	Kuroiwa et al. (1993c)
Ulva mutabilis		U (mt$^-$)	Bråten (1973)

Table 1 (continued)

Species	Inheritance of mtDNA[a]	Inheritance of cpDNA[a]	References
Volvox carteri	M	M	Adams et al. (1990); Kuroiwa et al. (1993b)
Phaeophyceae (brown algae)			
Dictyota sp.		M	Manton (1970)
Rhodophyta (red algae)			
Corallina sp.		M	Peel and Duckett (1975)
Gracilaria sp.		M	van der Meer (1978)
Muscopsida (Musci, mosses)			
Spagnum sp. (peat moss)		M	Manton (1970)
Sphaerocarpus donnelli (liverwort)	M	M	Diers (1967)
Pterophyta (ferns)			
Pteridium aquilinium (eagle fern)		M	Bell and Duckett (1976)
Pellaea hybrids (chelanthoid ferns)	M	M	Gastony and Yatskievych (1992)
Gymnosperms			
Calocedrus decurrens (incense cedar)	P	P	Neale et al. (1991)
Cryptomeria japonica (Japanese sugi)		P	Ohba et al. (1971)
Latrix decidua x *L. leptolepis* (European x Japanese larch)	M	P, rarely M	Szmidt et al. (1987); De Verno et al. (1993)
Picea hybrids (spruce)	M	P	Sutton et al. (1991); David and Keathley (1996)

Species			Reference
Pinus banksiana, *P. banksiana* x *P. contorta* (jack pine x lodgepole pine)	M	P	Dong et al. (1992); Dong and Wagner (1993)
Pinus monticola (western white pine)		P, B	White (1990)
Pinus taeda (loblolly pine)	M	P	Neale and Sedoroff (1989)
Pseudotsuga menziesii (Douglas-fir)	M	P	Neale et al. (1986); Aagaard et al. (1995)
Sequoia sempervirens (coast redwood)	P	P	Neale et al. (1989)
Angiosperms – dicotylodons			
Actinidia arguta x *A. deliciosa* hybrids (kiwifruit)		P	Cipriani et al. (1995); Matsunaga et al. 1996
Arabidopsis thaliana		M	Martinez et al. (1997)
Arachis hypogaea (peanut)		P	Grieshammer and Wynne (1990)
Beta vulgaris (beetroot)	M	M	Samoilov et al. (1986); Hagemann and Schröder (1989)
Brassica napus (rapeseed)	M, B	M	Ichikawa and Hirai (1983); Erickson and Kemble (1993)
Coffea canephora, *C. arabica* x *C. canephora* (coffee plant)		M	Lashermes et al. (1996)
Daucus carota sativus (carrot)	M	P, M	Boblenz et al. (1990); Steinborn et al. (1995)
Elais guinensis (oil palm)	M		Jack et al. (1995)
Epilobium hybrids (willow-herb)	M	M, rarely P	Schmitz and Kowalik (1987); Schmitz (1988)
Eucalyptus nitens (gum)		M	Byrne et al. (1993)
Glycine soja (soya bean)		M	Hatfield et al. (1985)
Helianthus hybrids (sunflower)	M	M	Rieseberg et al. (1994)
Lens culinaris (lentil)	M, B	M, B	Rajora and Mahon (1994, 1995)
Magnolia hybrids (lily tree)		M, P	Sewell et al. (1993)
Malus domestica (apple)	M	M	Ishikawa et al. (1992)

Table 1 (continued)

Species	Inheritance of mtDNA[a]	Inheritance of cpDNA[a]	References
Medicago sativa (alfalfa)	M	P, M and B	Schumann and Hancock (1989); Smith (1989); Masoud et al. (1990); Forsthoefel et al. (1992); Zhu et al. (1991, 1993)
Nicotiana plumbaginifolia, N. tabacum (tobacco)	M	M, trace B, P	Medgyesy et al. (1986); Horlow et al. (1990); Avni and Edelman (1991)
Oenothera hybrids (evening primrose)	P	B, M	Hachtel (1980); Brennicke and Schwemmle (1984); Chiu et al. (1988); Chiu and Sears (1993)
Oenothera erythrosepala, O. hookeri		B	Hageman and Schröder (1989)
Pelargonium zonale (geranium, zonal pelargonium)	Probably B	B, M	Metzlaff et al. (1981); Tilney-Bassett et al. (1992); Kuroiwa et al. (1993a)
Petunia hybrida (petunia)		M, rarely B	Derepas and Dulieu (1992)
Pharbitis hybrids (*Ipomea*, morning glory)	M	P, B	Hu et al. (1996); Hu and Hu (1996)
Phaseolus vulgaris (kidney bean)		M, B	Liu et al. (1997)
Pisum sativum (garden pea)	M	M	Liu and Hu (1995)
Populus hybrids (poplar)	M	M	Rajora et al. (1992); Rajora and Dancik (1992)
Quercus robur (common oak)	M	M	Dumolin et al. (1995)
Rosa species (wild rose)	M	M	Mastumoto et al. (1997)
Silene alba (white campion)		M	McCauley (1994)
Solanum tuberosum (potato)	M	M	Hagemann and Schröder (1989); Gounaris (1993)
Solanum nigrum (black nightshade)		M	Kavanagh et al. (1994)
Stellaria longipes (stitchwort)		M, P, B	Chong et al. (1994)
Angiosperms – monocotylodons			
Aegilops	M		Breiman (1987)

Allium hydris (onion)	M	M	Havey (1995)
Chlorophytum comosum (green lily)		M	Hagemann and Schröder (1989)
Festuca pratensis-Lolium perenne hybrids	M, P	M, P	Kiang et al. (1994)
Iris hybrids (Louisiana irises)		M, B, P	Cruzan et al. (1993)
Hordeum vulgare (barley)	M	M	Soliman et al. (1987); Hageman and Schröder (1989)
Hordeum x *Secale* hybrids	B	B	Soliman et al. (1987)
Musa acuminata (banana)	P	M	Faure et al. (1994)
Pennisetum americanum (fountain grass)		B	Krishna Rao and Koduru (1978)
Pennisetum glaucum (pearl millet)		M	Sujatha and Subrahmanyam (1991)
Secale cereale (rye)		B	Fröst et al. (1970)
Triticum aestivum (wheat) and polyploid hybrids	M	M	Vedel et al. (1981); Breiman (1987); Miyamura et al. (1987)
Triticum aestivum x *Secale cereale* (triticale)	M, trace B		Laser et al. (1997)
Zantedeschia hybrids (arum lily)		M, B, P	Yao et al. (1994)
Zea mays (sweat corn), *Z. mays* x *Z. perennis* hybrids	M	M	Conde et al. (1979); Hageman and Schröder (1989)

[a]B = biparental, M = maternal inheritance, P = paternal, U = uniparental inheritance. The predominate mode is given first in cases of pattern variability.

mechanisms within the organelle that ensure an ordered nonrandom segregation of mtDNA into emerging buds (Nunnari et al. 1997). Although part of the buds emerging from a zygote receive mtDNA from both parents, heteroplasmic mtDNA populations sort out very rapidly during vegetative mitotic growth in roughly 20 generations (Thomas and Wilkie 1968). Intracellular selection and relaxed mtDNA replication has been made responsible for the loss of one type of mtDNA (Birky 1994).

The most famous examples for a mixed plastid inheritance in higher plants are *Pelargonium zonale*, *Oenothera* species, and *Medicago sativa* (Table 1). Chloroplast inheritance has been studied in these species in reciprocal crosses with parents containing wild-type green plastids and parents harboring plastid mutations causing a white chloroplast pheno-

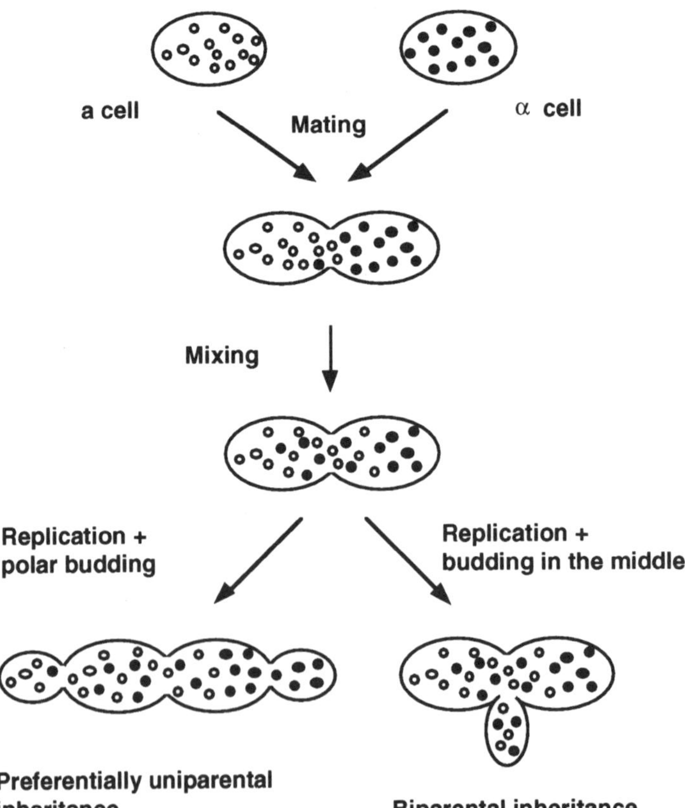

Fig. 2. Biparental inheritance of mitochondria and mtDNA segregation in *Saccharomyces cerevisiae*. Both mates contribute their mitochondria to the zygote. Budding in the diploid zygote can be predominantly polar and cells may arise having just one mitochondrial type, depending on the degree of mixing of mitochondria in the zygote prior to cell division. Buds with mixed mtDNA populations are formed when cell division occurs in the middle of the zygote from the zone of mitochondrial mixing. mtDNAs of the different parents are indicated by *open* and *filled* circles

type: progeny with either a green or a white color of leaves inherited plastids of only one parent. Green and variegated plants received chloroplasts from both parents which successively sorted out during plant development. *P. zonale* produces maternal, paternal, and biparental zygotes in varying proportions. *Oenothera* species produce only maternal and biparental but no parental zygotes. In contrast, *M. sativa* produces predominantly paternal and biparental zygotes (see references in Table 1). A similar segregation pattern reoccurs in crosses with a given female and varying male parents in *Pelargonium*, but variable patterns appear in crosses with a constant male and different females, indicating a maternal control in inheritance. Reciprocal crosses in *Oenothera* lead to clear reciprocal differences in the progenies with a distinct bias of maternal plastids. Reciprocal crosses of *Medicago* show a clear dominance (Tilney-Bassett 1994). Coincidently with biparental chloroplast inheritance, plastids and cpDNA were detected in sperm and egg cells of alfalfa (Zhu et al. 1991, 1993) and in generative cells of *P. zonale* (Nagata et al. 1997) and *Oenothera* species (Nakamura et al. 1992), but there is no consistent correlation among the absolute amount of cpDNA per generative cell and the strength of male plastid transmission.

β) Uniparental Maternal and Uniparental Paternal Inheritance

In the majority of plants and fungi, organelle genes are predominantly inherited from only one parent (Table 1). However, uniparental inheritance does not act via a single mode. A great variety of cellular and molecular mechanisms are found amongst different organisms. Blocks in organelle gene inheritance can act at different stages of the sexual cycle: at the prezygotic, fertilization, zygotic, and embryo stages. A single species may apply more than one method, especially in cases when mitochondria and plastids follow different modes of inheritance. A given aproach might not be completely effective on its own but sequential operation of distinct mechanisms can make organelle inheritance more rigid (Whatley 1982; Hagemann and Schröder 1989; Birky 1994, 1995).

Prezygotic mechanisms eliminate organelles or organelle genomes during gametogenesis. The most prominent mechanism found in some algae, a number of fungi, and generally in higher plants, employs male and female gametes of different size and content of organelles and, therefore, of organelle DNA. In these cases of *anisogamy* and *oogamy*, the macrogamete, the maternal parent, is larger than the microgamete, the paternal gamete. The zygote receives therefore an input of organelle genes that is biased towards one, normally the maternal parent. On the contrary, in isogamous forms, where gametes are not distinguished into male and female by size but into + or − mating types by physiological differences, the input bias is not determined by differences in cell volume or organelle contents. Organelles of one type still might be excluded from the zygote, e.g., by destruction: chloroplasts are degraded in gametes of the isogamous filamentous green algae *Temnogyra collinsii* (Birky 1995). Anisogamous forms also can choose such an eradicating way to exclude one type of organelle from the zygote. For example, in the syncytical green algae *Bryopsis maxima*, mitochondrial and chloroplast DNA vanish from the male gamete during differentiation (Kuroiwa et al. 1991). Similarly, organelle DNAs disappear at the tricellular pollen-grain stage during pollen development in wheat, *Lilium longiflorum*, and various other angiosperms (Miyamura et al. 1987; Sodmergen et al. 1992). In conifers, the structure of plastids will be disrupted and the maternal cpDNA destroyed in the egg cell whereas mitochondria migrate in aggregates to the perinuclear structure and become incorporated into the cytoplasma of the embryo (Owens and Morris 1990). Unequal distribution of organelles due to premitotic polarization of generative cells ensures in *Plumbago zeylanica* (leadwort) that the load of organelles in sperm cells is much greater than that in the egg (Russell 1987). However, most cases of nonrandom, polarized positioning of organelles during pollen development in angiosperms support their exclusion fromt the sperm cells (Hagemann and Schröder 1989).

During *fertilization*, organelles of the male gamete may fail to enter the female cell and are thus eliminated from the zygote. In higher basidiomycetes such as *Coprinus cinereus* (May and Taylor 1988), *Armillaria bulbosa* (Smith et al. 1990), and others (Table 1), nuclei invade and migrate through the mycelium of a compatible mating type, a property not given to mitochondria (Barroso and Labarère 1997). In *Ginkgo* and cycads, a single spermatozoid nucleus enters the egg during fertilization but the male cytoplasm is discarded just outside the egg cell near the point of nuclear entry (Whatley 1982). Similarly, in barley and spinach, only sperm nuclei merge with the egg cell (Wilms 1981; Mogensen 1988).

Zygotic deterministic mechanisms include selective degradation of organelles and organelle DNA after fusion. Chloroplasts of one parent desintegrate in zygotes of *Enteromorpha*, *Spirogyra*, *Volvox*, and other green algae (Whatley 1982; Kuroiwa et al. 1993c) and the male organelles in zygotes of the fern *Pteridium aquilinium* (Bell and Duckett 1976). mt and cpDNA degrade enzymatically in *Chlamydomonas reinhardtii* (Gilham 1994). Uniparental inheritance via partitioning is found in the green alga *Cylindrocystis*: two chloroplasts of each parent are present in the zygote that will not divide or fuse but will be distributed one by one into the four products of meiosis (Birky 1995). In multicellular plants, a localized accumulation of organelles in the unfertilized or fertilized egg can settle the elimination of one type of parental organelle from the embryo by partitioning them into extraembryonic cells during the early divisions of the zygote. Such defined separation mechanisms have been described in larch (Szmidt et al. 1987), pine (Willemse 1974), and alfalfa (Mogensen 1996).

In most instances, uniparental maternal and uniparental paternal inheritance show some degree of leakage and produce offspring with mixed organelles, or offspring with organelles of the opposite type arise next to the favored class of progeny (Baptista-Ferreira et al. 1983; Neale et al. 1986; Szmidt et al. 1987; Avni and Edelman 1919; Matsumoto and Fukumasa-Nakai 1996; Barroso and Labarère 1997; Laser et al. 1997). Other organisms where high numbers of progeny were analyzed seem not so amenable for leakage (for example *Aspergillus nidulans*, Coenen et al. 1996). It depends on its frequency in a population but also very much on sample sizes and methods used for analyses whether deviations from the predominant pattern of inheritance can be detected (Reboud and Zeyl 1994).

Organelle inheritance has been studied by genetic, cytological, and molecular techniques. Mutants, especially when of dominant character, serve very well to pursue organelles during reproduction in large numbers of offspring. Unluckily, such powerful genetic approaches are restricted to species where appropriate mutants are available. Marker phenotypes to trace mitochondria are respiration deficiencies (petite in bakers yeast, Ephrussi 1953; cytochrome *aa3* deficiency in *C. cinereus*, Baptista-Ferreira et al. 1983), mycelial growth retardation in the mould *Neurospora crassa* (poky and others, Griffiths et al. 1995), resistance to antibiotics in various fungi (e.g., chloramphenicol, erythromycin, oligomycin, Thraikill et al. 1980; Baptista-Ferreira et al. 1983; Coenen et al. 1996) and cytoplasmic male sterility (CMS) in many higher plants (Lonsdale 1987). Phenotypes that mark chloroplasts include extranuclear chlorphyll deficiencies (Ohba et al. 1971; Tilney-Bassett 1994) and resistance to herbicides and drugs (e.g., triazine, streptomycin, tentoxin, Horlow et al. 1990; Erickson and Kemble 1993; Kavanagh et al. 1994). Techniques of more general use, but restricted by sample numbers, are electron microscopy and immunogold labeling with anti-DNA antibodies, and fluorescence microscopy to detect presence or absence of organelle DNA by DAPI (4,6-dimidine-2-phenylindole) staining (Miyamura et al. 1987; Nakamura et al. 1992; Nagata et al. 1997). Plastids have been found in male gametes of 18–27% of analyzed angiosperms (Corriveau and Coleman 1988; Harris and Ingram 1991) but this does not inevitably denote paternal inheritance, as later-acting mechanisms may exclude them from the zygote or the offspring (see above). Absence of organelles is generally taken as an indication for uniparental inheritance, but this can be misleading: biparental chloroplast inheritance was described in rye and beans by genetic means but the cytological approach failed to detect plastids in male gametes (Fröst et al. 1970; Karas and Cass 1976; Liu et al. 1997). Moreover, DAPI staining may fail to detect DNA whilst molecular methods still can reveal its presence (Corriveau and Coleman 1991). Molecular techniques employ DNA-DNA hybridizations, RFLPs (restriction fragment length polymorphisms), PCR (polymerase chain reaction), and plasmids (Borkhardt and Olsen 1983; Silliker et al. 1996, Barroso and Labère 1997; Laser et al. 1997). Among molecular techniques, PCR is the most sensitive to detect substoichiometric amounts of organelle DNA, with a detection level that is comparable to the resolution obtained in genetic studies (Medgyesy et al. 1986; Coenen et al. 1996; Laser et al. 1997).

Frequency of leakage can be influenced by the nuclear genome of the host, or by the organelle genome, or by both (Sect. 2.b, c) – even the presence of plasmids in an organelle can change the mode of inheritance (Kawano et al. 1995). Frequencies in a population respond to intracellular selection. This becomes obvious when looking at organelle-encoded resistances in *S. cerevisiae*. When sensitive colonies are exposed to antibiotics, resistant colonies accumulate gradually over time, due not to induction of new mutations but to intracellular selection of substoichiometric mtDNAs (Birky 1994). Such a frequency selection is also known in higher plants. Chloroplasts were believed to be inherited strictly maternally in *Nicotiana abaccum*, but low-frequency paternal inheritance may occur as seen when selecting for resistance (Avni and Edelman 1991). Like mtDNA of *S. cerevisiae*, organelle DNAs of higher plants appoint rolling circle replication for proliferation (Backert et al. 1996). This gives an opportunity for uncontrolled multiplication of underrepresented DNA molecules and will finally lead to a new equilibrium state of organelle DNA (Albert et al. 1996). Unrecognized frequency selection can cause misinterpretation of results in crosses when organelle DNA of one parent is present in substoichiometric amounts in the other. Substoichiometric amounts of paternal mtDNA found in maternal mitochondria made it impossible to decide upon paternal or maternal inheritance of these sequences in intergenetic crosses between wheat and rye (Laser et al. 1997). A similar situation has been described in hexaploid wheats (Morere Le Paven et al. 1994) and in a study showing maternal inheritance of organellar DNA in carrots (Steinborn et al. 1995), making an earlier report on paternal cpDNA inheritance in carrots doubtful (Boblenz et al. 1990).

b) Genetic Control of Organelle Inheritance

In some instances, specific nuclear genes are known that control or contribute to organelle transmission (Sect. 2.c) but most of such functions await genetic and molecular identification. Plastid inheritance may be controlled by genes expressed in the maternal lineage as in *Pelargonium* (Tilney-Basett et al. 1992; Amoatey and Tilney-Basssett 1994), paternally as in *Petunia hybrida* (Derepas and Dulieu 1992), or by both the maternal and the paternal genotypes as in *Medicago sativa* (Smith 1989; Masoud et al. 1990). Differences between plastids are more decisive for the percentage outcome of biparental and maternal inheritance in *Oenothera* hybrids, although there is an additional influence of nuclear DNA (Chiu et al. 1988; Chiu and Sears 1993). Strain-dependent plastome/genome interactions are known in *Stellaria longipes* (Chong et al. 1994) and plastome/genome incompatibilities have been described in *Zantedeschia* hybrids (Yao et al. 1994). Maternal inheritance of mtDNA in *Brassica napus* is due to genes active in both parents (Erickson and Kemble 1993). Nuclear *Secale* genes are responsible for biparental inheritance in *Hordeum* x *Secale* hybrids (Soliman et al. 1987). Certain mtDNA mutations shift the frequency of paternal and maternal inheritance in *Chlamydomonas*, but environmental factors (blue light, nitrogen depletion) have also an influence (Beckers et al. 1991; Sears and VanWinckle-Swift 1994; Gloeckner and Beck 1995). Nuclear genotypes and mitochondrial haplotypes influence maintenance of mitochondria in mixed dikaryons of the basidiomycete *Agrocybe aegerita* (Barroso and Labarère 1997) and mitochondrial mixing in *S. cerevisiae* partly depends on the mitochondrial genotypes of the parent cells (Azpiroz and Butow 1993; Piskur 1997).

c) Nuclear Contribution to Organelle Inheritance

Due to the variability of mechanisms in which organisms verify uni- or biparental organelle inheritance, only a certain amount of conservation is to be expected in the participating functions. Recent studies have begun to identify nuclear components in the process of mitochondrial inheritance in *S. cerevisiae*. It is the ideal organism for analyses of nuclear-mitochondrial interactions since defects in nuclear genes affecting mitochondrial inheritance and maintenance of mtDNA are rarely lethal (Table 2). So far, investigations of functions in mitochondrial inheritance concentrated primarily on mitotic transfer, but it is believed that at least some of the discovered mitotic functions will also contribute to meiotic transfer (Berger and Yaffe 1996). Due to the yeast genome sequencing project, the list of identified functions is growing quickly and, in addition, first interactions between components are emerging. Some proteins involved in mitochondrial transmission are localized in the cytoplasm, others are directed to the mitochondria itself, often to the organelle membranes. Manifold links have now been established between the cell cytoskeleton and transfer of mitochondria into buds by protein and mutant characterization (Table 2). Evidence for an actin-dependent organelle movement leads to a model in which organelle-associated motor molecules utilize the energy of ATP binding and hydrolysis to drive along actin cables (Berger and Yaffe 1996; Simon and Pon 1996). Groups of other nuclear-encoded functions contributing to organelle transfer are responsible for mitochondrial protein import, correct protein folding, and nucleic acid metabolism. Mitochondrial inheritance mutants of these types are often thermosensitive and also show altered mitochondrial morphology and defects in the maintenance of the mtDNA (for details see Table 2).

Only a few nuclear functions influencing mitochondrial inheritance have been isolated from organisms other than *S. cerevisia*, namely the *clu1* homologue from *Dictyostelium discoideum* (Zhu et al. 1997), the *mdm10* homologue from *P. anserina* (Jamet-Vierny et al. 1997a), the *mdm12* homologue and a putative *cce1* homologue from *S. pombe* (Berger et al. 1997; White and Lilley 1997b), and the *mip1* homologue from *Schizophyllum commune* (Isaya et al. 1995), *C. cinereus* (Casselton et al. 1995), and *Coprinus bilanatus* (U. Kües and M. P. Challen, unpubl.). Where known, the cytological effects on mitochondrial inheritance and structure are the same as described in *S. cerevisiae* (Table 2).

The most prominent of all nuclear associations to organelle inheritance are in many fungi and algae the mating types. In the slime mold *Physarum polycephalum*, different forms of the multiallelic mating types locus *matA* can be ranked in a linear hierarchy to determine loss of mtDNA (Kawano et al. 1995). In the oomycete *Phytophthora infestans*, a certain type of mitochondria tends to cosegregrate with *A1*, one of the two possible mating types (Whittaker et al. 1994). Cells of the mating type *a2* in the heterobasidiomycete *Microbotryum violaceum (Ustilago violacea)* preferentially receive the mitochondria of the *a2* parent whereas *a1* cells retain either type of mitochondrion. This strongly suggests that mitochondrial selectivity is a function of the *a2* mating type allele or of a gene on the same chromosome (Wilch et al. 1992). An analogous link of mitochondrial inheritance and the *a2* mating type locus as in *M. violaceum* were not established in *U. maydis* (M. Bölker, pers. comm.). However, two genes, *lga2* and *rga2*, are localized in the

Table 2. Nuclear encoded functions affecting mitochondrial inheritance and maintenance of mtDNA in *Saccharomyces cerevisiae*[a]

Protein	Gene on chromosome	Null mutant	Function
			Proteins linked to cytoskeleton function
DEC1 (MDM20)	XV	Viable	• Null mutants fail to transport mitochondria into buds, lack actin cables • Interacts with CIN8, a kinesin-related protein involved in establishment and maintenance of the mitotic spindle
MDM1	XIII	Viable	• Inermediate filament protein, similar to mammalian vimentin and to mouse keratin • May form an intracellular framework to which mitochondria and spindle pole bodies are attached at distinct times in the cell cycle • Mutants are temperature-sensitive, have abnormal mitochondrial distibutions at non-permissive temperatures, are defective in transmission of mitochondria to buds
MMM1	XII	Viable	• Integral component of the mitochondrial outer membrane, essential for establishment and maintenance of mitochondrial shape and structure • Probably interacts with the cytoskeleton • Sequence similar to MDM12 • Mutants have problems in mitochondrial segregation; mitochondria accumulate at the bud site
TPM1	XIV	Viable	• Tropomyosin, coiled-coil protein localized to actin cables • Overproduction partially suppresses defect of OLE1 (MDM2) null mutant
TPM2	IX	Viable	• Tropomyosin isoform 2, coiled-coil protein • Overproduction partially suppresses defect of OLE1 (MDM2) null mutant
			Proteins involved in protein import and folding
ATM1 (MDY)	XIII	Lethal	• Required for mtDNA maintenance • Suspected to function in protein translocation, ABC transporter

HSP78	IV	Viable	• Mitochondrial heat shock protein, functions as a mitochondrial chaperone with HSP70 • Required for mitochondrial DNA integrity during extreme temperature stress
MGE1 (GRPE, YGE1)	XV	Lethal	• Required for maintenance of mitochondrial function • Participates in folding of proteins during mitochondrial import, cooperates with SSC1, homologue of *Escherichia coli* GrpE
MIP1 (YKL134C)	XI	Viable	• Mitochondrial intermediate peptidase (thiol-dependent metallopeptidase), required for second N-terminal cleavage of various proteins during mitochondrial import • Null mutant undergoes loss of functional mitochondrial genomes
PIM1(LON)	II	Viable	• La-like protease, may serve as chaperone for mitochondrial protein complex assembly and as protease for incorrectly assembled substrates • Null mutant loses mitochondrial genome function

Proteins involved in nucleic acid metabolism

ABF2 (HIM1, HM)	XIII	Viable	• HMG1 homologue, abundant DNA-binding protein of mitochondria, involved in differentiated organization of mtDNA nucleoids, required for mtDNA expression • Null mutant loses mtDNA when grown on glucose but not when grown on glycerol
CCE1 (MGT1)	XI	Viable	• Cruciform cutting endonuclease (endo X3 resolvase) • Involved in meiotic transmission of DNA, required for selective displacement of rho^+ by rho^- mtDNA in diploids mated from rho^+ and rho^- parents • Affects number of unresolved mitochondrial Holliday junctions and consequently the number of heritable units of mtDNA
MIP1 (YOR330C)	XV	Viable	• Mitochondrial DNA polymerase, possesses 3'–5' exonuclease activity
MSH1	VIII	Viable	• Mismatch repair enzyme, binds to single base mismatches • Involved in stabilization of the mitochondrial genome
MTF1	XIII	Viable	• Mitochondrial RNA polymerase specificity factor • Null mutant loses mtDNA

Table 2 (continued)

Protein	Gene on chromosome	Null mutant	Function
MRS2	XV	Viable	• Protein essential for splicing of mitochondrial group II introns and involved in a function unrelated to splicing • Essential for mitochondrial DNA maintenance at elevated temperatures
NHP6A (NHPA)	XVI	Viable	• Homologue to mammalian nonhistone protein HMG1, bends DNA • Suppressor of ABF2 mutations
NUC1	X	Viable	• Major nuclease of mitochondria with endonclease activity for DNA and RNA and 5' double-stranded DNA endonuclease activity • Homologous recombination and gene conversion are affected in null mutants
NUC2	XI	?	• Endo-exonuclease involved in induction of petite mutations
PIF1 (TST1)	XIII	Viable	• Mitochondrial DNA helicase (5' to 3' activity) with ATPase activity, homologous to the *Escherichia coli* single-strand binding protein SSB • Null mutant loses mtDNA at 36 °C • Mutation can be suppressed by extra copies of RIM1
RIM1	III	Viable	• Single-stranded zinc finger DNA binding protein required for replication of mtDNA
RIM2 (MRS12)	II	Viable	• Protein of mitochondrial carrier family, putative ADP/ATP translocase • Null mutation causes loss of mitochondrial DNA
RNA12 (PRP12, YME2)	XIII	Viable	• Integral inner mitochondrial membrane protein • Mutants have increased escape of DNA from mitochondria to nucleus • May be involved in methylation of pre-rRNA
RPO41	VI	Viable	• Core enzyme of mitochondrial RNA polymerase • Certain mutations lead to unstable mitochondrial genomes

Gene	Chromosome	Phenotype	Function
SSC1 (ENS1)	X	Lethal	• Mitochondrial heat shock protein HSP70 family, chaperone in protein folding during import • Cooperates with HSP78 in maintenance of mitochondrial function • Large subunit of mitochondrial endonuclease SceI that initiates homologous recombination by introduction of double-strand breaks

Other functions and unclarified molecular functions

Gene	Chromosome	Phenotype	Function
CLU1 (TIF31)	XIII	Viable	• Translation initiation factor EIF3, P135 subunit • Involved in dispersal and inheritance of mitochondria • Null mutant has clustered mitochondria, some mutants form petites
ERV1	VII	Lethal	• Essential for mitochondrial biogenesis, maintenance of mtDNA
ILV5	XII	Viable	• Ketol-acid reductoisomerase involved in valine and isoleucine biosynthesis • Null mutation lead to rho⁻ petite cells, affect on mtDNA is unrelated to branched-chain amino acid synthetase • Suppressor of ABF2 minus phenotype when slightly overexpressed
GCS1	IV	Viable	• GTPase-activating protein (GAP), involved in transition from stationary phase to G1 • Loss of mtDNA in *gcs1 sed1* double mutants
MDM10 (FUN37)	I	Viable	• Integral protein of outer mitochondrial membran • Depletion of MDM10 leads to giant mitochondria that do not enter the buds
MDM12	XV	Viable	• Integral membrane protein of the mitochondrial outer membrane • Null mutant has elevated frequency of loss of mtDNA (rho⁻ and rho0 formation)
MGM1	XV	Viable	• Mutations cause loss of mtDNA and altered mitochondrial morphology • Member of dynamin family of GTPases • Localized in the mitochondrial outer membrane • N-Terminal mitochondrial targeting domain has homology to the bacterial ribonuclease inhibitor barstar
MGM101 (MGM9)	X	Viable	• Mutation results in temperature sensitive loss of mtDNA • Basic protein with inidentified mitochondrial function

Table 2 (continued)

Protein	Gene on chromosome	Null mutant	Function
MHR1	?	?	• Mutations repress Endo.Sce1-induced gene conversion and cause UV sensitivity • Mutants are temperature-sensitive in DNA maintenance
OLE1 (MDM2)	VII	Viable	• Stearol-coA desaturase, Δ-9 fatty acid desaturase • Involved in regulation of mitochondrial inheritance and morphology • ts Mutants fial to transfer mitochondria into the growing bud
PPA2 (IPP2)	XIII	Viable	• Mitochondrial inorganic pyrophosphatase • mtDNA loss in null mutants
SHM1 (SHMT1)	II	Viable	• Mitochondrial serine (glycine) hydroxymethyltransferase • Suppressor of ABF2 null mutant when overexpressed
SOT1	?	?	• Mutations suppress defects of both MDM10 and MDM12 • Gene not yet molecularly identified
TYS1 (TTS1)	VII	Lethal	• Tyrosyl-tRNA synthetase • Complements MGM104 mutation that leads to temperature sensitive loss of mtDNA
YME1 (OSD1, YTA11)	XV1	Viable	• Potential ATP-dependent mitochondrial zinc-dependent protease, associated with the matrix side of the mitochondrial inner membrane • Likely to play a role in mitochondrial diversion or fusion • Mutants show increased escape of DNA from mitochondria to nucleus • Mutations lethal in petite mutants

[a]Most of this information has been collected from and original references can be found in the Yeast Protein Database maintained by J. I. Garrels at Proteome, Inc (http://quest7.proteome.com/YPDhome.html). Information for NUC2 is from Chow and Kunz (1991), for MHR1 from Ling et al. (1995), and for SOT1 from Berger et al. (1997).

unique DNA of the *a2* mating type but are missing in the *a1* mating type. These genes encode putative mitochondrial matrix proteins but it is not yet known whether they are involved in mitochondrial inheritance (Urban et al. 1996). *mip1* genes for mitochondrial intermediate peptidase have been shown to be directly linked to the *A* mating type loci in higher basidiomycetes (Casselton et al. 1995; Isaya et al. 1995, U. Kües and M. P. Challen, unpubl.). Similarly, in *Chlamydomonas* species, there is considerable evidence for control of organelle inheritance by the mating type genes or a closely gene(s). Genes connected to chloroplast transmission (functions permitting specific protection of organelles of one parent and defined destruction of the other) have been postulated to reside within the mating type mt^+ locus while those for mitochondrial inheritance are proposed to be present within the mt^- locus (Armbrust et al. 1995). A first characterization gene, *egy1*, present in seven to eight copies within the mating type region, seems to participate in m^- chloroplast degradation (Armbrust et al. 1993).

3 Recombination of Organelle Genomes

The circular chloroplast genomes of higher plants and algae are highly conserved in size (120–217 kb, as exception 400–450 kb are found in algae) and gene arrangements (Palmer 1991). Recombination of chloroplast genes is relatively rare in higher plants (Hoot and Palmer 1994; Kawata et al. 1997) but common in the lower eukaryote *Chlamydomonas* (Boudreau and Turmel 1995). In contrast, mitochondrial genomes in plants and in fungi are highly diverse in their size (17–180 kb in fungi, 16 to 220 kb in algae, up to more than 2000 kb in higher plants), in sequence contents and gene arrangement due to a high frequency of recombination (Schuster and Brennicke 1994; Leblanc et al. 1997). With a few exceptions, mitochondrial DNA in plants and fungi is reported to be circular although linear molecules of various sizes seem in many species to be the major form of DNA within the mitochondrion (Bendich 1993). Linear molecules are thought to be products of rolling circle replication of existing circular mtDNA molecules. Successive intramolecular recombination might circularize the DNA and this is made responsible for the 10–15% circular mtDNA molecules found in a cell (Maleszka et al. 1991; Bendich 1993; Backert et al. 1996).

a) Intraorganellar Recombination

High frequency of intramolecular recombination of mtDNA has been observed in many plants (e.g., in *Brassica*, maize, *Petunia*, and soybean, Palmer and Shields 1984; Conklin and Hanson 1993; Fauron et al. 1995;

Moeykens et al. 1995), and occasionally also in fungi (Barroso et al. 1992). Recombination mostly involves pairs of direct or inverted repeats sized between some 100 and 12 kbp. Recombinational events across repeated sequences in direct orientation are thought to lead to the generation of subgenomic DNA molecules (Fig. 3A). In contrast, recombination across inverted repeats results in DNA inversions. The higher the number of repeats within a given mtDNA, the more complex is the structure of the organelle genome: due to recombination via all possible repeats, it is composed of a multicircular population of master chromosome and subgenomic circles (Fauron et al. 1995). In contrast, most chloroplast genomes contain only two large inverted repeats (IR) of 5 to 80 kb which are separated by a large single-copy region (LSCR) and a small single-copy region (SSCR). Products of reciprocal exchange by a flip-flop mechanism within the IR differ only in orientation of the single-copy regions relative to each other (Palmer 1991; Fig. 3B).

The lack of conserved sequence motifs amongst large repeats in mtDNA of different species, even if closely related, suggests that recombination may be based on a homologous mechanism, rather than being site-specific (Conklin and Hanson 1993) – although, since the presence of repeats alone is not sufficient for high frequency recombination, other, perhaps site-specific, mechanisms might exist (Hanson and Folkerts 1992). So far, recombination via homologous repeated sequences has not been verified enzymatically. However, proteins related to RecA, pivotal for homologous recombination in *Escherichia coli*, have been detected in plant chloroplasts (Cerutti et al. 1993, Sect. 3.c). Moreover, gene replacement by homologous recombination has been observed in organelle transformation (Maliga et al. 1993).

In addition to the frequent recombination events that occur across large repeats, rare recombination of organelle DNAs has been observed that employs short repeats and sequences as small as a few bp (Table 3) up to a few hundred bp (André et al. 1992; Cosner et al. 1997). Specific recombination sequences, often closely linked or located within the large repeats, obviously act as recombination hot spots. Recombinational events of this type can cause large deletions, duplications, or inversions

---

Fig. 3A–C. Intraorganellar recombination. **A** Intramolecular recombination in the mtDNA of *Brassica campestris* using a pair of direct repeats. Recombination is reversible, leading to an equilibrium between the large master chromosome (218 kb) and the two smaller resolution products (135 and 83 kb, respectively). The schematic presentation follows the model of Palmer and Shields (1984). **B** Intramolecular recombination in the cpDNA of the brown alga *Dictyota dichotoma* at a pair of inverted repeats. Flip-flop isomerization leads to a mixture of cpDNA molecules differing only in the relative orientation of a region containing genes for photosystem I and II, indicated by *bold lines* marking their localization (Kuhsel and Kowallik 1987). **C** Recombination at small repeats (*R1* and *R2*) in master circles of plant mtDNA leads to formation of subcircles that can form new master circles by subsequent homologous recombination. (Small et al. 1989)

A

B

C

R1
R2
Progenitor
Master Circle
R1
R2
Recombination
at Repeat R1
R1
R2
A
Subcircles
B
R1
R2
Recombination
at Repeat R2
R1
R2
C
Subcircles
D
R1
R2
Homologous Recombination
at large Repeats
R1
R2
New Master
Circle (B + C)
R1
R2

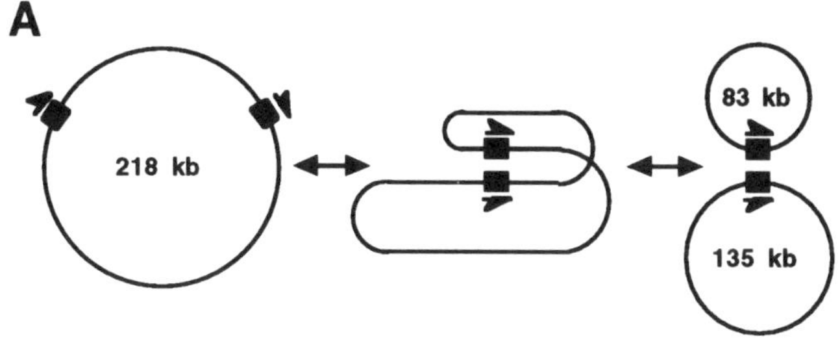

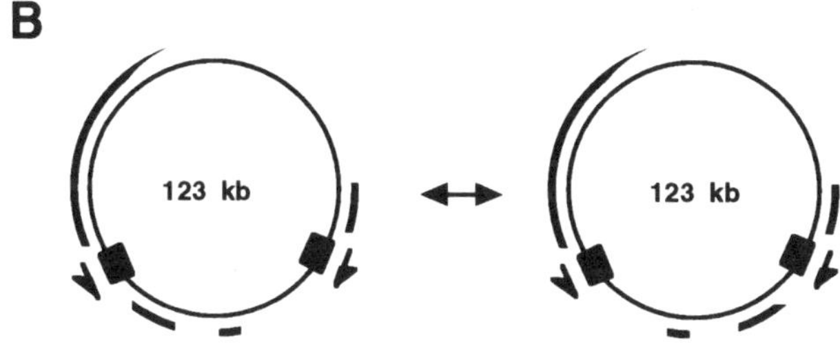

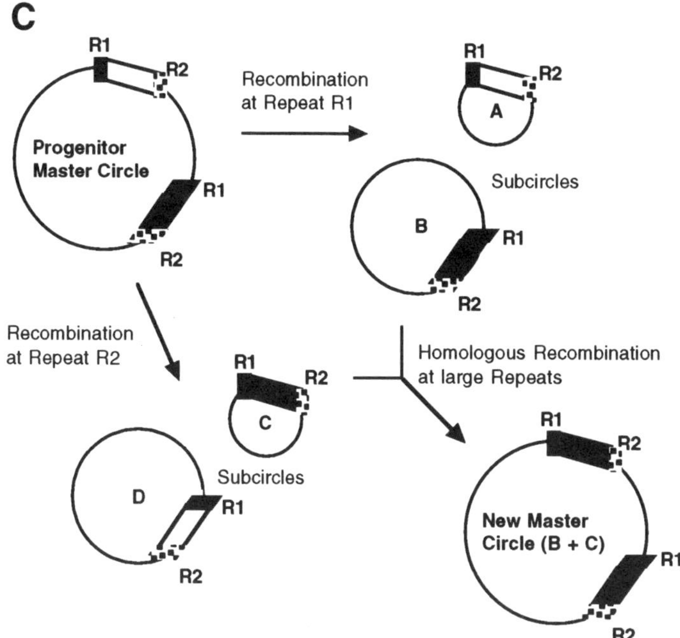

Fig. 3A–C.

in the DNA molecules and lead to mutant phenotypes which are characterized by impaired organelle function (Cosner et al. 1997; references in Table 3). Recombination between very short repeats in mtDNA results in molecules termed sublimons which are present in substoichiometric amounts relative to the main mitochondrial genome. When different substoichiometric intermediates participate via a common region in homologous recombination, new master circles are formed carrying the common region as a duplication (Small et al. 1989; Fig. 3C).

Small repeats serving as recombination hot spots are often either A+T- or G+C-rich, and often they contain stretches of only one base type (Table 3). Sometimes, sequences of such recombination hot spots have been found to be conserved between species (Moeykens et al. 1995; Johnson and Hattori 1996), but in many species there is no similarity amongst recombinogenic repeats. Moreover, within the same mt or cpDNA, different sequences may give rise to rearrangements (Table 3). Therefore, a distinct sequence per se is not the basis for recombination but rather its repetition within a given DNA.

Accumulation of short inverted and direct repeats within the borders of a 2.4-kb repeated element (Wendy) in cpDNA of *C. reinhardtii* suggests that some of the repeats found in organelle DNA may be the remnants of transposons or other mobile elements (Fan et al. 1995). Small repeats are often overrepresented in a genus, like the GC clusters in *S. cerevisiae* and MUSE (mitochondrial ultrashort elements) in *P. anserina*. They are preferentially located in the intergenic spacers and in the variable parts of the small and large rRNAs. It has been suggested that these short sequences are mobile elements that spread via gene conversion in *S. cerevisiae* and, at least in *P. anserina*, via a reverse transcriptase mechanism (Koll et al. 1996). In any case, RNA-processing sites and introns in organelle DNA appear to be favored substrates for sequence-specific rearrangements and sequence duplication (André et al. 1992). Homing (the movement of an intron to an intronless DNA) of a group I intron has been demonstrated in mitochondria of *S. cerevisiae* and in chloroplasts of *C. reinhardtii* (Lambowitz and Belfort 1993; Dürrenberger et al. 1996) and for mitochondrial *S. cerevisiae* and *P. anserina* group II introns (Sellem et al. 1993; Yang et al. 1996). The mobility of group I introns depends on double-strand breaks (DBS) introduced by an intron-encoded endonuclease at specific DNA sites, immediately followed by DBS repair via gene conversion using intron-containing DNA as the template for recombination. Homing of group II introns which encode reverse transcriptase-like enzymes occurs via reverse splicing of the generated cDNA (Guo et al. 1997).

Deletions and insertions in organelle DNA may be mediated by direct repeats via a type of illegitimate recombination, either by replication slippage (Sears et al. 1996), misalignment of repeated sequences during homologous recombination (Fan et al. 1995), or by topoisimerase-catalyzed DNA breakage and rejoining (Woelfle et al. 1993; Obernauerová and Subik 1994). Inverted repeats, on the other hand, may offer specific recombination signals due to secondary folding. Consistent with this, an endo-exonuclease from *N. crassa* has been shown in vitro to cleave such sequences in mtDNA at points of recombination (Almassan and Mishra 1991). Inverted repeats also might undergo homologous recombination: formation of head-to-head and/or tail-to-tail oligomers in cpDNA of rice

Table 3. Compilation of short invers and/or direct repeated sequences shown to be involved in mtDNA and cpDNA recombination

Species	mtDNA sequence	cpDNA sequence	Reference
Fungi		Not applicable	
Kluyveromyces lactis	CTAATATATAT		Hardy et al. (1989)
Neurospora crassa	CCCCGCCCC		Almasan and Mishra (1991)
Podospora anserina	GGCGCAAGCTC, GCCAGC/GTGT, GCCT/AGGTGTG/TC, GGCAAAA GGAATTTTAGGATT, TGGTTTTG, CTGATGCTGAAGGTTG/CTTTT, ATGGTGAAAACGGTT, CCTGG, TCGCTACAGACTGGTTCACC, AATGATGGGTA, GGTTGTTTT, ATAAATAAGATAAATG, TATAT, TGTGTTTTCTTAATGCAGAAAA, AAGGTTACT and others		Schulte et al. (1988); Silliker et al. (1996); Jamet-Vierny et al. (1997b)
Saccharomyces cerevisiae	GC-clusters of 20–50 bp, AATAATATAAA		de Zamaroczy and Bernadi (1986); Weiller et al. (1991)
Algae			
Chlamydomonas species		GACGTCCCCTTCGGGCAAATAAA, GGTACCACTGCCACTGGCGTCCT, CCTGTCAGTCGGGTAAAA, AACGAAGTGAAGGG, ATAAAGCTAAAACCT, GTAAAAAGTAAAAACAGA	Boynton et al. (1992); Boudreau and Turmel (1996)

Table 3 (continued)

Species	mtDNA sequence	cpDNA sequence	Reference
Mosses			
Marchantia polymorpha (liverwort)		CCCATTTTTTTATT	Zhou et al. (1988)
Angiosperms – dicotylodons			
Brassica oleracea (rapeseed)	AAAGGGGAAATAGAGGGGAAAG AGGAAAAAAAGAGGGG	13–18 bp T-rich blocks	Johnson and Hattori (1996); Bellaoui et al. (1998)
Crassulaceae		AAGTA/CT	van Ham et al. (1994)
Faopyrum (buckwheat)		TTCTATTTCTATCTAGA, AAACC(T)CC	Aii et al. (1997)
Glycine max (soybean)	CCCCTCCCC		Moeykens et al. (1995)
Mecicago sativa (alfalfa)		AGAAATT/AAAT, CAAAAA/GAAAAAAAT/GAAT, CAAAAAAAAAAA/TA/CAAA/TTT	Aldrich et al. (1988)
Nicotiana tabacum	ATCCC/ACTCC(CC)	17–18 bp and larger T-rich blocks	Kanzawa et al. (1994); Johnson and Hattori (1996); Goulding et al. 1996)
Oenothera species (evening primrose)	GGAAGCCGCC	AAAT/AG/CAT, ACAGAA, GATTAGATA	Mann and Brennicke (1986); vom Stein and Hachtel (1988); Nimzyk et al. (1993)
Petunia hybrida	GAAAAGGAAATCCTATTT		Conklin and Hanson (1993)
Pisum sativum (pea)		TTCCCCT, TTCTTTTCAATTTC, AATTCGTTTATA, 13–18 bp T-rich blocks	Johnson and Hattori (1996), Kawata et al. (1997)
Saxifraga hypnoides		ACTT/AC/TGATTT/CT	van Ham et al. (1994)

Angiosperms – monocotyledons			
Oryza sativa (rice)	TTCCCTC	CGGGTTCGATTCCC, TTTAGAATA	Hiratsuka et al. (1989); Kadowski et al. (1990);
		CTTTTTTTTAGAATAC, AGAAAAAAAT, TTACTTTTTTCA, AATAGAA/TAATGAG, AGGGTTTT	Kanno et al. (1993); Morton and Clegg (1993)
Triticum, Aegilops (wheat)	AGAATGTT, GTTTCGTT	CATTTTTTT, CTTTTTTATTC, ATTAT, GAAT, GAAAAAA, TGGT TCCTTTC, 13–18 bp T-rich blocks	Howe (1985); Ogihara et al. (1988, 1992); Mohr et al. (1993); Hartmann et al. (1994); Johnson and Hattori (1996)
Zea mays (maize)	CAGAG/ACAAAAGGGAAGGG, GCCCACCCCAG	TATAGAATACTTTTTT, ATTCTATATCTATAA	Hunt and Newton (1991); Morton and Clegg (1993); Marienfeld and Newton (1994)
Gymnosperms			
Pseudotsuga menziesii (Douglas-fir)		CATCTTTT	Tsai and Strauss (1989)

is thought to occur be homologous recombination between the stems of identical pairs of cruciforms located on different molecules (Kawata et al. 1997). Rayko and Goursot (1996) alternatively explain the formation of head-to-tail, head-to-head, and tail-to-tail mitochondrial genomes in petite mutants of *S. cerevisiae* as a consequence of recombination-directed rolling circle replication. According to this model, replicative strands invade homologous inverted nicked DNA of the same molecule at regions that are already replicated. Nascent strands subsequently cross-link and parental strands join end-to-end (the rocking circle model).

Most of the recombination processes described so far will occur in a DNA population within a single organelle or within the organelles of the same cell. However, recombination is also possible between organelle DNA from different parental origin, providing that at least some mixing of cytoplasm and organelle fusion is granted in a zygote or fusion product (Sect. 2). Recombination between parental genomes in most organisms is rare, but relatively frequent in chloroplasts of *Chlamydomonas* and in mitochondria of *S. cerevisiae* and a few other ascomycetes (Birky 1995). Homologous recombination between common regions in petite mtDNAs can restore wild-type respiration in *S. cerevisiae*. Interestingly, the recombinant mtDNA molecules carry duplicated regions similar to new master circles formed by recombination of sublimons in plant mitochondria (Evans and Clark-Walker 1985; Fig. 3C). The occasional ocurrence of recombinant DNA fragments in organelle genomes in the sexual progeny of other fungi and plants has been taken as a criterion for low biparental inheritance (Sect. 2). In most cases, the type of recombinational event is not clear, often perhaps due to low biparental transfer frequencies and to the complexity of genomes involved. However, high frequency of interparental recombination of organelle DNA has been observed in somatic hybrids of plants and fungi after overcoming sexual and vegetative incompatibility barriers by protoplasting. Such studies with somatic hybrids provide insight into mechanism of interparental recombination (Akagi et al. 1995; Kevei et al. 1997; Bellaoui et al. 1998).

b) Interorganellar Sequence Transfer

The transfer of DNA sequences between nuclear DNA and organelle genomes is a repeatedly occurring phenomenon during the evolution of eukaryotes (Brennicke et al. 1993). Transfer of cpDNA and mtDNA to the nucleus (Blanchard and Schmidt 1995) and transfer of nuclear and cpDNA to mitochondria (Schuster and Brennecke 1987) have been reported in higher plants. In contrast, mtDNA has been transferred to the nucleus in algae (Kobayashi et al. 1997) but in the mitochondria, foreign (promiscuous) DNA inserts were not found (Leblanc et al. 1997). In fungi, mitochondrial DNA differs from species to species by the nature and number of mitochondrial genes that were shifted to the nucleus (Paquin et al. 1997). It had been suggested that genetic information transfers between organelles via RNA and reverse transcription. Strong evidence for such a mode of transfer is found for mitochondrial genes whose nuclear localized counterparts resemble the edited mitochondrial transcripts (Wischmann and Schuster 1995). RNA-mediated transfer may also account for insertion of fully transcribed cpDNA segments into mtDNA (Schuster and Brennicke 1987). However, all possible cpDNA sequences have been found in mitochondrial genomes, although the composition of the transferred chloroplast sequences varies from plant to plant. Amongst the integrated fragments examined, nontranscribed DNA regions, segments that contain introns and segments that are transcribed in different directions, have been identified. All these cases argue against an RNA-mediated transfer of at least these chloroplast sequences (Nakazono and Hirai 1993; Zhen et al. 1997). Investigations of the borders of cpDNA integrated in rice mtDNA revealed no common sequence motifs or structures that might explain their mode of integration. Thus, integration might be nonhomologous or, alternatively, secondary deletions and/or rearrangements around the junction sites lead to the loss of such structures (Nakazono and Hirai 1993). Due to the dynamic and flexible character of mtDNA (Sect. 3.a), inserted cpDNA is constantly subjected to rearrangements in which short repeated sequences may play a role (Nakazono and Hirai 1993; Watanabe et al. 1994; Kubo et al. 1995). Some chloroplast sequences are found in more than one place in the mtDNA. These copies might either come from extensive mtDNA rearrangements including sequence duplication or, a second fragment of cpDNA has been inserted by homologous recombination via the first, earlier inserted, copy (Watanabe et al. 1994).

c) Proteins Involved in Organelle Recombination

The nature of nucleo-organellar interactions during organelle biogenesis has been extensively studied in the yeast *S. cerevisiae*. In fact, most of the proteins located in the mitochondria are nuclear encoded and many nuclear functions are necessary to maintain the integrity of the mitochondrial genome (Table 2). Many functions, such as the mitochondrial protein synthesis machinery (Myers et al. 1985), have an indirect effect on mtDNA maintenance. Other proteins are directly involved in DNA metabolism (Table 2): various endonucleases, more or less specific in their sequence requirements are known to initiate recombination by creating double-strand breaks in the DNA (Chow and Kunz 1991, Ezekiel and Zassenhaus 1994; Shibata et al. 1995). Homologous recombination as well as gene conversion is extremely reduced (nearly 50%) in mutants lacking the NUC1 endo/exonuclease. The important function of this enzyme in recombination appears to be the enlargement of heteroduplex tracts by its 5' exonuclease activity (Zassenhaus and Denninger 1994). In the course of recombination, Holliday junctions are created, and these were shown to be cut by the mitochondrial CCE1 (MGT1) junction-resolving enzyme (White and Lilley 1997a). Since recombination junction remain unresolved in * cce1* mutants, mtDNA molecules are linked together in branched structures, and this abolishes the normal rapid segregation of mtDNA in sexual crosses (Lockshon et al. 1995). Furthermore, various topoisomerases have been isolated from yeast mitochondria which may take part in illegitimate recombination (Ezekiel et al. 1994; Murthy and Patsupathy 1994; Obernauerová and Subik 1994; Tua et al. 1997). The mismatch repair protein MSH1 plays a role in homeologous recombination (Chi and Kolodner 1994).

Evidences for various topoisomerases also have been found in chloroplasts of *C. reinhardtii*, in the goose foot, *Chenopodium album*, and in the garden pea (Meissner et al. 1992; Woelfle et al. 1993; Zagariya and Sitailo 1995), and a gyrase is known to influence inheritance of cpDNA in the alga *Cyanidioschyzon merolae* (Itoh et al. 1997). Most interesting is the detection of RecA-like enzymes in the chloroplasts of *C. reinhardtii*, *Arabidopsis thaliana*, and *Pisum sativum*. RecA is a single-stranded DNA-binding protein that mediates strand displacement and pairing with homologous double-stranded DNA and such strand transfer activity has been found in pea chloroplasts (Cerutti and Jagendorf 1993). The nuclear encoded RecA homologue of *A. thaliana* has been shown to partially complement *recA* mutations in the bacterium *E. coli* (Pang et al. 1993), and dominant negative *recA* alleles of *E. coli* interfere with the activity of the RecA homologue in chloroplasts of *C. reinhardtii* transformants (Cerutti et al. 1995). All in all, the information on nucleocytoplasmic interactions in plants that influence the structural organization of plant organelle DNA is, however, limited. Nuclear genes that control changes in the organization of the mitochondrial genome are known to exist in wheat, maize, soybean, and *Arabidopsis* (Escote-Carlson et al. 1990; Hartmann et al. 1992; He et al. 1995; sakamoto et al. 1996), but it is not clear if these nuclear genes differentially influence recombination and/or replication of the various members of the multipartite mitochondrial genome.

Some insight has been gained in nuclear control of mitochondrial recombination in the fungus *P. anserina*, mostly due to the senescence phenomenon, the unavoidable arrest of vegetative growth which is correlated with structural rearrangements of the mtDNA (Jamet-Vierny et al. 1997b, c). It is estimated from a mutagenesis screen that at least 600 nuclear genes modulate the life span of the fungus and possibly also contribute to the defective structure of mtDNA molecules (Rossignol and Silar 1996; Silar et al. 1997b). The first isolated genes made it clear that, as in *S. cerevisiae*, many of these genes will not directly contribute to mtDNA recombination. Among the so far characterized genes that appear to be associated to specific mitochondrial DNA rearrangements is the long-known gene *grisea* that encodes a putative copper-activated transcription factor (Borghouts et al. 1997). In other cases, genes for the cytosolic ribosomal proteins S7, S12, and S19 have been identified, the first of which is a homologue to the S13 ribosomal protein of *S. cerevisiae* known to be involved in fidelity control of cytosolic translation (Dequard-Chablat and Sellem 1994; Silar et al. 1997a). A suppressor gene of the S12 mutation *AS1–4* encodes the mitochondrial outer membrane protein Tom70 that, by analogy in *S. cerevisiae*, is involved in the import of proteins from the cytosol into the mitochondria. Another suppressor is homologous to MDM10, which is also an outer membrane protein of *S. cerevisiae* (Jamet-Vierny et al. 1997a; Table 2). For more than 40 years, the *mt⁻* mating type was believed to influence the senescence process and, consequently, mitochondrial rearrangement. However, recent analysis revealed that a new locus (*rpm*) tightly linked to the mating type genes is responsible for certain mtDNA deletions and possibly in an allele-specific manner for the effect of the *mt⁻* haplotype on senescence (Contamine et al. 1996).

4 Plasmids

Natural plasmids have been frequently found in mitochondria of plants and fungi, but not in chloroplasts. Mitochondrial plasmids might be classified by their structure, by the presence or absence of homologies to the host mtDNA, and by the effects they might have on the host mtDNA. In fungi, most but not all plasmids are linear. In higher plants, circular plasmids may predominate. However, as shown in *N. crassa*, both circular and linear plasmids can occur in one species and even in one population of an organism. Most mitochondrial plasmids do not provide any known phenotypic advantage to their host but they encode functions (e.g., DNA- and RNA polymerase, reverse transcriptase) related to their own propagation. Some plasmids, however, have been implicated in fungal senescence, longevity, and rearrangements of the mitochondrial genome, resulting in mitochondrial malfunction and growth abnormalities (Griffiths 1995; Kawano et al. 1995).

a) Plasmid Inheritance and Horizontal Plasmid Transfer

As in organelles, there is a preference for maternal inheritance of mito-chondrial-associated plasmids (Griffiths 1995). However, plasmid Kalilo in *N. crassa*, Han-2 and Har-L in *N. intermedia*, pAL2-1 in *P. anserina* and Callan-a in *Epichloë typhina* are not or not always inherited along the maternal line. These plasmids are frequently lost during sexual propagation although the maternal mtDNA is maintained (Hermanns et al. 1994; Debets et al. 1995; Griffiths and Yang 1995; Chung et al. 1996). Occasional male transmission of mitochondrial plasmids (in frequencies ranging from 1–15%) has been documented in some crosses of *N. crassa*, *N. intermedia*, and *E. typhina* (May and Taylor 1989; Debets et al. 1995; Chung et al. 1996). In addition, a mitochondrion-associated linear plasmid in *Brassica* also tends to be paternally transmitted (Erickson and Kemble 1990). Plasmids may force their own transmission by overriding the normal pattern of mitochondrial inheritance, like plasmid mF in *P. polycephalum* that induces mitochondrial fusions (Kawano et al. 1995). The discovery of homologous mitochondrial plasmids in different *Neurospora* species also suggests inheritance of plasmids independently of their host mitochondrion (May and Taylor 1989). Similarly, mitochondrial plasmids in plants are conservative, and phylogenetic analysis supports their independent transfer (Palmer et al. 1983). Beside the usual mode of inheritance during the sexual cycle (vertical transfer), genetic information may be transmitted by horizontal gene transfer. It allows genes to invade a foreign genome, even if donor and receptor are not able to enter the reproductive cycle. In filamentous fungi, hyphal fusion (anastomosis) between genotypically dissimilar organisms offers the possibility of horizontal gene transfer. Transfer of mitochondrial plasmids has been reported in the genus *Neurospora* (Debets et al. 1994). Kempken (1995) demonstrated horizontal transfer of the mitochondrial *Ascobolus* plasmid pAI2 into the unrelated dung fungus *P. anserina*, but pAI2 has a low stability in the foreign host.

b) Plasmid Recombination

Various recombinative events have been observed in connection with mitochondrial plasmids. Certain plasmids are generated by recombination, others undergo intra- and interplasmid recombination and/or interact with the mitochondrial DNA. Amongst the circular plasmids released from and associated with rearrangements of the mitochondrial DNA are the well-studied senDNAs (α, β, γ, δ, ε, and θ) found in senescent cultures of *P. anserina*. The most common subcircle is the 2.5-kb p1DNA or α-senDNA that corresponds exactly to the first intron of the COX1 gene which encodes a protein with reverse transcriptase (Faßbender et al. 1994) and probably endonuclease activity (Sainsard-Chanet et al. 1994). The initial step for p1DNA liberation has been proposed to be an excision event mechanistically equivalent to the DNA-splicing procedure or alternatively, p1DNA might be produced by reverse transcription of the *cox1* pre-mRNA (Esser et al. 1980; Schmidt et al. 1990). Other scientists prefer a model where the intron transposes downstream from its own upstream exon, followed by a looping-out excision event of intrachromosomal tandem repeats of the intron (Sainsard-Chanet et al. 1994). Although neither of these models requires autoreplication for p1DNA propagation, there is strong evidence (presence of replication forks and the replication origin) that p1DNA, once liberated, is able to replicate autonomously (Wedde 1994). In contrast to p1DNA, other senDNAs are generated by progressive fragmentation of the wild-type mtDNA via intramolecular recombination. They have variable size and termini and do not arise as frequently. Interestingly, MUSE sequences (GGCGCAAGCTC, see Table 3) contribute in some cases to the release of β-senDNAs, either directly as the repeated sequence in which recombination occurs or recombination occurs through unrelated but closely linked direct repeats (Koll et al. 1996; Jamet-Vierny et al. 1997b, c).

p1DNA of *P. anserina* is able to reintegrate into the mtDNA at other sites than its original place by transposition via reverse splicing. Such integration events are proposed to contribute to site-specific deletion events in the mtDNA (Sainsard-Chanet et al. 1993; Sellem et al. 1993). Circular mitochondrial plasmids (Mauriceville, 3.6 kb; Varkud, 3.8 kb) with retroposon characteristics have been isolated from mitochondria of certain *Neurospora* strains that integrate via RNA intermediates into mtDNA and cause growth defects to the host. The plasmid-encoded reverse transcriptase initiates cDNA synthesis at the 3'-CCA end of the plasmid transcript that adopts a tRNA-like structure. Similarly, the reverse transcriptase recognizes 3' ends of tRNAs and produces the equivalent cDNA. Hybrid cDNAs are formed by template switching between the plasmid transcript and tRNAs. Homologous recombination between hybrid cDNAs and mtDNA sequences coding for tRNA accounts for integration of plasmid cDNA copies into the mtDNA (Chen and Lambowitz 1997; Chiang and Lambowitz 1997).

Mitochondrial linear plasmids have an ORF for a DNA polymerase, terminal inverted repeats (TIRs), and proteins associated at the 5' ends like some DNA viruses. The terminal repeats of linear plasmids have been implicated in integration events into mitochondrial DNA: a 60-bp deletion and a 13-bp insertion within the inverted repeat abolishes the ability of the 9-kb *Neurospora* plasmid Kalilo to integrate into mtDNA. Kalilo in its unmutated form integrates at seven distinct regions of the mtDNA. Integration always generates long inverted repeats of mtDNA flanking the ends of the inserted plasmid and, always, short imperfect palindromes of 5–18 bp are lost from each end of Kalilo. In contrast, the unrelated 7-kb plasmid Maranhar, found in other *Neurospora* isolates, integrates completely into mtDNA under formation of flanking inverted repeats of mtDNA. The integrative forms of both plasmids are thought to be single-stranded replication intermediates that have folded into a panhandle structure by pairing of the bases in the TIR sequence. There is no sequence homology between Maranhar and its mtDNA integration sites, but 5-bp regions of homology have been identified between the termini of Kalilo and its integration sites in the mtDNA, which might support cross-over and insertion of the plasmid (Griffiths 1995). In contrast to Kalilo and Maranhar, that induce fungal senescence by integration into mtDNA, the 8.4-kb *P. anserina* plasmid pAL2-1 confers longevitiy to its host. Integration of pAL2-1 is site-specific within the third intron of the apocytochrome b gene and is accompanied by the formation of long inverted repeats of mtDNA attached to the ends of the plasmid by a 15-bp AT linker (Hermanns and Osiewacz 1996). The plasmid integration mechanism in *Zea mays* is fundamentally different from those described so far. The mtDNA contains a 186-bp stretch of the 208-bp-long TIR of the linear plasmids S1 and S2 which integrate into mtDNA via homologous recombination through these regions. The mitochondrial genome linearizes in consequence of this mode of plasmid integration (Schardl et al. 1984). Similarly, the linear plasmid mF of *P. polycephalum* recombines with the host mtDNA via a common sequence of 479 bp. Due to this homologous recombination, two linear recombinant mtDNAs with the 205-bp plasmid TIR at one end are found in the mitochondrion along with the unrecombined plasmid and mtDNA. Common sequences between plasmid and mtDNA in these and other organisms (*Agaricus bisporus, Claviceps purpurea, Secale cereale*) comprise TIR sequences and parts of polymerase genes and are possibly relics of former plasmid integrations (Kawano et al. 1995).

Plasmids may also undergo intramolecular rearrangements by mechanisms similar to those observed in rearrangements of mtDNA (Sect. 3). A population of four circular related plasmids in rice mitochondria is thought to derive either from intramolecular recombination or from slipped mispairing during replication. Consistent with this observation, short repeats were found at the plasmid deletion/insertion sites (Miyata et al. 1995). Similarly, a set of a 26-bp imperfect direct repeat created a deletion in the circular S plasmid of date palm (Benslimane et al. 1996) and shortened forms of Kalilo might be generated by replication slippage (Vierula and Bertrand 1992). In other cases, different plasmids recombined with each other, as in teosinte (*Zea luxurians*), where the linear M1 plasmid obviously arose from a nonhomologous recombination event between two other linear plasmids (Grace et al. 1994). A new linear plasmid, Harbin-L, and a new small circular plasmid Harbin 0.9, were formed in a Chinese strain of *N. intermedia* by reciprocal recombination within 7-bp regions of identity between the circular plasmid Zhisi and the linear plasmid Harbin-1 (Griffiths and Yang 1995). Variant mitochondrial plasmids in *Vicia faba* arose from two succesive or simultaneous recombination events involving the two circular plasmids mtp1 and mtp2. One recombination event occurred within a 276-bp region identical in the parent plasmids, creating chimeric plasmids. The second type of recombination events causing sequence deletions was located in a region characterized by numerous direct repeats of a 31-bp motif. Interestingly, the presence of such mutated plasmids is nuclear-controlled at the level of either plasmid creation or replication (Flamand et al. 1993).

5 Conclusions

Classically, it was assumed that organelle DNA is maternally inherited. However, more and more examples for a preferential paternal inheritance are accumulating in the literature, overthrowing the strict dogma of maternal inheritance. In other cases, leakage in maternal inheritance enables paternal DNA to be transferred into the progeny. Fusion of the parental mitochondria or plastids during sexual crosses is a prerequisite for interparental recombination of organelle DNA but organellar recombination is generally rare during sexual crosses. In many organisms, organelle DNA frequently undergoes recombination without the introduction of foreign sequences. Various mechanisms contribute to rearrangements of the organelle genome. Most of the observed rearrangements do not affect metabolism and finess of the host organism. In other cases, rearrangements of organelle DNA, especially mtDNA, interferes, for instance, with the life span of some fungi and pollen sterility in plants. In general, mitochondrial genomes seem to be more accessible to recombination than the cpDNA. Recombination events may be either accidental or directed. An example for the latter is the homologous recombination along the large inverted repeats in chloroplast genomes, which is thought to contribute to the conservation and stability of the cpDNA (Palmer 1991).

Acknowledgments. We thank Prof. M. Bölker for sharing unpublished data, Dr. Robert P. Boulianne for discussions and critical comments on the manuscript. Work by U. K. was supported by a Glasstone Fellowship of the University of Oxford, by a grant of the Swiss Nationalfond (Grant 31-46'940.96) and by the ETH Zürich.

References

Aagaard JE, Vollmer SS, Sorensen FC, Strauss SH (1995) Mitochondrial DNA products among RAPD profiles are frequent and strongly differentiated between races of Douglas-fir. Mol Ecol 4:441–447

Adams CR, Stamer KA, Miller JK, McNally JG, Kirk MM, Kirk DL (1990) Patterns of organellar and nuclear inheritance among progeny of two geographically isolated strains of *Volvox carteri*. Curr Genet 18:141–153

Aii J, Kishima Y, Mikami T, Adachi T (1997) Expansion of the IR in the chloroplast genomes of buckwheat species is due to incorporation of an SSC sequence that could be mediated by an inversion. Curr Genet 31:276–279

Ainsworth AM, Beeching JR, Broxholme SJ, Hunt BA, Rayner ADM, Scard PT (1992) Complex outcome of reciprocal exchange of nuclear DNA between two members of the basidiomycete genus *Stereum*. J Gen Microbiol 138:1147–1157

Akagi H, Shimada H, Fujimara T (1995) High-frequency inter-parental recombination between mitochondrial genomes of rice cybrids. Curr Genet 29:58–65

Albert B, Godelle B, Atlan A, De Paepe R, Gouyon PH (1996) Dynamics of plant mitochondrial genome: model of a three-level selection process. Genetics 144:369–382

Alberts B, Bray D, Lewis J, Raff M, Roberts K, Watson JD (1994) Molecular biology of the cell, 3rd edn. Garland, New York

Aldrich J, Cherney BW, Merlin F, Christopherson L (1988) The role of insertions/deletions in the evolution of the intergenic region between *psbA* and *trnH* in the chloroplast genome. Curr Genet 14:137–146

Almasan A, Mishra NC (1991) Recombination by sequence repeats with formation of suppressive or residual mitochondrial DNA in *Neurospora*. Proc Natl Acad Sci USA 88:7684–7688

Amoatey HM, Tilney-Basett RAE (1994) A test of the complementary gene model for the control of biparental plastid inheritance in zonal pelargoniums. Heredity 72:69–77

André C, Levy A, Walbot V (1992) Small repeated sequences and the structure of plant mitochondrial genomes. Trends Genet 8:128–132

Armbrust EV, Ferris PJ, Goodenough UW (1993) A mating type-linked gene cluster expressed in *Chlamydomonas* zygotes participates in the uniparental inheritance of the chloroplast genome. Cell 74:801–811

Armbrust EV, Ibrahim A, Goodenough UW (1995) A mating type-linked mutation that disrupts the uniparental inheritance of chloroplast DNA also disrupts cell-size control *Chlamydomonas*. Mol Biol Cell 6:1807–1818

Avni A, Edelman M (1991) Direct selection for paternal inheritance of chloroplasts in sexual progeny of *Nicotiana*. Mol Gen Genet 225:273–277

Azpiroz R, Butow RA (1993) Patterns of mitochondrial sorting in yeast zygotes. Mol Biol Cell 4:21–36

Backert S, Dörfel P, Lurz R, Börner T (1996) Rolling-circle replication of mitochondrial DNA in the higher plant *Chenopodium album* (L.). Mol Cell Biol 16:6285–6294

Baptista-Ferreira JEC; Economou A, Casselton LA (1983) Mitochondrial genetics of *Coprinus*: recombination of mitochondrial genomes. Curr Genet 7:405–407

Barroso G, Labarère J (1997) Genetic evidence for nonrandom sorting of mitochondria in the basidiomycete *Agrocybe aegerita*. Appl Environ Microbiol 63:4686–4691

Barroso G, Moulinier T, Labarère J (1992) Involvement of a large inverted repeated sequence in a recombinational rearrangement of the mitochondrial genome of the higher fungus *Agrocybe aegerita*. Curr Genet 22:155–161

Beckers MC, Munaut C, Minet A, Matagne RF (1991) The fate of mitochondrial DNAs of mt^+ and mt^- origin in gametes and zygotes of *Chlamydomonas*. Curr Genet 20:239–243

Belcour L, Begel O (1977) Extranuclear recombination in *Aspergillus nidulans*: closely linked multiple chloramphenicol- and oligomycin-resistance loci. Mol Gen Genet 156:303–311

Bell PR, Duckett JG (1976) Gametogenesis and fertilization in *Pteridium*. Bot J Linn Soc 73:47–78

Bellaoui M, Martin-Canadell A, Pelletier G, Budar F (1998) Low-copy-number molecules are produced by recombination, actively maintained and can be amplified in the mitochondrial genome of Brassicaceae: relationship to reversion of the male sterile phenotype in some cybrids. Mol Gen Genet 257:177–185

Bendich A (1993) Reaching for the ring: the study of mitochondrial genome structure. Curr Genet 24:279–290

Benslimane AA, Rode A, Qutier F, Hartmann C (1996) Intramolecular recombination of a mitochondrial minicircular plasmid-like DNA of date-palm mediated by a set of short direct-repeat sequences. Curr Genet 26:535–541

Berger KH, Yaffe MP (1996) Mitochondrial distribution and inheritance. Experientia 52:1111–1116

Berger KH, Sogo LF, Yaffe MP (1997) Mdm12p, a component required for mitochondrial inheritance that is conserved between budding and fission yeast. J Cell Biol 136:545–553

Birky CW Jr (1994) Relaxed and stringent genomes: why cytoplasmic genes don't obey Mendel's laws. J Hered 85:355–365

Birky CW Jr (1995) Uniparental inheritance of mitochondrial and chloroplast genes: Mechanisms and evolution. Proc Natl Acad Sci USA 92:11331–11338

Blanchard JL, Schmidt GW (1995) Pervasive migration of organellar DNA to the nucleus in plants. J Mol Evol 41:397–406

Boblenz K, Nothnagel T, Metzlaff M (1990) Paternal inheritance of plastids in the genus *Daucus*. Mol Gen Genet 220:489–491

Borghouts C, Kimpel E, Owiewacz HD (1997) Mitochondrial DNA rearrangements of *Podospora anserina* are under the control of the nuclear gene *grisea*. Proc Natl Acad Sci USA 94:10768–10773

Borkhardt B, Olsen LW (1983) Paternal inheritance of the mitochondrial DNA in interspecific crosses of the aquatic fungus *Allomyces*. Curr Genet 7:403–405

Boudreau E, Turmel M (1995) Gene rearrangements in *Chlamydomonas* chloroplast DNAs are accounted for by inversions and by expansion/contraction of the inverted repeat. Plant Mol Biol 27:351–364

Boudreau E, Turmel M (1996) Extensive gene rearrangements in the chloroplast DNAs of *Chlamydomonas* species featuring multiple dispersed repeats. Mol Biol Evol 13:233–243

Boynton JE, Harris EH, Burkhardt BD, Lamerson PM, Gillman NW (1987) Transmission of mitochondrial and chloroplast genomes in crosses of *Chlamydomonas*. Proc Natl Acad Sci USA 84:2391–2395

Boynton JE, Gillham NW, Newmann SM, Harris EH (1992) Organelles genetics and transformation of *Chlamydomonas*. In: Herrmann RG (ed) Cell organelles. Springer, Vienna New York, pp 3–64

Brasier CM, Kirk SA (1986) Maternal inheritance of chloramphenicol tolerances in *Ophiostoma ulmi*. Trans Br Mycol Soc 87:460–462

Bråten T (1973) Autoradiographic evidence for the rapid disintegration of one chloroplast in the zygote of the green alga *Ulva mutabilis*. J Cell Sci 12:385–389

Breiman A (1987) Mitochondrial DNA diversity in the genera of *Triticum* and *Aegilops* revealed by Southern blot hybridisation. Theor Appl Genet 73:563–570

Brennicke A, Schwemmle B (1984) Inheritance of mitochondrial DNA in *Oenothera bertiana* and *Oenothera odorata* hybrids. Z Naturforsch 39c:191–192

Brennicke A, Grohmann L, Hiesel R, Knoop V, Schuster W (1993) The mitochondrial genome on its way to the nucleus: different stages of gene transfer in higher plants. FEBS Lett 325:140–145

Byrne M, Moran GF, Tibbits WN (1993) Restriction map and maternal inheritance of chloroplast DNA in *Eucalyptus nitens*. J Hered 84:218–220

Casselton LA, Asante-Owusu RN, Banham AH, Kingsnorth CS, Kües U, O'Shea SF, Pardo EH (1995) Mating type control of sexual development in *Coprinus cinereus*. Can J Bot 73: Suppl 1, S266–S272

Cerutti H, Jagendorf AT (1993) DNA strand-transfer activity in pea (*Pisum sativum* L.) chloroplasts. Plant Physiol 102:145–153

Cerutti H, Ibrahim Z, Jagendorf AT (1993) Treatment of pea (*Pisum sativum* L.) protoplasts with DNA-damaging agents induces a 39-kilodalton chloroplast protein immunologically related to *Escherichia coli* RecA. Plant Physiol 102:155–163

Cerutti H, Johnson AM, Boynton JE, Gillham NW (1995) Inhibition of chloroplast DNA recombination and repair by dominant negative mutants of *Escherichia coli* RecA. Mol Cell Biol 15:3003–3011

Chen B, Lambowitz AM (1997) *De novo* an DNA primer-mediated initiation of cDNA synthesis by the Mauriceville retroplasmid reverse transcriptase involve recognition of a 3' CCA sequence. J Mol Biol 271:311–332

Chi NW, Kolodner RD (1994) The effect of DNA mismatches on the ATPase activity of MSH1, a protein in yeast mitochondria that recognizes DNA mismatches. J Biol Chem 269:29993-29997

Chiang CC, Lambowitz A (1997) The Mauriceville retroplasmid reverse transcriptase initiates cDNA synthesis de novo at the 3' end of tRNAs. Mol Cell Biol 17:4256–4535

Chiu W-L, Sears BB (1993) Plastome-genome interactions affect plastid transmission in *Oenothera*. Genetics 133:998–997

Chiu W-L, Stubbe W, Sears BB (1988) Plastid inheritance in *Oenothera*: organelle genome modifies the extent of biparental plastid transmission. Curr Genet 13:181–189

Chong DKX, Chinnappa CC, Yeh FC, Chuong S (1994) Chloroplasst inheritance in the *Stellaria longipes* complex (Caryophyllaceae). Theor Appl Genet 88:614–617

Chow TY, Kunz BA (1991) Evidence that an endo-exonuclease controlled by the *NUC2* gene functions in the induction of petite mutations in *Saccharomyces cerevisiae*. Curr Genet 20:39–44

Chung K-R, Leuchtmann A, Schardl CL (1996) Inheritance of mitochondrial DNA and plasmids in the asomycetous fungus, *Epichloë typhina*. Genetics 142:259–265

Cipriani G, Testolin R, Morgante M (1995) Paternal inheritance of plastids in interspecific hybrids of the genus *Actinidia* revealed by PCR-amplification of chloroplast fragments. Mol Gen Genet 247:693–697

Coenen A, Croft JH, Slakhorst M, Debets F, Hoekstra R (1996) Mitochondrial inheritance in *Aspergillus nidulans*. Genet Res 67:93–100

Conde MF, Pring DR, Levings CS III (1979) Maternal inheritance of organelle DNAs in *Zea mays–Zea perennis* reciprocal crosses. J Hered 70:2–4

Conklin PL, Hanson MR (1993) A truncated recombination repeat in the mitochondrial genome of a *Petunia* CMS line. Curr Genet 23:477–482

Contamine V, Lecellier G, Belcour L, Picard M (1996) Premature death in *Podospora anserina*: sporadic accumulation of the deleted mitochondrial genome, translational parameters and innocuity of the mating types. Genetics 144:541–555

Corriveau JL, Coleman AW (1988) Rapid screening methods to detect potential biparental inheritance of plastid DNA and results for over 200 angiosperm species. Am J Bot 75:1443–1459

Corriveau JL, Coleman AW (1991) Monitoring by epifluorescence microscopy of organelle DNA fate during pollen development in five angiosperm species. Dev Biol 147:271–280

Cosner ME, Jansen RK, Palmer JD, Downie SR (1997) The highly rearranged chloroplast genome of *Trachelium caeruleum* (Campanulaceae): multiple inversions, inverted repeat expansion and contraction, transposition, insertions/deletions, and several repeat families. Curr Genet 31:419–429

Cruzan MB, Arnold ML, Carney SE, Wollenberg KR (1993) cpDNA inheritance in interspecific crosses and evolutionary inference in Lousiana irises. Am J Bot 80:344–350

David AJ, Keathley DE (1996) Inheritance of mitochondrial DNA in interspecific crosses of *Picea glauca* and *Picea omorika*. Can J For Res 26:428–432

Debets F, Yang X, Griffiths AJF (1994) Vegetative incompatibility in *Neurospora*: its effect on horizontal transfer of mitochondrial plasmids and senescence in natural populations. Curr Genet 26:113–119

Debets F, Yang X, Griffiths AJF (1995) The dynamics of mitochondrial plasmids in a Hawaiian population of *Neurospora intermedia*. Curr Genet 29:44–49

Dequard-Chablat M, Sellem CH (1994) The S12 ribosomal protein of *Podospora anserina* belongs to the S19 bacterial family and controls the mitochondrial genome integrity through cytoplasmic translation. J Biol Chem 269:14951–14956

Derepas A, Dulieu H (1992) Inheritance of the capacity to transfer plastids by the pollen parent in *Petunia hybrida* Hort. J Hered 83:6–10

DeVerno LL, Charest PJ, Bonen L (1993) Inheritance of mitochondrial DNA in the confier *Larix*. Theor Appl Genet 86:383–388

de Zamaroczy M, Bernardi G (1986) The GC clusters of the mitochondrial genome of yeastand their evolutionary origin. Gene 41:1–22

Diers LA (1967) Der Feinbau des Spermatozoids von *Sphaerocarpus donnellii* Aust. (*Hepaticaceae*). Planta 72:119–145

Dong J, Wagner DB (19939 Taxonomic and population differentiation of mitochondrial diversity in *Pinus banksiana* and *Pinus contorta*. Theor Appl Genet 86:573–578

Dong J, Wagner DB, Yanchuk AD, Carlson MR, Magnussen S, Wang XR, Szmitdt AE (1992) Paternal chloroplast DNA inheritance in *Pinus contorta* and *Pinus banksiana*: independence of parental species or cross direction. J Hered 83:419–422

Dumolin S, Demesure B, Petit RJ (1995) Inheritance of chloroplast and mitochondrial genomes in pedunculate oak investigated with an efficient PCR method. Theor Appl Genet 91:1253–1256

Dürrenberger F, Thompson AJ, Herrin DL, Rochaix J-D (1996) Double-strand break-induced recombination in *Chlamydomonas reinhardtii* chloroplasts. Nucleic Acids Res 24:3323–3331

Ephrussi B (1953) Nucleo-cytoplasmic relations in microorganisms. Clarendon, Oxford

Erickson L, Kemble R (1990) Paternal inheritance of mitochondria in rapeseed (*Brassica napus*). Mol Gen Genet 222:135–139

Erickson L, Kemble R (1993) The effect of genotype on pollen transmission of mitochondria in rapeseed (*Brassica napus*). Sex Plant Rep 6:33–39

Erickson L, Kemble R, Swanson E (1989) The *Brassica* mitochondrial plasmid can be sexually transmitted. Pollen transfer of a cytoplasmic genetic element. Mol Gen Genet 218:419–422

Escote-Carlson LJ, Gabay-Laughnan S, Laughnan JR (1990) Nuclear genotypes affects mitochondrial genome organization of CMS-S maize. Mol Gen Genet 223:457–464

Esser K, Tudzynski P, Stahl U, Kück U (1980) A model to explain senescence in the filamentous fungus *Podospora anserina* by the action of plasmid-like DNA. Mol Gen Genet 178:213–216

Evans RJ, Clark-Walker GD (1985) Elevated levels of petite formation in strains of *Saccharomyces cerevisiae* restored to respiratory competence. II. Organization of mitochondrial genomes in strains having high and moderate frequencies of petite mutant formation. Genetics 111:403–432

Ezekiel UR, Zassenhaus HP (1994) Evidence for a site-specific endonuclease in yeast mitochondria which recognizes the sequence 5'GCCGGC. Biochem Biophys Res Commun 201:208–214

Ezekiel UR, Towler EM, Wallis JW, Zassenhaus HP (1994) Evidence for a nucleotide-dependent topoisomerase activity from yeast mitochondria. Curr Genet 27:31–37

Fan W-H, Woelfle MA, Mosig G (1995) Two copies of a DNA element, Wendy, in the chloroplast chromosome of *Chlamydomonas reinhardtii* between rearranged gene clusters. Plant Mol Biol 29:63–80

Faßbender S, Bruhl KH, Ciriacy M, Kück U (1994) Reverse transcription activity of an intron encoded polypeptide. EMBO J 13:2075–2083

Faure S, Noyer J-L, Carreel F, Horry J-P, Bakry F, Lanaud C (1994) Maternal inheritance of chloroplast genome and paternal inheritance of mitochondrial genome in bananas (*Musa acuminata*). Curr Genet 25:265–269

Fauron C, Casper M, Gao Y, Moore B (1995) The maize mitochondrial genome: dynamic, yet functional. Trends Genet 11:228–235

Flamand M-C, Duc G, Goblet J-P, Hong L, Louis O, Briquet M, Boutry M (1993) Variant mitochondrial plasmids of broad bean arose by recombination and are controlled by the nuclear genome. Nucleic Acids Res 21:5468–5473

Förster H, Coffey MD (1990) Mating behaviour of *Phytohthora parasitica*: Evidence for sexual recombination in oospores using DNA restriction fragment length polymorphisms as genetic markers. Exp Mycol 14:351–359

Forsthoefel NR, Bohnert HJ, Smith SE (1992) Discordant inheritance of mitochondrial and plastid DNA in diverse alfalfa genotypes. J Hered 83:342–345

Fröst S, Vaivaras L, Carlbom C (1970) Reciprocal extrachromosomal inheritance in rye (Secale cereale L.). Hereditas 65:251–260

Fukuda M, Harada Y, Imahori S, Fukumasa-Nakai Y, Hayashi Y (1995) Inheritance of mitochondrial DNA in sexual crosses and protoplast cell fusions in *Lentinus edodes*. Curr Genet 27:550–554

Gastony GJ, Yatskievich G (1992) Maternal inheritance of the chloroplast and mitochondrial genomes in cheilanthoid ferns. Am J Bot 76:716–722

Gilham NW (1994) Organelle genes and genomes. Oxford University Press, New York

Gloeckner G, Beck CF (1995) Genes involved in light control of sexual differentitation in *Chlamydomonas reinhardtii*. Genetics 141:937–943

Goulding SE, Olmstead RG, Morden CW, Wolfe KH (1996) Ebb and flow of the chloroplast inverted repeat. Mol Gen Genet 252:195–206

Gounaris Y (1993) Comparison of restriction patterns of mitochondrial DNA from low and high sugar accumulating cultivars/seedlings. J Plant Physiol 141:423–427

Grace KS, Allen JO, Newton KJ (1994) R-type plasmids in mitochondria from a single source of *Zea luxurians* teosinte. Curr Genet 25:258–264

Grieshammer U, Wynne JC (1990) Mendelian and non-Mendelian inheritance of three isozymes in peanut (*Arachis hypogaea* L.). Peanut Sci 17:101–105

Griffiths AJF (1995) Natural plasmids of filamentous fungi. Microbiol Rev 59:673–685

Griffiths AJF, Yang X (1995) Recombination between heterologous linear and circular mitochondrial plasmids in the fungus *Neurospora*. Mol Gen Genet 249:25–36

Griffiths AJF, Collins RA, Nargang FE (1995) Mitochondrial genetics of *Neurospora*. In: Kück U (ed) The Mycota, vol II. Springer, Berlin Heidelberg New York, pp 93–108

Guo H, Zimmerly S, Perlman PS, Lambowitz AM (1997) Group II intron endonuclease use both RNA and protein subunits for recognition of specific sequences in double-stranded DNA. EMBO J 16:6835–6848

Hachtel W (1980) Maternal inheritance of chloroplast DNA in some *Oenothera* species. J Hered 71:191–194

Hagemann R, Schröder M-B (1989) The cytological basis of the plastid inheritance in angiosperms. Protoplasma 152:57–64

Hanson MR, Folkerts O (1992) Structure and function of the higher plant mitochondrial genomes. Int Rev Cytol 141:129–172

Hardy CM, Galeotti CL, Clark-Walker GD (1989) Deletions and rearrangements in *Kluyveromyces lactis* mitochondrial DNA. Curr Genet 16:419–427

Harris SA, Ingram R (1991) Chloroplast DNA and biosystematics: the effects of intra-specific diversity and plastid transmission. Taxon 40:393–412

Hartmann C, De Buyser J, Henry Y, Morère-Le Paven MC, Dyer T, Rode A (1992) Nuclear genes control changes in the organization of the mitochondrial genome in tissue cultures derived from immature embryos of wheat. Curr Genet 21:515–520

Hartmann C, Récipon H, Jubier M-F, Valon C, Delcher-Besin E, Henry Y, De Buyser J, Lejeune B, Rode A (1994) Mitochondrial DNA variability detected in a single wheat regenerant involves rare recombination event across a short repeat. Curr Genet 25:456–464

Hatfield PM, Shoemaker RC, Palmer RG (1985) Maternal inheritance of chloroplast DNA within the genus *Glycine*, subgenus *soja*. Heredity 76:373–374

Havey MJ (1995) Identification of cytoplasm using the polymerase chain reaction to aid in the extraction of maintainer lines from open-pollinated populations of onion. Theor Appl Genet 90:263–268

He S, Lynzik A, Mackenzie S (1995) Pollen fertility restoration by nuclear gene *Fr* in CMS bean: nuclear-directed alteration of a mitochondrial population. Genetics 139:955–962

Hermanns J, Osiewacz HD (1996) Induction of longevity by cytoplasmic transfer of a linear plasmid in *Podospora anserina*. Curr Genet 29:250–256

Hermanns J, Asseburg A, Osiewaczs HD (1994) Evidence for a life span-prolonging effect of a linear plasmid in a longevity mutant of *Podospora anserina*. Mol Gen Genet 243:297–307

Hintz WEA, Anderson LB, Horgen PA (1988) Nuclear migration and mitochondrial inheritance in the mushroom *Agaricus bitorquis*. Genetics 119:35–41

Hiratsuka J, Shimada H, Whittier R, Ishibashi T, Sakamoto M, Mori M, Kondo C, Hinji Y, Sun C-R, Meng B-Y, Li Y-Q, Kanno A, Nishizawa Y, Hirai A, Shinozaki K, Sugiura M (1989) The complete sequence of the rice (*Oryza sativa*) chloroplast genome: intermolecular recombination between distinct tRNA genes accounts for a major plastid DNA inversion during the evolution of the cereals. Mol Gen Genet 217:185–194

Hoot SB, Palmer JD (1994) Structural rearrangements, including parallel inversions, within the chloroplast genome of *Anemone* and related genera. J Mol Evol 38:274–281

Horlow C, Goujard J, Lepingle A, Missonier C, Bourgin J-P (1990) Transmission of paternal chloroplasts in tobacco (*Nicotiana tabacum*). Plant Cell Rep 9:249–252

Howe CJ (1985) The endpoints of an inversion in wheat chloroplast DNA are associated with short repeated sequences containing homology to *att*-lambda. Curr Genet 10:139–145

Hu ZM, Hu SY (1996) Changes of plastids, mitochondria and their DNA in the egg cell before and after fertilization in *Pharbitis*. Acta Bot Sin 38:257–261

Hu ZM, Hu SY, Zhang JZ (1996) Paternal inheritance of plastid DNA in genus *Pharbitis*. Acta Bot Sin 38:253–256

Hunt MD, Newton KJ (1991) The *NCS3* mutation: genetic evidence for the expression of ribosomal protein genes in *Zea mays* mitochondria. EMBO J 10:1045–1052

Ichikawa H, Hirai A (1983) Search for the female parent in the genesis of *Brassica napus* by chloroplast DNA restriction patterns. Jpn J Genet 58:419–424

Isaya G, Sakati WR, Rollins RA, Shen GP, Hanson LC, Ullrich RC, Novotny CP (1995) Mammalian mitochondrial intermediate peptidase: structure/function analysis of a new homologue from *Schizophyllum commune* and relationship to thimet oligopeptidase. Genomics 28:450–461

Ishikawa S, Kato S, Imakawa S, Mikami T, Shimamoto Y (1992) Organelle DNA polymorphism in apple cultivars and rootstocks. Theor Appl Genet 83:963–967

Itoh R, Takahashi H, Toda K, Kuroiwa H, Kuroiwa T (1997) DNA gyrase involvement in chloroplast-nucleoid division in *Cyanidioschyzon merolae*. Eur J Cell Biol 73:252–258

Jack PL, Dimitrijevic TAF, Mayes S (1995) Assessment of nuclear, mitochondrial and chloroplast RFLP markers in oil palm (*Elaeis guieensis* Jacq.). Theor Appl Genet 1995:643–649

Jamet-Vierny C, Contamine V, Boulay J, Zickler D, Picard M (1997a) Mutations in genes encoding the mitochondrial outer membrane proteins Tom70 and Mdm10 of *Podospora anserina* modify the spectrum of mitochondrial DNA rearrangements associated with cellular death. Mol Cell Biol 17:6359–6366

Jamet-Vierny C, Boulay J, Briand J-F (1997b) Intramolecular cross-overs generate deleted mitochondrial DNA molecules in *Podospora anserina*. Curr Genet 31:162–170

Jamet-Vierny C, Boulay J, Begel O, Silar P (1997c) Contribution of various classes of defective mitochondrial DNA molecules to senescence in *Podospora anserina*. Curr Genet 31:171–178

Jin T, Horgen PA (1994) Uniparental mitochondrial transmission in the cultivated button mushroom, *Agaricus bisporus*. Appl Environ Microbiol 60:4456–4460

Johnson DA, Hattori J (1996) Analysis of a hotspot for deletion formation within the intron of the chloroplast *trnI* gene. Genome 39:999–1005

Kadowski K-I, Suzuki T, Kazama S (1990) A chimeric gene containing the 5' portion of *atp6* is associated with cytoplasmic male-sterility of rice. Mol Gen Genet 224:10–16

Kanazawa A, Tsutsumi N, Hirai A (1994) Reversible changes in the composition of the population of mtDNAs during dedifferentiation and regeneration in tobacco. Genetics 138:865–879

Kanno A, Hirai A (1993) A transcription map of the chloroplast genome from rice (*Oryza sativa*). Curr Genet 23:166–174

Kanno A, Watanabe N, Nakamurda I, Hirai A (1993) Variations in chloroplast DNA from rice (*Oryza sativa*): differences between deletions mediated by short direct-repeat sequences within a single species. Theoret Appl Genet 86:579–584

Karas I, Cass DD (1976) Ultrastructural aspects of sperm cell formation in rye: evidence for cell plate involvement in generative cell division. Phytomorphology 26:36–45

Kavanagh TA, O'Driscoll KM, McCabe PF, Dix PJ (1994) Mutations conferring lincomycin, spectinomycin, and strepotmycin resistance in *Solanum nigrum* are located in three different chloroplast genes. Mol Gen Genet 242:675–680

Kawano S, Takano H, Kuroiwa T (1995) Sexuality of mitochondria: fusion, recombination, and plasmids. Int Rev Cytol 161:49–110

Kawata M,. Harada T, Shimamoto Y, Oono K, Takaiwa F (1997) Short inverted repeats function as hotspots of intramolecular recombination giving rise to oligomers of deleted plastid DNAs (ptDNAs). Curr Genet 31:179–184

Kempken F (1995) Horizontal transfer of a mitochondrial plasmid. Mol Gen Genet 248:89–94

Kevei F, Tóth B, Coenen A, Hamari Z, Varga J, Croft JH (1997) Recombination of mitochondrial DNAs following transmission of mitochondria among incompatible strains of black *Aspergilli*. Mol Gen Genet 254:379–388

Kiang A-S, Connolly V, McConnell DJ, Kavanagh TA (1994) Paternal inheritance of mitochondria and chloroplasts in *Festuca pratensis-Lolium perenne* intergenic hybrids. Theor Appl Genet 87:681–688

Kobayashi Y, Knoop V, Fukuzawa H, Brennecke A, Ohyama K (1997) Interorganellar gene transfer in bryophytes: the functional *nad7* gene is nuclear encoded in *Marchantia polymorpha*. Mol Gen Genet 256:589–592

Koll F, Boulay J, Belcour L, d'Aubenton-Carafa Y (1996) Contribution of ultra-short invasive elements to the evolution of the mitochondrial genome in the genus *Podospora*. Nucleic Acids Res 24:1734–1741

Krishna Rao M, Koduru PRK (1978) Inheritance of genetic male sterility in *Pennisetum americanum* (1.) Leeke Pearlmillet. Euphytica 27:777–783

Kubo T, Yanai Y, Konoshita T, Mikami T (1995) The chloroplast *trnP-trnW-petG* gene cluster in the mitochondrial genomes of *Beta vulgaris, B. trigyna* and *B. webbiana*: evolutionary aspects. Curr Genet 27:285–289

Kuhsel M, Kowallik KV (1987) The plastome of brown alga, *Dictyota dichotoma*. II. Location of structural genes coding for ribosomal RNAs, the large subunit of ribulose-1,5-bisphosphate carboxylase/oxygenase and for polypeptides of photosystems I and II. Mol Gen Genet 207:361–368

Kuroiwa T, Kawano S, Watanabe M, Hori T (1991) Preferential digestion of chloroplast DNA in male gametangia during the late stage of gametogenesis in the anisogamous *Bryopsis maxima*. Protoplasma 163:102–113

Kuroiwa T, Kawazu T, Uchida H, Ohta T, Kuroiwa H (1993a) Direct evidence of plastid DNA and mitochondrial DNA in sperm cells in relation to biparental inheritance of organelle DNA in *Pelargonium zonale* by fluorescence/electron microscopy. Eur J Cell Biol 62:307–313

Kuroiwa H, Nozaki H, Kuroiwa T (1993b) Preferential digestion of chloroplast nuclei in sperms before and during ferilization in *Volvox carteri*. Cytologia 58:281–291

Kuroiwa T, Uchida H, Hori T, Maegawa M (1993c) Preferential disappearance of chlorophyll preceding digestion of male-derived chloroplast DNA in young zygotes of *Monostroma latissiumum* by a DAPI-epifluorescence microscopy and electron microscopy. Cytologia 58:331–336

Lambowity AM; Belfort M (1993) Introns as mobile genetic elements. Annu Rev Biochem 62:587–622

Laser B, Mohr S, Odenbach W, Oettler G, Kück U (1997) Parental and novel copies of the mitochondrial *orf25* gene in the hybrid crop-plant triticale: predominant transcriptional expression of the maternal gene copy. Curr Genet 32:337–347

Lashermes P, Cros J, Combes MC, Trouslot P, Anthony F, Hamon S, Charrier A (1996) Inheritance and restriction fragment length polymorphism of chloroplast DNA in the genus *Coffea* L. Theor Appl Genet 93:626–632

Leblanc C, Richard O, Kloareg B, Viehmann S, Zetsche K, Boyen C (1997) Origin and evolution of mitochondria: what have we learnt from red algae? Curr Genet 31:193–207

Lee RW, Langille B, Lemieux C, Boer PH (1990) Inheritance of mitochondrial and chloroplast genome markers in backcrosses of *Chlamydomonas eugametos* x *Chlamydomonas moewusii* hybrids. Curr Genet 17:73–76

Lee SB, Taylor JW (1993) Uniparental inheritance and replacement of mitochondrial DNA in *Neurospora tetrasperma*. Genetics 134:1063–1075

Ling F, Makishima F, Morishima N, Shibata T (1995) A nuclear mutation defective in mitochondrial recombination in yeast. EMBO J 14:4090–4101

Liu XL, Hu SY (1995) Studies of generative cell development in *Pisum sativum* – with special reference to cytological basis of plastid maternal inheritance. Acta Bot Sin 37:749–753

Liu XL, Hu SY, Wang S (1997) Cytological basis of plastid inheritance in *Phaeseolus vulgaris*: Investigations of plastids, mitochondria and their DNA nucleiodes during pollen development. Acta Bot Sin 39:106–110

Lockshon D, Zweifel SG, Freeman-Cook LL, Lorimer HE, Brewer BJ, Fangman WL (1995) A role for recombination junctions in the segregation of mitochondrial DNA in yeast. Cell 81:947–955

Lonsdale D (1987) Cytoplasmic male sterility: a molecular perspective. Plant Physiol Biochem 25:265–272

Maleszka R, Skelly PJ, Clark-Walker GD (1991) Rolling circle replication of DNA in yeast mitochondria. EMBO J 10:3923–3929

Maliga P, Carrer H, Kanevski I, Staub J, Svab Z (1993) Plastid engineering in land plants: a conservative genome is open to change. Philos Trans R Soc Lond B 342:203–208

Manna E, Brennicke A (1986) Primary and secondary structure of 26S ribosomal RNA of *Oenothera* mitochondria. Mol Gen Genet 203:377–381

Mannella CA, Pittenger TH; Lambowitz AM (1979) Transmission of mitochondrial deoxyribonucleic acid in *Neurospora crassa* sexual crosses. J Bacteriol 137:1449–1451

Manton I (1970) Plant spermatozoids. In: Bacetti B (ed) Comparative spermatology. Academic Press, New York, pp 143–158

Marienfeld JR, Newton KJ (1994) The maize *NCS2* abnormal growth mutant has a chimeric *nad4-nad7* mitochondrial gene and is associated with reduced Complex I function. Genetics 138:855–863

Martin FN (1989) Maternal inheritance of mitochondrial DNA in sexual crosses of *Pythium sylvaticum*. Curr Genet 16:373–374

Martinez P, Lopez C, Roldan M, Sabater B, Martin M (1997) Plastid DNA of five ecotypes of *Arabidopsis thaliana*: sequence of *ndhG* gene and maternal inheritance. Plant Sci 123:113–122

Masoud SA, Johnson LB, Sorensen EL (1990) High transmission of paternal plastid DNA in alfalfa plants demonstrated by restriction fragment polymorphic analysis. Theor Appl Genet 79:49–55

Matsumoto S, Wakita H, Fukui H (1997) Molecular classification of wild roses using organelle DNA probes. Sci Hortic 68:191–196

Matsumoto T, Fukumasa-Nakai Y (1996) Mitochondrial inheritance in sexual crosses of *Pleurotus ostreatus*. Curr Genet 30:549–552

Matsunaga S, Sakai A, Kawano S, Kuroiwa T (1996) Cytological analysis of the mature pollen of *Actinidia deliciosa* (Kiwifruit). Cytologia 61:337–341

May G, Taylor JW (1988) Patterns of mating and mitochondrial DNA inheritance in the agaric basidiomycete *Coprinus cinereus*. Genetics 118:213–220

May G, Taylor JW (1989) Independent transfer of mitochondrial plasmids in *Neurospora crassa*. Nature 359:320–322

McCauley DE (1994) Contrasting the distribution of chloroplast DNA and allozyme polymorphism among local population of *Silene alba*: implications for studies of gene flow in plants. Proc Natl Acad Sci USA 91:8127–8131

Medgyesy P, Pày A, Màrton L (1986) Transmission of paternal chloroplasts in *Nicotiana*. Mol Gen Genet 204:195–198

Meissner K, Dörfel P, Börner T (1992) Topoisomerase activity in mitochondrial lysates of a higher plant (*Chenopodium album* L.). Biochem Int 27:1119–1125

Metzlaff M, Börner T, Hagemann R (1981) Variations of chloroplast DNAs in the genus *Pelargonium* and their biparental inheritance. Theor Appl Genet 60:37–41

Milgroom MC, Lipari SE (1993) Maternal inheritance and diversity of mitochondrial DNA in the chestnut blight fungus, *Cryphonectria parasitica*. Phytopathology 83:563–567

Mirfakhrai M, Tanaka Y, Yanagisawa K (1990) Evidence for mitochondrial DNA polymorphism and uniparental inheritance in the cellular slime mould *Polyspondylium pallidum*: effect of intraspecies mating on mitochondrial DNA transmission. Genetics 124:607–613

Miyamura S, Kuroiwa T, Nagata T (1987) Disappearance of plastid and mitochondrial nucleotids during the formation of generative cells of higher plants revealed by fluorescence microscopy. Protoplasma 141:149–159

Miyata S, Kanazawa A, Tsutsumi N, Sano Y, Hirai A (1995) Mitochondrial plasmid-like DNAs of the B1 family in the genus *Oryza*: sequence heterogeneity and evolution. Jpn J Genet 70:675–685

Mogensen HL (1988) Exclusion of male mitochondria and plastids during syngamy as a basis for maternal inheritance. Proc Natl Acad Sci USA 85:2594–2597

Mogensen HL (1996) The hows and whys of cytoplasmic inheritance in seed plants. Am J Bot 83:383–404

Mohr S, Schulte-Kappert E, Odenbach W, Oettler G, Kück U (1993) Mitochondrial DNA of cytoplasmic male-sterile *Triticum timopheevi*: rearrangements of upstream sequences of the *atp6* and *orf25* genes. Theor Appl Genet 86:259–268

Morere Le Paven MC, De Buyser J, Henry Y, Hartmann C, Rhode A (1994) Unusual inheritance of the mitochondrial genome organization in the progeny of reciprocal crosses between alloplasmic hexaploid wheat regenerants. Theor Appl Genet 89:572–576

Morton BR; Clegg MT (1993) A chloroplast DNA mutational hotspot and gene conversion in a noncoding region near *rbcL* in the grass family (Poaceae). Curr Genet 24:357–365

Moeykens CA, Mackenzie SA, Shoemaker RC (1995) Mitochondrial genome diversity in soybean: repeats and rearrangements. Plant Mol Biol 29:245–254

Murthy V, Pasupathy K (1994) Evidence for the presence of topoisomerase-like activity in mitochondria of *Saccharomyces cerevisiae*. Biochem Biophys Res Commun 198:387–392

Myers AM, Pape LK, Tzagoloff A (1985) Mitochondrial protein synthesis is required for maintenance of intact mitochondrial genomes in *Saccharomyces cerevisiae*. EMBO J 4:2087–2092

Nagata N, Sodmergen TS, Saito C, Sakai A, Kuroiwa H, Kuroiwa T (1997) Preferential degradation of plastid DNA with preservation of mitochondrial DNA in the sperm cells of *Pelargonium zonale* during pollen development. Protoplasma 197:217–219

Nakamura S, Ikehara T, Uchida H, Suzuki T, Sodmergen TS (1992) Fluorescence microscopy of plastid nucleoids and a survey of nuclease C in higher plants with respect to mode of plastid inheritance. Protoplasma 169:68–74

Nakazono M, Hirai A (1993) Identifikation of the entire set of transferred chloroplast DNA sequences in the mitochondrial genome of rice. Mol Gen Genet 236:341–346

Neale DB, Sederoff RR (1989) Paternal inheritance of chloroplast DNA and maternal inheritance of mitochondrial DNA in loblolly pine. Theor Appl Genet 77:212–216

Neale DB, Wheeler NC, Allard RW (1986) Paternal inheritance of chloroplast DNA in Douglas-fir. Can J For Res 6:1152–1154

Neale DB, Marshall KA, Sederoff RR (1989) Chloroplast and mitochondrial DNA are paternally inherited in *Sequioa sempervirens* D. Don. Endl. Proc Natl Acad Sci USA 86:9347–9349

Neale DB, Marshall KA, Harry DE (1991) Inheritance of chloroplast and mitochondrial DNA in incense-cedar (*Calocedrus decrurrens*). Can J For Res 21:717–720

Nimzyk R, Schöndorf T, Hachtel W (1993) In-frame length mutations associated with short tandem repeats are located in unassigned open reading frames of *Oenothera* chloroplast DNA. Curr Genet 23:265–270

Nunnari J, Marshall WF, Straight A, Murray A, Sedat JW, Walter P (1997) Mitochondrial transmission during mating in *Saccharomyces cerevisiae* is determined by mitochondrial fusion and fission and the intramitochondrial segregation of mitochondrial DNA. Mol Biol Cell 8:1233–1242

Obernauerová M, Subik J (1994) Energy-dependent mitochondrial mutagenicity of antibacterial ofloxacin and its recombinogenic activity in yeast. Curr Genet 26:281–284

Ogihara Y, Terachi T, Sasakuma T (1988) Intramolecular recombination of the chloroplast genome mediated by short direct-repeat sequences in wheat species. Proc Natl Acad Sci USA 85:8573–8577

Ogihara Y, Terachi T, Sasakuma T (1992) Structural analysis of length mutations in a hot-spot region of wheat chloroplast DNAs. Curr Genet 22:251–258

Ohba K, Iwakawa M, Okada Y, Murai M (1971) Paternal transmission of a plastid anomaly in some reciprocal crosses of sugi, *Cryptomeria japonica*. D. Don. Silvae Genet 20:101–107

Owens JN, Morris SJ (1990) Cytological basis for cytoplasmic inheritance in *Pseudotsuga menziesii*. Pollen tube and archegonial development. Am J Bot 77:433–445

Palmer JD (1991) Plastid chromosomes: structure and evolution. In: Bogorad L, Vasil IK (eds) The molecular biology of plastids, vol 7A. Academic Press, San Diego, California, pp 5–53

Palmer JD, Shields CR (1984) Tripartite structure of the *Brassica campestris* mitochondrial genome. Nature 307:437–440

Palmer JD, Shields CR, Cohen DB, Orton TJ (1983) An unusual mitochondrial DNA plasmid in the genus *Brassica*. Nature 301:725–728

Pang Q, Hays JB, Rajagopal I (1993) Two cDNAs from the plant *Arabidopsis thaliana* that partially restore recombination proficiency and DNA-damage resistance to *E. coli* mutants lacking recombination-intermediate-resolution activities. Nucleic Acids Res 21:1647–1653

Paquin B, Laforest MJ, Forget L, Roewer I, Wang Z, Longcore J, Lang BF (1997) The fungal mitochondrial genome project: evolution of fungal mitochondrial genomes and their gene expression. Curr Genet 31:380–395

Peel MC, Duckett JG (1975) Studies of spermatogenesis in the Rhodophyta. In: Duckett JG, Racey PA (eds) Biology of the male gamete. Linnean Society of London, pp 1–13

Piskur J (1997) The transmission disadvantage of yeast mitochondrial intergenic mutants is eliminated in the *mgt1 (cce1)* background. J Bacteriol 179:5614–5617

Rajora OP, Dancik BP (1992) Chloroplast inheritance in *Populus*. Theor Appl Genet 84:280–285

Rajora OP, Mahon JD (1994) Inheritance of mitochondrial DNA in lentil (*Lens culinaris* Medik.). Theor Appl Genet 89:206–210

Rajora OP, Mahon JD (1995) Paternal plastid DNA can be inherited in lentil. Theor Appl Genet 90:543–649

Rajora OP, Barrett JW, Dancik BP, Strobeck C (1992) Maternal transmission of mitochondrial DNA in interspecific hybrids of *Populus*. Curr Genet 22:141–145

Rayko E, Goursot R (1996) Amphimeric mitochondrial genomes of petite mutants of yeast. II. A model for the amplification of amphimeric mitochondrial petite DNA. Curr Genet 30:135–144

Reboud X, Zeyl C (1994) Organelle inheritance in plants. Heredity 72:132–140

Rieseberg LH, van Fossen C, Arias D, Carter RL (1994) Cytoplasmic male sterility in sunflower: origin, inheritance, and frequency in natural populations. J Hered 85:233–238

Rossignol M, Silar P (1996) Genes that control longevity in *Podospora anserina*. Mech Ageing Dev 90:183–193

Russel SD (1987) Quantitative cytology of the egg and central cell of *Plumbago zeylanica* and its impact on cytoplasmid inheritance pattern. Theor Appl Genet 74:693–699

Sainsard-Chanet A, Begel O, Belcour L (1993) DNA deletion of mitochondrial introns is correlated with the process of senescence in *Podospora anserina*. J Mol Biol 234:1–7

Saindard-Chanet A, Begel O, Belcour L (1994) DNA double-strand break *in vivo* at the 3' extremity of exons located upstream of group II introns. Senesence and circular DNA introns in *Podospora anserina*. J Mol Biol 242:630–643

Sakamoto W, Kondo H, Murata M, Motoyoshi F (1996) Altered mitochondrial gene expression in a maternal distorted leaf mutant of *Arabidopsis* induced by *chloroplast mutator*. Plant Cell 8:1377–1390

Samoilov AM, Komarnitskit IK, Gleba YY, Peretyatko VG (1986) Analysis of chloroplast and mitochondrial DNA of some varieties and species of the genus *Beta* L. in connection with the study of the genetic control of cytoplasmic male sterility. Genetika 22:26–29

Schardl CL, Lonsdale SM, Pring DR, Rose KR (1984) Linearization of maize mitochondrial chromosomes by recombination with linear episomes. Nature 310:292–296

Schmidt U, Riederer B, Mörl M, Schmelzer C, Stahl U (1990) Self-splicing of the mobile group II intron of the filamentous fungus *Podospora anserina* (COI I1) in vitro. EMBO J 9:2289–2298

Schmitz UK (1988) Molecular analysis of mitochondrial DNA and its inheritance in *Epilobium*. Curr Genet 13:411–415

Schmitz UK, Kowallik KV (1987) Why are plastids maternally inherited in *Epilobium*? Plant Sci 53:139–145

Schulte E, Kück U, Esser K (1988) Multipartite structure of mitochondrial DNA in a fungal longlife mutant. Mol Gen Genet 211:342–349

Schumann CM, Hancock JF (1989) Paternal inheritance of plastids in *Medicago sativa*. Theor Appl Genet 78:863–866

Schuster W, Brennicke A (1987) Plastid, nuclear and reverse transcriptase sequences in the mitochondrial geome of *Oenothera*: is genetic information transferred between organelles via RNA? EMBO J 6:2857–2863

Schuster W, Brennicke A (1994) The plant mitochondrial genome: Physical structure, information content, RNA editing, and gene migration to the nucleus. Annu Rev Plant Physiol Plant Mol Biol 45:61–78

Sears BB, VanWinkle-Swift KP (1994) The salvage/turnover/repair (STOR) model for uniparental inheritance in *Chlamydomonas*: DNA as a source of sustenance. J Hered 85:366–376

Sears BB, Stoike LL, Chiu W-L (1996) Proliferation of direct repeats near the *Oenothera* chloroplast DNA origin of replication. Mol Biol Evol 13:850–863

Sellem CH, Lecellier G, Belcour L (1993) Transposition of group II intron. Nature 366:176–178

Sewell MM, Qiu YL, Parks CR, Chase MW (1993) Genetic evidence for trace paternal transmission of plastids in *Liriodendron* and *Magnolia* (Magnoliaceae). Am J Bot 80:854–858

Shibata T, Nakagawa K, Morishima N (1995) Multi-site-specific endonucleases and the initiation of homologous genetic recombination in yeast. Adv Biophys 31:77–91

Silar P, Koll F, Rossignol M (1997a) Cytosolic ribosomal mutations that abolish accumulation of circular intron in the mitochondria without preventing senescence of *Podospora anserina*. Genetics 145:697–705

Silar P, Vierny C, Gagny B, Rossignol M, Haedens V (1997b) Genetic analysis of two cellular degenerations in the filamentous fungus *Podospora anserina*. C R Seances Soc Biol Fil 191:563–577

Silliker ME, Collins OR (1988) Non-mendelian inheritance of mitochondrial DNA and ribosomal DNA in the myxomycete, *Didymium iridis*. Mol Gen Genet 213:370–378

Silliker ME, Liotta MR, Cummings DJ (1996) Elimination of mitochondrial mutations by sexual reproduction: two *Podospora anserina* mitochondrial mutants yield only wild-type progeny when mated. Curr Genet 30:318–324

Simon VR, Pon LA (1996) Actin-based organelle movement. Experientia 52:1117–1122

Small I, Suffolk R, Leaver CJ (1989) Evolution of plant mitochondrial genomes via sub-stoichiometric intermediates. Cell 58:69–76

Smith ML, Duchesne LC, Bruhn JN, Anderson JB (1990) Mitochondrial genetics in a natural population of the plant pathogen *Armillaria*. Genetics 126:575–582

Smith SE (1989) Influence of parental genotypes on plastid inheritance in *Medicago sativa*. J Hered 80:214–217

Sodmergen TS, Kawano S, Nakamura S, Tano S, Kuroiwa T (1992) Behavior of organelle nuclei (nucleoids) in generative and vegetative cells during maturation of pollen in *Lillium longiflorum* and *Pelargonium zonale*. Protoplasma 168:73–81

SolimanK, Fedak G, Allard RW (1987) Inheritance of organelle DNA in barley and *Hordeum* x *Secale* intergeneric hybrids. Genome 29:867–872

Specht CA, Novotny CP, Ullrich C (1992) Mitochondrial DNA of *Schizophyllum commune*: restriction map, genetic map, and mode of inheritance. Curr Genet 22:129–134

Sriprakash KS, Batum C (1981) Segregation and transmission of mitochondrial markers in fusion products of the asporogenous yeast *Torulopsis glabrata*. Curr Genet 4:73–80

Steinborn R, Linke B, Nothnagel T, Börner T (1995) Inheritance of chloroplast and mito-chondrial DNA in alloplasmic forms of the genus *Daucus*. Theor Appl Genet 91:632–638

Sujatha M, Subrahmanyam NC (1991) Characterization of nuclear gene controlled yellow stripe mutant of *Pennisetum glaucum* (L.) R. Br. Plant Sci 73:55–64

Sutton BCS, Flanagan DJ, Gawley JR, Newton CH, Lester DT, El-Kassaby YA (1991) Inheritance of chloroplast and mitochondrial DNA in *Picea* and composition of hybrids from introgression zones. Theor Appl Genet 82:242–248

Szmidt AE, Alden T, Hällgren J (1987) Paternal inheritance of chloroplast DNA in *Larix*. Plant Mol Biol 9:59–64

Thomas DY, Wilkie D (1968) Recombination of mitochondrial drug-resistance factors in *Saccharomyces cerevisiae*. Biochem Biophys Res Commun 30:368–372

Thrailkill KM, Birky CW Jr, Lückemann G, Wolf K (1980) Intracellular population genet-ics: evidence for random drift of mitochondrial gene frequencies in *Saccharomyces cerevisiae* and *Schizosaccharomyces pombe*. Genetics 96:237–262

Tilney-Bassett RAE (1994) Nuclear controls of chloroplast inheritance in higher plants. J Hered 85:347–354

Tilney-Bassett RAE, Almouslem AB, Amoatey HM (1992) Complementary genes control biparental plastid inheritance in *Pelargonium*. Theor Appl Genet 85:317–324

Tsai CH, Strauss SH (1989) Dispersed repetitive sequences in the chloroplast genome of Douglas-fir. Curr Genet 16:211–218

Tua A, Wang J, Kulpa V, Wernette CM (1997) Mitochondrial topoisomerase I of *Saccharomyces cerevisiae*. Biochimie 79:341–350

Urban M, Kahmann R, Bölker M (1996) The biallelic *a* mating type locus of *Ustilago maydis*: remnants of an additional pheromone gene indicate evolution from a multiallelic ancestor. Mol Gen Genet 250:414–420

van der Meer JP (1978) Genetics of *Gracilaria* sp. (Rhodophycaceae, Gigartinales) marine red algae. III. Non-mendelian gene transmission. Phycologia 17:314–318

van Ham RCHJ, 't Hart H, Mes THM, Sandbrink JM (1994) Molecular evolution of non-coding regions of the chloroplast genome in the Crassulaceae and related species. Curr Genet 25:558–566

van Horn R, Clay K (1995) Mitochondrial DNA variation in the fungus *Atkinsonella hypoxylon* infecting sympatric *Danthonia* grasses. Evolution 49:360–371

Vedel F, Quetier F, Cauderon Y, Doshba F, Doussinault G (1981) Studies on maternal inheritance in polyploid wheats with cytoplasmic DNAs as genetic markers. Theor Appl Genet 59:239–245

Vierula PJ, Bertrand H (1992) A deletion derivative of the *kalilo* senescence plasmid forms hairpin and duplex DNA structures in the mitochondria of *Neurospora*. Mol Gen Genet 234:361–368

vom Stein J, Hachtel W (1988) Deletions/insertions, short inverted repeats, sequences resembling *att*-lamda and frame shift mutated open reading frames are involved in chloroplast DNA differences in the genus *Oenothera* subsection Munzia. Mol Gen Genet 213:513–518

Watanabe N, Nakazono M, Kanno A, Tsutsumi N, Hirai A (1994) Evolutionary variations in DNA sequences transferred from chloroplast genomes to mitochondrial genomes in the Gramineae. Curr Genet 26:512–518

Wedde M (1994) Replikation des mobilen Intron (p1DNA) in Mitochondria von *Podospora anserina*. Bibliotheca Mycologica 156. J Cramer, Berlin, 93 pp

Weiller GF, Bruckner H, Kim SH, Pratje E, Schweyen RJ (1991) A GC cluster repeat is a hotspot for mit⁻ macro-deletions in yeast mitochondrial DNA. Mol Gen Genet 226:233–240

Whatley JM (1982) Ultrastructure of plastid inheritance: green algae to angiosperms. Biol Rev 57:527–569

White EE (1990) Chloroplast DNA in *Pinus monticola* 2. Survey of within-species variability and detection of heteroplasmic individuals. Theor Appl Genet 79:251–255

White MF, Lilley DM (1997a) The resolving enzyme CCE1 of yeast opens the structure of the four-way DNA junction. J Mol Biol 266:122–134

White MF, Lilley DMJ (1997b) Characterization of a Holliday junction-resolving enzyme from *Schizosaccharomyces pombe*. Mol Cell Biol 17:6465–6471

Whittaker SL, Assinder SJ, Shaw DS (1994) Inheritance of mitochondrial DNA in *Phytophthora infestans*. Mycol Res 98:569–575

Wilch G, Ward S, Castle A (1992) Transmission of mitochondrial DNA in *Ustilago violacea*. Curr Genet 22:135–140

Willemse MTM (1974) Megagametogenesis and formation of neocytoplasm in *Pinus sylvestris* L. In: Linskens HF (ed) Fertilization in higher plants. North-Holland, Amsterdam, pp 97–102

Wilms HJ (1981) Pollen tube penetration and fertilization in spinach. Acta Bot Neerl 30:101–122

Wischmann C, Schuster W (1995) Transfer of *rps10* from the mitochondrion to the nucleus in *Arabidopsis thaliana*: evidence for RNA-mediated transfer and exon shuffling at the integration site. FEBS Lett 374:152–156

Woelfle MA, Thompson RJ, Mosig G (1993) Roles of novobiocin-sensitive topoisomerases in chloroplast DNA replication in *Chlamydomonas reinhardtii*. Nucleic Acids Res 21:4231–4238

Yang J, Zimmerly S, Perlman PS, Lambowitz AM (1996) Efficient integration of an intron RNA into double-stranded DNA by reverse splicing. Nature 381:332–335

Yao JL, Cohen D, Rowland RE (1994) Plastid inheritance and plastome-genome incompatibility in interspecific hybrids of *Zantedeschia* (Araceae). Theor Appl Genet 88:255–260

Zagariya AM, Sitailo LA (1995) The influence of antibiotics and antitumor agents on the relaxation activity of *Pisum sativum* leaf chloroplast topoisomerase I. Arch Biochem Biophys 320:177–181

Zassenhaus HP, Denniger G (1994) Analysis of the role of the NUC1 endo/exonuclease in yeast mitochondrial DNA recombination. Curr Genet 25:142–149

Zheng D, Nielsen BL, Daniell H (1997) A 7.5-kbp region of the maize (T cytoplasm) mitochondrial genome contains a chloroplast-like *trn*I (CAT) pseudo gene and many short segments homologous to chloroplast and other known genes. Curr Genet 32:125–131

Zhou DX, Massenet O, Quigley F, Marion MJ, Moneger F, Huber R; Mache R (1988) Characterization of a large inversion in the spinach chloroplast genome relative to *Marchantia*: a possible transposon-mediated origin. Curr Genet 13:433–439

Zhu Q, Hulen D, Liu T, Clarke M (1997) The *cluA*-mutant of *Dictyostelium* identifies a novel class of proteins required for dispersion of mitochondria. Proc Natl Acad Sci 94:7308–7313

Zhu T, Mogensen HL, Smith SE (1991) Quantitative cytology of the alfalfa generative cell and its relation to male plastid inheritance patterns in three genotypes. Theor Appl Genet 81:21–26

Zhu T, Mogensen HL, Smith SE (1993) Quantitative, three-dimensional analysis of alfalfa egg cells in two genotypes: Implications for biparental plastid inheritance. Planta 190:143–150

Zinn AR, Pohlman JK, Perlman PP, Butow RA (1987) Kinetic and segregational analysis of mitochondrial DNA recombination in yeast. Plasmid 17:248–256

Heike Röhr
Prof. Dr. Ulf Stahl
Institut für Biotechnologie
Fachgebiet Mikrobiologie und Genetik
Technische Universität Berlin
Gustav-Meyer-Allee 25
D-13355 Berlin, Germany

Dr. Ursula Kües
Mikrobiologisches Institut
ETH-Zentrum
Schmelzbergstraße 7
CH-8092 Zürich, Switzerland

Edited by
K. Esser

Mutation:
Nuclear and Plastomic Transformation
of Higher Plants Using Microprojectile Bombardment

Christer Jansson and Pirkko Mäenpää

1 Introduction

Biolistic transformation is the dominant method for construction of transgenic cereals. Although the biolistic approach can be optimized for transformation of any plant (and non-plant) species, the efficiency of generated transformants is significantly lower compared to *Agrobacterium*-based transformations. Protocols for *Agrobacterium*-mediated transformation of grasses have been successfully developed during the past years. However, biolistic transformation continues to be of major importance for gene technological advances in cereals and other higher plants.

In plants, there are three different types of genetic materials: the nuclear genome, the plastid genome (the plastome) and the mitochondrial genome. Many structures and functions in plant metabolism, such as those pertaining to photosynthetic and respiratory electron transport, depend on interactions between the nucleus and the organellar genomes. Methods for stable transformation of the chloroplast genome in higher plants have recently been established. However, as yet, no methods for manipulation of the mitochondrial genome have been reported.

In this chapter, we will discuss the recent advances in biolistic transformation of higher plants, including the construction of stable nuclear and plastomic mutants.

2 Biolostic Transformation – Methodology

The general procedure for biolistic transformation is outlined in Fig. 1 and a schematic representation of the steps is shown in Fig. 2. The detailed protocol varies depending on organism and tissue, and examples can be found in several references. Both commercial and home-built biolistic devices are in use. Microprocetiles are commonly made of gold, although tungsten or platinum particles are also sometimes in use. Rupture discs for various pressures are available. As for *Agrobacterium*-based transformation, a large number of reporter genes and selectable

DNA preparation

Microprojectile preparation

Rinse and suspend microprojectiles in water

Resuspend microprojectiles in DNA solution

Vortex suspension

Resuspend DNA/microprojectile preparation in EtOH

Bombardment

Set the distances between rupture disc and macrocarrier,
between the macrocarrier and the stopping screen,
and between the stopping screen and the sample

Insert macrocarrier

Resuspend DNA/microcarrier precipitate and dispense onto
the macrocarrier surface

Insert the stopping screen

Insert sample on petri dish

Evacuate chamber

Activate Helium pressure

Fig. 1. Outline showing a general protocol for biolistic transformation

marker genes can be employed in biolistic transformation. These include
nptII, bar, gfp, luc, gus and regulatory genes for anthocyanin synthesis
(Pang et al. 1996; Sheen et al. 1995; for a review see McElroy and Brettell
1994).

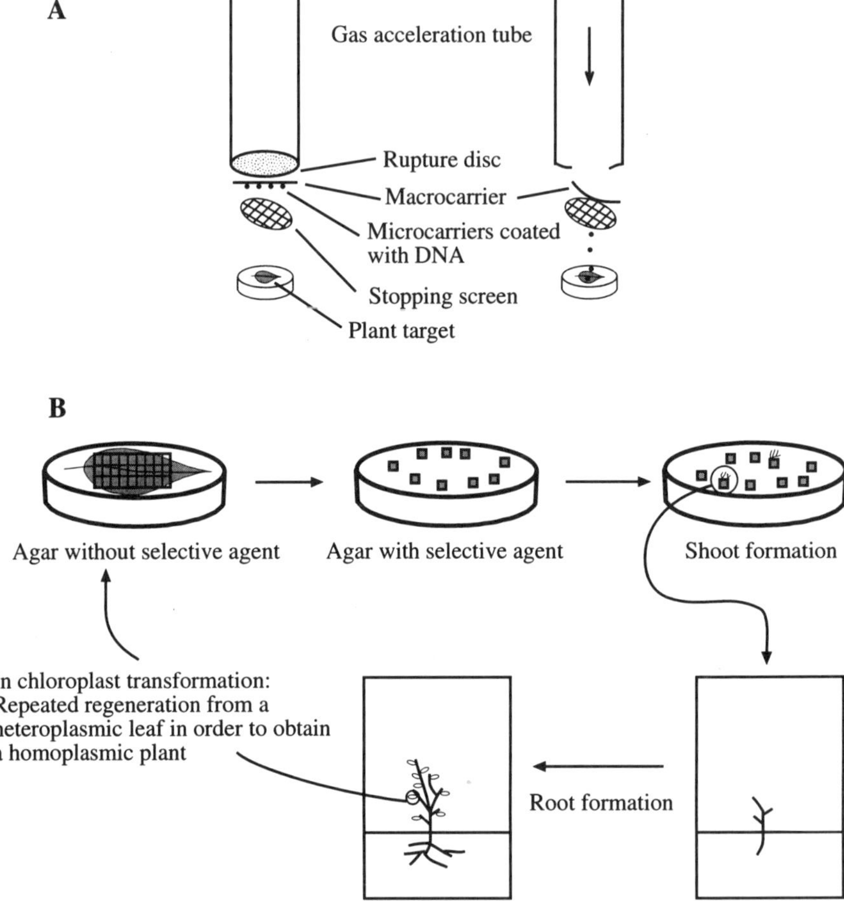

Fig. 2A, B. Schematic representation of biolistic transformation of plant tissue

3 Biolistic Transformation – Pros and Cons

Next to the *Agrobacterium*-mediated protocol, microprojectile bombardment has become the most utilized method for genetic transformation of plants (AhlGoy and Duesing, 1995). Until around 1995, success
with *Agrobacterium*-mediated transformation was limited to decotyle-
donous plants (Songstad et al. 1995). Thus, typically, biolistic transformation is chosen for plants that have been shown to be recalcitrant to
Agrobacterium-based transformation, primarily cereals and other
monocotyledons, but also various dicotyledonous plants. A large number of transgenic plants have now been constructed with the particle
bombardment procedure, including the important crops maize
(Gordon-Kamm et al. 1990), rice (Christou et al. 1991), wheat (Vasil et al.

1992), barely (Wan and Lemaux 1994; Ritala et al. 1994), soybean (McCabe et al. 1988), cotton (Finer and McMullen 1990) and cucumber (Schulze et al. 1995), as well as other economically important plants, e.g. white spruce (Ellis et al. 1993). The success rate of *Agrobacterium*-based transformation of cereals is rapidly increasing and transgenic cereal plants produced by this method now include rice (Hiei et al. 1994), maize (Ishida et al. 1996), wheat (Cheng et al. 1997) and barley (Tingay et al. 1997). However, biolistic transformation still remains a very important and powerful method for construction of transgenic cereals. Another economically important monocotyledonous crop where biolistic transformation has proven successful is turfgrass. The turfgrass is the second largest seed market in the US, surpassed only by hybrid seed corn (Lee 1996). Biolistic transformation of turfgrass has been reported for several species (Hartman 1994; Lee 1996). Also for several dicotyledonous crops, microprojectile bombardment still surpasses *Agrobacterium*-based protocols as a method for transformation. Examples are certain *Eucalyptus* species (Serrano et al. 1996) and sunflower (Hunold et al. 1995).

One problem often encountered with biolistic transformation is the integration, at one and the same locus, of linked, multiple copies of the transgenes, frequently including rearranged vector fragments (Gordon-Kamm et al. 1990; Register et al. 1994; Vasil et al. 1992; Wan and Lemaux 1994; for a review see Hansen and Chilton 1996). Since the multiple copies of the inserted transgenes are linked, they cannot be segregated in a subsequent breeding program. This phenomenon is observed also for *Agrobacterium*-based transformation. However, the integration pattern for transgenes using *Agrobacterium* is quite different from that using microprojectile bombardment (Chilton 1993), and the copy number of inserted transgenes via *Agrobacterium*-based transformation is often far less than using biolistic transformation (Hansen and Chilton 1996). The two major problems associated with integration of multiple copies of transgenes are often referred to as gene silencing (or transgene inactivation) and cosuppression (Finnegan and McElroy 1994; Hansen and Chilton 1996). In cosuppression, not only expression of the transgene but also of the corresponding endogenous gene(s) is suppressed. The mechanisms of gene silencing and cosuppression are unknown, but most likely they involve interaction between transgenes and between a transgene and related host genes, e.g. homologous recombination (Finnegan and McElroy 1994; Hansen and Chilton 1996).

Since gene silencing most commonly is associated with insertion of multiple copies of the transgene, methods to reduce the copy number of the integrated transgene would be of great value (Finnegan and McElroy 1994). Several approaches in this direction have been made with varying success. An elegant, novel biolistic transformation method that reduces the copy number of inserted transgenes was developed by Hansen and

Chilton (1996), who termed their method agrolistic transformation. In agrolistic transformation, the vector construction contains the left and right border sequences of the *Agrobacterium tumefaciens* Ti-plasmid flanking the transgene. Furthermore, the *Agrobacterium* virulence genes *virD1* and *virD2* are codelivered with the transgene. In *Agrobacterium* infection, the virulence genes are required for excision of the T-strands from the Ti-plasmid and the border sequences are necessary for integration of the T-strands into the plant host genome (for a review on *Agrobacterium* infection see Zambryski 1992). Using this technique, which combines the efficiency of the biolistic DNA delivery approach with the precision of *Agrobacterium*-based transformation, Hansen and Chilton (1996) reported successful transformation of tobacco and maize with a significantly reduced number of inserted transgene copies.

4 Biolistic Transformation of Plastid Genomes

a) The Plastid Genome of Higher Plants

Highly coordinated operations of the nuclear and plastid genomes are prerequisites for the biogenesis of plant chloroplasts. In fully developed chloroplasts, gene products encoded by the nucleus as well as by the plastid are necessary.

Plastid genomes from 12 species have been sequenced (see Rochaix 1997). The plastome typically exists as a circular DNA molecule with a size between 50 and 300 kb. The plastome sequence of green algae and higher plants reveals about 120 genes (Sugiura 1996). These genes can be divided into three groups. The first and largest group consists of genes for plastid genome function, such as genes encoding subunits of RNA polymerase, tRNAs and ribosomal proteins. The second group is formed by genes encoding subunits for components of the photosynthetic machinery. It is typical for photosynthetic multisubunit enzyme complexes, such as photosystem (PS) I, PS II, Rubisco, the ATP synthase and the cytochrome b/f complex, in that they contain subunits, some of which are encoded by the nuclear genome and by the plastome. The third and smallest group of plastome genes consists of open reading frames (*orf*), the role of which, at least in higher plants, is more or less unknown. The plastid genome of non-green algae expresses more variation (Reith and Munholland 1995).

b) Higher Plant Chloroplast Transformation Technology

Chloroplast transformation was first established in the green unicellular alga *Chlamydomonas reinhardtii* using the biolistic technique (Boynton et al. 1988). The method was soon introduced to higher plants and to-bacco chloroplast transformation was reported by Svab et al. (1990). Since then, chloroplast transformation has provided a powerful tool for studies of chloroplast gene function in both *Chlamydomonas* and to-bacco. In addition to the biolistic method, tobacco chloroplast transfor-mation is reported also from PEG-treated protoplasts (Golds et al. 1993), but so far the biolistic method has been more in use. Until now, tobacco is the only plant specimen that has enabled the technically demanding chloroplast transformation. Before the development of chloroplast transformation, genetic engineering of many genes with a central role in oxygenic autotrophy was possible only in cyanobacteria (see Nixon and Jansson 1996).

In the biolistic-mediated higher plant chloroplast transformation, DNA-coated tungsten or gold particles are delivered to leaf chloroplasts. While in chloroplasts, the transformation DNA is released and inte-grated into the plastome by homologous recombination, thus allowing targeted DNA surgery including insertions, deletions and site-specific mutagenesis. New plants are regenerated under selective pressure using standard tissue culture techniques from the transformed cells through the callus state. The plants obtained from this first regeneration are het-eroplasmic – they contain both transformed and wild-type plastome copies. A repetition of the selective regeneration starting from leaves of the newly regenerated plants is required for homoplasmicity, that is the state where the plant carries only transformed copies of the plastome (Maliga 1993).

Selectable markers for tobacco chloroplast transformation confer re-sistance to spectinomycin, based on mutant alleles of the 16S rRNA gene (Svab et al. 1990) or the bacterial *aadA* (amino glycoside adenyl trans-ferase) gene (Svab and Maliga 1993). There are additionl potential mark-ers conferring resistance to kanamycin (Carrer et al. 1993) or lincomycin (Moll et al. 1990). The most versatile marker at the moment is the *aadA* gene. When this is used, tobacco cells containing transformed plastomes can be identified by their green colour on a white background of un-transformed tissue culture cells.

c) Current Achievements of Higher Plant Chloroplast Transformation

Higher plant chloroplast transformation has already now demonstrated its potential by providing insight into the role of certain plastome genes and the regulation of plastid gene expression in vivo. Using tobacco chloroplast transformation, an in vivo system to study splicing in chlo-roplasts was developed (Bock and Maliga 1995). A light-responsible to-bacco promoter was studied in vivo by using a fused reporter gene con-struct that was introduced into the tobacco chloroplast genome by targeted insertion (Allison and Maliga 1995). The results reveal the complex architecture of the promoter and pointed to DNA-binding proteins as possible mediators in transcriptional regulation of the *psbD* gene encoding the D2 polypeptide of PS II. A similar study revealed control of accumulation of the D1 polypeptide of PS II by regulatory elements through the untranslated region of the *psbA* mRNA (Staub and Maliga 1993). A further study indicates that initiation of D1 translation in tobacco is controlled via the *psbA* 5' untranslated region (UTR; Staub and Maliga 1994).

The plastome encodes several *orf*s and genes, the disruption of which by chloroplast genetic engineering will be important in finding out their specific roles in higher plants. Homoplastic plants with plastid genes disrupted have been obtained, provided the selection and regeneration steps are performed on a sucrose-containing medium (Kanevski and Maliga 1994). If the disrupted gene is essential for plant growth and development, the homoplastic state is not reached and the plant remains heteroplasmic despite repeated regeneration cycles.

The *sprA* gene disruption has resulted in homoplasmic transplastomic mutants, but the role of the gene product remains to be clarified (Sugita et al. 1998). Plastomes of higher plants contain several genes with homology to the mitochondrial NADH dehydrogenase subunit genes. Some of the *ndh* genes of tobacco plastome have been disrupted (Rochaix 1997 and references therein). The results from preliminary characterization of the mutants are in line with the proposed role of the *ndh* genes in the chlororespiratory chain. Homoplasmic deletion of the *rpoB* gene from tobacco plastids resulted in strongly reduced levels in plastid gene transcripts involved in photosynthesis, but plastid transcripts of the protein synthesis machinery remained at levels close to those in the wild type (Allison et al. 1996). Thus it was inferred that the residual RNA polymerase activity in the plastids of the homoplasmic mutants must be the result of a plastid-localized, nuclear-encoded enzmye. This conclusion receives support from elsewhere, i.e. the small plastome of the parasitic, non-photosynthetic *Epifagus virginiana* lacks *E. coli*-like RNA polymerase genes, and is still transcribed (Wolfe et al. 1992).

There is an RNA-editing system in the mitochondria and chloroplasts of higher plants, resulting in RNA sequences different from the corresponding DNA (Maier et al. 1996). The spinach, but not the tobacco, plastome-encoded *psbF* mRNA contains one of the chloroplast-editing sites. At protein levels the gene products of spinach and tobacco have 100% identity. By plastome gene displacement, the tobacco *psbF* was replaced by the spinach gene. The heterologous editing site in *psbF* mRNA remained unmodified in tobacco, demonstrating that the ability to edit individual sites is not necessary maintained in another specimen – even if they are both dicotyledons. The transplastomic tobacco with the spinach *psbF* gene had an obvious phenotype that revealed editing of the *psbF* mRNA as a necessary processing step. This represents the first conclusive proof for the biological significance of plant organellar RNA editing (Bock et al. 1994).

d) Technical Problems and Future Perspectives of Higher Plant Chloroplast Genome Transformation

Plant chloroplast transformation is technically difficult when compared to *Chlamydomonas* chloroplast transformation. The difficulty is based on several reasons: first, the genetic information in the plastome is in many locations desely packed, or, if there appears to be a non-informative sequence, it might no be characterized well enough to help estimating if the insertion of a selectable marker is possible without side effects. Second, in a single tobacco target cell the content of up to 10 000 plastome copies, combined with the multicellular organization of plants, lowers the efficiency of selection for transformants as compared to the more simple organization of *Chlamydomonas* (see Rochaix 1995). Third, only plant species with tissue culture techniques optimized for plant regeneration can be directly considered as candidates for chloroplast transformation.

During the selection period, the transforming DNA should end up in all the plastome copies in plastids thorough the plant. The selection takes place at the same time with the regeneration of the plant from callus. The whole process is time-consuming, and the conditions should remain strictly optimal. In addition, a drawback in utilization of higher plant chloroplast transformation is that it is not possible to apply it directly to economically important plants. However, the potential of higher plant plastid transformation is now becoming realized and, clearly, in spite of problems, it is not possible to replace recombinant DNA studies of higher plant plastome by *Chlamydomonas* or by cyanobacteria, especially not when tissue-specific expression of genes or molecular mechanisms of plant stress tolerance is the aim of the study.

One possible means to generate a more simple system for plastid transformation is reduction of the plastome copy number per cell. This reduction can be achieved by treatment with chloroplast DNA synthesis inhibitors nalidixic acid and novobiocin (Ye and Sayre 1990). Also nuclear-reduced chloroplast number mutants (arc) of *Arabidopsis* (Pyke and Leech 1992) may provide potential starting material for plastid transformation because of the extremely low number of chloroplasts per cell. In a cell with possibly only one chloroplast, the transforming DNA has only to be introduced into all the plastome copies in the single chloroplast rather than in over 100 separate chloroplasts per cell, as is the case in tobacco. It can be expected that chloroplast transformation in the future will help to broaden our understanding of higher plant chloroplast biochemistry as well as enabling plastome manipulation in commercially important ways. Concerning higher plants, a topic of particular interest is how chloroplast metabolism is regulated by environmental conditions.

References

Ahl Goy P, Duesing JH (1995) From pots to plots: genetically modified plants on trial. Bio/Technology 13:454–458

Allison LA, Maliga P (1995) Light-responsive and transcription-enhancing elements regulate the plastid *psbD* core promoter. EMBO J 14:3721–3730

Allison LA, Simon LD, Maliga P (1996) Deletion of *rpoB* reveals a second distinct transcription system in plastids of higher plants. EMBO J 15:2802–2809

Bock R, Maliga P (1995) Correct splitting of a group II intron from a chimeric reporter gene transcript in tobacco plastids. Nucleic Acid Res 23:2544–2547

Bock R, Kössel H, Maliga P (1994) Introduction of a heterologous editing site into the tobacco plastid genome: the lack of RNA editing leads to a mutant phenotype. EMBO J 13:4623–4628

Boynton JE, Gilham NW, Harris EH Hosler JP, Johnson AM, Jones AR, Randolph-Anderson BL, Robertson D, Klein TM, Shark KB, Sanford JC (1988) Chloroplast transformation in *Chlamydomonas* with high velocity microprojectiles. Science 240:1534–1538

Carrer H, Hockenberry TN, Svab Z, Maliga P (1993) Kanamycin resistance as a selectable marker for plastid transformation in tobacco. Mol Gen Genet 241:49–56

Cheng M, Fry JE, Pang S, Zhou H, Hironaka CM, Duncan DR, Conner TW, Wan Y (1997) Genetic transformation of wheat mediated by *Agrobacterium tumefaciens*. Plant Physiol 115:971–980

Chilton M-D (1993) *Agrobacterium* gene transfer: progress on a "poor man's vector" for maize. Proc Natl Acad Sci USA 90:3119–3120

Christou P, Ford TL, Kofron M (1991) Production of transgenic rice (*Ozyza sativa* L.) plants from agronomically important indica and japonica varieties via electric discharge particle acceleration of exogenous DNA into immature zygotic embryos. Bio/Technology 9:957–962

Ellis DD, McCabe DE, McInnis S, Ramachandran R, Russel DR, Walace KM, Martinell BJ, Roberts DR, Raffla KF, McCown BH (1993) Stable transformation of *Picea glauca* by particle acceleration. Bio/Technology 11:84–89

Finer JJ, McMullen MD (1990) Transformation of cotton (*Gossypium hirsutum* L.) via particle bombardment. Plant Cell Rep 8:586–589

Finnegan J, McElrroy D (1994) Transgenic inactivation: plants fight back! Bio/Technology 12:883–888

Golds T, Maliga P, Koop H-U (1993) Stable plastid transformation in PEG-treated protoplasts of *Nicotiana tabacum*. Biotechnology 11:95–97

Gordon-Kamm WJ, Spencer TM, Mangano ML, Adams TR, Daines RJ, Start WG, O'Brien JV, Chambers SA, Adams WR, Willets NG, Rice TB, Mackey CJ, Krueger RW, Kausch AP, Lemaux PG (1990) Transformation of maize cells and regeneration of fertile transgenic plants. Plant Cell 2:603–618

Hansen G, Chilton M-D (1996) "Agrolistic" transformation of plant cells: Integration of T-strands generated *in planta*. Proc Natl Acad Sci USA 93:14978–14983

Hartman CL, Lee L, Day PR, Tumer NE (1994) Herbicide resistance turfgrass (*Agrostis palustris* Huds.) by biolistic transformation. Bio/Technology 12:919–923

Hiei Y, Ohta S, Komari T, Kumashiro T (1994) Efficient transformation of rice (*Oryza sativa* L.) mediated by *Agrobacterium* and sequence analysis of the boundaries of the T-DNA. Plant J 6:271–282

Hunold R, Burrus M, Bronner R, Duret J-P, Hahne G (1995) Transient gene expression in sunflower (*Helianthus annuus* L.) following microprojectile bombardment. Plant Sci 105:95–109

Ishida Y, Saito H, Ohta S, Hiei Y, Komari T, Kumashiro T (1996) High efficiency transformation of maize (*Zea mays* L.) mediated by *Agrobacterium tumefaciens*. Nat Biotechnol 14:745–750

Jansson C, Mäenpää P (1997) Site-directed mutangesis for structure-function analysis of the Photosystem II reaction center protein D1. In: Esser K (ed) Progress in Botany, vol 58. Springer, Berlin Heidelberg New York, pp 352–367

Kanevski I, Malinga P (1994) Relocation of the plastid *rbcL* gene to the nucleus yields functional ribulose-1,5 bisphosphate carboxylase in tobacco chloroplasts. Proc Natl Acad Sci USA 91:1969–1973

Lee L (1996) Turfgrass biotechnology. Plant Sci 115:1–8

Maier RM, Zeltz P, Kössel H, Bonnard G, Gualberto JM, Grienenberger JM (1996) RNA editing in plant mitochondria and chloroplasts. Plant Mol Biol 32:343–365

Maliga P, Carrer H, Kanevski I, Staub J, Svab Z (1993) Plastid engineering in land plants: a conservative genome is open to change. Philos Trans R Soc Lond B 342:203–208

McCabe DE, Swain WF, Martinell BJ, Christou P (1988) Stable transformation of soybean (*Glycine max*) by particle acceleration. Bio/Technology 6:923–926

Moll B, Poslby L, Maliga P (1990) Streptomycin and lincomycin resistances are selective plastid markers in cultured *Nicotiana* cells. Mol Gen Genet 221:245–250

Nixon P, Jansson C (1996) Cyanobacterial transformation and gene regulation. In: Andersson B, Salter H, Barber J (eds) Molecular genetics of photosynthesis. Oxford Univ Press, Oxford, pp 197–224

Pang S-Z, DeBoer DL, Wan Y, Ye G, Layton JG, Neher MK, Armstrong CL, Fry JE, Hinchee AW, Fromm ME (1996) An improved green fluorescence protein gene as a vital marker in plants. Plant Physiol 113:893–900

Pyke KA, Leech RM (1992) Chloroplast division and expansion is radically altered by nuclear mutations in *Arabidopsis thaliana*. Plant Physiol 99:1005–1008

Register JC III, Peterson DJ, Bell PJ, Bullock WP, Evans IJ, Frame B, Greenlands AJ, Higgs NS, Jepson I, Jiao S, Lewnau JL, Sillick JM, Wilson HM (1994) Structure and function of selectable and non-selectable transgenes in maize after introduction by particle bombardment. Plant Mol Biol 25:951–961

Reith M, Munholland J (1995) Complete nucleotide sequence of the *Phorphyra purpurea* chloroplast genome. Plant Mol Biol Rep 13:333–342

Ritala A, Aspegren K, Kurten U, Salmenkallio-Marttila M, Mannonen L, Hannus R, Kauppinen V, Teeri TH, Enari TM (1994) Fertile transgenic barley to particle bombardment of immature embryos. Plant Mol Biol 24:317–325

Rochaix J-D (1995) *Chlamydomonas reinhardtii* as the photosynthetic yeast. Annu Rev Genet 29:209–230

Rochaix J-D (1997) Chloroplast reverse genetics: new insights into the function of plastid genes. Trends Plant Sci 2:419–425

Schulze J, Balko C, Zellner B, Koprek T, Hänsch R, Nerlich A, Mendel RR (1995) Biolistic transformation of cucumber using embryogenic suspension cultures: long-term expression of reporter genes. Plant Sci 112:197–206

Serrano L, Rochange F, Semblat JP, Marque C, Teulières C, Boudet A-M (1996) Genetic transformation of *Eucalyptus globulus* through biolistics: complementary development of procedures for organogenesis from zygotic embryos and stable transformation of corresponding proliferating tissues. J Exp Bot 47:285–290

Sheen J, Hwang S, Niwa Y, Kobayashi H, Galbraith DW (1995) Green-fluorescent protein as a new vital marker in plant cells. Plant J 8:777–784

Shinozaki K, Ohme M, Tanaka M, Wakasugi T, Hayashida N, Matsubayashi T, Zaita N, Chunwongse J, Obokata J, Yamaguchi-Shinozaki K, Ohto C, Torazawa K, Meng BY, Sugita M, Deno H, Kamogashira T, Yamada K, Kusuda J, Takaiwa F, Kato A, Tohdoh N, Shimada H, Sugiura M (1986) The complete nucleotide sequence of the tobacco chloroplast genome. EMBO J 5:2043–2049

Songstad DD, Somers DA, Griesbach RJ (1995) Advances in alternative DNA delivery techniques. Plant Cell Tissue Org Cult 40:1–15

Staub JM, Maliga P (1993) Accumulation of D1 polypeptide in tobacco plastids is regulated via the untranslated region of the *psbA* mRNA. EMBO J 12:601–606

Staub JM, Maliga P (1994) Translation of *psbA* mRNA is regulated by light via the 5' untranslated region in tobacco plastids. Plant J 6:547–553

Sugita M, Svab Z, Maliga P, Sugiura M (1997) Targeted deletion of sprA from the tobacco plastid genome indicates that the encoded small RNA is not essential for pre-16S rRNA maturation in plastids. Mol Gen Genet 257:23–27

Sugiura M (1996) Structure and replication of chloroplast DNA. In: Andersson B, Salter AH, Barber J (eds) Molecular genetics of photosynthesis. Oxford University Press, Oxford, pp 58–74

Svab Z, Maliga P (1993) High-frequency plastid transformation in tobacco by selection for a chimeric aadA gene. Proc Natl Acad Sci USA 90:913–917

Svab Z, Hajdukiewitz P, Maliga P (1990) Stable transformation of plastids in higher plants. Proc Natl Acad Sci USA 87:8526–8530

Tingay S, McEllroy D, Kalla R, Fieg S, Wang M, Thornton S, Brettell R (1997) *Agrobacterium tumefaciens*-mediated barley transformation. Plant J 11:1369–1376

Vasil V, Castillo AM, Fromm ME, Vasil IK (1992) Herbicide resistant fertile transgenic wheat plants obtained by microprojectile bombardment of regenerable embryogenic callus. Bio/Technology 10:667–674

Wan Y, Lemaux PG (1994) Generation of large numbers of independantly transformed fertile barley plants. Plant Physiol 104:37–48

Wolfe KH, Morden CW, Palmer JD (1992) Function and evolution of a minimal plastid genome from a nonphotosynthetic parasitic plant. Proc Natl Acad Sci USA 89:10648–10652

Ye JS, Sayre RT (1990) Reduction of chloroplast DNA content in *Solanum nigrum* suspension cells by treatment with chloroplast DNA inhibitors. Plant Physiol 94:1477–1483

Zambryski P (1992) Chronicles from the *Agrobacterium*-plant cell-DNA transfer story. Annu Rev Plant Physiol Mol Biol 43:465–490

Prof. Dr. Christer Jansson
University of Stockholm
Department of Biochemistry
Svante Arrhenius Väg
S-10691 Stockholm, Sweden

Dr. Pirkko Mäenpää
University of Turku
Department of Biology
BioCity A 6th floor
Tykistökatu 6
FIN-20520 Turku, Finland

Edited by
K. Esser

Extranuclear Inheritance:
Genetics and Biogenesis of Mitochondria

By Thomas Lisowsky, Karlheinz Esser, Torsten Stein, Elke Pratje, and Georg Michaelis

1 Introduction

This chapter is a continuation of our previous articles in this series (Bauerfeind et al. 1997; Riemen et al. 1993). Research activities and progress in the field of mitochondrial genetics, mitochondrial biogenesis and nuclear-mitochondrial interactions refer to the complete sequence of new mitochondrial genomes, transcription of mitochondrial DNA, processing, editing, and stability of mitochondrial RNA, the import of proteins and RNA, and the characterization of many nuclear genes required for mitochondrial biogenesis. Three topics have been selected to be reviewed here. (1) The complete sequence of the mitochondrial genome of *Arabidopsis thaliana* will be discussed in the section on mitochondrial genomes. The mitochondrial DNA of the flagellate *Reclinomonas americana* may represent a very ancient type of mitochondrial genome and is much more eubacterial-like than any other mitochondrial DNA analyzed so far. (2) For several years, laboratories have been searching for genes encoding plant organelle RNA polymerases. Recently, the sequence of the first genes were published and are reviewed here. (3) The identification of a nuclear gene restoring cytoplasmic male sterility is another remarkable result that will be discussed.

For additional topics the reader is referred to the following reviews or articles on mitochondrial inheritance in filamentous fungi (Griffiths 1996) or mammals (Lightowlers et al. 1997), red algae and evolution of mtDNA (Leblanc et al. 1997), evolution of mitochondrial introns and exons (Laroche et al. 1997), migration of nucleic acids between chloroplasts, mitochondria and the nucleus (Thorsness and Weber 1996), structure of the mitochondrial DNA (Backert et al. 1997; Bendich 1996), genome organization (Janska and Woloszynska 1997), mitochondrial DNA maintenance in vertebrates (Shadel and Clayton 1997), regulation of organelle biogenesis (Nunnari and Walter 1996), RNA editing and translation of partial edited transcripts (Blanc et al. 1996; Hanson 1996), scrambled ribosomal RNA genes in green algae (Nedelcu 1997), protein import into fungal and mammalian mitochondria (Haucke and Schatz 1997; Neupert 1997; Pfanner et al. 1997; Ryan et al. 1997; Stuart and Neupert 1996; Suzuki et al. 1997; several articles in the Journal of Bioenergetics and Biomembranes 29, pp 3–54, 1997) and into plant mitochondria (Braun and Schmitz 1997; Silva Filho et al. 1996; Whelan and Glaser 1997), ageing (Gershon 1997; Ozawa 1997a; Wallace 1997), and mitochondria and apoptosis (Ozawa 1997b).

2 Mitochondrial Genomes

a) *Arabidopsis thaliana*

The complete DNA sequences of mitochondrial (mt) genomes from the red alga *Chondrus crispus*, the green algae *Chlamydomonas reinhardtii* and *Prototheca wickerhamii* and the liverwort *Marchantia polymorpha* were reviewed in previous volumes of Progress in Botany (Bauerfeind et al. 1997; Riemen et al. 1993). In the meantime, the first complete mt genome of a higher plant was determined (Unseld et al. 1997). The 367-kb mt genome of *Arabidopsis thaliana* is of medium size compared to the values of higher plant mtDNA: 180 kb for *Brassica hirta* to 2400 kb for muskmelon. The *Arabidopsis* mtDNA is about twice as large as that of *Marchantia polymorpha* although the gene content is similar (Table 1). The 57 genes of *Arabidopsis* mtDNA are arranged on both strands with a calculated gene density of about one gene per 8 kb.

A. *thaliana* mtDNA encodes 3 rRNA genes for 26S, 18S and 5S rRNA. The 5S rRNA gene is described for land plants and the green alga *Prototheca wickerhamii*, but is absent from all animal mt genomes analyzed so far. The 22 predicted tRNA genes of *Arabidopsis* are not sufficient to read all codons. The tRNAs for six amino acids have to be imported from the cytosol. Import of tRNAs into mitochondria is known from several flowering plants. The number and types of imported tRNAs vary considerably between different plant species (Kumar et al. 1996). In *Marchantia*, all except one (the isoleucine tRNA$_{AAU}$) are mitochondrially encoded (Akashi et al. 1996), whereas in the green alga *Chlamydomonas reinhardtii* only three tRNA genes are present on the mt genome.

The protein encoding information of *Arabidopsis* mtDNA is similar to that of the liverwort (Table 1). Differences include (1) nad7, a pseudogene in *Marchantia* but a functional gene in *Arabidopsis*; (2) rps14 and rps19, which seem to be pseudogenes in *Arabidopsis*; (3) genes encoding subunits of the succinate dehydrogenase complex which are absent from the higher plant mt genome, and (4) genes for ribosomal proteins, 7 of which are encoded by *Arabidopsis* mtDNA and 16 by *Marchantia* mtDNA.

Twenty three introns are found in protein coding genes of *Arabidopsis*. Eighteen of these introns are *cis*-splicing, and five introns in nad1, nad2 and nad5 are of the *trans*-splicing type. The evolution of *trans*-splicing plant introns was analyzed by Malek et al. (1997). The nad2 and nad5 genes of the ferns *Asplenium nidus* and *Marsilea drummondii* contain *cis*-splicing introns in contrast to their counterpart in higher plants. Only one intron of *Arabidopsis* carries a maturase-like reading frame. In *Marchantia* 10 open reading frames are present in 32 introns.

Mitochondrial DNA of higher plants contains repeats and sequences originating from either plastids or the nucleus. In *Arabidopsis* two large

Table 1. Comparison of the coding information in mitochondrial DNA from various organisms

	Reclinomonas americana	Chondrus crispus	Chlamydomonas reinhardtii	Prototheka wickerhamii	Marchantia polymorpha	Arabidopsis thaliana	Homo sapiens
Transcription and translation							
Ribosomal RNAs rrn 5	+	–	–	+	+	+	–
rrn18	+	+	+	+	+	+	+
rrn26	+	+	+	+	+	+	+
tRNAs	26	23	3	26	29	22	22
Rnase P RNA	+	–	–	–	–	–	–
Ribosomal proteins							
Small subunit (rps)	12	3	–	10	12	4	–
Large subunit (rpl)	15	1	–	3	4	3	–
Elongation factor (tufA)	+	–	–	–	–	–	–
RNA polymerase (rpo A,B,C,D)	4	–	–	–	–	–	–
Respiration							
Complex I							
nad1	+	+	+	+	+	+	+
nad2	+	+	+	+	+	+	+
nad3	+	+	–	+	+	+	+
nad4	+	+	+	+	+	+	+
nad4L	+	+	–	+	+	+	+
nad5	+	+	+	+	+	+	+
nad6	+	+	+	+	+	+	+

Table 1 (continued)

	Reclinomonas americana	*Chondrus crispus*	*Chlamydomonas reinhardtii*	*Prototheka wickerhamii*	*Marchantia polymorpha*	*Arabidopsis thaliana*	*Homo sapiens*
nad7	+	−	−	+	−	+	−
nad8	+	−	−	−	−	−	−
nad9	+	−	−	+	+	+	−
nad10	+	−	−	−	−	−	−
nad11	+	−	−	−	−	−	−
Complex II							
sdh2	+	+	−	−	−	−	−
sdh3	+	+	−	−	+	−	−
sdh4	+	+	−	−	+	−	−
Complex III							
cob	+	+	+	+	+	+	+
Complex IV							
cox1	+	+	+	+	+	+	+
cox2	+	+	−	+	+	+	+
cox3	+	+	−	+	+	+	+
Complex V							
atp1	+	−	−	+	+	+	−
atp3	+	−	−	−	−	−	−
atp6	+	+	−	+	+	+	+
atp8	+	−	−	−	−	−	+
atp9	−	+	−	+	+	+	−

Protein maturation,
assembly, transport

Cytochrome c biogenesis	4	–	–	–	3	3	–
Cytochrome oxidase assembly	1	–	–	–	–	–	–
Protein transport (sec Y)	1	–	–	–	–	–	–
Intronic orfs	–	–	–	2	10	1	–
orfs with unknown function	5	3	1	3	3	3	–
Number of introns	1	1	–	5	32	23	–
Number of protein coding genes	67	21	8	36	37	27	13
Number of total genes	92	46	13	65	67	57	37
Genome size (in kb)	69	26	16	55	187	367	17

Data were taken from the following references: *Reclinomonas americana* (Lang et al. 1997), *Chondrus crispus* (Leblanc et al. 1995), *Chlamydomonas reinhardtii* (Boer and Gray 1988; Michaelis et al. 1990; Vahrenholz et al. 1993), *Prototheka wickerhamii* (Wolff et al. 1994), *Marchantia polymorpha* (Oda et al. 1992), *Arabidopsis thaliana* (Unseld et al. 1997); *Homo sapiens* (Anderson et al. 1981). Intact genes are represented by (+), missing genes by (–).

identical direct repeated sequences seem to be responsible for frequent recombination and the occurrence of two subgenomic circles. These 2 large and 144 small repeats contribute to about 7% of the entire mt genome.

Sequences with homology to plastid DNA comprise about 1% of the *Arabidopsis* mt genome. These sequences include fragments from the psbD, rbcL and ndhB genes. Plastid sequences in mitochondria are thought to be non-functional with the exception of the tRNA genes. However, the utilization of a chloroplast-derived sequence was reported to act as promoter element for the nad9 gene in rice mitochondria (Nakazono et al. 1996).

About 4% of the *Arabidopsis* mt genome seems to originate from nuclear DNA. These sequences show homology to retrotransposons like *copia, gypsy* and *LINE* (Knoop et al. 1996). So far, the identified *Arabidopsis* genes can account for only 10% of the complete mt genome. Putative open reading frames, introns, duplications and integrated sequences from plastid or nuclear origin represent about 30% of the mtDNA, leaving 60% with unknown functions. Open reading frames or similarities to known nucleotide or protein sequences were undetectable in these 60% of the mt genome.

Although the complete DNA sequence of the *Arabidopsis* mt genome has been determined, its final coding information has to be verified by comparison with cDNA sequences because of extensive RNA editing. In trypanosomes the specificity for RNA editing is provided by guide RNAs (gRNA). Evidence for gRNAs in plant mitochondria is lacking so far, but as soon as the edited *Arabidopsis* mt sequences are known, a search for gRNA sequences can be started in the complete mt genome. On the other hand, in chloroplasts of the hornwort *Anthoceros formosa* editing sites seem to be selected by the two-dimensional RNA structure (Yoshinaga et al. 1997). In this moss, a sequence complementary to every editing site exists within the respective RNA molecules, which suggests that mispairing acts as a signal for editing. It is not known whether such a mechanism exists in higher plant mitochondria.

The biochemical mechanism of editing differs between mitochondria of higher plants and trypanosomes (reviewed by Bauerfeind et al. 1997; Gray 1996; Stuart et al. 1997). In plants, RNA editing involves a deamination from cytosine to uridine and to some extent uridine to cytosine changes, whereas in trypanosomes uridines are inserted or deleted. An enzyme complex which catalyzes the editing reaction has been isolated from trypanosomes (Rusché et al. 1997). This enzyme complex consists of an endo- and exonuclease, terminal uridylyl transferase and RNA ligase.

b) *Reclinomonas americana*

The complete DNA sequence of the mt genome from the heterotrophic flagellate *Reclinomonas americana* revealed several surprising features (Lang et al. 1997). The 69-kb mtDNA of this freshwater protozoon is much more eubacterial-like than any other mt genome analyzed so far and seems to resemble the endosymbiotic progenitor. It contains 97 genes, significantly more than the 67 mt genes in *Marchantia polymorpha*, 57 in *Arabidopsis thaliana*, 37 in human or 13 in *Chlamydomonas reinhardtii* (Table 1).

In addition to the protein-coding genes previously found in various mtDNAs, 18 genes are unique to the *Reclinomonas* mt genome. These new genes encode additional 9 proteins of the mitochondrial ribosome, 1 subunit of the ATP synthetase (*atp*3) and a new component with a function for cytochrome c biogenesis. Genes for cytochrome oxidase assembly (*cox*11), protein transport (*sec*Y), a translation factor (*tuf*A), and four components for an eubacterial-like RNA polymerase (*rpo*A–D) are the first examples described for a mt genome.

The presence of bacterial-like *rpo* genes is of special interest because such genes were thought to be lost early in evolution. Only nuclear-encoded bacteriophage-like RNA polymerases could be identified in mitochondria so far (see Sect. 3 on mitochondrial transcription below).

All bacteria contain a Sec-dependent protein export pathway, but such a pathway seems to be absent from mitochondria analyzed so far. Sec homologous genes were undetectable in the complete genome of *Saccharomyces cerevisiae* (Glick and von Heijne 1996), suggesting a loss of this original endosymbiotic trait. In yeast, the nuclear gene PET1402/OXA1 functions in protein export from the mitochondrial matrix across the mitochondrial inner membrane (Bauer et al. 1994; Bonnefoy et al. 1994; He and Fox 1997; Hell et al. 1997; Herrmann et al. 1997; Meyer et al. 1997a). Mutations in the PET1402/OXA1 gene induce the accumulation of the precursor of cytochrome oxidase subunit II (Cox2) in the mitochondrial matrix (He and Fox 1997; Hell et al. 1997) and affect the assembly and activity of the ATP synthetase (Altamura et al. 1996) and cytochrome bc_1 (Meyer et al. 1997b) complexes. A homologous nuclear gene was recently identified from *Arabidopsis thaliana* (Hamel et al. 1997), suggesting a similar pathway in higher plant mitochondria. Whether components like the inner membrane peptidase Imp1/Imp2 and the small Som1 protein of yeast are directly involved in a general export pathway has to be shown (Bauerfeind et al. 1998; Esser et al. 1996; Nunnari et al. 1993; Pratje et al. 1994, 1997). The presence of a secY gene in *Reclinomonas* mitochondria supports the ancient trait for this flagellate.

The translation factor tufA is another example for the eubacterial-like character of *Reclinomonas* mtDNA. A tufA gene is not present in the mtDNA of any other organism studied so far. The *Reclinomonas* gene maps upstream of the rps10 gene as in *E. coli*. A gene organization comparable to that of *E. coli* includes several mitochondrial *Reclinomonas* gene clusters, such as genes encoding subunits of the NADH dehydrogenase, succinate dehydrogenase, or cytochrome c biogenesis.

Other eubacterial characters refer to the possible ribosome binding sites at the 5' end of *Reclinomonas* mitochondrial mRNAs, the presence of a prokaryotic-type gene encoding the RNA component of RNase P, and the absence of RNA editing.

The 26 tRNA species of *Reclinomonas* have the potential to read all codons used with the exception of the ACN (threonine) codon family.

In summary, the mt genome of *Reclinomonas* seems to be the most original studied so far and its gene content, organization and expression resemble more closely the ancestral protomitochondrial genome than any other mt genome characterized at present.

3 Mitochondrial Transcription

Mitochondria contain an unusual transcription apparatus (Clayton 1996). The core enzyme for RNA polymerase in most of the investigated eukaryotic organisms is a large single subunit protein of the bacteriophage SP6/T7-type (Masters et al. 1987). In contrast to the bacteriophage enzymes, the mitochondrial RNA polymerases depend on additional factors for correct initiation of RNA synthesis at the promoter (Schinkel et al. 1987; Lisowsky and Michaelis 1988; Shadel and Clayton 1993, 1995). All these proteins are encoded by nuclear genes and imported into mitochondria after translation of the respective messenger RNAs in the cytosol. The number and type of additional transcription factors (Tracy and Stern 1995) that are necessary for regulated expression of specific mitochondrial genes are still open questions. These factors are also important for the problem of species specificity of mitochondrial transcription (Fisher et al. 1989; Bauerfeind et al. 1997).

In the most recent review in this series on mitochondrial transcription (Bauerfeind et al. 1997), the major point was the universal picture that emerged for the mitochondrial transcription apparatus of the eukaryotic cell (Clayton 1996). During the past year, decisive progress has been made concerning elucidation of the evolution of the peculiar mitochondrial transcription system with the bacteriophage-type core enzyme of RNA polymerase that exists in most of the present eukaryotic species (Cermakian et al. 1996; Gray and Lang 1998).

Another important finding is the identification of different genes for bacteriophage-type core enzymes in the nuclear genome of the higher plant *Arabidopsis thaliana* (Hedtke et al. 1997) and experimental verification that the respective proteins are imported into either mitochondria or chloroplasts (Gray and Lang 1998).

a) Evolution of the Mitochondrial Transcription Apparatus

One highlight of the latest research on the evolution of mitochondria is the finding that the mitochondrial genome of the protozoon *Reclinomonas americana* (see Sect. 2) closely resembles the ancient situation of the former endosymbiont that later developed into the present mitochondrial organelles (Lang et al. 1997). This is the first time that genes have been identified on a mitochondrial genome that encode eubacterial transcription components typical for present prokaryotes like *Escherichia coli* (Palmer 1997). Up to now all the other eukaryotes investigated have had a nuclear-encoded RNA polymerase of the bacteriophage-type as core enzyme for mitochondrial gene expression (Cermakian et al. 1996). This allows us to draw a new detailed model for the evolution of the mitochondrial transcription apparatus in the eukaryotic

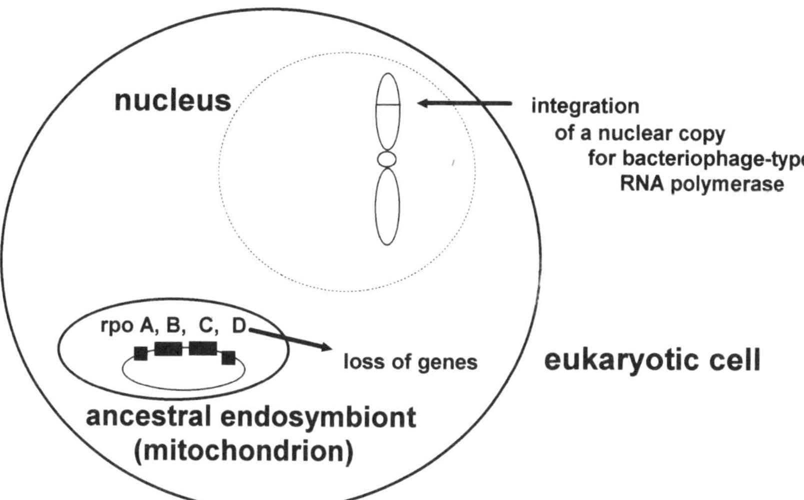

Fig. 1. Model for the evolution of the mitochondrial transcription apparatus, from an eubacterial transcription machinery to a bacteriophage-type single subunit core enzyme. In the ancient endosymbiosis between the eukaryotic cell and the endosymbiont which developed into mitochondria, the genome of the ancestral endosymbiont encoded typical eubacterial genes (*rpoA, B, C* and *D*) for transcription of its own DNA. During evolution, these genes were completely eliminated from nearly all existing mitochondrial genomes after the integration of the gene for a bacteriophage-type RNA polymerase into one of the chromosomes of the nucleus. The source of the gene copy for this enzyme is still unknown today. The protozoon *Reclinomonas americana* is the first living representative that still exhibits the ancestral proto-mitochondrial genome organization and encodes functional subunits of the eubacterial RNA polymerase on its residual mitochondrial genome (Lang et al. 1997)

cell. According to this model, the ancestral eubacterial organism that represents the endosymbiont for the evolution of present mitochondria brought the eubacterial transcription machinery into the eukaryotic cell. After integration of a gene copy for the bacteriophage-type RNA polymerase subunit into the nuclear genome, this enzyme took over the transcription of the organelle genes. After this event, the eubacterial gene copies were eliminated from the mitochondrial genome. The origin of the gene copy for the bacteriophage-type enzyme is still an open question. It could have been introduced into the cell by the endosymbiont or by other independent genetic elements that arrived in the eukaryotic cell at a later time point. Independent of the origin of this gene, the respective reading frame had to obtain an extension coding for the mitochondrial targeting sequence which directs the protein to the organelles. The major consequences for the latest model of mitochondrial evolution is shown in Fig. 1.

b) Bacteriophage-Type Core Enzymes in Mitochondria and Plastids of Higher Plants

The second, very important finding is the identification of two nuclear copies for bacteriophage-type core enzymes in the higher plant *Arabidopsis thaliana* (Hedtke et al. 1997). The genomic sequences of the genes demonstrate that the two genes have nearly identical intron-exon structures and therefore have been created by gene duplication. The related proteins also exhibit a high degree of identical amino acids over the entire length of the proteins with the exception of the amino terminus. One of the proteins has a targeting sequence typical for mitochondria and the other leader sequence is characteristic for plastids. The different locations of the two polymerases have been proven experimentally by in vitro import studies with isolated organelles (Hedtke et al. 1997). It is of special interest that just recently a third gene for a bacteriophage-type core enzyme has been identified in *Arabidopsis thaliana* (A. Weihe, pers. comm.; Sanchez and Schuster, GeneBank Acc. No. AJ001037). The amino terminal extension of this protein gives no conclusive evidence whether this polymerase is imported into mitochondria, plastids or even both organelles. The latest data for organelle transcription in the higher plant *Arabidopsis thaliana* are summarized in Fig. 2. The different core enzymes in higher plants could be used for tissue- and development-specific expression of the genetic information in the organelles (Stern et al. 1997).

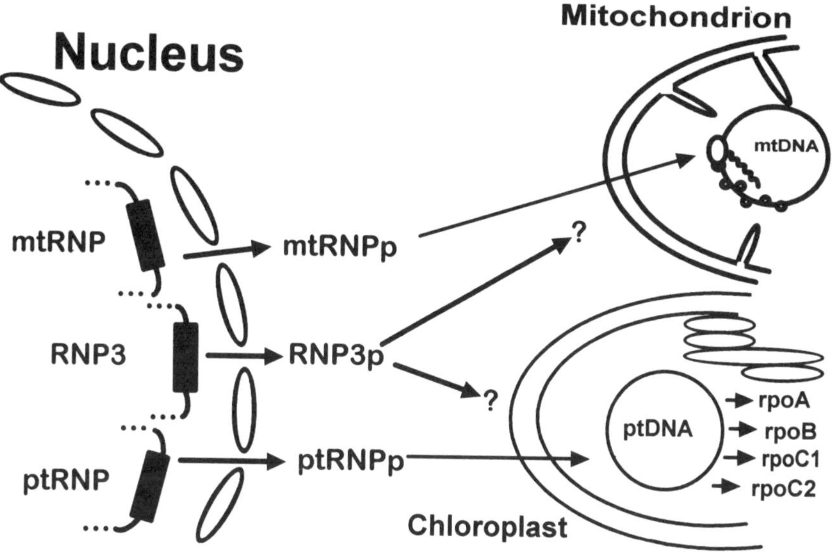

Fig. 2. Genes for bacteriophage-type RNA polymerases and eubacterial subunits for transcription in the higher plant *Arabidopsis thaliana*. In the nucleus of this species the latest research identified two genes encoding bacteriophage-type core enzymes for RNA polymerase (Hedtke et al. 1997). One gene encodes the mitochondrial RNA polymerase (*mtRNPp*), another a plastid version of this enzyme (*ptRNPp*). A third gene copy (*RNP3*) of a bacteriophage-type RNA polymerase has just been identified (Weihe, pers. comm.; Sanchez and Schuster, GeneBank Acc. No. AJ001037) and the related gene product may be imported into mitochondria or plastids or even both organelles. The *question marks* (?) indicate that experimental evidence for the precise localization of RNP3p is still missing. The nuclear enzymes all have to be transcribed in the nucleus, translated into protein (p) in the cytosol and imported into the specific organelles. The plastid genome contains four genes (*rpoA, B, C1, C2*) which code for eubacterial subunits of RNA polymerase and which give this organelle a limited autonomy for transcription of plastid DNA (Stern et al. 1997)

c) Structural Features of the Core Enzymes

The growing number of complete nuclear gene sequences for the mitochondrial core enzymes from different species from plants (Hedtke et al. 1997) to humans (Tiranti et al. 1997) allows a first detailed comparison to identify common elements and differences important for function (see Fig. 3). All the currently available protein sequences of the core enzymes demonstrate the presence of eight conserved domains that are also found in the bacteriophage SP6/T7 RNA polymerases. The amino terminal regions, on the other hand, exhibit extraordinarily high degrees of length and sequence variations. The first amino acids of the organelle enzymes represent the targeting sequences. There is first evidence that

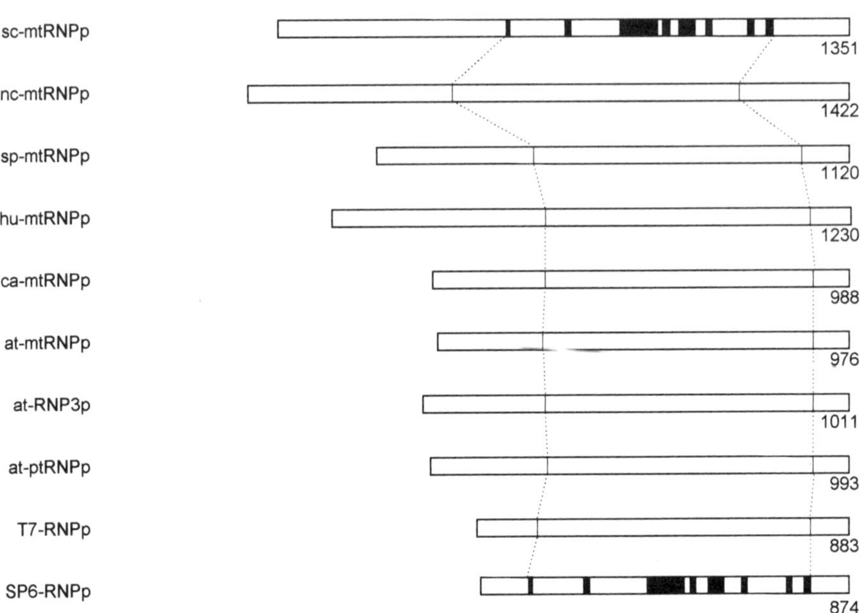

Fig. 3. Domain structure and special features of the bacteriophage-type RNA polymerase enzymes (RNPp) from species of which the complete DNA sequence of the genes and the derived amino acid sequences are now available. The enzymes for mitochondria (mt) or plastids (pt) have been aligned with the single subunit proteins for bacteriophages SP6 or T7. Eight conserved sequence blocks found in all enzymes are marked by *black boxes*. *Dashed lines* mark the position of the borders of the highly conserved sequences in the different proteins. The lengths of the proteins are given by the *numbers at the end of the related box*. The amino-terminal parts of the proteins exhibit a very high degree of sequence and length variations. **sc** *Saccharomyces cerevisiae*; **nc** *Neurospora crassa*; **sp** *Schizosaccharomyes pombe*; **hu** human; **ca** *Chenopodium album*; **at** *Arabidopsis thaliana*; **RNP3** third RNA polymerase of the bacteriophage-type found in *Arabidopsis thaliana*; **T7** bacteriophage T7; **SP6** bacteriophage SP6

the leader sequences in these enzymes are longer than for other mitochondrial proteins. For the yeast enzyme a leader sequence of more than 100 amino acid residues is predicted. The same is true for the core enzymes from *Arabidopsis thaliana* (A. Weihe, pers. comm.), but the presumptive long leader sequences alone cannot explain the high variations in these regions, like for example the extension in the up to now largest protein of *Neurospora crassa* with 1422 amino acid residues (Chen et al. 1996). This protein is more than 400 amino acids longer than the plant enzymes. The fact that also domains with no sequence similarities at all exist in these regions hints at a special functional role of the amino terminus. Even after the presumable cleavage of a larger leader sequence and import into the organelles there are still long amino terminal extensions present which have no sequence similarities with other core enzymes. The species-specific differences and variations in these regions could be important for species-specific expression of different promoter elements on the mitochondrial or plastid genomes and could play a critical role in the interactions with the related specificity factors for transcription (Bauerfeind et al. 1997).

The alignment for the conserved regions of the bacteriophage core enzymes allows the grouping into three classes represented by bacteriophage, linear plasmid and organelle (mitochondrial and plastid) enzymes. A new mutation in the yeast *RPO41* gene for the mitochondrial core enzyme gives additional functional evidence for this grouping (see Fig. 4). Determination of the changed amino acid that is responsible for the mutant phenotype (Lisowsky et al. 1996) identifies a serine residue next to a conserved histidine. This special pair of amino acids is found only in mitochondrial or plastid core enzymes for RNA polymerase but not in the linear plasmid or bacteriophage enzymes. The two mutants of yeast mitochondrial RNA polymerase give thereby valuable insights into the functional similarities and differences of the core enzymes. Up to now there are no other mutants for the organelle enzymes, as it proved to be very difficult to manipulate the DNA sequences of the related reading frames in vitro because special regions are very sensitive to errors occurring upon replication. This phenomenon seems to be a general characteristic of all the gene sequences (Lisowsky et al. 1996; D. A. Clayton, D. Lonsdale, A. Weihe, pers. comm.).

d) Perspectives and New Problems

The surprising new insights into the evolution of the organelle transcription system raises the new question concerning the origin of the bacteriophage-type core enzyme. It will be necessary to investigate further ancient protozoa in order to reconstruct in more detail the different steps during the adaptation of the eubacterial transcription machinery to the bacteriophage core enzyme. For the core enzyme one of the important functional aspects is the involvement of the long amino-terminal extension and its interaction with possible additional transcription factors. In this respect, it is necessary to identify mitochondrial specificity factors for transcription from species other than yeasts (Carrodeguas et al. 1997). The different forms of the bacteriophage-type core enzymes that exist in higher plants hint at a possible role of regulation of organelle gene expression in response to differentiation and development. This problem can now be addressed experimentally in *Arabidopsis*. The close relation of the mitochondrial and plastid transcription machinery in higher plants also raises the question of regulatory mechanisms in the cross-talk between mitochondria, plastids and the nucleus.

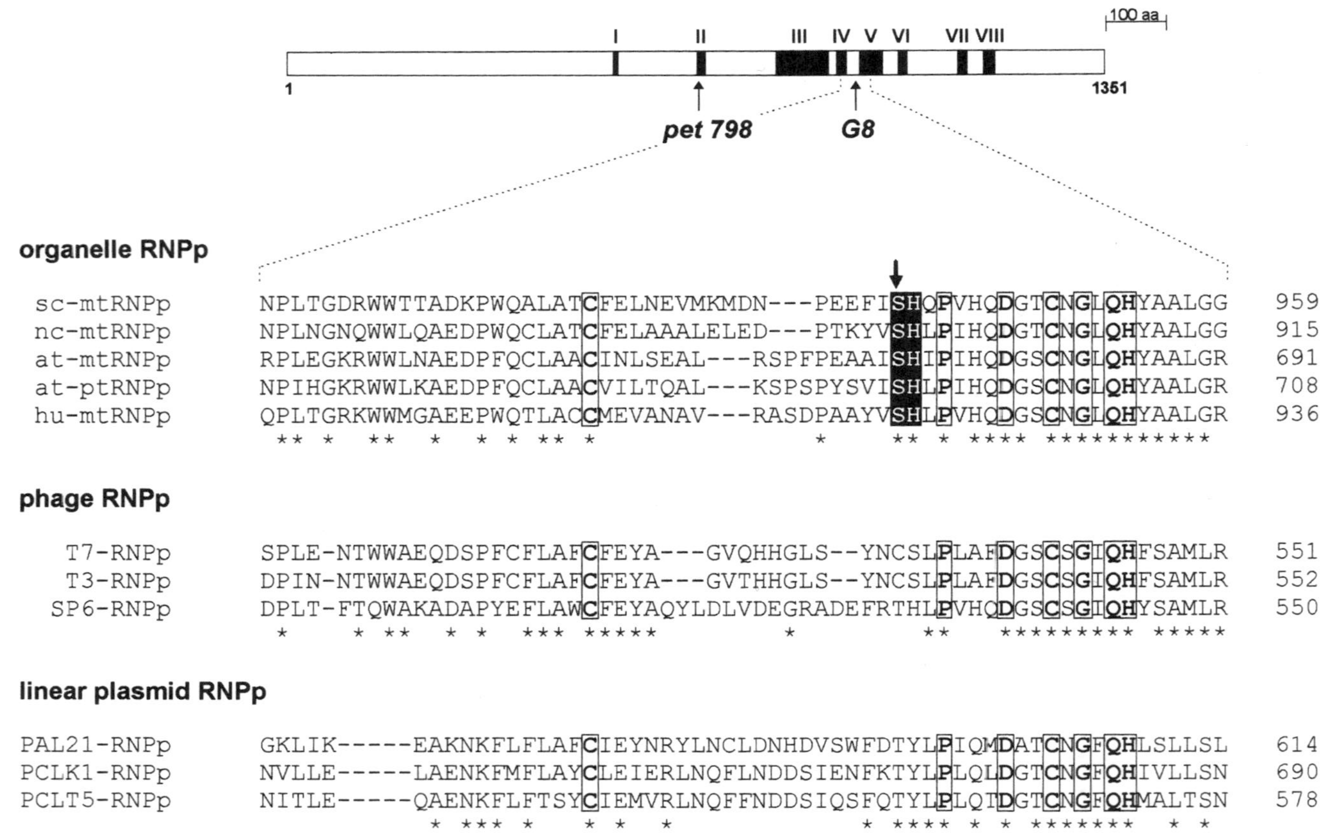

I   II   III IV V VI   VII VIII
100 aa
1
1351
pet 798
G8
organelle RNPp
sc-mtRNPp   NPLTGDRWWTTADKPWQALATCFELNEVMKMDN---PEEFISHQPVHQDGTCNGLQHYAALGG   959
nc-mtRNPp   NPLNGNQWWLQAEDPWQCLATCFELAAALELED---PTKYVSHLPIHQDGTCNGLQHYAALGG   915
at-mtRNPp   RPLEGKRWWLNAEDPFQCLAACINLSEAL---RSPFPEAAISHIPIHQDGSCNGLQHYAALGR   691
at-ptRNPp   NPIHGKRWWLKAEDPFQCLAACVILTQAL---KSPSPYSVISHLPIHQDGSCNGLQHYAALGR   708
hu-mtRNPp   QPLTGRKWWMGAEEPWQTLACCMEVANAV---RASDPAAYVSHLPVHQDGSCNGLQHYAALGR   936
            ** *   **   *   * * ** *   *   ** * **** * ***********
phage RNPp
T7-RNPp   SPLE-NTWWAEQDSPFCFLAFCFEYA---GVQHHGLS--YNCSLPLAFDGSCSGIQHFSAMLR   551
T3-RNPp   DPIN-NTWWAEQDSPFCFLAFCFEYA---GVTHHGLS--YNCSLPLAFDGSCSGIQHFSAMLR   552
SP6-RNPp   DPLT-FTQWAKADAPYEFLAWCFEYAQYLDLVDEGRADEFRTHLPVHQDGSCSGIQHYSAMLR   550
            *   * **   * *   *** *****   *   ** ********** *****
linear plasmid RNPp
PAL21-RNPp GKLIK-----EAKNKFLFLAFCIEYNRYLNCLDNHDVSWFDTYLPIQMDATCNGFQHLSLLSL   614
PCLK1-RNPp NVLLE-----LAENKFMFLAYCLEIERLNQFLNDDSIENFKTYLPLQLDGTCNGFQHIVLLSN   690
PCLT5-RNPp NITLE-----QAENKFLFTSYCIEMVRLNQFFNDDSIQSFQTYLPLQTDGTCNGFQHMALTSN   578
              * *** *   * *   *   * **** * * * ******   * *

4 Nuclear Restorer Genes of Cytoplasmic Male Sterility (CMS)

Cytoplasmic male sterility (CMS) is a defect that prevents the production of functional pollen without interfering with female fertility. The trait has been reported to occur in about 140 plant species (Wise et al. 1996) and is used for hybrid seed production of several cultivated plants. CMS is maternally inherited and results from the expression of unusual or aberrant mitochondrial genes. Different CMS systems in *Zea mays*, *Sorghum*, *Brassica*, *Petunia*, *Nicotiana*, *Plantago*, and *Helianthus* are fairly well characterized. The mechanisms of CMS and pollen fertility restoration by nuclear restorer genes offer the possibility to investigate the role of nuclear genes in mitochondrial gene expression during a special differentiation step. In this section recent progress in the characterization of nuclear CMS restorer genes will be summarized.

CMS and restoration of the fertile phenotype by nuclear genes is best characterized for the maize T-cytoplasm (Cui et al. 1996). Mitochondrial DNA of sterile plants contains the chimeric gene T-urf13, which is the result of a recombination event and encodes a protein that is localized in the mitochondrial inner membrane and is correlated with male sterility. Male sterile plants accumulate T-urf13 transcripts. Restoration can be achieved by mutations in the T-urf13 region or defined alleles of nuclear Rf genes (restorers of fertility). Using transposon tagging, Wise et al. (1996) generated mutated Rf1 alleles and analyzed their effect on T-urf13 transcripts. The authors concluded that the Rf1 gene product is not involved in transcriptional regulation, but alters T-urf13 RNA processing, in this way leading to *trans*-inhibition of translation of the truncated RNA. The tagging assay also resulted in the cloning of the first nuclear restorer gene, Rf2 (Cui et al. 1996), which codes for a putative mitochondrial aldehyde dehydrogenase. As a possible mechanism of fertility restoration by this enzyme, the authors discuss a detoxification reaction,

---

Fig. 4. Three distinct classes of bacteriophage-type RNA polymerases. The *upper part of the figure* represents the yeast mitochondrial RNA polymerase. *Black boxes* mark the eight conserved domains and *arrows* indicate the position of the mutation *pet798* and *G8*. The mutation *pet798* is located in a conserved domain whereas the recently described *G8* mutation is of special interest as it identifies a functionally important serine residue next to a conserved histidine (*black arrow* and *black box*). These two amino acids are conserved only in mitochondrial or plastid RNA polymerase core enzymes. Comparison of the domains and the conserved amino acids allows the classification of the bacteriophage-type enzymes into three groups. The RNA polymerases (*RNPp*) from organelles [mitochondria (mt) or plastids (pt)], phages and linear plasmids are aligned for the functionally important and conserved domains IV and V. sc *Saccharomyces cerevisiae*; nc *Neurospora crassa*; hu human; at *Arabidopsis thaliana*; T7 bacteriophage T7; T3 bacteriophage T3; SP6 bacteriophage SP6; PAL21 linear mitochondrial plasmid from *Podospora anserina*; PCLK1 linear mitochondrial plasmid from *Claviceps purpurea*; PCLT5 linear mitochondrial plasmid from *Claviceps purpurea*. For a complete list of the respective sequence accession numbers see Lisowsky et al. (1996)

a role in energy metabolism, or a direct or indirect interaction with the T-urf13 protein. The Rf8 gene, a recently described third restorer, has the ability to replace Rf1 partially and, like Rf1, leads to accumulation of an aberrant T-urf13 transcript (Dill et al. 1997). Another weak restorer identified in this work is also associated with slightly changed T-urf13 RNAs and was designated Rf*, because its allelic identity was not tested. Transcript mapping identified the conserved motif 5'-CNACNNU-3', included in the 5' end of all Rf1, Rf8, and Rf* associated T-urf13-transcripts. This motif might be important for T-urf13 expression and is also conserved in Rf3-associated orf107 transcripts of CMS *Sorghum*. The analysis of the Rf genes reveals, that cms-T restoration in maize can be mediated by multiple unlinked nuclear genes that directly or indirectly cause posttranscriptional modifications of T-urf13 transcripts.

Recent work on fertility restoration has also been published for another CMS system of maize, cms-S (Zabala et al. 1997), for sterile *Sorghum* (Tang et al. 1996; Howad and Kempken 1997), the Polima (pol) CMS of *Brassica napus* (Jean et al. 1997; Singh et al. 1996), and the Ogura CMS of *Brassica* (Bellaoui et al. 1997). The work of de Haan et al. (1997) demonstrates that CMS and restorer genes are not an invention of plant breeders but occur in nature. The authors analyzed genetically the frequency and mode of action of nuclear restorer genes in natural populations of *Plantago lanceolata*. Several dominant and recessive acting restorer genes for three different CMS types illustrate the complex situation found in natural populations of this plant species.

References

Akashi K, Sakurai K, Hirayama J, Fukujawa H, Okyama K (1996) Occurrence of nuclear encoded tRNA[Ile] in mitochondria of the liverwort *Marchantia polymorpha*. Curr Genet 30:181–185

Altamura N, Capitanio N, Bonnefoy N, Papa S, Dujardin G (1996) The *Saccharomyces cerevisiae OXA1* gene is required for the correct assembly of the cytochrome c oxidase and oligomycin-sensitive ATP synthase. FEBS Lett 382:111–115

Anderson S, Bankier AT, Barrell BG, De Bruijn MHL, Coulson AR, Drouin J, Eperon IC, Nierlich DP, Roe BA, Sanger F, Schreier PH, Smith AJH, Staden R, Young IG (1981) Sequence and organization of the human mitochondrial genome. Nature 290:457–465

Backert S, Nielsen BL, Börner T (1997) The mystery of the rings: structure and replication of mitochondrial genomes from higher plants. Trends Plant Sci 2:477–483

Bauer M, Behrens M, Esser K, Michaelis G, Pratje E (1994) *PET1402*, a nuclear gene required for proteolytic processing of cytochrome oxidase subunit 2 in yeast. Mol Gen Genet 245:272–278

Bauerfeind M, Esser K, Lisowsky T, Pratje E, Stein G, Stein T, Michaelis G (1997) Extranuclear inheritance: genetics and biogenesis of mitochondria. Prog Bot 58:386–409

Bauerfeind M, Esser K, Michaelis G (1998) The *Saccharomyces cerevisiae SOM1* gene: heterologous complementation studies, homologues in other organisms, and a mitochondrial inner membrane association of the small gene product. Mol Gen Genet 257:635–640

Bellaoui M, Pelletier G, Budar F (1997) The steady-state level of mRNA from the Ogura cytoplasmic male sterility locus in Brassica cybrids is determined post-transcriptionally by its 3'region. EMBO J 16:5057–5068

Bendich AJ (1996) Structural analysis of mitochondrial DNA molecules from fungi and plants using moving pictures and pulse-field gel electrophoresis. J Mol Biol 255:564–588

Blanc V, Jordana X, Litvak S, Araya A (1996) Control of gene expression by base deamination: the case of RNA editing in wheat mitochondria. Biochimie 78:511–517

Boer PH, Gray MW (1988) Scrambled ribosomal RNA gene pieces in *Chlamydomonas reinhardtii* mitochondrial DNA. Cell 55:399–411

Bonnefoy N, Chalvet F, Hamel P, Slonimski PP, Dujardin G (1994) *OXA1*, a *Saccharomyces cerevisiae* nuclear gene whose sequence is conserved from prokaryotes to eukaryotes controls cytochrome oxidase biogenesis. J Mol Biol 239:201–212

Braun HP, Schmitz UK (1997) The mitochondrial processing peptidase. Int J Biochem Cell Biol 29:1043–1045

Carrodeguas JA, Yun S, Shadel GS, Clayton DA, Bogenhagen DF (1997) Functional conservation of yeast mtTFB despite extensive sequence divergence. Gene Expr 6:219–230

Cermakian N, Ikeda TM, Cedergren R, Gray MW (1996) Sequences homologous to yeast mitochondrial and bacteriophage T3 and T7 RNA polymerases are widespread throughout the eukaryotic lineage. Nucleic Acids Res 24:648–654

Chen B, Kubelik AR, Mohr S, Breitenberger S (1996) Cloning and characterization of the *Neurospora crassa cyt*-5 gene. J Biol Chem 271:6573–6544

Clayton DA (1996) Mitochondrial DNA gets the drift. Nat Genet 14:123–125

Cui X, Wise RP, Schnable PS (1996) The ref2 nuclear restorer gene of male-sterile T-cytoplasm maize. Science 227:1334–1336

De Haan AA, Koelewijn HP, Hundscheid MPJ, Van Damme JMM (1997) The dynamics of gynodioecy in *Plantago lanceolata* L. II. Mode of action and frequencies of restorer alleles. Genetics 147:1317–1328

Dill CL, Wise RP, Schnable PS (1997) Rf8 and Rf* mediate unique T-urf13-transcript accumulation, revealing a conserved motif associated with RNA processing and restoration of pollen fertility in T-cytoplasm maize. Genetics 147:1367–1379

Esser K, Pratje E, Michaelis G (1996) SOM1, a small new gene required for mitochondrial inner membrane peptidase function in *Saccharomyces cerevisiae*. Mol Gen Genet 252:437–445

Fisher RP, Parisi MA, Clayton DA (1989) Flexible recognition of rapidly evolving promoter sequences by mitochondrial transcription factor 1. Genes Dev 3:2202–2217

Gershon D (1997) More on mitochondria and senescence. Bioessays 19:533–534

Glick BS, von Heijne G (1996) *Saccharomyces cerevisiae* mitochondria lack a bacterial-type Sec machinery. Prot Sci 5:2651–2652

Gray MW (1996) RNA editing in plant organelles: A fertile field. Proc Natl Acad Sci USA 93:8157–8159

Gray MW, Lang BF (1998) Transcription in chloroplasts and mitochondria: a tale of two polymerases. Trends Microbiol 6:1–3

Griffiths AJF (1996) Mitochondrial inheritance in filamentous fungi. J Genetics 75:403–414

Hamel P, Sakamoto W, Wintz H, Dujardin G (1997) Functional complementation of an oxaľ yeast mutation identifies an *Arabidopsis thaliana* cDNA involved in the assembly of respiratory complexes. Plant J 12:1319–1327

Hanson MR (1996) Protein products of incompletely edited transcripts are detected in plant mitochondria. Plant Cell 8:1–3

Haucke V, Schatz G (1997) Import of proteins into mitochondria and chloroplasts. Trends Cell Biol 7:103–106

He S, Fox TD (1997) Membrane translocation of mitochondrially coded Cox2p: distinct requirements for export of N and C termini and dependence on the conserved protein Oxa1p. Mol Biol Cell 8:1449–1460

Hedtke B, Börner T, Weihe A (1997) Mitochondrial and chloroplast phage-type RNA polymerases in *Arabidopsis*. Science 277:809–811

Hell K, Herrmann JM, Pratje E, Neupert W, Stuart RA (1997) Oxa1p mediates the export of the N- and C-termini of pCoxII from the mitochondrial matrix to the intermembrane space. FEBS Lett 418:367–370

Herrmann JH, Neupert W, Stuart RA (1997) Insertion into the mitochondrial inner membrane of a polytopic protein, the nuclear-encoded Oxa1p. EMBO J 16:2217–2226

Howad W, Kempken F (1997) Cell type-specific loss of atp6 RNA editing in cytoplasmic male sterile *Sorghum bicolor*. Proc Natl Acad Sci 94:11090–11095

Janska H, Woloszynska M (1997) The dynamic nature of plant mitochondrial genome organization. Acta Biochim Pol 44:239–250

Jean M, Brown GG, Landry BS (1997) Genetic mapping of nuclear fertility restorer genes for the 'Polima' cytoplasmic male sterility in canola (*Brassica napus* L.) using DNA markers. Theor Appl Genet 95:321–328

Knoop V, Unseld M, Marienfeld J, Brandt P, Sünkel S, Ullrich H, Brennicke A (1996) Copia, gypsy- and LINE-like retrotransposon fragments in the mitochondrial genome of *Arabidopsis thaliana*. Genetics 142:579–585

Kumar R, Maréchal-Dronard L, Akama K, Small J (1996) Striking differences in mitochondrial tRNA import between different plant species. Mol Gen Genet 252:404–411

Lang BF, Burger G, O'Kelly CJ, Cedergren R, Golding BG, Lemieux C, Sankoff D, Turmel M, Gray MW (1997) An ancestral mitochondrial DNA resembling a eubacterial genome in miniature. Nature 387:493–497

Laroche J, Li P, Maggia L, Bousquet J (1997) Molecular evolution of angiosperm mitochondrial introns and exons. Proc Natl Acad Sci 94:5722–5727

Leblanc C, Boyen C, Richard O, Bonnard G, Grienenberger JM, Kloareg B (1995) Complete sequence of the mitochondrial DNA of the rhodophyte *Chondrus crispus* (Gigartinales). Gene content and genome organization. J Mol Biol 250:484–495

Leblanc C, Richard O, Kloareg B, Viehmann S, Zetsche K, Boyen C (1997) Origin and evolution of mitochondria: what have we learnt from red algae? Curr Genet 31:193–207

Lightowlers RN, Chinnery PF, Turnbull DM, Howell N (1997) Mammalian mitochondrial genetics: heredity, heteroplasmy and disease. Trends Genet 13:450–455

Lisowsky T, Michaelis G (1988) A nuclear gene essential for mitochondrial replication suppresses a defect of mitochondrial transcription in *Saccharomyces cerevisiae*. Mol Gen Genet 214:218–223

Lisowsky T, Stein T, Michaelis G, Guan M-X, Chen XJ, Clark-Walker DG (1996) A new point mutation in the nuclear gene of yeast mitochondrial RNA polymerase *RPO41* identifies a functionally important amino acid residue in a protein region conserved among mitochondrial core enzymes. Curr Genet 30:389–395

Malek O, Brennicke A, Knoop V (1997) Evolution of *trans*-splicing plant mitochondrial introns in pre-permian times. Proc Natl Acad Sci USA 94:553–558

Masters BS, Stohl LL, Clayton DA (1987) Yeast mitochondrial RNA polymerase is homologous to those encoded by bacteriophages *T3* and *T7*. Cell 51:89–99

Meyer W, Bauer M, Pratje E (1997a) A mutation in cytochrome oxidase subunit 2 restores respiration of the mutant *pet ts1402*. Curr Genet 31:401–407

Meyer W, Boemer U, Pratje E (1997b) Mitochondrial inner membrane bound Pet1402 protein is rapidly imported into mitochondria and affects the integrity of the cytochrome oxidase and ubiquinol-cytochrome c oxidoreductase complexes. Biol Chem 378:1373–1379

Michaelis G, Vahrenholz C, Pratje E (1990) Mitochondrial DNA of *Chlamydomonas reinhardtii*: the gene for apocytochrome b and the complete functional map of the 15.8 kb DNA. Mol Gen Genet 223:211–216

Nakazono M, Nishiwaki S, Tsutsumi N, Hirai A (1996) A chloroplast-derived sequence is utilized as a source of promoter sequences for the gene for subunit 9 of NADH dehydrogenase (nad 9) in rice mitochondria. Mol Gen Genet 252:371–378

Nedelcu AM (1997) Fragmented and scrambled mitochondrial ribosomal RNA coding regions among green algae: a model for their origin and evolution. Mol Biol Evol 14:506–517

Neupert W (1997) Protein import into mitochondria. Annu Rev Biochem 66:863–917

Nunnari J, Walter P (1996) Regulation of organelle biogenesis. Cell 84:389–394

Nunnari J, Fox TD, Walter P (1993) A mitochondrial protease with two catalytic subunits of nonoverlapping specificities. Science 262:1997–2004

Oda K, Yamato K, Ohta E, Nakamura Y, Takemura M, Nozato N, Akashi K, Kanegae T, Ogura Y, Kohchi T, Ohyama K (1992) Gene organization deduced from the complete sequence of liverwort *Marchantia polymorpha* mitochondrial DNA. J Mol Biol 223:1–7

Ozawa T (1997a) Genetic and functional changes in mitochondria associated with aging. Physiol Rev 77:425–464

Ozawa T (1997b) Oxidative damage and fragmentation of mitochondrial DNA in cellular apoptosis. Biosci Rep 17:237–250

Palmer JD (1997) The mitochondrion that time forgot. Nature 387:454–455

Pfanner N, Craig EA, Hönlinger A (1997) Mitochondrial preprotein translocase. Annu Rev Cell Dev Biol 13:25–51

Pratje E, Esser K, Michaelis G (1994) The mitochondrial inner membrane peptidase: In: Heijne G von (ed) Signal peptidases. Landes Company, Austin, pp 105–112

Pratje E, Esser K, Bauerfeind M, Meyer W, Bauer M, Michaelis G (1997) The mitochondrial inner membrane peptidase of yeast and genes affecting its function. In: Hopsu-Havu VK, Jaervinen M, Kirschke H (eds) Proteolysis in cell functions, IOS Press, Amsterdam, pp 348–354

Riemen G, Lisowsky T, Maggouta F, Michaelis G, Pratje E (1993) Extranuclear inheritance: mitochondrial genetics. Prog Bot 54:318–333

Rusché LN, Cruz-Reyes J, Piller KJ, Sollner-Webb B (1997) Purification of a functional enzymatic editing complex from *Trypanosoma brucei* mitochondria. EMBO J 16:4069–4081

Ryan MT, Naylor DJ, Hoj PB, Clark MS, Hoogenraad NJ (1997) The role of molecular chaperones in mitochondrial protein import and folding. Int Rev Cytol Surv Cell Biol 174:127–193

Schinkel AH, Groot-Koerkamp MJA, Touw EPW, Tabak HF (1987) Specificity factor of yeast mitochondrial RNA polymerase. Purification and interaction with core RNA polymerase. J Biol Chem 262:12785–12791

Shadel GS, Clayton DA (1993) Mitochondrial transcription initiation. J Biol Chem 268:16083–16086

Shadel GS, Clayton DA (1995) A *Saccharomyces cerevisiae* mitochondrial transcription factor, sc-mtTFB, shares features with sigma factors but is functionally distinct. Mol Cell Biol 15:2101–2108

Shadel GS, Clayton DA (1997) Mitochondrial DNA maintenance in vertebrates. Annu Rev Biochem 66:409–435

Silva Filho MC, Chaumont F, Leterme S, Boutry M (1996) Mitochondrial and chloroplast targeting sequences in tandem modify protein import specificity in plant organelles. Plant Mol Biol 30:764–780

Singh M, Hamel N, Menassa R, Li XQ, Young B, Jean M, Landry BS, Brown GG (1996) Nuclear genes associated with a single Brassica CMS restorer locus influence transcripts of three different mitochondrial gene regions. Genetics 143:505–516

Stern DB, Higgs DC, Yang J (1997) Transcription and translation in chloroplasts. Trends Plant Sci 2:308–315

Stuart K, Allen TE, Heidmann S, Seiwert SD (1997) RNA editing in kinetoplastid protozoa. Microbiol Mol Biol Rev 61:105–120

Stuart RA, Neupert W (1996) Topogenesis of inner membrane proteins of mitochondria. TIBS 21:261–267

Suzuki CK, Rep M, van Dijl JM, Suda K, Grivell LA, Schatz G (1997) ATP-dependent proteases that also chaperone protein biogenesis. TIBS 22:118–123

Tang HV, Pring DR, Shaw LC, Salazar RA, Muza FR, Yan B, Schertz KF (1996) Transcript processing internal to a mitochondrial open reading frame is correlated with fertility restoration in male-sterile sorghum. Plant J 10:123–133

Thorsness PE, Weber ER (1996) Escape and migration of nucleic acids between chloroplasts, mitochondria, and the nucleus. Int Rev Cytol Surv Cell Biol 165:207–234

Tiranti V, Savoia A, Forti F, Dapolito MF, Centra M, Racchi M, Zeviani M (1997) Identification of the gene encoding the human mitochondrial RNA polymerase (H MTRPOL) by cyberscreening of the expressed sequence tags database. Human Mol Genet 6:615–625

Tracy RL, Stern DB (1995) Mitochondrial transcription initiation: promoter structures and RNA polymerases. Curr Genet 28:205–216

Unseld M, Marienfeld JR, Brandt P, Brennicke A (1997) The mitochondrial genome of *Arabidopsis thaliana* contains 57 genes in 366,924 nucleotides. Nat Genet 15:57–61

Vahrenholz C, Riemen G, Pratje E, Dujon B, Michaelis G (1993) Mitochondrial DNA of *Chlamydomonas reinhardtii*: the structure of the ends of the linear 15.8 kb genome suggests mechanisms for DNA replication. Curr Genet 24:241–247

Wallace DC (1997) Mitochondrial DNA in aging and disease. Sci Am 277:40–47

Whelan J, Glaser E (1997) Protein import into plant mitochondria. Plant Mol Biol 33:771–789

Wise RP, Dill CL, Schnable PS (1996) Mutator-induced mutations of the rf1 nuclear fertility restorer of T-cytoplasm maize alter the accumulation of T-urf13 mitochondrial transcripts. Genetics 143:1383–1394

Wolff G, Plante I, Lang BF, Kück U, Burger G (1994) Complete sequence of the mitochondrial DNA of the chlorophyte alga *Prototheca wickerhamii*. J Mol Biol 237:75–86

Yoshinaga K, Kakehi T, Shima Y, Iinuma H, Masuzawa T, Ueno M (1997) Extensive RNA editing and possible double-stranded structures determining editing sites in the atpB transcripts of hornwort chloroplasts. Nucleic Acids Res 25:4830–4834

Zabala G, Gabay-Laughnan S, Laughnan JR (1997) The nuclear gene Rf3 affects the expression of the mitochondrial chimeric sequence R implicated in S-type male sterility in maize. Genetics 147:847–860

Thomas Lisowsky
Karlheinz Esser
Torsten Stein
Georg Michaelis
Botanisches Institut
der Universität Düsseldorf
Universitätsstraße 1
D-40225 Düsseldorf, Germany

Elke Pratje
Institut für Allgemeine Botanik
Universität Hamburg
Ohnhorststraße 18
D-22609 Hamburg, Germany

Edited by
K. Esser

Genetics of Phytopathology:
Phytopathogenic Bacteria

By Holger Jahr, Rainer Bahro, and Rudolf Eichenlaub

1 Introduction

In this chapter we will cover the exopolysaccharides (EPS) and plant cell wall-degrading enzymes of phytopathogenic bacteria as a follow up on our previous review (Bahro et al. 1996). Historically, the excretion of these components by phytopathogenic bacteria has been discussed as an important factor in disease development. It seems quite obvious that a vascular pathogen, producing large amounts of highly viscous EPS, can block xylem vessels and thus cause infected plants to wilt. On the other hand, extracellular bacterial enzymes attacking the plant cell wall may lead to maceration of plant tissue and result in soft rot symptoms in diseased plants.

With the advent of molecular genetics in plant pathology it became possible to address the question whether EPS and extracellular enzymes are really crucial pathogenicity factors. Genetic studies, especially the isolation of mutants affecting the biosynthesis of these components, showed that this seems to be the case only in some systems, while in other cases the disease is the result of various synergistically acting pathogenicity factors. In the sections below, EPS of some selected bacteria will be addressed, followed by the description of some cell wall-degrading enzymes for which involvement in pathogenicity has been demonstrated.

2 Extracellular Polysaccharides (EPS)

a) General Remarks

Exopolysaccharides (EPS) are known to be excreted by many genera of bacteria as part of the extracellular matrix (Sutherland 1988; Leigh and Coplin 1992). EPS are present either in form of a capsule covalently bound to the bacterial cell envelope or as a loosely associated slime, which can be released into the environment. Due to their special properties, EPS are thought to enable bacteria to colonize special habitats

(Ferris and Beveridge 1985; Costerton et al. 1987). Virtually all plant-pathogenic and associative bacteria produce large amounts of EPS in culture and also in their host plant, and thus it was suggested that production of EPS is an essential factor in pathogenic or symbiotic interactions of bacteria with plants (Van Alfen 1989; Coplin and Cook 1990; Leigh and Coplin 1992, Denny 1995).

In this review the role of the EPS in the symbiotic *Rhizobium*-legumes and the tumor-generating *Agrobacterium*-plant interaction will not be addressed. For these topics the reader is referred to reviews by Leigh Walker (1994) and Denny (1995), respectively.

Chemically, EPS molecules are either homo- or heteropolymers with a complex structure. Most of these macromolecules are acidic heteropolysaccharides with high molecular weights consisting of repeating units containing neutral sugars and uronic acids. In addition, the long sugar chains formed by these repeating units are usually branched. Common side groups, linked to the sugar chains by ester bonds are pyruvate, succinate, and acetate (Leigh and Coplin 1992; Denny 1995). Synthesis of EPS is regulated and affected by environmental factors and growth stage. Due to the hydrophilic and anionic properties, EPS determine multiple and important biological functions. EPS may generate a water-saturated matrix around the bacteria as a protection against desiccation, detergents, and hydrophobic polypeptide antibiotics due to a prevention of contact with the hydrophobic molecules of the cell membrane (Leigh and Coplin 1992). Furthermore, it enables free-living bacteria to adhere to abiotic or biological surfaces, where higher concentrations of nutrients may be found (Costerton et al. 1987).

The acidic EPS produced by many plant pathogens can act as an ion exchanger concentrating minerals and nutrients, or binding toxic compounds (Norberg and Persson 1984; Coplin and Majerczak 1990). EPS may also promote infection and colonization and enhance survival of bacteria within the host tissue, which is necessary for the development of disease symptoms (Tharaud et al. 1994; Bermpohl et al. 1996; Saile et al. 1997). Thus, EPS production is often positively correlated with virulence of phytopathogenic bacteria and EPS⁻ mutants of several plant pathogens have been shown to exhibit a reduced virulence. Capsules and slime may prevent recognition of the pathogen by the plant defence system and block agglutinins or lectins and detoxify phytoalexins or reactive oxygen species (Bradshow-Rouse et al. 1981; Romeiro et al. 1981; Young and Sequeira 1986; Kiraly et al. 1997).

The biosynthetic pathways for EPS are very complex and bacteria often produce more than one kind of EPS. The genes for EPS production are generally clustered and in some cases EPS synthesis seems to be coordinated with the expression of other virulence factors. Expression of EPS synthesis genes is modulated both positively and negatively by a complex, hierarchical regulatory network consisting of cell density-dependent autoinduction, two-component regulatory systems, and transcriptional activators.

In the following paragraphs we will focus on well-studied, vascular plant pathogens, like *Ralstonia (Pseudomonas) solanacearum, Erwinia* sp., *Xanthomonas campestris* pv. *campestris* and *Clavibacter michiganensis* subsp. *michiganensis*. In these systems sufficient data have been accumulated to allow discussion of possible functions of the EPS in pathogenesis.

b) *Erwinia amylovora*

The Gram-negative bacterium *Erwinia amylovora* causes fire blight on apple and pear trees and other rosaceous plants (Aldwinckle and Beer 1979). *E. amylovora* produces the homopolysaccharide levan and the acidic heteropolysaccharide amylovoran (Steinberger and Beer 1988; Ayers et al. 1979), with the formation of amylovoran being required for pathogenicity (Bennet and Billing 1978; Bellemann and Geider 1992; Bernhard et al. 1993b; Tharaud et al. 1994). Amylovoran is composed of a pentasaccharide repeating unit of four differently linked galactose molecules and a glucuronic acid residue which are decorated with pyruvate and acetate groups (Smith et al. 1990; Nimtz et al. 1996). The enzymes for amylovoran synthesis are encoded by the *ams* (*amy*lovoran synthesis) operon (Bernhard et al. 1993b) consisting of 12 genes (*ams*A-L). The operon is expressed from a single RcsA/B-regulated promoter in front of *ams*G (Burgert and Geider 1995). RcsA and RcsB are transcriptional activator proteins of the LuxR type (Coleman et al. 1990; Stout et al. 1991; Bereswill and Geider 1997). RcsB and another protein RcsC constitute a two-component regulatory system (Bernhard et al. 1990; Stout 1994; Bereswill and Geider 1997). The regulation of EPS synthesis by the *rcs* genes is a common system among enterobacteria (Chatterjee et al. 1990; Leigh and Coplin 1992). The current model described below is based on regulation of colanic acid synthesis in *E. coli* (Gottesman and Stout 1991).

RcsC is suggested to be an environmental sensor protein of the ITR class located in the cytoplasmatic membrane which activates RcsB by phosphorylation (Stout and Gottesman 1990; Parkinson and Kofoid 1992; Stout 1994). In the phosphorylated state, RcsB forms a dimer and activates transcription of the *ams* genes. Regardless of the state of phosphorylation, RcsB can alternatively form a heterodimer with RcsA, also activating transcription. Since RcsA is subjected to rapid degradation by *lon* protease (Eastgate et al. 1995), expression of the *ams* genes depends either on RcsA concentration or signal-mediated phosphorylation of RcsB by RcsC. Accordingly, mutation in the *lon* gene results in an overproduction of amylovoran (Eastgate et al. 1995) while mutations in the genes of the *ams* operon and the regulatory genes *rcs*A or *rcs*B cause reduction of amylovoran production and these mutants are less virulent (Bernhard et al. 1990; Chatterjee et al. 1990; Coleman et al. 1990; Bernhard et al. 1993b; Tharaud et al. 1994; Bereswill and Geider 1997).

In the pathogenic interaction, the capsule of amylovoran may create a favourable microenvironment and protects the bacteria from recognition and thereby prevents immobilization by plant defence (Romeiro et al. 1981; Metzger et al. 1994; Tharaud et al. 1994). Also, amylovoran may cause plugging of xylem vessels and disrupt water flow and thus cause wilting of diseased plants (Huang and Goodman 1976).

In addition to the acidic amylovoran, *E. amylovora* produces levan, a neutral homopolymer composed of β-2,6-linked fructose, synthesized by extracellular levansucrase (Gross et al. 1992). Levansucrase is constitutively expressed and inactivation of its gene (*lsc*) by transposon mutagenesis had no significant effect on virulence (Geier and Geider 1993).

c) *Erwinia stewartii*

Erwinia stewartii is a pathogen of maize causing Stewart's wilt and blight and is transmitted by the corn flea beetle. The bacteria colonize the vascular system, are able to invade the intercellular spaces of leaves and produce water-soaked lesions. The EPS produced by *E. stewartii*, stewartan, consists of glucuronic acid, galactose and glucose in a ratio of 1:3:3 (Bernhard et al. 1996). Stewartan is an important virulence factor, since EPS-deficient mutants of *E. stewartii* fail to cause wilting and water soaking. However, these mutants can still induce local necrotic lesions (Dolph et al. 1988; Braun 1990; Coplin and Majerczak 1990).

The *cps* gene cluster required for stewartan synthesis has been cloned (Dolph et al. 1988) and sequence data suggest that it has an organization similar to the *ams* gene cluster of *E. amylovora* (Bernhard et al. 1993a, 1996). Furthermore, the synthesis of stewartan is regulated similarly to that of *E. amylovora*, involving the two-component RcsC-RcsB system and RcsA (Poetter and Coplin 1991; Stout 1994). In addition, N-acyl-homoserine lactones (acyl-HSLs) are engaged in regulation of EPS synthesis. Inability to produce acyl-HSLs results in loss of virulence (Beck von Bodman and Farrand 1995).

Bernhard and coworkers (1996) demonstrated that the ability to synthesize either amylovoran or stewartan can be transferred to different *Erwinia* species by cosmids carrying the *ams* (cosmid pEA109) or the *cps* (cosmid pES2144) gene cluster. *E. amylovora amsA-amsF* mutants carrying pES2144 synthesize stewartan, but this EPS did not substitute for amylovoran in pathogenesis, since merodiploids were avirulent on pear slices and apple seedlings. In contrast, mutants in *cps*B-*cps*D of *E. stewartii*, carrying pEA109, produce amylovoran and have regained virulence on corn seedlings. This indicates that in the interaction of *E. amylovora* with the host plant the structure and composition of the EPS is a crucial factor.

d) *Ralstonia solanacearum*

Ralstonia (Pseudomonas) solanacearum (Yabuuchi et al. 1995) causes bacterial wilt in more than 200 plant species. Agronomically important hosts are tomato, potato, tobacco, peanut and banana (Buddenhagen and Kelman 1964). Soilborne wild-type *R. solanacearum* efficiently enter host plants via roots, penetrate the vascular system and systemically colonize the xylem (Hayward 1991). The major EPS of virulent strains is EPS I, a > 1000-kDa acidic, unbranched heteropolymer, composed of a repeating trimeric unit of N-acetylgalactosamine, 2-N-acetylgalactos-aminuronic acid, and 2-N-acetylbacillosamine decorated with 3-hydroxybutyric acid (Orgambide et al. 1991; Schell et al. 1993). Trans-poson-induced mutants defective in EPS I production were always found to be affected in virulence (Denny et al. 1988; Cook and Sequeira 1991; Denny and Baek 1991; Kao et al. 1992). Possibly, EPS prevent adherence of bacteria to the plant cell wall and thereby propagate the ability for colonization (Young and Sequeira 1986; Saile et al. 1997).

Production of EPS I requires the 18-kb *eps* gene cluster encoding several membrane-associated and soluble enzymes involved in biosynthesis and export of EPS I (Denny and Baek 1991; Schell et al. 1993). Furthermore, synthesis of EPS I involves genes of the *ops* gene cluster (lipopolysaccharide biosynthesis) required for synthesis of some common sugar precursors for both EPS I and lipopolysaccharides (Cook and Sequeira 1991; Kao and Sequeira 1991). A single promoter directing transcription of at least seven genes of the *eps* gene cluster (Huang and Schell 1995) is regulated by PhcA, a LysR-type transcriptional activator (Brumbley and Denny 1990; Schell 1993). Mutations in *phcA* reduce production of EPS I and decrease virulence (Brumbley and Denny 1990; Orgambide et al. 1991; Brumbley et al. 1993).

The network regulating EPS I expression includes, besides PhcA, at least three two-component systems (PhcS/R, VsrA/D and VsrB/C), acyl-HSLs acting as autoinducers, the cell density-dependent autoregulator 3-hydroxy palmitic acid methyl ester (3-OH PAME) and another, unusual regulator protein called XpsR. All these systems need to be activated for production of EPS I (Huang et al. 1995; Clough et al. 1997a, b; Flavier et al. 1997a, b).

Autoinduction is the ability of bacteria to respond to extracellular concentrations of autoinducers in order to express specific genes coordinately within a population. This type of synchronized activation, which usually occurs only when a population reaches a high cell density within a confined location, is also termed quorum sensing (Fuqua et al. 1994, 1996; Sitnikov et al. 1995). A novel autoregulator, 3-OH PAME, is synthesized by PhcB at high cell densities and controls virulence in *R. solanacearum*. Mutants in *phcB* that cannot synthesize 3-OH PAME do not produce EPS I and, as a consequence, are not virulent on tomato plants. 3-OH PAME modulates PhcA-regulated genes via the PhcS/R two-component regulatory system which represses expression and/or function of *phcA* (Clough et al. 1997a, b; Flavier et al. 1997a, b).

In its active state PhcA regulates expression of extracellular enzymes, production of acyl-HSLs and cell motility. In the regulation of EPS production PhcA and the two-component regulatory system VsrA/D control the production of XpsR (Schell et al. 1994). XpsR is a unique basic regulatory protein that appears to interact with VsrC, the response regulator of the two-component system VsrB/C to activate the transcription at the *eps* promoter (Huang et al. 1995). Thus, XpsR and PhcA are the central components in a multifactorial regulatory cascade. In addition, two acyl-HSLs can be detected at high cell density in extracts of wild-type *R. solanacearum* (Flavier et al. 1997b; Shaw et al. 1997) but it remains unclear at what stage these putative autoinducers act in the regulatory network.

Based on the complex regulation of EPS and other virulence factors, it was suggested that *R. solanacearum* can switch between two specialized phenotypes controlled by PhcA (Denny et al. 1994; Clough et al. 1997a). At cell densities below 10^7 cfu/ml the bacteria show a phenotype comparable to a *phcA* mutant, being low in virulence, motile, and producing only small quantities of EPS and other virulence factors. This phenotype might well be sufficient for survival in soil or plant debris, a saprophytic existence, and/or early parasitic activities. Above a titre of 10^8 cfu/ml the phenotype switches to high virulence. The bacteria produce large amounts of EPS and cell wall-degrading enzymes which may facilitate colonization of host tissue, induce release of nutrients by the diseased plant and may allow them to avoid host defence reactions.

e) *Xanthomonas campestris* pv. *campestris*

Xanthomonas campestris pv. *campestris* (*Xcc*) causes black rot of crucifers (Williams 1980) on all members of the cabbage family. The characteristic symptom of the infection process is vein blackening, a reaction of the plant to the occlusion and disorganization of the vascular system by the bacteria (Sutton and Williams 1970). Chlorosis and wilting are the subsequent results of the induced water stress. Virulence of *Xcc* is correlated to the ability to produce EPS, since EPS is proposed to mask the bacteria, preventing induction of plant defence reactions (Ramirez et al. 1988; Newman et al. 1994). *Xcc* produces the EPS xanthan with pentasaccharide repeating units consisting of glucose, mannose and glucuronic acid in a molar ratio of 2:2:1, decorated with acetate and pyruvate side groups (Jannson et al. 1975). Xanthan biosynthesis is well characterized; both the biosynthetic genes and the biochemical activity of the gene products are known (reviewed by Coplin and Cook 1990). A large gene cluster (*xps/gum*B-M) consisting of at least 12 genes is required for xanthan biosynthesis (Barrere et al. 1986; Harding et al. 1987, 1993).

The production of EPS and other virulence factors, for example extracellular enzymes, is positively controlled by at least two two-component regulatory systems (Osbourn et al. 1990; Tang et al. 1991). A regulatory gene cluster, *rpf*A-H, was found to have coordinated effects on expression of EPS biosynthesis genes and other virulence factors (Tang et al. 1991; Dow and Daniels 1994). Mutations in any of these genes result in significant reduction of EPS, extracellular enzymes and virulence. One of these genes, *rpf*C encodes a fused two-component sensor-regulator protein of the ITR class (Tang et al. 1991; Parkinson and Kofoid 1992). Evidence for a negative regulator gene which seems to balance the system has also been reported (Tang et al. 1990). In addition, it was determined recently that EPS production is controlled by diffusible, low-molecular-weight substances (DF and DSF) of unknown chemical structure, which were found in the culture supernatant. These factors are part of two distinct signal transduction pathways in *Xcc* that cross-talk in regulation of EPS biosynthesis (Poplawsky et al. 1998).

f) *Clavibacter michiganensis* subsp. *michiganensis*

Clavibacter michiganensis subsp. *michiganensis (Cmm)* is a Gram-positive bacterium causing bacterial wilt and canker of tomato (see recent review by Metzler et al. 1997). EPS produced by these bacteria is an acidic, high-molecular-weight heteropolysaccharide composed of a tetrasaccharide repeating unit with fucose, glucose and galactose in a molar ratio of 2:1:1, decorated with acetate and pyruvate residues (Eichenlaub et al. 1991; Van den Bulk et al. 1991; Bermpohl et al. 1996). It has been proposed that EPS may induce wilting by blocking xylem vessels, interfering with water transport within the plant (Van den Bulk et al. 1989, 1991). However, histological studies of infected tomato plants presented no indication of occlusion of xylem vessels (Wallis 1977); instead, tissue maceration occurred, indicating that extracellular enzymes may be relevant for virulence (Benhamou 1991).

EPS$^-$ mutants producing only 10% of the amount of EPS of the wild type are not altered in virulence (Bermpohl et al. 1996). Genes responsible for wilting of infected plants were located on two plasmids, encoding an endoglucanase and a putative extracellular protease (Meletzus et al. 1993; Dreier et al. 1997). Thus, in this system EPS does not seem to be directly involved in the establishment of disease symptoms, but at most may cause some water stress accompanied by a slight reduction of biomass of infected plants. However, in some isolates of *Cmm* a correlation was found between variations in the sugar composition of EPS and the inability to colonize the host, indicating that the chemical structure of the EPS may play a role in host recognition and early stages of infection (Bermpohl et al. 1996).

3 Plant Cell Wall-Degrading Enzymes

In addition to the excretion of exopolysaccharides by the wilt-inducing
bacteria of the genera *Clavibacter, Erwinia, Ralstonia* and *Xanthomonas*,
secretion of enzymes capable of attacking the plant cell wall occurs. The
following sections will discuss to what extent these enzymes constitute
essential pathogenicity factors and what is known so far about their secretion, function in pathogenesis and regulation.

a) Secretion of Bacterial Extracellular Enzymes

An effective secretory system is required to bring cell wall-degrading
enzymes in contact with their substrate. The secretion of extracellular
virulence factors by plant pathogenic bacteria has been reviewed by
Salmond (1994) and Finlay and Falkow (1997). Extracellular enzymes are
generally secreted by two distinct pathways. Type I-proteins (e.g. proteases in *E. chrysanthemi*) are secreted by a Sec-independent, but ATP-
dependent, one-step pathway, functionally similar to hemolysin secretion in *E. coli* (Pirhonen et al. 1993), while the type II mechanism is a
two-step, Sec- and ATP-dependent, *out*-cluster-mediated pathway (Salmond and Reeves 1993; Reeves et al. 1993). The *out* gene cluster of
E. carotovora subsp. *carotovora* encodes the proteins of the general secretory pathway (GSP) that is required for pectinase and cellulase secretion (Thomas et al. 1997). A two-step mechanism has been shown for the
endoglucanase of *Ralstonia solanacearum* as a consequence of an unusual two-component leader sequence (Huang and Schell 1990, 1992).
E. carotovora and *E. chrysanthemi* cellulases, pectate lyases, pectin
methyl esterases and polygalacturonases, as well as amylases, cellulases,
polygalacturonases and proteases of *X. campestris* are examples for type
II secretion (Salmond 1994). Moreover, most pectate lyases (PelADE) of
E. chrysanthemi are secreted by a type II pathway (Barras et al. 1994).
Interestingly, Palomaki and Saarilahti (1995, 1997) reported that the C-
terminus of the *E. carotovora* subsp. *carotovora* polygalacturonase
(PehA) is directly involved in a conformation-sensitive structure essen-
tial for secretion.

b) Cellulolytic Enzymes

It is well established that the hydrolysis of cellulose requires a consor-
tium of microbial enzymes (Gilbert and Hazlewood 1993; Béguin and
Aubert 1994). Studies on saphrophytic bacteria such as some *Pseudomonas* species (Gilbert et al. 1988; Hazlewood et al. 1992) or *Cellvibrio*
(Millward-Sadler et al. 1995) have indicated that cellulases from aerobic

bacteria may not associate, but bind individually to the substrate, in contrast to the "cellulosomes" known from *Clostridium thermocellum* (one of the few bacterial cellulase producers of commercial interest; see Bayer et al. 1994).

Endoglucanases (1,4-β-D-glucan glucanohydrolase, EC 3.2.1.4) cleave β-glucosidic bonds in crystalline regions of cellulose, creating sites for exoglucanase activity. The phytopathogenic genus *Clavibacter* (recently reviewed by Metzler et al. 1997) produces endoglucanases, but so far only one endoglucanase of the potato ring rot inducing *C. m.* subsp. *sepedonicus* (De Boer and Slack 1984; Bishop and Slack 1987) has been partially purified and characterized (Baer and Gudmestad 1995). Laine et al. (1996) propose an important role of the endoglucanase in virulence, while the study of Baer and Gudmestad (1995) did not support such a function, suggesting that involvement in pathogenicity may not be an absolute necessity. Endoglucanase genes are also not required for the wilting disease induced by *R. solanacearum*; however, they reduce the time required for disease development and plant death (Roberts et al. 1988).

In the plant pathogen *E. chrysanthemi* two glucanases are secreted: the poorly expressed CelY and the recently characterized endoglucanase EGZ (Boyer et al. 1984; Aymeric et al. 1988; Bortoli-German et al. 1995; Brun et al. 1995, 1997). However, the phytopathogenic relevance remains unclear. In the soft rot causing enterobacterial phytopathogen *E. carotovora* subsp. *carotovora*, involvement of a cellulase in pathogenesis was shown for CelV, the major secreted β-1,4-endoglucanase, since *celV* mutants exhibit a significantly reduced virulence in a potato disc assay (Walker et al. 1993).

Exoglucanases (syn: cellobiohydrolase; 1,4-β-D-glucan cellobiohydrolase, EC 3.2.1.91.) cleave cellulose chains at the non-reducing end, liberating cellobiose (Coughlan and Mayer 1992) and are regarded as key enzymes for the degradation of the native substrate. So far there is no indication of phytopathogenic relevance of these enzymes.

c) Xylanolytic Enzymes

β-1,4-Endoxylanases (1,4-β-D-xylan xylohydrolase; EC 3.2.1.8) cleave the internal glycosidic linkage of the heteroxylan backbone. Hydrolysis of the substrate is not random, and hydrolysis of bonds depends on the length and extent of branching or the presence of substituents in the substrate (Reilly 1981). Keen et al. (1996) showed by marker exchange mutagenesis of the *xyn*A gene of *E. chrysanthemi* that a complete loss of endoxylanase activity does not affect virulence on corn leaves. Although numerous xylanolytic enzymes with different properties have been described in recent years (Jeffries 1996; Sunna and Antranikian 1997), their role in phytopathogenicity remains unclear.

d) Pectolytic Enzymes

Pectins generally consisting of linear polygalacturonate chains interspersed with highly branched rhamnogalacturonan regions and "smooth" parts exclusively composed of α-1,4-linked D-galacturonic acid are important structural constituents of plant cell walls. Some of these galacturonates are modified by methyl-esterification at the carboxyl group or by acetyl-esterification on the hydroxyl groups at the C-2 or C-3 position. The extent of modification varies greatly, depending on pectin type. The depolymerization of pectin is limited by its extent of methylation, because pectate lyases exhibit a reduced activity on highly methylated substrates (Tardy et al. 1997). Thus, pectin methyl esterases facilitate pectate lyase action by removing the methyl groups in pectin (Shevchik and Hugouvieux-Cotte-Pattat 1997).

Plant pathogenic bacteria differ widely in their capacities to synthesize pectic enzymes. Soft rotting *E. chrysanthemi* and *E. carotovora* secrete many pectolytic enzymes, such as endo-polygalacturonase (Peh, EC 3.2.1.15), pectin esterase (EC 3.1.1.11) (Shevchik and Hugouvieux-Cotte-Pattat 1997) and exo-poly-α-D-galacturonosidase (EC 3.2.1.82) (Collmer and Keen 1986; Collmer et al. 1991). These enzymes have pH optima below pH 6 and are inhibited by bivalent cations. Lyases, such as pectate lyase (PL, Pel; EC 4.2.2.2) and pectin lyase (EC 4.2.2.10) have an activity optimum at basic pH values and need bivalent cations for their activity. The Pels can be divided into four families (Chatterjee et al. 1995), including extracellular PelADE, extracellular PelBC, periplasmatic Pels and Pel3 of *E. carotovora* (Liao et al. 1996). The enzymes enable *Erwinia* to macerate plant tissues (Pérombelon and Kelman 1980). Host-specific pathovars of *P. syringae* have a weaker, variable pectolytic activity (Bauer and Collmer 1997), causing necrosis rather than rot.

E. chrysanthemi, causing soft rot on various plants (Agrios 1988), produces an exo-polygalcturonase PehX (He and Collmer 1990), eight pectate lyases (PelA-D,L,Z,X), the exopectate lyase PelX (Brooks et al. 1990), five major endopectate lyases (PelA-E) (Tardy et al. 1997) and a set of secondary endo-pectate lyases (PelL, PelZ), together with some pectin esterases (Pissavin et al. 1996). Among the major pectate lyases, PelA, D and E play the most important role in pathogenicity, and PemA (pectin methyl esterase A) and PaeY (pectin acyl esterase Y) seem to be necessary especially during the initial invasion of plants (Boccara and Chatain 1989; Shevchik and Hugouvieux-Cotte-Pattat 1997). The *pel*A-E-D-*paeY*-*pem*A gene cluster can, therefore, be considered as a virulence locus of phytopathogenic *E. chrysanthemi* (Beaulieu et al. 1992; Shevchik and Hugouvieux-Cotte-Pattat 1997).

Production of pectolytic enzymes appears to be vital for the virulence of both *E. chrysanthemi* and *E. carotovora*, as demonstrated by isolation of avirulent mutants that were pleiotropically affected in either extracellular enzyme production or secretion (*out* mutants) (Murata et al. 1991; Pirhonen et al. 1993; Liu et al. 1992; Jones et al. 1993). A mutation of single or multiple *pel* genes leads to reduced virulence (Boccara et al. 1988; Boccara and Chatain 1989; Beaulieu et al. 1992; Kelemu and Collmer 1993). In addition, the involvement of a cellulase was demonstrated (Walker et al. 1993). The *pel* genes are regulated by autoinducers in a cell density-dependent manner (Jones et al. 1993; Pirhonen et al. 1993). PelB, a novel pectate lyase, showing 93% identical amino acids to Pel3 of *E. chrysanthemi*, is calcium-independent, and a member of a new Pel family in *E. carotovora*.

Other plant pathogens like *Pseudomonas viridiflava* (Liao et al. 1988), *Xanthomonas campestris* pv. *vesicatoria* (Beaulieu et al. 1991), *Bacillus subtilis* (Nasser et al. 1990) and *P. syringae* pv. *lachrymans* (Collmer et al. 1991; Bauer and Collmer 1997) produce only a single pectate lyase. The pectic enzymes of *Pseudomonas fluorescence* and *P. viridiflava* causing postharvest soft rot have been partially characterized (Liao et al. 1992), while little is known about pectic enzymes of *Pseudomonas syringae* (Bauer and Collmer 1997). The pectate lyase PelS of *P. syringae* pv. *lachrymans*, the only pectic enzyme of this bacterium, is not involved in pathogenicity (Collmer et al. 1991).

Nevertheless, pectate lyase seems to represent a key pathogenicity factor in soft rot disease caused by *P. viridiflava* (Liao et al. 1992, 1994), since transposon-induced mutants of the *pel* locus were unable to induce soft rot in potato tuber slices. In the plant pathogen *Burkholderia (Pseudomonas) cepacia*, a polygalacturonase is discussed as being responsible for maceration of plant tissue (Gonzalez et al. 1997). *PehA* mutants of *R. solanacearum*, the most important wilt-inducing pseudomonad, show a reduced virulence, indicating a putative role of the polygalacturonase in wilting. For the formation of virulence symptoms, however, extracellular endoglucanase and EPS I are required additionally (Allen et al. 1997).

e) Regulatory Aspects

As already described for EPS synthesis, several physiological processes in bacteria are regulated by autoinducers (reviewed by Salmond 1994). For phytopathogenic *Erwinia* (Jones et al. 1993; Pirhonen et al. 1993; Beck von Bodman and Farrand 1995) it was shown that the synthesis of extracellular enzymes is regulated by N-acyl-homoserine lactones. In addition to the biosynthesis of EPS, the expression of extracellular enzymes (PglA, Pme, Egl) in *R. solanacearum* is regulated by PhcA and the products of the *phc*BSR operon (Clough et al. 1997a, b).

In *E. carotovora* subsp. *carotovora*, recent studies by Flego et al. (1997) implicate a role for calcium in virulence control. High concentrations of calcium in plant tissue were shown to decrease maceration. This is probably due to a Ca^{2+}-dependent pectate stabilization, leading to a structurally enhanced wall integrity (McGuire and Kelman 1986; Carpita and Gibeaut 1993). In addition, production of PehA, an important virulence factor, during early stages of infection by *E. carotovora* subsp. *carotovora* (Saarilahti et al. 1992; Saarilahti 1993) is repressed by Ca^{2+}, resulting in a reduced virulence of the pathogen (Flego et al. 1997). Besides Ca^{2+} cations, other ions may also trigger differential expression of *pel* genes, since in *E. chrysanthemi* the expression of a *pel*D::*uid*A fusion protein was stimulated by the presence of iron chelators (Masclaux et al. 1996).

4 Concluding Remarks

An interesting aspect of the regulation of EPS genes and those for extra-cellular enzymes is the involvement of autoregulators which mediate a cell-density-dependent synchronous expression of genes required for pathogenicity. It is to be expected that in the future more such autoregulators will be detected and their exact functions in the regulatory cascade will be elucidated. In addition, it needs to be clarified how bacterial cells actually measure the parameter "cell density".

Two-component regulatory networks seem to be involved in a growing number of expression systems and the complex interaction of the various protein components needs to be addressed by future work. Also it seems possible that not all components engaged have been identified so far.

With respect to the phytophathogenic aspects of this review, it seems to emerge that in bacteria causing vascular infection resulting in wilting of the host plant in most cases EPS together with enzymes and other pathogenicity factors act synergistically. However, in soft rot-inducing bacteria the plant cell wall-degrading enzymes seem to be the principal cause of the disease.

References

Agrios GN (1988) Plant pathology, 3rd edn. Academic Press, London

Aldwinckle HS, Beer SV (1979) Fire blight and its control. Hortic Rev 1:423–475

Allen C, Gay J, Simon-Buela L (1997) A regulatory locus, *peh*SR, controls polygalacturonase production and other virulence functions in *Ralstonia solanacearum* Mol Plant Microbe Interact 10:1054–1064

Ayers AR, Ayers SB, Goodman R (1979) Extracellular exopolysaccharide of *Erwinia amylovora*: a correlation with virulence. Appl Environ Microbiol 38:659–666

Aymeric J-L, Guiseppi A, Pascal M-C, Chippaux M (1988) Mapping and regulation of the *cel* genes in *Erwinia chrysanthemi*. Mol Gen Genet 211:95–101

Baer D, Gudmestad NC (1995) In vitro cellulolytic activity of the plant pathogen *Clavibacter michiganensis* subsp. *sepedonicus*. Can J Microbiol 41:877–888

Bahro R, Dreier J, Eichenlaub R (1996) Genetics of phytopathogenic bacteria. Springer, Berlin Heidelberg New York. Prog Bot 58:410–427

Barras F, van Gijsegem F, Chatterjee AK (1994) Extracellular enzymes and soft-rot pathogenesis of soft-rot *Erwinia*. Annu Rev Phytopathol 32:201–234

Barrere GC, Barber CE, Daniels MJ (1986) Molecular cloning of genes involved in the biosynthesis of the extracellular polysaccharide xanthan by *Xanthomonas campestris* pv. *campestris*. Int J Biol Macromol 8:372–374

Bauer WD, Collmer A (1997) Molecular cloning, characterization, and mutagenesis of a *pel* gene from *Pseudomonas syringae* pv. *lachrymans* encoding a member of the *Erwinia chrysanthemi* PelADE family of pectate lyases. Mol Plant Microbe Interact 10:369–379

Bayer EA, Morag E, Lamed R (1994) The cellulosome-a treasure-trove for biotechnology. Trends Biotechnol 12 (9):379–386

Beaulieu C, Minsavage GV, Canteros BI, Stall RE (1991) Biochemical and genetic analysis of a pectate lyase gene from *Xanthomonas campestris* pv. *campestris*. Mol Plant Microbe Interact 4:446–451

Beaulieu C, Boccara M, Van Gijsegem F (1992) Pathogenic behavior of pectinase-defective *Erwinia chrysanthemi* mutants on different plants. Mol Plant Microbe Interact 6:197–202

Beck von Bodman S, Farrand SK (1995) Capsular polysaccharide biosynthesis and pathogenicity in *Erwinia stewartii* require induction by an N-acylhomoserine lactone autoinducer. J Bacteriol 177:5000–5008

Béguin P, Aubert JP (1994) The biological degradation of cellulose. FEMS Microbiol Rev 13(1):25–58

Bellemann P, Geider K (1992) Localization of transposon insertions in pathogenicity mutants of *Erwinia amylovora* and their biochemical characterization. J Gen Microbiol 138:931–940

Benhamou N (1991) Cell surface interactions between tomato and *Clavibacter michiganense* subsp. *michiganense*: location of some polysaccharides and hydroxyproline-rich glycoproteins in infected host leaf tissues. Physiol Mol Plant Pathol 38:15–38

Bennet RA, Billing E (1978) Capsulation and virulence in *Erwinia amylovora*. Ann Appl Biol 89:41–45

Bereswill S, Geider K (1997) Characterization of *rcs*B gene from *Erwinia amylovora* and its influence on exopolysaccharide synthesis and virulence of th fire blight pathogen. J Bacteriol 179:1354–1361

Bermpohl A, Dreier J, Bahro R, Eichenlaub R (1996) Exopolysaccharides in the pathogenic interactions of *Clavibacter michiganensis* subsp. *michiganensis* with tomato plants. Microbial Res 151:391–399

Bernhard F, Poetter K, Geider K, Coplin DL (1990) The *rcs*A gene of *Erwinia amylovora*: identification, nucleotide sequence, and regulation of exopolysaccharide biosynthesis. Mol Plant Microbe Interact 3:429–437

Bernhard F, Belleman P, Burgert P, Coplin DL, Geider K (1993a) Heterologous expression of EPS-genes in *Erwinia amylovora* and *Erwinia stewartii*. Acta Hortic 338:265–272

Bernhard F, Coplin DL; Geider K (1993b) A gene cluster for amylovoran synthesis in *Erwinia amylovora*: characterization and relationship to *cps* genes in *Erwinia stewartii*. Mol Gen Genet 239:158–168

Bernhard F, Schullerus D, Bellemann P, Nimtz M, Coplin DL, Geider K (1996) Genetic transfer of amylovoran and stewartan synthesis between *Erwinia amylovora* and *Erwinia stewartii*. Microbiology 142:1087–1096

Bishop AL, Slack SA (1987) Effect of cultivar, inoculum dose, and strain of *Clavibacter michiganense* subsp. *sepedonicum* on symptom development in potatoes. Phytopathology 77:1085–1098

Boccara M, Chatain V (1989) Regulation and role in pathogenicity of *Erwinia chrysanthemi* 3937 pectin methylesterase. J Bacteriol 171(7):4085–4087

Boccara M, Diolez A, Rouve M, Kotoujansky A (1988) The role of individual pectate lyases of *Erwinia chrysanthemi* strain 3937 in pathogenicity on *Saintpaulia* plants. Physiol Mol Pathol 33:95–104

Bortoli-German I, Haiech J, Chippaux M, Barras F (1995) Informational suppression to investigate structural functional and evolutionary aspects of the *Erwinia chrysanthemi* cellulase EGZ. Mol Biol 246:82–94

Boyer MH, Chambost JP, Magnan M, Cattanéo J (1984) Carboxymethyl-cellulase from *Erwinia chrysanthemi*. I. Production and regulation of extracellular carboxymethly-cellulase. J Biotechnol 1:229–239

Bradshaw-Rouse JJ, Whatley MA, Coplin DL, Woods A, Sequeira L, Kelman A (1981) Agglutination of strains of *Erwinia stewartii* with a corn agglutinin: correlation with extracellular polysaccharide production and pathogenicity. Appl Environ Microbiol 42:344–350

Braun EJ (1990) Colonization of resistant and susceptible maize plants by *Erwinia stewartii* strains differing in exopolysaccharide production. Physiol Mol Plant Pathol 36:363–379

Braun EJ, Rodrigues CA (1993) Purification and properties of an endoxylanase from a corn stalk rot strain of *Erwinia chrysanthemi*. Phytopathology 83:332–338

Brooks AD, He SY, Gold S, Keen NT, Collmer A, Hutchenson SW (1990) Molecular cloning of the structural gene for exopolygalacturonate lyase from *Erwinia chrysanthemi* EC16 and characterization of the enzyme product. J Bacteriol 172:6950–6958

Brumbley SM, Denny TP (1990) Cloning of *phcA* from wild-type *Pseudomonas solanacearum*, a gene that when mutated alters expression of multiple traits that contribute to virulence. J Bacteriol 172:5677–5685

Brumbley SM, Carney BF, Denny TP (1993) Phenotype conversion in *Pseudomonas solanacearum* due to spontaneous inactivation of PhcA, a putative LysR transcriptional activator. J Bacteriol 175:5477–5487

Brun E, Gans P, Marion D, Barras F (1995) Overproduction, purification and characterization of the cellulose-binding domain of the *Erwinia chrysanthemi* secreted endoglucanase EGZ. Eur J Biochem 231:142–148

Brun E, Moriaud F, Gans P, Blackledge MJ, Barras F, Marion D (1997) Solution structure of the cellulose-binding domain of the endoglucanase Z secreted by *Erwinia chrysanthemi*. Biochemistry 36:16074–16086

Buddenhagen I, Kelman A (1964) Biological and physiological aspects of bacterial wilt caused by *Pseudomonas solanacearum*. Annu Rev Phytopathol 2:203–230

Burgert P, Geider K (1995) Molecular analysis of the *ams*-operon required for exopolysaccharide synthesis in *Erwinia amylovora*. Mol Microbiol 15:917–933

Carpita NC, Gibeaut DM (1993) Structural models of primary cell walls in flowering plants: consistency of molecular structure with the physical properties of the walls during growth. Plant J 3:1–30

Chatterjee A, Chun W, Chatterjee AK (1990) Isolation and characterization of an *rcsA*-like gene of *Erwinia amylovora* that activates extracellular polysaccharide production in *Erwinia* species, *Escherichia coli*, and *Salmonella typhimurium*. Mol Plant Microbe Interact 3:144–148

Chatterjee A, Liu Y, Chatterjee AK (1995) Nucleotide sequence of a pectate lyase structural gene, *pel*1 of *Erwinia carotovora* subsp. *carotovora* strain 71 and structural relationship of *pel*1 with other *pel* Genes of *Erwinia* species. Mol Plant Microbe Interact 8:92–95

Clough SJ, Flavier AB, Schell MA, Denny TP (1997a) Differential expression of virulence genes and motility in *Ralstonia (Pseudomonas) solanacearum* during exponential growth. Appl Environ Microbiol 63:844–850

Clough SJ, Lee KE, Schell MA, Denny TP (1997b) A two-component system in *Ralstonia (Pseudomonas) solanacearum* modulates production of *phc*A-regulated virulence factors in response to 3-hydroxypalmitic acid methyl ester. J Bacteriol 179:3639–3648

Coleman M, Pearce R, Hitchin F, Busfield F, Mansfield JW, Roberts IS (1990) Molecular cloning, expression and nucleotide sequence of the *rcsA* gene of *Erwinia amylovora*, encoding a positive regulator of capsule expression: evidence for a family of related capsule activator proteins. J Gen Microbiol 136:1799–1806

Collmer A, Keen NT (1986) The role of pectic enzymes in plant pathogenesis. Annu Rev Phytopathol 24:383–409

Collmer A, Bauer DW, He SY, Lindeberg M, Kelemus S, Rodriguez-Palenzuala P, Burr T, Chatterjee AK (1991) Pectic enzyme production and bacterial pathogenicity. In: Hennecke H, Verma DPS (eds) Advances in molecular genetics of plant-microbe interactions, vol I. Kluwer, Dordrecht

Cook D, Sequeira L (1991) Genetic and biochemical characterization of a gene cluster from *Pseudomonas solanacearum* required for extracellular polysaccharide production and for virulence. J Bacteriol 173:1654–1662

Coplin DL, Cook D (1990) Molecular genetics of extracellular polysaccharide biosynthesis in vascular phytopathogenic bacteria. Mol Plant Microbe Interact 3:271–279

Coplin DL, Majerczak DR (1990) Extracellular polysaccharide genes in *Erwinia stewartii*: directed mutagenesis and complementation analysis. Mol Plant Microbe Interact 3:286–292

Costerton JW, Cheng KJ, Geesey GG, Ladd TI, Nickel JC (1987) Bacterial biofilms in nature and disease. Annu Rev Microbiol 41:435–464

Coughlan MP, Mayer F (1992) The cellulose-decomposing bacteria and their enzyme systems. In: Balows A, Truper HG, Dworkin M, Harder W, Schleifer K-H (eds) The prokaryotes 2. Edition, vol I., Springer, New York, pp 460–516

De Boer SH, Slack SA (1984) Current status and prospects for detecting and controlling bacterial ring rot of potatoes in North America. Plant Dis 68:841–844

Denny TP (1995) Involvement of bacterial polysaccharides in plant pathogenesis. Annu Rev Phytopathol 33:173–197

Denny TP, Baek SR (1991) Genetic evidence that extracellular polysaccharide is a virulence factor of *Pseudomonas solanacearum*. Mol Plant Microbe Intract 4:198–206

Denny TP, Makini FW, Brumbley SM (1988) Characterization of *Pseudomonas solanacearum* Tn5 mutants deficient in extracellular polysaccharide. Mol Plant Microbe Interact 1:215–223

Denny TP, Brumbley SM, Carney BF, Clough SJ, Schell MA (1994) Phenotype conversion of *Pseudomonas solanacearum*. In: Hayward AC, Hartman GL (eds) Bacterial wilt: the disease and its causative agent, *Pseudomonas solanacearum*. CAB International, Wallingford, pp 137–143

Dolph PJ, Majerczak DR, Coplin DL (1988) Characterization of a gene cluster for exopolysaccharide biosynthesis and virulence in *Erwinia stewartii*. J Bacteriol 170:865–871

Dow JM, Daniels MJ (1994) Pathogenicity determinants and global regulation of pathogenicity in *Xanthomonas campestris* pv. *campestris*. In: Dangl J (ed) Bacterial pathogenesis of plants and animals. Current topics in microbiology and immunology, vol 192. Springer, Berlin Heidelberg New York, pp 29–41

Dreier J, Meletzus D, Eichenlaub R (1997) Characterization of the plasmid encoded virulence region *pat*-1 of phytopathogenic *Clavibacter michiganensis* subsp. *michiganensis*. Mol Plant Microbe Interact 10:195–206

Eastgate JA, Taylor N, Coleman MJ, Healy B, Thompson L, Roberts IS (1995) Cloning, expression and characterization of the *lon* gene of *Erwinia amylovora*: evidence for a heat shock response. J Bacteriol 177:932–937

Eichenlaub R, Bermpohl A, Meletzus D (1991) Genetic and physiological aspects of the pathogenic interaction of *Clavibacter michiganense* subsp. *michiganense* with the host plant. In: Hennecke H, Verma DPS (eds) Advances in molecular genetics of plant-microbe interactions, vol 1. Kluwer, Dordrecht, pp 99–102

Ferris GG, Beveridge TJ (1985) Functions of bacterial cell surface structures. BioScience 35:172–177

Finlay BB, Falkow S (1997) Common themes in microbial pathogenicity revisited. Microbiol Mol Biol Rev 61:136–169

Flavier AB, Clough SJ, Schell MA, Denny TP (1997a) Identification of 3-hydroxypalmitic acid methyl ester as a novel autoregulator controlling virulence in *Ralstonia solanacearum*. Mol Microbiol 26:251–259

Flavier AB,Ganova-Raeva LM, Schell MA, Denny TP (1997b) Hierarchical autoinduction in *Ralstonia solanacearum*: control of acyl-homoserine lactone production by a novel autoregulatory system responsive to 3-hydroxypalmitic acid methy ester. J Bacteriol 179:7089–7097

Flego D, Pirhonen M, Saarilahti, Palva TK, Palva ET (1997) Control of virulence gene expression by plant calcium in the phytopathogen *Erwinia carotovora*. Mol Microbiol 25:831–838

Fuqua WC, Winans SC, Greenberg EP (1994) Quorum sensing in bacteria: the LuxR–LuxI family of cell density-responsive transcriptional regulators. J Bacteriol 176:269–275

Fuqua WC, Winans SC, Greenberg EP (1996) Census and consensus in bacterial ecosystems: the LuxR–LuxI family of quorum-sensing transcriptional regulators. Annu Rev Microbiol 50:727–751

Geier G, Geider K (1993) Characterization and influence on virulence of levansucrase gene from the fire blight pathogen *Erwinia amylovora*. Physiol Mol Plant Pathol 42:387–404

Gilbert HJ, Hazlewood GP (1993) Bacterial cellulases and xylanases. J Gen Microbiol 139:187–194

Gilbert HJ, Sullivan DA, Jenkins G, Kellet LE, Minton NP, Hall J (1988) Molecular cloning of multiple xylanase genes from *Pseudomonas fluorescens* subsp. *cellulosa*. J Gen Microbiol 134:3239–3247

Gonzales FC, Pettit EA, Valadez VA, Provin EM (1997) Mobilization, cloning, and sequence determination of a plasmid-encoded polygalacturonase from a phytopathogenic *Burkholderia (Pseudomonas) cepacia*. Mol Plant Microbe Interact 10:840–851

Gottesman S, Stout V (1991) Regulation of capsular polysaccharide synthesis in *Escherichia coli*. Mol Microbiol 5:1599–1606

Gross M, Geier G, Rudolph K, Geider K (1992) Levan and levansucrase synthesized by the fireblight pathogen *Erwinia amylovora*. Physiol Mol Plant Pathol 40:371–381

Harding NE, Cleary JM, Cabanas DK, Rosen IG, Kang KS (1987) Genetic and physical analysis of a cluster of genes essential for xanthan gum biosynthesis in *Xanthomonas campestris*. J Bacteriol 169:2854–2861

Harding NE, Raffo S, Raimondi A, Cleary JM, Ielpi L (1993) Identification, genetic and biochemical analysis of genes involved in synthesis of sugar nucleotide precursors of xanthan gum. J Gen Microbiol 139:447–457

Hayward AC (1991) Biology and epidemiology of bacterial wilt caused by *Pseudomonas solanacearum*. Annu Rev Phytopathol 29:65–87

Hazlewood GP, Laurie JI, Ferreira LM, Gilbert HJ (1992) *Pseudomonas fluorescens* subsp. *cellulosa*: an alternative model for bacterial cellulase. J Appl Bacteriol 72:244–251

He SY, Collmer A (1990) Molecular cloning, nucleotide sequence, and marker exchange mutagenesis of the exo-poly-alpha-D-galacturnosidase-encoding *pehX* gene of *Erwinia chrysanthemi* EC16. J Bacteriol 172:4988–4995

Huang P, Goodman RN (1976) Ultrastructural modifications in apple stems induced by *Erwinia amylovora* and the fire blight toxin. Phytopathology 66:269–276

Huang JZ, Schell MA (1990) Evidence that extracellular export of the endoglucanase encoded by *egl* of *Pseudomonas solanacearum* occurs by a two-step process involving a lipoprotein intermediate. J Biol Chem 265:11628–11632

Huang JZ, Schell MA (1992) Role of the two-component leader sequence and mature amino acid sequences in extracellular export of endoglucanase EGL from *Pseudomonas solanacearum*. J Bacteriol 174(4):1314–1323

Huang J, Schell MA (1995) Molecular characterization of the *eps* gene cluster of *Pseudomonas solanacearum* and its transcriptional regulation at a single promoter. Mol Microbiol 16:977–989

Huang J, Carney BF Denny TP, Weissinger AK, Schell MA (1995) A complex network regulates expression of *eps* and other virulence genes of *Pseudomonas solanacearum*. J Bacteriol 177:1259–1267\

Jannson PE, Kenne L, Lindberg B (1975) Structure of the extracellular polysaccharide from *Xanthomonas campestris*. Carbohydr Res 45:275–282

Jeffries TW (1996) Biochemistry and genetics of microbial xylanases. Environ Biotechnol 7:337–342

Jones S, Yu B, Bainton NJ, Birdsall M, Bycroft BW, Chabra SR, Cox AJR, Golby P, Reeves PJ, Stephens S, Winson MK, Salmond GPC, Steward GSAB, Williams P (1993) The *lux* autoinducer regulates the production of exoenzyme virulence determinants in *Erwinia carotovora* and *Pseudomonas aeruginosa*. EMBO J 12:2477–2482

Kao CC, Sequeira L (1991) A gene cluster required for coordinated biosynthesis of lipopolysaccharide and extracellular polysaccharide also affects the virulence of *Pseudomonas solanacearum*. J Bacteriol 173:7841–7847

Kao CC, Barlow L, Sequeira L (1992) Extracellular polysaccharide is required for wild-type virulence of *Pseudomonas solanacearum*. J Bacteriol 174:1068–1071

Keen NT, Boyd C, Henrissat B (1996) Cloning and characterization of a xylanase gene from corn strains of *Erwinia chrysanthemi*. Mol Plant Microbe Interact 9:651–657

Kelemu S, Collmer A (1993) *Erwinia chrysanthemi* EC16 produces a second set of plant-inducible pectate lyase isozymes. Appl Environ Microbiol 59:1756–1761

Kiraly Z, El-Zahaby HM, Klement Z (1997) Role of extracellular polysaccharides (EPS) slime of plant pathogenic bacteria in protecting cells to reactive oxygen species. J Phytopathol 145:59–68

Laine MJ, Nakhei H, Dreier J, Lehtilä K, Meletzus D, Eichenlaub R, Metzler M (1996) Stable transformation of the gram-positive phytopathogenic bacterium *Clavibacter michiganensis* subsp. *sepedonicus* with several cloning vectors. Appl Environ Microbiol 65:1500–1506

Leigh JA, Coplin DL (1992) Exopolysaccharides in plant-bacterial interactions. Annu Rev Microbiol 46:307–346

Leigh JA, Walker GC (1994) Exopolysaccharides of *Rhizobium*: synthesis regulation and symbiotic function. Trends Genet 10:63–67

Liao C-H (1989) Analysis of pectate lyases produced by soft rot bacteria associated with spoilage of vegetables. Appl Environ Microbiol 55:1677–1683

Liao C-H, Hung H-Y, Chatterjee AK (1988) An extracellular pectate lyase is the pathogenicity factor of the soft-rotting bacterium *Pseudomonas viridiflava*. Mol Plant Microbe Interact 1:199–206

Liao C-H, Sasaki G, Nagahashi G, Hicks KB (1992) Cloning and characterization of a pectate lyase gene from the soft-rotting bacterium *Pseudomonas viridiflava*. Mol Plant Microbe Interact 4:301–308

Liao C-H, McCallus DE, Fett WF (1994) Molecular characterization of two gene loci required for production of the key pathogenicity factor pectate lyase in *Pseudomonas viridiflava*. Mol Plant Microbe Interact 7:391–400

Liao C-H, Gaffney TD, Bradley SP, Wong L-JC (1996) Cloning of a pectate lyase gene from *Xanthomonas campestris* pv. *malvacearum* and comparison of its sequence relationship with *pel* genes of soft-rot *Erwinia* and *Pseudomonas*. Mol Plant Microbe Interact 9:14–21

Liu Y, Murata H, Chatterjee A, Chatterjee AK (1992) Characterization of a novel regulatory gene *aepA* that controls extracellular enzyme production in the phytopathogenic bacterium *Erwinia carotovora* subsp. *carotovora*. Mol Plant Microbe Interact 6:299–308

Masclaux C, Hugouvieux-Cotte-Pattat, Expert D (1996) Iron is a triggering factor for differential expression of *Erwinia chrysanthemi* strain 3937 pectate lyases in pathogenesis of african violets. Mol Plant Microbe Interact 9:198–205

McGuire RG, Kelman A (1986) Calcium in potato tuber cell walls in relation to tissue maceration by *Erwinia carotovora* pv. *atroseptica*. Phytopathology 76:400–406

Meletzus D, Bermpohl A, Dreier J, Eichenlaub R (1993) Evidence for plasmid-encoded virulence factors in the phytopathogenic bacterium *Clavibacter michiganensis* subsp. *michiganensis*. J Bacteriol 175:2131–2136

Metzger M, Bellemann P, Burgert P, Geider K (1994) Genetics of galactose metabolism of *Erwinia amylovora* and its influence on polysaccharide synthesis and virulence of the fire blight pathogen. J Bacteriol 176:450–459

Metzler MC, Laine MJ, De Boer SH (1997) The status of molecular biological research on the plant pathogenic genus *Clavibacter*. FEMS Microbiol Lett 150:1–8

Millward-Sadler SJ, Davidson K, Hazlewood GP Black GW, Gilbert HJ, Clarke JH (1995) Novel cellulose-binding domains. NodB homologues and conserved modular architecture in xylanases from the aerobic soil bacteria *Pseudomonas fluorescens* subsp. *cellulosa* and *Cellvibrio mixtus*. Biochem J 312:39–48

Murata H, McEvoy JL, Chatterjee A, Collmer A, Chatterjee AK (1991) Molecular cloning of an *aep*A gene that activates production of extracellular pectolytic, cellulolytic, and proteolytic enzymes in *Erwinia carotovora* subsp. *carotovora*. Mol Plant Microbe Interact 4:239–246

Nasser W, Chalet F, Robert-Baudouy J (1990) Purification and characterization of extracellular pectate lyases from *Bacillus subtilis*. Biochimie 72:689–695

Newman MA, Conrads-Strauch J, Scofield G, Daniels MJ, Dow JM (1994) Defense-related gene induction in *Brassica campestris* in response to defined mutants of *Xanthomonas campestris* with altered pathogenicity. Mol Plant Microbe Interact 7:553–562

Nimtz M, Mort A, Domke T, Wray V, Zhang Y, Qiu F, Choplin D, Geider K (1996) Structure of amylovoran, the capsular exopolysaccharde from the fire blight pathogen *Erwinia amylovora*. Carbohydr Res 287:59–76

Norberg AB, Persson H (1984) Accumulation of heavy-metal ions by *Zoogloea ramigera*. Biotechnol Bioeng 26:239–246

Orgambide G, Montrozier H, Servin P, Roussel J, Trigalet-Demery D, Trigalet A (1991) High heterogeneity of the exopolysaccharides of *Pseudomonas solanacearum* strain GMI 1000 and the complete structure of the major polysaccharide. J Biol Chem 266:8312–8321

Osbourn AE, Clarke BR, Stevens BJH, Daniels MJ (1990) Use of oligonucleotide probes to identify members of two-component regulatory systems in *Xanthomonas campestris* pathovar *campestris*. Mol Gen Genet 222:145–151

Palomaki T, Saarilahti HT (1995) The extreme C-terminus is required for secretion of both the native polygalacturonase (PehA) and PehA/Bla hybrid proteins in *Erwinia carotovora* ssp. *carotovora*. Mol Microbiol 17:449–459

Palomaki T, Saarilahti HT (1997) Isolation and characterization of new C-terminal substitution mutations affecting secretion of polygalacturonase in *Erwinia carotovora* ssp. *carotovora*. FEBS Lett 400:122–126

Parkinson JS, Kofoid EC (1992) Communication modules in bacterial signalling proteins. Annu Rev Genet 26:71–112

Pérombelon MCM, Kelman A (1980) Ecology of the soft-rot *Erwinias*. Annu Rev Phytopathol 18:361–387

Pirhonen M, Flego D, Heikinheimo R, Palva ET (1993) A small diffusible signal molecule is responsible for the global control of virulence and exoenzyme production in the plant pathogen *Erwinia carotovora*. EMBO J 12:2467–2476

Pissavin C, Robert-Baudouy J, Hugouvieux-Cotte-Pattat N (1996) Regulation of pelZ, a gene of the pelB-pelC cluster encoding a new pectate lyase of *Erwinia chrysanthemi* 3937. J Bacteriol 178:7187–7196

Poetter K, Coplin DL (1991) Structural and functional analysis of the rcsA gene from *Erwinia stewartii*. Mol Gen Genet 229:155–160

Poplawsky AR, Chum W, Slater H, Daniels MJ, Dow JM (1998) Synthesis of extracellular polysaccharide, extracellular enzymes, and xanthomonadin in *Xanthomonas campes-*

tris: evidence for the involvement of two intercellular regulatory signals. Mol Plant Microbe Interact 11:68–70

Ramirez ME, Fucikovsky L, Garcia-Jimenez F, Quintero R, Galindo E (1988) Xanthan gum production by altered pathogenicity variants of *Xanthomonas campestris*. Appl Microbiol Biotechnol 29:5–10

Reeves PJ, Whitcombe D, Wharam S, Gibson M, Allison G, Bunce N, Barallon R, Douglas P, Mulholland V, Stevens S et al. (1993) Molecular cloning and characterization of 13 out genes from *Erwinia carotovora* subspecies *carotovora*: genes encoding members of a general secretion pathway (GSP) widespread in gram-negative bacteria. Mol Microbiol 8:443–456

Reilly PJ (1981) Xylanases: structure and function. In: Hollaender AE (ed) Trends in biology of fermentation for fuels and chemicals. Plenum Press, New York, pp 111–129

Roberts DP, Denny TP, Schell MA (1988) Cloning of the egl gene of *Pseudomonas solanacearum* and analysis of its role in phytopathogenicity. J Bacteriol 170:1445–1451

Romeiro R, Karr A, Goodman R (1981) Isolation of a factor from apple that agglutinates *Erwinia amylovora*. Plant Physiol 68:772–777

Saarilahti HT (1993) Characterization of polygalacturonases. Gene 124:145–147

Saarilahti HT, Pirhonen M, Karlsson M-B, Flego D, Palva ET (1992) Expression of *pehA-bla* gene fusions in *Erwinia carotovora* subsp. *carotovora* and isolation of regulatory mutants affecting polygalacturonase production. Mol Gen Genet 234:81–88

Saile E, McGarvey JA, Schell MA, Denny TP (1997) Role of extracellular polysaccharide and endoglucanase in root invasion and colonization of tomato plants by *Ralstonia solanacearum*. Phytopathology 87:1264–1271

Salmond GPC (1994) Secretion of extracellular virulence factors by plant pathogenic bacteria. Annu Rev Phytopathol 32:181–200

Salmond GPC, Reeves PJ (1993) Membrane traffic wardens and protein secretion in gram-negative bacteria. Trends Biochem Sci 18:7–12

Schell MA (1993) Molecular biology of the LysR family of transcriptional regulators. Annu Rev Microbiol 47:597–626

Schell MA, Denny TP, Clough SJ, Huang J (1993) Further characterization of genes encoding extracellular polysaccharide of *Pseudomonas solanacearum* and their regulation. In: Nester EW, Verma DPS (eds) Advances in molecular genetics of plant-microbe interactions, vol 2. Kluwer, Dordrecht, pp 231–239

Schell MA, Denny TP, Huang J (1994) VsrA, a second two-component system regulating virulence genes of *Pseudomonas solacearum*. Mol Microbiol 11:489–500

Shaw PD, Ping G, Daly SL, Cha C, Cronan JE, Rinehart KL, Farrand SK (1997) Detecting and characterizing *N*-acyl-homoserine lactone signal molecules by thin-layer chromatography. Proc Natl Acad Sci USA 94:6036–6041

Shevchik VE, Hugouvieux-Cotte-Pattat N (1997) Identification of a bacterial pectin acetyl esterase in *Erwinia chrysanthemi* 3937. Mol Microbiol 24:1285–1301

Sitnikow DM, Schineller JB, Baldwin TO (1995) Transcriptional regulation of bioluminescence genes from *Vibrio fischeri*. Mol Microbiol 17:801–812

Smith ARW, Rastall RA, Rees NH, Hignett RC (1990) Structure of the extracellular polysaccharide of *Erwinia amylovora*: a preliminary report. Acta Hortic 273:211–219

Steinberger EM, Beer SV (1988) Creation and complementation of pathogenicity mutants of *Erwinia amylovora*. Mol Plant Microbe Interact 1:135–144

Stout V (1994) Regulation of capsule synthesis includes interactions of the RcsC/RcsB regulatory pair. Res Microbiol 145:389–392

Stout V, Gottesman S (1990) RcsB and RcsC: a two-component regulator of capsule synthesis in *Escherichia coli*. J Bacteriol 172:659–669

Stout V, Torres-Cabassa A, Maurizi MR Gutnick D, Gottesman S (1991) RcsA, an unstable positive regulator of capsular polysaccharide biosynthesis. J Bacteriol 173:1738–1747

Sunna A, Antranikian G (1997) Xylanolytic enzymes from fungi and bacteria. Crit Rev Biotechnol 17:39–67

Sutherland IW (1988) Bacterial surface polysaccharides: structure and function. Int Rev Cytol 113:187–231

Sutton JC, Williams PH (1970) Relation of xylem plugging to black rot lesion development in cabbage. Can J Bot 48:391–401

Tang JL, Gough CL, Daniels MJ (1990) Cloning of genes involved in negative regulation of production of extracellular enzymes and polysaccharide of *Xanthomonas campestris* pathovar *campestris*. Mol Gen Genet 222:157–160

Tang JL, Liu YN, Barber CE, Dow JM, Wootton JC, Daniels MJ (1991) Genetic and molecular analysis of a cluster of *rpf* genes involved in positive regulation of synthesis of extracellular enyzmes and polysaccharide in *Xanthomonas campestris* pathovar *campestris*. Mol Gen Genet 226:409–417

Tardy F, Nasser W, Robert-Baudouy J, Hugouvieux-Cotte-Pattat N (1997) Comparative analysis of the five major *Erwinia chrysanthemi* pectate lyases: enzyme characteristics and potential inhibitors. J Bacteriol 179:2503–2511

Tharaud M, Menggad M, Paulin JP, Laurent J (1994) Virulence, growth, and surface characteristics of *Erwinia amylovora* mutants with altered pathogenicity. Microbiology 140:659–669

Thomas JD, Reeves PJ, Salmond GP (1997) The general secretion pathway of *Erwinia carotovora* subsp. *carotovora*: analysis of the membrane topology of OutC and OutF. Microbiology 143:713–720

Van Alfen NK (1989) Reassessment of plant wilt toxins. Annu Rev Phytopathol 27:533–550

Van den Bulk RW, Löffler HJM, Dons JJM (1989) Effect of phytotoxic compounds produced by *Clavibacter michiganensis* subsp. *michiganensis* on resistent and susceptible tomato plants. Neth J Plant Pathol 95:107–117

Van den Bulk RW, Zevenhuizen LPTM, Cordewener JHG, Dons JJM (1991) Characterization of the extracellular polysaccharide produced by *Clavibacter michiganensis* subsp. *michiganensis*. Phytopathology 81:619–623

Walker DS, Reeves PJ, Salmond GPC (1993) The major secreted cellulase, CelV, of *Erwinia carotovora* subsp. *carotovora* is an important soft rot virulence factor. Mol Plant Microbe Interact 7:425–431

Williams PH (1980) Black rot, a continuing threat to world crucifers. Plant Dis 64:736–742

Wallis FM (1977) Ultrastructural histopathology of tomato plants infected with *Corynebacterium michiganense*. Physiol Plant Pathol 11:333–342

Yabuuchi E, Kosako Y, Yano I, Hotta H, Nishiuchi Y (1995) Transfer of two *Burkholderia* and and *Alcaligenes* species to *Ralstonia* gen. nov.: proposal of *Ralstonia pickettii* (Ralston, Palleroni and Doudoroff 1973) com. nov., *Ralstonia solanacearum* (Smith 1896) comb. nov. and *Ralstonia eutropha* (Davis 1969) comb. nov.. Microbiol Immunol 39:897–904

Young DH, Sequeira L (1986) Binding of *Pseudomonas solanacearum* fimbriae to tobacco leaf cell walls and its inhibition by bacterial extracellular polysaccharides. Physiol Mol Plant Pathol 28:393–402

Dipl. Biol. Holger Jahr
Dipl. Biol. Rainer Bahro
Prof. Dr. Rudolf Eichenlaub
Lehrstuhl Gentechnologie/Mikrobiologie
Universität Bielefeld
Universitätsstraße 25
D-33501 Bielefeld, Germany

Edited by
K. Esser

Plant Breeding:
Male Sterility in Higher Plants –
Fundamentals and Applications

By Frank Kempken and Daryl Pring

1 Introduction

Under normal conditions, plants undergo a life cycle that consits of an alternating vegetative sporophytic generation and a much-reduced sexual gametophytic generation. During these cycles, seeds germinate, the mature plant organs differentiate, and finally vegetative growth terminates in flower formation, which in principle leads to sepals, petals, anthers, and carpels. Male sterile mutants which cannot produce fertile pollen or functional anthers can often be observed in higher plant species. In this chapter we differentiate between nuclear and cytoplasmic-nuclear male sterility (CMS); the latter is particularly useful for production of hybrid seed, which is the main application of CMS. To date, F_1 hybrid varieties are produced in most agricultural and horticultural crops. The successful exploitation of heterosis requires a simple and reliable system to produce female parents and perform crosses for the production of hybrid seed. Without a CMS system, male floral organs must be removed mechanically, which is usually not economical nor practical. While nuclear male sterility is based solely on mutations which occur in nuclear genes, CMS is maternally inherited and based on changes in mitochondrial gene expression as influenced by nuclear genes. Importantly, the CMS phenotype may be corrected by nuclear fertility restoration (RF) genes. In the first section, we discuss anther and pollen development and present recent molecular data as well. In the second part, some of the most important CMS systems are presented. Finally, approaches to genetically engineering male sterility in higher plants are discussed. Genetically engineered male sterility may be applied to any crop, including those crops where CMS systems are not available or are unreliable. This area thus has a tremendous potential in plant breeding.

Some aspects of this review have been covered by other reviews in the past years (e.g. Hanson et al. 1988; Pring and Lonsdale 1989; Hanson and Folkerts 1992; Pring et al. 1993; Horn et al. 1994; Kück and Wricke 1995; Maier et al. 1996).

Progress in Botany, Vol. 60
© Springer-Verlag Berlin Heidelberg 1999

2 Male Sterility in Natural and Breeding Systems

This section deals with spontaneously occurring male sterility and with male sterility introduced by breeding or crop improvement, excluding the introduction of male sterility by genetic engineering.

a) Anther Development and Nuclear Male Sterility

The development of anthers is part of the genetic program to produce flowers. Flower development has been extensively studied in recent years, focusing on *Antirrhinum majus* and particularly on *Arabidopsis thaliana* (reviewed in Theißen and Saedler 1995; Weigel 1995; Riechmann and Meyerowitz 1997). These investigations led to the ABC model of flower development to explain the formation of sepals, petals, anthers, and carpels in the four whorls (Weigel and Meyerowitz 1994). The A activity in the first whorl generates sepals, and, together with the B activity in the second whorl, petals are formed. B and C activities are required to generate anthers in the third whorl, and C activity leads to carpel formation in the fourth whorl. These activities are due to the spatially regulated activity of specific transcription factors encoded by organ identity genes. In *Arabidopsis thaliana* three organ identity genes control anther formation, i.e., *apetala 3* (AP3), *pistilata* (PI) and *agamous* (AG). The corresponding genes have also been cloned from *Antirrhinum majus (deficiens, globosa,* and *plena*). AP3 and PI are both required for B activity and their gene products form a heterodimer, while AG mediates the C activity (Coen 1992, Schwarz-Sommer et al. 1992, Tröbner et al. 1992). The spatial pattern, i.e., the restriction of activity of AG to the inner whorl, is controlled by cadastral genes, which, in turn, are activated by floral meristem genes.

While the action of the genes controlling the formation of organs is not fully understood, even less is known about the genetic network downstream of the B/C organ identity genes. A generalized overview of anther development was presented by Goldberg et al. (1993). There, anther development is divided in two phases. In phase I, the morphology of the anthers is established, as well as cell and tissue differentiation. This phase ends with meiosis occurring in microspore mother cells. In phase II pollen grains differentiate and the anther enlarges to its final size. This phase terminates when dehiscence and pollen release occur (Goldberg et al. 1993).

The anther cell types which carry out specific tasks (Esau 1977) can be traced to specific floral meristem layers. A number of genes have been isolated which control various steps in this complex developmental pattern (see Sect. 2). The isolation of anther-, pollen- or tapetum-specific

Table 1. Examples of pollen development genes identified in *Arabidopsis thaliana*

Gene name	Gene function	Mutant phenotype	Reference
Antherless	Male sexual organs	Anthers absent	Chaudhury et al. (1992)
ap3	Organ identity gene	Stamens absent	Bowman et al. (1989)
axr1	Male sexual organs	Short filaments	Estelle and Somerville (1987)
bcp1	Pollen development, active in pollen and tapetum	Tapetum mutant: dead shriveled pollen, microspore mutant: one- to-one-segregation of dead-to-viable pollen	H. Xu et al. (1995)
ms (male sterility) 2	Fatty acid reductase?	Pollenless	Aarts et al. (1993, 1997)
ms1, 7, 8	Microsporogenesis	Postmeiotic, microspore become aberrant after being released from tetrads	Chaudhury (1993)
ms3, 4, 5, 15	Microsporogenesis	Premeiotic or meiotic arrest of microspore development	Chaudhury (1993)
msH	Dehiscence	Anthers fail to dehiscence	Dawson et al. (1993)
pistilatat	Organ identity gene	Stamens absent	Bowman et al. (1989)
pop1	Pollen function	Lack of tryphine	Preuss et al. (1993)
Quartet	Pollen separation	Outer wall of meiotic products of pollen mother cell are fused	Preuss et al. (1994)
Sidecar pollen	Required for normal cell division pattern	High degree of abortive and extra-celled pollen	Chen and McCormick (1996)
Stud	Male-specific cytokinesis after telophase II of meiosis	No wall formation in tetranucleate microspore, giant pollen	Hulskamp et al. (1997)
Tetraspore	Meiotic cytokinesis	Microspore nuclei remain in same cytoplasm	Spielman et al. (1997)

promoters was a major step toward engineered male sterility (see Sect. 3.a) and a more detailed study of the dehiscence process. This requires breakage at a special anther structure, the stomium, at dehiscence (Bonner and Dickinson 1989). In a recent paper (Beals and Goldberg 1997), the differentiation of the stomium and adjacent cells (circular cell cluster), and the roles that these regions play in the anther dehiscence process, are analyzed and led to the conclusion that anther dehiscence depends on the presence of a functional stomium region and that these cells can specifically be ablated by using the barnase/barstar genes (Beals and Goldberg 1997). Data from Koltunow et al. (1990) and Mariani et al. (1990) show that dehiscence occurs normally even if no viable pollen is present. Therefore, a signal from the flower, as discussed by Beals and Goldberg (1997), may be necessary for coordination of pollen developmental and the dehiscence program.

Mutations in any gene involved in the differentiation processes described above may lead to male sterility (reviewed in: Chaudhury 1993). For example, there are more than 60 genes that affect male fertility in maize or rice (Albertsen and Palmer 1979; Albertsen and Phillips 1981; Kaul 1988; Hu and Rutger 1992). Recent efforts using *Arabidopsis thaliana* have led to the cloning and molecular characterization of a number of male sterility genes. Table 1 lists some examples of these genes from *Arabidopsis thaliana*. However, similar genes are also known from maize, i.e. the recessive mutant *antherless* (Weijer 1952; Kaul 1988) or cotton. There, a dominant mutant (*ms4*) has malformed anthers (Allison and Fisher 1964). These mutants fall into two groups, i.e., mutants that perturb male sexual organs (e.g., *pistillata*) and mutants that impair pollen development (e.g., *male sterility 2*). According to Chaudhury (1993), the latter group may be split into (1) premeiotic and meiotic mutants, (2) postmeiotic mutants, (3) pollen release mutants, and (4) pollen function mutants (see Table 1 for examples). Further analysis of these and other nuclear mutations affecting anther and pollen development will be useful to elucidate the genetic networks which control these developmental processes.

b) Cytoplasmic-Nuclear Male Sterility (CMS)

Substantial evidence indicates that many plants where CMS occurs are characterized by unusual mitochondrial genes or open reading frames. Most of these clearly resulted from recombination and/or duplication events, often involving sequences of defined normal, constitutive mtDNA genes (Table 2). The manifestation of CMS can occur prior to (sporophytic action) or following (gametophytic action) meiosis, but both processes terminate the male gametophytic generation, i.e., aborted pollen.

Table 2. Chimeric mitochondrial genes. Chimeric mitochondrial genes or open reading frames suspected as causal of CMS in higher plants are often positioned 5' or 3' to normal genes, and may be driven by promoters of the 5' gene, or derived from other genes. The listed promoters have been determined in some examples, or are probable in other examples

| | | | Constituents | | |
Promoter	5' Gene	Gene/orf	5'	3'	3' Feature/gene
atp6		Maize T-*urf13*	*3', rrn26*	*rrn26*	*orf25*
S-1		Maize *orf355*	R1 plasmid		
atp9		Maize *atp6-C*	*atp9*, unique		
atp6		Maize *cox2-C*	*atp6*		
		Maize *atp9-C*			Unique
		Sorghum 9E *cox1*		Novel	
Novel		Sorghum *orf107*	*atp9*	*orf79*	
Novel, *Zmatp6*		Sorghum *orf130*			
atp6	*atp6*	Rice *orf79*	*cox1*	*orf107*	
		Brassica *orf224*	*orfB, rps3*		*t element, atp6*
		Brassica *orf222*	*orfB, rps3*		*nad5C, orf139*
atp6	*atp6*	Brassica *orf263*	*nad5*		
trnfM	*trnfM*	Radish *orf138*			*orfB*
atp1	*atp1*	Sunflower *orf522*	*orfB*		
atp9		Petunia *pcf*	*atp9, cox2*		*nad3, rps12*
cox1		Wheat *orf256*	*cox1*		*cox1*

References: Maize T-*urf13*: Dewey et al. (1986); maize *orf355*: Zabala et al. 1997; maize *atp6-C, cox2-C and atp9-C*: Dewey et al. (1991); sorghum 9E*cox1*: Bailey-Serres et al. (1986); sorghum *orf107*: Tang et al. (1996b) and *orf130*: Tang et al. (1996a); rice *orf79*: Iwabuchi et al. (1993; Akagi et al. (1994); *Brassica orf224*: Singh and Brown (1991): Handa and Nakajima (1992); *Brassica orf 222*: L'Homme et al. (1997); *Brassica orf 263*: Landgren et al. (1996); radish *orf138*: Bonhomme et al. (1992); Krishnasamy and Makaroff (1993); Krishnasamy et al. (1994); sunflower *orfH522*: Köhler et al. (1991); Laver et al. (1919); petunia *pcf*: Hanson et al. (1995); wheat *orf256*: Rathburn and Hedgcoth (1991); Song and Hedgcoth (1994a).

Mitochondrial gene expression can be regulated by factors associated with transcription, *cis-* and *trans-*transcript splicing, site specific cleavage, RNA editing, and translation. In addition, mtDNA gene expression has been shown to be regulated temporally or spatially. The current status of understanding the roles of fertility restoration (RF) genes invokes possible roles in several of these regulatory mechanisms. Still, many considerations of CMS represent an enigma, including the apparent tissue-specific deleterious effect of CMS-associated genes or orfs and why CMS-associated structures are not deleterious in non-reproductive tissues. In this section we present overviews about some of the best-characterized CMS systems.

α) Maize

CMS in maize (*Zea mays*, L) has been extensively studied. There are three major sources of CMS in maize, the T, S and C groups, differentiated by the genetics of fertility restoration and a variety of other criteria (Duvick 1965; Laughnan and Gabay-Laughnan 1983).

T Cytoplasm. The T, or Texas, source of CMS in maize is probably the best understood example of CMS in higher plants, in terms of its probable mtDNA lesion and the nature of required RF genes. A major factor in elucidating this system is the impact T cytoplasm has on maize production, and the resultant susceptibility to two fungal pathogens and their host-specific toxins, resulting in disease epidemics in the US (reviewed in Pring and Lonsdale 1989; Levings and Siedow 1992; Levings 1993; Ward 1995).

CMS in this example is associated with compelling evidence implicating the mitochondrial gene T-*urf13*, a 345-bp chimeric open reading frame consisting of sequences found 3' to *rrn26*, sequences internal to *rrn26*, and sequences of unknown origin (Dewey et al. 1986). Important leads in implicating T-*urf13* in CMS and a role in disease toxin sensitivity were obtained through analyses of male-fertile disease toxin-resistant mutants regenerated from tissue culture. Most of these mutants are characterized by deletion of T-*urf13* (Rottmann et al. 1987; Wise et al. 1987a), but a particularly interesting mutant, designated T-4, exhibits a truncation of T-*urf13* resulting from a frame shift and premature stop codon (Wise et al. 1987a).

T-*urf13* encodes a 13-kDa polypeptide, URF13 (Dewey et al. 1987; Wise et al. 1987b). URF13 and its interactions with the mitochondrion and the fungal toxins has been extensively examined. URF13 is associated with the inner mitochondrial membrane (Dewey et al. 1987), and oligomers bind to toxins of the fungal pathogens, leading to the formation of hydrophillic pores (Braun et al. 1990). URF13 has three membrane-spanning alpha helices, and oligomers undergo a conformational change in the presence of the fungal toxins, allowing a rapid permealization of the inner mitochondrial membrane (Siedow et al. 1995; reviewed in: Rhoads et al. 1995; Levings et al. 1995).

Indeed, in transgenic tobacco, URF13 caused toxin sensitivity, but failed to cause CMS. This may indicate that URF13 is required for CMS but not sufficient on its own (Chaumont et al. 1995). However, these experiments were not fully conclusive and further efforts may be necessary.

The restoration of fertility in T cytoplasm maize requires the action of two genes, *Rf1* and *Rf2*. Critically, action of the gene *Rf1* results in the appearance of a novel 1.6 kb T-*urf13* transcript (Dewey et al. 1986) within a 5' terminus within T-*urf13* (Dewey et al. 1987; Kennell and Pring 1989). Although the novel 1.6 kb transcript represents only a minor part of the T-*urf13* transcripts in maize lines with *Rf1*, action of the gene is accompanied by a 80% reduction in abundance of URF13 in leaf tissues (Dewey et al. 1987).

This may indicate that the effect of *Rf1* is not restricted to T-*urf13* processing, but may influence a yet unknown transcript as well. Dill et al. (1997) raise the possibility that the novel transcripts may interfere with translation.

Evidence that *Rf1* indeed confers fertility restoration and the novel transcript, as opposed to linkage of the two traits, was obtained through tagging the gene with the Mutator transposon family; resultant *rf1-m* lines were male-sterile and did not have the 1.6-kb transcript (Wise et al. 1996). Recently, two additional partial restorer genes, *Rf8* and *Rf**, were shown to exhibit similar affects, but generating transcripts with termini within T-*urf13* that occur 5' to the transcript terminus conferred by *Rf1* (Dill et al. 1997).

A major advance in understanding CMS and the nature of RF genes was the successful isolation and characterization of *Rf2*, the second gene required for fertility restoration in T-cytoplasm maize. A series of transposon-induced mutants of *Rf2* were identified (Schnable and Wise 1994) and Mu-hybridizing clones were analyzed. Sequencing revealed an open reading frame that probably encodes an aldehyde dehydrogenase (ALDH; Cui et al. 1996). Cui et al. (1996) hypothesized that this gene may have a function independent of fertility restoration, and could play a role in oxidation of fatty acids or in detoxifying acetaldehyde. Alternatively, the protein may interact with T-*urf13* in some manner, or modify components of the inner mitochondrial membrane.

S Cytoplasm. The USDA, or S, source of CMS in maize represents a complex example of mtDNA recombinational events that can be associated with the expression of CMS, and the appearance of spontaneous male-fertile revertants. Instability, expressed as fertile individuals, has been painstakingly traced to cytoplasmic revertants to male fertility, or the appearance of new nuclear restorer genes (reviewed in Laughnan and Gabay-Laughnan 1983; Gabay-Laughnan et al. 1995).

Identification and characterization of a series of spontaneous cytoplasmic revertants identified a region of S mtDNA that is consistently associated with reversion. Sequences of episomes characteristic of S cytoplasm are involved in recombination with the principle mtDNA genome, generating linearized molecules with S episomes at their ends (Schardl et al. 1984). Most spontaneous fertile revertants are characterized by the loss of the episomes and of linear mtDNA configurations (Schardl et al. 1985), but analyses of revertants in the Wf9 background showed that this class of revertants retained the episomes and linear configurations (Escote et al. 1985). The S episomes share terminal invert repeat sequences which are represented within the principle mtDNA genome and are the sites of recombination with the S episomes (Schardl et al. 1984, 1985). One of these sites, δ (Schardl et al. 1984), is flanked by sequences designated R (Houchins et al. 1986; Zabala et al. 1997).

Examinations of the Wf9 revertants identified rearranged mtDNA sequences in the δ region at the termini of linear chromosomes (Small et al. 1988; Zabala et al. 1997). Sequencing of the progenitor region in S cytoplasm showed two chimeric open reading frames, *orf355* and the 3' *orf77* (Zabala et al. 1997). *Orf355* is composed of sequences highly similar to part of the R sequences in N cytoplasm and sequences of unknown origin. *Orf77* is composed in part of coding sequences of *atp9* and sequences found 5' to *orf25* in maize. Transcription of this region is in part probably driven by a promoter within the inverted repeat, characteristic of the S and R episomes.

Fertility restoration is exacted by the single dominant gene *Rf3* (Laughnan and Gabay-Laughnan 1983; Gabay-Laughnan et al. 1995; Kamps et al. 1996). Since this is a gametophytic system, wherein action of *Rf3* is manifested at the haploid, pollen stage, Zabala et al. (1977) critically examined expression of *orf355–orf77* in pollen. Transcripts of this region were reduced in abundance in pollen from heterozygous *Rf3rf3* plants, and importantly, further reduced in homozygous *Rf3Rf3* lines. Since these effects could also be observed in immature ear tissues, action of *Rf3* is evident in sporophytic and gametophytic tissues. Although effects associated with RF genes in nonreproductive tissues have been observed in many CMS examples, this is the first case wherein gene action demonstrated at the haploid stage is also expressed in diploid tissues. Since alteration of transcripts in the presence of *Rf3* occurs in sporophytic and gametophytic tissues, a pollen-specific deleterious effect of expression of the open reading frames was invoked (Zabala et al. 1997).

C Cytoplasm. The Charrua, or C source of CMS in maize, is characterized by three novel chimeric mtDNA rearrangements involving the genes *atp6*, *atp9*, and *cox2* (Dewey et al. 1991).

The *atp6* gene in C cytoplasm (*atp6*-C) is characterized by sequences of a novel 5' extension to the conserved core *atp6* sequence, which replaced an amino extension characteristic of T cytoplasm mtDNA. The 5' terminus of this long open reading frame includes amino-terminal sequences derived from *atp9*. The "missing" *atp6* 5' extension characteristic of T cytoplasm is found as part of the extended open reading of C cytoplasm *cox2* (*cox2*-C), replacing *cox2* sequences characteristics of T cytoplasm. The third unusual gene in C mtDNA is *atp9*. The sequence is identical to that of N cytoplasm, but sequences flanking the gene are distinct from those of N cytoplasm. The 5' sequences in C cytoplasm are present in T cytoplasm but not transcribed, and are not present in N or S cytoplasm mtDNAs. Thus, the gene can be considered as transposed from its normal context. Fertility restoration did not affect transcription of any of these genes, nor the size or abundance of COX2, in etiolated seedlings (Dewey et al. 1991).

β) Sorghum

At least seven distinguishable groups of sorghum male-sterile cytoplasms have been differentiated in the US, and additional sources from India have been described. The mtDNAs of many of these can be distinguished (G.-W. Xu et al. 1995; Pring et al. 1995), but to date only two cytoplasms are known to be associated with unusual genes/orfs.

The 9E male-sterile cytoplasm exhibits the only currently known variant of a normal mtDNA gene that could possibly be involved in cms. In this cytoplasm *cox1* differs from that of the milo cytoplasm by recombinational events that resulted in the loss of carboxy terminal sequences and replacement with sequences of unknown origin, generating a longer open reading frame (Bailey-Serres et al. 1986). The resultant open reading frame is longer than that of milo, and accordingly, the resultant gene product is larger. It is not known whether fertility restoration alters expression of the 9E *cox1*.

The IS1112C, or A3 source of cms, is characterized by several unusual features (Pring et al. 1995). In this case two chimeric mitochondrial open reading frames have been shown to be specific for the A3 cytoplasm.

Orf265/130 is a chimeric open reading frame consisting of sequences similar to those 5' to maize T-*urf13* and *atp6*, sorghum *urf209*, sequences found 5' to sorghum *atp6–2*, and sequences of unknowing origin (Tang et al. 1996a). Two promoters were identified, one corresponding to a maize T-*urf13/atp6* promoter (Yan and Pring 1997). An in-frame start codon 3' to sequences found 5' to *atp6–2* defines *orf130*, which has one silent RNA editing site (Tang et al. 1996a). No differences in transcript patterns were observed in plants restored to fertility.

Orf107 is a chimeric orf consisting of sequences derived from sorghum *atp9*, sequences of unknown origin, and sequences similar to those of *orf79* (Tang et al. 1996b), a chimeric orf associated with cms in rice (Iwabuchi et al. 1993; Akagi et al. 1994). Transcription is driven by three promoters of unknown origin (Yan and Pring 1997). Fertility restoration (Tang et al. 1996b) or action of the gene *Rf3*, one of two genes required for restoration (H. V. Tang et al., submitted), results in a novel nucleolytic cleavage event within the open reading frame. The 3' product, a 380-nt transcript, accumulates at high abundance, while the 5' processing product is degraded. Since a trace, basal level of transcript processing is presenet in lines that are *rf3rf3*, it was postulated that the *Rf3* gene regulates processing activity (H. V. Tang et al., submitted).

In male-sterile plants, *orf107* transcripts are highly edited at three positions, one shared with *atp9*, and a fourth site is infrequently edited. About 75% of whole-length transcripts are cleaved by action of *Rf3*. Examinations of RNA editing in the remaining whole-length transcripts revealed highly frequent editing at the site shared with *atp9*, but low levels of editing at the two sites that are edited at high frequency in male-sterile plants (D. Pring et al., submitted). A dependence of editing for action of the transcript processing activity was hypothesized.

Uniquely, RNA editing of *atp6* has been shown to be related to the restoration of male fertility in this cytoplasm. RNA editing in different tissues of lines carrying the A3 cytoplasm was compared. Whereas RNA editing of *atp9*, *nad4*, and *nad3–rps12* transcripts in anthers was similar to that of etiolated shoots, mitochondrial *atp6* RNA editing was strongly reduced in anthers of the A3Tx398 and A3Tx7000 male-sterile lines (Howad and Kempken 1997a, b, and unpublished data). Transcripts of *atp6* in wheat anthers, and selected plastid transcripts in sorghum anthers showed normal RNA editing, which indicate that the reduction in editing frequency of mitochondrial *atp6* is specific for this male-sterile sorghum cytoplasm. Restoration of fertility in F_1 and F_2 lines correlates with an increase in RNA editing of *atp6* transcripts. These data suggest that loss of *atp6* RNA editing efficiency might contribute to or cause cms in sorghum. Interestingly, the data from anther *atp6* RNA editing are correlated with aspects of fertility restoration of the A3 cytoplasm, wherein the role of the second gene required for restoration, *Rf4*, is unknown (H. V. Tang et al., submitted). This may suggest a functional correlation. However, since *orf107* transcript processing is not pollen-specific, a yet unidentified anther-specific factor may be involved.

Altered editing of *atp6* transcripts is also associated with cms in rice (Iwabuchi et al. 1993) and in a transgenic approach, unedited ATP9 protein caused cms in tobacco (Hernould et al. 1993; see Sect. 3).

γ) *Brassica*

The biology, genetics, and characterization of sequences associated with CMS in *Brassica* species has recently been reviewed (Makaroff 1995). Four sources of CMS, the Ogura, *pol*, *nap*, and *tour* cytoplasms, have been characterized. In each case a novel open reading frame has been identified, 5' or 3' to genes which can be regarded as normal mitochondrial genes.

Ogura Cytoplasm. Male sterility induced by the Ogura cytoplasm of *Raphanus*, radish, is associated with the mitochondrial gene *orf138* (Bonhomme et al. 1992; Krishnasamy et al. 1994). *Orf138* is restricted to the Ogura cytoplasm (Bonhomme et al. 1991; Krishnasamy and Makaroff

1993), although sequences hybridizing to an *orf138* probe are present in normal cytoplasm radish and several *Brassica* species (Krishnasamy and Makaroff 1993). Like many examples of CMS-associated genes, *orf138* is cotranscribed with a 3' gene, which in this case is a gene designated *orfB* (Bonhomme et al. 1992; Krishnasamy et al. 1994). Transcripts of *orf138* are not edited (Krishnasamy and Makaroff 1994). *Orf138* encodes a 19–20-kDa polypeptide, which is membrane-associated, and may occur in multimers (Grelon et al. 1994; Krishnasamy and Makaroff 1994). ORF138 is readily detected in roots and leaves, and anthers, petals, sepals and ovaries of sterile radish flower buds contain about equally abundant levels of the protein (Krishnasamy and Makaroff 1994), indicating that the product is not deleterious in these somatic tissues. Fertility restoration has no effect on *orf138–orfB* transcript patterns (Bonhomme et al. 1992; Krishnasamy and Makaroff 1994), but the abundance of ORF138 is dramatically reduced in flowers and leaves of restored plants (Krishnasamy and Makaroff 1994).

polima, or pol Cytoplasm. The *polima,* or *pol* source of CMS in *Brassica* includes an mtDNA configuration including the open reading frame *orf224* (Singh and Brown 1991; Handa and Nakajima 1992; Handa et al. 1995). The open reading frame is chimeric, consisting of 58 codons from *orfB*, 43-bp derived frome exon 1 of *rps3* (Hada et al. 1995), and sequences of unknown origin. Transcripts of *orf224* are edited at a single site within the *rps3* sequence-derived region (Stahl et al. 1994; Handa et al. 1995). *Orf224* is positioned 5' to the cotranscribed *atp6*. Fertility restoration by action of *Rfp1* results in the appearance of novel transcripts (Singh and Brown 1991; Witt et al. 1991), which were shown to be processed with 5' termini within *orf224*, releasing monocistronic *atp6* transcripts (Singh and Brown 1991, 1993). The processing activity is less efficient in floral tissue than in seedling tissues (Singh and Brown 1993). Two restorer genes, *Rfp1* and *Rfp2*, have identical effects on transcription (Singh and Brown 1991), and are probably alleles (Jean et al. 1997).

napus, or nap Cytoplasm. An extraordinary observation has been put forward regarding the nature of sequences associated with male sterility of the *pol* and *nap* (*B. napus*) cytoplasms. The *nap* cytoplasm does not induce male sterility in most nuclear backgrounds, but is male-sterile in selected backgrounds (L'Homme and Brown 1993; L'Homme et al. 1997). Sequences of *orf224* were detected in *nap* cytoplasm plants (L'Homme and Brown 1993), and subsequent isolation and sequencing of *nap* mtDNA revealed the chimeric open reading frame *orf222*, with remarkable similarity to *pol orf224* (L'Homme et al. 1997).

Sequences of orf222 are 85% identical to that of *orf224*, including 5' sequences of *orfB*. Translation of *orf222* revealed 79% similarity to *orf224*. Similar to the *pol* configuration, *orf222* is co-transcribed with other genes located 3', in this case the trans-spliced exon c of *nad5* and an open reading frame designated *orf139*. Fertility restoration influences transcripts of this region.

tour, or B. tournefortii Cytoplasm. A chimeric mtDNA orf associated with CMS in *Brassica* carrying the *B. tournefortii* cytoplasm, *orf263*, has been described (Landgren et al. 1996). *Orf263* is positioned 5' to *atp6*, and consists of sequences highly similar to the 5' end of exon a of *nad5*, sequences found 3' to *apt9* in *Raphanus*, and sequences of unknown origin. Transcripts spanning *atp6* and *orf263* were not observed in parental *B. tournefortii*, but were apparent in two male-sterile lines carrying the *tour* cytoplasm. A 29-kDa-polypeptide, which may correspond to the *orf263* gene product, was detected in male-sterile lines with the *tour* cytoplasm, but not present in *B. tournefortii*, consistent with the differential transcription patterns.

δ) *Phaseolus*

CMS in common bean derived from the cytoplasm of the line G08063 has several unusual features. Investigations of spontaneous revertants to male fertility and of fertility restoration by the gene *Fr* indicated that mtDNA was rearranged in both cases (Mackenzie et al. 1988), which was shown to be a deletion (Mackenzie and Chase, 1990). This possible mechanism of fertility restoration is unprecedented in higher plants. Transcripts characteristic of the deleted region were lost in revertants or lines restored to fertility by *Fr* (Mackenzie and Chase 1990).

The deleted region, designated *pvs*, is 3736 bp, and includes several open reading frames, including *orf239* (Chase and Ortega 1992; Johns et al. 1992). Sequences of *orf239* are of unknown origin. The source of this male-sterile cytoplasm, G08063, is male-fertile, and thus carries the requisite RF gene; this gene, designated *Fr2*, does not cause mtDNA deletions (Mackenzie 1991), nor alters transcripts of the *pvs* region (Mackenzie and Chase 1990; Chase 1994).

Antibody to part of ORF239 indicated that *orf239* is expressed at very low levels in most tissues, but the polypeptide could be detected in ovules, mitochondria of pollen mother cells, and the callose layer and developing primary cell walls of microspores (Abad et al. 1995). Tobacco plants transformed with *orf239* were characterized by presence of ORF239 in the cell wall of aberrant developing microspores, and all male-sterile plants among the transgenic progeny carried detectable ORF239 (He et al. 1996).

ε) Petunia

CMS in petunia (reviewed in: Conley and Hanson 1995; Hanson et al. 1995) is characterized by the presence of the chimeric orf *pcf* (petunia CMS associated *fused* gene) which consists of portions of mitochondrial respiratory complex genes (*atp9* and *cox2*) and sequences of unknown origin (Young and Hanson 1987). The *pcf* gene is positioned 5' to, and cotranscribed with *nad3–rps12*; transcription is driven by three initiation sites (Hanson et al. 1988; Rasmussen and Hanson 1989; Pruitt and Hanson 1991). Transcription of *pcf* appears to be regulated in a tissue-specific manner, since transcript abundance in anthers is about four to five times higher than in leaves (Young and Hanson 1987).

The petunia *Rf* restorer gene reduces the abundance of the smallest *pcf* transcript in anthers (Pruitt and Hanson 1991). *Pcf* is translated into a 25 kDa peptide designated URF-S. Since URF-S does not contain ATP9 or COX2 sequences, it apparently results from protein processing (Nivison and Hanson 1989; Nivison et al. 1994). Tissue-print immunoblots revealed URF-S to be abundant in sterile plants, but absent in fertile isonuclear lines. Near-isonuclear plants restored to fertility exhibit low levels of URF-S (Conley et al. 1991). In experiments with transgenic plants CMS was not induced by the presence of *pcf*. This may suggest that expression of *pcf* early in meiosis may be critical for the action of the URF-S protein (Wintz et al. 1995). Alternatively, the presence of URF-S may be necessary, but not sufficient, to cause CMS.

ξ) Sunflower

CMS in *H. annuus* carrying the *H. petiolaris*, or PET1 cytoplasm is associated with the mitochondrial chimeric open reading frame *orfH522*, which is cotranscribed with the 5' gene *atp1* (Köhler et al. 1991; Laver et al. 1991). *OrfH522* is chimeric, including ten codons duplicated from *orfB* (Köhler et al. 1991; Laver et al. 1991).

RNA editing of transcripts includes two C-to-U modifications, changing two predicted amino acids, and fertility restoration has no effect on editing frequency (Laver et al. 1991). Fertility restoration also has no affect on *atp1–orfH522* transcript patterns in seedling tissues (Köhler et al. 1991; Laver et al. 1991).

A novel 15–16-kDa polypeptide was identified in polypeptide synthesis patterns from isolated mitochondria, and abundance of the protein showed little or no variation in mitochondria isolated from seedlings of restored lines (Horn et al. 1991; Laver et al. 1991). The protein was shown to be the *orfH522* product (Horn et al. 1996; Moneger et al. 1994).

Expression of *orfH522*, uniquely among CMS-related genes, exhibits tissue-specific regulation associated with fertility restoration. The abundance of transcripts encoding *orfH522* and *atp1* was shown to be reduced in florets of restored lines, associated with a decrease in abundance of the gene product (Moneger et al. 1994). In situ hybridization demonstrated that the reduction in abundance of transcripts localized to meiotic cells (Smart et al. 1994).

η) Other CMS Systems

A chimeric mtDNA open reading frame, *orf256*, is associated with CMS in *Triticum aestivum* carrying the *T. timopheevi* cytoplasm (Rathburn and Hedgcoth 1991; Song and Hedgcoth 1994a). The open reading frame consists of sequences duplicated from the coding and 5' sequences of *cox1*, and sequences of unknown origin. *Orf256* is weakly transcribed in *T. timopheevi*, but has abundant transcripts in the *T. aestivum* background. Antibody to the polypeptide predicted by *orf256* detected a novel 7 kDa protein in male-sterile lines, but not in lines restored to fertility (Song and Hedgcoth 1994b).

Nuclear-controlled RNA processing events are associated with the expression of the *T. timopheevi* male-sterile cytoplasm. The gene regions coding for *atp1* and *apt9* exhibit identical DNA sequences in wheat, rye, and the intergenic hybrid triticale (*x Tritico secale* Wittmack). However, cotranscripts containing these genes show different sizes depending on the nuclear genotype (Laser and Kück 1995), which probably results from differential processing All wheat lines contain a 2.6-kb *atp1–atp9* mRNA. In rye and five triticale lines an additional 2.35 kb-mRNA is present. The processing event occurs more frequently in those triticale lines which carry the *Triticum timopheevi* cytoplasm. This may indicate a better interaction of *trans*-acting factors from rye with the cytoplasm from *T. timpheevi*. Interestingly, this cytoplasm causes CMS in alloplasmic wheat, but not in triticale.

MtDNA of the male-sterile Chinsurah Boro II cytoplasm of rice is associated with rearrangements 3' to one of two copies of *atp6*, including sequences similar to *cox1* (Iwabuchi et al. 1993), which was shown to include an orf designated *orf79* (Akagi et al. 1994). The amino terminus of the predicted *orf79* gene product is derived from the *cox1* sequence, and the carboxy terminus is highly similar to that predicted for sorghum *orf107* (see Sect. 2.b.β). Fertility restoration by the gene *Rf-1* is associated with an unusual endonucleolytic transcript processing activity, which apparently processes an *atp6–orf79* cotranscript into a monocistronic *atp6* transcript and a shorter transcript that may include *orf79* (Iwabuchi et al. 1993; Akagi et al. 1994).

A role of RNA editing was invoked in the restoration of fertility in this system. Examination of cDNAs from part of B-*atp6* indicated that editing frequency in the non-processed co-transcript was less frequent than in the processed, monocistronic B-*atp6* transcript resulting from action of *Rf-1* (Iwabuchi et al. 1993). These observations are consistent with the postulated effects of fertility restoration on editing of *atp6* in anthers of sorghum (see Sect. 2.b.β).

c) Fertility Restoration Genes in CMS

α) Enigmas of CMS

Why is the Deleterious Effect of CMS-Associated Genes or orfs Tissue-specific? One of the enigmas of current knowledge of CMS is the question of why expression of these genes is not deleterious in nonreproductive cells. In only one case is there suggestive evidence that the gene product, a priori, indeed may be toxic. The frequency with which maize T-*urf13* is spontaneously deleted in maize callus tissue culture was forwarded as indicating that expression may confer a competitive disadvantage (Pring et al. 1988), and toxicity of URF13 expressed in bacteria, and insect cell cultures or larvae, led to the conclusion that the protein was toxic in a variety of organism (Levings 1993).

Expression of a cms-related, deleterious, gene at only microsporongenesis has been observed in one case, which could represent a rational explanation for deleterious events in reproductive tissues. In this example, common bean ORF239, the protein is primarily expressed in microspore cell walls, the callose layer, and mitochondria of pollen mother cells (Abad et al. 1995).

Several examples indicate that RF genes can exert marked effects in expression of ordinary or cms-related genes in reproductive tissues. The depressed frequency of *atp6*-specific editing in anthers of male-sterile sorghum is returned to normal levels by fertility restoration (Howard and Kempken 1997b). RF genes have also been shown to confer reduction in the abundance of radish ORF138 (Krishnasamy and Makaroff 1994) and decreased abundance of sunflower *orfH522* transcripts (Smart et al. 1994) and the gene product (Moneger et al. 1994).

Another explanation is the assumption that a high demand for energy exists specifically during early pollen development, which is based on the dramatic increase in mitochondria number observed in tapetal and microspore mother cells of T-cytoplasm maize during microsporogenesis (Warmke and Lee 1978; Lee and Warmke 1979). A putative deleterious effect of a cms-related gene may not be sufficient to cause lesions in nonreproductive tissues, while energy demand in tapetal and microspore mother cells may represent a stress condition that allows a deleterious effect. This model is similar to what is believed to happen in certain mitochondrially inherited human diseases (Grossman and Shoubridge 1996).

Are RF Genes Linked to Other Genes Influencing Mitochondrial Expression, or Alleles? The identification of the gene product of maize *Rf2*, ALDH, has ramifications in understanding the nature of *Rf* genes. Critically, the line carrying the reference, nonrestoring *rf2* allele displayed a transcript the same size as lines with *Rf2* (Cui et al. 1996), and a mitochondrial protein was detected in this line with antibodies to the *Rf2* protein (P.S. Schnable, pers. comm.). These observations indicate the distinct possibility of a metabolic role of ALDH in microsporogenesis in lines carrying T-*urf13*. The abundance of URF13 is reduced 80% in lines carrying *Rf1* (Dewey et al. 1987), but presence of *Rf2*, the possible ALDH, is required for restoration. The recessive, nonrestoring *rf2* allele may encode a defective ALDH, unable to ameliorate damage associated with URF13, even in lines with the *Rf1*-reduced abundance of URF13. The recessive *rf2* allele may allow the accumulation of toxic levels of acetaldehyde and alcohol, which could be critical to tapetal cells during microsporogenesis (Cui et al. 1996; Levings 1996).

The possibility that nuclear genes conferring the restoration of male fertility may be linked to genes influencing expression of normal genes, or may represent alleles, has recently emerged. Two of the examples reflect alterations in transcript patterns, probably through nucleolytic processing.

The possibility of linkage of RF genes and genes affecting transcription of mtDNA genes was first raised by Makaroff (1995), who observed that many lines that restored fertility to the Ogura male-sterile radish cytoplasm also conferred *atp1* transcript alterations.

Compelling evidence for such a linkage was observed by Singh et al. (1996) regarding the linkage of *Rfp1*, which restores fertility of the *pol* cytoplasm, and a gene designated *Mmt*, which influences transcript modification of two other mtDNA regions. In this case the recessive *rfp1* allele, or a tightly linked *Mmt*, modified transcripts of *nad4* and a gene involved in cytochrome c biogenesis. Linkage of the sorghum gene *Rf3*, which regulates nucleolytic processing within sorghum *orf107* (Tang et al. 1996b), and a gene designated *Mmt1*, which similarly confers enhanced processing of transcripts 5' to sorghum *urf209* (Tang et al. 1996a), has recently been described (H. V. Tang et al., submitted). Interestingly, a trace basal level of processing was observed in *rf3rf3Mmt1Mmt* lines; that the two activities are separable was obtained through the identification of *rf3rf3Mmt1Mmt1* lines (H. V. Tang et al., submitted). Recently, four RF genes in *Phaseolus* were assigned to the same linkage group, and two were shown to be allelic with a gene designated *Fr2*; the fourth gene, *Fr*, is postulated to have been derived by mutation (Jia et al. 1997).

In the sorghum (H. V. Tang et al., submitted), *Brassica* (Singh et al. 1996) and *Phaseolus* (Jia et al. 1997) examples, characteristics of the systems under study raised the possibility of allelism, or evolution of the RF genes from a common ancestor.

The RF gene *Rf3* is tightly linked to *Mmt1* in IS1112C sorghum. The line Tx7000 is *rf3rf3Mmt1Mmt1*, and is thus a maintainer line that does process transcripts 5' to *urf209*. If these genes are alleles, the Tx7000 allele could be considered a normal gene, conferring transcript processing 5' to *urf209* as part of a transcript maturation process. The *Rf3* allele is infrequent, whereas *Mmt1* occurs frequently among sorghum lines examined to date, indicating that the IS1112C allele could have evolved such that it acquired the additional capability to process *orf107* transcripts (Tang et al. 1996b; H. V. Tang et al., submitted). The absence of *orf107* in lines other than IS1112C might be explained by the lack of selective pressure for modification of the RF gene in these lines. Thus, with respect to *orf107* processing they are maintainers of CMS.

β) New Clues to an Old Hypothesis

Although CMS has been extensively studied, the mechanisms of pollen abortion still represent a number of enigmas. Several recent developments in the biology of CMS in maize are relevant to the modeling of mechanisms of CMS. Based on observations of the effects of host-specific fungal toxins on T cytoplasm maize mitochondria, Flavell hypothesized that an anther-specific substance could mimic the effect of the host-specific toxins, interacting with "structures altered by cytoplasmically inherited mutations" (Flavell 1974). As discussed above, the specific interaction of the host-specific toxins with URF13 and the resultant permealization of the inner mitochondrial membrane is well documented (Levings and Siedow 1992; Levings 1993; Rhoads et al. 1995; Levings et al. 1995). A tapetum-specific synthesis of a compound that mimics the action of the fungal toxins has been invoked (Cui et al. 1996; Levings 1996), based on the observation that maize *Rf2* may encode an aldehyde dehydrogenase (ALDH; Cui et al. 1996). Interestingly, high ALDH expression levels have recently been shown to accompany pollen development in tobacco (op den Camp and Kuhlemeier 1997; Tadege and Kuhlemeier 1997). Cumulatively, these observations are consistent with the possibility that an anther-specific substance may interact with URF13, causing mitochondrial membrane disruption, invoking a possible role of ALDH in ameloriation of these events (Cui et al. 1996). A candidate for such an anther-specific substance, which affects respiration of cms T mitochondria and *E. coli* expressing URF13, may have been identified (Levings 1996).

Actions of *Rf1* and *Rf2* result in decreased URF13 abundance (Dewey et al. 1987) and probable enhanced repair capabilities (Cui et al. 1996), allowing diploid tapetal cells to function in microsporogenesis. It is interesting to note that one gene decreases abundance of the deleterious mtDNA gene product and that the second gene possibly ameloriates the effects of the mtDNA gene product. Neither reduced URF13 abundance nor apparent functional ALDH activity alone is sufficient to restore normal tapetal function, in that lines that are *Rf1Rf1rf2rf2* or *rf1rf1Rf2Rf2* are sterile. It is important to note that restoration is sporophytic, wherein the genotype of the sporophyte, and not the gamete, determines pollen viability. Plants carrying *Rf1rf1Rf2rf2* shed 100% viable pollen, in spite of the fact that gametes can be recessive for one or both genes, such as pollen carrying *Rf1rf2*, *rf1Rf2*, and *rf1rf2*. Thus, a potentially deleterious substance invoked for T cytoplasm maize has no effect after the second meiotic division, when haploids are present, implicating the diploid tapetum, microspore mother cell, or dyads as primary sites of action.

An extension of Flavell's (1974) hypothesis to cms systems with a gametophytic mode of restoration invokes a postmeiotic synthesis and appearance of a deleterious substance(s) at the microspore or pollen stage. In gametophytic systems such as S cytoplasm maize (Gabay-Laughnan et al. 1995), A3 cytoplasm sorghum (Tang et al., submitted) and the Bo rice cytoplasm (Iwabuchi et al. 1993), segregation for sterility-fertility occurs at the haploid, pollen level. In the maize and rice single-gene examples, RF effects are manifested by transcription aberrations of cms-associated orfs (Zabala et al. 1997; Iwabuchi et al. 1993; Akagi et al. 1994), which mimic the effect of maize *Rf1* (Dewey et al. 1987; Kennel and Pring 1989) and may be associated with a reduction in abundance of the putative gene products. The sorghum A3 example similarly indicates a transcriptional aberration associated with one RF gene, but a function for the second required gene has not been determined (H. V. Tang et al., submitted), and may involve anther-specific loss of RNA editing (Howad and Kempken 1997b).

3 Artificial Male Sterility in Transgenic Plants

In recent years, many strategies have been published to genetically engineer male sterility. These approaches are particularly useful where CMS systems are not available. Some of the most promising approaches are discussed below.

a) Tapetum/Pollen-Specific Gene Expression

Pollen ablation may be caused by anther specific expression of ribonucleases. Mariani et al. (1990, 1992) made use of Barnase, a gene for an extracellular RNase from *Bacillus amyloliquefaciens*. This bacterium uses the Barnase as a defense system against competing bacteria. *B. amyloliquefaciens* also expresses Barstar, a specific inhibitor of Barnase.

The Barnase gene was fused to a tapetum-specific promoter (TA29) from tobacco and this *RNase* was specifically expressed in tapetal cells of transformants. As a consequence these cells were ablated and no pollen was produced. More importantly, a second construct using the same promoter but fused to the Barstar inhibitor, did not cause tapetum ablation. Instead, this construct inhibited Barnase activity in a dominant way and thus behaved as a restorer gene. The tapetum specific promoter TA29 is correctly regulated in many monocots and dicots (Mariani et al. 1992) and this system is therefore useful in a wide variety of plants.

In a similar strategy the Osg6B promoter from rice fused to an endo-β-1,3-glucanase was employed. This gene is pathogenesis-related and was previously identified from soybean. The construct was transformed into tobacco, and during formation of tetrads it was expressed in the tapetum cells, leading to a significant reduction of the number of fertile pollen (Tsuchiya et al. 1995). In addition, a number of other approaches include the use of another glucanase (Worrall et al. 1992), antisense inhibition of flavonoid biosynthesis in tapetum cells (van der Meer et al.

1992), and an engineered increase in auxin activity in the anther (Spena et al. 1992).

Yet another strategy is based on induction of male sterility by tapetum-specific deacetylation of externally applied nontoxic N-acetyl-L-phosphinothricin (Kriete et al. 1996). The N-acetyl-L-ornithine deacetylase from *E. coli* was fused to the tobacco TA29 tapetum-specific promoter. Application of externally applied N-acetyl-L-phosphinothricin led to empty anthers without pollen. This system is easy to handle and may be applied to many different crops. Since male sterility is induced only in the presence of N-acetyl-L-phosphinothricin, no fertility restoration is required in the F_1.

b) A Peptide from a CMS-Related orf May Cause Male Sterility in Transgenic Tobacco

In CMS common bean, a novel 3736-bp mitochondrial DNA sequence of unknown origin, designated *pvs*, is present in male-sterility lines (Chase and Ortega 1992; Johns et al. 1992). Within the sequence is an open reading frame, *orf239* which is translated into a 27-kDa polypeptide that is present in reproduction tissues only. Interestingly the protein is found to localize to the callose layer and primary cell wall of developing pollen, although it is of mitochondrial origin (Abad et al. 1995). It is believed that the ORF239 protein interferes with normal pollen development and thus causes CMS in common bean. In a transgenic approach, the *orf239* sequence was transferred into tobacco. Several transformants exhibited a semisterile or male-sterile phenotype, regardless of whether or not a mitochondrial targeting sequence was used (He et al. 1996). These experiments demonstrate the potential use of CMS-specific orfs to generate male sterility. However, many open questions remain to be answered, such as how a mitochondrial encoded protein is transported to the callose layer.

c) Targeting of Unedited ATP9 into Plant Mitochondria

An approach toward male sterility might involve the use of unedited proteins targeted to plant mitochondria (Hernould et al. 1993). RNA editing is an important processing step for almost all mitochondrial transcripts of higher plants and results in predicted amino acid sequences that are conserved in many mitochondrial genes in animals and fungi (Covello and Gray 1989; Gualberto et al. 1980; Hiesel et al. 1989; reviewed in: Pring et al. 1993; Maier et al. 1996; see also Sect. 2.b.β).

A French group succeeded in targeting an unedited ATP9 protein into mitochondria of transgenic tobacco (Hernould et al. 1993). They fused the genomic, unedited sequence of *atp9* from wheat to a yeast *coxIV* import sequence, and the construct was introduced into tobacco. In wheat, RNA editing occurs at six positions in the *atp9* transcript (Begu et al. 1990; Nowak and Kück 1990). About 50% of the transformants were either semi or fully male sterile. In the remaining transformants, which were male-fertile, no expression of the construct was observed. It is assumed that the unedited ATP9 protein competes with edited ATP9 for assembly in the ATPase F_0 complex. Alternatively, one may speculate that the imported unedited ATP9 may insert into the inner membrane and form proton channels which disrupt the H^+ gradient (Hernould et al. 1993).

Recently it was demonstrated that an antisense *atp9* sequence may be used to restore fertility in plants expressing unedited ATP9 protein. F_1 plants produced up to 80% of normal pollen levels, and seed yield was 2.4 g of seed per plant compared to 3 g found for fertile plants (Zabaleta et al. 1996). These results demonstrate the potential usefulness of the technique to generate male-sterile plants which regain fertility in the F_1.

4 Conclusions

The analysis of anther development is yet in its beginning. However, as an important step toward understanding this developmental process, the genes which control flower formation were identified, and, particularly in *Arabidopsis thaliana*, a number of mutants have been identified. This will allow a molecular approach into anther and pollen development.

As shown above, recent developments in cytoplasmic-nuclear male sterility have led to a much better understanding of how the formation of chimeric orfs may occur and contribute to CMS. However, CMS may have other causes as well, e.g. reduced transcript editing of normal, constitutive genes (Howad and Kempken 1997b).

A unique feature of CMS is that expression of the trait is influenced by nuclear RF genes, and the current state of knowledge reveals several possible leads to the possible origins of these genes. One of the most potentially important observations, derived from the maize *Rf2* data, is that an RF gene may represent a common wild-type gene, and that the non-restoring allele may be defective in some manner. Alternatively, the possibility that a mutation in a transcript processing gene may have coevolved with chimeric orf formation, thus generating a restorer allele, is indicated by several examples of possible alleslism of RF genes.

While until recent years, breeders were forced to use naturally occurring CMS systems, the invention of gene technology now allows other, more targeted approaches as well. In this review we have listed some of the major advances in this field. Particularly strategies such as the tapetum-specific deacetylation of externally applied non-toxic N-acetyl-L-phosphinothricin (Kriete et al. 1996) appear to be very tempting, since there fertility restoration is not necessary because sterility occurs only after appropriate chemical treatment. It has yet to be seen which of these approaches will lead to large scale applications in plant breeding.

Acknowledgments. The authors thank Mrs. Alexandra Wakenhut for help with typing the manuscript. The laboratory work of the authors was funded by Cooperative Investigations, U. S. Department of Agriculture-Agriculture Research Service and Institute of Food and Agricultural Sciences, University of Florida to D.R.P. and a grant from the Deutsche Forschungsgemeinschaft (Ke409/3-3) To F. K.

References

Aarts MGM, Dirkse WG, Stiekema WJ, Pereira A (1993) Transposon tagging of a male sterility gene in *Arabidopsis*. Nature 363:715–717

Aarts MGM, Hodge R, Kalantidis K, Florack D, Wilson ZA, Mulligan BJ, Stiekeman WJ, Scott R, Pereira A (1997) The *Arabidopsis* MALE STERILITY 2 protein shares similarity with reductases in elongation/condensation complexes. Plant J 12:615–623

Abad AR, Mehrtens BJ, Mackenzie SA (1995) Specific expression in reproductive tissues and fate of a mitochondrial sterility-associated protein in cytoplasmic male-sterile bean. Plant Cell 7:271–285

Akagi H, Sakamoto M, Shinjyo C, Shimada H, Fujimura T (1994) A unique sequence located downstream from the rice mitochondrial *atp6* may cause male sterility. Curr Genet 25:52–58

Albertsen MC, Palmer RG (1979) A comparative light- and electron-microscopic study of microsporogenesis in male sterile (*msl*) and male fertile soybeans. Am J Bot 66:253–265

Albertsen MC, Phillips RL (1981) Developmental cytology of 13 genetic male sterile loci in maize. Can J Genet Cytol 23:195–208

Allison DC, Fisher WD (1964) A dominant gene for male sterility in upland cotton. Crop Sci 4:548–549

Bailey-Serres J, Hanson DK, Fox TD, Leaver CJ (1986) Mitochondrial genome rearrangement leads to extension and relocation of the cytochrome c oxidase subunit I gene in sorghum. Cell 47:567–576

Beals TP, Goldberg RB (1997) A novel cell ablation strategy blocks tobacco anther dehiscence. Plant Cell 9:1527–1545

Begu D, Graves PV, Domec C, Arselin G, Litvak S, Araya A (1990) RNA editing of wheat mitochondrial ATP synthase subunit 9: direct protein and cDNA sequencing. Plant Cell 2:1283–1290

Bonhomme S, Budar R, Ferault J, Pelletier G (1991) A 2.5 kb NcoI fragment of Ogura radish mitochondrial DNA is correlated with cytoplasmic male sterility in *Brassica* cybrids. Curr Genet 19:121–127

Bonhomme S, Budar F, Lancelin D, Small I, Defrance MC, Pelletier G (1992) Sequence and transcript analysis of the Nco2.5 Ogura-specific fragment correlated with cytoplasmic male-sterility in *Brassica* cybrids. Mol Gen Genet 235:340–348

Bonner LJ, Dickinson HG (1989) Anther dehiscence in *Lycopersicon esculetum*. I. Structural aspects. New Phytol 113:97–115

Bowman JL, Smyth DR, Meyerowith EM (1989) Genetic directing flower development in *Arabidopsis*. Plant Cell 1:37–52

Braun CJ, Siedow JN, Levings CS III (1990) Fungal toxins bind to the URF-13 protein in maize mitochondria and *Escherichia coli*. Plant Cell 2:153–161

Chase CD (1994) Expression of CMS-unique and flanking mitochondrial DNNA sequences in Phaseolus vulgaris L. Curr Genet 25:245–251

Chase CD, Ortega VM (1992) Organization of ATPA coding and 3' flanking sequences associated with cytoplasmic male sterility in *Phaseolus vulgaris* L. Curr Genet 22:147–153

Chaudhury AM (1993) Nuclear genes controlling male fertility. Plant Cell 5:1277–1283

Chaudhury AM, Craig S, Farell L, Bloemer K, Dennis ES (1992) Genetic control of male fertility in higher plants. Aust J Plant Physiol 19:419–425

Chaumont F, Bernier B, Buxant R, Williams ME, Levings CS III, Boutry M (1995) Targeting the maize T-*urf13* product into tobacco mitochondria confers methomyl sensitivity to mitochondrial respiration. Proc Natl Acad Sci USA 92:1167–1171

Chen YC, McCormick S (1996) *sidecar* pollen, an *Arabidopsis thaliana* male gametophytic mutant with abberant cell divisions during pollen development. Development Suppl 122:3243–3253

Coen ES (1992) Flower development. Curr Opin Cell Biol 4:929–933

Conley CA, Hanson MR (1995) How do alterations in plant mitochondrial genomes disrupt pollen development? J Bioenerg Biomemb 27:447–457

Conley CA, Nivison HT, Wilson RK, Hanson MR (1991) Localization of mitochondrial proteins in tissue prints of cytoplasmic male sterile (cms) and fertile *Petunia* lines. J Cell Biol 115:300A

Covellos PS, Gray MW (1989) RNA editing in plant mitochondria. Nature 341:662–666

Cui X, Wise RP, Schnable PS (1996) The rf2 nuclear restorer gene of male-sterile T-cytoplasm maize. Science 272:1334–1336

Dawson J, Wilson ZA, Aarts MGM, Braithwaite A, Briarty LGB, Mulligan BJ (1993) Microspore and pollen development in six male-sterile mutants of *Arabidopsis thaliana*. Can J Bot 71:629–638

Dewey RE, Levings CS III, Timothy DH (1986) Novel recombinations in the maize mitochondrial genome produce a unique transcriptional unit in the Texas male-sterile cytosplasm. Cell 44:439–449

Dewey RE, Timothy DH, Levings CS III (1987) A unique mitochondrial protein associated with cytoplasmic male sterility in the T cytoplasm of maize. Proc Natl Acad Sci USA 84:5374–5378

Dewey RE, Timothy DH, Levings CS III (1991) Chimeric mitochondrial genes expressed in the C male-sterile cytoplasm of maize. Curr Genet 20:475–482

Dill CL, Wise RP, Schnable PS (1997) Rf8 and Rf* mediate unique T-urf13-transcript accumulation, revealing a conserved motif associated with RNA processing and restoration of fertility restoration in T-cytoplasm maize. Genetics 147:1367–1379

Duvick DN (1965) Cytoplasmic pollen sterility in corn. Adv Genet 13:1–56

Esau K (1977) Anatomy of Seed Plants.Wiley, New York

Escote LJ, Gabay-Laughnan SJ, Laughnan JR (1985) Cytoplasmic reversion to fertility in cms-S maize need not involve loss of linear mitochondrial plasmids. Plasmid 14:264–267

Estelle MA, Somerville CR (1987) Auxin-resistant mutants of *Arabidopsis* with an altered morphology. Mol Gen Genet 206:200–206

Flavell R (1974) A model for the mechanism of cytoplasmic male sterility in plants, with special reference to maize. Plant Sci Lett 3:259–263

Gabay-Laughnan S, Zabala G, Laughnan JR (1995) S-type cytoplasmic male sterility in maize. In: Levings CS III, Vasil IK (eds) The molecular biology of plant mitochondria. Kluwer, Dordrecht, pp 395–432

Goldberg R, Beals TP, Sanders PM (1993) Anther development: basic principles and practical applications. Plant Cell 5:1217–1229

Grelon M, Budar F, Bonhomme S, Pelletier G (1994) Ogura cytoplasmic male-sterility (CMS)-associated orf138 is translated into a mitochondrial membrane polypeptide in male-sterile *Brassica* cybrids. Mol Gen Genet 243:540–547

Grossmann LI, Shoubridge EA (1996) Mitochondrial genetics and human disease. Bioessays 18:983–991

Gualberto JM, Lamattina L, Bonnard G, Weil JH, Grienenberger JM (1989) RNA editing in wheat mitochondria results in the conservation of protein sequences. Nature 341:660–662

Handa H, Nakajima K (1992) Different organization and altered transcription of the mitochondrial atp6 gene in the male-sterile cytoplasm of rapeseed (*Brassica napus* L.). Curr Genet 21:153–159

Handa H, Gualberto JM, Grienenberger JM (1995) Characterization of the mitochondrial orfB and its derivative orf224 a chimeric open reading frame specific to one mitochondrial genome of the Polima male-sterile cytoplasm in rapeseed (*Brassica napus* L.). Curr Genet 28:546–552

Hanson MR, Younr EG, Rothenberg M (1988) Sequence and expression of a fused mitochondrial gene, associated with *Petuni* cytoplasmic male sterility, compared with normal mitochondrial genes in fertile and sterile plants. Philos Trans R Soc Lond 319:199–208

Hanson MR, Folkerts O (1992) Structure and function of the higher plant mitochondrial genome. Int Rev Cytol 141:129–172

Hanson MR, Nivison HT, Conley MR (1995) Cytoplasmic male sterility in *Petunia*. In: Levings CS III, Vasil IK (eds) The Molecular Biology of Plant Mitochondria. Kluwer, Boston, pp 497–514

He S, Abad AR, Gelvin SB, Mackenzie S (1996) A cytoplasmic male sterility-associated mitochondrial protein causes pollen disruption in transgenic tobacco. Proc Natl Acad Sci USA 93:11763–11768

Hernould M, Shuharsono S, Litvak S, Araya A, Mouras A (1993) Male sterility induction in transgenic tobacco plants with an unedited *apt9* mitochondrial gene from wheat. Proc Natl Acad Sci USA 90:2370–2374

Hiesel R, Wissinger B, Schuster W, Brennicke A (1989) RNA editing in plant mitochondria. Science 246:1632–1634

Horn R, Kohler RH, Zetsche K (1991) A mitochondrial 16 kDa protein is associated with cytoplasmic male sterility in sunflower. Plant Mol Biol 17:29–36

Horn R, Hahn V, Friedt W (1994) Recombination: Effects on structure and function of the mitochondrial genome. Prog Bot 55:219–235

Horn R, Hustedt JEG, Horstmeyer A, Hahnen A, Zetsche K, Friedt W (1996) The CMS-associated 16-kDa protein encoded by orfH522 in the PET1 cytoplasm is also present in other male-sterile cytoplasms of sunflower. Plant Mol Biol 30:523–538

Houchins JP, Ginsburg H, Rohrbaugh M, Dale RMK, Schardl CL, Hodge TP, Lonsdale DM (1986) DNA sequence analysis of a 5.27-kiolobase repeat occurring adjacent to the regions of S-episome homology in maize mitochondria. EMBO J 5:2781–2788

Howad W, Kempken F (1997a) Sequence analysis and transcript processing of the mitochondrial *nad3–rps12* genes from *Sorghum bicolor*. Plant Sci 129:65–68

Howad W, Kempken F (1997b) Cell-type specific loss of *atp6* RNA editing in cytoplasmic male sterile *Sorghum bicolor*. Proc Natl Acad Sci USA 94:11090–11095

Hu J, Rutger JN (1992) Pollen characteristics and genetics of induced and spontaneous genetic male-sterile mutants in rice. Plant Breed 109:97–107

Hulskamp M, Parekh NS, Grini P, Schneitz K, Zimmermann I, Lolle SJ, Pruitt RE (1997) The STUD gene is required for male-specific cytokinesis after telophase II of meiosis in *Arabidopsis thaliana*. Dev Biol 187:114–124

Iwabuchi M, Kyozuka J, Shimamoto K (1993) Processing followed by complete editing of an altered mitochondrial atp6 RNA restores fertility of cytoplasmic male sterile rice. EMBO J 12:1437–1446

Jean M, Brown GG, Landry BS (1997) Genetic mapping of nuclear fertility restorer genes for the 'Polima' cytoplasmic male sterility in canola (*Brassica napus* L.) using DNA markers. Theor Appl Genet 95:321–328

Jia MH, He S, Vanhouten W, Mackenzie S (1997) Nuclear fertility restorer genes map to the same linkage group in cytoplasmic male-sterile bean. Theor Appl Genet 95:205–210

Johns C, Lu M, Lyznik A, Mackenzie S (1992) A mitochondrial DNA sequence is associated with abnormal pollen development in cytoplasmic male sterile bean plants. Plant Cell 4:435–449

Kamps TL, McCarthy DR, Chase CD (1996) Gametophytic genetics in *Zea mays* L. dominance of a restoration-of-fertility allele (Rf3) in diploid pollen. Genetics 142:1001–1007

Kaul MLH (1988) Male sterility in higher plants. Monographs on theoretical and applied genetics, vol 10. Springer, Berlin Heidelberg New York

Kennell JC, Pring DR (1989) Initiation and processing of atp6, T-urf13 and ORF221 transcripts from mitochondria of T cytoplasm maize. Mol Gen Genet 216:16–24

Köhler RH, Horn R, Lössl A, Zetsche K (1991) Cytoplasmic male sterility in sunflower is correlated with the co-transcription of a new open reading frame with the atpA gene. Mol Gen Genet 227:369–376

Koltunow AM, Truettner J, Cox KH, Wallroth M, Goldberg RB (1990) Different temporal and spatial gene expression patterns occur during anther development. Plant Cell 2:1201–1224

Kriete G, Niehaus K, Perlick AM, Puhler A, Broer I (1996) Male sterility in transgenic tobacco plants induced by tapetum-specific deacetylation of the externally applied non-toxic compound N-acetyl-L-phosphinothricin. Plant J 9:809–818

Krishnasamy S, Makaroff CA (1993) Characterization of the radish mitochondrial orfB locus: possible relationship with male sterility in Ogura radish. Curr Genet 24:156–163

Krishnasamy S, Makaroff CA (1994) Organ-specific reduction in the abundance of a mitochondrial protein accompanies fertility restoration in cytoplasmic male-sterile radish. Plant Mol Biol 26:935–946

Krishnasamy S, Grant RA, Makaroff CA (1994) Subunit 6 of the Fo-ATP synthase complex from cytoplasmic male-sterile radish: RNA editing and NH2-terminal protein sequencing. Plant Mol Biol 24:129–141

Kück U, Wricke G (1995) Genetic mechanisms for hybrid breeding. Advances in plant breeding 18. Berlin: Blackwell Wissenschafts-Verlag

L'Homme Y, Brown GG (1993) Organizational differences between cytoplasmic male-sterile and male-fertile *Brassica* mitochondrial genomes are confined to a single transposed locus. Nucleic Acids Res 21:1903–1909

L'Homme Y, Stahl RJ, Li X-Q Hameed A, Brown GG (1997) Brassica nap cytoplasmic male sterility is associated with expression of a mtDNA region containing a chimeric gene similar to the pol CMS-associated orf224 gene. Curr Genet 31:325–335

Landgren M, Zetterstrand M, Sundberg E, Glimelius K (1996) Alloplasmic male-sterility *Brassica* lines containing *B. tournefortii* mitochondria express an ORF 3 of the *atp6* gene and a 32 kDA protein. Off. Plant Mol Biol 32:879–890

Laughnan JR, Gabay-Laughnan S (1983) Cytoplasmic male sterility in maize. Annu Rev Genet 17:27–48

Laser B, Kück U (1995) The mitochondrial *atpA/atp9* co-transcript in wheat and triticale: RNA processing depends on the nuclear genotype. Curr Genet 29:50–57

Laver HK, Reynolds SJ, Monegar F, Leaver CJ (1991) Mitochondrial genome organization and expression associated with cytoplasmic male sterility in sunflower (Helianthus annuus). Plant J 1:185–193

Lee S-LJ, Warmke HE (1979) Organelle size and number in fertile and T-cytoplasmic male-sterile corn. Am J Bot 66:141–148

Levings CS III (1993) Thoughts on cytoplasmic male sterility in cms-T maize. The Plant Cell 5:1285–1290

Levings CS III (1996) Infertility treatment: A nuclear restorer gene in maize. Science 272:1279–1280

Levings CS III, Siedow JN (1992) Molecular basis of disease susceptibility in the Texas cytoplasm of maize. Plant Mol Biol 19:135–147

Levings CS III, Rhoads DM, Siedow JN (1995) Molecular interactions of *Bipolaris maydis* T-toxin and maize. Can J Bot 73:S483–S489

Mackenzie SA, Pring DR, Bassett M, Chase C (1988) Mitochondrial DNA rearrangement associated with fertility restoration and cytoplasmic reversion to fertility in cytoplasmic male sterile Phaseolus vulgaris L. Proc Natl Acad Sci USA 85:2714–2717

Mackenzie SA, Chase CD (1990) Fertility restoration is associated with loss of a portion of the mitochondrial genome in cytoplasmic male-sterile common bean. Plant Cell 2:905–912

Mackenzie SA (1991) Identification of a sterility-inducing cytoplasm in a fertile accession line of *Phaseolus vulgaris* L. Genetics 127:411–416

Maier RM, Zeltz P, Kössel H, Bonnard G, Gualberto JM, Grienenberger JM (1996) RNA editing in plant mitochondria and chloroplasts. Plant Mol Biol 32:343–365

Makaroff CA (1995) Cytoplasmic male sterility in *Brassica* species. In: Levings CS III, Vasil IK (eds) The molecular biology of plant mitochondria. Kluwer, Boston, pp 515–555

Mariani C, De Beuckeleer M, Truettner J, Leemans J, Goldberg RB (1990) Induction of male sterility in plants by a chimaeric ribonuclease gene. Nature 347:737–741

Mariani C, Gossele V, De Beuckeleer M, De Block M, Goldberg RB, De Greef W, Leemans J (1992) A chimaeric ribonuclease-inhibitor gene restores fertility to male-sterile plants. Nature 357:384–387

Moneger F, Smart CJ, Leaver CJ (1994) Nuclear restoration of cytoplasmic male sterility in sunflower is associated with the tissue-specific regulation of a novel mitochondrial gene. EMBO J 13:8–17

Nivison HT, Hanson MR (1989) Identification of a mitochondrial protein associated with cytoplasmic male sterility in *Petunia*. Plant Cell 1:1121–1130

Nivison HT, Sutton CA, Wilson RK, Hanson MR (1994) Sequencing, processing, and localization of the petunia CMS-associated mitochondrial protein. Plant J 5:613–623

Nowak C, Kück U (1990) RNA editing of the mitochondrial atp9 transcript from wheat. Nucleic Acids Res 18:7164

Op den Camp RGL, Kuhlemeier C (1997) Aldehyde dehydrogenase in tobacco pollen. Plant Mol Biol 35:355–365

Preuss D, Lemieux B, Yen G, Davis RW (1993) A conditional sterile mutation eliminates surface components from *Arabidopsis* pollen and disrupts cell signaling during fertilization. Genes Dev 7:974–985

Preuss D, Rhee SY, Davis RW (1994) Tetrad analysis possible in *Arabidopsis* with mutation of the QUARTET (QRT) genes. Science 264:1458–1460

Pring DR, Lonsdale DM (1989) Cytoplasmic male sterility and maternal inheritance of disease susceptibility in maize. Annu Rev Phytopathol 27:483–502

Pring DR, Gengbach BG, Wise RP (1988) Recombination is associated with polymorphism of the mitochondrial genomes of maize and sorghum. Phil Trans R Soc Lond [Biol] 319:187–198

Pring DR Brennicke A, Schuster W (1993) RNA editing gives a new meaning to the genetic information in mitochondria and chloroplasts. Plant Mol Biol 21:1163–1170

Pring DR, Tang HV, Schertz KF (1995) Cytoplasmic male sterility and organelle DNAs of sorghum. In: Levings CS III, Vasil IK (eds) The molecular biology of plant mitochondria. Kluwer, pp 461–495

Pruitt KD, Hanson MR (1991) Transcription of the *Petunia* mitochondrial CMS-associated *pcf* locus in male sterile and fertility-restored lines. Mol Gen Genet 227:348–355

Rasmussen J, Hanson MR (1989) A NADH dehydrogenase subunit gene is co-transcribed with the abnormal petunia mitochondrial gene associated with cytoplasmic male sterility. Mol Gen Genet 215:332–336

Rathburn HB, Hedgcoth C (1991) A chimeric open reading frame in the 5' flanking region of cox1 mitochondrial DNA from cytoplasmic male-sterile wheat. Plant Mol Biol 16:909–912

Rhoads DM, Levings CS III, Siedow JN (1995) URF13, a ligand-gated, pore-forming receptor for T-toxin in the inner mitochondrial membrane of cms-T mitochondria. J Bioenerg Biomembr 27:437–445

Riechmann JL, Meyerowitz EM (1997) MADS domain proteins in plant development. Biol Chem 378:1079–1101

Rottman WH, Brears T, Hodge TP, Lonsdale DM (1987) A mitochondrial gene is lost via homologous recombination during reversion of CMS T maize to fertility. EMBO J 6:1541–1546

Schardl CL, Lonsdale DM, Pring DR, Rose KR (1984) Linearization of maize mitochondrial chromosomes by recombination with linear episomes. Nature 310:292–296

Schardl CL, Pring DR, Fauron CM-R, Lonsdale DM (1985) Mitochondrial DNA rearrangements resulting in fertile revertants of S-type male sterile maize. Cell 43:361–368

Schnable PS, Wise RP (1994) Recovery of heritable, transposon-induced, mutant alleles of the *rf 2* nuclear restorer of T-cytoplasm maize. Genetics 136:1171–1185

Schwarz-Sommer Z, Hue I, Huijser P, Flor PJ, Hansen R, Tetens F, Lönnig WE, Saedler H, Sommer H (1992) Characterization of the *Antirrhinum* floral homeotic MADS-box gene *deficiens*: evidence for DNA binding and autoregulation of its persistent expression throughout flower development. EMBO J 11:251–263

Siedow JN, Rhoads DM, Ward GC, Levings CS III (1995) The relationship between the mitochondrial gene T-*urf13* and fungal pathotoxin sensitivity in maize. Biochim Biophys Acta 1271:235–240

Singh M, Brown GG (1991) Suppression of cytoplasmic male sterility by nuclear genes alters expression of a novel mitochondrial gene region. The Plant Cell 3:1349–1362

Singh M, Brown GG (1993) Characterization of expression of a mitochondrial gene region associated with the *Brassica* Polima CMS: developmental influences. Curr Genet 24:316–322

Singh M, Hamel N, Menassa R, Li X-Q, Young B, Jean M, Landry B, Brown GG (1996) Nuclear genes associated with a single *Brassica* CMS restorer locus influence transcripts of three different mitochondrial gene regions. Genetics 143:505–516

Small ID, Earle ED, Escote-Carlson LJ, Gabay-Laughnan JR, Leaver CJ (1988) A comparison of cytoplasmic revertants to fertility from different CMS-S maize sources. Theor Appl Genet 76:609–618

Smart CJ, Moneger F, Leaver CJ (1994) Cell-specific regulation of gene expression in mitochondria during anther development in sunflower. Plant Cell 6:811–825

Song J, Hedgcoth C (1994a) Influence of nuclear background and transcription on a chimeric gene (orf256) and cox1 in fertile and cytoplasmic male sterile wheats. Genome 37:203–209

Song J, Hedgcoth C (1994b) A chimeric gene (orf256) is expressed as protein only in cytoplasmic male-sterile lines of wheat. Plant Mol Biol 26:535–539

Spena A, Estruch JJ, Prensen E, Nacken W, Van Onckelen H, Sommer H (1992) Antherspecific expression of the *rolB* gene of *Agrobacterium rhizogenes* increases IAA content in anthers and alters anther development in whole flower growth. R Appl Genet 84:520–527

Spielman M, Preuss D, Li FL, Browne W, Scott R, Dickinson H (1997) TETRASPORE is required for male meiotic cytokinesis in *Arabidopsis thaliana*. Development Suppl 124:2645–2657

Stahl R, Sun S, L'Homme Y, Ketela T, Brown GG (1994) RNA editing of transcripts of a chimeric mitochondrial gene associated with cytoplasmic male-sterility in *Brassica*. Nucleic Acids Res 22:2109–2113

Tadege M, Kuhlemeier C (1997) Aerobic fermentation during tobacco pollen development. Plant Mol Biol 35:343–354

Tang HV, Pring DR, Muza RF, Yan B (1996a) Sorghum mitochondrial orf25 and a related chimeric configuration of a male-sterile cytoplasm. Curr Genet 29:265–274

Tang HV, Pring DR, Shaw LC, Salazar FA, Muza FR, Yan B, Schertz KF (1996b) Transcript processing internal to a mitochondrial open reading frame is correlated with fertility restoration in male-sterile sorghum. The Plant J 10:123–133

Theißen G, Saedler H (1995) MADS-box genes in plant ontogeny and phylogeny: Haeckel's biogenetic law revisited. Curr Opin Gene Dev 5:628–639

Tröbner W, Ramirez L, Motte P, Hue I, Huijser P, Lönnig WE, Saedler H, Sommer H, Schwarz-Sommer Z (1992) GLOBOSA: A homeotic gene which interacts with DEFICIENS in the control of *Antirrhinum* floral organogenesis. EMBO J 11:4693–4704

Tsuchiya T, Toriyama K, Yoshikawa M, Ejiri S, Hinata K (1995) Tapetum-specific expression of the gene for an endo-beta-1,3-glucanase causes male sterility in transgenic tobacco. Plant Cell Physiol 36:487–494

Van der Meer IM, Stam ME, van Tunen AJ, Mol JNM Stuitje AR (1992) Antisense inhibition of flavonoid biosynthesis in petunia anthers results in male sterility. Plant Cell 4:253–262

Ward GC (1995) The Texas male-sterile cytoplasm of maize. In: Levings CS III, Vasil IK (eds) The molecular biology of plant mitochondria. Kluwer, Dordrecht, pp 433–459

Warmke HE, Lee S-LJ (1978) Pollen abortion in T cytoplasmic male-sterile corn (*Zea mays*): a suggested mechanism. Science 200:561–563

Weigel D (1995) The genetic of flower development: From floral induction to ovule morphogenesis. Annu Rev Genet 29:19–39

Weigel D, Meyerowitz EM (1994) The ABCs of floral homeotic genes. Cell 78:203–209

Weijer J (1952) A catalogue of genetic maize types together with a maize bibliography. Bibl Genet 14:189–425

Wintz H, Chen HC, Sutton Ca, Conley CA, Cobb A, Ruth D, Hanson MR (1995) Expression of the CMS-associated *urfS* sequence in transgenic petunia and tobacco. Plant Mol Biol 28:83–92

Wise RP, Pring DR, Gengenbach BG (1987a) Mutation to male fertility and toxin intensitivity in T-cytoplasm maize is associated with a frameshift in a mitochondrial open reading frame. Proc Natl Acad Sci USA 84:2858–2862

Wise RP, Fliss AE Jr, Pring DR, Gengenbach GB (1987b) urf13-T of T cytoplasm maize mitochondria encodes a 13,000 kD polypeptide. Plant Mol Biol 9:121–126

Wise RP, Dill CL, Schnable PS (1996) *Mutator*-induced mutations of the *rf1* nuclear fertility restorer of T-cytoplasm maize alter the accumulation of T-*urf13* mitochondrial transcripts. Genetics 143:1383–1394

Witt U, Hansen S, Albaum M, Abel WO (1991) Molecular analyses of the CMS-inducing Polima cytoplasm in *Brassica napus* L. Curr Genet 19:323–327

Worrall D, Hird DL, Hodge R, Paul W, Draper J, Scott R (1992) Premature dissolution of the microsporocyte callose wall causes male sterility in transgenic tobacco. Plant Cell 4, 759–771

Xu G-W, Cui Y-X, Schertz KF, Hart GE (1995) Isolation of mitochondrial DNA sequences that distinguish male-sterility-inducing cytoplasms in *Sorghum bicolor* (L.) Moench. Theor Appl Genet 90:1180–1187

Xu H, Know RB, Taylor PE, Singh MB (1995) *Bcp1*, a gene required for male fertility in *Arabidopsis*. Proc Natl Acad Sci USA 92:2106–2110

Yan B, Pring DR (1997) Transcriptional initiation sites in sorghum mitochondrial DNA indicate conserved and variable features. Curr Genet 32:287–295

Young EG, Hanson MR (1987) A fused mitochondrial gene is associated with cytoplasmic male sterility is developmentally regulated. Cell 50:41–49
Zabala G, Gabay-Laughnan S, Laughnan JR (1997) The nuclear gene Rf3 affects the expression of the mitochondrial chimeric sequence R implicated in S-type male sterility in maize. Genetics 147:847–860
Zabaleta E, Mouras A, Hernould M, Araya S, Araya A (1996) Transgenic male-sterile plant induced by an unedited *atp9* gene is restored to fertility by inhibiting its expression with antisense RNA. Proc Natl Acad Sci USA 93:11259–11263

Dr. Frank Kempken
Lehrstuhl für Allgemeine Botanik
Ruhr-Universität Bochum
D-44780 Bochum, Germany

Dr. Daryl Pring
Crop Genetics and Environment Research Unit,
USDA-ARS
Department of Plant Pathology
1453 Fifield Hall
University of Florida
Gainesville, Florida 32611, USA

Edited by
K. Esser

Plant Breeding:
Genetic Mapping in Woody Crops

By Eva Zyprian

1 Introduction

The recent establishment of genetic maps in woody plants reviewed here will focus on the major fruit-producing horticultural perennials. These are citrus, apple, grape and others, whose products can be produced in high numbers, stored and shipped, and which have therefore acquired great economical importance in modern agriculture. In general, these species have been under traditional breeding for a long time. They have been chosen for genetic mapping in order to accelerate breeding progress.

A second group of woody plants currently studied comprises forest trees. Some of them are of high economic relevance or natural wood constituents threatened by diseases. A short summarizing description of the recent achievements on their genetic maps will be given at the end of this chapter.

The fruit-producing perennials share characteristics that make them recalcitrant to classical genetic analyses and hence they have been analytically neglected for a long time. Genetic maps for woody fruit crops have been published only recently, in contrast to the major herbaceous crops of the world, whose molecular maps have been elaborated since about 1986 (as compiled in Graner and Wenzel, 1992). The difficulties encountered in the woody species result from long generation cycles (long phases of juvenility), high heterozygosity levels (outbreeding systems) and inbreeding depression excluding the generation of homozygous material. Polyploidy or chilling requirement may cause additional problems (Weeden, 1994). High contents of phenolic compounds and polysaccharides in the tissues can hamper DNA extraction and analysis. These facts render analyses difficult and time-consuming. Fortunately, some of these species have nevertheless been judged to be worth investing the effort to construct genetic maps, hoping that the information revealed will also be useful for related, but economically not quite so important species, enabling their genetic analysis exploiting synteny relationships. The range of plants being studied with modern molecular techniques is rapidly expanding.

The availability of genetic maps is of enormous value to plant breeders, enabling them to select the most suitable parental types in crosses to obtain the best combination of genes in a shorter time frame. The maps will help to understand and better exploit gene action (additive, dominant, epistatic interaction of genes or pleiotropy) (Allen 1994) and are the prerequisite to gaining information about QTLs (quantitative trait loci). These are genetic loci that contribute to the control of a specific trait under polygenic determination. In this case, several genes affect the phenotype of one trait. The individual contributing genes may exert a strong or weak influence. In addition, environmental factors can affect the corresponding phenotype to a certain extent. Hence the traits under polygenic control do not follow a clear scheme of presence or absence, but show a continuous variation over many phenotypic levels. Agronomically important traits like yield, growth type or vigour, fruit size or seedlessness, but also resistance to diseases or to abiotic stresses are frequently controlled by such QTLs. Breeders have a great interest in their localization and analysis in order to learn about their individual contributory effects on a specific trait and their best composition. The resolution of the individual loci contributing to a complex characteristic allows them to be treated and analyzed individually as Mendelian factors rather than using quantitative genetics. This concept was first developed with a molecular marker (RFLP) map of tomato and the quantitative traits for fruit mass, soluble solid concentration and pH (Paterson et al. 1988). Moreover, the mapping of major QTLs carries the potential to develop selection schemes within breeding programs, where the phenotypic trait may be scored at a very early development stage employing molecular markers genetically linked to the QTL in question (marker-assisted selection, MAS). The advent of molecular marker techniques, especially those based on PCR (polymerase chain reaction) technology enabled the genetic mapping in woody plants, where classical genetic methods can be applied only with difficulty.

Genetic maps will allow the breeder to follow chromosomes or fragments thereof, or individual markers, from initial breeding to cultivar release and through generations of cultivar development. Within the traditional cultivars of a crop, the genotypes may be evaluated by going back through previous generations determining the allele combinations breeders have been consistently selecting for or against (Shoemaker et al. 1994). In addition to the exploitation for screening procedures in breeding schemes as by MAS, the knowledge of the location of monogenic factors and QTLs for important traits opens experimental strategies for their molecular analysis (by positional cloning and physical mapping) and understanding of the functions involved. This aspect is of increasing importance regarding the upcoming methods of gene transfer in the improvement of cultivars by modern breeding by biotechnology, especially in woody fruit species (e.g. Puite and Schaart 1996; Yao et

1996; Guitérrez et al. 1997; Pena et al. 1997; Oliviera et al. 1996 for review).

2 The Plants Under Consideration

Although there seems a plentitude of plants providing us with fruit products within the woody species, only some have already been analyzed, at least for a few varieties of the corresponding species. Regarding, e.g. the Rosaceae family in the temperate area (Table 1), there are many fruit (and some timber) producers. From all these it is **apple** and **peach** where the efforts to construct genetic maps were first justified and maps are published corresponding to the species with the highest world production.

Another important group is the genus *Citrus* (*Rutaceae*) including the oranges (*Citrus sinensis*, world production 1994: 59 x 10^6 t), mandarins (*C. deliciosa*), satsumas (*C. unshiu*) and tangerines (*C. tangerina*) clementines (mandarin hybrids), grapefruit (*Citrus* x *paradisi*), shaddocks (*Citrus maxima*), lemon (*Citrus limon*), citron (*Citrus medica*), lime (*Citrus aurantiifolia*), bitter orange (*Citrus aurantium*), *Citrus myrtifolia* and *Citrus bergamina*. A close relative is *Poncirus trifoliata*, frequently used as rootstock for varieties of the *Citrus* species and *Citrus*-interspecific hybrids (e.g. tangelos, chironjjas, citranges). Substantial mapping work has been conducted here.

Similarly, we find a number of blueberries in the genus **Vaccinium** (Ericaceae) including *Vaccinium myrtillus* (bilberry), *Vaccinium corymbosum* (highbush blueberry), *Vaccinium vitis-idaea* (cowberry), *Vaccinium oxycoccos* and *Vaccinium macrocarpon* (cranberry). The

Table 1. Rosaceae fruit producers and their world production. (Data from Franke 1997)

Rosaceae subfamiliy Pomoideae		World production
Apple	*Malus* x *domestica*	49 x 10^6 t
Pear	*Pyrus communis*	11.2 x 10^6 t
Quince	*Cydonia oblonga*	–
Rosaceae subfamily Prunoideae		
Peach	*Prunus persica*	11 x 10^6 t
Plum	*Prunus domestica*	7.3 x 10^6 t
Apricot	*Prunus armeniaca*	2.4 x 10^6 t
Almond	*Prunus dulcis*	1.3 x 10^6 t
Sweet Cherry	*Prunus avium*	1.2 x 10^6 t
Sour Cherry	*Prunus cerasus*	1 x 10^6 t
–	*Prunus salicina*	–
–	*Prunus serotina*	–

tetraploid highbush blueberries and the cranberries are cultured plants in the US. Diploid wild relatives of **blueberries** have been chosen for genetic analyses.

The **grapevine** (*Vitis* sp. L., Vitaceae) also is a plant currently under genetic mapping. Grapes have been in culture for thousands of years. Viticulture started in the Neolithic era (6000–5000 *B.C.*) in Transcaucasia, from where it spread (Mullins et al. 1992). Grape-growing areas exist nowadays all around the world (Europe, North America, South America, Australasia), employing a variety of scion- and rootstock cultivars (estimated total number about 9000; E. Dettweiler, pers. comm.), in each case selected for and adapted to the local climatic conditions.

The total world production of grape berries is estimated at 60 x 10^6 t (plus 11 x 10^6 t of raisins). Besides the production of wine, other products like raisins, juice, table fruits, liquors etc. are obtained. Most vineyards are planted with vegetatively propagated traditional material (*Vitis vinifera* usually as scion) containing subtle gene combinations. However, the need for chemical protection from (mainly fungal) diseases raises a problem concerning the pollution of the environment with toxic compounds (Mullins et al. 1992). Breeding for disease resistance is thus one of the modern tasks. The desire to accelerate breeding for improved grapevine varieties resulted in the effort to construct a first molecular map for two grapevine varieties (Lodhi et al. 1995). Mapping in grapevine is currently being continued for other varieties.

3 Mapping and Molecular Markers

Traditional Mendelian genetics follows the observation of phenotypes through generations of crosses (F_2) or backcrosses (BC), where segregating markers can be followed. For mapping purposes backcrosses of a heterozygous F_1 to one recurrent (homozygous) parent (Aa x aa) are usually designed as testcross permitting the observation of segregation, linkage and recombination of markers from the heterozygous plant. This analysis reveals the inheritance of a marker corresponding to a genetic locus specifying a certain trait. If a multitude of loci is known, they can be localized to chromosomes aggregating those that segregate in common ("linked", recombination frequency < 50%) rather than independently (recombination frequency = 50%, as the segregation of independent factors will be 1:1).

The loci on one chromosome constitute a linkage group. The distance between two linked loci can be estimated from the amount of recombination (the recombination fraction) between them, caused by unlinking through meiotic crossovers in the heterozygous state. The basic logic of mapping in the first approach is quite simple: the further apart two loci are on one chromosome, the higher the chance that they will be separated by a crossover event in meiosis; the closer the loci are located to each other, the smaller their chance of separation. If at least three loci are analyzed (three-factor cross), the relative order of the three markers can be deduced.

However, this simple logic is complicated in reality by the occurrence of multiple crossovers (double crossovers in three marker crosses obscuring the recombination events), by interference (one crossover in a genomic region affecting the frequency of crossovers in neighbouring regions) and by the fact that the recombination frequency is not constant thoughout the physical length of an individual chromosome. The problem of multiple crossovers can be dealt with by including accordingly modified statistic functions in the mapping calculation programs (e.g. Kosambi 1994), while the problem of uneven recombination frequency can only be resolved by analyzing a high number of loci along a chromosome (high density mapping) and finally determining the real physical distances by physical mapping and molecular analyses (cloning, restriction analysis and sequencing). Mapping by recombination analysis can provide only rough estimates of distances as it is a functional mapping. However, this technique is extremely useful for research and breeding (Kaudewitz 1992).

Recombination analysis therefore requires knowledge about the segregation of a multitude of individual genetic loci scoreable as phenotypes. Especially in the woody crops, it is very difficult to obtain this data by traditional observation. Isozymes have been frequently employed as biochemical markers but are not numerous enough for detailed mapping. The follow up of a morphological trait like fruit colour, for example, will require waiting for the first fruit set after a controlled cross, which may last several years due to the long juvenile stage. The scoring of the agronomically important traits that represent quantitative characteristics requires even long-term observations under field conditions over many years, and at several places, in case environmental effects are also involved. For this reason, the use of molecular markers has been of rapid development through the past years in the analysis of crops. Their application has enabled the genetic analysis of perennials since about 1990.

Molecular markers exist in several variants depending on the experimental way in which they are produced. However, all of them represent a very small piece of the whole genome and can be followed as individual Mendelian factors in testcrosses. They provide the basis for genetic mapping studies by recombination analysis. Furthermore, they can be used in breeding selection schemes as "indicators" for a characteristic trait, if they are genetically tightly linked to this desired trait (MAS, see above). All molecular markers are based on the analysis of differences in DNA sequences and have the intrinsic advantage of being independent of the developmental stage, physiological conditions or environmental influences. In addition, molecular markers can be obtained in practically unlimited numbers and therefore meet the requirement for genetic mapping.

The currently most used types of molecular markers are compiled here.

RFLP (restriction fragment length polymorphisms) markers (Botstein et al. 1980) have proven very useful for the establishment of molecular maps. Individual RFLP markers are inherited as codominant markers. However, this technique has two drawbacks: first, even single- or low-copy probes obtained from expressed genes can sometimes give rather little information and detect less polymorphisms than expected, as the plants show little variation within coding genes that are under selective pressure for functionality (Cai et al. 1994). Genomic regions that (in the majority) do not encode genetic functions that are expressed will tolerate more mutations/alterations detectable as polymorphisms. Second, the RFLP procedure requires the preparation of substantial amounts of genomic DNA (5–10 µg/reaction) for the hybridization procedure and the rather expensive (radioactive or non-radioactive) labeling of a high number of probes. Both drawbacks are resolved by more recent techniques relying on PCR technology.

The **PCR reaction** (Saiki et al. 1988) was developed in numerous applications including molecular marker generation for mapping of plants. Currently there are three major variants in use:

RAPD (random amplified polymorphic DNA) markers are obtained by offering single primers of arbitrary sequence in amplification reactions on genomic DNA. No prior sequence information is necessary. Two genotypes differing in RAPD markers will have different DNA sequences at the primer annealing sites or insertions/deletions between two primer binding sites flanking an amplicon. RAPD markers are inherited as dominant markers and can be used to construct genetic maps in many species (Welsh and McClelland 1990; Williams et al. 1990, 1993; Tingey and del Tufo 1993). They are simply scored as absence or presence of the band of a particular size produced from a certain primer.

RAPD-PCR has been widely used in molecular mapping, as it requires only small amounts of DNA (about 20 ng/reaction), is quickly performed and the products are scored on simple agarose gels. The frequency of polymophisms detected is satisfactory (Tingey and del Tufo 1993). The availability of different primers is practically unlimited. Arbitrary dekamer oligos with GC contents of 50% and higher are commonly used. Recent refinement of the method indicated that also longer arbitrary primers (17- to 24-mers) can be successfully employed to reveal polymorphisms (Ye et al. 1996). Nevertheless, RAPD markers exhibit sensitivity to experimental parameters and hence the conditions for RAPD analysis have to be thoroughly adjusted (Büscher et al. 1993). The ease with which RAPD markers can be obtained, however, largely outweighs these drawbacks.

Cloning and sequencing of informative RAPD bands will allow the design of longer, specific primers amplifying the marker of interest with improved experimental stability (**SCAR** markers, sequence-characterized amplified region) (Paran and Michelmore 1993).

STMS (sequence-tagged microsatellite markers, also called **SSR** for simple sequence repeat) are PCR-based markers that enjoy wide use in molecular mapping. However, for this application, substantial prior work is necessary. Clones of genomic DNA containing microsatellite repeats must be generated and sequenced. Microsatellites are DNA stretches of di- or trinucleotide repeats scattered in the genomes. They exhibit an extremely high level of length polymorphisms. This polymorphisms can be detected by amplifying the repeats from specific primer pairs flanking them in neighbouring conserved regions. The polymorphic products differ by few or more bases and are resolved with sequencing qualtiy separation techniques. STMS proved very useful, e.g. for the differentiation of cultivars that show low levels of polymorphisms with other marker techniques and are now employed in mapping studies. Their mode of inheritance is generally codominant, rendering them excellent analytical tools (Cregan et al. 1994).

AFLP (amplified fragment length polymorphisms) markers are based on the selective amplification of subsets of restriction fragments obtained by cutting the genomic DNA with two different enzymes (a frequently cutting and a less frequently cutting one, producing different fragment ends, at least in a part of the mixture of fragments). The restriction fragments are ligated to short adaptors (of known sequence) and then amplified in two steps with primers annealing to the adaptor sequence plus one, two or three nucleotides specified by the primer (+N, +NN or +NNN) that need to be present as complementary bases in the adjacent genomic DNA (Zabeau and Vos 1992). In this way, specific subsets of the restriction fragments are amplified and resolved by gel electrophoresis (in sequencing quality). The clear and multiple bands (50 to 100 obtained per sample) may even be scored as homo- or heterozygous due to different band intensity. This system has a high resolving power and yields codominant (and dominant) markers. AFLP data have not yet been published for the plants described in the following sections. However, it is such a powerful technique, that applications are to be expected very soon, also for fruit crops.

Molecular marker types and mapping strategies have been reviewed (e.g. Jones et al. 1997). Further variations of marker-generating methods exist (nicely compiled in Staub et al. 1996) and automation of analysis is an important aspect for the future.

In order to calculate genetic distances, the segregation data over a test population will be transferred into a digital (0/1) matrix. The observed segregation has to be tested for "goodness of fit" to the expected ratios. This is usually accomplished by chi-square calculations. For the further processing of data several computer programs are available. They calculate the distances between loci in a linkage group from recombination fractions. Typically, two-point analyses are the first step to give an estimation about linked markers. If linkage groups with more than two markers are defined, they can be used to deduce the relative map order by three- and n-point analyses. One of the programs very often used for this purpose is MAPMAKER (e.g. MAPMAKER/Exp 3.0, Lander et al. 1987). It uses the method of maximum likelihood for the statistical calculations (Chakravarti et al. 1991). The critical thresholds are defined by LOD scores, representing the ratio of the likelihood for linkages versus the likelihood for non-linkage (the odds ratio). The LOD score is given as a decadic logarithm, meaning that an LOD score of three, e.g. will mean a 1000:1 ratio of the probability for linkage versus the probability of non-linkage (Risch 1992).

MAPMAKER, however, is only capable of using data sets from a single marker segregation type (1:1, 1:2:1 or 3:1). If mixed segregation types are observed during the analysis, other calculating programs are employed, e.g. LINKAGE.1 (Suiter et al. 1983), or JoinMap (Stam 1993).

Once a genetic map is constructed, the phenotypes governed by QTLs can be superimposed with their segregation patterns from the mapping population, and cosegregating molecular markers can be identified. For instance, the program MAPMAKER/QTL 1.1 (Paterson et al. 1988) will then perform an interval mapping looking for statistically significant associations between the phenotype and a genetic area defined by intervals between and at the map positions, where molecular markers have been localized (Lander and Botstein 1989). This approach was frequently used with success but can cause problems, if more than one QTL per linkage group is present. Therefore, some variations and refinements of the mapping test statistics exist. The interested reader is referred to e.g. Knapp et al. (1990), Zeng (1993), Zeng (1994), Jansen and Stam (1994), Churchill and Doerge (1994).

4 Mapping Strategies and Results

a) Mapping in Backcrosses or the F$_2$ Generation

α) *Citrus*

The genetic analysis of *Citrus* is a very elegant example to demonstrate the development of molecular marker maps. Its analysis started with two publications in 1992 (Durham et al. 1992; Jarrell et al. 1992), both showing first RFLP maps in this genus. The map was extended with RAPDs 2 years later (Cai et al. 1994) and was recently refined by the inclusion of STMS markers (Kijas et al. 1997). The study published in 1994 (Cai et al. 1994) already announced the localization of a cDNA clone with a cold-responsive gene on the map. Furthermore, some RAPD markers cosegregating with resistance to citrus tristeza virus were converted into SCAR markers (Deng et al. 1997).

Citrus is a diploid plant with a low chromosome number ($n = 9$) and a small genome ($1C = 0.62$ pg) but a high degree of apomixis is added to the usual problems encountered in woody perennials. Inbreeding is almost impossible, demanding mapping strategies starting from heterozygous material. However, interspecific and intergeneric hybridization is easily accomplished and used in *Citrus* breeding, as exemplified by *Citrus* x *Poncirus* hybrids. *Poncirus trifoliata* (trifoliate orange) is a non-edible, morphologically distinct relative of *Citrus* with agronomically important traits like cold hardiness, resistance to tristeza virus, *Phytophthora* root rot and the *Citrus* nematode (*Tylenchulus semipenetrans*). Several *Citrus* x *Poncirus* hybrids are therefore used as rootstocks for the cultured *Citrus* species and varieties. For these reasons, a linkage map was constructed from a *Citrus* x *Poncirus* F_2 population (Jarrell et al. 1992). The F_2 was generated by hybridizing a Sacaton citrumelo originating from *C. paradisi* x *P. trifoliata* with pollen from Troyer citrange, which is derived from *C. sinensis* x *P. trifoliata.*

Linkage analysis of the *Citrus* x *Poncirus* F_2 led to a first core map of 31 loci in 9 linkage groups. Seven additional 1:1 segregating markers could be joined to the core map. Finally, 38 out of 46 loci could be placed on the map in 10 linkage groups spanning in total 351 cM. Two loci were found duplicated, one on two different linkage groups and one duplicated within the same linkage group. The size range of the linkage groups was 101 to 3 cM (Jarrell et al. 1992).

This *Citrus* map was improved by the inclusion of STMS markers (Kijas et al. 1997). Fourteen different primer pairs flanking microsatellite repeats were developed from the DNA of a hybrid of rangpur lime (*Citrus* x *limonia* Osb.) and trifoliate orange [*Poncirus trifoliata.* (L.) Raf.]. These were checked for amplification of length-polymorphic products within the population of plants used before (Jarrell et al. 1992). Seven STMS loci could be assigned to the map, rendering the largest linkage group of *Citrus* 74 cM in length. Interestingly, the STMS loci are located at the termini of different linkage groups. This indicates their spreading widely thoughout the genome. It probably does not mean that STMS are non-randomly distributed at chromosome ends, as only few markers constitute the *Citrus* map as yet. Segregation analysis of the STMS indicated a substantial amount of distorted segregation, as observed with other markers before. Nevertheless, these markers proved very useful for genetic analysis, even in such a wide intergeneric cross.

Published at the same time as the Jarrell study, *Citrus* maps were constructed from another two backcross populations, one intergeneric backcross of *Citrus grandis* (L.) Osb. cv. Tong Dee x *Poncirus trifoliata* (L.) Raf. cv. Pomeroy using *C. grandis* as the recurrent female (C x P), and a second interspecific backcross obtained from *C. reticulata* Blanco cv. Clementine x *C.x paradisi* Macf. cv. Duncan crossed back to *C. reticulata* as the recurrent male parent (C x C) (Durham et al. 1992). In these two BC populations most isozyme and RFLP data followed the 1:1 segregation ratio as expected. In addition, 1:2:1 and 1:1:1:1 ratios for a minority of markers occurred due to some heterozygous loci in the recurrent parents.

Finally, 52 loci were arranged in 11 linkage groups in the C x P population with a total length of 553 cM. The distance between the markers ranged from 0 to 32.2 cM (average 13.5 cM). For the C x C population 32 loci were obtained in 8 linkage groups with a total map length of 314 cM and a marker distance range from 0 to 32.7 (average 11.6 cM).

Interestingly, in this study, the maps elaborated could be compared. Overall, they exhibit conservation of locus order indicating similar genome organization. This is supported by the fact that most *Citrus* species are interfertile and interspecific hybrids are usually fertile. However, some differences may account for the high ratio of distorted markers observed and the affected loci were different in the two populations.

This citrus map (Durham et al. 1992) was extended with RAPD markers.

These experiments yielded 373 polymorphic bands from 60 BC_1 plants of which 266 could be unequivocally scored for linkage analysis. The segregating bands were grouped into three types: 146 band of type aa x aA (band present in the F_1 but not in *C. grandis*), 48 bands of type Aa x Aa (band present in F_1 and in *C. grandis*) and 72 bands of type Aa x aa (band present in *C. grandis* but absent in the F_1). The 31 of the 48 bands of type Aa x Aa segregated 3:1 as was to be expected for a dominant marker; 128 out of 186 band of both other types conformed to Mendelian ratios with a segregation of 1:1. Thus again, the remaining bands (ca. 40%) exhibited distorted segregation. This phenomenon seems to be common to *Citrus* crosses and may be due to some very early selection. These markers were included in the linkage analysis and 77 of the 107 distorted loci could be placed on the map. A high percentage of them clustered on linkage group I and most of the others clustered within linkage groups II, III, IV and VIII. These areas correspond to regions with skewed RFLP markers in the earlier studies, supporting the hypothesis that loci with distorted segregation might be linked to genes exposed to direct selection.

The study showed that the recurrent parent *C. grandis* was highly heterozygous (120 out of 266 RAPD markers). Use of two-point and multiple-point analysis for linkage and recombination of the aa x Aa type markers assorted 122 of the 146 RAPDs into nine linkage groups at a LOD level > 3. Within the remaining 24 markers were another four linked couples. Integration with recorded RFLP and isozyme markers led to a map of eight linkage groups (by merging some of the formerly established smaller ones). The total length of this map is 1192 cM with a mean distance of 7.5 cM between the markers. Individual linkage groups ranged in length from 29.8 to 243.8 cM. This map is estimated to cover 70 to 80% of the *Citrus* genome.

This refined map was used to localize cDNA clones with genes responsive to cold acclimation as RFLP markers. Two loci were mapped to group IV. Both were found to be linked to each other with a recombination frequency of 13.6%. A third locus mapped to linkage group IX (Cai et al. 1994). Very recently, RAPD markers cosegregating with resistance to citrus tristeza virus (*Ctv*) from *Poncirus trifoliata* were identified (Gmitter et al. 1996). This resistance gene follows single dominant behaviour. Bulked segregant analysis (BSA) of resistant and susceptible pools from the cross used before (Durham et al. 1992; Cai et al. 1994) with additional 280 dekamer primers identified 12 more dominant polymorphic RAPDs cosegregating with CTV resistance (Deng et al. 1997).

The principle of this method (Michelmore et al. 1991) is to combine DNA from individuals representing the phenotypic extreme classes of a segregating population (bulked segregant) and then looking for markers unique to one or the other class. In a screening for disease resistance, e.g. DNA from the susceptible and resistant segregants would be pooled and analyzed for markers that are specific to either one pool. If such markers can be identified, they will be verified by checking their cosegregation with the analyzed trait over a large segregation population. The identification of markers in linkage with desired traits can be accelerated in this way, by randomizing the appearance of non-linked segregating markers in the background of the pool.

Seven of the correlating RAPD products could be successfully converted into SCAR markers. All seven loci were placed on the map of the *Ctv* region of *Poncirus trifoliata*. This region thus became densely covered with molecular markers (Deng et al. 1997).

β) *Prunus*

Of all members of the genus *Prunus*, peach (*Prunus persica*) is the best characterized. Peach is a diploid plant with $n = 8$ chromosomes and a very small genome (0.55 pg/2C, about twice the size of *Arabidopsis thaliana*). Although many morphological and isozyme markers are described, only little information on linkages was available until the construction of a genetic map in *P. persica* was initiated. The aim is not only to understand peach genetics, but also to provide information that will help to perform mapping in other *Prunus* species based on synteny relationships (Chaparro et al. 1994). In contrast to other woody crops discussed here, peach is a self-pollinating species with a low level of genetic variation. Segregating populations are therefore created by selfing F_1s of wide crosses.

The segregation of markers was followed in nine F_2 populations, mapping one of them (NC174RL x Pillar). This resulted in a linkage map of 15 groups covering 83 RAPDs, one isozyme and four morphological markers. The linkage groups span about 396 cM with an average density of 4.8 cM; 16 linkage groups are to be expected from the chromosome number, two for each chromosome representing each of the homologs. Homology could be established between six of the linkage groups (1a–1b, 2a–2b and 3a–3b linkage groups). An additional linkage group was deduced from another cross (Marsun x White Glory F_2 family).

Additionally to map construction, the authors here undertook an approach to identify markers in linkage with 4 isozyme/morphological markers. The red leaf locus is incompletey dominant and present as *Gr/Gr* in NC174RL and *gr/gr* (green leaf) in Pillar. Four markers in linkage to *Gr* and six linked to the *gr* allele were identified in bulked segregant analyses. Screening for markers linked to *Mdh1-1/Mdh1-1* versus *Mdh1-2/Mdh1-2* in Marsun x White Glory identified three markers linked to the *Mdh1-1* allele (Chaparro et al. 1994).

γ) *Vaccinium*

In the work published on blueberry, mapping has been the long-term goal to identify molecular markers linked to genes that control winter dormancy and cold hardiness (Rowland and Levi 1994).

Blueberries require chilling during the dormancy period for vegetative and floral budbreak in spring. The breeding programs are targeted towards the creation of highbush blueberry varieties (*Vaccinuim corymbosum* L.) with low chilling requirements. This character would render them suitable for culture in the southern US, because they would be early-ripening.

V. corymbosum is a tetraploid, and huge progeny numbers would be necessary to follow segregation ratios. Therefore the authors constructed test populations from crosses of interspecific hybrids with wild diploid species. An interspecific hybrid US388 from the *V. darrowi* clone Fla4B (evergreen, low-chilling lowbush) x *V. elliottii* clone Knight (decidous, moderate high-chilling highbush) was used. (The diploid *V. darrowi* has been extensively used as donor of the low-chilling requirement in blueberry breeding, making use of plants that occasionally have unreduced gametes).

Diploid blueberries are essential self-sterile (probably due to inbreeding depression) and it is difficult to obtain F_2 or backcross material. Thus, the interspecific F_1 hybrid US388 was crossed to another plant of the same species as one of the original parents, the *V. darrowi* US795. This design is equivalent to a testcross (backcross) and dominant heterozygous markers from the F_1 hybrid should segregate 1:1 in this arrangement. About 40 resulting progeny plants of this population were used to construct a first linkage map for blueberry with RAPD markers.

The linkage groups obtained ranged from 19 to 165 cM with a total map length of 954 cM. The distance between adjacent markers varied from 3 to 30 cM (average distance 16 cM). Although in this case the number of linkage groups was found to correlate with the number of chromosomes ($n = 12$), it seems likely, that due to the still limited data, not all chromosomes are defined, and that small linkage groups will be fused as soon as more markers are tested. However, these data on *Vaccinium* creating the first linkage map provide the basis for further investigation.

b) Mapping with the Double Pseudotestcross Stragetgy

As described in the Introduction, many species of fruit crops are highly heterozygous trees and this fact prohibits the efficient scoring of genetic markers in backcrosses or classical F_2 populations. However, this can be of advantage when a different mapping strategy is applied, the double pseudotestcross (Weeden 1994). In this strategy, two highly heterozygous individuals are crossed. It is assumed that in many cases a locus *Aa* present in a parental type I will be homozygous recessive (*aa*) in the second parent II and vice versa. In the cross I x II, the marker *A* (and many other heterozygous markers of parent I) will therefore segregate 1:1. The same holds for parent II with markers in the arrangement *Bb* crossed to *bb* in parent I. This cross will therefore yield a substantial amount of markers segregating 1:1 from parent I and 1:1 from parent II. This segregation corresponds formally to the situation in a classical test-cross (backcross with homozygous recessive type) and has been named pseudotestcross. As both parental types are homozygous recessive at a number of loci, it is a double pseudotestcross. Observing the segregation of the markers from parent I and parent II separately will result in two linkage maps, one for each parent. Homologous linkage groups between both parental types may be identified on the basis of dominant markers segregating 3:1 or codominant markers segregating 1:2:1, that are heterozygous in both parents. Markers homozygous in either parent will not segregate in this arrangement (and hence be useless for mapping by recombination analysis). This strategy works very well with dominant RAPD markers and codominant RFLPs or isozyme markers (Hemmat et al. 1994; Weeden 1994; Lodhi et al. 1995).

α) Apple

The double pseudotestcross strategy was successfully applied to construct a molecular linkage map in apple (Hemmat et al. 1994). Apple (*Malus* x *domestica* Borkh.) has a high chromosome number ($2n = 34$), is highly heterozygous due to outbreeding and self-incompatibility and represents one of the economically most important fruit trees.

Fifty six trees from the cross of Rome Beauty x White Angel were analyzed. Rome Beauty represents a common apple cultivar, while White Angel is a crabapple of unknown origin, presumed to carry *Malus seiboldii* in its parentage and to be heterozygous for a dominant gene encoding resistance to powdery mildew (*Podosphaera leucotricha*). Rome Beauty is highly susceptible to this pathogen. Isozymes, RFLPs and RAPD markers were employed and scored according to the double pseudotestcross for linkage analysis. After statistical verification of the segregation ratios and processing of the data, two linkage maps for either parent were constructed. After this basic alignment had been done, additional software was used to determine multipoint recombination frequencies and refine the map. The two maps are based on 409 markers in total, 34 of which were allozyme polymorphisms, 8 RFLPs and the remaining one RAPD markers; 268 heterozygous markers originated from White Angel and 180 from Rome Beauty; 39 markers seemed to be heterozygous in both parental types. RAPD markers appeared highly polymorpic and most useful in this analysis.

The map of White Angel contained 253 markers arranged in 24 linkage groups spanning 950 cM and 156 markers from Rome Beauty could be placed into 21 linkage groups. For 15 linkage groups of White Angel the homologs from Rome Beauty could be identified. RAPD markers produced from different primers were found clustered in some regions (several markers in a distance of about 10 cM) (Hemmat et al. 1994).

In this study only about 10% of the RAPDs appeared present in both parents. This fact is of advantage for the double pseudotestcross strategy but may cause problems once RAPDs in correlation with agronomically important traits (QTLs or monogenic factors) are identified. The transferability of such valuable markers from one cross to another may be limited. For this reason and for the better resolution of segregation/recombination data with codominant markers, the development of STMS markers was recently initiated (Guilford et al. 1997) and was applied to apple cultivar differentiation. Ten of the newly developed STMS were checked for segregation in a cross of Royal Gala x A172-2 and followed simple Mendelian inheritance. For a diploid species, no more than two alleles would be expected. However, 25% of the STMS resulted in more than two amplification products and may be biallelic. Such complex banding patterns are discussed to reflect ancestral chromosome duplication events or may demonstrate allopolyploidy. It is to be expected that these STMS markers will soon be integrated in the available apple molecular linkage map.

β) Grapevine

In grapevine, the urgent need to differentiate the many cultivars has led to the development of molecular markers (RFLPs, RAPDs and recently STMS) since about 1990. However, the first report on genetic mapping was published only in 1995 (Lodhi et al. 1995) and was performed with the double pseudotestcross strategy (Weeden 1994). Sixty plants from an interspecific hybrid population obtained from Cayuga White x Aurore were used for mapping based on RAPD markers. The parental types both represent complex hybrids (Cayuga White is a hybrid of *Vitis vinifera, V. labrusca* and *V. rupestris,* whereas Aurore is a hybrid of *V. vinifera, V rupestris* and *V. aestivalis*). The F_1 population segregates for disease resistance and many other traits. Primers producing at least two polymorphic bands were selected from a prescreening of parental types and used to follow segregation of the amplification products.

Previous isozyme and RFLP loci were also included in the study of segregation.

As a result of this first mapping in grape, there is a Cayuga White map of 214 markers over 1196 cM in 20 linkage groups and the map of Aurore with 225 markers arranged in 22 linkage groups, overall spanning 1477 cM. The number of linkage groups could be reduced to that corresponding to the chromosome number ($n = 19$), by the observation that one linkage group from one parent was homologous to more than one linkage group of the second parent. In total, ten of the linkage groups could be assigned to their homolog. The maps did not exhibit similar lengths in recombination units, due to an uneven saturation with markers. In both linkage maps, clustering of markers was observed in several cases. The recombination frequencies on comparison of both parental maps were not the same: Aurore had a higher recombination frequency in linkage group VII, while Cayuga White had a higher recombination frequency on linkage group IV. Hence medium density maps were constructed. From the information on the genome size of grapevine (4.75 to 5×10^6 bp, Lohdi et al. 1995) it can be calculated that 1 cM in grapevine approximates 300 kb, which is a distance suitable for further analytical steps like map-based cloning.

The question of transferability of RAPDs from one population to another was recently addressed (Lodhi et al. 1997). Hybridization studies with RAPD bands to genomic DNA from different grapevines and to RAPD amplification profiles of the corresponding primer were performed. The data indicate that RAPDs can originate from repetitive as well as single-copy sequences in the genome. In addition, evidence for internal priming of smaller RAPD products within larger RAPD amplification products (by the same dekamer) during reamplification was obtained.

In order to test transferability of RAPD-generated markers from one population to another, 140 RAPD bands produced from the parental types of the mapping population outlined above (Cayuga White x Aurore) were scored in the parental types of a second population derived from Horizon x Illinois 547-1. Horizon is of the same origin as Cayuga White (a sibling) and therefore both populations are genetically related. Illionois 547-1 has been derived from a cross between *V. rupestris* and *V. cinerea*; 100 of the bands were common to both populations and 47 were polymorphic in at least one population. Ten bands were polymorphic only in Aurore x Cayuga White and 22 in Horizon x Illinois 547-1. These data, in conjunction with results from genomic hybridizations, indicate that at least a portion of markers developed in a cross can be transferred, with caution, to a second cross (Lohdi et al. 1997). However, a similar study with genetically unrelated populations has to be undertaken to learn more about the wider use of such markers. The recent development of STMS markers in grapevine (Thomas et al. 1993a, 1994; Bowers et al. 1996) will also be useful to address this question.

c) Mapping in Forest Trees

A second group of woody plants under molecular mapping is represented by the forest trees. Some of them are of high economic relevance, as they are used for the industrial production of paper and related products from their wood and include fast-growing species such as *Eucalyptus* (Myrtaceae) (Grattapaglia et al. 1996; Verhaegen and Plomion 1996). They may also represent major constituents of the natural forests like the pines with their huge genomes (Kamm et al. 1996) or Norway spruce (*Picea abies*). Some of these natural timber producers are threatened by specific diseases where resistance gene mapping and analysis is the most important current analytical goal, as it is the case in sugar pine (*Pinus lambertiana* Dougl.; Devey et al. 1995) or poplar (*Populus* sp.; Salicaceae, Villar et al. 1996).

The recent achievements in forest trees are briefly summarized.

α) Angiosperms

In the **angiospermous** trees, the fast-growing *Eucalyptus* has deserved the most attention. A molecular map has been derived for *E. nitens* with a combination of RFLP, isozyme and RAPD markers (Byrne et al. 1995). *Eucalyptus grandis* and *E. urophylla* were mapped with segregating RAPDs (Verhaegen and Plomion 1996). Mapping in conjuction with the identification of QTLs affecting the vegetative propagation qualtities was done employing the pseudotestcross strategy in a cross of *E. grandis* x *E. urophylla* (Grattapaglia et al. 1995). Some QTLs affecting growth and wood quality were identified in *E. grandis* (Grattapaglia et al. 1996). Very recently, genomic regions involved in the expression of plant height and leaf area in seedlings of *E. nitens* could also be identified (Byrne et al. 1997).

The development of marker techniques has also begun in members of the Fagaceae family. In *Quercus* microsatellites were characterized and five STMS-tagged loci proved useful as segregating markers in a cross of *Q. petraea* x *Q. robur* (Steinkellner et al. 1997). In *Betula alleghaniensis* (yellow birch, a tree of economic relevance in Canada) RAPD markers generated by 11-mer oligonucleotide primers produced polymorphic, dominant markers that could be followed in intraspecific crosses (Roy et al. 1992). This is the first step towards mapping.

In the Salicaceae family the genus *Populus* represents a model plant for forest tree biology. This genus comprises species with rather fast sexual propagation (in the greenhouse), a small genome (2C/1.2pg and $2n = 38$), ease of transformation to transgenic plants, ease in obtaining fertile interspecific hybrids, and easy vegetative propagation (Bradshaw et al. 1994). This genus therefore represents an exception to all the other plants discussed here. The genetic heterogeneity of the species allows many polymorphic markers to be followed in F_2 generations of interspecific crosses. RFLP and RAPD markers were employed in conjunction with specific markers obtained from sequencing the ends of RFLP probes and designing longer, specific primer pairs for amplification (sequence-tagged sites, STS). Analysis resulted in a preliminary map of 35 linkage groups with a marker density sufficiently high to permit QTL analysis (Bradshaw et al. 1994). Further application of RAPD markers in *Populus* allowed tagging of a genomic region with markers linked to resistance to poplar rust (*Melampsora larici-populina*) (Villar et al. 1996).

β) Gymnosperms

The **gymnospermous** forest trees have also recently attracted much attention, as they are ecologically and economically important (e.g. timer production). Many studies have therefore been initiated in the Pinaceae family. The trees of this family exhibit huge genome sizes, large portions of repetitive DNA and a rather constant chromosome number. They have one advantage, however: in contrast to the angiosperms, the megagametophytic tissue is large. This tissue arises in the seed by mitotic divisions from a single meiotic division product (Nelson et al. 1993; Devey et al. 1995). Hence it is haploid and identical to the maternal genetic contribution to a seed. This haploid tissue suffices for marker analyses and is extremely useful for genetic analyses (avoiding the problems associated with dominance). Megagametophytes from a single tree can be directly used for analysis, if the tree is heterozygous for the loci under study.

Megagametophytes were thus used for the analysis of slash pine (*Pinus elliottii* Engelm. var. *elliottii*) with $n = 12$, RAPD markers yielding 13 linkage groups and 9 pairs of markers. Thus, a first partial linkage map was obtained for this species (Nelson et al. 1993). Slash pine was also used in a cross with longleaf pine (*Pinus palustris*) to construct parent-specific maps from the resulting F_1. In this study, 18 linkage groups and 3 marker pairs were identified in longleaf pine and 13 groups and 6 marker pairs in slash pine. The aim of this investigation is mainly to address the problem of delayed growth in longleaf pine caused by extended juvenile growth (Kubisiak et al. 1995). In *Pinus pinaster* (maritime pine) with the impressive genome size of 2.4 x 10^{10} bp and $n = 12$, megagametophytes from a single interracial F_1 hybrid were used to contruct 12 major linkage groups with RAPD markers. In this case it was shown that many of the RAPDs contained highly repetitive sequences (Plomion et al. 1995). In sugar pine (*Pinus lambertina*) markers are identified that are tightly linked to a single dominant gene for resistance to white pine blister rust (Devey et al. 1995). The development of microsatellite markers also started within the pines (Echt and May-Marquardt 1997). In *Pinus taeda*, a single dominant resistance gene conferring resistance to fusiform rust disease has been identified as a genomic region (Wilcox et al. 1996).

In the genus *Picea* there has been a study developing a first linkage map based on 72 megagametophytes from a single tree of *Picea abies* (Norway spruce) with RAPDs. The map contained 17 major linkage groups covering 165 markers and 3584 cM. This species has $n = 12$ chromosomes and a genome of about 4 x 10^6 bp/C. The mapping should provide the basis for ecological and population genetic studies on this important forest tree (Binelli and Bucci 1994).

5 Summary and Conclusions

This compilation of recently published work on fruit crops and forest trees shows that there has been a rapid "take off" in the construction of genetic maps based on linkage analysis of molecular markers, starting from scratch in about 1990. As such molecular analyses are costly, they have so far been applied only to a few species of high economic (or ecologic) relevance. However, costs will be reduced with progress in technology and upcoming automation efforts. Equally important, the basic knowledge obtained may facilitate analysis of other related fruit crops exploiting synteny relationships. It will bring tremendous advantages for modern plant breeding.

Over the past years, first genetic maps have been obtained for *Citrus* (and *Poncirus*), *Prunus* and *Vaccinium* by following the segregation of molecular markers in F_2 populations. For strongly heterozygous fruit crops like apply (*Malus*) and grape (*Vitis*), the double pseudotestcross strategy is employed. Molecular maps are currently available for two cultivars of each of these species. Regarding forest trees, *Eucalyptus*, *Quercus* and *Populus* within the angiospermous genera and *Pinus* and *Picea* within the gymnosperms have derserved attention. In the latter case of gymnosperms, the special advantage of the large haploid megagametophyte permits mapping from a single tree.

The molcular maps constructed and published until now are mainly based on RAPD markers but usually include some few RFLP and isozyme data. STMS markers are developing and SCAR markers are useful in specific cases, where some association between a (converted) RAPD marker and an agronomically important trait has been established. RAPDs carry the intrinsic disadvantages of a dominant marker system, that is largely outweighed by the ease in obtaining them. An application of RAPDs in the double pseudotestcross format is especially useful.

In many cases of the reports reviewed here, distorted segregation of markers (RFLPs and RAPDs) has been described, most clearly exemplified in the *Citrus* studies. Such segregation deviating from the expected Mendelian ratios is frequently hypothesized as indicating very early (pre- or postzygotic) selective pressure prohibiting the regular segregation of markers linked to the genes affected (e.g. Durham et al. 1992). The appearance of distortion may depend on the compatibility of the parental types and hence be influenced by the width of such a testcross. Nevertheless, markers with distortion were usually included in the mapping calculations and could mostly be placed on the maps.

Another phenomenon is clustering of markers in certain regions. In the application of RAPDs this might be explained by the fact that a substantial amount of RAPDs is derived from repetitive DNA or may show internal priming sites. In this case, a smaller and a larger amplicon generated with the same primer will map to the same location. However, clustering is a phenomenon common to all marker types and can only be resolved by increasing the marker numbers and using a variety of marker types.

Although all these analyses started only a few years ago, there are already results surpassing the basic map construction. In *Citrus* three loci with genes responsive to cold acclimation could be mapped (Cai et al. 1994). Markers cosegregating with resistance to citrus tristeza virus (*Ctv*) were identified in the *Poncirus* genome (Gmitter et al. 1996; Zeng et al. 1997). Similarly, molecular markers linked to morphological and isozymatic markers have been identified in *Prunus* (Chaparro et al. 1994). In grapevine, the crucial question of transferability of molecular markers from one population to another has been addressed (Lodhi et al. 1997) and lends support to optimistic assessments for future work.

Also in the forest trees, markers labeling segregation of a single resistance gene have been identified. The analysis here is mainly aimed at the development of markers linked to QTLs of complex characteristics, and the first maps provide the basis for these studies.

All this newly available knowledge will be used for further molecular analyses and exploited in plant breeding for agronomically important traits, e.g. the characterization of disease resistance genes. It will even serve in better management of precious forest trees. It is to be expected that molecular mapping studies will be initiated soon on more species, including the minor crops. Increased knowledge on resistance and complex desired characteristics will enable improved agriculture, horticulture, viticulture and forestry in the near future.

References

Allen FL (1994) Usefulness of plant genome mapping to plant breeding. In: Gresshoff PM (ed) Plant genome analysis. CRC Press, Boca Raton, pp 11–18

Becker H (1993) Pflanzenzüchtung. Verlag Eugen Ulmer, Stuttgart

Binelli G, Bucci G (1994) A genetic linkage map for *Picea abies* Krst., based on RAPD markers as a tool in population genetics. Theor Appl Genet 88:283–288

Botstein D, White RL, Skolnick MH, Davis RW (1980) Construction of a linkage map in man using restriction fragment length polymorphisms. Am J Hum Genet 32:314–331

Bowers JE, Dangl GS, Vignani R, Meredith CP (1996) Isolation and characterization of new polymorphic simple sequence repeat loci in grape (*Vitis vinifera* L.). Genome 39:628–633

Bradshaw HD, Villar M, Watson BD, Otto KG, Stewart S, Stettler RF (1994) Molecular genetics of growth and development in *Populus*. III. A genetic linkage map of a hybrid poplar composed of RFLP, STS and RAPD markers. Theor Appl Genet 89:167–178

Büscher N, Zyprian E, Blaich R (1993) Identification of grapevine cultivars by DNA analyses: pitfalls of random amplified polymorphic DNA techniques using 10mer primers. Vitis 32:187–188

Byrne M, Murrell JC, Allen B, Moran GF (1995) An integrated genetic linkage map for eucalypts using RFLP, RAPD and isozyme markers. Theor Appl Genet 91:869–875

Byrne M, Murrel JC, Owen JV, Kriedemann P, Williams ER, Moran GF (1997) Identification and mode of action of quantitative trait loci affecting seedling height and leaf area in *Eucalyptus nitens*. Theor Appl Genet 94:674–681

Cai Q, Guy CL, Moore GA (1994) Extension of the linkage map in *Citrus* using random amplified polymorphic DNA (RAPD) markers and RFLP mapping of cold-acclimation-responsive loci. Theor Appl Genet 89:606–614

Chakravarti A, Lasher LK, Reefer JE (1991) A maximum likelihood method for estimating genome length using genetic linkage data. Genetics 128:175–182

Chaparro JX, Wener DJ, O'Malley D, Sederoff RR (1994) Targeted mapping and linkage analysis of morphological, isozyme, and RAPD markers in peach. Theor App Genet 87:805–815

Churchill GA, Doerge RW (1994) Empirical threshold values for quantitative trait mapping. Genetics 139:963–971

Cregan PB, Akkaya MS, Bhagwat AA, Lavi U, Rongwen J (1994) Length polymorphisms of simple sequence repeat (SSR) DNA as molecular markers in plants. In: Gresshoff PM (ed) Plant genome analysis. CRC Press, Boca Raton, pp 47–56

Deng Z, Huang S, Xiao S, Gmitter G Jr (1997) Development and characterization of SCAR markers linked to the citrus tristeza virus resistance gene from *Poncirus trifoliata*. Genome 40:697–704

Devey ME, Delfino-Mix A, Kinloch B Jr, Neale DB (1995) Random amplified polymorphic DNA markers tightly linked to a gene for resistance to white pine blister rust in sugar pine. Proc Natl Acad Sci USA 92:2066–2070

Durham RE, Liou PC, Gmitter FG Jr, Moore GA (1992) Linkage of restriction fragment length polymorphisms and isozymes in *Citrus*. Theor Appl Genet 84:39–48

Echt CS, May-Marquardt P (1997) Survey of microsatellites in pine. Genome 40:9–17

Franke W (1997) Nutzpflanzenkunde, 6th ed. Georg Thieme Verlag, Stuttgart

Gmitter FG Jr, Xiao SY, Huang S, Hu XH, Garnsey SM, Deng Z (1996) A localized linkage map of the citrus tristeza virus resistance region. Theor Appl Genet 92:688–695

Graner A, Wenzel G (1992) Towards an understanding of the genome – new molecular markers increase the efficiency of plant breeding. Agro-Food-Industry Hi-Tech 3:18–23

Grattapaglia D, Bertolucci FK, Sederoff RR (1995) Genetic mapping of QTLs controlling vegetative propagation in *Eucalyptus grandis* and *E. urophylla* using a pseudotest cross strategy and RAPD markers. Theor Appl Genet 90:933–947

Grattapaglia D, Bertolucci FLG, Penchel R, Sederoff RR (1996) Genetic mapping of quantitative trait loci controlling growth and wood quality traits in *Eucalyptus grandis* using a maternal half sib family and RAPD markers. Genetics 144:1205–1214

Guilford P, Prakash S, Zhu JM, Rikkerink E, Gardiner S, Bassett H, Forster R (1997) Microsatellites in *Malus* x *domestica* (apple): abundance, polymorphism and cultivar identification. Theor Appl Genet 94:249–254

Guitérrez MA, Luth ED, Moore GA (1997) Factors affecting *Agrobacterium*-mediated transformation in *Citrus* and producing sour orange (*Citrus aurantium* L.) plants expressing the coat protein gene of the citrus tristeza virus. Plant Cell Rep 16:745–753

Hemmat M, Weeden NF, Manganaris AG, Lawson DN (1994) Molecular marker linkage map for apple. J Hered 85:4–11

Jansen RC, Stam P (1994) High resolution of quantitative traits into multiple loci via interval mapping. Genetics 136:1447–1455

Jarrell DC, Roose ML, Traugh SN, Kupper RS (1992) A genetic map of citrus based on the segregation of isozymes and RFLPs in an intergeneric cross. Theor Appl Genet 84:49–56

Jones N, Ougham H, Thomas H (1997) Markers and mapping: we are all geneticists now. New Phytol 137:165–177

Kamm A, Doudrick RL, Heslop-Harrison JS, Schnidt T (1996) The genomic and physical organization of *Ty1-copia*-like sequences as a component of large genomes in *Pinus elliottii* var. *elliottii* and other gymnosperms. Proc Natl Acad Sci USA 93:2708–2713

Kaudewitz F (1992) Genetik. Verlag Eugen Ulmer, Stuttgart

Kijas JMH, Thomas MR, Fowler JCS, Roose ML (1997) Integration of trinucleotide microsatellites into a linkage map of *Citrus*. Theor Appl Genet 94:701–706

Knapp SJ, Bridges WC Jr, Birkes D (1990) Mapping quantitative trait loci using molecular linkage maps. Theor Appl Genet 79:583–592

Kosambi DD (1944) The estimation of map distances from recombination values. Ann Eugenet 12:172–175

Kubisiak TL, Nelson CD, Nance WL, Stine M (1995) RAPD linkage mapping in a longleaf pine x slash pine F1 family. Theor Appl Genet 90:1119–1127

Lander ES, Botstein D (1989) Mapping Mendelian factors underlying quantitative traits using RFLP linkage maps. Genetics 121:185–199

Lander E, Green P, Abrahamson J, Barlow A, Daley M, Lincoln S, Newburg L (1987) MAPMAKER: an interactive computer package for constructing primary genetic linkage maps of experimental and natural populations. Genomics 1:174–181

Levi A, Rowland L, Hartung JS (1993) Production of reliable randomly amplified polymorphic DNA (RAPD) markers from DNA of woody plants. HortScience 28:1188–1190

Lodhi MA, Daly MJ, Ye G-N, Weeden NF, Reisch BI (1995) A molecular marker based linkage map of *Vitis*. Genome 38:786–794

Lodhi MA, Weeden NF, Reisch BI (1997) Characterization of RAPD markers in *Vitis*. Vitis 36:133–140

Michelmore RW, Paran I, Kesseli RV (1991) Identification of markers linked to disease-resistance genes by bulked segregant analysis: a rapid method to detect markers in specific genomic regions by using segregating populations. Proc Natl Acad Sci USA 88:9828–9832

Mullins MG, Bouquet A, Williams LE (1992) Biology of the grapevine. Cambridge University Press, Cambridge

Nelson CD, Nance WL, Doudrick RL (1993) A partial genetic linkage map of slash pine (*Pinus elliottii* Engelm. var. *eliottii*) based on random amplified polymorphic DNA. Theor Appl Genet 87:145–151

Oliviera MM, Miguel CM, Raquel MH (1996) Transformation studies in woody fruit species. Plant Tissue Cult Biotechnol 2(2):76–93

Paran I, Michelmore RW (1993) Development of reliable PCR-based markers linked to downy mildew resistance genes in lettuce. Theor Appl Genet 85:985–993

Paterson A, Lander E, Lincoln S, Hewitt J, Peterson S, Taksley S (1988) Resolution of quantitative traits into Mendelian factors using a complete RFLP linkage map. Nature 335:721–726

Pena L, Cervera M, Juárez J, Navarro A, Pina JA, Navarro L (1997) Genetic transformation of lime (*Citrus aurantifolia* Swing.): factors affecting transformation and regeneration. Plant Cell Rep 16:731–737

Plomion C, Bahrmann N, Durel C-E, O'Malley DM (1995) Genomic mapping in *Pinus pinaster* (maritime pine) using RAPD and protein markers. Heredity 74:661–668

Risch N (1992) Genetic linkage: interpreting Lod scores. Science 255:803–804

Rowland LJ, Lewi A (1994) RAPD-based genetic linkage map of blueberry derived from a cross between diploid species (*Vaccinium darrowi* and *V. elliottii*). Theor Appl Genet 87:863–868

Roy A, Fracaria N, MacKay J, Bousquet J (1992) Segregating random amplified polymorphic DNAs (RAPDs) in *Betula alleghamiensis*. Theor Appl Genet 85:173–180

Saiki RK, Gelfand DH, Stoffel S, Scharf SJ, Higuchi R, Horn GT, Mullis KB, Erlich HA (1988) Primer-directed enzymatic amplification of DNA with a thermostable DNA polymerase. Science 239:487–491

Shoemaker RC, Lorenzen LL, Diers BW, Olson TC (1994) Genome mapping and agriculture. In: Gresshoff PM (ed) Plant genome analysis. CRC Press, Boca Raton, pp 1–10

Southern EM (1975) Detection of specific sequences among DNA fragments separated by gel electrophoresis. J Mol Biol 98:503–517

Stam P (1993) Construction of integrated generic linkage maps by means of a new computer package: JoinMap. Plant J 3:739–744

Staub JE, Serquen F, Gupta M (1996) Genetic markers, map construction, and their application in plant breeding. HortScience 31:729–740

Steinkellner H, Fluch S, Turetschek E, Lexer C, Streiff R, Kremer A, Burg K, Gloessl J (1997) Identification and characterization of (GA-CT)n-microsatellite loci from *Quercus petraea*. Plant Mol Biol 33:1093–1096

Suiter KA, Wendel JF, Case JS (1983) LINKAGE-1: a PASCAL computer program for the detection and analysis of genetic linkage. J Hered 74:203–204

Thomas MR, Cain P, Matsumoto S, Scott NS (1993a) Microsatellite sequences in grapevine for mapping and fingerprinting. In: Hayashi T et al. (eds) Techniques on gene diagnosis and breeding in fruit trees. Fruit Tree Research station, Ministry of Agriculture, Forestry and Fisheries, Fujimoto, Tsukuba, Ibaraki, 305, Japan

Thomas MR, Matsumoto S, Cain P, Scott NS (1993b) Repetitive DNA of grapevine: classes present and sequences suitable for cultivar identification. Theor Appl Genet 86:173–180

Thomas MR, Cain P, Scott NS (1994) DNA typring of grapevines: A universal methodology and database for describing cultivars and evaluating genetic relatedness. Plant Mol Biol 25:939–949

Tingey SV, del Tufo JP (1993) Genetic analysis with random amplified polymorphic DNA markers. Plant Physiol 101:349–352

Verhaegen D, Plomion C (1996) Genetic mapping in *Eucalyptus urophylla* and *Eucalyptus grandis* using RAPD markers. Genome 39:1051–1061

Villar M, Lefèvre F, Bradshaw HD Jr, Teissier du Cros E (1996) Molecular genetics of rust resistance in poplars (*Melampsora larici-populina* Kleb/*Populus* sp.) by bulked segregant analysis in a 2 x 2 factoral mating design. Genetics 143:531–536

Weeden NF (1994) Approaches to mapping in horticultural crops. In: Gresshoff PM (ed) Plant genome analysis. CRC Press, Boca Raton, pp 57–68

Welsh J, McClelland M (1990) Fingerprinting genomes using PCR with arbitrary primers. Nucleic Acids Res 18:7213–7218

Wilcox PL, Amerson HV, Kuhlman EG, Liu BH, O'Malley DM, Sederoff RR (1996) Detection of a major gene for resistance to fusiform rust disease in loblolly pine by genomic mapping. Proc Natl Acad Sci USA 93:3859–3864

Williams JGK, Kubelik AR, Livak KJ, Rafalski JAS, Tingey SV (1990) DNA polymorphisms amplified by arbitrary primers are useful as genetic markers. Nucleic Acids Res 18:6531–6535

Williams JGK, Hanafey MK, Rafalski JA, Tingey SV (1993) Genetic analysis using random amplified polymorphic DNA markers. Meth Enzymol 218:704–740

Yao JL, Wu JH, Gleave AP, Morris BAm (1996) Transformation of citrus embryogenic cells using particle bombardment and production of transgenic embryos. Plant Sci 113:175–183

Ye GN, Hemmat M, Lohdi MA, Weeden NF, Reisch BI (1996) Long primers for RAPD-mapping and fingerprinting of grape and pear. BioTechniques 20:368–371

Zabeau M, Vos P (1992) Selective restriction fragment amplification: a general method for DNA fingerpringing. European patent application 92402629.7, No 0 534 858 A1

Zeng Z-B (1993) Theoretical basis for separation of multiple linked gene effects in mapping quantitative trait loci. Proc Natl Acad Sci USA 90:10972–10976

Zeng Z-B (1994) Precision mapping of quantiative trait loci. Genetics 136:1457–1468

Dr. rer. nat. habil. Eva Zyprian
Bundesanstalt für Züchtungsforschung
an Kulturpflanzen
Institut für Rebenzüchtung Geilweilerhof
D-76833 Siebeldingen, Germany

Edited by
K. Esser

Cell Biology and Physiology

Plant Water Relations

By Rainer Lösch

Since the last general overview on plant water relations in Prog Bot 56 (Lösch 1995), another nearly 4000 publications on this topic have appeared. Only a small fraction of them can be mentioned in the following. The present report does not cover aspects of water relations of poikilohydric plants and of stomatal patchiness (Prog Bot 59: Hartung et al. 1997; Beyschlag and Eckstein 1997), habitat water relations, and aspects of water relations during germination and growth. A new textbook in German on plant-water relations gives a comprehensive overview of the whole topic (Lösch 1998).

1 Cell and Tissue Water Status and Its Influence upon Metabolism

Water in cells and tissues is more or less strongly bound by hydration forces or for osmotic reasons. A new equation for calculating osmotic potential of a solution based on a molecular model of the water structure is proposed by Cochrane (1994). Michel and Radcliffe (1995) developed a computer programme that calculates osmotic potentials and related parameters in the range between 0 and 40 °C for solutions of manitol, PEG 8000, NaCl, KCl and sucrose. Osmotic water flow across porous, semipermeable membranes is treated from a point of view of irreversible thermodynamics by Comper (1994). An overview on the characteristics of osmotically active solutes is given by Niu et al. (1997), detailed studies on barley epidermis cell solutes and their effect on the local water potential relations have been made by Fricke et al. (1994, 1995; see also Fricke 1997). Oparka (1994) reviews what is known about plasmolysis against the background of modern insights into the interactions between the plasmamembrane and the adjacent apoplastic and symplastic compartments.

Up to more than one third of the tissue water content can be bound as hydration shells to membranes and macromolecules. Rascio et al. (1997) propose a relatively simple procedure to estimate the particular tissue affinity for strongly bound water. Overview articles by Chapman (1994) and Jendrasiak (1996) describe the interaction between water and biomembrane components. For lipid headgroups in membranes a progressive hydration induces a changed protrusion of choline dipoles relative to the plain of the membrane and increases their mobilities. The latter results from a shift in the lipid phase transition temperature (Ulrich and Watts 1994a, b). This again is influenced by the ratio of saturated to desaturated fatty acids. An acclimational shift of this ratio to a higher amount of saturated bonds brings about a sustained growth of potato cells under low water potential (Leone et al. 1996). An altered membrane hydration and/or a changed phospholipid packing density are accompanied by a changed light-scattering behavior of membranes (White et al. 1996). The changed energetic status due to the degree of hydration, bending, Van der Waals interactions, and structural irregularities when model membranes undergo a hexagonal-lamellar-hexagonal transition is analyzed by Kozlov et al. (1994). The reentrant transition, lamellar-hexagonal, appears to be driven by a delicate balance between the hydration energy in the lamellar and the bending energy in the hexagonal phase. The energy of voids in hexagonal interstices defines the relevant energy scale and temperature range. Hydrodynamics of interactions between monolayers in bilayers or under stacked membrane conditions are treated by Seifert and Langer (1994). Hydration repulsion decays exponentially when bilayers are less than ~ 7 Å apart (Gleeson et al. 1994). Soluble sugars reduce the increase in the gel-to-fluid phase transition temperature of palmitoyl-oleoyl-phosphatidyl-choline under dehydration (Koster et al. 1994). They keep the membrane structure undisturbed by their interaction with phosphatidyl headgroups of phospholipids (Hoekstra et al. 1997). The glycosylated hydroquinone arbutin also depresses the gel-to-liquid crystalline transition temperature, possibly by inhibition of phospholipase effects on membranes under advancing dehydration (Oliver et al. 1996). Higher amounts of cholesterol stabilize membranes composed of different phospholipid components regulating their motional freedom and the degrees of hydrophobicity (studies on the effects of high hydrostatic pressure on membranes by Bernsdorff et al. 1997).

Water crosses bilayers on account of the lateral dynamics of the phospholipids; large-volume water translocation is facilitated by water channels (Haines 1994). This family of membrane channel proteins is called aquaporins and is present in both the plasmalemma and the tonoplast membrane. Aquaporins can make up nearly one fifth of all the integral plasma membrane protein (Johansson et al. 1996). Many aquaporins are Hg- and Zn-sensitive (Daniels et al. 1996; Tazawa et al. 1996). Probably, a membrane is composed by aquaporine-rich domains and lipid domains (Henzler and Steudle 1995). The changed membrane permeabilities under steep osmotic gradients do not result from a changed domain distribution but from localized effects on membrane structure (Svetek et al. 1994). According to a model of Welling et al. (1996), the minimum effective pore radius of a water channel should be 1.78 Å, a distance that fits the aquaporin-1 channel.

Cell enlargement and growth of organs depend closely on cell respective tissue water relations that interact with cell wall enlargement. The dynamics of both osmohydraulic forces and structural development tend to keep a turgor homeostasis of enlarging cells. Pritchard (1994), in a Tansley review, describes these processes for the case of root cell expansion. Tomos and Pritchard (1994) distinguish two phases of cell expansion: (1) an acceleration phase in which the cell walls loosen, (2) a second decelerating phase when the walls become rigid. Enzymes that transiently disconnect the wall hemicellulose macromolecules (xyloglucan-endotransglycosylases) are most active during this period. The sustained turgor pressure under an osmoregulatory solute accumulation (Kutschera and Köhler 1994) enlarges the cell volume. In the decelerating phase, the walls stiffen so that even a nearly unchanged turgor does not bring about further cell enlargement (Kutschera 1996). Auxin increases elastic wall properties, and polysaccharide hydrolases and transferases are active during the first phase. These wall-extending enzymes are called expansins. They catalyze the slippage between cellulose microfibrils and the polysaccharide matrix of the wall (McQueen-Mason and Cosgrove 1994, 1995). Their molecular biology has been analyzed in detail by Kieliszewski and Lamport (1994). Expansins become activated by an increase in the apoplastic proton concentration (Mizuno and Katou 1996; Cho and Kende 1997). Auxin induces growth by a mechanism independent of this cell-wall acidification and therefore not by the expansin mechanism of rearrangement of covalent interfibrillar bonds (Kutschera 1994). It increases the elastic and the plastic wall extensibilities irrespective whether or not a turgor-driven cell enlargement occurs, but does not bring about a breakdown of cell-wall material or a significant change in the incorporation of amino acids into the cell-wall protein fraction and of glucose into the polysaccharides (Edelmann and Kutschera 1993; Edelmann and Köhler 1995). Extension growth is resumed therefore in an overproportial degree after periods of water stress-dependent low turgor on account of this "stored growth" (Serpe and Matthews 1994; Hohl and Schopfer 1995). Possibly, the operation point of expansins is particularly the stretch-resistant load-bearing cellulose/xyloglucan network, whereas the auxin-dependent growth mechanism affects particularly the compression-resistent pectic polysaccharide network. Passioura (1994) includes such molecular processes into an extended Lockhart model of cell expansion growth in order to interpret the variable nature of yield threshold and extensibility in the relevant equation. Under undisturbed growth conditions, both wall enlargement mechanisms should work synergistically. McCann and Roberts (1994), distinguishing these different structural networks, emphasize that the apoplastic continuum is, in fact, a mosaic of domains with heterogenous properties concerning wall extension growth and functions. Also at the integrational level of organs, growth occurs in a tissue-specific way. This results particularly from the (turgor-related) mechanical impedance put up by peripheral tissue in response to the volume increment of tissues in the center, e.g., of a shoot. This is the very background of the epidermal growth-control hypothesis of Kutschera (e.g., Kutschera 1995; see also Peters and Tomos 1996). Cell volume extension ceases during the deceleration phase (Tomos and Pritchard 1994). Water stress-dependent turgor reduction brings about a still reversible increase in wall stiffening (Hohl and Schopfer 1995), but probably entails wall rearrangements if cell maturation is approached. Intensive growth brings about a local decrease in water potential (Boyer 1993; Maruyama and Boyer 1994) and in this way – additionally to the active osmotic adjustment of the growing cells – strongly directs the water potential gradient to the growing plant parts.

Osmoregulation under water stress can lower water and osmotic potentials by several bars. This is enough in some cases to keep the turgor potential high, irrespective of a high water demand for transpiration (olive plants: Rieger 1995). More often, however, decreased transpiration rates under stress stabilize the plant water status more efficiently than osmotic adjustment (*Thuja occidentalis*: Edwards and Dixon 1995). The degree of osmotic adjustment is rather versatile, depending, e.g., on the rate of development of the water deficits (Basnayake et al. 1996), the population-specific adjustments of tissue elasticity (Chimenti and Hall 1994; Fan et al. 1994), and, most important, the genetic predisposition for an increased production of solutes under stress (e.g., Tschaplinski and Tuskan 1994; Basnayake et al. 1995). In tropical maize lines there are few, if any, links between the ability for efficient osmotic adjustment and yield characters (Guei and Wassom 1993). Solutes accumulated during osmoregulation comprise various carbohydrates (glucose in *Triticum*: Kameli and Losel 1994; sucrose and hexoses in *Festuca arundinacea*: Spollen and Nelson 1994; sorbitol in *Malus domestica*: Wang et al. 1995, 1996; mannitol and malate in *Fraxinus excelsior*: Peltier and Marigo 1996; Peltier et al. 1994). Also inorganic ions can contribute to osmoregulation (e.g., Rascio et al. 1994), but mostly to a lesser degree than organic osmolytes. The compatible solutes proline and quarternary ammonium compounds are increased considerably as an osmoregulatory response to water stress by upregulated gene expression (proline biosynthesis: Verbruggen et al. 1993; Savoure et al. 1997). The beneficial effect of an increased proline level relies not only on its function as a solute and as a protection molecule for dehydration-sensitive macromolecular surfaces, but also on its ability to reduce the generation of free radicals (Alia et al. 1993). Betaines and related quarternary ammonium compounds function similarly to proline as compatible solutes, particularly if long-term water stress (caused also by NaCl effects) prevails (Köhl 1996, 1997a, b). Also these compounds keep the steady-state metabolism functional, e.g., by strongly stabilizing the oxygen-evolving photosystem II complex (Papageorgiou and Murata 1995; Allakhverdiev et al. 1996). Glycinebetaine (GB) can be taken up from external applications (Makela et al. 1996). Such artificially supplied GB can, however, exert also undesired effects upon protein synthesis (e.g., downregulation of synthesis of the large Rubisco subunit) and proline accummulation (Gibon et al. 1997). Key enzymes of GB synthesis (genes encoding betaine aldehyde dehydrogenase: Wood et al. 1996) are not completely specific (Vojtechova et al. 1997), and biosynthetic pathways originating from the choline metabolism may end in different compatible solutes, provided that specific enzymes and substrates for intermediary metabolic steps are available (Rivoal and Hanson 1994b). Often, an osmoregulatory change in the amount of cell and tissue solutes is accompanied by changed wall elasticities. This also improves the plant resistance to water deficits (Neumann 1995). Fan et al. (1994) studied the relative contributions of both components to turgor maintenance in several tree species. The modulus of elasticity undergoes changes also during the ontogeny of a plant (Onwona-Agyeman et al. 1995).

Osmoregulation and osmoprotection can alleviate the effects of water and turgor loss to a certain degree. Nevertheless, very often water stress disturbes metabolic processes and even aggravates so drastically that it results in drought damage. Noninvasive methods to monitor the plant water status, are, among others, electrical resistance measurements of tissues and organs (Borchert 1994), in-situ NMR spectroscopy (Van As et al. 1994; Koeckenberger et al. 1997), and fluorescence measurements both at leaf level and remotely sensed by τ-LIDAR (Cerovic et al. 1996). Ontogenetic changes in plant morphology under drought comprise decreased shoot/root ratios, an increased length of very deep-reaching roots, and decreased specific leaf areas (Van Splunder et al. 1996: *Salix* spp.), sometimes an increased leaf thickness and degradation of chlorophyll (Bussotti et al. 1995: *Fagus silvatica*), and in *Triticum* flag leaves an increased stomatal frequency and a higher number of bulliform cells accompanied by a smaller amount of intermediate veins (Zagdanska and Kozdoj 1994).

Often, drought stress is more efficiently avoided than tolerated by plants (general systematic overviews on plant drought responses are given by Belhassen et al. 1995; Monneveux and Belhassen 1996). Certain peanut cultivars, for instance (Rucker et al. 1995), develop much more extended roots when soil dries out, in this way creating better survival conditions, (likewise population-specific differences of rooting depths are found in *Armeria* on sandy or salt marsh soils: Köhl 1996). Similarly, under drought, beech provenances from Sicily allocate more biomass into the roots than provenances from less drought-affected sites in Central Italy. Their leaf-specific conductivity and percent loss in hydraulic conductivity is smaller. The whole syndrome of developmental responses of beech to a reduced water supply results in a delay of detrimental drought effects (Tognetti et al. 1995). Particularly under extended, but mild to moderate water stress, it is more the whole-plant response that offsets the suboptimal conditions than osmoregulatory biochemical effects based on stress-responsive gene expressions at the cellular level (Blum 1996). Tardieu (1996) points out that the (at present very popular) efforts to select crop cultivars for better resistance against cell dehydration may not be the best strategy for improving drought resistance in crops. This holds true certainly with even more justification concerning attempts to improve drought tolerance by genetic engineering of plant water stress responses. Most often, this will be a rather awckward operation that affects the fine-tuned cascades of plant stress responses.

Cellular desiccation has influence on many metabolic processes, often by direct up or downregulation of gene expression (Bartels et al. 1996). Osborne and Boubriak (1994) review the mechanisms by which specialized cells keep their DNA macromolecules functional even under water shortage. Obviously, in seed embryos two DNA forms exist with respect to hydration amounts. They possess different conformational structures, base sequences, and specific binding proteins. Only DNA from desiccation-tolerant cells retains its functional integrity, and the changes between this and the desiccation-sensitive DNA form determine whether or not a certain seed maturation or germination stage shows poikilohydric or homoiohydric behavior. Seeds, seedlings, and adult plants under water stress produce by specific gene expression drought-stress proteins that are called dehydrins, in developing seeds often also LEA proteins – see Prog Bot 57, 21 – (e.g., Cheng et al. 1993; Finch-Savage et al. 1994; Wood and Goldsbrough 1997). Dehydrin production is tissue-specific and depends upon the particular ontogenetic stage of the plant (Han and Kermode 1996; Han et al. 1997). Different dehydration stresses (drought, cold, NaCl, also exogenous ABA supply) can induce the expression of different dehydrins. Drought hardening is often related to an intensified expression of dehydrins (Stanca et al. 1996). There are also great differences in the ability of provenances and crop cultivars to express such stress proteins (Ristic et al. 1996). Dehydrins stabilize macromolecular structures with detergent- or chaperone-like properties (Close 1996). There are reports on enzyme-like effects of drought-stress proteins, particularly if functioning as protein kinases (e.g., Lopez et al. 1994; Jonak et al. 1996). Some of these stress proteins seem to be extremely destruction-resistant (Pelah et al. 1995: boiling stability of a water-stress-responsive protein in aspen; species-specific expression of this protein in *Populus*: Pelah et al. 1997). Stress proteins may interact also in various ways with antioxidant enzymes that protect the cellular metabolism from a burst of reactive radicals under water shortage (Burdon 1993). In order to detect stress-responsive protein production, often hormonally deficient or responsive mutants (the latter involved in signal transduction pathways) are used experimentally (Vartanian 1996). Selection for higher desiccation tolerance of crop cultivars identifies quantitative trait loci, QTLs, as molecular markers of the genome that are linked with plant drought responses (Lebreton et al. 1995; Quarrie 1996; Agrama and Moussa 1996; Mian et al. 1996; Tuinstra et al. 1996).

A changed expression of structural and functional proteins controls degree and metabolic cascades of the photosynthetic carbon assimilation. Under moderate water stress, only the apoproteins of the light-harvesting complex II become decreased due to reduced gene transcription. Under severe water stress, both chlorophyll and apoproteins decrease (Hao et al. 1996). The photochemical efficiency of photosystem II (PS II) of *Quercus ilex* and *Q. pubescens* as expressed by fluorescence changes is seriously depressed when predawn leaf water potential is lower than -4 MPa (Methy et al. 1996), a situation only seldom experienced by adult trees on the growing sites, more often by seedlings with a much less developed root system. The amount of the key enzyme of the Calvin cycle, Rubisco, remains relatively unaffected as long as water stress is moderate and exerts its influence by preference on stomatal conductance for CO_2 (Dreesmann et al. 1994, see also Lal and Edwards 1996). Rubisco activity depression by severe water stress is mostly reversible, as is the case also with stress-dependent reductions of chlorophyll and protein contents (Castrillo and Trujillo (1994). Using plants with antisense Rubisco-DNA and measuring photosynthesis reduction, Rubisco and 1,5-bisphosphate (RuBP) levels under drought, Gunasekera and Berkowitz (1993) conclude that stress effects on an enzymatic step involved in RuBP regeneration must be responsible for a decreased photosynthesis. Within the sucrose production pathway an accumulation of fructose-2,6-bisphosphate occurs under water stress (Reddy 1996). A comparative study of activity depression and recuperation of the CO_2-fixing or accumulating enzymes carbonic anhydrase, PEP-carboxylase, and Rubisco, when leaf water potential becomes de- and increased was performed by Prakash and Rao (1996) – an increase in carbonic anhydrase activity as an assimilatory response to drought-increased ABA is reported by Popova et al. (1996). Leaf internal CO_2 concentration (c_i) under drought is influenced by the different sensitivities to water stress of the stomatal control of CO_2 uptake and the photosynthetic assimilation process in the chloroplast. Upon water shortage, c_i is first reduced for stomatal limitation of CO_2 uptake, than c_i increases due to nonstomatal inhibition of photosynthesis. The species-specific minimal value of the quotient c_i/c_a [c_a = leaf external CO_2 concentration] and the water potential level when this quotient is reached are related together (Brodribb 1996). Assimilates available for export and storage are strongly reduced under a more severe water stress (Marur et al. 1996). Rape photosynthetic adaptation to drought is accompanied by a changed spectral reflexion index that again depends on leaf area and plant nitrogen status (C. R. Jensen et al. 1996).

CAM is generally assumed to be a particular water-saving form of carbon assimilation. However, succulent C_3 and CAM plants with similar growth forms coexisting under the same environmental conditions may not always differ in their water use efficiency, according to comparative measurements by Eller and Ferrari (1997) with the Namibian *Cotyledon orbiculata* (CAM) and *Othonna opima* (C_3). Plants where CAM is inducible by NaCl or drought show modulative adaptations to the water stress situation at various integration levels (Lüttge 1993): (1) turgor and gas exchange regulation at the whole-plant level, (2) metabolic adjustment at the cellular level, (3) adapted transport proteins at the membrane and (4) macromolecular level and (5) altered gene expression patterns. Particularly facultative CAM plants (studies within the Crassulacean genus *Sedum*: Lee and Kim 1994; Conti and Smirnoff 1994; Castillo 1996) are very versatile in the drought responses of their carbon metabolism. Besides *Mesembryanthemum crystallinum*, the plant most renowned for its CAM induction ability (Lüttge 1993) and representative of the Mesembryanthemaceae, also within the Portulacaceae genera with inducible CAM exist (*Talinum* with a maximum of 24% diel gain during the night: Guerere et al. 1996). The shift from full CAM to CAM idling in *Xerosicyos danguyi* (Cucurbitaceae) occurs when water potential drops steeply from –0.3 to –1.2 MPa at approx. 50% relative water content. This is accompanied by a strong ABA increase, bringing about stomatal closure and a higher availability of PEP-carboxylase, the K_m of which is significantly decreased (Bastide et al. 1993).

Leaf mitochondrial respiration exhibits a different sensitivity to chronic and acute water stress, respectively (Collier and Cummins 1996). Under reduced water potentials the reduced demand for shoot growth energy, fueled by mitochondrial respiration, is compensated by an increased demand for maintenance energy (Collier and Cummins 1993). The latter is, in part, provided by the alternative respiration pathway. In roots, water stress stimulates transiently both respiration pathways.

An increased mobilization of nitrogen compounds occurs in tomato plants under water stress. It covers the nitrogen demand for a premature development of reproductive organs. Cytosolic glutamine synthetase gene expression is involved in this stress reponse (Bauer et al. 1997). Nitrate reductase is inhibited by water stress due to the action of the higher levels of free oxygen radicals that stimulate the hydrolysis of this protein (Kenis et al. 1994).

Elevated levels of active oxygen under drought stress result from an increase in one-electron transitions of oxygen due to excess energy gain not used up in the photosynthetic process. Within the intermediate metabolic sequences the oxygen radicals create other free radicals so that radical-scavenging molecules and enzymatic processes become overstrained, and a control of radical activities is lost under continued severe stress (Elstner and Osswald 1993; Smirnoff 1993). At the end, structural decompartmentalization brings about cell death. An important source of active oxygen is a disrupted mitochondrial electron transport chain with an enhanced leaching of oxygen radicals from mitochondria to the surroundings (Leprince et al. 1993). Lipid peroxidation increases severalfold under such conditions in sensitive species (Olsson et al. 1996). Activity of the glutathione system, which removes a lot of dangerous radicals, becomes strongly increased when a moderate water stress is prolonged (Sgherri and Navari-Izzo 1995; Zhang and Kirkham 1996). Cellular uptake of reduced glutathione is channeled by a specific transporter in the cell membrane that has a preference for oxidized glutathione (Jamai et al. 1996). In the poikilohydric *Boea hygroscopica*, glutathione protects particularly sulfhydryl groups of the thylakoid proteins so that they remain in the reduced form despite a higher radical density in the dehydration-affected cells (Navari-Izzo et al. 1997). Some of the isoenzyme forms of manganese-superoxide dismutase which is involved in enzymatic radical scavenging processes are expressed only under abscisic acid (ABA) influence or under osmotic treatment (e.g., Sod3.2, –3.3 and –3.4), Sod3.1-expression is independent from such an induction (Zhu and Scandalios 1994).

Current knowledge of ABA action and signalling is summarized by Giraudat et al. (1994) The importance of this phytohormone in root-shoot signalling of drought stress with the consequence of stomatal closure is beyond any doubts. ABA is required further in order to induce the gene expression of some stress-responsive proteins (Griffiths and Bray 1996). In roots, ABA concentration increases upon a drop in turgor. Its effect there prolongs the time when root growth under moderate water stress is still possible (Griffiths et al. 1996; Munns and Sharp 1993). Leaf and shoot growth, by contrast, cease then under the influence of ABA. Under such conditions, ABA accumulation in floral organs reaches half the concentrations found in vegetative plant parts, but reduces also here flower and fruit development (Westgate et al. 1996). When shoot turgor is kept (artificially) high, this ABA accumulation in floral organs and its developmental consequences do not happen, irrespective of a high shoot ABA supply from the root zone. Leaves metabolize xylem-delivered ABA rather quickly (half lifetimes about 1 h) and control by this way the duration of its effects on stomatal apertures and leaf growth (Jia et al. 1996). A markedly reduced turgor prolongs the lifetime of ABA in leaf tissues (Zhang et al. 1997a) as it does the increase in xylem pH (Jia and Zhang 1997). Under both conditions, reexportation of ABA from leaves to the shoot phloem is strongly reduced. Possibly, a rise in xylem sap pH, if this occurs under a decreased water potential, can by influencing the ABA turnover rate, control degree and duration of stomatal closure under low water potential, not only of the root but also the shoots and leaves.

2 Root Water Uptake and Water Movement Through the Plant

Root zone water reserves stem often from different sources and show a considerable degree of patchiness. The relative contribution of different soil layers to the plant's water supply were studied by analysis of stable isotope relations (Thorburn and Ehleringer 1995) or after application of radioactive tracers, e.g., by Thornburn and Walker (1994: utilization of stream water, groundwater or floodwater, respectively, by *Eucalyptus camaldulensis* populations), by Bishop and Dambrine (1995: exploitation of soil layers for water by Scots pine and Norway spruce root systems), by Dawson and Pate (1996), and Farrington et al. (1996; both: seasonal exploitation of soil water resources, of Australian phreatophytes in the first, of *Eucalyptus marginata* in the second paper). Foraging of roots for water and nutrients in a heterogenous soil is a well-known phenomenon. It has been described in detail in several studies. Only few of them may be cited for example purposes: palm root system in the Lamto savanna/Ivory coast (Mordelet et al. 1996), below-ground competition between shortgrass prairie *Bouteloua* tussocks (Hook and Lauenroth 1994), fine-root dynamics of a Michigan mixed hardwood forest (Pregitzer et al. 1993), of a *Eucalyptus globulus* plantation under different irrigation/fertilization treatments (Katterer et al. 1995), of corn (and other cereals) rooting patterns depending on soil penetration resistances, water and nutrient availability (Lipiec et al. 1993; Watt et al. 1996). Zwieniecki and Newton (1995) describe how elongating roots exploit rock fissures as small as 100 µm. By X-ray computer tomography, Grose et al. (1996) demonstrated that water contents around the roots of wheat, radish, and cotton were of large enough range to provide heterogenous sites in the rhizosphere which favor or restrict the existence there of different rhizosphere fungi.

Water diffuses from bulk soil to the root surface along concentration (respective water potential) gradients. Kage and Ehlers (1996) calculated that no transport limitation for field crop water uptake will occur if rooting densities are higher than 0.1 cm^{-3}. Root-soil air gaps may offer a high resistance barrier for the soil-root water passage in arid habitats. North and Nobel (1997) showed that this really is the case in field-grown *Agave deserti*. However, they emphazise that these gaps beneficially maintain a higher root water potential in the early stages of drought. Later, they prevent a root-soil water loss under a water potential gradient directed to the dry soil. Soil sheets around young roots intensify this protection against water leakage from plant to soil (Huang et al. 1993). A quickly reversible drought-dependent loss of root hydraulic conductivity of epiphytic cacti serves the same purpose (North and Nobel 1994). A good water uptake resumption after relief of drought depends upon the ability of near-surface roots to become functional rapidly (Wraith et al. 1995). In Brassicaceae and related families, short, tuberized, hairless roots are produced under drought stress. They withstand a prolonged drought period and give rise to a new functional root system upon rehydration. Endogenous ABA and auxin play a promotive role in this drought rhizogenesis (Vartanian et al. 1994). Shrubs of arid lands are well adapted to maintain high nutrient uptake capacities even under very low soil water potentials (*Artemisia tridentata*: Matzner and Richards 1996).

When soil water reserves become depleted, it is essential that plant roots can follow the retreating moisture front by growth. A moderate water stress (0.15 MPa) stimulates *Pinus pinaster* root growth despite a turgor reduction from 0.5 to 0.4 MPa. This growth under lowered water potential profits from wall loosening of the meristematic cells even under reduced turgor. At higher stresses, cell expansion ceases for reasons of complete loss of turgor (Triboulot et al. 1995). The more intense cell-wall loosening at low water potentials results from increased xyloglucan endotransglycolase activity (Wu et al. 1994, 1996) under the influence of higher ABA levels in the roots (Sharp et al. 1994). ABA is synthesized in water-stressed maize roots and accumulated by preference by young primary roots (Zhang and Tardieu 1996). A sufficient ascorbate supply decreasing cell-wall-bound peroxidase activities favors root elongation growth (Del Cordoba-Pedregosa et al. 1996); elevated root zone CO_2 concentrations are beneficial for intensive root growth (Ferris and Taylor 1994). Root orientation during the elongation growth is both hydro- and gravitropic (Takahasi 1994; Nakamoto 1995). The root cap is the sensory site for the environmental stimuli. The expansion of the mucilagenous root tip of maize was studied in a series of papers by McCully and co-workers (Selaey et al. 1995; McCully and Selaey 1996; McCully and Boyer 1997). Unlike the expanded blob on water-surrounded root tips in hydroponic experiments, the mucilage in a soil is a dry coating over the tip of which soil particles adhere. Upon water supply, it increases in a swelling type of kinetics, with no change in carbon concentration between tip and periphery. At its surface a condensed gel phase seals the imbibed mucilage that has no water-retaining capacity. It slows root tip desiccation only by pulling together soil particles and increasing the capillary forces between them. It functions as a lubricant, facilitating root extension into the soil. Bret-Harte and Silk (1994) emphasize that the carbohydrate demand at the root tip is higher than the possible sucrose supply only by symplasmic diffusion (root tip plasmodesmata structure and composition: Turner 1994).

Only a minor amount of the water taken up into the roots passes the root tips, the greater amount enters the plant behind them, across the rhizodermis with its various cell types. The root tip bypass interface to the soil is not controlled by a suberized Casparian band. It makes up only 0.031% of the total surface area of a maize root endodermis system (Steudle et al. 1993). Dolan (1996) gives an overview of the ontogeny of root epidermis patterning, Peters and Farquhar (1996) review what is known about root hairs. If an exodermis exists with suberized cells, as in onion roots, a cell dimorphism is found with long and short cells (Peterson and Waite 1996). The latter function as passage cells, having less dense suberin lamellae and thin cell walls (Peterson and Enstone 1996). Exodermal passage cells have a cytoplasmic structure, suggesting an active role in ion uptake. The hydraulic conductance of the root cell membranes depends on their (P- and N-)nutritional status and is influenced by temperature effects on membrane fluidity (Carvajal et al. 1996).

Radial water flow across the cortex and the endodermis is best characterized by the "composite transport model of the root" (Steudle 1994) that includes both apoplastic and symplastic transport. The prevalence of the latter is emphasized by Varney et al. (1993) based on fluorescent tracer studies. With a pressure-clamp technique, Magnani et al. (1996) determine the (symplastic) cell-to-cell conductance as 43% of the whole-root conductance. Drying/rewetting treatments of *Opuntia* roots indicate hydration-dependent variable apoplastic conductivities of the cortex (North and Nobel 1996); the overall hydraulic conductivity of *Opuntia* roots is drastically lowered by low temperatures (Cui and Nobel 1994). Whole-root hydraulic conductivities and their components (review: Huang and Nobel 1994) were studied comparatively in shallow vs. deep lateral roots of *Gutierrezia* (Wan et al. 1994) and of Australian Proteaceae roots (Pate et al. 1995), in *Gossypium* roots (Yang and Grantz 1996), and in Norway spruce (Hallgren et al. 1994). The latter study pays particular attention to the effects of either hydrostatic or osmotic forcing of water through the roots when quantifying the hydraulic properties. The different forces influence the relative contributions of apoplastic and symplastic water flow to the overall conductance. Osmotic gradients cause a much smaller water flow than hydrostatic ones (Steudle and Frensch 1996).

After symplasmic passage of the endodermis, water and solutes are released to the xylem. Cations and anions are transported from xylem parenchyma cells to the vessel apoplast through various channels, activated at different membrane potentials (Wegner and Raschke 1994). The simultaneous flow of the differently charged ions is probably driven by electrochemical gradients that are built up by the ion exchange processes along the radial pathways across the root.

Xylem vessels are fully functional only if forming pipes of dead cells, then offering the least transport resistance for the water flow through the plant and not impeded by remnants of degenerated protoplasm in the cell lumina (Mapfumo et al. 1993; Wang et al. 1994). Their ontogenetic beginning as living cells is, however, of great importance for the future function: developing vessels accumulate the highest potassium concentrations of all root tissues (up to 190 mM: McCully 1994). This guarantees full turgescence by osmotic water attraction even in competition with other cells (root sieve tube osmotic potentials: −1.4 MPa: Pritchard 1996). Embedded into a parenchyma without any voids, they are embolism-free starting points for the water columns under tension, higher up in the shoot xylem. Similarly, an uninterrupted hydraulic continuity must exist between living and dead cells in the shoot xylem. The cambium layer probably plays an important role under this aspect. Eventually, even an ultrafiltration from phloem to xylem following the water potential gradient occurs there (Münch water) so that the outermost xylem vessels show higher than average water potentials. Milburn (1996) indicates this mechanism of xylem water supply in order to explain the only small xylem tensions measured by Zimmermann with a pressure probe and offered in evidence for his challenge of the cohesion theory (Zimmermann et al. 1994; see Prog Bot 56, 66). Possibly, it is simply a tension release by aspiration from outside that occurs in the attempts to measure directly with a pressure probe the negative xylem pressure when the vessels become pierced by the pressure probe and no living plasma seals the lesion (as is the case in living cells), possibly the values reported by the Zimmermann group are meaningful. Anyway, the pressure probe data are in clear conflict with the coherent body of evidence in favor of the cohesion theory (Steudle 1995; Richter 1997). The occurrence of reasonable (negative) water potential values in the xylem is proven by measurements with methodically quite different approaches (psychrometry, pressure bomb, osmotic methods, determination of equal xylem embolism thresholds under pressure and tension: Sperry et al. 1996), and it agrees with water potential thermodynamics. The cohesion forces between water molecules and the physical laws of capillarity permit also from a theroretical point of view the low water potential values that are reported a hundred times in the ecological and physiological literature. Experimental determinations of moderate tensions sustained maximally by water in glass capillaries (Smith 1994) fall short of considerably more negative water potential thresholds of water column rupture in real plant axes under centrifugal tension (Holbrook et al. 1995). Also an additional compensating pressure theory (Canny 1995) need not be called in, that demands positive pressures from the surroundings upon the tracheary elements. By this compensation pressure xylem water potentials remain, according to this hypothesis, at levels safe from cavitation.

For acropetal water transport through the xylem, the cohesion theory of sap ascent in plants (Oertli 1993) requires a hydraulic continuum of the water colums in vessels even under strong tensions. It does not require that xylem vessels remain free from embolism all their lifetime. The appreciation of xylem embolism as a catastrophic event has been abandoned thoroughly in the meantime. Rather, it became clear that every species has a certain cavitation water potential threshold (e.g., oaks: Higgs and Wood 1995; *Curatella*: Sobrado 1996; *Ochroma* and

Pseudobombax: Machado and Tyree 1994). Only beyond the threshold range does an exponential increase of the vessel embolism occur and become dangerous ("runaway embolisms"). This threshold is determined by hydraulic respective acoustic "vulnerability curves" (methods of determination: Kolb et al. 1996; Jackson and Grace 1996). Under normal habitat conditions, minimal water potentials just approach this threshold in the worst case: if they became more negative, at least leaf and twig losses would occur or the plant would even die. Hydraulic vulnerability thresholds therefore sometimes determine the range of occurrence of a taxon (beech ecotypes: Borghetti et al. 1993; *Acer grandidentatum*: Alder et al. 1996; *Larrea tridentata*: Pockman and Sperry 1997) or even of vegetational units (riparian cottonwoods: Tyree et al. 1994b). However, leaf petioles, shoot branches, stems, and roots can have different embolism sensitivities (Hacke and Sauter 1996a), the roots being as a rule the most sensitive plant axes (Sperry and Saliendra 1994; Sperry and Ikeda 1997).

It is quite normal that during a vegetation period the majority of the vessels ceases functioning for reasons of drought embolism. Repair is rare, but can occur (Edwards et al. 1994) when during a rainy and nearly transpirationless period stem water potentials are raised to nearly zero and, additionally, root pressure pushes water upwards. This is often the case in moist-tropical environments, and lianas with their wide vessels and long axes profit from such a repair mechanism (Cochard et al. 1994). Much more commonly, frost-induced embolisms of plants in winter-cold climates (Sperry et al. 1994) become repaired after thawing (Tognetti and Borghetti 1994; Hacke and Sauter 1996b). With winter temperatures around 0 °C, *Juglans* xylem experiences slight tensions at positive, slight pressures at negative temperatures (Ameglio et al. 1995). In the latter case partial repair of the xylem embolisms occurs. Strongly increased cambium activity in spring (beech: Magnani and Borghetti 1995) and the production of new tracheary elements thereafter compensates for embolisms that end the operational lifetime of the individual vessels or tracheids. With respect to time, they function like any other cellular structure and metabolic process: sensitive for disturbance and failure in a metastabile steady state. Replacement of functionless vessels by new ones is particularly important in ring-porous trees that by winterfrost embolism lose $\geq 90\%$ of the hydraulic conductivity developed during a vegetation period (Sperry et al. 1994).

Sensitive stomatal regulation prevents in many plants water potentials from approaching the critical vulnerability threshold (e.g., oaks: Lo Gullo et al. 1995; Tyree and Cochard 1996; Cochard et al. 1996). When stem water potentials approach the high vulnerability threshold of the tropical hemiepiphyte *Clusia uvitana*, imminent damage is avoided by switching to the water-saving CAM under drought (Zotz et al. 1994). There are, however, many species with poor stomatal regulation during the day so that their water potential drops to very low values. As a rule, their vulnerability threshold is very low for reasons of the particular anatomical structure of the xylem elements (small pores in the pit membranes of mostly also small vessels: Hargrave et al. 1994). Once the vulnerability threshold is approached, leaf loss occurs (*Juglans*: Tyree et al. 1993) reducing the transpiring surface and alleviating by this feedback the drop of water potential (Tyree and Cochard 1996). Shoot water uptake at prestress levels can occur then only after sufficient new vessels are produced or after repair of embolisms (roses: Van Doorn and Suiro 1996). The coexistence of drought-deciduous and evergreen trees in a tropical dry forest can be related to the different sensitivity to embolism of the two life-forms (Sobrado 1993). Comparative studies on embolism threshold and habitat growth conditions were carried out with several Californian chaparral species (Kolb and Davis 1994; Jarbeau et al. 1995; Langan et al. 1997), but even species within one genus can differ drastically in their susceptibility to embolism (Van Doorn and Jones 1994) so that they are restricted to habitats of different drought stress probabilities. Xylem-invading parasites (e.g., the nematode *Bursaphelenchus xylophilus* causing pine wilt disease) exagerate the danger of embolism in an advanced stage of infection (Ikeda 1996).

Roughness of the internal surface of vessel walls (Roth 1996), shape of perforation plates (Schulte and Castle 1993a, b) and interlacing of vessels (MacFall and Johnson 1994) have evolved during the cormophyte phylogeny from a primitive to a very efficient status. This xylem evolution occurred in several parallel progressions under the selection pressure to increase efficiency and safety of hydraulic architecture (Baas and Wheeler 1996). The Silurian/Devonian Rhyniales still lacked a developed endodermis and an intraxylary parenchyma, but they were already free from gas spaces in the xylem, that was centered in a protostele (Raven 1994). An optimal relationship between water flow tension and conductivity has probably been a goal of xylem evolution since then. The central position of conducting tissues is sufficient in slender plants. In wider plant axes the displacement of vessels to the shoot periphery (attempted by the actinostele, fully realized in the siphonostele) is the better solution (Roth et al. 1994a, b; Roth and Mosbrugger 1996). Similarly, vessel diameter and wall structure evolution were under selection pressure from the danger of embolism (Tyree et al. 1994a).

Mostly by anatomical characters determined are species-specific hydraulic parameters that include the water flux per unit pressure gradient (K_h), $K_l = K_h$ (leaf area)$^{-1}$, $K_s = K_h$ (conductive stem cross-section area)$^{-1}$ and the Huber value [(conductive stem cross-section area)(leaf area)$^{-1}$)]. K_h of *Eucalyptus* lignotubers is half that of the stem (Myers 1995), K_l and Huber values of hemi-epiphytic *Ficus* species are smaller than those of free-standing fig trees (Patino et al. 1995). In Scots pine, K_l decreases with age (Mencuccini and Grace 1996), lupines with a comparatively smaller root system have a K_h three to five times higher than wheat (Gallardo et al. 1996). Loss of nearly one third of the conducting sapwood area due to stem bending does not influence negatively the hydraulic conductivity of *Pinus taeda* (Fredericksen et al. 1994). Probably, this high tolerance to function loss of parts of the xylem results from the involvement in acropetal water transport of most of the stem cross-section area of loblolly pine: the inner sapwood still carries one third of the sap flux density measured in the outer parts of the xylem (Phillips et al. 1996). A versatile reactivation of older vessels for water transport upon embolism blockade of earlywood vessels permits even in ring-porous oaks a certain amount of crown water supply (Granier et al. 1994). Infrared thermography is a means to visualize such differential sap flow activities (Anfodillo et al. 1993). By the more conventional dye-infusion method, Larson et al. (1994) demonstrated strongly sectored hydraulic pathways in the stems of *Thuja occidentalis*.

Velocity (Barrett et al. 1995; Cohen and Li 1996; Dye et al. 1996 – all focused upon method validation) and amount of stem sap flow is measured by heat pulse and heat balance methods, respectively. Edwards et al. (1997) propose a unified nomenclature for sap flow measurements, Smith and Allen (1996) give an overview of the (commercially) available heat balance methods, Hatton et al. (1995) suggest optimal sampling strategies when heat compensation methods are applied. Improvements of existing approaches, methodical comparisons, special constructions or applications are described, among others, by Chandra et al. (1994), Gutierrez et al. (1994a), Weibel and De Vos (1994), Peressotti and Ham (1996), Köstner et al. (1996) and Herzog et al. (1997). Cienciala and Lindroth (1995) and Lindroth et al. (1995) determined by heat balance measurements the water expenditure in *Salix viminalis* energy plant forests as 2–4 kg H_2O_2 tree^{-1}. Jimenez et al. (1996) calculated from heat balance measurements the transpiration of a *Laurus azorica* canopy. The value of 636 mm a^{-1} is equivalent to 65% of the potential laurel forest evapotranspiration or 80% of the annual precipitation in the investigated area.

Surprisingly, an appreciable shoot water flow occurs also in completely submerged aquatic plants or parts of them (*Lobelia, Sparganium, Myriophyllum, Mentha aquatica*), probably supplying nutrients from the root zone to distal plant parts (Sytsma and Anderson 1993; Pedersen and Sand-Jensen 1997). It is driven probably by root pressure and results in a clearly detectable guttation from the youngest leaves (Pedersen 1993, 1994).

The xylem sap is a strongly diluted solution of inorganic ions (Engels and Marschner 1993; Dambrine et al. 1995), amino acids (Andersen et al. 1993; Schneider et al. 1994; Ohtake et al. 1996), organic acids, sugars (Andersen et al. 1995a), phytohormones (Else et al. 1994; De Kock et al. 1994; Motosugi et al. 1996), and even macromolecules (Munns et al. 1993, Polle and Glavac 1993; Zimmermann et al. 1994). The concentrations of these substances vary diurnally (Andersen et al. 1995b; Urrestarazu et al. 1996a) and seasonally (Rennenberg et al. 1994; Schill et al. 1996). Root exudates often differ in the concentration of their contents from what is transported in intact plant stems (Schurr and Schulze 1995). As a consequence of these differences and changes in time of xylem sap constituents (and velocity, e.g., volume per cross-section) also its pH value shows diurnal patterns of variation (Urrestarazu et al. 1996b) that might function as a signal between different regions alongside the xylem pathway. With respect to drought signalling (A. B. Jensen et al. 1996), this occurs by a pH influence on ABA sequestration (Wilkinson and Davies 1997; Jia and Zhang 1997). ABA in the sap can rise more than ten times as a drought situation progresses (Jackson et al. 1995: *Pinus, Picea*; Jokhan et al. 1996: *Ricinus*). Correia and Pereira (1994) report a 100-fold ABA concentration increase in *Lupinus albus* leaf xylem upon drought – by contrast, Triboulot et al. (1996) found no significant ABA increase under drought in *Quercus* twigs. When soil becomes dry (or more compacted: Mulholland et al. 1996a, b), somewhat higher amounts of ABA are synthesized in roots, and the xylem sap concentration of the phytohormone becomes increased with this new ABA and with phloem-derived ABA. This results from the much reduced ABA catabolism in roots under drought (Liang et al. 1997). The higher ABA concentrations delivered to the leaves influence stomatal apertures, shoot elongation, and leaf extension growth. Under the influence of ABA, cell expansion is maintained in roots, but inhibited in shoots (Munns and Sharp 1993). Another group of root-borne phytohormones distributed in the shoot via the xylem stream are cytokinins. Their xylem sap concentrations are highest in the morning; loading of cytokinins into the sap depends upon the transpiration rate and is (in *Urtica*) influence by the plant's nitrogen status (Beck and Wagner 1994). The phytohormonal long-distance effects interact with hydraulic (Chazen and Neumann 1994) and electrical signals (Rekha et al. 1996), and may be modified by mycorrhizal influences (Auge et al. 1994; Ebel et al. 1994). Mansfield and De Silva (1994) review comprehensively the influences upon stomatal regulation of the hormonal signals from roots.

3 Stomatal Biology, Transpiration and Evapotranspiration

Osmoregulatory ion transport across guard cell plasmalemma and tonoplast (concerning the latter only a few studies can be mentioned, e.g. Amodeo et al. 1994; Ward and Schroeder 1994; tonoplast-bound ATPase: Willmer et al. 1994) provides the solute changes that bring about changed osmotic and turgor relations of guard cells (Meidner and Edwards 1996). These cell water paramenters control stomatal apertures, which, finally, constitute the very cause of short-term variable leaf conductances for the water vapor transport from the plant to the atmosphere. The ion transport occurs through transmembrane channels the gating of which is controlled by membrane polarization. Potassium channels (general overview: Hedrich and Dietrich 1996) exist in the guard cell plasmalemma and are either outward- (e.g., Lemtiri-Chlieh 1996; Miedema and Assmann 1996) or inward-rectifying (e.g., Blatt and Thiel 1994; Kelly et al. 1995; Grabov and Blatt 1997). There are differences, too, in channel-specific conductances (Wu 1995). Under depolarization of the *Vicia faba* guard cell plasma membrane approx. 500 K^+ efflux channels per protoplast are activated (Ilan et al. 1994). The conductance of hyperpolarization-activated channels (influx) becomes increased with decrease of the apoplastic pH (Ilan et al. 1996). A low concentration of ATP is required for activation of the inward directed K^+ channels (Wu and Assmann 1995). As a rule, functional aspects of these potassium channels were studied by patch clamp techniques. Attention was particularly paid to possible interactions between the channel protein complexes (Hoth et al. 1997) and phytohormones (auxin: Thiel et al. 1993; Dunleavy and Ladley 1995; ABA: Schwartz et al. 1994; Lemtiri-Chlieh 1996; Li and Assmann 1996).

Aspects of the signal transduction pathway of ABA in guard cells that brings about stomatal closure were investigated in several studies, but a concise picture of the whole process does not yet exist. From the continued ability of ABA-injected, still viable *Commelina* stomata to open, Anderson et al. (1994) conclude that the primary ABA reception site must be on the extracellular side of the plasma membrane. Then, a phospholipase-linked G-protein (Guanosin triphosphate-binding protein) must be involved in the signal transduction cascade (Lee et al. 1994; Wu and Assmann 1994). Mori and Muto (1997) determined the molecular weight of such an ABA-responsive protein to be 48 kDa. Extracellular Ca^{2+} seems to be required for its activation. Probably the ABA-responsive kinase together with a Ca^{2+}-dependent activator kinase constitute the membrane-linked system that increases the gating of K^+-effluxes and the subsequent stomatal closure. The conductance of inward rectifying potassium channels is probably affected by higher cytosolic ABA concentrations (Schwartz et al. 1994). Here, too, protein kinases must be operative (Li and Assmann 1996). It seems to be evident that ABA influences different K^+ channels from both sides of the plasma membrane. The situation is complicated further by events at the tonoplast membrane. Here, cell-internal ABA modifies channel conductances, the gating of which is dependent upon the cytosolic potassium concentration (MacRobbie 1995). The effect of the latter upon the K^+-current of outward-rectifying channels was investigated by Lemtiri-Chlieh (1996).

Calcium involved in these partly interacting sequences of metabolic events is not only taken up from outside via Ca^{2+}-channels, but also liberated from storage compounds (Lemtiri-Chlieh and MacRobbie 1994).

Upon ABA treatment of *Vicia faba* protoplasts, the guard cell levels of inositoltrisphosphate (IP_3) increase strongly (Y. Lee et al. 1996c). An IP_3 interaction with the endoplasmatic reticulum is accompanied by a cytosolic Ca^{2+} accumulation. Also the calmodulin-Ca^{2+} system is involved in such shifts in the cytosolic calcium status (Nejidat 1995; Cousson et al. 1995) that control the membrane channel gatings. On a protein basis, calmodulin is three- to sevenfold more abundant in guard cell than in mesophyll cell protoplasts, and many guard-cell proteins bind calmodulin in a Ca^{2+}-dependent manner (Cotelle et al. 1996). From the apoplastic Ca^{2+} pool the ions can enter the guard cell interior by a specific stretch-activated channel and using the incomplete specificity of K^+-influx channels. Across the tonoplast voltage-gated Ca^{2+} release channels exist (Allen and Sander 1994, 1995, 1996), which are affected in a feed-forward manner by the cytosolic $[Ca^{2+}]$ and in a feed-back manner by dephosphorylation processes in the guard cell cytoplasm. Influencing probably the cytosolic Ca^{2+} homeostasis, also cytokinins have an influence on the guard-cell metabolism (Darjania et al. 1994). Oxidative substances at low concentrations promote stomatal closure (McAinsh et al. 1996) by their effects upon the cytosolic free Ca^{2+} (at higher concentrations they destroy the membranes). Increases in cytosolic Ca^{2+} occur under an elevated CO_2 supply to the guard cells, suggesting that calcium may act as a second messanger in the CO_2 signal transduction pathway (Webb et al. 1996).

Changed leaf-internal CO_2 concentrations modify also the malate concentration in guard cells. This again effects the voltage-dependent properties of the anion-release channel of the plasma membrane and may therefore be one step in the CO_2-sensing mechanism of guard cells (Hedrich et al. 1994). This channel is activated by cytoplasmic ATP in a pH-dependent manner. The activation process involves the interaction of four ATP-binding sites (Schulz-Lessdorf et al. 1996). Fast and slow gating modes of this channel are transformable (Dietrich and Hedrich 1994), its ability for anion discrimination is low (Schmidt and Schroeder 1994). The anion release brings about a membrane depolarization (Schwartz et al. 1995) and in this way influences the conductivity also of other transmembrane channels.

The apoplastic supply of osmoregulatory solutes to the guard cells is to some extent decoupled from the overall xylem sap solute concentrations. If K^+ supply falls to rather low values under zero K^+ fertilization stomatal apertures are not much affected, probably for reasons of K^+ substitution by other solutes (Ridolfi et al. 1994). On the other hand, Ca^{2+} concentrations in the region adjacent to the guard cells do not increase higher than 0.05 mol m^{-3} ($[K^+]$ = 50–75 mol m^{-3}) even if much higher calcium amounts (concentrations up to 4 mol m^{-3}) are delivered to the leaf in the xylem sap (De Silva et al. 1996). Particularly calcicole plants must possess very efficient mechanisms in order to keep the high calcium load of xylem sap away from access to the guard cells; otherwise an – in nature nonexistent – malfunction of the stomata would occur (De Silva and Mansfield 1994).

While under natural conditions the degree of nutrient supply is a long-term characteristic of the growing place, plant water supply, and the atmospheric environment undergo medium-to-short-term changes. Stomatal responses to them change in their intensity, but not quality during the ontogenetic development (Tewolde et al. 1993; Yu et al. 1996). Different water supply to the leaves may influence stomatal apertures by direct hydraulic transmission of the xylem water status to the epidermis (Fuchs and Livingston 1996), but, as a rule, an involvement of an ABA-mediated drought signal transduction can be detected. Stomatal response characteristics to ABA are more sensitive in the beginning of a drought stress cycle (Peng and Weyers 1994) than at the end, when water stress is very severe (Correia and Pereira 1995). Greenhouse-grown plants respond to water deficits and ABA treatment less sensitively than outdoor-grown individuals (Hirasawa et al. 1995). Superimposed on the stomatal responses to plant drought stress are responses to VPD_{LA}, the water vapor pressure difference between leaf and the ambient air (Gucci et al. 1996). The normally very strong VPD_{LA} responses of stomata (Comstock and Ehleringer 1993) become even intensified when wind reduces the leaf boundary layer (Aphalo and Jarvis 1993; Gutierrez et al. 1994b). They are reduced, but not absent, if the whole canopy conductance under a changed bulk air VPD_{LA} is studied (Pitacco and Gallinaro 1996: assessment by micrometeorological methods of canopy partitioning between latent and sensible heat under a varied VPD_{LA}). In rare cases the stomatal VPD_{LA} response is so drastic that with an increase of VPD_{LA} at higher rates even a decrease in steady-state transpiration occurs ("feed-forward responses of stomata", critical verification: Franks et al. 1997). It is still a matter of debate whether this response essentially occurs to a changed humidity situation or a changed transpirational water loss (Assmann and Gershenson 1991; Monteith 1995; Bunce 1997). This has implications for either the hydropassive or hydroactive nature of the underlying mechanism. There are indications that ABA is involved in stomatal responses to humidity under an unchanged bulk leaf water potential (Bunce 1997; Kerstiens 1997), indicating a hydroactive process. Stomata of the CAM plant *Plectranthus marrubioides* respond to altered humidity in the dark phase (Herppich 1997). This indicates that the mechanism is functional also in highly CO_2 responsive stomata. Leaf carbon gain becomes reduced in this case, however, only under very high VPD_{LA} values. In C_3 plants (*Vicia faba*) the relative stomatal sensitivity to VPD_{LA} decreases with both leaf temperature and CO_2. $[CO_2]$ during growth is more important than the actual $[CO_2]$ at the time of measurements (Wilson and Bunce 1997). This acclimation of the stomatal CO_2 response to elevated CO_2 – and in a manner similar to the temperature response to the prevailing growth temperature – was found in several studies (e.g., Chen et al. 1995; Santrucek and Sage 1996) indicating adaptational trends to keep an homeostasis in carbon gain. The irradiance responses of *Phaseolus vulgaris* stomata are more sensitive under drought than under well-watered conditions, the closing response upon (experimental) darkness always being nearly twice as rapid as the opening process under light (Barradas et al. 1994). Irradiance gain during previous days can influence the sensitivity and velocity of stomatal movements (Blom-Zandstra et al. 1995). A light stimulus that induces stomatal pore opening in epidermal strips from *Mesembryanthemum crystallinum* leaves in the C_3 mode of photosynthesis is ineffective when CAM has been induced (Mawson and Zaugg 1994). Also a light-dependent zeaxanthin formation in guard cell chloroplasts does not work in this case. Possibly, there is a regulatory link between the altered xanthophyll cycle in the CAM mode of ice plant leaves and the shift from diurnal to nocturnal stomatal opening (Tallman et al. 1997). Only the blue light responsive sensor is operative in this facultative CAM species. Parvathi and Raghavendra (1995) emphasize that this plasmamembrane redox system functions as a third bioenergetic process that provides the energy required for guard cell ion uptake, besides respiration and photosynthesis. ABA inhibits the polarizing transmembrane proton transport effected by the plasmalemma blue light absorption (Goh et al. 1996). The blue light absorption also exerts regulating effects upon the respiratory electron flow in guard cells via either the cytochrome or the alternative pathway (Mawson 1994).

From the differentiated responsiveness of stomata to the influencing factors diurnal courses of leaf conductances and transpiration result, the shapes of which mirror the gradual change or the short-term fluctuations of the microclimatic situation on the one hand, and are partly decoupled from the daily march of evaporation on the other. Measurements of diurnal courses of transpiration together with values of plant water potential, carbon gain, nutrient turnover, and with microclimate and soil water data can characterize the ecological situation of the particular habitat or the ecophysiological peculiarities of individual species (e.g., Meinzer et al. 1996; *Acacia koa* stands; Gallego et al. 1994: *Quercus pyrenaica* forests; Koch et al. 1994: rainforest canopy surface gas exchange). Many relevant studies derive from measured data models that describe patterns of stomatal behavior and/or attempt an upscaling from leaf level transpiration to canopy water loss (mostly using Penman-Monteith-type calculations, see Prog Bot 55, 87–89). As examples for the first type of generalization of actual data with the aim of prognostic purposes, Pitman may be cited (1996: tropical evergreen dipterocarp *Hopea*), also Baille et al. (1996: roses under greenhouse conditions), or Sala and Tenhunen (*Quercus ilex*). Successful attempts for upscaling and prognoses of evapotranspiration were reported, among others, by Kowalik et al. (1997: Tuscanian mountains beech forests), Aschan et al. (1997: Canary Islands laurel forests), Granier and Breda (1996: sessile oak forest) and David et al. (1997: *Eucalyptus* plantation). Transpiration data as measured by porometry or heat balance techniques were compared in several studies with results of micrometeorological approaches (Bowen ratio, eddy correlation, or similar techniques). Agreement of results obtained independently by the different methods serves as a mutual validation of the measurement approaches (e.g., Loustau et al. 1996; Berbigier et al. 1996: *Pinus pinaster* forest in Portugal; Herbst et al. 1996: maize field in northern Germany, Busch et al. 1996; Ammann et al. 1996: triticale field). Such procedures proved to be useful also when effects upon the stand hydrology had to be assessed of successional replacements of dominating plant species (e.g., *Quercus virginiana* savanna invasion by *Juniperus ashei*: Owens 1996; *Cladium* repression by spread of *Typha* in Florida Everglades: Koch and Rawlik 1993).

References

Agrama HAS, Moussa ME (1996) Mapping QTLs in breeding for drought tolerance in maize (*Zea mays* L.) Euphytica 91:89–97

Alder NN, Sperry JS, Pockman WT (1996) Root and stem xylem embolism, stomatal conductance, and leaf turgor in *Acer grandidentatum* populations along a soil moisture gradient. Oecologia 105:293–301

Alia P, Saradhi P, Mohanty P (1993) Proline in relation to free radical production in seedlings of *Brassica juncea* raised under sodium chloride stress. Plant Soil 155/156:497–500

Allakhverdiev SI, Feyziev YM, Ahmed A, Hayashi H, Aliev JA, Klimov VV, Murata N, Carpentier R (1996) Stabilization of oxygen evolution and primary electron transport reactions in photosystem II against heat stress with glycinebetaine and sucrose. J Photochem Photobiol B 34:149–157

Allen GJ, Sanders D (1994) Two voltage-gated, calcium release channels coreside in the vacuolar membrane of broad bean guard cells. Plant Cell 6:685–694

Allen GJ, Sanders D (1995) Calcineurin, a type 2B protein phosphatase, modulates the Ca^{2+}-permeable slow vacuolar ion channel of stomatal guard cells. Plant Cell 7:1473–1483

Allen GJ, Sanders D (1996) Control of ionic currents in guard cell vacuoles by cytosolic and luminal calcium. Plant J 10:1055–1069

Ameglio T, Cruiziat P, Beraud S (1995) Tension/pressure alternation in walnut xylem sap during winter: effect on hydraulic conductivity of twigs. C R Acad Sci [III] 318:351–357

Ammann C, Meixner FX, Busch J, Lösch R (1996) CO_2 and H_2O gas exchange of a triticale field. II. Micrometeorological flux studies and comparison with upscaling from porometry. Phys Chem Earth 21:151–155

Amodeo G, Escobar A, Zeiger E (1994) A cationic channel in the guard cell tonoplast of *Allium cepa*. Plant Physiol 105:999–1006

Andersen PC, Brodbeck BV, Mizell RF (1993) Diurnal variation of amino acids and organic acids in xylem fluid from *Lagerstroemia indica*: an endogenous circadian rhythm. Physiol Plant 89:783–790

Andersen PC, Brodbeck BV, Mizell RF (1995a) Water stress- and nutrient solution-mediated changes in water relations and amino acids, organic acids and sugars in xylem fluid of *Prunus salicina* and *Lagerstroemia indica*. J Am Soc Hortic Sci 120:36–42

Andersen PC, Brodbeck BV, Mizell RF (1995b) Diurnal variation in tension, osmolarity, and the composition of nitrogen and carbon assimilates in xylem fluid of *Prunus persica, Vitis* hybrid, and *Pyrus communis.* J Am Soc Hortic Sci 120:600–606

Anderson BE, Ward JM (1994) Evidence for an extracellular reception site for abscisic acid in *Commelina* guard cells. Plant Physiol 104:1177–1183

Anfodillo T, Sigalotti GB, Tomasi M, Semenzato P, Valentini R (1993) Applications of a thermal imaging technique in the study of the ascent of sap in woody species. Plant Cell Environ 16:997–1001

Aphalo PJ, Jarvis PG (1993) The boundary layer and the apparent responses of stomatal conductance to wind speed and to the mole fractions of CO_2 and water vapour in the air. Plant Cell Environ 16:771–783

Armstrong J, Armstrong W, Armstrong IB, Pittaway GR (1996) Senescence, and phytotoxin, insects, fungal and mechanical damage: factors reducing convective gas-flows in *Phragmites australis.* Aquat Bot 54:211–226

Aschan G, Lösch R, Jiménez MS, Morales D (1997) Energiebilanz von Waldbeständen in nicht-idealem Gelände – Abschätzungen auf der Grundlage von standörtlicher Klimaerfassung und flankierenden Gaswechselmessungen am Beispiel eines Lorbeerwaldbestandes auf Teneriffa. EcoSys (Kiel) Suppl 20:145–160

Assmann SM, Gershenson A (1991) The kinetics of stomatal responses to VPD in *Vicia faba*: electrophysical and water relations models. Plant Cell Environ 14:455–466

Auge RM, Duan X, Ebel RC, Stodola AJW (1994) Nonhydraulic signalling of soil drying in mycorrhizal maize. Planta 193:74–82

Baas P, Wheeler EA (1996) Parallelism and reversibility in xylem evolution: a review. Int Assoc Wood Anat J 17:351–364

Baille M, Romero-Aranda R, Baille A (1996) Stomatal conductance of rose whole plants in greenhouse conditons: Analysis and modelling. J Hortic Sci 71:957–970

Barradas VL, Jones HG, Clark JA (1994) Stomatal responses to changing irradiance in *Phaseolus vulgaris* L. J Exp Bot 45:931–936

Barrett DJ, Hatton TJ, Ash JE, Ball MC (1995) Evaluation of the heat pulse velocity technique for measurement of sap flow in rainforest and eucalypt forest species of southeastern Australia. Plant Cell Environ 18:463–469

Bartels D, Furini A, Ingram J, Salamini F (1996) Responses of plants to dehydration stress: a molecular analysis. Plant Growth Regul 20:111–118

Basnayake J, Cooper M, Ludlow MM, Henzell RG, Snell PJ (1995) Inheritance of osmotic adjustment to water stress in three grain sorghum crosses. Theor Appl Genet 90:675–682

Basnayake J, Cooper M, Henzell RG, Ludlow MM (1996) Influence of rate of development of water deficit on the expression of maximum osmotic adjustment and desiccation tolerance in three grain sorghum lines. Field Crops Res 49:65–76

Bastide B, Sipes D, Hann J, Ting IP (1993) Effect of severe water stress on aspects of crassulacean acid metabolism in *Xerosicyos.* Plant Physiol 103:1089–1096

Bauer D, Biehler K, Fock H, Carrayol E, Hirel B, Migge A, Becker TW (1997) A role for cytosolic glutamine synthetase in the remobilization of leaf nitrogen during water stress in tomato. Physiol Plant 99:241–248

Beck E, Wagner BM (1994) Quantification of the daily cytokinin transport from the root to the shoot of *Urtica dioica* L. Bot Acta 107:342–348

Belhassen E, This D, Monneveux P (1995) Adaptation génétique a secheresse. Cah Agric 4:251–261

Berbigier P, Bonnefond JM, Loustau D, Ferreira MI, Davis JS, Pereira JS (1996) Transpiration of a 64-year-old maritime pine stand in Portugal. II. Evapotranspiration and canopy stomatal conductance measured by an eddy covariance technique. Oecologia 107:43–52

Bernsdorff C, Wolf A, Winter R, Gratton E (1997) Effect of hydrostatic pressure on water penetration and rotational dynamics in phospholipid-cholesterol bilayers. Biophys J 72:1264–1277

Beyschlag W, Eckstein J (1997) Stomatal patchiness. Prog Bot 59:283–298

Bishop K, Dambrine E (1995) Localization of tree water uptake in Scots pine and Norway spruce with hydrological tracers. Can J For Res 25:286–297

Blatt MR, Thiel G (1994) K^+ channels of stomatal guard cells: bimodal control of the K^+ inward-rectifier evoked by auxin. Plant J 5:55–68

Blom-Zandstra M, Pot CS, Maas FM, Schapendonk AHCM (1995) Effects of different light treatments on the nocturnal transpiration and dynamics of stomatal closure of two rose cultivars. Sci Hortic 61:252–262

Blum A (1996) Crop responses to drought and the interpretation of adaptation. Plant Growth Regul 20:135–148

Borchert R (1994) Electric resistance as a measure of tree water status during seasonal drought in a tropical dry forest in Costa Rica. Tree Physiol 14:299–312

Borghetti M, Leonardi S, Raschi A, Snyderman D, Tognetti R (1993) Ecotypic variation of xylem embolism, phenological traits, growth parameters and allozyme characteristics in *Fagus sylvatica*. Funct Ecol 7:713–720

Boyer JS (1993) Temperature and growth-induced water potential. Plant Cell Environ 16:1099–1106

Bret-Harte MS, Silk WK (1994) Nonvascular, symplasmic diffusion of sucrose cannot satisfy the carbon demands of growth in the primary root tip of *Zea mays* L. Plant Physiol 105:19–33

Brodribb T (1996) Dynamics of changing intercellular CO_2 concentration (c_i) during drought and determination of minimum functional c_i. Plant Physiol 111:179–185

Bunce JA (1996) Does transpiration control stomatal responses to water vapour pressure deficit? Plant Cell Environ 20:131–135

Burdon RH (1993) Stress proteins in plants. Bot J Scotl 46:463–475

Busch J, Lösch R, Meixner FX, Ammann C (1996) CO_2 and H_2O_2 gas exchange of a tricicale field. I. Leaf level porometry and upscaling to canopy level. Phys Chem Earth 21:143–149

Bussotti F, Bottacci A, Bartolesi A, Grossoni P, Tani C (1995) Morpho-anatomical alterations in leaves collected from beech trees *Fagus sylvatica* L. in conditions of natural water stress. Environ Exp Bot 35:201–213

Canny MJ (1995) A new theory for the ascent of sap-cohesion supported by tissue pressure. Ann Bot 75:343–357

Carvajal M, Cooke DT, Clarkson DT (1996) Plasma membrane fluidity and hydraulic conductance in wheat roots: interactions between root temperature and nitrate or phosphate deprivation. Plant Cell Environ 19:1110–1174

Castillo FJ (1996) Antioxidative protection in the inducible CAM plant *Sedum album* L. following the imposition of severe water stress and recovery. Oecologia 107:469–477

Castrillo M, Trujillo I (1994) Ribulose-1,5-bisphosphate carboxylase activity and chlorophyll and protein contents in two cultivars of French bean plants under water stress and rewatering. Photosynthetica 30:175–181

Cerovic ZG, Goulas Y, Gorbunov M, Briantais J-M, Camenen L, Moya I (1996) Fluorosensing of water stress in plants: diurnal changes of the mean lifetime and yield of chlorophyll fluorescence, measured simultaneously and at a distance with a τ-LIDAR and a modified PAM-fluorimeter, in maize, sugar beet, and kalanchoe. Remote Sens Environ 58:311–321

Chandra S, Lindsey PA, Bassuk NL (1994) A gauge to measure the mass flow rate of water in trees. Plant Cell Environ 17:867–874

Chapman D (1994) The role of water in biomembrane structures. J Food Eng 22:367–380

Chazen O, Neumann PM (1994) Hydraulic signals from the roots and rapid cell-wall hardening in growing maize (*Zea mays* L.) leaves are primary responses to polyethylene glycol-induced water deficits. Plant Physiol 104:1385–1392

Chen XM, Begonia GB, Hesketh JD (1995) Soybean stomatal acclimation to long-term exposure to CO_2-enriched atmospheres. Photosynthetica 31:51–57

Cheng Y, Weng J, Joshi CP, Nguyen HT (1993) Dehydration stress-induced changes in translatable RNAs in sorghum. Crop Sci 33:1397–1400

Chimenti CA, Hall AJ (1994) Responses to water stress of apoplastic water fraction and bulk modulus of elasticity in sunflower (*Helianthus annuus* L.) genotypes of contrasting capacity for osmotic adjustment. Plant Soil 166:101–107

Cho H-T, Kende H (1997) Expansins in deep-water rice internodes. Plant Physiol 113:1137–1143

Cienciala E, Lindroth A (1995) Gas-exchange and sap flow measurements of *Salix viminalis* trees in short-rotation forest. I. Transpiration and sap flow. Trees 9:289–294

Close TJ (1996) Dehydrins: emergence of a biochemical role of a family of plant dehydration proteins. Physiol Plant 97:795–803

Cochard H, Ewers FW, Tyree MT (1994) Water relations of a tropical vine-like bamboo (*Rhipidocladum racemiflorum*): root pressures, vulnerability to cavitation and seasonal changes in embolism. J Exp Bot 45:1085–1089

Cochard H, Breda N, Granier A (1996) Whole tree hydraulic conductance and water loss regulation in *Quercus* during drought: evidence for stomatal control of embolism? Ann Sci For 53:197–206

Cochrane TT (1994) A new equation for calculating osmotic potential. Plant Cell Environ 17:427–433

Cohen Y, Li Y (1996) Validating sap flow measurement in field-grown sunflower and corn. J Exp Bot 47:1699–1707

Collier DE, Cummins WR (1993) Effect of osmotic stress on the respiratory properties of shoots and roots of *Arnica alpina*. Can J Bot 71:1102–1108

Collier DE, Cummins WR (1996) The rate of development of water deficits affects *Saxifraga cernua* leaf respiration. Physiol Plant 96:291–297

Comper WD (1994) The thermodynamic and hydrodynamic properties of macromolecules that influence the hydrodynamics of porous systems J Theor Biol 168:421–427

Comstock J, Ehleringer J (1993) Stomatal response to humidity in common bean (*Phaseolus vulgaris*): implications for maximum transpiration rate, water-use efficiency and productivity. Aust J Plant Physiol 20:669–691

Conti S, Smirnoff N (1994) Rapid triggering of malate accumulation in the C_3/CAM intermediate plant *Sedum telephium*: relationship with water status and phosphoenolpyruvate carboxylase. J Exp Bot 45:1613–1621

Correia MJ, Pereira JS (1994) Abscisic aid in apoplastic sap can account for the restriction in leaf conductance of white lupins during moderate soil drying and after rewatering. Plant Cell Environ 17:845–852

Correia MJ, Pereira JS (1995) The control of leaf conductance of white lupin by xylem ABA concentration decreases with the severity of water deficits. J Exp Bot 46:101–110

Cotelle V, Forestier C, Vavasseur A (1996) A reassessment of the intervention of calmodulin in the regulation of stomatal movement. Physiol Plant 98:619–628

Cousson A, Cotelle V, Vavasseur A (1996) Induction of stomatal closure by vanadate or a light/dark transition involves Ca^{2+}-calmodulin-dependent protein phosphorylation. Plant Physiol 109:491–497

Cui M, Nobel PS (1994) Water budgets and root hydraulic conductivity of opuntias shifted to low temperatures. Int J Plant Sci 155:167–172

Dambrine E, Martin F, Carisey N, Grainer A, Hallgren J-E, Bishop K (1995) Xylem sap composition: a tool for investigating mineral uptake and cycling in adult spruce. Plant Soil 168/169:233–241

Daniels MJ, Chaumont F, Mirkov TE, Chrispeels MJ (1996) Characterization of a new vacuolar membrane aquaporin sensitive to mercury at a unique site. Plant Cell 8:587–599

Darjania L, Curvetto N, Delmastro S (1994) Cytokinins can elicit cytosolic calcium concentration changes in guard cell protoplasts of *Vicia faba*: presumable linking with cAMP. Biocell 18:59–75

David TS, Ferreira MI, David JS, Pereira JS (1997) Transpiration from a mature *Eucalyptus globulus* plantation in Portugal during a spring-summer period of progressively higher water deficit. Oecologia 110:153–159

Dawson TE, Pate JS (1996) Seasonal water uptake and movement in root systems of Australian phreatophytic plants of dimorphic root morphology: a stable isotope investigation. Oecologia 107:13–20

De Kock M, Theron KI, Swart P, Weiler EW, Bellstedt DU (1994) Cytokinins in the xylem sap of the dioecious fynbos shrub, *Leucadendron rubrum* Burm. f.: Seasonal fluctuations and their possible interaction with morphological characteristics as expressed in the two sexes. New Phytol 127:749–759

De Silva DLR, Mansfield TA (1994) The stomatal physiology of calcium delivered in the xylem sap. Proc R Soc Lond [Biol] 257:81–85

De Silva DLR, Honour SJ, Mansfield TA (1996) Estimations of apoplastic concentrations of K^+ and Ca^{2+} in the vicinity of stomatal guard cells. New Phytol 134:463–469

Del Cordoba-Pedregosa C, Gonzalez-Reyes JA, Del Sagrario Canadillas M, Navas P, Cordoba F (1996) Role of apoplastic and cell-wall peroxidases on the stimulation of root elongation by ascorbate. Plant Physiol 112:1119–1125

Dietrich P, Hedrich R (1994) Interconversion of fast and slow gating modes of GCAC1, a guard cell anion channel. Planta 195:301–304

Dolan L (1996) Pattern in the root epidermis: an interplay of diffusible signals and cellular geometry. Ann Bot 77:547–553

Dreesmann DC, Harn C, Daie J (1994) Expression of genes encoding Rubsico in sugarbeet (*Beta vulgaris* L.) plants subjected to gradual desiccation. Plant Cell Physiol 35:645–653

Dunleavy PJ, Ladley PD (1995) Stomatal responses of *Vicia faba* L. to indole acetic acid and abscisic acid. J Exp Bot 46:95–100

Dye PJ, Soko S, Poulter AG (1996) Evaluation of the heat pulse velocity method for measuring sap flow in *Pinus patula*. J Exp Bot 47:975–981

Ebel RC, Stodola AJW, Duan X, Auge RM (1994) Non-hydraulic root-to-shoot signalling in mycorrhizal and non-mycorrhizal sorghum exposed to partial soil drying or root severing. New Phytol 127:495–505

Edelmann HG, Köhler K (1995) Auxin increases elastic wall-properties in rye coleoptiles: implications for the mechanism of wall loosening. Physiol Plant 93:85–92

Edelmann HG, Kutschera U (1993) Tissue pressure and cell-wall metabolism in auxin-mediated growth of sunflower hypocotyls. J Plant Physiol 142:467–473

Edwards DR, Dixon MA (1995) Mechanisms of drought response in *Thuja occidentalis* L. I. Water stress conditioning and osmotic adjustment. Tree Physiol 15:121–127

Edwards WRN, Jarvis PG, Grace J, Moncrieff JB (1994) Reversing cavitation in tracheids of *Pinus sylvestris* L. under negative water potentials. Plant Cell Environ 17:389–397

Edwards WRN, Becker P, Cermak J (1997) A unified nomenclature for sap flow measurements. Tree Physiol 17:65–67

Eller BM, Ferrari S (1997) Water use efficiency of two succulents with contrasting CO_2 fixation pathways. Plant Cell Environ 20:93–100

Else MA, Davies WJ, Whitford PN, Hall KC, Jackson MB (1994) Concentrations of abscisic acid and other solutes in xylem sap from root systems of tomato and castor-oil plants are distorted by wounding and variable sap flow rates. J Exp Bot 45:317–323

Elstner EF, Osswald W (1993) Mechanisms of oxygen activation during plant stress. Proc R Soc Edinb [B] 102:131–154

Engels C, Marschner H (1993) Influence of the form of nitrogen supply on root uptake and translocation of cations in the xylem exudate of maize (*Zea mays* L.). J Exp Bot 44:1695–1701

Fan S, Blake TJ, Blumwald E (1994) The relative contribution of elastic and osmotic adjustment to turgor maintenance of woody species. Physiol Plant 90:408–413

Farrington P, Turner JV, Gailitis V (1996) Tracing water uptake by jarrah (*Eucalyptus marginata*) trees using natural abundances of deuterium. Trees 11:9–15

Ferris R, Taylor G (1994) Increased root growth in elevated CO_2: a biophysical analysis of root cell elongation. J Exp Bot 45:1603–1612

Finch-Savage WE, Pramanik SK, Bewley JD (1994) The expression of dehydrin proteins in desiccation-sensitive (recalcitrant) seeds of temperate trees. Planta 193:478–485

Franks PJ, Cowan IR, Farquhar GD (1997) The apparent feedforward response of stomata to air vapour pressure deficit: Information revealed by different experimental procedures with two rainforest trees. Plant Cell Environ 20:142–145

Fredericksen TS, Hedden RL, Williams SA (1994) Effect of stem bending on hydraulic conductivity and wood strength of loblolly pine. Can J For Res 24:442–446

Fricke W (1997) Cell turgor, osmotic pressure and water potential in the upper epidermis of barley leaves in relation to cell location and in response to NaCl and air humidity. J Exp Bot 48:45–58

Fricke W, Leigh RA, Tomos AD (1994) Epidermal solute concentrations and osmolality in barley leaves studied at the single-cell level: changes along the leaf blade, during leaf ageing and NaCl stress. Planta 192:317–323

Fricke W, Hinde PS, Leigh RA, Tomos AD (1995) Vacuolar solutes in the upper epidermis of barley leaves: intercellular differences follow patterns. Planta 196:40–49

Fuchs EE, Livingston NJ (1996) Hydraulic control of stomatal conductance in Douglas fir [*Pseudotsuga menziesii* (Mirb.) Franco] and alder [*Alnus rubra* (Bong)] seedlings. Plant Cell Environ 19:1091–1098

Gallardo M, Eastham J, Gregory PJ, Turner NC (1996) A comparison of plant hydraulic conductances in wheat and lupins. J Exp Bot 47:233–239

Gallego HA, Rico M, Moreno G, Santa Regina I (1994) Leaf water potential and stomatal conductance in *Quercus pyrenaica* Willd. forests: vertical gradients and response to environmental factors. Tree Physiol 14:1039–1047

Gibon Y, Bessieres MA, Larher F (1997) Is glycine betaine a non-compatible solute in higher plants that do not accumulate it? Plant Cell Environ 20:329–340

Giraudat J, Parcy F, Bertauche N, Gosti F, Leung J, Morris P-C, Bouvier-Durand M, Vartanian N (1994) Current advances in abscisic acid action and signalling. Plant Mol Biol 26:1557–1577

Gleeson JT, Erramilli S, Gruner SM (1994) Freezing and melting water in lamellar structures. Biophys J 67:706–712

Goh C-H, Kinoshita T, Oku T, Shimazaki K-I (1996) Inhibition of blue light-dependent H^+ pumping by abscisic acid in *Vicia* guard cell protoplasts. Plant Physiol 111:433–440

Grabov A, Blatt MR (1997) Parallel control of the inward-rectifier K^+ channel by cytosolic free Ca^{2+} and pH in *Vicia* guard cells. Planta 201:84–95

Granier A, Breda N (1996) Modelling canopy conductance and stand transpiration of an oak forest from sap flow measurements. Ann Sci For 53:537–546

Granier A, Anfodillo T, Sabatti M, Chochard H, Dreyer E, Tomasi M, Valentini R, Breda N (1994) Axial and radial water flow in the trunks of oak trees: a quantitative and qualitative analysis. Tree Physiol 14:1383–1396

Griffiths A, Bray EA (1996) Shoot induction of ABA-requiring genes in response to soil drying. J Exp Bot 47:1525–1531

Griffiths A, Parry AD, Jones HG, Tomos AD (1996) Abscisic acid and tugor pressure regulation in tomato roots. J Plant Physiol 149:372–376

Grose MJ, Gilligan CA, Spencer D, Goddard BVD (1996) Spatial heterogeneity of soil water around single roots: use of CT-scanning to predict fungal growth in the rhizosphere. New Phytol 133:261–272

Gucci R, Massai R, Xiloyannis C, Flore JA (1996) The effect of drought and vapour pressure deficit on gas exchange of young kiwifruit (*Actinidia deliciosa* var. *deliciosa*) vines. Ann Bot 77:605–613

Guei RG, Wassom CE (1993) Genetics of osmotic adjustment in breeding maize for drought tolerance. Heredity 71:436–441

Guerere I, Tezara W, Herrera C, Fernandez MD, Herrera A (1996) Recycling of CO_2 during induction of CAM by drought in *Talinum paniculatum* (Portulacaceae). Physiol Plant 98:471–476

Gunasekera D, Berkowitz GA (1993) Use of transgenic plants with ribulose-1,5-bisphosphate carboxylase/oxygenase antisense DNA to evaluate the rate limitation of photosynthesis under water stress. Plant Physiol 103:629–635

Gutierrez MV, Harrington RA, Meinzer FC, Fownes JH (1994a) The effect of environmentally induced stem temperature gradients on transpiration estimates from the heat balance method in two tropical woody species. Tree Physiol 14:179–190

Gutierrez MV, Meinzer FC, Grantz DA (1994b) Regulation of transpiration in coffee hedgerows: covariation of environmental variables and apparent responses of stomata to wind and humidity. Plant Cell Environ 17:1305–1313

Hacke U, Sauter JJ (1996a) Drought-induced xylem dysfunction in petioles, branches, and roots of *Populus balsamifera* L. and *Alnus glutinosa* (L.) Gaertn. Plant Physiol 111:413–417

Hacke U, Sauter JJ (1996b) Xylem dysfunction during winter and recovery of hydraulic conductivity in diffuse-porous and ring-porous trees. Oecologia 105:435–439

Haines TH (1994) Water transport across biological membranes. FEBS Lett 346:115–122

Hallgren SW, Ruedinger M, Steudle E (1994) Root hydraulic properties of spruce measured with the pressure probe. Plant Soil 167:91–98

Han BD, Kermode AR (1996) Dehydrin-like proteins in castor bean seeds and seedlings are differentially produced in response to ABA and water-deficit-related stresses. J Exp Bot 47:933–939

Han BD, Hughes W, Galau GA, Bewley JD, Kermode AR (1997) Changes in late-embryogenesis-abundant (LEA) messenger RNAs and dehydrins during maturation and premature drying of *Ricinus communis* L. seeds. Planta 201:27–35

Hao L-M, Wang H-L, Wen J-Q, Liang H-G (1996) Effects of water stress on light-harvesting complex II (LHCII) and expression of a gene encoding LHCII in *Zea mays*: J Plant Physiol 149:30–34

Hargrave KR, Kolb KJ, Ewers FW, Davis SD (1994) Conduit diameter and drought-induced embolism in *Salvia mellifera* Greene (Labiatae). New Phytol 126:695–705

Hartung W, Schiller P, Dietz K-J (1997) Physiology of poikilohydric plants. Prog Bot 59:299–327

Hatton TJ, Moore SJ, Reece PH (1995) Estimating stand transpiration in a *Eucalyptus populnea* woodland with the heat pulse method: measurement errors and sampling strategies. Tree Physiol 15:219–227

Hedrich R, Dietrich P (1996) Plant K^+ channels: similarity and diversity. Bot Acta 109:94–101

Hedrich R, Marten I, Lohse G, Dietrich P, Winter H, Lohaus G, Heldt H-W (1994) Malate-sensitive anion channels enable guard cells to sense changes in the ambient CO_2 concentration. Plant J 6:741–748

Henzler T, Steudle E (1995) Reversible closing of water channels in *Chara* internodes provides evidence for a composite transport model of the plasma membrane. J Exp Bot 46:199–209

Herbst M, Kappen L, Thamm F, Vanselow R (1996) Simultaneous measurements of transpiration, soil evaporation and total evaporation in a maize field in northern Germany. J Exp Bot 47:1957–1962

Herppich WB (1997) Stomatal responses to changes in air humidity are not necessarily linked to nocturnal CO_2 uptake in the CAM plant *Plectranthus marrubioides* Benth. (Lamiaceae). Plant Cell Environ 20:393–399

Herzog KM, Thum R, Zweifel R, Hasler R (1997) Heat balance measurements: to quantify sap flow in thin stems only? Agric For Meteorol 83:75–94

Higgs KH, Wood V (1995) Drought susceptibility and xylem dysfunction in seedlings of four European oak species. Ann Sci For 52:507–513

Hirasawa T, Wakabayashi K, Touya S, Ishihara K (1995) Stomatal responses to water deficits and abscisic acid in leaves of sunflower plants (*Helianthus annuus* L.) grown under different conditions. Plant Cell Physiol 36:955–964

Hoekstra FA, Wolkers WF, Buitink J, Goloyina EA, Crowe JH, Crouwe LM (1997) Membrane stabilization in the dry state. Comp Biochem Physiol [A] 117:335–341

Hohl M, Schopfer P (1995) Rheological analysis of viscoelastic cell wall changes in maize coleoptiles as affected by auxin and osmotic stress. Physiol Plant 94:499–505

Holbrook NM, Burns MJ, Field CB (1995) Negative xylem pressures in plants: a test of the balancing pressure technique. Science 270:1193–1194

Hook PB, Lauenroth WK (1994) Root system response of a perennial bunchgrass to neighbourhood-scale soil water heterogeneity. Funct Ecol 8:738–745

Hoth S, Dreyer I, Dietrich P, Becker D, Mueller-Roeber B, Hedrich R (1997) Molecular basis of plant specific acid activation of K^+ uptake channels. Proc Natl Acad Sci USA 94:4806–4810

Huang B, Nobel PS (1994) Root hydraulic conductivity and its components with emphasis on desert succulents. Agron J 86:767–774

Huang B, North GB, Nobel PS (1993) Soil sheaths, photosynthate distribution to roots, and rhizosphere water relations for *Opuntia ficus-indica*. Int J Plant Sci 154:425–431

Ikeda T (1996) Xylem dysfunction in *Bursaphelenchus xylophilus*-infected *Pinus thunbergii* in relation to xylem cavitation and water status. Ann Phytopathol Soc Jpn 62:554–558

Ilan N, Schwartz A, Moran N (1994) External pH effects on the depolarization-activated K channels in guard cell protoplasts of *Vicia faba*. J Gen Physiol 103:807–831

Ilan N, Schwartz A, Moran N (1996) External protons enhance the activity of the hyperpolarization-activated K channels in guard cell protoplasts of *Vicia faba*. J Membr Biol 154:169–181

Jackson GE, Grace J (1996) Field measurements of xylem cavitation: are acoustic emissions useful? J Exp Bot 47:1643–1650

Jackson GE, Irvine J, Grace J, Khalil AAM (1995) Abscisic acid concentration and fluxes in droughted conifer saplings. Plant Cell Environ 18:13–22

Jamai A, Rommasini R, Martinoia E, Delrot S (1996) Characterization of glutathione uptake in broad bean leaf protoplasts. Plant Physiol 111:1145–1152

Jarbeau JA, Ewers FW, Davis SD (1995) The mechanism of water-stress-induced embolism in two species of chaparral shrubs. Plant Cell Environ 18:189–196

Jendrasiak G (1996) The hydration of phospholipids and its biological significance. J Nutr Biochem 7:598–609

Jensen AB, Busk PK, Figueras M, Alba MM, Peracchia G, Messeguer R, Goday A, Pages M (1996) Drought signal transduction in plants. Plant Growth Regul 20:105–110

Jensen CR, Mogensen VO, Mortensen G, Andersen MN, Schjoerring JK, Thage JH, Koribidis J (1996) Leaf photosynthesis and drought adaptation in field-grown oilseed rape. Aust J Plant Physiol 23:631–644

Jia W, Zhang J (1997) Comparison of exportation and metabolism of xylem-delivered ABA in maize leaves at different water status and xylem sap pH. Plant Growth Regul 21:43–49

Jia W, Zhang J, Zhang D-P (1996) Metabolism of xylem-delivered ABA in relation to ABA flux and concentration in leaves of maize and *Commelina communis*. J Exp Bot 47:1085–1091

Jiménez MS, Cermak J, Kucera J, Morales D (1996) Laurel forests in Tenerife, Canary Islands: the annual course of sap flow in *Laurus* trees and stand. J Hydrol 183:307–321

Johansson I, Larsson C, Ek B, Kjellbom P (1996) The major integral proteins of spinach leaf plasma membranes are putative aquaporins and are phosphorylated in response to Ca^{2+} and apoplastic water potential. Plant Cell 8:1181–1191

Jokhan AD, Else MA, Jackson MB (1996) Delivery rates of abscisic acid in xylem sap of *Ricinus communis* L. plants subjected to part-drying of the soil. J Exp Bot 47:1595–1599

Jonak C, Kiegerl S, Ligterink W, Barker PJ, Huskisson NS, Hirt H (1996) Stress signalling in plants: a mitogen-activated protein kinase pathway is activated by cold and drought. Proc Natl Acad Sci USA 93:11274–11279

Kage H, Ehlers W (1996) Does transport of water to roots limit water uptake of field crops? Z Pflanzenernähr Bodenkd 159:583–590

Kameli A, Losel DM (1994) Contribution of carbohydrates and other solutes to osmotic adjustment in wheat leaves under water stress. J Plant Physiol 145:363–366

Katterer T, Fabiao A, Madeira M, Ribeiro C, Steen E (1995) Fine-root dynamics, soil moisture and soil carbon content in a *Eucalyptus globulus* plantation under different irrigation and fertilisation regimes. For Ecol Manage 74:1–12

Kelly WB, Esser JE, Schroeder JI (1995) Effects of cytosolic calcium and limited, possible dual, effects of G protein modulators on guard cell inward potassium channels. Plant J 8:479–489

Kenis JD, Rouby MB, Edelman MO, Silvente ST (1994) Inhibition of nitrate reductase by water stress and oxygen in detached oat leaves: a possible mechanism of action. J Plant Physiol 144:735–739

Kerstiens G (1997) In vivo manipulation of cuticular water permeance and its effect on stomatal response to air humidity. New Phytol 137:473–480

Kieliszewski MJ, Lamport DTA (1994) Extensin: repetitive motifs, functional sites, post-translational codes, and phylogeny. Plant J 5:157–172

Koch GW, Amthor JS, Goulden ML (1994) Diurnal patterns of leaf photosynthesis, conductance and water potential at the top of a lowland rain forest canopy in Cameroon: measurements from the Radeau des Chimes. Tree Physiol 14:347–360

Koch MS, Rawlik PS (1993) Transpiration and stomatal conductance of two wetland macrophytes (*Cladium jamaicense* and *Typha domingensis*) in the subtropical Everglades. Am J Bot 80:1146–1154

Koeckenberger W, Pope JM, Xia Y, Jeffrey KR, Komor E, Callaghan PT (1997) A non-invasive measurement of phloem and xylem water flow in castor bean seedlings by nuclear magnetic resonance microimaging. Planta 201:53–63

Köhl KI (1996) Population-specific traits and their implication for the evolution of a drought-adapted ecotype in *Armeria maritima*. Bot Acta 109:206–215

Köhl KI (1997a) The effect of NaCl on growth, dry matter allocation and ion uptake in salt marsh and inland populations of *Armeria maritima*. New Phytol 135:213–225

Köhl KI (1997b) NaCl homoeostasis as a factor for the survival of the evergreen halophyte *Armeria maritima* (Mill.) Willd. under salt stress in winter. Plant Cell Environ 20:1253–1263

Kolb KJ, Davis SD (1994) Drought tolerance and xylem embolism in co-occurring species of coastal sage and chaparral. Ecology 75:648–659

Kolb KJ, Sperry JS, Lamont BB (1996) A method for measuring xylem hydraulic conductance and embolism in entire root and shoot systems. J Exp Bot 47:1805–1810

Koster KL, Webb MS, Bryant G, Lynch DV (1994) Interactions between soluble sugars and POPC (1-palmitoyl-2-oleoylphosphatidylcholine) during dehydration: vitrification of sugars alters the phase behavior of the phospholipid. Biochim Biophys Acta 1193:143–150

Köstner B, Biron P, Siegwolf R, Granier A (1996) Estimates of water flux and canopy conductance of Scots pine at the tree level utilizing different xylem sap flow methods. Theor Appl Climatol 53:105–113

Kowalik P, Borghetti M, Borselli L, Magnani F, Sanesi G, Tognetti R (1997) Diurnal water relations of beech (*Fagus sylvatica* L.) trees in the mountains of Italy. Agric For Meteorol 84:11–23

Kozlov MM, Leikin S, Rand RP (1994) Bending, hydration and interstitial energies quantitatively account for the hexagonal-lamellar-hexagonal reentrant phase transition in dioleoylphosphatidyle thanolamine. Biophys J 67:1603–1611

Kutschera U (1994) Tansley Review no 66. The current status of the acid-growth hypothesis. New Phytol 126:549–569

Kutschera U (1995) Tissue pressure and cell turgor in axial plant organs: implications for the organismal theory of multicellularity. J Plant Physiol 146:126–132

Kutschera U (1996) Cessation of cell elongation in rye coleoptiles is accompanied by a loss of cell-wall plasticity. J Exp Bot 45:591–595

Kutschera U, Köhler K (1994) Cell elongation, turgor and osmotic pressure in developing sunflower hypocotyls. J Exp Bot 45:591–595

Lal A, Edwards GE (1996) Analysis of inhibition of photosynthesis under water stress in the C4 species *Amaranthus cruentus* and *Zea mays*: electron transport, CO_2 fixation and carboxylation capacity. Aust J Plant Physiol 23:403–412

Langan SJ, Ewers FW, Davis SD (1997) Xylem dysfunction caused by water stress and freezing in two species of co-occurring chaparral shrubs. Plant Cell Environ 20:425–437

Larson DW, Doubt J, Matthes-Sears U (1994) Radially sectored hydraulic pathways in the xylem of *Thuja occidentalis* as revealed by the use of dyes. Int J Plant Sci 155:569–582

Lebreton C, Lazic-Jancic V, Steed A, Peekic S, Quarrie SA (1995) Identification of QTL for drought responses in maize and their use in testing causal relationships between traits. J Exp Bot 46:853–865

Lee J, Yi H, Lee S, Lee Y (1994) Involvement of G-protein in signal transduction process of abscisic acid-induced stomatal closure. J Plant Biol 37:429–434

Lee KS, Kim J-H (1994) Changes in crassulacean acid metabolism (CAM) of *Sedum* plants with special reference to soil moisture conditions. J Plant Biol 37:9–15

Lee Y, Choi YB, Suh S, Lee J, Assmann SM, Joe CO, Kelleher JF, Crain RC (1996) Abscisic acid-induced phosphoinositide turnover in guard cell protoplasts of *Vicia faba*. Plant Physiol 110:987–996

Lemtiri-Chlieh F (1996) Effects of internal K^+ and ABA on the voltage- and time-dependence of the outward K^+-rectifier in *Vicia* guard cells. J Membr Biol 153:105–116

Lemtiri-Chlieh F, MacRobbie EAC (1994) Role of calcium in the modulation of *Vicia* guard cell potassium channels by abscisic acid: a patch-clamp study. J Membr Biol 137:99–107

Leone A, Costa A, Grillo S, Tucci M, Horvath I, Vigh L (1996) Acclimation to low water potential determines changes in membrane fatty acid composition and fluidity in potato cells. Plant Cell Environ 19:1103–1109

Leprince O, Hendry GAF, Atherton NM (1994) Free radical processes induced by desiccation in germinating maize: the relationship with respiration and loss of desiccation tolerance. Proc R Soc Edinb [B] 102:211–218

Li J, Assmann SM (1996) An abscisic acid-activated and calcium-independent protein kinase from guard cells of fava bean. Plant Cell 8:2359–2368

Liang J, Zhang J, Wong MH (1997) How do roots control xylem sap ABA concentration in response to soil drying? Plant Cell Physiol 38:10–16

Lindroth A, Cermak J, Kucera J, Cienciala E, Eckersten H (1995) Sap flow by the heat balance method applied to small size *Salix* trees in a short-rotation forest. Biomass Bioenergy 8:7–15

Lipiec J, Ishioka T, Hatano R, Sakuma T (1993) Effects of soil structural discontinuity on root and shoot growth and water use of maize. Plant Soil 157:65–74

Lo Gullo MA, Salleo S, Piaceri EC, Rosso R (1995) Relations between vulnerability to xylem embolism and xylem conduit dimensions in young trees of *Quercus cerris*. Plant Cell Environ 18:661–669

Lopez F, Vansuyt G, Fourcroy P, Casse-Delbart F (1994) Accumulation of a 22-kDa protein and its mRNA in the leaves of *Raphanus sativus* in response to salt stress or water deficit. Physiol Plant 91:605–614

Lösch R (1995) Plant water relations. Prog Bot 56:56–96

Lösch R (1998) Wasserhaushalt der Pflanzen. Quelle und Meyer, Wiesbaden

Loustau D, Berbigier P, Roumagnac P, Arruda-Pacheco C, David JS, Ferreira MI, Pereira JS, Tavares R (1996) Transpiration of a 64-year-old maritime pine stand in Portugal. I. Seasonal course of water flux through maritime pine. Oecologia 107:33–42

Lüttge U (1993) The role of Crassulacean acid metabolism (CAM) in the adaptation of plants to salinity. New Phytol 125:59–71

MacFall JS, Johnson GA (1994) The architecture of plant vasculature and transport as seen with magnetic resonance microscopy. Can J Bot 72:1561–1573

Machado J-L, Tyree MT (1994) Patterns of hydraulic architecture and water relations of two tropical canopy trees with contrasting leaf phenologies: *Ochroma pyramidale* and *Pseudobombax septenatum*. Tree Physiol 14:219–240

MacRobbie EAC (1995) ABA-induced ion efflux in stomatal guard cells: multiple actions of ABA inside and outside the cell. Plant J 7:565–576

Magnani F, Borghetti M (1995) Interpretation of seasonal changes of xylem embolism and plant hydraulic resistance in *Fagus sylvatica*. Plant Cell Environ 18:689–696

Magnani F, Centritto M, Grace J (1996) Measurement of apoplasmic and cell-to-cell components of root hydraulic conductance by a pressure-clamp technique. Planta 199:296–306

Makela P, Peltonen-Sainio P, Jokinen K, Peehu E, Setala H, Hinkhanen R, Somersalo S (1996) Uptake and translocation of foliar-applied glycinebetaine in crop plants. Plant Sci 121:221–230

Mansfield TA, De Silva DLR (1994) Sensory systems in the roots of plants and their role in controlling stomatal function in the leaves. Physiol Chem Phys Med NMR 26:89–99

Mapfumo E, Aspinall D, Hancock T, Sedgley M (1993) Xylem development in relation to water uptake by roots of grapevine (*Vitis vinifera* L.) New Phytol 125:93–99

Marur CJ, Mazzafera P, Magalhaes AC (1996) Carbon assimilation and export in leaves of cotton plants under water deficit. Rev Bras Fisiol Veg 8:181–186

Maruyama S, Boyer JS (1994) Auxin action on growth in intact plants: threshold turgor is regulated. Planta 193:44–50

Matzner SL, Richards JH (1996) Sagebrush (*Artemisia tridentata* Nutt.) roots maintain nutrient uptake capacity under water stress. J Exp Bot 47:1045–1056

Mawson BT (1994) Cyanide-resistant, alternative pathway respiration in guard cell protoplasts of *Vicia faba*. Can J Bot 72:150–156

Mawson BT, Zaugg MZ (1994) Modulation of light-dependent stomatal opening in isolated epidermis following induction of crassulacean acid metabolism in *Mesembryanthemum crystallinum* L. J Plant Physiol 144:740–746

McAinsh MR, Clayton H, Mansfield TA, Hetherington AM (1996) Changes in stomatal behavior and guard cell cytosolic free calcium in response to oxidative stress. Plant Physiol 111:1031–1042

McCann MC, Roberts K (1994) Changes in cell wall architecture during cell elongation. J Exp Bot 45:1683–1691

McCully ME (1994) Accumulation of high levels of potassium in the developing xylem elements in roots of soybean and some other dicotyledons. Protoplasma 183:116–125

McCully ME, Boyer JS (1997) The expansion of maize root-cap mucilage during hydration. III. Changes in water potential and water content. Physiol Plant 99:169–177

McCully ME, Sealey LJ (1996) The expansion of maize root-cap mucilage during hydration. II. Observations on soil-grown roots by cryo-scanning electron microscopy. Physiol Plant 97:454–462

Mcqueen-Mason SJ, Cosgrove DJ (1994) Disruption of hydrogen bonding between plant cell wall polymers by proteins that induce wall extension. Proc Natl Acad Sci USA 91:6574–6578

Mcqueen-Mason SJ, Cosgrove DJ (1995) Expansin mode of action on cell walls: analysis of wall hydrolysis, stress relaxation, and binding. Plant Physiol 107:87–100

Meidner H, Edwards M (1996) Osmotic and turgor pressures of guard cells. Plant Cell Environ 19:503

Meinzer FC, Fownes JH, Harrington RA (1996) Growth indices and stomatal control of transpiration in *Acacia koa* stands planted at different densities. Tree Physiol 16:607–615

Mencuccini M, Grace J (1996) Developmental patterns of above-ground hydraulic conductance in a Scots pine (*Pinus slyvestris* L.) age sequence. Plant Cell Environ 19:939–948

Methy M, Damesin C, Rambal S (1996) Drought and photosystem II activity in two Mediterranean oaks. Ann Sci For 53:255–262

Mia MAR, Bailey MA, Ashley DA, Wells R, Carter TE Jr, Parrott WA, Boerma HR (1996) Molecular markers associated with water use efficiency and leaf ash in soybean. Crop Sci 36:1252–1257

Michel BE, Radcliffe D (1995) A computer program relating solute potential to solute composition for five solutes. Agron J 87:126–130

Miedema H, Assmann SM (1996) A membrane-delimited effect of internal pH on the K^+ outward rectifier of *Vicia faba* guard cells. J Membr Biol 154:227–237

Milburn JA (1996) Sap ascent in vascular plants: challengers to the cohesion theory ignore the significance of immature xylem and the recycling of Munch water. Ann Bot 78:399–407

Mizuno A, Katou K (1996) Regulation of plant elongation growth by surface and xylem proton pumps. J Plant Res 109:85–91

Monneveux P, Belhassen E (1996) The diversity of drought adaptation in the wide. Plant Growth Regul 20:85–92

Monteith JL (1995) A reinterpretation of stomatal responses to humidity. Plant Cell Environ 18:357–364

Mordelet P, Barot S, Abbadie L (1996) Root foraging strategies and soil patchiness in a humid savanna. Plant Soil 182:171–176

Mori IC, Muto S (1997) Abscisic acid activates a 48-kilodalton protein kinase in guard cell protoplasts. Plant Physiol 113:833–839

Motosugi H, Nishijima T, Hiehata N, Koshioka M, Sugiura A (1996) Endogenous gibberellins in the xylem exudate from apple trees. Biosci Biotechnol Biochem 60:1500–1502

Mulholland BJ, Black CR, Taylor IB, Roberts JA, Lenton JR (1996a) Effect of soil compaction on barley (*Hordeum vulgare* L.) growth. I. Possible role for ABA as a root-sourced chemical signal. J Exp Bot 47:539–549

Mulholland BJ, Taylor IB, Black CR, Roberts JA (1996b) Effect of soil compaction on barley (*Hordeum vulgare* L.) growth. II. Are increased xylem sap ABA concentration involved in maintaining leaf expansion in compacted soils? J Exp Bot 47:551–556

Munns R, Sharp RE (1993) Involvement of abscisic acid in controlling plant growth in soils of low water potential. Aust J Plant Physiol 20:425–437

Munns R, Passioura JB, Milborrow BV, James RA, Close TJ (1993) Stored xylem sap from wheat and barley in drying soil contains a transpiration inhibitor with a large molecular size. Plant Cell Environ 16:867–872

Myers BA (1995) The influence of the lignotuber on hydraulic conductance and leaf conductance in *Eucalyptus behriana* seedlings. Aust J Plant Physiol 22:857–863

Nakamoto T (1995) Gravitropic reaction of primary seminal roots of *Zea mays* L. influenced by temperature and soil water potential. J Plant Res 108:71–75

Navari-Izzo F, Meneguzzo S, Loggini B, Vazzana C, Sgherri CLM (1997) The role of the gluathione system during dehydration of *Boea hygroscopica*. Physiol Plant 99:23–30

Nejidat A (1995) Possible involvement of calmodulin in the regulation of ATPase activity in guard cells. Physiol Plant 94:411–414

Neumann PM (1995) The role of cell wall adjustment in plant resistance to water deficits. Crop Sci 35:1258–1266

Niu DK, Wang M, Wang YF (1997) Plant cellular osmotica. Acta Biotheor (Leiden) 45:161–169

North GB, Nobel PS (1994) Changes in root hydraulic conductivity for two tropical epiphytic cacti as soil moisture variees. Am J Bot 81:46–53

North GB, Nobel PS (1996) Radial hydraulic conductivity of individual root tissues of *Opuntia ficus-indica* (L.) Ann Bot 77:133–142

North GB, Nobel PS (1997) Root-soil contact for the desert succulent *Agave deserti* in wet and drying soil. New Phytol 135:21–29

Oertli JJ (1993) The ascent of sap in trees. Vierteljahresschr Naturforsch Ges Zürich 138:169–190

Ohtake N, Nishiwaki T, Mizukoshi K, Minagawa R, Takahashi Y, Chinushi T, Ohyama T (1995) Amino acid composition in xylem sap of soybean related to the evaluation of N_2 fixation by the relative ureide method. Soil Sci Plant Nutr 41:95–102

Oliver AE, Crowe LM, De Araujo PS, Fisk E, Crowe JH (1996) Arbutin inhibits PLA_2 in partially hydrated model systems. Biochim Biophys Acta 1302:69–78

Olsson M, Nilsson K, Liljenberg C, Hendry GAF (1996) Drought stress in seedlings: lipid metabolism and lipid peroxidation during recovery from drought in *Lotus corniculatus* and *Cerastium fontanum*. Physiol Plant 96:577–584

Onwona-Agyeman S, Morioka N, Kondo M, Kitagawa K (1995) Seasonal changes in the modulus of elasticity of living branches of three coniferous species. Ecol Res 10:199–206

Oparka KJ (1994) Plasmolysis: new insights into an old process. New Phytol 126:571–591

Osborne DJ, Boubriak II (1994) DNA and desiccation tolerance. Seed Sci Res 4:175–185

Owens MK (1996) The role of leaf and canopy-level gas exchange in the replacement of *Quercus virginiana* (Fagaceae) by *Juniperus ashei* (Cupressaceae) in semiarid savannas. Am J Bot 83:617–623

Papageorgiou GC, Murata N (1995) The unusually strong stabilizing effects of glycine betaine on the structure and function of the oxygen-evolving Photosystem II complex. Photosynth Res 44:243–252

Parvathi K, Raghavendra AS (1995) Bioenergetic processes in guard cells related to stomatal function. Physiol Plant 93:146–154

Passioura JB (1994) The physical chemistry of the primary cell wall: implications for the control of expansion rate. J Exp Bot 45:1675–1682

Pate JS, Jeschke WD, Alyward MJ (1995) Hydraulic architecture and xylem structure of the dimorphic root system of Southwest Australian species of Proteaceae. J Exp Bot 46:907–915

Patino S, Tyree MT, Herre EA (1995) Comparison of hydraulic architecture of woody plants of differing phylogeny and growth form with special reference to free-standing and hemi-epiphytic *Ficus* species from Panama. New Phytol 129:125–135

Pedersen O (1993) Long-distance water transport in aquatic plants. Plant Physiol 103:1369–1375

Pedersen O (1994) Acropetal water transport in submerged plants. Bot Acta 107:61–65

Pedersen O, Sand-Jensen K (1997) Transpiration does not control growth and nutrient supply in the amphibious plant *Mentha aquatica*. Plant Cell Environ 20:117–123

Pelah D, Shoseyov O, Altman A (1995) Characterization of BspA, a major boiling-stable, water-stress-responsive protein in aspen (*Populus tremula*). Tree Physiol 15:673–678

Pelah D, Wang W, Altman A, Shoseyov O, Bartels D (1997) Differential accumulation of water stress-related proteins, sucrose synthase and soluble sugars in *Populus* species that differ in their water stress response. Physiol Plant 99:153–159

Peltier J-P, Marigo G (1996) Adjustment processes in ash and water stress situation. C R Acad Sci [III] 319:425–429

Peltier J-P, Agasse F, De Bock F, Marigo G (1994) Osmotic adjustment in ash and water stress situation. C R Acad Sci [III] 317:679–684

Peng Z-Y, Weyers JDB (1994) Stomatal sensitivity to abscisic acid following water deficit stress. J Exp Bot 45:835–845

Peressotti A, Ham JM (1996) A dual-heater gauge for measuring sap flow with an improved heat-balance method. Agron J 88:149–155

Peters RL, Farquhar ML (1996) Roots hairs: specialized tubular cells extending root surfaces. Bot Rev 62:1–40

Peters WS, Tomos AD (1996) The history of tissue tension. Ann Bot 77:657–665

Peterson CA, Enstone DE (1996) Functions of passage cells in the endodermis and exodermis of roots. Physiol Plant 97:592–598

Peterson CA, Waite JL (1996) The effect of suberin lamellae on the vitality and symplasmic permeability of the onion root exodermis. Can J Bot 74:1220–1226

Phillips N, Oren N, Zimmermann R (1996) Radial patterns of xylem sap flow in non-, diffuse- and ring-porous tree species. Plant Cell Environ 19:983–990

Pitacco A, Gallinaro N (1996) Micrometeorological assessment of sensitivity of canopy resistance to vapour pressure deficit in a Mediterranean oak forest. Ann Sci For 53:513–520

Pitman JI (1996) Ecophysiology of tropical dry evergreen forest, Thailand: measured and modelled stomatal conductance of *Hopea ferrea*, a dominant canopy emergent. J Appl Ecol 33:1366–1378

Pockman WT, Sperry JS (1997) Freezing-induced xylem cavitation and the northern limit of *Larrea tridentata*. Oecologia 109:19–27

Polle A, Glavac V (1993) Seasonal changes in the axial distribution of peroxidase activity in the xylem sap of beech (*Fagus sylvatica* L.) trees. Tree Physiol 13:409–413

Popova LP, Tsonev TD, Lazova GN, Stoinova ZG (1996) Drought- and ABA-induced changes in photosynthesis of barley plants. Physiol Plant 96:623–629

Prakash KR, Rao VS (1996) The altered activities of carbonic anhydrase, phosphoenolpyruvate carboxylase and ribulose bisphosphate carboxylase due to water stress and after its relief. J Environ Biol 17:39–42

Pregitzer KS, Hendrick RL, Fogel RT (1993) The demography of fine roots in response to patches of water and nitrogen. New Phytol 125:575–580

Pritchard J (1994) The control of cell expansion in roots. New Phytol 127:3–26

Pritchard J (1996) Aphid stylectomy reveals an osmotic step between sieve tube and cortical cells in barley roots. J Exp Bot 47:1519–1524

Quarrie SA (1996) New molecular tools to improve the efficiency of breeding for increased drought resistance. Plant Growth Regul 20:167–178

Rascio A, Platani C, Scalfati G, Tonti A, Di Fonzo N (1994) The accumulation of solutes and water binding strength in durum wheat. Physiol Plant 90:715–721

Rascio A, Russo M, Di Fonzo N (1997) Simplified procedure for estimating tissue affinity for bound water in durum wheat. Crop Sci 37:275–277

Raven JA (1994) Physiological analysis of aspects of the functioning of vascular tissue in early plants. Bot J Scotl 47:49–64

Reddy AR (1996) Fructose 2,6-bisphosphate-modulated photosynthesis in sorghum leaves grown under low water regimes. Phytochemistry 43:319–322

Rekha G, Sudarshana L, Prasad TG, Kulkarni MJ, Sashidar VR (1996) Slower-chemical or faster-electrical signalling under stress in plants: is it the hare and tortoise story of a slower signal winning the race? Curr Sci 71:284–289

Rennenberg H, Schupp R, Glavac V, Jochheim H (1994) Xylem sap composition for beech (*Fagus sylvatica* L.) trees: seasonal changes in the axial distribution of sulfur compounds. Tree Physiol 14:541–548

Richter H (1997) Water relations of plants in the field: some comments on the measurement of selected parameters. J Exp Bot 48:1–7

Ridolfi M, Garrec JP, Louguet P, Laffray D (1994) Effects of potassium and calcium deficiencies on stomatal functioning in intact leaves of *Vicia faba*. Can J Bot 72:1835–1842

Rieger M (1995) Offsetting effects of reduced root hydraulic conductivity and osmotic adjustment following drought. Tree Physiol 15:379–385

Ristic Z, Williams G, Yang G, Martin B, Fullerton S (1996) Dehydration, damage to cellular membranes, and heat-shock proteins in maize hybrids from different climates. J Plant Physiol 149:424–432

Rivoal J, Hanson AD (1994b) Choline-O-sulfate biosynthesis in plants: identification and partial characterization of a salinity-inducible choline sulfotransferase from species of *Limonium* (Plumbaginaceae). Plant Physiol 106:1187–1193

Roth A (1996) Water transport in xylem conduits with ring thickenings. Plant Cell Environ 19:622–629

Roth A, Mosbrugger B (1996) Numerical studies of water conduction in land plants: evolution of early stele types. Paleobiology 22:411–421

Roth A, Mosbrugger V, Neugebauer HJ (1994a) Efficiency and evolution of water transport systems in higher plants: a modelling approach. I. The earliest land plants. Philos Trans R Soc Lond [Biol] 345:137–152

Roth A, Mosbrugger V, Neugebauer HJ (1994b) Efficiency and evolution of water transport systems in higher plants: a modelling approach. II. Stelar evolution. Philos Trans R Soc Lond [Biol] 345:153–162

Rucker KS, Kvien CK, Holbrook CC, Hook JE (1995) Identification of peanut genotypes with improved drought-avoidance traits. Peanut Sci 22:14–18

Sala A, Tenhunen JD (1996) Simulations of canopy net photosynthesis and transpiration in *Quercus ilex* L. under the influence of seasonal drought. Agric For Meteorol 78:203–222

Santrucek J, Sage RF (1996) Acclimation of stomatal conductance to a CO_2-enriched atmosphere and elevated temperature in *Chenopodium album*. Aust J Plant Physiol 23:467–478

Savoure A, Hua X-J, Bertauche N, Van Montagu M, Verbruggen N (1997) Abscisic acid-independent and abscisic acid-dependent regulation of proline biosynthesis following cold and osmotic stresses in *Arabidopsis thaliana*. Mol Gen Genet 254:104–109

Schill V, Hartung W, Orthen B, Weisenseel MH (1996) The xylem sap of maple (*Acer platanoides*) trees – sap obtained by a novel mthod shows changes with season and height. J Exp Bot 47:128–133

Schmidt C, Schroeder JI (1994) Anion selectivity of slow anion channels in the plasma membrane of guard cells: large nitrate permeability. Plant Physiol 106:383–391

Schneider A, Kreuzwieser J, Schupp R, Sauter JJ, Rennenberg H (1994) Thiol and amino acid composition of the xylem sap of poplar trees (*Populus x canadensis* robusta). Can J Bot 72:347–351

Schulte PJ, Castle AL (1993a) Water flow through vessel perforation plates: the effects of plate angle and thickness for *Liriodendron tulipifera*. J Exp Bot 44:1143–1148

Schulte PJ, Castle AL (1993b) Water flow through vessel perforation plates: a fluid mechanical approach. J Exp Bot 44:1135–1142

Schulz-Lessdorf B, Lohse G, Hedrich R (1996) GCAC1 recognizes the pH gradient across the plasma membrane: a pH-sensitive and ATP-dependent anion channel links guard cell membrane potential to acid and energy metabolism. Plant J 10:993–1004

Schurr U, Schulze E-D (1995) The concentration of xylem sap constituents in root exudate, and in sap from intact, transpiring castor bean plants (*Ricinus communis* L.). Plant Cell Environ 18:409–420

Schwartz A, Wu W-H, Tucker EB, Assmann SM (1994) Inhibition of inward K^+ channels and stomatal response by abscisic acids: an intracellular locus of phytohormone action. Proc Natl Acad Sci USA 91:4019–4023

Schwartz A, Ilan N, Schwarz M, Scheaffer J, Assmann SM, Schroeder JI (1995) Anion channel blockers inhibit S-type anion channels and abscisic acid responses in guard cells. Plant Physiol 109:651–658

Sealey LJ, McCully ME, Canny MJ (1995) The expansion of maize root-cap mucilage during hydration: I. Kinetics. Physiol Plant 93:38–46

Seifert U, Langer SA (1994) Hydrodynamics of membranes: the bilayer aspect and adhesion. Biophys Chem 49:13–22

Serpe MD, Matthews MA (1994) Changes in cell wall yielding and stored growth in *Begonia argenteo-guttata* L. leaves during the development of water deficits. Plant Cell Physiol 35:619–626

Sgherri CLM, Navari-Izzo F (1995) Sunflower seedlings subjected to increasing water deficit stress: oxidative stress and defence mechanisms. Physiol Plant 93:25–30

Sharp RE, Wu Y, Voetberg GS, Saab IN, Lenoble ME (1994) Confirmation that abscisic acid accumulation is required for maize primary root elongation at low water potentials. J Exp Bot 45:1743–1751

Smirnoff N (1993) Tansley Review no 52. The role of active oxygen in the response of plants to water deficit and desiccation. New Phytol 125:27–58

Smith A (1994) Xylem transport and the negative pressures sustainable by water. Ann Bot 74:647–651

Smith DM, Allen SJ (1996) Measurement of sap flow in plant stems. J Exp Bot 47:1833–1844

Sobrado MA (1993) Trade-off between water transport efficiency and leaf life-span in a tropical dry forest. Oecologia 96:19–23

Sobrado MA (1996) Embolism vulnerability of an evergreen tree. Biol Plant 38:297–301

Sperry JS, Ikeda T (1997) Xylem cavitation in roots and stems of Douglas-fir and white fir. Tree Physiol 17:275–280

Sperry JS, Saliendra NZ (1994) Intra- and inter-plant variation in xylem cavitation in *Betula occidentalis*. Plant Cell Environ 17:1233–1241

Sperry JS, Nichols KL, Sullivan JEM, Eastlack SE (1994) Xylem embolism in ring-porous, diffuse-porous, and coniferous tress of northern Utah and interior Alaska. Ecology 75:1736–1752

Sperry JS, Saliendra NZ, Pockman WT, Cockhard H, Cruiziat P, Davis SD, Ewers FW, Tyree MT (1996) New evidence for large negative xylem pressures and their measurement by the pressure chamber method. Plant Cell Environ 19:427–436

Spollen WG, Nelson CJ (1994) Response of fructan to water deficit in growing leaves of tall fescue. Plant Physiol 106:329–336

Stanca AM, Crosatti C, Grossi M, Lacerenza NG, Rizza F, Cattivelli L (1996) Molecular adaptation of barley to cold and drought conditions. Euphytica 92:215–219

Steudle E (1994) Water transport across roots. Plant Soil 167:79–90

Steudle E (1995) Trees under tension. Nature 378:663–664

Steudle E, Frensch J (1996) Water transport in plants: role of the apoplast. Plant Soil 187:67–79

Steudle E, Murrmann M, Peterson CA (1993) Transport of water and solutes across maize roots modified by puncturing the endodermis. Plant Physiol 103:335–349

Svetek J, Schara M, Nemec M, Nothnagel EA (1994) Osmolality effects on membrane transport properties and membrane domain structure measured in maize root tissue in situ by spin probe EPR. Acta Pharm 44:297–307

Sytsma MD, Anderson LWJ (1993) Transpiration by an emergent macrophyte: source of water and implications for nutrient supply. Hydrobiologia 271:97–108

Takahashi H (1994) Hydrotropism and its interaction with gravitropism in roots. Plant Soil 165:301–308

Tallman G, Zhu J, Mawson BT, Amodeo G, Nouhi Z, Levy K, Zeiger E (1997) Induction of CAM in *Mesembryanthemum crystallinum* abolishes the stomatal response to blue light and light-dependent zeaxanthin formation in guard cell chloroplasts. Plant Cell Physiol 38:236–242

Tardieu F (1996) Drought perception by plants. Do cells of droughted plants experience water stress? Plant Growth Regul 20:93–104

Tazawa M, Asai K, Iwasaki N (1996) Characteristics of Hg- and Zn-sensitive water channels in the plasma membrane of *Chara* cells. Bot Acta 109:388–396

Tewolde H, Dobrenz AK, Voigt RL (1993) Seasonal trends in leaf photosynthesis and stomatal conductance of drought stressed and nonstressed pearl millet as associated to vapor pressure deficit. Photosynth Res 38:41–49

Thiel G, Blatt MR, Fricker MD, White IR, Millner P (1993) Modulation of K^+ channels in *Vicia* stomatal guard cells by peptide homologs to the auxin-binding protein C terminus. Proc Natl Acad Sci USA 90:11493–11497

Thorburn PJ, Ehleringer JR (1995) Root water uptake of field-growing plants indicated by measurements of natural-abundance deuterium. Plant Soil 177:225–233

Thorburn PJ, Walker GR (1994) Variation in stream water uptake by *Eucalyptus camaldulensis* with differing access to stream water. Oecologia 100:293–301

Tognetti R, Borghetti M (1994) Formation and seasonal occurrence of xylem embolism in *Alnus cordata*. Tree Physiol 14:241–250

Tognetti R, Johnson JD, Michelozzi M (1995) The response of European beech (*Fagus sylvatica* L.) seedlings from two Italian populations to drought and recovery. Trees 9:348–354

Tomos D, Pritchard J (1994) Biophysical and biochemical control of cell expansion in roots and leaves. J Exp Bot 45:1721–1731

Triboulot MB, Pritchard J, Tomos D (1995) Stimulation and inhibition of pine root growth by osmotic stress. New Phytol 130:169–175

Triboulot MB, Fauveau ML, Breda N, Label P, Dreyer E (1996) Stomatal conductance and xylem-sap abscisic acid (ABA) in adult oak trees during a gradually imposed drought. Ann Sci For 53:207–220

Tschaplinski TJ, Tuskan GA (1994) Water-stress tolerance of black and eastern cottonwood clones and four hybrid progeny: II Metabolites and inorganic ions that constitute osmotic adjustment. Can J For Res 24:681–687

Tuinstra MR, Grote EM, Goldsbrough PB, Ejeta G (1996) Identification of quantitative trait loci associated with pre-flowering drought tolerance in sorghum. Crop Sci 36:1337–1344

Turner A, Wells B, Roberts K (1994) Plasmodesmata of maize root tips: structure and composition. J Cell Sci 107:3351–33612

Tyree MT, Cochard H (1996) Summer and winter embolism in oak: impact on water relations. Ann Sci For 53:173–180

Tyree MT, Cochard H, Cruiziat P, Sinclair B, Ameglio T (1993) Drought-induced leaf shedding in walnut: evidence for vulnerability segmentation. Plant Cell Environ 16:879–882

Tyree MT, Davis SD, Cochard H (1994a) Biophysical perspectives of xylem evolution: is there a tradeoff of hydraulic efficacy for vulnerability to dysfunction? Int Assoc Wood Anat J 15:335–360

Tyree MT, Kolb KJ, Rood SB, Patino S (1994b) Vulnerability to drought-induced cavitation of riparian cottonwoods in Alberta: a possible factor in the decline of the ecosystem? Tree Physiol 14:455–466

Ulrich AS, Watts A (1994a) Molecular response of the lipid headgroup to bilayer hydration monitored by ^{2}H-NMR. Biophys J 66:1441–1449

Ulrich AS, Watts A (1994b) Lipid headgroup hydration studied by ^{2}H-NMR: a link between spectroscopy and thermodynamics. Biophys Chem 49:39–50

Urrestarazu M, Sanchez A, Lorente FA, Guzman M (1996a) Chronophysiologic rhythm model for daily ionic variation of xylematic exudates in tomato plants. Commun Soil Sci Plant Anal 27:1843–1858

Urrestarazu M, Sanchez A, Lorente FA, Guzman M (1996b) A daily rhythmic model for pH and volume from xylem sap of tomato plants. Commun Soil Sci Plant Anal 27:1859–1874

Van As H, Reinders JEA, De Jager PA, Van De Sanden PACM, Schaafsma TJ (1994) In situ plant water balance studies using a portable NMR spectrometer. J Exp Bot 45:61–67

Van Doorn WG, Jones RB (1994) Ultrasonic acoustic emissions from excised stems of two *Thryptomene* species. Physiol Plant 92:431–436

Van Doorn WG, Suiro V (1996) Relationship between cavitation and water uptake in rose stems. Physiol Plant 96:305–311

Van Splunder I, Voesenek LACJ, Coops H, De Vries XJA, Bloom CWPM (1996) Morphological responses of seedlings of four species of Salicaceae to drought. Can J Bot 74:1988–1995

Varney GT, McCully ME, Canny MJ (1993) Sites of entry of water into the symplast of maize roots. New Phytol 125:733–741

Vartanian N (1996) Mutants as tools to understand cellular and molecular drought tolerance mechanisms. Plant Growth Regul 20:125–134

Vartanian N, Marcotte L, Giraudat J (1994) Drought rhizogenesis in *Arabidopsis thaliana*. Plant Physiol 104:761–767

Verbruggen N, Villarroel R, Van Montagu M (1993) Osmoregulation of a pyrroline-5-carboxylate reductase gene in *Arabidopsis thaliana*. Plant Physiol 103:771–781

Vojtechova M, Hanson AD, Munoz-Clares RA (1997) Betaine-aldehyde dehydrogenase from amaranth leaves efficiently catalyzes the NAD-dependent oxidation of dimethylsulfoniopropionaldehyde to dimethylsulfoniopropionate. Arch Biochem Biophys 337:81–88

Wan C, Sosebee RE, McMichael BL (1994) Hydraulic properties of shallow vs. deep lateral roots in a semiarid shrub. Am Midl Nat 131:120–127

Wang XL, McCully ME, Canny MJ (1994) The branch roots of *Zea*. IV. The maturation and openness of xylem conduits in first-order branches of soil-grown roots. New Phytol 126:21–29

Wang Z, Quebedeaux B, Stutte GW (1995) Osmotic adjustment: effect of water stress on carbohydrates in leaves, stems and roots of apple. Aust J Plant Physiol 22:747–754

Wang Z, Quebedeaux B, Stutte GW (1996) Partitioning of [^{14}C]glucose into sorbitol and other carbohydrates in apple under water stress. Aust J Plant Physiol 23:245–251

Ward JM, Schroeder JI (1994) Calcium-activated K^+ channels and calcium-induced calcium release by slow vacuolar ion channels in guard cell vacuoles implicated in the control of stomatal closure. Plant Cell 6:669–683

Watt M, Van Der Weele CM, McCully ME, Canny MJ (1996) Effects of local variations in soil moisture and hydrophobic deposits and dye diffusion in corn roots. Bot Acta 109:492–501

Webb AAR, Mcainsh MR, Mansfield TA, Hetherington AM (1996) Carbon dioxide induces increases in guard cell cytosolic free calcium. Plant J 9:297–304

Wegner LH, Raschke K (1994) Ion channels in the xylem parenchyma of barley roots: a procedure to isolate protoplasts from this tissue and a patch-clamp exploration of salt passageways into xylem vessels. Plant Physiol 105:799–813

Weibel FP, De Vos JA (1994) Transpiration measurements on apple trees with an improved stem heat balance method. Plant Soil 166:203–219

Welling DJ, Welling PA, Welling LW (1996) Filled pore approximation: a theoretical framework for solute-solvent coupling in narrow water channels. Am J Physiol 270:C1246–C1254

Westgate ME, Passioura JB, Munns R (1996) Water status and ABA content of floral organs in drought-stressed wheat. Aust Plant Physiol 23:763–772

White G, Pencer J, Nickel BG, Wood JM, Hallen FR (1996) Optical changes in unilamellar vesicles experiencing osmotic stress. Biophys J 71:2701–2715

Wilkinson S, Davies WJ (1997) Xylem sap pH increase: a drought signal received at the apoplastic face of the guard cell that involves the suppression of saturable abscisic acid uptake by the epidermal symplast. Plant Physiol 113:559–573

Willmer CM, Grammatikopoulos G, Lasceve G, Vavasseur A (1995) Characterization of the vacuolar-type H$^+$-ATPase from guard cell protoplasts of *Commelina*. J Exp Bot 46:383–389

Wilson KB, Bunce JA (1997) Effects of carbon dioxide concentration on the interactive effects of temperature and water vapour on stomatal conductance in soybean. Plant Cell Environ 20:230–238

Wood AJ, Goldsbrough PB (1997) Characterization and expression of dehydrins in water-stressed *Sorghum bicolor*. Physiol Plant 99:144–152

Wood AJ, Saneoka H, Rhodes D, Joly RJ, Goldsbrough PB (1996) Betaine aldehyde dehydrogenase in sorghum. Molecular cloning and expression of two related genes. Plant Physiol 110:1301–1308

Wraith JM, Baker JM, Blake TK (1995) Water uptake resumption following soil drought: a comparison among four barley genotypes. J Exp Bot 46:873–880

Wu W-H (1995) A novel cation channel in *Vicia faba* guard-cell plasma membrane. Acta Phytophysiol Sin 21:347–354

Wu W-H, Assmann SM (1994) A membrane-delimited pathway of G-protein regulation of the guard-cell inward K$^+$ channel. Proc Natl Acad Sci USA 91:6310–6314

Wu W-H, Assmann SM (1995) Is ATP required for K$^+$ channel activation in *Vicia* guard cells? Plant Physiol 107:101–109

Wu Y, Spollen WG, Sharp RE (1994) Root growth maintenance at low water potentials: increased activity of xyloglucan endotransglyosylase and its possible regulation by abscisic acid. Plant Physiol 106:607–615

Wu Y, Sharp RE, Durachko DM, Cosgrove DJ (1996) Growth maintenance of the maize primary root at low water potentials involves increases in cell-wall extension properties, expansion activity, and wall susceptibility to expansions. Plant Physiol 111:765–772

Yamada M, Sato S (1996) Effect of hypoxia on nucleoli in excised root tips of *Vicia faba*: immunoelectron microscopy using anti-DNA antibodies. Cytologia 61:403–410

Yang S, Grantz DA (1996) Root hydraulic conductance in pima cotton: comparison of reverse flow, transpiration, and root pressurization. Crop Sci 36:1580–1589

Yu G-R, Nakayama K, Lu H-Q (1996) Responses of stomatal conductance in field-grown maize leaves to certain environmental factors over a long term. J Agric Meteorol 52:311–320

Zagdanska B, Kozdoj J (1994) Water stress-induced changes in morphology and anatomy of flag leaf of spring wheat. Acta Soc Bot Pol 63:61–66

Zhang J, Kirkham MB (1996) Antioxidant responses to drought in sunflower and sorghum seedlings. New Physiol 132:361–373

Zhang J, Tardieu F (1996) Relative contribution of apices and mature tissues to ABA synthesis in droughted maize root systems. Plant Cell Physiol 37:598–605

Zhang J, Jia W, Zhang D-P (1997) Effect of leaf water status and xylem pH on metabolism of xylem-transported abscisic acid. Plant Growth Regul 21:51–58

                                                                233

Zhu D, Scandalios JG (1994) Differential accumulation of manganese-superoxide dismutase transcripts in maize in response to abscisic acid and high osmoticum. Plant Physiol 106:173–178

Zimmermann U, Zhu JJ, Meinzer FC, Goldstein G, Schneider H, Zimmermann G, Benkert R, Thuermer F, Melcher P, Webb D, Haase A (1994) High molecular weight organic compounds in the xylem sap of mangroves: Implications for long-distance water transport. Bot Acta 107:218–229

Zotz G, Tyree MT, Cochard H (1994) Hydraulic architecture, water relations and vulnerability to cavitation of *Clusia uvitana* Pittier: a C_3-CAM tropical hemiepiphyte. New Phytol 127:287–295

Zwieniecki MA, Newton M (1995) Roots growing in rock fissures: their morphological adaptation. Plant Soil 172:181–187

Prof. Dr. Rainer Lösch
Abt. Geobotanik
Heinrich-Heine-Universität
Universitätsstraße 1/26.13
D-40225 Düsseldorf, Germany

Edited by
U. Lüttge

Dynamics of Nutrient Transport from the Root to the Shoot

By Ulrich Schurr

1 Why Study Dynamics of Nutrient Transport?

In 1990, Tanner and Beevers stated that transpiration was of minor importance for nutrient transport, since they could find no differences in growth of plants when grown at very different air humidities and therefore very different transpiration rates. This paper provoked a considerable amount of response (e.g. Smith 1991), which clearly showed the need for a critical reevaluation of processes involved in nutrient transport in the xylem sap.

Considerable interaction takes place between the nutrient transport in the xylem and water transport, since nutrients are often dragged passively with the mass flow of water. The mode of water transport in the xylem is presently under debate (e.g. Zimmermann et al. 1994, 1995a; Steudle 1995), a topic which is clearly beyond the scope of this chapter. However, it is obvious that changes in the knowledge of water transport in the xylem might influence the approach to nutrient transport in the xylem.

Nutrient transport inside the plant is a very dynamic process, which plays a crucial role in supply of the parts of the plant with basic substance and therefore has significant importance not only for academic, but also for practical reasons. Nutrient transport – obviously not only in the xylem, but also in the phloem – and the local processes of ion transport, no matter whether active or passive, are the mechanisms underlying the distribution of nutrients in the plant. Ions which are metabolised (e.g. most of the inorganic anions) usually have to be transported to the site of assimilation and, last but not least, nutrient fluxes themselves have considerable regulatory impact on, e.g. nutrient uptake and assimilation (Pitman 1988; Marschner et al. 1996).

This chapter can obviously not cover all the mentioned aspects in detail, especially since the entire field is currently undergoing significant changes, but will focus on recent work on dynamics of nutrient transport from the root to the shoot and the processes, which determine this dynamic behaviour in intact plants. The review should be seen as a momentary and subjective impression of an active area of research.

Progress in Botany, Vol. 60
© Springer-Verlag Berlin Heidelberg 1999

2 Appropriate Methods

Adequate methods to sample xylem sap are crucial to study the dynamic variations of xylem sap composition and nutrient fluxes. Since the currently available techniques have recently been reviewed (Schurr 1998), this chapter will cover only some basic comparison of the techniques relevant for the topic of nutrient dynamics. Since we are mainly interested in the nutrient dynamics under natural conditions, the basic requirement of a technique is not to interfere with the forces and processes that govern nutrient fluxes in intact plants.

Common techniques like root pressure exudation and the Scholander bomb method include action destructive to the plant and thus contain the risk of massive interference with the relevant processes determining nutrient fluxes in intact plants. This is not to say that these methods provide no relevant information, but one must always keep in mind their limitations, and calculations of ion balances based on these are questionable.

a) Root Pressure Exudation

One main difference between root pressure exudates and xylem sap from intact plants is that the flux rates at which these solutions have been sampled differ significantly. Usually, the exudation rates are in the range of a few microliter per minute, while the fluxes in intact plants across the same cross-section can be higher by up to 100 times (Schurr and Schulze 1995). This obviously causes differences in the dilution of the ions loaded into the xylem by active transport processes in the root. Application of pneumatic pressure to the root system can increase the flux to transpiration-like rates, but even though the concentrations in the exudates under these conditions are in the range of saps from intact plants, the relative abundance of various ions is changing (Schurr and Schulze 1995), due to obvious impact of the sampling method on other processes that influence xylem loading (see below).

b) Scholander Pressure Method

The classical Scholander bomb technique has been criticised over the years for many reasons (Berger et al. 1994; Zimmermann et al. 1994), especially for the risk of variation in the composition of the sap in the xylem after the abscission of the piece of plant by local water and ion fluxes (Schurr 1998). Although it is clear that results obtained with this technique have to be interpreted with care, in many cases, especially in field studies, it is the only method which can be applied. Then, however, it needs careful controls and should be evaluated against a more defined technique (McDonald and Davies 1996).

c) Xylem Pressure Probe

This most refined method for sampling of xylem sap is based on the xylem pressure probe (Balling and Zimmermann 1990), which allows individual xylem vessels to be punctured by a glass capillary. The capillary is pushed into the tissue until a negative pressure value indicates positioning in a functional xylem vessel. For sap sampling the negative pressure is compensated by application of pressure to the root system (Zimmermann et al. 1995b). The technique has its strength in mechanistical approaches to xylem transport of nutrients, especially if very small sample volumes can be handled (Bazzanella et al. 1997) and when localising tracer techniques are applied (Kuhn et al. 1995; Schröder et al. 1996). However, its application in studies on ion balances is limited by the susceptibility of the xylem to cavitation during sampling. Additionally, since significant differences in individual xylem vessels have been detected by this method (Marienfeld et al. 1996), it is necessary to collect very many samples to obtain the mean values relevant for calculation of ion balances of entire plant organs.

d) Root Pressure Chamber

This technique allows xylem sap to be sampled continuously from intact, transpiring plants, which is an essential prerequisite to studying dynamics of nutrient transport in the xylem (Schurr et al. 1998). However, it is obviously a laboratory method, since it requires sealing entire root systems inside a pressure chamber (McDonald and Davies 1996).

Xylem sap is gained by compensation of the tension inside the xylem by application of pneumatic pressure at the root system. This can be done over a period of several days without significant alteration of gas exchange, growth and phloem transport (Schurr et al. 1998b). The controlling device regulates the oxygen partial pressure in the pressure chamber additionally to adjusting the pressure itself. The limitation of the system to soil-grown plants was eliminated recently with the development of the so-called spray-pressure chamber (Schurr et al. 1998a), in which nutrient solution is supplied by nozzles to roots hanging freely in the gaseous atmosphere (aeroponic culture). The root-pressure technique is presently the only method which allows continuous sampling of xylem sap from an intact, transpiring plant.

From this short description of the methods, it is clear that each has its drawbacks in applicability and that there is a continuing need for technical development and comparison of the techniques. However, we have a suitable range of methods in our hands now for sampling xylem sap to study the dynamics of nutrient fluxes in intact plants.

3 Temporal and Spatial Variation of Sap Composition and Nutrient Fluxes in the Xylem

Strong diurnal variations in xylem sap composition of cations, anions, amino acids and in pH have been reported in sunflower, castor bean and poplar in experiments using the root-pressure chamber (Gollan et al. 1992; Schurr et al. 1992, 1998a; Schurr and Schulze 1995, 1996). In all cases studied, the diurnal variations were much larger than the trend of the mean values with, e.g. stress treatments. Even in early papers, where we had not been aware of the intensity of diurnal changes, the variation of the concentrations in individual samples around the mean values can be attributed to diurnal changes (Gollan et al. 1992). This is clear, for example, from comparison of the range of diurnal variation in pH with the range of variation in drought stress treatments (Schurr and Schulze 1996). Diurnal changes in xylem sap composition have also been found with different sampling methods and different species (Andersen and Brodbeck 1989; Andersen et al. 1993, 1995) including even tropical trees (Barker and Becker 1995).

In many other cases, diurnal variations in exudation rate were recorded (Passioura and Munns 1984; Fiscus 1986; Delhon et al. 1995b). Especially under conditions in which water is readily available for the root, this is indicative of a diurnal variation in nutrient transport into the xylem. However, the absolute values of the measured concentrations have to be handled with care, since it is clear that detachment of the shoot can considerably interfere with the processes determining nutrient transport into the xylem (Schurr 1998).

Diurnal variations were observed in all analysed substances (Schurr and Schulze 1995). Three categories of substances have been suggested: In one category the diurnal course of concentration varied inversely with transpiration rate over the whole day. A second group showed high concentrations during the night and lower concentrations during the day, but concentrations varied during the light period despite constant transpiration rates. The concentrations of the third category changed proportionally to transpiration, thus the dilution effect of transpirational flux was overridden by other processes (Schurr and Schulze 1995). However, recent experiments show that amplitude and diurnal pattern of individual nutrients are very dependent on the availability of the individual nutrient for the root system (Schurr et al. 1998b).

Mass fluxes of nutrients are dependent on the relation between concentration in the xylem sap and flux rate of the xylem. However, it has to be kept in mind that additional pathways for ion transport can contribute to the distribution of ions in the plant. Ion transport in the xylem has some similarities with a liquid chromatographic system, since there is a "mobile phase" containing ion species, which can be exchanged with the surrounding "stationary phase" (Wolterbeek et al. 1984; Wolterbeek 1987; Senden et al. 1992, 1994). The composition of the exchanging compartments is rather variable, and their exchange capacity can be varied by physicochemical and metabolic means, since it depends on passive and active exchange processes along the xylem. These properties of xylem transport have the potential to alter diurnal variations of mass transport in the transpiration stream by lateral exchange; however, systematic experiments still have to be done.

Mass flux-driven nutrient transport in the xylem can be calculated as the product of transpirational mass flux and concentration of xylem sap. Since we are interested in the nutrient fluxes in intact plants, we need to use methods which give concentration and transpiration values as they exist in intact plants. Appropriate methods for xylem sap sampling have already been described, but one has additionally to keep in mind that estimation of transpiration rate in standard gas exchange systems may not be relevant, since the conditions inside a gas exchange cuvette can be considerably different from freely transpiring plants. For example, when gas exchange systems with intensive mass fluxes of gas or fans inside the curvette are used, the boundary layer at the leaf surface is significantly smaller than in an undisturbed situation. This is intended when stomatal properties are to be studied, but it is obviously not the best way to quantitatively determine the relevant water fluxes for ion transport in undisturbed plants. It is not sufficient to simply lower the flux rate or the turbulent flow at the leaf surface, sine this can have equally disturbing impact on the driving forces of transpiration by altering the water concentration around the leaf. Probably the easiest way is to weigh the plants carefully to obtain the entire shoot transpiration rate. Alternatively techniques need to be developed to measure and localise water loss in undisturbed conditions for leaves and parts of leaves, in order to obtain the data sets needed for calculations of localised ion fluxes.

Dynamics of concentrations and fluxes of substances in the xylem have been modelled on the basis of their interactions (Fiscus 1986). This has been done extensively for whole plants in recent years, especially for the transport of abscisic acid, which is involved in root-shoot signalling in drought-stressed plants (Davies and Zhang 1991). Models have been established which predict that the concentration and the flux of abscisic acid will be constant throughout the day due to the inversely proportional action of release in the root and water flux (Tardieu and Davies 1993; Jarvis and Davies 1997). This prediction seems to be true under certain conditions in the field; however, the relations between release and dilution do not seem to be tightly controlled, since significant increase in abscisic acid during the day has been observed in *Ricinus* during well-watered and drought-stressed conditions (Schurr and Schulze 1996). It is not astonishing that different drought conditions result in different dynamics of abscisic acid concentration and flux in the xylem, since many parameters play a role in the availability of abscisic acid for transport in the xylem including passive physiochemical processes (Hartung and Slovik 1991; Slovik et al. 1995) as well as metabolic control (Hartung et al. 1998; Wilkinson and Davies 1997).

4 Processes Potentially Involved in Temporal Variation

It is far from clear which processes contribute under which conditions to the temporal variation of xylem sap. However, it is possible to list a number of processes for which it is likely that they might contribute at least under certain conditions. This section emphasis the view that temporal and spatial variations are more likely to be normal than a constant xylem sap composition and that the plant has to cope with them, since they may have impact on leaf physiology via the apoplast concentration not only for stomatal control (Ruiz et al. 1993; Hartung et al. 1998), but also for nutrient fluxes.

a) Availability to the Root

Nutrient availability can vary in time and space due to many different processes in the soil (Nye and Tinker 1977; Cameron and Haynes 1986; Marschner 1995). However, in the context of this chapter, we are mainly interested in changes in nutrient availability that could be related to diurnally fluctuating nutrient fluxes in the xylem. Since these patterns of nutrient transport occur in soil and nutrient culture, it seems unlikely that they are due to changes in nutrient availability for uptake into the plant. Thus, nutrient availability is more likely to be involved in the amplitude of diurnal nutrient dynamics than in the pattern itself (Schurr et al. 1998a).

b) Uptake

Nutrient transport systems (Clarkson and Lüttge 1991) and systems involved in nitrate and ammonium uptake in roots have recently been reviewed in this series (Peuke and Kaiser 1996). Though the molecular approach has given important additional information since then (Tsay et al. 1993; Touraine and Glass 1997 and references therein), this review will concentrate on the physiological importance of uptake for the root-to-shoot supply and its possible role in the dynamics.

In many cases, nutrient uptake has been reported to vary diurnally (Clement et al. 1978; Hanson 1980; Delhon et al. 1995a, b) and measurement systems with high temporal resolution are available (Hatch et al. 1986; Macduff et al. 1997). The diurnal course of, e.g. nitrate uptake is rather variable. Macduff et al. (1997) found the highest net nitrate uptake during the late light period. Steingröver et al. (1986a, b) obtained a maximum of nitrate uptake during the first half of the dark period and Delhon et al. (1995a) report on a continuous decline in nitrate uptake immediately after the end of the light period. This is a situation which corresponds well to the variable diurnal courses of xylem sap composition and nutrient fluxes in the xylem sap (Schurr et al. 1998a) and may be related to differences in plant species, nutrient conditions (Steingröver et al. 1982; Macduff and Wild 1988) or even to the dual role of nitrate as key metabolite for nitrogen assimilation (Marschner 1995) and as an important osmoticum in the vacuole of the leaf (Steingröver et al. 1982, 1986b; Bloom-Zandstra and Lange 1985, Stulen 1985; Veen and Kleinendorst 1985). The major variation seems to be in the influx component of nitrate, while the efflux remains rather constant (Pearson et al. 1981; Delhon et al. 1995a).

Nutrient uptake into the root is due to active and passive processes. Since active processes are energy-dependent, energy limitation could be relevant to diurnal variations of nutrient uptake and therefore indirectly to nutrient transport to the shoot. Close correlations were observed between the diurnal course of import into the root system via the phloem (Rideout and Raper 1994) as well as for root respiration rate (Hansen 1980) and root nitrate uptake. Under permanent light, nitrate uptake did not decline (Delhon et al. 1995a). This is astonishing on the basis of the estimation of energy consumption of nutrient uptake to be only 5% of the entire carbon catabolism in nitrate-supplied plants (Bloom et al. 1992). This is significantly less than the 15% attributed to nitrate assimilation in the root system. However, nitrate uptake might be even less competitive for carbohydrates than nitrate reduction in comparison with root growth (Radin et al. 1978). At present it is not clear, if nitrate uptake is limited by the supply of carbon skeletons or energy per se, since external feeding of carbohydrates gave contradictory results (Touraine et al. 1988; Delhon et al. 1996b).

Nevertheless, nutrient uptake is dependent on a functional phloem import into the root system, since girdling and lowering of CO_2 concentration at the shoot prevented the increase in nitrate uptake when light was switched on (Delhon et al. 1996b). This supports findings in maize seedlings grown at low CO_2 concentration (Pace et al. 1990), where the effect was less dramatic, probably due to continued supply of carbon skeletons from the endosperm. This adverse effect could be overcome by addition of glucose to the rooting medium. Two hours after detachment of the shoot, root uptake rates for potassium, ammonium, nitrate and chloride declined in *Hordeum vulgare* (Bloom and Caldwell 1988). Clement et al. (1978) already noted that uptake rates into the root declined when they reduced the carbon status of the shoot by various means. It is not likely that nitrate uptake is subjected to a cascade of signals from the shoot, since defoliation, which also caused a decline in carbon availability to the root system, increased uptake rate of nitrate (Macduff and Jackson 1992) in order to sustain regrowth of leaves.

Close interactions between a variety of ions have been observed (Lee and Drew 1989; Jackson and Volk 1995; Macduff et al. 1997). Especially potassium and nitrate seem to be closely linked in their uptake (Macduff et al. 1997). They show the same diurnal variation and a similar amplitude during the day. Since nitrate and potassium also exhibit a strong correlation during the diurnal course of the xylem sap composition (Schurr and Schulze 1995), this is clearly a hint that either uptake plays a crucial role for the diurnal variation in nutrient transport or that similar processes, but with an opposite direction, are active in nutrient uptake at the root cortex and in xylem loading. Nevertheless, the amount of potassium taken up is less than the amount of nitrate, which indicates a significant proportion of potassium being cycled inside the plant (see below).

Regulation of nitrate uptake depends on a series of signals from inside the root and from the shoot. Nitrate uptake has been found to decline in response to accumulation of nitrate in the root system (Siddiqi et al. 1990; Mattsson et al. 1993; Imsande and Touraine 1994), while other studies indicate a demand-driven control for nutrients by the shoot (Engels et al. 1992; Engels and Marschner 1993, Imsande and Touraine 1994). However, the relation between diurnal variation of growth rate (Schurr 1997) and nutrient uptake or transport towards the shoot still needs to be established. Since the capacity of the shoot for nutrient assimilation varies with daytime, it is sensible to suggest that the diurnal variations in nutrient uptake (e.g. Macduff et al. 1997) are linked to the supply-demand status inside the plant. This is shown for phosphate (Clarkson 1988; Pitman 1988), sulphur (Rennenberg and Herschbach 1995; Lappartient and Touraine 1996) and most elaborately, for nitrate uptake (Imsande and Touraine 1994). Steingröver et al. (1986a, b) proposed a close link between the concentration of nitrate in the vacuoles of leaf blades with nitrate uptake on the basis of their studies with spinach. However, it has not been tested yet if nitrate uptake is controlled by nitrate concentration or nitrate flux into the leaf, as has been proposed for nitrate reductase (Shaner and Boyer 1976; Barneix et al. 1984; Ingemarson 1987). Studies with barley mutants deficient in their nitrate reductase (King et al. 1993) show an inhibition of $^{13}NO_3^-$ uptake, which hints at a signal directly from the nitrate pool itself, because these plants have low organic nitrogen status, indicating nitrogen limitation. This signal or the controlled process cannot prevent a massive accumulation of nitrate in the shoots, as has been observed in tobacco mutants and transformants with reduced nitrate reductase activity (Scheible et al. 1997).

The nature of the signal(s) from the shoot is not clear yet. Malate (Touraine et al. 1992) and amino acids transported in the phloem have been supposed in the case of nitrate (Clarkson 1985, Cooper et al. 1986; Cooper and Clarkson 1989; Muller and Touraine 1992; Imsande and Touraine 1994; Barneix and Causin 1996). External application of amino acids has been shown to reduce nitrate uptake rate in a series of plant species (Doddema and Otten 1979; Breteler and Arnozis 1985; Lee et al. 1992), with the influx system being affected by the treatment (Muller et al. 1995). However, the effectiveness of amino acids on the uptake during the diurnal course has not been thoroughly studied. Malate and other carboxylases have been proposed in the context of the Dijkhoorn-Ben Zioni hypothesis as being transported to the root to maintain charge balance when nitrate is reduced in the shoot (Ben Zioni et al. 1971). While a number of studies have calculated a considerable amount of carboxylastes transported towards the root from the shoot (e.g. Touraine et al. 1988), others indicate that most of the carboxylates remain in the shoot, which would diminish their potential role for controlling nitrate uptake (Peuke et al. 1996). Again, this might be dependent on the relative contribution of nitrate and organic acids to the osmotic status of the leaf (Blom-Zandstra and Lange 1985). External application of malate has not been found to release the inhibition of nitrate uptake during darkness (Delhon et al. 1996a), while it increased nitrate uptake during the light period (Muller and Touraine 1992; Delhon et al. 1996a). On the other hand, glucose enhanced nitrate uptake when malate did not and vice versa, which again hints at a multilayer signal. Since such signals will be transmitted via the phloem, dynamic changes in phloem transport (Rideout and Raper 1994; Schurr et al. 1998b) to the root might have significant impact on nutrient uptake at the root system, and even the decline of nitrate uptake in response to a low humidity-high transpiration treatment has been explained on the basis of the lack of signals from the shoot under these conditions (Brewitz et al. 1996).

c)   Transport Through the Root and Xylem Loading

It is difficult in the intact plant to distinguish between the direct effects of xylem loading in the stele and the impact of transport of the nutrient towards the loading site. It is beyond the scope of this chapter to list all the evidence for the different passage pathways of the individual nutrients towards the stele. It should be made clear, however, that since the individual ions take different routes towards the xylem (Läuchli 1976; Clarkson 1988), they are clearly subjected to different processes of storage and release in the root system. Nutrients that are assimilated by the plant are additionally subjected to a diurnally varying metabolic activity of the root, which may have impact on the amounts delivered to the sites of xylem loading. Delivery of nutrients to the xylem-loading sites also depends on the abundance of other ions. For example, it has been suggested that a high availability of potassium at the root can favour nitrate throughput to the xylem over local nitrate reduction in the root itself (Förster and Jeschke 1993).

Water flux through the root system can have significant influence on the partitioning of nutrients between the different compartments in the root. Delhon et al. (1995b) suggested that the reason for the accumulation of nitrate and ^{15}N compounds in the root during the night is a change in partitioning of nitrate between export into the xylem and assimilation/storage in the root system due to the decline of the transpiration flux. Intermittent storage and delayed export could even explain observations that during the light period more nitrate can be reduced than has been taken up (Rufty et al. 1984). Exchange of ions with compartments parallel to the pathway towards the xylem is altered by transpirational flux (Jeschke 1984). Compartmental analysis using ^{15}N nitrate identified at least two compartments (probably cytoplasm and vacuole) in parallel to the pathway towards the xylem (Devienne et al. 1994a, b). Even interactions between the nutritional status due to a single ion (sulphate) and the delivery of nitrogen to the shoot have been attributed to variations of lateral fluxes in the root in response to changes in water conductivity of the root system (Karmoker et al. 1991; Carvajal et al. 1995), plasma membrane fluidity (Carvajal et al. 1996a) and water channel function (Carvajal et al. 1996b). These changes might also be linked to the diurnal variation of the reflection coefficients of the entire root system with transpiration rate in intact plants (Zhu et al. 1995, Schneider et al. 1997). These would be congruent with a higher degree of unstirred layers in the root system during the night, which could favour nutrient uptake into the cell over nutrient transport to the shoot – a yet unproved, but interesting, prespective for the change in nutrient supply to the xylem loading sites, at least for those ions transported in the apoplast. The buffering effect and possibly intermittent storage in the root depends on the release characteristics when transpiration starts off again.

Xylem loading is another crucial step by which the plant can regulate transport towards the shoot (Engels and Marschner 1992). This is most clearly shown by mutants, which have sufficient uptake of, e.g. phosphate into the root, but lack the possibility to transport it to the shoot (Poirier et al. 1991). Analysis of ^{15}N nitrate during light and dark periods also indicate that the decline in transport towards the shoot is stronger than the decline in nitrate uptake, which causes an accumulation of nitrate in the root, while the shoot showed no diurnal variation in the same experiments (Delhon et al. 1995a). Similar indications exist from experiments in which additional sulphur was given to the shoots by gaseous supply. Demand-driven control acts on sulphur transport between roots and shoots by affecting xylem loading (Herschbach et al. 1995) and root uptake (Lappartient and Touraine 1996).

Xylem-loading studies have often used perfused roots (Clarkson and Hanson 1986; Lacan and Durand 1996), since experiments at the level of membrane transport have been hampered until recently by the lack of methods to isolate the specific cells that fulfil this role in the root. Since the root cortex consists of a much higher number of cells than the stele, protoplast isolations from whole roots contained only a minor share of relevant cells. Xylem parenchyma cells isolated by progressive digestion of barley roots manifested three cation-specific rectifying functions and one anion-specific transport function (Wegner and Raschke 1994). Besides one inward-rectifying channel (KIRC), two outward-rectifying channels were characterised (Wegner et al. 1994, De Boer and Wegner 1997) and shown to be influenced by different cytoplasmic and apoplastic effectors. Protoplasts isolated from the stele or the cortex separately show distinct features of their ion transport system (Roberts and Tester 1995), with the predominantly inward-rectifying properties of the cortex cells favouring uptake from the apoplast, while the outward-rectifying channels in the stele might be responsible for xylem loading (Roberts and Tester 1997). On the basis of these analyses, the different channels were proposed to be involved in the control of root-to-shoot transport of nutrients (De Boer and Wegner 1997).

d) Lateral Exchange on the Transport Pathway to the Shoot

Lateral exchange of nutrients is due to loading or unloading processes along the xylem pathway to the shoot. It involves uptake into living cells including, e.g. surrounding parenchyma and phloem as well as the ion exchange characteristics of mainly the cell walls of the xylem strands and the surrounding tissue (Wolterbeek et al. 1984; Wolterbeek 1987). The latter phenomenon has not gained much attentin for "classical" nutrients, while significant effects have been shown for the transport of heavy metals towards the shoot (Senden et al. 1992, 1994 and references therein). The exchange capacity of the cell wall is large (Richter and Dainty 1989a, b) and the retention of calcium and even potassium relative to water transport is significant and of importance during dynamic changes of xylem sap composition (Marienfeld et al. 1996; Schröder et al. 1996). In conjunction with metabolically controlled changes in organic acid concentration in the xylem sap for complexation of ions (Wolterbeek et al. 1984; Wolterbeek 1987) and pH variations (Gerendas and Schurr 1998), sorption-desorption processes could play an important role.

Unloading along the xylem transport pathway provides the surrounding tissues with nutrients and has the potential to buffer transport in the xylem by intermittent storage (Jeschke and Pate 1995). Perfusion experiments have been performed showing a homeostatic action of the surrounding tissue on the pH (Clarkson and Hanson 1986). In these experiments, root systems were detached and perfused with solutions of rather different composition. The pH at the outlet of the perfused tissue remained rather constant throughout the experiment even when very acidic or alkaline solutions were supplied. This was explained by the action of proton pumps, which has to be questioned on the basis of physiochemical analysis of the xylem sap (Gerendas and Schurr 1998). The relevance of this homeostatic function in intact plants is doubtful, since significant variations of pH have been found in the xylem sap of intact plants (Gollan et al. 1992; Schurr and Schulze 1995, 1996). One problem of perfusion systems might be the dysfunction of the phloem, since significant phloem-xylem and xylem-phloem transfer of ions and reduced nitrogen have been analysed in carbon/ion models (Jeschke and Pate 1991; Jeschke et al. 1996). This transfer can occur either via parenchymatic cells or via specialised transfer or contact cells (Sauter 1982; van der Shoot 1989).

5 Nutrient Fluxes on the Whole Plant Level

Nutrient cycling plays important roles in the plant (Marschner et al. 1996), including nutrition-related as well as regulatory functions. Thus, a range of different methods has been used to study this topic.

Modelling of ion balances on the basis of cabon and nitrogen fluxes (Jeschke and Pate 1991; Jeschke et al. 1994a) has been very successful on the level of the individual leaf (Jeschke and Pate 1992), for quantification of the distribution fluxes in whole plants at different stresses and nutrient regimes (e.g. Peuke and Jeschke 1993; Jeschke et al. 1996) and even for the quantitative analysis of fluxes between host and parasite plants (Jeschke et al. 1994a, b; Tennakoon and Pate 1997). However, these models provide net fluxes of carbon, nitrogen and nutrients over a period of several days (usually over a 9-day period, Jeschke and Pate 1991), which are the results of obviously much more dynamic and ample fluxes in xylem and phloem (see above). It is this diurnal time scale of fluxes which needs to be to analysed in order to understand regulation processes on the whole-plant level which integrate over the different organs and their function.

Tracer studies using radioactive or stabile isotopes can be used to determine fluxes within intact plants (e.g. Delhon et al. 1995a; Clarkson et al. 1996; Schurr et al. 1998a). The temporal resolution of this approach depends mainly on the analytical technique and the intensity of label (Clarkson et al. 1996). Very high temporal resolutions can be obtained by short-lived isotopes like ^{13}N and ^{11}C, but these require cyclotron access. Stable isotopes can be used for labelling fluxes as well as in some cases for localising and even imaging of ion distribution (Massiot et al. 1994; Kuhn et al. 1995; Schröder et al. 1996; Gojon et al. 1996). Analysis of tracer velocities includes the isotopic dilution, which can be significant if the unlabelled species is present in large amounts in the relevant compartments. To reduce this problem, ^{15}NO$_3$-tracer studies for nutrient transport inside the whole plant often use nitrate-depleted plants (e.g. Brewitz et al. 1996) and relatively long (several hours') labelling periods (Cooper and Clarkson 1989; Delhon et al. 1995a; Brewitz et al. 1996). Since retranslocation towards the root system can be considerable during this period, split-root systems are used in which label is given to only one part of the root system (Cooper and Clarkson 1989; Delhon et al. 1995a; Brewitz et al. 1996). Tracer in the non-labelled root system (receiver roots) can then be treated as retranslocation to the root system from the shoot. However, this approach is limited to those ions and substances that can be labelled isotopically, analysis is rather complex and relatively expensive, and splitting the root system can disturb the behaviour of the plant.

Root pressure chamber and transpiration measurements of the shoot allow a quasicontinuous analysis of the transport in the xylem with a high temporal resolution (Schurr 1998; Schurr et al. 1998a). This method additionally allows several ion species and non-ionic substances (amino acids, organic acids, xenobiotica) to be monitored simultaneously at the same temporal resolution. Spatial resolution can be obtained by opening several sampling sites along the plant architecture. A combination of this approach with nutrient uptake experiments and tracer methods has a high potential for further understanding the dynamics of nutrient transport on a whole-plant level.

6 Prospects

Nutrient transport form the root to the shoot is a dynamic and complex process which is important for whole plant fluxes of nutrients and signals. The methodological developments during recent years – especially the non-destructive techniques – provide the basis for temporal and also spatial resolution sufficient to study actual fluxes of nutrients and signals and will lead to a better understanding of the reasons and consequences of the dynamic processes involved. The combination of the existing and newly developed techniques with molecular approaches has the potential to identify key points of control in plant nutrition with relevance for basic and applied research.

Acknowledgements. The author is grateful to Anne Krapp for critical reading of the manuscript and the Deutsche Forschungsgemeinschaft (DFG) for the generous support within the SFB 199 TP A9 and TP C1 and the research project Apoplast of Higher Plants.

References

Andersen PC, Brodbeck BV (1989) Diurnal and temporal changes in the chemical profile of xylem exudate from *Vitis rotundifolia*. Physiol Plant 75:63–70

Andersen PC, Brodbeck BV, Mizell RF (1993) Diurnal variation of amino acids and organic acids in xylem fluid from *Lagerstroemia indica*: an endogenous circadian rhythm. Physiol Plant 89:783–790

Andersen PC, Brodbeck BV, Mizell RF (1995) Diurnal variation in tension, osmolality, and the composition of nitrogen and carbon assimilates in xylem fluid of *Prunus persica. Vitis* hybird, and *Pyrus communis*. J Am Soc Hortic Sci 120:600–606

Balling A, Zimmermann U (1990) Comparative measurements of the xylem pressure of *Nicotiana* plants by means of the pressure bomb and pressure probe. Planta 182:325–338

Barker M, Becker P (1995) Sap flow rate and sap nutrient content of a tropical rain forest canopy species, *Dryobalanops aromatica*, in Brunei. Selbyana 16:201–211

Barneix AJ, Causin HF (1996) The central role of amino acids on nitrogen utilisation and plant growth. J Plant Physiol 149:358–362

Barneix AJ, James DM, Watson EF, Hewitt EJ (1984) Some effects of nitrate abundance and starvation on metabolism and accumulation of nitrogen in barley (*Hordeum vulgare* L. cv. Sonja). Planta 162:469–476

Bazzanella A, Lochmann H, Mainka A, Bächmann K (1997) Determination of inorganic anions, carboxylic acids and amino acids in plant matrices by capillary electrophoresis. Chromatographia 45:59–62

Ben Zioni A, Vaadia Y, Lips SH (1971) Nitrate uptake by roots as regulated by nitrate reduction products of the shoot. Physiol Plant 24:288–290

Berger A, Oren R, Schulze ED (1994) Element concentrations in the xylem sap of *Picea abies* (L.) Karst. seedlings extracted by various methods under different environmental conditions. Tree Physiol 14:111–128

Bloom AJ, Caldwell MM (1988) Root excision decreases nutrient absorption and gas fluxes. Plant Physiol 87:794–796

Bloom AJ, Sukrapanna SE, Warner RH (1992) Root respiration associated with ammonium and nitrate absorption and assimilation in barley. Plant Physiol 99:1294–1301

Blom-Zandstra M, Lange EM (1985) The role of nitrate in the osmoregulation of lettuce (*Lactuca sativa* L.) grown at different light intensities. J Exp Bot 36:1043–1052

Breteler H, Arnozis P (1985) Effect of amino compounds on nitrate utilisation by roots of dwarf bean. Phytochemistry 24:653–658

Brewitz E, Larsson CM, Larsson M (1996) Responses of nitrate assimilation and N translocation in tomato (*Lycopersicon esculentum* Mill.) to reduced ambient air humidity. J Exp Bot 47:855–861

Cameron KC, Haynes RJ (1986) Retention and movement of nitrogen in soils. In: Haynes RJ (ed) Mineral nitrogen in the plant-soil-system. Academic Press, Orlando

Carjaval M, Cooke DT, Clarkson DT (1995) The effect of nutrient deprivation on the biochemical and biophysical properties of wheat root plasma membranes and their relation to root hydraulic conductivity. J Exp Bot 46:51–58

Carjaval M, Cooke DT, Clarkson DT (1996a) Plasma membrane fluidity and hydraulic conductance in wheat roots: interaction between root temperature and nitrate or phosphate deprivation. Plant Cell Eniviron 19:1110–1114

Carjaval M, Cooke DT, Clarkson DT (1996b) Responses of wheat plants to nutrient deprivation may involve the regulation of water-channel function. Planta 199:372–381

Clarkson DT (1985) Regulation of the absorption and release of nitrate by plant cells: a review of current ideas and methodology. In: Lambers H, Neeteson J, Stulen I (eds) Physiological, ecological and applied aspects of nitrogen metabolism in higher plants. Nijhoff and Junk, Den Haag, 158–179

Clarkson DT (1988) Movement of ions across roots. In: Baker DA, Hall JL (eds) Solute transport in plant cells and tissues. Longman, New York, pp 251–303

Clarkson DT, Hanson JB (1986) Proton fluxes and the activity of a stelar proton pump in onion roots. J Exp Bot 37:1136–1150

Clarkson DT, Lüttge U (1991) Mineral nutrition: inducible and repressible nutrient transport systems. Prog Bot 52:61–83

Clarkson DT, Gojon A, Saker LR, Woersema PK, Purves JV, Tillard P, Arnold GM, Paans AJM, Vaalsburg W, Stulen I (1996) Nitrate and ammonium influxes in soybean (*Glycine max*) roots: direct comparison of ^{13}N and ^{15}N tracing. Plant Cell Environ 19:859–868

Clement CR, Hopper MJ, Jones LHP, Leafe EL (1978) The uptake of nitrate by *Lolium perenne* from flowing nutrient solution. II. Effect of light, defoliation, and relationship to CO_2 flux. J Exp Bot 29:1173–1183

Cooper HD, Clarkson DT (1989) Cycling of amino-nitrogen and other nutrients between shoots and roots in cereals – a possible mechanism integrating shoot and root in regulation of nutrient uptake. J Exp Bot 40:753–762

Cooper HD, Clarkson DT, Johnson M, Whiteway J, Loughman BC (1986) Cycling of amino-nitrogen between shoots and roots in wheat seedlings. Plant Soil 91:319–322

Davies WJ, Zhang J (1991) Root signals and the regulation of growth and development of plants in drying soil. Annu Rev Plant Mol Biol 42:55–76

De Boer AH, Wegner LH (1997) Regulatory mechanisms of ion channels in the xylem parenchyma. J Exp Bot 48:441–449

Delhon P, Gojon A, Tillard P, Passama L (1995a) Diurnal regulation of NO_3^- uptake in soybean plants. I. Changes in NO_3^- influx, efflux, and N utilisation in the plant during the day/night cycle. J Exp Bot 46:1585–1594

Delhon P, Gojon A, Tillard P, Passama L (1995b) Diurnal regulation of NO_3^- uptake in soybean plants. II. Relationship with accumulation of NO_3^- and asparagine in the roots. J Exp Bot 46:1595–1602

Delhon P, Gojon A, Tillard P, Passama L (1996a) Diurnal regulation of NO_3^- uptake in soybean plants. III. Implication on the Dijkshoorn-Ben Zioni model in relation with the diurnal changes in NO_3^- assimilation. J Exp Bot 47:885–892

Delhon P, Gojon A, Tillard P, Passama L (1996b) Diurnal regulation of NO_3^- uptake in soybean plants. IV. Dependence on current photosynthesis and sugar availability to the roots. J Exp Bot 47:893–900

Devienne F, Mary B, Lamaze T (1994a) Nitrate transport in intact wheat roots. I. Estimation of cellular fluxes and NO_3^- distribution using compartmental analysis from data of $^{15}NO_3^-$ efflux. J Exp Bot 45:667–676

Devienne F, Mary B, Lamaze T (1994b) Nitrate transport in intact wheat roots. II. Long-term effects of NO_3^- concentration in the nutrient solution on NO_3^- unidirectional fluxes and distribution within the tissues. J Exp Bot 45:677–684

Doddema H, Otten H (1979) Uptake of nitrate by mutants of *Arabidopsis thaliana*, disturbed in uptake or reduction of nitrate. III. Regulation. Physiol Plant 45:339–346

Engels C, Marschner H (1992) Adaptation of potassium transport into the shoot of maize (*Zea mays*) to shoot demand: evidence for xylem loading as a regulatory step. Physiol Plant 86:263–268

Engels C, Marschner H (1993) Influence of the form of nitrogen supply on root uptake and translocation of cations in the xylem exudate of maize (*Zea mays* L.). J Exp Bot 44:1695–1701

Engels C, Münkle L, Marschner H (1992) Effect of root zone temperature and shoot demand on uptake and xylem transport of macronutrients in maize (*Zea mays* L.). J Exp Bot 43:537–547

Fiscus EL (1986) Diurnal changes in volume and solute transport coefficients in *Phaseolus vulgaris* roots. Plant Physiol 80:752–759

Förster JC, Jeschke WD (1993) Effects of potassium withdrawal on nitrate transport and the contribution of the root to nitrate reduction in the whole plant. J Plant Physiol 141:322–328

Gerendas J, Schurr U (1988) Physicochemical aspects of ion relations and pH regulation in the apoplast – a quantitative approach. J Exp Bot (submitted)

Gojon A, Grignon A, Tillard P, Massiot P, Levebvre F, Thellier M, Ripoll C (1996) Imaging and microanalysis of ^{14}N and ^{15}N by SIMS microscopy in yeast and plant samples. Cell Mol Biol 42:351–360

Gollan T, Schurr U, Schulze ED (1992) Stomatal response to drying soil in relation to changes in the xylem sap composition of *Helianthus annuus*. I. The concentration of cations, anions, amino acids and the pH of the xylem sap. Plant Cell Environ 15:551–559

Hansen GK (1980) Diurnal variation of root respiration rates and nitrate uptake as influenced by nitrogen supply. Physiol Plant 48:421–427

Hartung W, Slovik S (1991) Physiochemical properties of plant growth regulators and plant tissues determine their distribution and redistribution: stomatal regulation by abscisic acid in leaves. New Phytol 119:361–382

Hartung W, Wilkinson S, Davies WJ (1998) Factors that regulate abscisic acid concentrations at the primary site of action at the guard cell. J Exp Bot (in press)

Hatch DJ, Hopper MJ, Dhanos MS (1986) Measurement of ammonium ions in the flowing solution culture and diurnal variation in uptake by *Lolium perenne* L. J Exp Bot 7:589–596

Herschbach C, De Kok LJ, Rennenberg H (1995) Net uptake of sulphate and its transport to the shoot in spinach plants fumigated with H_2S or SO_2: does atmospheric sulphur affect the "inter-organ" regulation of sulphur nutrition? Bot Acta 108:41–46

Imsande J, Touraine B (1994) N demand and the regulation of nitrate uptake. Plant Physiol 105:3–7.

Ingermarsson B (1987) Nitrogen utilisation in *Lemna*. I. Relations between net nitrate flux, nitrate reduction, and in vivo activity and stability of nitrate reductase. Plant Physiol 85:856–859

Jackson WA, Volk RJ (1995) Attributes of the nitrogen uptake systems of maize (*Zea mays* L.): maximal suppression by exposure to both nitrate and ammonium. New Phytol 130:327–335

Jarvis AJ, Davies WJ (1997) Whole plant ABA flux and the regulatin of water loss in *Cedrella odorata*. Plant Cell Environ 20:521–527

Jeschke WD (1984) Effects of transpiration on potassium and sodium fluxes in root cells and the regulation of ion distribution between roots and shoots of barley plants. J Plant Physiol 117:267–285

Jeschke WD, Pate JS (1991) Modelling of the partitioning, assimilation and storage of nitrate within root and shoot organs of castor beans (*Ricinus communis* L.). J Exp Bot 42:1091–1103

Jeschke WD, Pate JS (1992) Temporal patterns of uptake, flow and utilisation of nitrate, reduced nitrogen and carbon in a leaf of salt-treated castor bean (*Ricinus communis* L.). J Exp Bot 43:393–402

Jescke WD, Pate JS (1995) Mineral nutrition and transport in the xylem and phloem of *Banksia prionotes* (Protaceae), a tree with dimorphic root morphology. J Exp Bot 46:895–905

Jeschke WD, Räth N, Bäumel P, Czygan FC, Proksch P (1994a) Modelling of the flows and partitioning of carbon and nitrogen in the holoparasite *Cuscuta refluxa* Roxb. and its host *Lupinus albus* L. I. Methods for estimating flows. J Exp Bot 45:791–800

Jeschke WD, Bäumel P, Räth N, Czygan FC, Proksch P (1994b) Modelling of the flows and partitioning of carbon and nitrogen in the holoparasite *Cuscuta refluxa* Roxb. and its host *Lupinus albus* L. II. Flows between host and parasite and within the parasite host. J Exp Bot 45:801–812

Jeschke WD, Peuke A, Kirkby EA, Pate JS, Hartung W (1996) Effects of P deficiency on the uptake, flows and utilisation of C, N and H_2O within intact plants of *Ricinus communis* L. J Exp Bot 47:1737–1754

Karmoker JL, Clarkson DT, Saker LR, Rooney JM, Purveys JV (1991) Sulphate deprivation depresses the transport of nitrogen to the xylem and the hydraulic conductivity of barley (*Hordeum vulgare* L.) roots. Planta 185:269–278

King BJ, Siddiqi MY, Ruth TJ, Warner RL, Glass ADM (1993) Feedback control of nitrate influx in barley roots by nitrate, nitrite and ammonium. Plant Physiol 102:1279–1286

Kuhn A, Bauch J, Schröder WH (1995) Monitoring uptake and contents of Mg, Ca and K in Norway spruce as influenced by pH and Al using microprobe analysis and stable isotope labelling. Plant Soil 168/169:135–150

Lacan D, Durand M (1996) Na^+-K^+-exchange at the xylem/symplast boundary. Plant Physiol 110:705–711

Lappartient AG, Touraine B (1996) Demand-driven control of root ATP sulfurylase activity and SO_4^{2-} uptake in intact canola. Plant Physiol 111:147–157

Läuchli A (1976) Apoplastic transport in plants. In: Lüttge U, Mitman MG (eds) Encyclopedia of plant physiology, vol 2B. Springer, Berlin Heidelberg New York, pp 3–34

Lee RB, Drew MC (1989) Rapid, reversible inhibition of nitrate influx in barley by ammonium. J Exp Bot 40:741–752

Lee RB, Purves JV, Ratcliffe RG, Saker LR (1992) Nitrogen assimilation and the control of ammonium and nitrate absorption by maize roots. J Exp Bot 43:1385–1396

Macduff JH, Wild J (1988) Changes in NO_3^- and K^+ uptake by four species in flowing nutrient solution culture in response to increased irradiance. Physiol Plant 74:251–256

Macduff JH, Jackson SB (1992) Influx and efflux of nitrate and ammonium in Italian ryegrass and white clover roots: comparison between effects of darkness and defoliation. J Exp Bot 43:525–535

Macduff JH, Bakken AK, Dhanoa MS (1997) An analysis of the physiological basis of commonality between diurnal patterns of NH_4^+, NO_3^- and K^+-uptake by *Phleum pratense* and *Festuca pratensis*. J Exp Bot 48:1691–1701

Marienfeld S, Zhu JJ, Koyro HW, Schröder W, Zimmermann U (1996) Direkte Probeentnahme aus dem Xylem: Vergleich mit konventionellen Proben. Proceedings of the German Botanical Society, Düsseldorf, Abstr. 325

Marschner H (1995) Mineral nutrition of higher plants. Academic Press, Harcourt Brace, New York

Marschner H, Kirkby EA, Cakmak I (1996) Effect of mineral nutritional status on shoot-root partitioning of photoassimilates and cycling of mineral nutrients. J Exp Bot 47:1255–1263

Massiot P, Sommer F, Gojon A, Grignon N, Lefebvre F, Thellier M, Ripoll C (1994) Comparison of three different methods of analytical imaging of the stable isotopes of nitrogen for application to plant studies. J Trace Microprobe Techn 12:103–122

Mattson M, Lundborg T, Larsson CM (1993) Nitrogen utilisation in N-limited barley during vegetative and generative growth. IV. Translocation and remobilisation of nitrogen. J Exp Bot 44:537–546

McDonald AJS, Davies WJ (1996) Keeping in touch: responses of whole plant to deficits in water and nitrogen supply. Adv Bot Res 22:229–300

Muller B, Touraine B (1992) Inhibition of NO_3^- uptake by various phloem-translocated amino acids in soybean seedlings. J Exp Bot 43:617–623

Muller B, Tillard P, Touraine B (1995) Nitrate fluxes in soybean seedling roots and their responses to amino acids: an approach using ^{15}N. Plant Cell Environ 18:1267–1279

Nye PH, Tinker PB (1977) Solute movement in the soil-root system. Blackwell, Oxford, p 342

Pace GM, Volk RJ, Jackson WA (1990) Nitrate reduction in reponse to CO_2-limited photosynthesis. Relationships to carbohydrate supply and nitrate reductase activity in maize seedlings. Plant Physiol 92:286–292

Passioura JB, Munns R (1984) Hydraulic resistance of plants. II. Effect of rooting medium and time of day, in barley and lupin. Aust J Plant Physiol 11:341–350

Pearson CJ, Volk RJ, Jackson WA (1981) Daily changes in nitrate influx, efflux and metabolism in maize and pearl millet. Planta 152:319–324

Peuke AD, Jeschke WD (1993) The uptake and flow of C, N and ions between roots and shoots in *Richinus communis* L. 1. Growth with ammonium or nitrate as a nitrogen source. J Exp Bot 44:1167–1176

Peuke AD, Kaiser WM (1996) Nitrate or ammonium uptake and transport, and rapid regulation of nitrate reduction in higher plants. Prog Bot 57:93–113

Peuke AD, Glaab J, Kaiser WM, Jeschke WD (1996) The uptake and flow of C, N and ions between roots and shoots in *Ricinus communis* L. IV. Flow and metabolism of inorganic nitrogen and malate depending on nitrogen nutrition and salt treatment. J Exp Bot 47:377–385

Pitman MG (1988) Whole plants. In: Baker DA, Hall JL (eds) Solute transport in plant cells and tissues. Longman, New York, pp 346–391

Poirier Y, Thoma S, Sommerville C, Schiefelbein J (1991) A mutant of *Arabidopsis thaliana* deficient in xylem loading. Plant Physiol 97:1087–1093

Radin JW, Parker LL, Sell CR (1978) Partitioning of sugars between growth and nitrate reduction in cotton roots. Plant Physiol 62:550–553

Rennenberg H, Herschbach C (1995) Sulphur nutrition of trees: a comparison of spruce (*Picea abies* L.) and beech (*Fagus sylvatica* L.). Z Pflanzenernahr Bodenkd 158:513–517

Richter C, Dainty J (1989a) Ion behaviour in plant cell walls. I. Characterisation of *Sphagnum russowii* cell wall ion exchanger. Can J Bot 67:451–459

Richter C, Dainty J (1989b) Ion behaviour in plant cell walls. II. Measurements of the Donnan free space, anion exclusion space, anion-exchange capacity, and cation exchange capacity in delignified *Sphagnum russowii* cell walls. Can J Bot 67:460–465

Rideout JW, Raper CD (1994) Diurnal changes in net uptake of nitrate are associated with changes in estimated export of carbohydrates to roots. Int J Plant Sci 155:173–179

Roberts SK, Tester M (1995) Inward and outward K^+-selective currents in the plasma membrane of protoplasts from maize root cortex and stele. Plant J 8:811–825

Roberts SK, Tester M (1997) Permeation of Ca^{2+} and monovalent cations through an outward rectifying channel in maize root cells. J Exp Bot 48:839–846

Rufty TW, Israel DW, Volk RJ (1984) Assimilation of $^{15}NO_3^-$ taken up by plants in the light and in the dark. Plant Physiol 76:769–775

Ruiz LP, Atkinson CJ, Mansfield TA (1993) Calcium in the xylem and its influence on the behaviour of stomata. Philos Trans R Soc Lond Biol 341:67–74

Sauter JJ (1982) Transport in Markstrahlen. Ber Dtsch Bot Ges 95:593–618

Scheible WR, Lauerer M, Schulze ED, Caboche M, Stitt M (1997) Accumulation of nitrate acts as a signal to regulate shoot-root allocation in tobacco. Plant J 11:671–691

Schneider H, Zhu JJ, Zimmermann U (1997) Xylem and cell turgor pressure probe measurements in intact roots of glycophytes: transpiration induces a change in the radial and cellular reflection coefficients. Plant Cell Environ 20:221–229

Schröder WH, Zhu JJ, Schneider H, Thürmer F, Zimmermann U, Marienfeld S (1996) Visualisierung von K-, Mg- und Ca-Tracern in einzelnen Xylemelementen. Proceedings of the German Botanical Society, Düsseldorf, Abstr. 324

Schurr U (1997) Growth physiology: approaches to a spatially and temporarily varying problem. Prog Bot 59:355–373

Schurr U (1998) Xylem sap sampling – new aspects and techniques in an old topic. Trends in Plant Sci 3:293–298

Schurr U, Schulze ED (1995) The concentration of xylem sap constituents in root exudate, and in sap from intact, transpiring castor bean plants (*Ricinus communis* L.). Plant Cell Environ 18:409–420

Schurr U, Schulze ED (1996) Effect of drought on nutrient transport and ABA transport in *Ricinus communis*. Plant Cell Environ 19:665–674

Schurr U, Gollan T, Schulze ED (1992) Stomatal response to drying soil in relation to changes in the xylem sap composition of *Helium annuus*. II. Stomatal sensitivity to abscisic acid imported from the xylem sap. Plant Cell Environ 15:561–567

Schurr U, Herdel K, Schmidt P (1998a) Nutrient status affects diurnal variation of nutrient transport in the xylem. Plant Cell Environ (in preparation)

Schurr U, Jahnke S, Schulze ED (1998b) Carbon transport and partitioning in *Ricinus communis* L.: impact of local changes in water potential and phloem sampling. J Exp Bot (submitted)

Senden MHMN, Van Paassen FJM, Van der Meer AJGM, Wolterbeek HAT (1992) Cadmium-citric acid-xylem wall interactions in tomato plants. Plant Cell Environ 15:71–79

Senden MHMN, Van der Meer AJGM, Verburg TG, Wolterbeek HAT (1994) Effects of cadmium on the behaviour of citric acid in isolated tomato cell walls. J Exp Bot 45:597–606

Shaner DL, Boyer JS (1976) Nitrate reductase activity in maize (*Zea mays* L.) leaves. Plant Physiol 58:499–504

Siddiqi MY, Glass ADM, Ruth TJ, Rufty TW (1990) Studies on the uptake of nitrate in barley. I. Kinetics of $^{13}NO_3^-$ influx. Plant Physiol 93:1426–1432

Slovik S, Daeter W, Hartung W (1995) Compartmental redistribution and long-distance transport of abscisic acid (ABA) in plants as influenced by environmental changes in the rhizosphere. A biomathematical model. J Exp Bot 46:881–894

Smith JAC (1991) Ion transport and the transpiration stream. Bot Acta 104:416–421.

Steingröver E, Oosterhuis R, Wieringa F (1982) Effect of light treatment and nutrition on nitrate accumulation in spinach (*Spinacea oleracea* L.). Z Pflanzenphysiol 107:97–102

Steingröver E, Siesling J, Ratering P (1986a) Daily change in uptake, reduction and storage of nitrate in spinach grown at low light intensity. Physiol Plant 66:550–556

Steingröver E, Siesling J, Ratering P (1986b) Effect of one night with "low light" on uptake, reduction and storage of nitrate in spinach. Physiol Plant 66:557–562

Steudle E (1995) Trees under tension. Nature 378:663–664

Stulen I (1985) Interactions between nitrogen and carbon metabolism in a whole plant context. In: Lambers H, Neeteson J, Stulen I (eds) Physiological, ecological and applied aspects of nitrogen metabolism in higher plants. Nijhoff and Junk, Den Haag, pp 234–257

Tanner W, Beevers H (1990) Does transpiration have an essential function in long-distance transport in plants? Plant Cell Environ 13:745–750

Tardieu F, Davies WJ (1993) Integration of hydraulic and chemical signalling and the control of stomatal conductance and water status of droughted plants. Plant Cell Environ 16:341–349

Tennakoon KU, Pate JS (1997) Xylem fluxes of fixed N through nodules of the legume *Acacia littorea* and haustoria of an associated N-dependent root hemiparasite *Olax phyllanthi*. J Exp Bot 48:1061–1069

Touraine B, Glass ADM (1977) NO_3^- and ClO_3^- fluxes in the chl-5 mutant of *Arabidopsis thaliana*. Plant Physiol 114:137–144

Touraine B, Grignon N, Grignon C (1988) Charge balance in NO_3^- fed soybean. Estimation of K^+ and carboxylate recirculation. Plant Physiol 88:605–612

Touraine B, Muller B, Grignon C (1992) Effect of phloem-translocated malate on nitrate uptake by roots of intact soybean plants. Plant Physiol 99:1118–1123

Tsay YF, Schröder JI, Feldmann A, Crawford NM (1993) The herbicide sensitivity gene *CHL1* of *Arabidopsis* encodes a nitrate-inducible nitrate transporter. Cell 72:705–713

Van der Shoot C (1989) Determinants of xylem-phloem transfer in tomato. PhD Thesis University of Delft

Veen BW, Kleinendorst A (1985) Nitrate accumulation and osmotic regulation in Italian ryegrass (*Lolium multiflorum* Lam.) J Exp Bot 36:211–218

Wegner LH, Raschke K (1994) Ion channels in the xylem parenchyma or barley roots: a procedure to isolate protoplasts from this tissue and a patch-clamp exploration of salt passageways into xylem vessels. Plant Physiol 105:799–813

Wegner LH, De Boer AH, Raschke K (1994) Properties of K^+ inward rectifier in the plasma membrane of xylem parenchyma cells from barley roots: effects of TEA^+, Ca^{2+}, Ba^{2+}, and La^{3+}. J Membr Biol 142:363–379

Wilkinson S, Davies WJ (1997) Xylem sap pH increase: a drought signal received at the apoplastic face of the guard cell which involves the suppression of saturable ABA uptake by the epidermal symplast. Plant Physiol 113:559–573

Wolterbeek HT (1987) Cation exchange in isolated cell walls of tomato. I. Cd^{2+} and Rb^+ exchange in adsorption experiments. Plant Cell Environ 10:39–44

Wolterbeek HT, Van Luiupen J, De Bruin M (1984) Non-steady state xylem transport of fifteen elements into the tomato leaf as measured by gamma-ray spectroscopy: a model. Physiol Plant 61:559–606

Zhu JJ, Zimmermann U, Thürmer F, Haase A (1995) Xylem pressure response in maize roots subjected to osmotic stress: determination of radial coefficients by use of the xylem pressure probe. Plant Cell Environ 18:906–912

Zimmermann U, Meinzer FC, Benkert R, Zhu JJ, Schneider H, Goldstein G, Kuchenbrod E, Haase A (1994) Xylem water transport: is the available evidence consistent with the cohesion theory? Plant Cell Environ 17:1169–1181

Zimmermann U, Meinzer F, Bentrup FW (1995a) How does water ascend in tall trees and other vascular plants? An Bot 76:545–551

Zimmermann U, Bentrup FW, Haase A (1995b) Xylem pressure, flow and sap composition of trees determined by means of the xylem pressure probe. In: Terazawa M (ed) Xylem sap utilisation. Hokaido University Press, Sapporo, pp 59–70

Dr. Ulrich Schurr
Universität Heidelberg
Botanisches Institut
Im Neuenheimer Feld 360
D-69120 Heidelberg, Germany

Edited by
U. Lüttge

Photosynthesis. Carbon Metabolism:
In and Beyond the Chloroplast

By Grahame J. Kelly

> *"It is far more probable that starch is only elaborated within the cell when the supply of nutriment is in excess of the cell requirements, and that most of the assimilated products never pass through the stage of starch at all. (Our experiments) .... point to the somewhat unexpected conclusion that, at any rate in the leaves of the Tropaeolum, cane sugar is the first sugar to be synthesised ..."* (Brown and Morris 1893).

1 Introduction

Views concerning the central sugars of photosynthesis have varied. Starch was popular in the late 1800s after Sachs (1862) found that light promoted its synthesis in Chlorophyllkörnern. The possibility that the flow of photosynthetically fixed carbon could bypass starch was not seriously entertained until the turn of the century, (see above quotation), after which cane sugar (the disaccharide sucrose) gradually achieved the reputation of being the end product of photosynthesis in the green leaf cell; but simplistic views seldom hold true in the biological world. The anticipation (see Walker 1997) and characterization (Heldt and Rapley 1970) of the chloroplast's phosphate translocator ultimately gave glyceraldehyde (as glyceraldehyde-3-P) the distinction of being *the* product of photosynthesis within the chloroplast. Still, glyceraldehyde-3-P (GAP) never accumulates in quantity, whereas sucrose (which is synthesized from GAP in the cytosol) does. Additionally, sucrose is a principal form in which photosynthetic product is translocated to other parts of the plant. Therefore, the triose phosphates (GAP and its partner dihydroxyacetone-P) and sucrose currently share the title of central sugars of photosynthesis; one in, and the other beyond, the chloroplast. Recent research on photosynthetic carbon metabolism reflects this distinction: more and more attention is being devoted to what happens beyond the chloroplast, while research on intrachloroplast events has been directed at the perennially popular enzyme Rubisco, and at the mechanism chloroplasts use to dissipate whatever portion of harvested light energy cannot be used to convert CO_2, via GAP, to sucrose. These topics are reviewed below, covering the literature that has appeared since our last chapter in Volume 58 of Progress in Botany.

Progress in Botany, Vol. 60
© Springer-Verlag Berlin Heidelberg 1999

2 From CO_2 to GAP

a) Ribulose-Bisphosphate Carboxylase/Oxygenase (Rubisco)

The Calvin cylcle's CO_2 fixing enzyme, Rubisco, has continued to receive dedicated attention. Its reputation for being the major determinant of photosynthetic rate in nonstressed sunlit plants in the field has been extended to C_4 plants (Furbank et al. 1996). It also contributes to the establishment of the lower photosynthetic rates of stressed plants: stresses such as low water availability (Kanechi et al. 1996), rain treatment (Ishibashi et al. 1996), oxidative stress (Desimone et al. 1996; Stieger and Feller 1997; Ishida et al. 1997), and ultraviolet radiation (Ferreira et al. 1996; Greenberg et al. 1996; Allen et al. 1997) all cause Rubisco to be physically damaged and/or reduced in amount.

Research on the regulation of Rubisco has centered on the activity-enhancing protein Rubisco activase and the inhibitory sugar-P carboxyarabinitol-1-P (CA1P). Rubisco activase is a relatively heat-sensitive protein (Crafts-Brandner et al. 1997; Eckardt and Portis 1997) that slowly (Woodrow et al. 1996) promotes dissociation of inhibitory sugar-Ps from Rubisco, permitting the latter to display its full activity. Exactly how it does this is obscure (Salvucci and Ogren 1996), although there are hints that it first forms an oligomer reminiscent of the better-known microfilament-protein actin (Lilley and Portis 1997). Another intriguing parallel has been noted by Komatsu et al. (1996), who discovered an amino acid sequence homology between the activase and a gibberellin-binding protein. Whatever its mechanism, it is clear from antisense-gene experiments that the activase is vital for photosynthesis, the only point of contention is whether or not it is present in leaves in great excess (Mate et al. 1996; Eckardt et al. 1997).

Little progress, and some odd results, have appeared in relation to the CA1P inhibition of Rubisco. This sugar-P is presumably synthesized at night when it binds to any CO_2-plus-Mg^{2+}-activated Rubisco and stops it from working. However, this regulatory event occurs only in some plant species, e.g., French bean and tobacco, and perhaps in the green alga *Dunaliella tertiolecta* (MacIntyre et al. 1997), but not in others, e.g., spinach and wheat. In French bean leaves it is unclear whether it is restricted to chloroplasts (Moore et al. 1995) or whether some exists in unexpected locations such as leaf veins (Anwaruzzaman et al. 1996). Finally, its mode of synthesis remains unclear: it is quite possibly derived from fructose-1,6-P_2 via the seldom-met sugar hamamelose (Andralojc et al. 1996; Martindale et al. 1997), but it is annoying that a kinase needed to phosphorylate carboxyarabinitol to CA1P has not yet been detected. In addition, to complicate matters, recent research by the Rothamsted group has raised a suspicion that there exists another tight-binding inhibitor of Rubisco that, unlike CA1P, is present by day as well as by night (Keys et al. 1995; Parry et al. 1997), but there is so far no clue concerning the identity of this inhibitor.

The possibility that the *oxygenase* activity of Rubisco (and, consequently, photorespiration: Sect. 5) might be eliminated continues to be entertained. It has been argued that the oxygenase activity is inevitable in normal air, but the existence of certain aldolases that catalyze reactions involving similar chemistry but which nevertheless display no tendency to catalyze a similar oxygenase reaction bolster hopes that molecular biologists may some day create an oxygenase-free Rubisco (Hixon et al. 1996; Wildner et al. 1996). Slight success in this direction has been reported by Kostov et al. (1997) for a cyanobacterial Rubisco, but no recent efforts have been reported for higher plant Rubiscos; they all possess very similar ratios of carboxylase to oxygenase activities that are about double those of the cyanobacterial Rubiscos (Balaguer et al. 1996; Laisk and Loreto 1996; Uemura et al. 1996).

b) Remainder of the Calvin Cycle

Almost all attention to the remainder of the Calvin cycle has been focused on the mechanistic details by which certain enzymes undergo light-mediated activation, and on the existence of multienzyme complexes. Light-mediated activation involves the reduction of disulfide bonds by reduced thioredoxin. The cysteine residues between which the disulfide bonds are formed and broken have been more precisely identified in the higher plant and *Chlamydomonas reinhardtii* variants of glyceraldehyde-3-P (GAP) dehydrogenase, fructose-1,6-bisphosphatase (FBPase), and sedoheptulose-1,7-bisphosphatase (SBPase) (Anderson et al. 1996a; Jacquot et al. 1997a; Li et al. 1997). One of the cysteine residues is missing from a cyanobacterial GAP dehydrogenase, and indeed this enzyme probably is not light-activated (Tamoi et al. 1996). Reduced thioredoxin has attracted more attention because this small protein occurs not only in the chloroplast in two forms that participate in light-mediated enzyme activation (Reche et al. 1997), but also in mitochondria in two forms, one of which may have an as yet undefined role in photosynthesis, and in seeds, where it reduces the disulfide bonds of storage proteins prior to their mobilization during germination (Konrad et al. 1996; Lozano et al. 1996; Jacquot et al. 1997b). It may even have a role in protecting stressed plants by substituting for the enzyme dehydroascorbate reductase (Morell et al. 1997). Finally, it has attained importance in fatty acid biosynthesis in that it has been shown to participate in the light-mediated activation of the chloroplast's acetyl-CoA carboxylase (Sasaki et al. 1997).

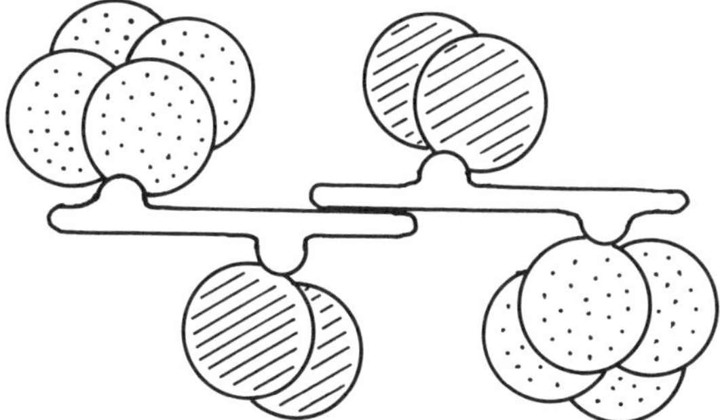

Fig. 1. An adaption of the model of Wedel et al. (1997) for the association of two molecules of GAP dehydrogenase (*dotted*, four-subunit enzyme), two molecules of Ru5P kinase (*cross-hatched*, two-subunit enzyme), and two molecules of a small linking protein (*thin, unshaded*). It is proposed that the enzymes are reductively activated (by reduced thioredoxin) while associated as shown, but that photosynthetically generated NADPH induces dissociation of the enzymes from the dimeric linking protein; the free, reduced enzymes are the fully active forms that participate in photosynthesis

Increased attention is being given to protein-protein interactions in the chloroplast. The simplest is aggregation of one enzyme, as evidenced by the recent demonstrations of the association of four molecules of either GAP dehydrogenase (Baalmann et al. 1995) or FBPase (Grotjohann 1997) into relatively inactive oligomers. Alternatively, Calvin cycle enzymes may associate with each other or with other proteins. Examples include an association between FBPase and thioredoxin that appears to regulate the FBPase activity in a nonreductive fashion (Häberlein and Vogeler 1995), the association of a major or minor portion of each of six Calvin cycle enzymes with thylakoid membranes (Anderson et al. 1996b), and associations of GAP dehydrogenase with glycerate-3-P kinase (Wang et al. 1996) and with ribulose-5-P (Ru5P) kinase (Avilan et al. 1997; Wedel et al. 1997). These protein-protein interactions may be expected in the chloroplast stroma, where the concentration of proteins is quite high, and they may benefit the Calvin cycle's operation by channeling intermediates from one enzyme to the next (Harris and Königer 1997); evidence for such chanelling in the *oxidative* pentose phosphate pathway in root nodules and in yeast has recently been presented (Debnam et al. 1997). Of special interest is the interpretation of Wedel et al. (1997) that there exists in chloroplasts a small protein with the specific task of mediating the tantalizing (Lebreton et al. 1997) GAP dehydrogenase/Ru5P kinase association and participating in the light-mediated activations of these enzymes (Fig. 1). Given that other Calvin cycle enzyme associations have been reported (see above, and our previous reviews), one wonders whether similar small proteins also exist. This one must be important if, as concluded by Fridlyand et al. (1997), GAP dehydrogenase has a part in determining the rate of CO_2-

saturated photosynthesis (cf. Rubisco is the major determinant with air levels of CO_2).

c) The C_4 Pathway

Photosynthetic carbon metabolism begins with the carboxylation of ribulose-1,5-P_2 (RuBP) in C_3 plants, but in C_4 plants, and in plants with crassulacean acid metabolism (CAM), preceding metabolic events designed to concentrate and/or indirectly store the CO_2 destined for RuBP carboxylation occur. The majority of these events occur beyond the chloroplast.

In C_4 plants, the well-known C_4 pathway uses the enzyme phosphoenolpyruvate (PEP) carboxylase in mesophyll cells to collect CO_2 (as HCO_3^-) and form oxaloacetate which, after being reduced to malate, is either directly or in a roundabout metabolic way transported to bundle-sheath cells and decarboxylated, thereby feeding CO_2 to the Calvin cycle, which is restricted to these cells (but see Castrillo et al. 1997). This CO_2 pump greatly suppresses photorespiration and increases water use efficiency. Its evolution was encouraged about 7 million years ago by a drop in the level of atmospheric CO_2; consequently, terrestrial C_4 plants have existed only during the last 2% of the period when vascular plants have been present on land (Cerling et al. 1997). However, a submersed monocot (*Hydrilla verticillata*) that displays C_4-type photosynthesis has been identified as a possible ancestor of terrestrial C_4 plants, and may have existed up to 100 million years ago (Magnin et al. 1997).

Several reports have emphasized the importance and properties of component enzymes of the C_4 pathway. The most studied is PEP carboxylase, partly because of its fascinating regulation via a phosphorylation (that activates it) catalyzed by another enzyme appropriately named PEP-carboxylase kinase (reviewed by Vidal and Chollet 1997). The mRNA for this kinase increases when plants are illuminated (Hartwell et al. 1996) via an as yet only partly elucidated signaling system that involves Ca^{2+} and glycerate-3-P (Giglioli-Guivarc'h et al. 1996). A similar, but not identical, system potentially operates in the cells of C_3-plant leaves (Li et al. 1996; Smith et al. 1996; but see Leport et al. 1996), and in other organs including germinating wheat seeds (Osuna et al. 1996), banana fruit (Law and Plaxton 1997), and soybean root nodules, in which photosynthate supply, rather than light, triggers the signaling system (Zhang and Chollet 1997a). Finally, Du et al. (1997) have added stomatal guard cells to the list of cells in which PEP carboxylase has a central role and is regulated by phosphorylation. In all of these cases, phosphorylation of the carboxylase seems to activate it primarily by reducing its susceptibility to inhibition by malate, although altered sensitivities to other regulatory metabolites could be important (Gao and Woo 1996); these would include the recently reported inhibitors phosphate and sulfate (Salahas et al. 1997) and, interestingly, shikimic acid, which is synthesized from PEP via a metabolic pathway that does not involve PEP carboxylation (Colombo et al. 1996).

PEP carboxylation produces oxaloacetate which is reduced to malate by a malate dehydrogenase shown to be essential for normal C_4 photosynthesis (Trevanion et al. 1997), and which undergoes light-mediated activation via reduction of two disulfide bonds per subunit (Issakidis et al. 1996). The malate is then processed by one of three possible metabolic routes (that designate the three subtypes of C_4 plants), each of which may vary subtly between plant species (Meister et al. 1996) and have special requirements such as an active 2-oxoglutarate/malate translocator (Taniguchi and Sugiyama 1997) or reduced capacity for respiration in bundle-sheath mitochondria (Agostino et al. 1996). Ultimately, CO_2 is released to the Calvins cycle's Rubisco, and the remaining 3-carbon pyruvate is returned to the mesophyll cells, where it is reconverted to PEP in the chloroplast by the remarkable enzyme pyruvate, P_i dikinase (PPDK). Molecular biologists managed to introduce this C_4 enzyme, and a functionally similar bacterial enzyme, into C_3 plants, but nothing spectacular happened (Ishimaru et al. 1997; Panstruga et al. 1997). PPDK produces, in addition to PEP, a good amount of AMP, which is returned to the ADP/ATP pool by adenylate kinase, an enzyme that is abundant in mesophyll chloroplasts where it seems to assemble into an inactive storage form overnight (Wild et al. 1997).

Further reports confirm that when the C_4 pathway concentrates CO_2 into bundle-sheath cells, photorespiration (Sect. 5) is largely, although not totally, inhibited, and although back diffusion of CO_2 is greater than earlier believed, the mesophyll cell's PEP carboxylase recaptures it so that the overall pathway remains effective (Dai et al. 1996; He and Edwards 1996; Saliendra et al. 1996; Lacuesta et al. 1997). However, without the C_4 pathway, C_4 plants can photorespire (Devi et al. 1996, Lacuesta et al. 1997).

Most C_4 plants are stressed by cool temperatures and, like all plants, by severe water stress. Sensitivity to cool temperatures appears to be due to the impairment of critical enzymes such as PEP carboxylase and PPDK, but a few species possess cold-resistant forms of these enzymes (Usami et al. 1995; Simon 1996; Matsuba et al. 1997) so that in cool climates C_4 plants, while rare, can be found (Beale et al. 1996). In the case of the enzyme PPDK, it was found that substitution of only 0.3% of the amino acid residues could confer cold tolerance (Ohta et al. 1996). PPDK is also implicated in water-stressed C_4 plants whose photosynthesis is decreased not only by a reduced supply of CO_2 to Rubisco (Lal and Edwards 1996), but also by a dramatic, over nine-fold drop in the PPDK activity (Du et al. 1996).

Characteristics of C_4 photosynthesis have again been reported in two aquatic plants, the leafless sedge *Eleocharis vivipara* (Ueno 1996) and the monocot *Hydrilla verticillata* (Spencer et al. 1996; Magnin et al. 1997), and in the ears of C_3 cereals and green tomato fruits that refix respiratory CO_2 (Bort et al. 1996; Xu et al. 1997). Of particular note is the predominant role being attributed to PPDK in the lemmas, paleae, and immature seeds of rice (Sadimantara et al. 1996; Imaizumi et al. 1997).

d) Crassulacean Acid Metabolism (CAM)

Like C_4 plants, CAM plants employ considerable metabolism beyond the chloroplast prior to RuBP carboxylation. Events within the cytosol and vacuole, and on the vacuolar membrane (tonoplast) are especially notable. In summary, PEP carboxylation in the cytosol at night produces malate that is delivered to and stored in the vacuole, from where it is dispensed back to the cytosol next day and decarboxylated, thereby providing CO_2 for Rubisco behind closed stomates. Stomate closure by day can help to protect CAM plants during drought (Güerere et al. 1996), but not if conditions become too extreme (Castillo 1996; Roberts et al. 1996), and indeed the expected superior water use efficiency of CAM plants (Helliker and Martin 1997) has recently been thrown into doubt (Eller and Ferrari 1997). It is also worth noting that CAM is not restricted to dry, sunny habitats: it is common among rainforest epiphytes (Zotz and Ziegler 1997), while some uncommon examples include species from the shaded rainforest understory, including *Aechmea magdalenae* (Skillman and Winter 1997; Zotz and Ziegler 1997). Two extreme examples are the tiny *Blossfeldia liliputana*, a unique poikilohydric cactus (Barthlott and Porembski 1996), and the brown marine macroalga *Ectocarpus siliculosus*, where a CAM-like mechanism helps to overcome the limiting CO_2 availability in seawater (Schmid and Dring 1996).

The mechanism of CAM in conventional CAM plants has been further investigated. The upside-down behavior of CAM-plant stomates (open at night and closed by day) has been correlated with a loss of the sensitivity of stomatal guard cells to blue light (Tallman et al. 1997) which is detected by the guard-cell chloroplasts of other plants (Quiñones et al. 1996). CO_2 that enters stomates at night is captured by a PEP carboxylase that is activated via a phosphorylation; in this case a circadian oscillator, rather than light (as in C_3 and C_4 plants), initiates the signaling system that culminates in enzyme phosphorylation (Hartwell et al. 1996). By day the PEP carboxylase is deactivated, more so in those species that display CAM more strongly (Borland and Griffiths 1997). Nocturnal PEP carboxylation leads to the production of the large quantity of malic acid that accumulates in the vacuole (Fig. 2); an accumulation of citric acid is also detected in some species, but does not seem to be a central component of CAM (Herppich et al. 1995; Borland et al. 1996). Malate is moved into the vacuole by a recently characterized tonoplast transporter (Steiger et al. 1997; Fig. 2); accumulation is powered by a tonoplast H^+-ATPase (Zhigang et al. 1996) that is possibly regulated by a H^+-

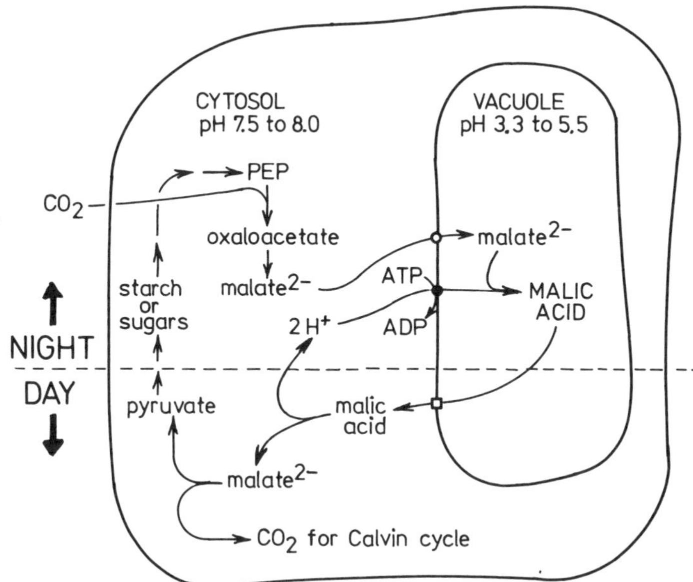

Fig. 2. The role of the tonoplast during CAM. By night, active transport of malate anions into the vacuole is driven by an ATPase that delivers H^+ into the same vacuole. Thus it is the acid form (malic acid) that accumulates. Next day, this malic acid diffuses passively (via a channel protein?) into the cytosol, where it first dissociates (returning the H^+, which is presumably accomodated by the buffering capacity of cytosolic metabolism), and then undergoes the decarboxylation that supplies CO_2 for the Calvin cycle and pyruvate for synthesis of the carbohydrate reserve needed to generate the next night's supply of PEP. (After Lüttge et al. 1995)

pyrophosphatase found on the same membrane (Lüttge et al. 1995). Next day, malic acid moves passively across the tonoplast and into the cytosol (Fig. 2). This process is notably sensitive to high or low temperature, indicating that the biophysical state of the membrane is important for its normal operation (Lüttge et al. 1995; Grams et al. 1996; Grams et al. 1997). Once in the cytosol, the malate is decarboxylated by one of three possible enzymes; one of these (PEP carboxykinase) is similar to PEP carboxylase in that its activity is potentially regulated by phosphorylation (Walker and Leegood 1996). The remaining 3C-pieces are converted gluconeogenically to stores of chloroplast starch or extrachloroplast sugars (Borland 1996; Christopher and Holtum 1996) from which PEP is derived the following night. It was thought that the storage form (starch or sugars) accumulated by a particular CAM species correlated with its particular decarboxylation enzyme, but it is now clear that this rule is invalid (Christopher and Holtum 1996).

3 The Multifaceted Fate of GAP

a) Transitory Starch in Chloroplasts

The conclusion of Brown and Morris (1893) that "... starch is only elaborated in the cell when the supply of nutriment is in excess ..." (see intro-
ductory quotation) has been confirmed for peach, a species that synthe-
sizes sorbitol rather than cane sugar (sucrose) as the predominant translocated carbohydrate (Escobar-Gutiérrez and Gaudillère 1997), and for
spinach and soybean deprived of nitrogen (Robinson 1997). However,
no new details concerning the biochemistry of the synthesis of this chloroplast starch by day and its mobilization by night have been reported,
except for further complications concerning the role of phosphorylase in
the mobilization process: C_4 plant leaves contain even more nonchloro-
plastic forms of this enzyme than do C_3 plant leaves (Shatters and West
1996; Venkaiah and Kumar 1996), and Duwenig et al. (1997a) have indications that the (chloro)plastidic form might be more important for
starch synthesis in fruits and seeds than for starch degradation in leaves.
Thus, it seems to be even more certain that the nocturnal mobilization of
starch in chloroplasts is hydrolytic (not phosphorylytic), producing
maltose and glucose that are exported to the cytosol (Rost et al. 1996).

b) Three Notes About Chloroplast Lipids

The major role of the chloroplast in *mature* leaves is to act as a factory that manufactures
soluble sugars (sucrose, or a sugar alcohol such as sorbitol, or one of the small raffinose
type of oligosaccharides; Moore et al. 1997) for export to metabolic or storage sinks; but
in *immature* leaves the chloroplast implements several other biosynthetic capabilities
whereby the molecular building blocks of the proteins and membranes needed by a
growing leaf are generated from Calvin cycle precursors (Fig. 3; Herbers and Sonnewald
1996). Three interesting pieces of information concerning lipid synthesis in chloroplasts
have appeared:

1. The initial enzyme of fatty acid synthesis, acetyl-CoA carboxylase, is similar to several Calvin cycle enzymes in that it undergoes a reductive light-mediated activation
   (Sasaki et al. 1997).
2. The enzymes that synthesize long-chain fatty acids from acetyl-CoA are organized
   into a multienzyme complex adjacent to the thylakoid, thereby permitting substrate
   channeling and the consequent increase in rate of fatty acid synthesis for a given
   chloroplast content of each substrate (Roughan and Ohlrogge 1996; Roughan 1997).
3. Isopentenyl-P_2, the precursor of chloroplast isoprenoids such as the carotenoids and
   the phytol tail of chlorophyll, is not synthesized through the classical pathway (from
   acetate, and via mevalonate), but rather through a different pathway in which one
   molecule each of GAP and pyruvate are its precursors; in the 5-carbon sugar de-
   oxyxylulose is an intermediate in this pathway (Schwender et al. 1996; Arigoni et al.
   1997; Lichtenthaler et al. 1997; Fig. 3).

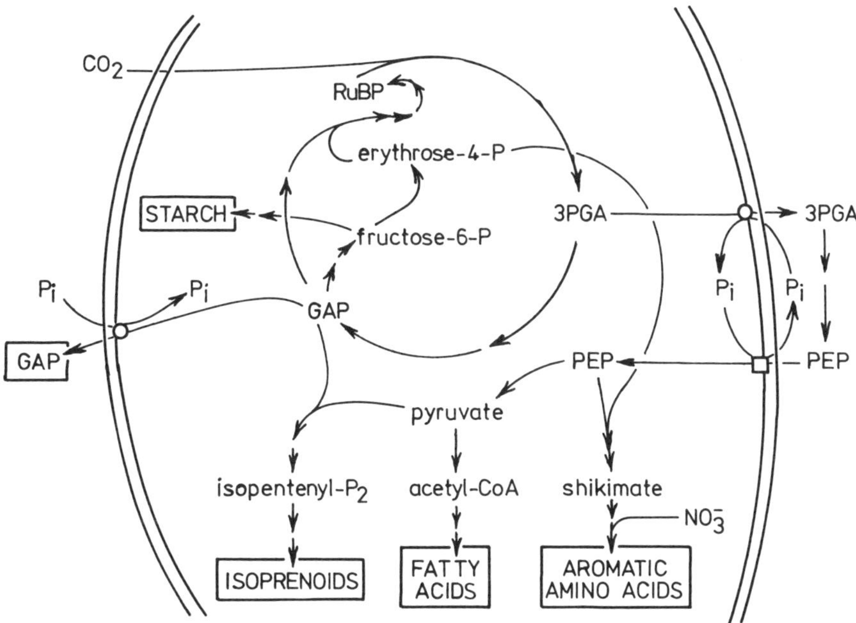

Fig. 3. Five products of photosynthesis (*boxed items*) in developing leaves. Based on research covered in our earlier reviews, and on Herbers and Sonnewald (1996), Fischer et al. (1997), and Lichtenthaler et al. (1997). *3PGA* Glycerate-3-P

c) Export from the Chloroplast

The now-classical phosphate translocator that exports net photosynthetic product, as GAP, from the chloroplast (in exchange for an incoming P_i; Fig. 3) is clearly not alone. It is now more certain that a separate, distinct envelope protein that favors the transport of glucose-6-P is present: such a transporter was identified on *Capsicum annuum* fruit chromoplasts (Quick and Neuhaus 1996) and on the chloroplasts of a CAM plant (Neuhaus and Schulte 1996; Kore-eda and Kanai 1997), where, presumably, it could participate in the nocturnal conversion of chloroplast starch to cytosolic PEP. In addition, a third phosphate translocator that exchanges PEP and P_i has been confirmed for chloroplasts and other types of plastids; it could be important for the synthesis of aromatic amino acids (Fischer et al. 1997; Fig. 3), and is almost certainly central to the C_4 pathway (Aoki and Kanai 1995). Finally, the presence of an ATP/ADP translocator has been more firmly established (Neuhaus et al. 1997), and the probability that there is a transporter for maltose has been raised (Rost et al. 1996). More surprises may be imminent, given the results of studies with other types of plastids (see below, Sect. 4.b).

d) Cytosolic Sucrose Synthesis and its Regulation

When GAP arrives in the cytosol of leaf cells, it is converted to soluble sugars ready for export from the leaf. Synthesis of the sugar sucrose has been by far the most intensively studied. Two regulatory enzymes have received recent attention: cytosolic FBPase that forms fructose-6-P from the FBP generated from two GAPs, and sucrose-P synthase (SPS) that forms sucrose-P from two fructose-6-Ps (one of which is first converted to UDP-glucose). The FBPase is potentially regulated by a reductive process (Anderson et al. 1997) similar to that of its chloroplast counterpart (Sect. 2.b), but the value of this is not yet established. Its activity is also controlled through inhibition by the regulatory sugar-P fructose-2,6-P_2 (Kruger and Scott 1995), especially during water stress (Reddy 1996). Plants containing low amounts of the FBPase clearly made less sucrose (Micallef and Sharkey 1996; Zrenner et al. 1996), but in one case (genetically engineered potato), plant growth was unaffected because photosynthate partitioning was appropriately adjusted, i.e., chloroplasts made more starch and then, by exporting hexose or hexose-P at night, the cytosolic FBPase was not required (Zrenner et al. 1996).

The well-established principal regulatory mechanism for SPS, viz. activation by dephosphorylation and inactivation by phosphorylation, catalyzed by a SPS phosphatase and a SPS kinase (respectively), has been reviewed (Huber and Huber 1996). The activity of the SPS phosphatase is, in turn, modulated by a circadian-rhythm-based signaling system (Jones and Ort 1997). An additional level of complexity was reported by Toroser and Huber (1997), who found that osmotic stress somehow activates a second form of SPS kinase that attaches a second phosphate group to the already phosphorylated (and therefore inactive) SPS. This, like total dephosphorylation, revives the enzyme's activity, in this case perhaps to promote the formation of sucrose for osmoregulatory purposes. There is also one report that SPS itself might occur in more than one form in leaves (Reimholz et al. 1997), and another that it forms a complex with the following (and final) enzyme of sucrose synthesis, viz. sucrose-P phosphatase (Echeverria et al. 1997). Interest in SPS will continue because there is evidence that it is a principal determinant of plant production: tomato plants genetically engineered to contain extra SPS photosynthesized faster (Galtier et al. 1995), although they tended to produce sweeter fruit rather than more fruit (Laporte et al. 1997).

A new topic, introduced in our last review, is the mechanism by which soluble sugars (that accumulate in actively photosynthesizing cells when the ability of the rest of the plant to use them is exceeded) somehow initiate a feedback inhibition of photosynthesis. Recent clear examples of the effect were wheat and sugar beet leaves in which the mRNA transcripts for Rubisco, FBPase, SBPase, and glycerate-3-P kinase were reduced when the leaves were fed 1–2% sucrose or glucose (Jones et al. 1996; Lee and Daie 1997). The current hypothesis is that the genes for these Calvin cycle enzymes are downregulated by a molecular signaling system that begins with the traditional glycolytic enzyme hexokinase and its substrates glucose and/or fructose, comparable mechanisms are known from animal and yeast cells (Jang and Sheen 1997; Smeekens and Rook 1997). Using transgenic *Arabidopsis* plants, Jang et al. (1997) have further strong evidence that hexokinase is the sugar sensor that links sugar-related metabolic activities to photosynthesis rates in plants, but other research implies that there is some explaining to be done:

1. Strand et al. (1997) found that the correlation between accumulation of sugars and downregulation of genes in *Arabidopsis* broke down when leaves developed at a low temperature (5 °C). Under this temperature the well-known accumulation of sugars as a component of cold tolerance took place, and was supported by an unrestricted photosynthesis. Similarly, spinach leaves (Martindale and Leegood 1997) and wheat and rye (Gray et al. 1996) acclimated to low temperatures developed increased capacities for photosynthesis, in the latter case this was deemed necessary to compensate for the slower metabolism at lower temperatures and therefore the greater difficulty in using harvested light energy at these temperatures (Huner et al. 1996). It seems that the success of this adaptation mechanism may not be total (Savitch et al. 1997).

2. Sugar beet, tobacco, and *Flaveria bidentis* supplied with sucrose in culture systems showed no evidence for photosynthetic repression (Kovtun and Daie 1995; Furbank et al. 1997).

3. More than 90% of photosynthetically generated hexoses (glucose and fructose) was compartmented into the vacuole of tobacco, snapdragon, and parsley leaves (Moore et al. 1997), whereas the hexokinase-based sugar-sensing mechanism would be expected to be in the cytosol. Possibly the concentration of hexoses in the cytosol increases at the end of the day (Moore et al. 1997) after invertase hydrolyzes the vacuolar store of sucrose (Scholes et al. 1996) and the resultant hexoses are moved across the tonoplast. The accumulation of hexoses in the cytosol appears to be possible: Schaffer and Petreikov (1997) measured 30 mM fructose in the cytosol of young tomato-fruit cells.

A new angle to the hexokinase story might develop if plant scientists follow up the report that rabbit erythrocyte hexokinase is inactivated by dehydroascorbate (Fiorani et al. 1996). Ascorbate and its oxidation products (e.g., monodehydroascorbate) have central roles in the responses of plants to stress.

4 Distant Fates of Photosynthetic Product

a) Translocation

The majority of photosynthate produced by mature leaves is exported as soluble sugars to metabolic or storage sink tissues via the phloem. A concept for the loading of sucrose into the phloem, driven by a H^+-ATPase and a H^+/sucrose symporter (see Fig. 2 in our last review), has received more experimental support (Stadler et al. 1995; Kühn et al. 1996, 1997), although it seems there is some flexibility as to whether the H^+/sucrose symporter is on the sieve-tube membrane (e.g., tobacco, potato, and tomato) or on the membrane of its associated companion cell (e.g., *Plantago major*). The presence of the symporter implies that sucrose is loaded from an apoplastic source; however, its arrival at this source, at least in monocots, seems to require good symplastic (i.e., plasmodesmatal) connections between preceeding parenchyma cells (Evert et al. 1996; Russin et al. 1996).

b) Storage

The ultimate fates of translocated sugars include metabolism to form the proteins, membranes, and cell walls of metabolic sinks (such as growing shoots and roots); conversion to storage materials such as protein, starch, fructans, and oils in fruits, seeds, and modified organs; direct storage as soluble sugars; and respiration to power these processes. The synthesis of storage starch and (briefly) storage sucrose and oils is reviewed below.

Starch, like photosynthesis, is located in plastids; storage starch is located in amyloplasts (cf. transitory starch in chloroplasts, Sect. 3.a). Its biosynthesis begins with translocated sucrose, which is cleaved by one or

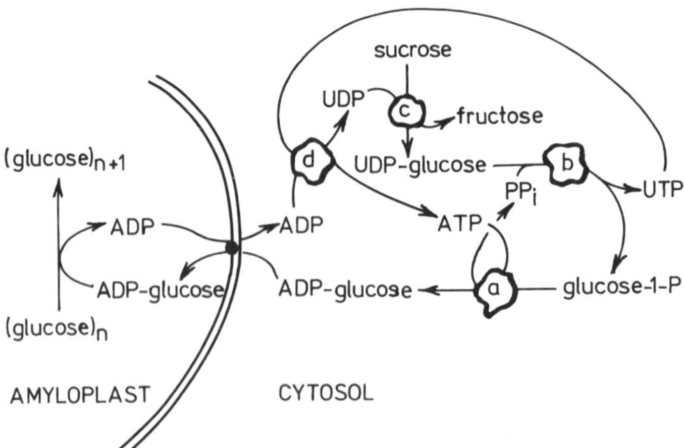

Fig. 4. Proposed pathway for the conversion of sucrose to starch in starch-storing tissues. The sum of the reactions is: sucrose + (glucose)$_n$ → fructose + (glucose)$_{n+1}$. Enzymes: *a* ADP-glucose pyrophosphorylase; *b* UDP-glucose pyrophosphorylase; *c* sucrose synthase; *d* nucleoside-P$_2$ kinase. (After Kleczkowski 1996)

other of the isozymes (Guerin and Carbonero 1997) of the rather inappropriately named sucrose synthase (Fig. 4). This enzyme, which is thought to be a reliable marker for sink tissues (Déjardin et al. 1997), is yet another example of an enzyme whose activity is regulated by a phosphorylation/dephosphorylation mechanism; in this case, phosphorylation increases activity (Huber et al. 1996; Zhang and Chollet 1997b). The UDP-glucose generated by sucrose synthase has, according to recent reports, one surprise fate: conversion *in the cytosol* to ADP-glucose. This is a surprise because the enzyme required to do this, ADP-glucose pyrophosphorylase, has been traditionally thought of as the principal regulatory enzyme of starch synthesis *within plastids*, but now Denyer et al. (1996) and Thorbjørnsen et al. (1996) report that the majority of the ADP-glucose pyrophosphorylase in the cells of maize and barley endosperm is in the cytosol rather than in the amyloplast, and Rudi et al. (1997) report that this cytosolic enzyme is not allosterically regulated by glycerate-3-P and Pi like its chloroplast counterpart.

The second group of authors have been quick to point out (Thorbjørnsen et al. 1996) that there still seems to be enough of this enzyme in barley amyloplasts to accommodate starch synthesis, possibly from imported glucose-6-P (Hill and Smith 1995; Emes and Neuhaus 1997; Möhlmann et al. 1997), and indeed another study with transgenic potato indicated that there is plenty of the enzyme in starch-storing tissues (Sweetlove et al. 1996). However, indications that isolated amyloplasts can synthesize starch from exogenously supplied ADP-glucose and that they possess a translocator that recognizes this nucleotide sugar (Emes and Neuhaus 1997; Möhlmann et al. 1997; Naeem et al. 1997) support the possibility that at least some amyloplast starch is generated from the ADP-glucose synthesized in the cytosol (Kleczkowski 1996; Fig. 4). The fructose shown in Fig. 4 need not be wasted. It could be phosphorylated by hexokinase and then isomerized to glucose-6-P, which could be imported into the amyloplast for starch synthesis. Perhaps this could occur in one step by using the enzyme phosphorylase, which is thought by Duwenig et al. (1997a) to have a role in starch formation. The P_i released by this enzyme would need to be exported to properly balance all reactions, thus the design of the amyloplastic glucose-6-P translocator such that it preferentially exports P_i to the cytosol (when P_i is counterexchanged with glucose-6-P) (Neuhaus and Maaß 1996) makes it eminently suitable for this role.

The mobilization of stored starch has received little attention. Amylases (Rost et al. 1996) rather than phosphorylases probably catalyze the initial reactions. As noted above, plastid phosphorylase is believed to be more important for starch synthesis, while a role for the cytosolic enzyme is still unclear. Remarkably, when the activity of this cytosolic enzyme in the potato plant was reduced by antisense technology, the plants appeared normal but gave double the yield of tubers (Duwenig et al.

1997b). Incidentally, another remarkable aspect of phosphorylase is its potential longevity: active enzyme has been extracted from *Nelumbo nucifera* (lotus) seeds estimated to be more than 500 years old (Minami-kawa et al. 1995).

Besides starch, sink tissues may also store sucrose or oils. Sucrose storage does not seem to involve the straightforward uptake and accu-mulation of translocated sucrose. Rather, the sucrose is first hydrolyzed by an apoplastic invertase and, in sugar-cane stem, resynthesis of su-crose by sucrose-P synthase occurs (Zhu et al. 1997). In tomato fruit, glucose released by the invertase appears to be pumped into cells by a cell membrane transporter powered by a H^+-ATPase (Brown et al. 1997). Tissues that store oil do so in modified plastids. In castor-bean en-dosperm, these plastids were found to possess a unique translocator that imports malate (the substrate for fatty acid synthesis) in exchange for an exported Pi (released when ATP is hydrolyzed to drive the acetyl-CoA carboxylase reaction) (Eastmond et al. 1997).

5 Photorespiration

Photorespiration (Leegood et al. 1995) consists of a cycle of reactions initiated by the oxygenase activity of Rubisco and interwoven with the Calvin cycle in such a way that, as clarified in our last review, it achieves nothing other than the consumption of photosynthetically generated ATP and NADPH when the CO_2 concentration drops to its compensation point, as might occur after leaf stomates shut on water-stressed plants. Theoretically, it could also occur if the oxygen evolved during intensive photosynthesis became trapped in the leaf (Tolbert et al. 1995), but Li-geza et al. (1997) believe that the normal leaf is too well ventilated for this to happen.

The fact that photorespiration consumes ATP and NADPH implies that it could be another approach whereby excess harvested light energy could be dissipated. This concept, which has had its ups and downs over the past years, is again on an "up" (Osmond et al. 1997). Firstly, experi-ments with the leaves of wheat, *Chenopodium bonus-henricus*, and pea all strongly supported it (Biehler and Fock 1996; Heber et al. 1996b; Park et al. 1996), and partial support came from another study with French bean leaves (Daniel 1997). Secondly, when the photorespiration of to-bacco leaves was retarded by first identifying the rate-limiting enzyme (it turned out to be the chloroplast's glutamine synthetase) and then using antisense technology to reduce the amount of this enzyme in the leaf, the leaves' susceptibility to high-light-induced photoinhibition was notably increased (Kozaki and Takeba 1996).

Apart from an interesting report that glycolate may be oxidized in chloroplasts by a glycolate-quinone oxidoreductase system (Goyal and Tolbert 1996) rather than is "usual" (i.e., in peroxisomes by glycolate dehydrogenase), most recent enzyme studies have centered on the mitochondrial glycine decarboxylase complex that converts two molecules of glycine to serine, CO_2 and NH_4^+; also, an NAD^+ is reduced. The reassimilation of this NH_4^+, and another subsequently released from the serine, is the role of the rate-limiting glutamine synthase mentioned above. However, the glycine decarboxylase itself appears to be only a little less important with respect to rate limitation (Wingler et al. 1997), especially in the palisade layer of mesophyll cells, which has less of the enzyme compared to the spongy cells (Guinel and Ireland 1996). In the mitochondrion, the enzyme is intimately associated with the pyruvate dehydrogenase complex (Bourguignon et al. 1996) that produces acetyl-CoA for the Krebs cycle, and much of the NADH it produces may be oxidized through the alternative cyanide-resistant electron transport system (Igamberdiev et al. 1997). Finally, there is good evidence that plants obtain most of their serine by courtesy of this enzyme, rather than others that can synthesize serine from glycine and formate (Prabhu et al. 1996).

6 Conclusions and Prospects

The literature on photosynthetic carbon metabolism from the past 2 years has contained a few exciting items, although (in our opinion) not as many as in some previous 2-year periods. Nevertheless, there are several indications that the future will not be dull. Firstly, there are new views of chloroplasts. Evolutionary relics of these organelles have been discovered in protozoan parasites of humans, such as *Toxoplasma gondii* (McFadden et al. 1996). In plants such as *Chenopodium album* there is a vertical gradient of chloroplast abundance, with the greatest number of chloroplasts per mesophyll cell in the uppermost fully expanded leaves (Yamasaki et al. 1996); and, most amazingly, within the cell the chloroplasts can no longer be considered as discrete, relatively independent organelles. Molecular biologists have used their fascinating new tool (green fluorescent protein) to rediscover a structural feature first reported in the 1960s, viz. chloroplasts are sometimes interconnected by thin tubules through which not only small molecules, but even proteins, may pass from the stroma of one chloroplast to the stroma of another (Köhler et al. 1997).

Secondly, the aim of linking photosynthesis research to increased plant productivity (an aim that has been entertained for many decades) has been addressed from both traditional and unexpected quarters. One of the latter was an experiment showing that the productivity of wheat plants was reduced by 25% when they were grown in space (on board the space shuttle Discovery; Tripathy et al. 1996). Meanwhile, back on Earth, two groups have attained productively *increases* of ca. 22% by using advanced plant biotechnology and plant breeding and selection procedures, but the increases appeared to be due to factors such as the improved acquisition and maintenance of photosynthetic leaves rather than to improvements in photosynthetic carbon metabolism (Black et al.

1995; Lawlor 1995; Medrano et al. 1995). Finally, the ancient discipline of philosophy has been applied to photosynthesis research by raising the more fundamental question of just what good increased plant productivity would do. While possible benefits in the sectors of socioeconomics and sustainable agriculture (Lawlor 1995) are undeniable, Walker (1995) has emphasized that these benefits could do little more than buy a little time, while more robust solutions to the central facets of the human predicament (including the increasing population of *Homo sapiens*) are sought. Meanwhile, he implies (Walker 1995) that the simple objective of satisfying scientific curiosity is as legitimate a reason for researching photosynthesis as entertainment is a reason for funding sporting events and operatic performances. We hope that the imagination of photosynthesis researchers will in the future be allowed, at least sometimes, to wander unhindered, so that a cornucopia of exciting items on photosynthetic carbon metabolism will appear in our next reviews.

Acknowledgments. We are most grateful to Agatha and Helen for their ever-cheerful assistance.

References

Agostino A, Heldt HW, Hatch MD (1996) Mitochondrial respiration in relation to photosynthetic C$_4$ acid decarboxylation in C4 species. Aust J Plant Physiol 23:1–7

Allen DJ, McKee IF, Farage PK, Baker NR (1997) Analysis of limitations to CO$_2$ assimilation on exposure of leaves of two *Brassica napus* cultivars to UV-B. Plant Cell Environ 20:633–640

Anderson LE, Huppe HC, Li AD, Stevens FJ (1996a) Identification of a potential redox-sensitive interdomain disulfide in the sedoheptulose bisphosphatase of *Chlamydomonas reinhardtii*. Plant J 10:553–560

Anderson LE, Gibbons JT, Wang X (1996b) Distribution of ten enzymes of carbon metabolism in pea (Pisum sativum) chloroplasts. Int J Plant Sci 157:525–538

Anderson LE, Li AD, Nehrlich SC, Hill MH, Stevens FJ (1997) The cytosolic fructose bisphosphatase of *Brassica napus* contains a new potential regulatory disulfide and is redox-sensitive. Plant Sci 128:23–30

Andralojc PJ, Keys AJ, Martindale W, Dawson WG, Parry MAJ (1996) Conversion of D-hamamelose into 2-carboxy-D-arabinitol and 2-carboxy-D-arabinitol 1-phosphate in leaves of *Phaseolus vulgaris* L. J Biol Chem 271:26803–26809

Anwaruzzaman, Nakano Y, Yokota A (1996) Different location in dark-adapted leaves of *Phaseolus vulgaris* of ribulose-1,5-bisphosphate carboxylase/oxygenase and 2-carboxyarabinitol 1-phosphate. FEBS Lett 388:223–227

Aoki N, Kanai R (1995) The role of phosphoenolpyruvate in proton/pyruvate cotransport into mesophyll chloroplasts of maize. Plant Cell Physiol 36:187–189

Arigoni D, Sagner S, Latzel C, Eisenreich W, Bacher A, Zenk MH (1997) Terpenoid biosynthesis from 1-deoxy-D-xylulose in higher plants by intramolecular skeletal rearrangement. Proc Natl Acad Sci USA 94:10600–10605

Avilan L, Gontero B, Lebreton S, Ricard J (1997) Memory and imprinting effects in multienzyme complexes. Eur J Biochem 246:78–84

Baalmann E, Backhausen JE, Rak C, Vetter S, Scheibe R (1995) Reductive modification and nonreductive activation of purified spinach chloroplast NADP-dependent glyceraldehyde-3-phosphate dehydrogenase. Arch Biochem Biophys 324:201–208

Balaguer L, Afif D, Dizengremel P, Dreyer E (1996) Specifity factor of ribulose bisphosphate carboxylase/oxygenase of *Quercus robur*. Plant Physiol Biochem 34:879–883

Barthlott W, Porembski ST (1996) Ecology and morphology of *Blossfeldia liliputana* (Cactaceae): a poikilohydric and almost astomate succulent. Bot Acta 109:161–166

Beale CV, Bint DA, Long SP (1996) Leaf photosynthesis in the C_4-grass *Miscanthus x giganteus*, growing in the cool temperate climate of southern England. J Exp Bot 47:267–273

Biehler K, Fock H (1996) Evidence for the contribution of the Mehler-peroxidase reaction in dissipating excess electrons in drought-stressed wheat. Plant Physiol 112:265–272

Black CC, Tu Z-P, Counce PA, Yao P-F, Angelov MN (1995) An integration of photosynthetic traits and mechanisms that can increase crop photosynthesis and grain production. Photosynth Res 46:169–175

Borland AM (1996) A model for the partitioning of photosynthetically fixed carbon during the C_3-CAM transition in *Sedum telephium*. New Phytol 134:433–444

Borland AM, Griffiths H (1997) A comparative study on the regulation of C_3 and C_4 carboxylation processes in the constitutive crassulacean acid metabolism (CAM) plant *Kalanchoë daigremontiana* and the C_3-CAM intermediate *Clusia minor*. Planta 201:368–378

Borland AM, Griffiths H, Maxwell C, Fordham MC, Broadmeadow MSJ (1996) CAM induction in *Clusia minor* L. during the transition from wet to dry season in Trinidad: the role of organic acid speciation and decarboxylation. Plant Cell Environ 19:655–664

Bort J, Brown RH, Araus JL (1996) Refixation of respiratory CO_2 in the ears of C_3 cereals. J Exp Bot 47:1567–1575

Bourguignon J, Merand V, Rawsthorne S, Forest E, Douce R (1996) Glycine decarboxylase and pyruvate dehydrogenase complexes share the same dihydrolipoamide dehydrogenase in pea leaf mitochondria: evidence from mass spectrometry and primary-structure analysis. Biochem J 313:229–234

Brown HT, Morris GH (1893) A contribution to the chemistry and physiology of foilage leaves. J Chem Soc 63:604–677

Brown MM, Hall JL, Ho LC (1997) Sugar uptake by protoplasts isolated from tomato fruit tissues during various stages of fruit growth. Physiol Plant 101:533–539

Castillo FJ (1996) Antioxidative protection in the inducible CAM plant *Sedum album* L. following the imposition of severe water stress and recovery. Oecologia 107:469–477

Castrillo M, Aso P, Longart M, Vermehren A (1997) *In situ* immunofluorescent localisation of ribulose-1,5-bisphosphate carboxylase/oxygenase in mesophyll of C_4 dicotyledonous plants. Photosynthetica 33:39–50

Cerling TE, Harris JM, MacFadden BJ, Leakey MG, Quade J, Eisenmann V, Ehleringer JR (1997) Global vegetation change through the miocene/pliocene boundary. Nature 389:153–158

Christopher JT, Holtum JAM (1996) Patterns of carbon partitioning in leaves of crassulacean acid metabolism species during deacidification. Plant Physiol 112:393–399

Colombo SL, Pairoba CF, Andreo CS (1996) Inhibitory effect of shikimic acid on PEP carboxylase activity. Plant Cell Physiol 37:870–872

Crafts-Brandner SJ, van de Loo FJ, Salvucci ME (1997) The two forms of ribulose-1,5-bisphosphate carboxylase/oxygenase activase differ in sensitivity to elevated temperature. Plant Physiol 114:439–444

Dai Z, Ku MSB, Edwards GE (1996) Oxygen sensitivity of photosynthesis and photorespiration in different photosynthetic types in the genus *Flaveria*. Planta 198:563–571

Daniel E (1997) The temperature dependence of photoinhibition in leaves of *Phaseolus vulgaris* (L.), Influence of CO_2 and O_2 concentrations. Plant Sci 124:1–8

Debnam PM, Shearer G, Blackwood L, Kohl DH (1997) Evidence for channeling of intermediates in the oxidative pentose phosphate pathway by soybean and pea nodule extracts, yeast extracts, and purified yeast enzymes. Eur J Biochem 246:283–290

Déjardin A, Rochat C, Wuillème S, Boutin J-P (1997) Contribution of sucrose synthase, ADP-glucose pyrophosphorylase and starch synthase to starch synthesis in developing pea seeds. Plant Cell Environ 20:1421–1430

Denyer K, Dunlap F, Thorbjørnsen T, Keeling P, Smith AM (1996) The major form of ADP-glucose pyrophosphorylase in maize endosperm is extra-plastidal. Plant Physiol 112:779–785

Desimone M, Henke A, Wagner E (1996) Oxidative stress induces partial degradation of the large subunit of ribulose-1,5-bisphosphate carboxylase/oxygenase in isolated chloroplasts of barley. Plant Physiol 111:789–796

Devi MT, Rajagopalan AV, Raghavendra AS (1996) Purification and properties of glycolate oxidase from plants with different photosynthetic pathways: distinctness of C_4 enzyme from that of a C_3 species and a C_3–C_4 intermediate. Photosynth Res 47:231–238

Du YC, Kawamitsu Y, Nose A, Hiyane S, Murayama S, Wasano K, Uchida Y (1996) Effects of water stress on carbon exchange rate and activities of photosynthetic enzymes in leaves of sugarcane (*Saccharum* sp.). Aust J Plant Physiol 23:719–726

Du Z, Aghoram K, Outlaw WH (1997) *In vivo* phosphorylation of phospho*enol*pyruvate carboxylase in guard cells of *Vicia faba* L. is enhanced by fusicoccin and suppressed by abscisic acid. Arch Biochem Biophys 337:345–350

Duwenig E, Steup M, Kossmann J (1997a) Induction of genes encoding plastidic phosphorylase from spinach (*Spinacia oleracea* L.) and potato (*Solanum tuberosum* L.) by exogenously supplied carbohydrates in excised leaf discs. Planta 203:111–120

Duwenig E, Steup M, Willmitzer L, Kossmann J (1997b) Antisense inhibition of cytosolic phosphorylase in potato plants (*Solanum tuberosum* L.) affects tuber sprouting and flower formation with only little impact on carbohydrate metabolism. Plant J 12:323–333

Eastmond PJ, Dennis DT, Rawsthorne S (1997) Evidence that a malate/inorganic phosphate exchange translocator imports carbon across the leucoplast envelope for fatty acid synthesis in developing castor seed endosperm. Plant Physiol 114:851–856

Echeverria E, Salvucci ME, Gonzalez P, Paris G, Salerno G (1997) Physical and kinetic evidence for an association between sucrose-phosphate synthase and sucrose-phosphate phosphatase. Plant Physiol 115:223–227

Eckardt NA, Portis AR (1997) Heat denaturation profiles of ribulose-1,5-bisphosphate carboxylase/oxygenase (rubisco) and rubisco activase and the inability of rubisco activase to restore activity of heat-denatured rubisco. Plant Physiol 113:243–248

Eckardt NA, Snyder GW, Portis AR, Ogren WL (1997) Growth and photosynthesis under high and low irradiance of *Arabidopsis thaliana* antisense mutants with reduced ribulose-1,5-bisphosphate carboxylase/oxygenase activase content. Plant Physiol 113:575–586

Eller BM, Ferrari S (1997) Water use efficiency of two succulents with contrasting CO_2 fixation pathways. Plant Cell Environ 20:93–100

Emes MJ, Neuhaus HE (1997) Metabolism and transport in non-photosynthetic plastids. J Exp Bot 48:1995–2005

Escobar-Gutiérrez AJ, Gaudillère J-P (1997) Carbon partitioning in source leaves of peach, a sorbitol-synthesizing species, is modified by photosynthetic rate. Physiol Plant 100:353–360

Evert RF, Russin WA, Botha CEJ (1996) Distribution and frequency of plasmodesmata in relation to photoassimilate pathways and phloem loading in the barley leaf. Planta 198:572–579

Ferreira RMB, Franco E, Teixeira ARN (1996) Covalent dimerization of ribulose bisphosphate carboxylase subunits by UV radiation. Biochem J 318:227–234

Fiorani M, Saltarelle R, De Sanctis R, Palma F, Ceccaroli P, Stocchi V (1996) Role of de-hydroascorbate in rabbit erythrocyte hexokinase inactivation induced by ascorbic acid/Fe(II). Arch Biochem Biophys 334:357–361

Fischer K, Kammerer B, Gutensohn M, Arbinger B, Weber A, Häusler RE, Flügge U-F (1997) A new class of plastidic phosphate translocators: a putative link between primary and secondary metabolism by the phosphoenolpyruvate/phosphate antiporter. Plant Cell 9:453–462

Fridlyand LE, Backhausen JE, Holtgrefe S, Kitzmann C, Scheibe R (1997) Quantitative evaluation of the rate of 3-phosphoglycerate reduction in chloroplasts. Plant Cell Physiol 38:1177–1186

Furbank RT, Chitty JA, von Caemmerer S, Jenkins CLD (1996) Antisense RNA inhibition of *RbcS* gene expression reduces rubisco level and photosynthesis in the C_4 plant *Flaveria bidentis*. Plant Physiol 111:725–734

Furbank RT, Pritchard J, Jenkins CLD (1997) Effects of exogenous sucrose feeding on photosynthesis in the C_3 plant tobacco and the C_4 plant *Flaveria bidentis*. Aust J Plant Physiol 24:291–299

Galtier N, Foyer CH, Murchie E, Alred R, Quick P, Voelker TA, Thépenier C, Lascève G, Betsche T (1995) Effects of light and atmospheric carbon dioxide enrichment on photosynthesis and carbon partitioning in the leaves of tomato (*Lycopersicon esculentum* L.) plants over-expressing sucrose phosphate synthase. J Exp Bot 46:1335–1344

Gao Y, Woo KC (1996) Regulation of phospho*enol*pyruvate carboxylase in *Zea mays* by protein phosphorylation and metabolites and their roles in photosynthesis. Aust J Plant Physiol 23:25–32

Giglioli-Guivarc'h N, Pierre J-N, Brown S, Chollet R, Vidal J, Gadal P (1996) The light-dependent transduction pathway controlling the regulatory phosphorylation of C4 phospho*enol*pyruvate carboxylase in protoplasts from *Digitaria sanguinalis*. Plant Cell 8:573–586

Goyal A, Tolbert NE (1996) Association of glycolate oxidation with photosynthetic electron transport in plant and algal chloroplasts. Proc Natl Acad Sci USA 93:3319–3324

Grams TEE, Beck F, Lüttge U (1996) Generation of rhythmic and arrhythmic behaviour of crassulacean acid metabolism in *Kalanchoe daigremontiana* under continuous light by varying the irradiance or temperature: measurements in vivo and model stimulations. Planta 198:110–117

Grams TEE, Borland AM, Roberts A, Griffiths H, Beck F, Lüttge U (1997) On the mechanism of reinitiation of endogenous crassulacean acid metabolism rhythm by temperature changes. Plant Physiol 113:1309–1317

Gray GR, Savitch LV, Ivanov AG, Huner NPA (1996) Photosystem II excitation pressure and development of resistance to photoinhibition. Plant Physiol 110:61–71

Greenberg BM, Wilson MI, Gerhardt KE, Wilson KE (1996) Morphological and physiological responses of *Brassica napus* to ultraviolet-B radiation: photomodification of ribulose-1,5-bisphosphate carboxylase/oxygenase and potential acclimation processes. J Plant Physiol 148:78–85

Grotjohann N (1997) Activation of chloroplast fructose-1,6-bisphosphatase of *Pisum sativum* by mole mass change. Bot Acta 110:323–327

Güerere I, Tezara W, Herrera C, Fernández MD, Herrera A (1996) Recycling of CO_2 during induction of CAM by drought in *Talinum paniculatum* (Portulacaceae). Physiol Plant 98:471–476

Guerin J, Carbonero P (1997) The spatial distribution of sucrose synthase isozymes in barley. Plant Physiol 114:55–62

Guinel FC, Ireland RJ (1996) Immunogold localization and quantitative distribution of two proteins of the glycine decarboxylase complex in developing leaflets of pea (Pisum sativum L.). Int J Plant Sci 157:539–545

Häberlein I, Vogeler B (1995) Completion of the thioredoxin reaction mechanism: kinetic evidence for protein complexes between thioredoxin and fructose 1,6-bisphosphatase. Biochim Biophys Acta 1253:169–174

Harris GC, Königer M (1997) The 'high' concentrations of enzymes within the chloroplast. Photosynth Res 54:5–23

Hartwell J, Smith LH, Wilkins MB, Jenkins GI, Nimmo HG (1996) Higher plant phosphoenolpyruvate carboxylase kinase is regulated at the level of translatable mRNA in response to light or circadian rhythm. Plant J 10:1071–1078

He D, Edwards GE (1996) Estimation of diffusive resistance of bundle sheath cells to CO_2 from modeling of C_4 photosynthesis. Photosynth Res 49:195–208

Heber U, Bligny R, Streb P, Douce R (1996) Photorespiration is essential for the protection of the photosynthetic apparatus of C3 plants against photoinactivation under sunlight. Bot Acta 109:307–315

Heldt HW, Rapley L (1970) Specific transport of inorganic phosphate, 3-P-glycerate and dihydroxyacetone-P, and of dicarboxylates across the inner membrane of spinach chloroplasts. FEBS Lett 10:143–148

Helliker BR, Martin CE (1997) Comparative water-use efficiences of three species of *Peperomia* (Piperaceae) having different photosynthetic pathways. J Plant Physiol 150:259–263

Herbers K, Sonnewald U (1996) Manipulating metabolic partitioning in transgenic plants. Trends Biotechnol 14:198–205

Herppich M, von Willert DJ, Herppich WB (1995) Diurnal rhythm in citric acid content preceded the onset of nighttime malic acid accumulation during metabolic changes form C_3 to CAM in salt-stressed plants of *Mesembryanthemum crystallinum*. J Plant Physiol 147:38–42

Hill LM, Smith AM (1995) Coupled movements of glucose 6-phosphate and triose phosphate through the envelopes of plastids from developing embryos of pea (*Pisum sativum* L.). J Plant Physiol 146:411–417

Hixon M, Sinerius G, Schneider A, Walter C, Fessner W-D, Schloss JV (1996) Quo vadis photorespiration: a tale of two aldolases. FEBS Lett 392:281–284

Huber SC, Huber JL (1996) Role and regulation of sucrose-phosphate synthase in higher plants. Annu Rev Plant Physiol Plant Mol Biol 47:431–444

Huber SC, Huber JL, Liao P-C, Gage DA, McMichael RW, Chourey PS, Hannah LC, Koch K (1996) Phosphorylation of serine-15 of maize leaf sucrose synthase. Plant Physiol 112:793–802

Huner NPA, Maxwell DP, Gray GR, Savitch LV, Krol M, Ivanov AG, Falk S (1996) Sensing environmental temperature change through imbalances between energy supply and energy consumption: redox state of photosystem II. Physiol Plant 98:358–364

Igamberdiev AU, Bykova NV, Gardeström P (1997) Involvement of cyanide-resistant and rotenone-intensitive pathways of mitochondrial electron transport during oxidation of glycine in higher plants. FEBS Lett 412:265–269

Imaizumi N, Samejima M, Ishihara K (1997) Characteristics of photosynthetic carbon metabolism of spikelets in rice. Photosynth Res 52:75–82

Ishibashi M, Usuda H, Terashima I (1996) The loss of ribulose-1,5-bisphosphate carboxylase/oxygenase caused by 24-hour rain treatment fully explains the decrease in the photosynthetic rate in bean leaves. Plant Physiol 111:635–640

Ishida H, Nishimori Y, Sugisawa M, Makino A, Mae T (1997) The large subunit of ribulose-1,5-bisphosphate carboxylase/oxygenase is fragmented into 37-kDa and 16-kDa polypeptides by active oxygen in the lysates of chloroplasts from primary leaves of wheat. Plant Cell Physiol 38:471–479

Ishimaru K, Ichikawa H, Matsuoka M, Ohsugi R (1997) Analysis of C4 maize pyruvate, orthophosphate dikinase expressed in C3 transgenic *Arabidopsis* plants. Plant Sci 129:57–64

Issakidis E, Lemaire M, Decottignies P, Jacquot J-P, Migniac-Maslow M (1996) Direct evidence for the different roles of the N- and C-terminal regulatory disulfides of sorghum leaf NADP-malate dehydrogenase in its activation by reduced thioredoxin. FEBS Lett 392:121–124

Jacquot J-P, Lopez-Jaramillo J, Miginiac-Maslow M, Lemaire S, Cherfils J, Chueca A, Lopez-Gorge J (1997a) Cysteine-153 is required for redox regulation of pea chloroplast fructose-1,6-bisphosphatase. FEBS Lett 401:143–147

Jacquot J-P, Lancelin J-M, Meyer Y (1997b) Thioredoxins: structure and function in plant cells. New Phytol 136:543–570

Jang J-C, Sheen J (1997) Sugar sensing in higher plants. Trends Plant Sci 2:208–214

Jang J-C, León P, Zhou L, Sheen J (1997) Hexokinase as a sugar sensor in higher plants. Plant Cell 9:5–19

Jones TL, Ort DR (1997) Circadian regulation of sucrose phosphate synthase activity in tomato by protein phosphatase activity. Plant Physiol 113:1167–1175

Jones PG, Lloyd JC, Raines CA (1996) Glucose feeding of intact wheat plants represses the expression of a number of Calvin cycle genes. Plant Cell Environ 19:231–236

Kanechi M, Uchida N, Yasuda T, Yamaguchi T (1996) Non-stomatal inhibition associated with inactivation of rubisco in dehydrated coffee leaves under unshaded and shaded conditions. Plant Cell Physiol 37:455–460

Kelly GJ, Latzko E (1996) Photosynthesis. Carbon metabolism: the carbon metabolisms of unstressed and stressed plants. Progress Bot 58:187–220

Keys AJ, Major I, Parry MAJ (1995) Is there another player in the game of rubisco regulation? J Exp Bot 46:1245–1251

Kleczkowski LA (1996) Back to the drawing board: redefining starch synthesis in cereals. Trends Plant Sci 1:363–364

Köhler RH, Cao J, Zipfel WR, Webb WW, Hanson MR (1997) Exchange of protein molecules through connections between higher plant plastids. Science 276:2039–2042

Komatsu S, Masuda T, Hirano H (1996) Rice gibberellin-binding phosphoprotein structurally related to ribulose-1,5-bisphosphate carboxylase/oxygenase activase. FEBS Lett 384:167–171

Konrad A, Banze M, Follmann H (1996) Mitochondria of plant leaves contain two thioredoxins. Completion of the thioredoxin profile of higher plants. J Plant Physiol 149:317–321

Kore-eda S, Kanai R (1997) Induction of glucose 6-phosphate transport activity in chloroplasts of *Mesembryanthemum crystallinum* by the C_3-CAM transition. Plant Cell Physiol 38:895–901

Kostov RV, Small CL, McFadden BA (1997) Mutations in a sequence near the N-terminus of the small subunit alter the CO_2/O_2 specificity factor for ribulose bisphosphate carboxylase/oxygenase. Photosynth Res 54:127–134

Kovtun Y, Daie J (1995) End-product control of carbon metabolism in culture-grown sugar beet plants. Plant Physiol 108:1647–1656

Kozaki A, Takeba G (1996) Photorespiration protects C3 plants from photooxidation. Nature 384:557–560

Kruger NJ, Scott P (1995) Integration of cytosolic and plastidic carbon metabolism by fructose 2,6-bisphosphate. J Exp Bot 46:1325–1333

Kühn C, Quick WP, Schulz A, Riesmeier JW, Sonnewald U, Frommer WB (1996) Companion cell-specific inhibition of the potato sucrose transporter SUT1. Plant Cell Environ 19:1115–1123

Kühn C, Franceschi VR, Schulz A, Lemoine R, Frommer WB (1997) Macromolecular trafficking indicated by localization and turnover of sucrose transporters in enucleate sieve elements. Science 275:1298–1300

Lacuesta M, Dever LV, Muñoz-Rueda A, Lea PJ (1997) A study of photorespiratory ammonia production in the C_4 plant *Amaranthus edulis*, using mutants with altered photosynthetic capacities. Physiol Plant 99:447–455

Laisk A, Loreto F (1996) Determining photosynthetic parameters from leaf CO_2 exchange and chlorophyll fluorescence. Plant Physiol 110:903–912

Lal A, Edwards GE (1996) Analysis of inhibition of photosynthesis under water stress in the C4 species *Amaranthus cruentus* and *Zea mays*: electron tansport, CO_2 fixation and carboxylation capacity. Aust J Plant Physiol 23:403–412

Laporte MM, Galagan JA, Shapiro JA, Boersig MR, Shewmaker CK, Sharkey TD (1997) Sucrose-phosphate synthase activity, and yield analysis of tomato plants transformed with maize sucrose-phosphate synthase. Planta 203:253–259

Law RD, Plaxton WC (1997) Regulatory phosphorylation of banana fruit phospho*enol*pyruvate carboxylase by a copurifying phospho*enol*pyruvate carboxylasekinase. Eur J Biochem 247:642–651

Lawlor DW (1995) Photosynthesis, productivity and environment. J Exp Bot 46:1449–1461

Lebreton S, Gontero B, Avilan L, Ricard J (1997) Information transfer in multienzyme complexes. 1. Thermodynamics of conformational constaints and memory effects in the bienzyme glyceraldehyde-3-phosphate-dehydrogenase-phosphoribulokinase complex of *Chlamydomonas reinhardtii* chloroplasts. Eur J Biochem 250:286–295

Lee S-J, Daie J (1997) End-product repression of genes involving carbon metabolism in photosynthetically active leaves of sugarbeet. Plant Cell Physiol 38:887–894

Leegood RC, Lea PJ, Adcock MD, Häusler RE (1995) The regulation and control of photorespiration. J Exp Bot 46:1397–1414

Leport L, Kandlbinder A, Baur B, Kaiser WM (1996) Diurnal modulation of phosphoenolpyruvate carboxylation in pea leaves and roots as related to tissue malate concentrations and to the nitrogen source. Planta 198:495–501

Li B, Zhang X-Q, Chollet R (1996) Phospho*enol*pyruvate carboxylase kinase in tobacco leaves is activated by light in a similar but not identical way as in maize. Plant Physiol 111:497–505

Li AD, Stevens FJ, Huppe HC, Kersanach R, Anderson LE (1997) *Chlamydomonas reinhardtii* NADP-linked glyceraldehyde-3-phosphate dehydrogenase contains the cysteine residues identified as potentially domain-locking in the higher plant enzyme and is light activated. Photosynth Res 51:167–177

Lichtenthaler HK, Schwender J, Disch A, Rohmer M (1997) Biosynthesis of isoprenoids in higher plant chloroplasts proceeds via a mevalonate-independent pathway. FEBS Lett 400:271–274

Ligeza A, Tikhonov AN, Subczynski WK (1997) In situ measurements of oxygen production and consumption using paramagnetic fusinite particles injected into a bean leaf. Biochim Biophys Acta 1319:133–137

Lilley RMcC, Portis AR (1997) ATP hydrolysis activity and polymerization state of ribulose-1,5-bisphosphate carboxylase oxygenase activase. Plant Physiol 114:605–613

Lozano RM, Wong JH, Yee BC, Peters A, Kobrehel K, Buchanan BB (1996) New evidence for a role for thioredoxin *h* in germination and seedling development. Planta 200:100–106

Lüttge U, Fischer-Schliebs E, Ratajczak R, Kramer D, Berndt D, Kluge M (1995) Functioning of the tonoplast in vacuolar C-storage and remobilization in crassulacean acid metabolism. J Exp Bot 46:1377–1388

MacIntyre HL, Sharkey TD, Geider RJ (1997) Activation and deactivation of ribulose-1,5-bisphosphate carboxylase/oxygenase (rubisco) in three marine microalgae. Photosynth Res 51:93–106

Magnin NC, Cooley BA, Reiskind JB, Bowes G (1997) Regulation and localization of key enzymes during the induction of Kranz-less, C_4-type photosynthesis in *Hydrilla verticillata*. Plant Physiol 115:1681–1689

Martindale W, Leegood RC (1997) Acclimation of photosynthesis to low temperature in *Spinacia oleraceae* L. 1. Effects of acclimation on CO_2 assimilation and carbon partitioning. J Exp Bot 48:1865–1872

Martindale W, Parry MAJ, Andralojc PJ, Keys AJ (1997) Synthesis of 2'-carboxy-D-arabinitol-1-phosphate in french bean (*Phaseolus vulgaris* L.): a search for precursors. J Exp Bot 48:9–14

Mate CJ, von Caemmerer S, Evans JR, Hudson GS, Andrews TJ (1996) The relationship between CO_2-assimilation rate, rubisco carbamylation and rubisco activase content in activase-deficient transgenic tobacco suggests a simple model of activase action. Planta 198:604–613

Matsuba K, Imaizumi N, Kaneko S, Samejima M, Ohsugi R (1997) Photosynthetic responses to temperature of phosphoenolpyruvate carboxykinase type C_4 species differing in cold sensitivity. Plant Cell Environ 20:268–274

McFadden GI, Reith ME, Munholland J, Lang-Unnasch N (1996) Plastid in human parasites. Nature 381:482

Medrano H, Keys AJ, Lawlor DW, Parry MAJ, Azcon-Bieto J, Delgado E (1995) Improving plant production by selection for survival at low CO_2 concentrations. J Exp Bot 46:1389–1396

Meister M, Agostino A, Hatch MD (1996) The roles of malate and aspartate in C_4 photosynthetic metabolism of *Flaveria bidentis* (L.). Planta 199:262–269

Micallef BJ, Sharkey TD (1996) Genetic and physiological characterization of *Flaveria linearis* plants having a reduced activity of cytosolic fructose-1,6-bisphosphatase. Plant Cell Environ 19:1–9

Minamikawa T, Maeda Y, Xu B-M (1995) Ancient lotus seeds preserve activity of starch phosphorylase. Naturwissenschaften 82:375–376

Möhlmann T, Tjaden J, Henrichs G, Quick WP, Häusler R, Neuhaus HE (1997) ADP-glucose drives starch synthesis in isolated maize endosperm amyloplasts: characterization of starch synthesis and transport properties across the amyloplast envelope. Biochem J 324:503–509

Moore BD, Sharkey TD, Seemann JR (1995) Intracellular localization of CA1P and CA1P phosphatase activity in leaves of *Phaseolus vulgaris* L. Photosynth Res 45:219–224

Moore BD, Palmquist DE, Seemann JR (1997) Influence of plant growth at high CO_2 concentrations on leaf content of ribulose-1,5-bisphosphate carboxylase/oxygenase and intracellular distribution of soluble carbohydrates in tobacco, snapdragon and parsley. Plant Physiol 115:241–248

Morell S, Follmann H, De Tullio M, Häberlein I (1997) Dehydroascorbate and dehydroascorbate reductase are phantom indicators of oxidative stress in plants. FEBS Lett 414:567–570

Naeem M, Tetlow IJ, Emes MJ (1997) Starch synthesis in amyloplasts purified from developing potato tubers. Plant J 11:1095–1103

Neuhaus H-E, Maaß U (1996) Unidirectional transport of orthophosphate across the envelope of isolated cauliflower-bud amyloplasts. Planta 198:542–548

Neuhaus HE, Schulte N (1996) Starch degradation in chloroplasts isolated from C_3 or CAM (crassulacean acid metabolism)-induced *Mesembryanthemum crystallinum* L. Biochem J 318:945–953

Neuhaus HE, Thom E, Möhlmann T, Steup M, Kampfenkel K (1997) Characterization of a novel eukaryotic ATP/ADP translocator located in the plastid envelope of *Arabidopsis thaliana* L. Plant J 11:73–82

Ohta S, Usami S, Ueki J, Kumashiro T, Komari T, Burnell JN (1996) Identification of the amino acid residues responsible for cold tolerance in *Flaveria brownii* pyruvate, orthophosphate dikinase. FEBS Lett 396:152–156

Osmond B, Badger M, Maxwell K, Björkman O, Leegood R (1997) Too many photons: photorespiration, photoinhibition and photooxidation. Trends Plant Sci 2:119–121

Osuna L, González M-C, Cejudo FJ, Vidal J, Echevarría C (1996) In vivo and vitro phosphorylation of the phospho*enol*pyruvate carboxylase from wheat seeds during germination. Plant Physiol 111:551–558

Panstruga R, Hippe-Sanwald S, Lee Y-K, Lataster M, Lipka V, Fischer R, Liao YC, Häusler RE, Kreuzaler F, Hirsch H-J (1997) Expression and chloroplast-targeting of active phosphoenolpyruvate synthetase from *Escherichia coli* in *Solanum tuberosum*. Plant Sci 127:191–205

Park Y-I, Chow WS, Osmond CB, Anderson JM (1996) Electron transport to oxygen mitigates against the photoinactivation of photosystem II *in vivo*. Photosynth Res 50:23–32

Parry MAJ, Andralojc PJ, Parmar S, Keys AJ, Habash D, Paul MJ, Alred R, Quick WP, Servaites JC (1997) Regulation of rubisco by inhibitors in the light. Plant Cell Environ 20:528–534

Prabhu V, Chatson KB, Abrams GD, King J (1996) ^{13}C nuclear magnetic resonance detection of interactions of serine hydroxymethyltransferase with C1-tetrahydrofolate synthase and glycine decarboxylase complex activities in Arabidopsis. Plant Physiol 112:207–216

Quick WP, Neuhaus HE (1996) Evidence of two types of phosphate translocators in sweet-pepper (*Capsicum annuum* L.) fruit chromoplasts. Biochem J 320:7–10

Quiñones MA, Lu Z, Zeiger E (1996) Close correspondence between the action spectra for the blue light responses of the guard cell and coleoptile chloroplasts, and the spectra for blue light-dependent stomatal opening and coleoptile phototropism. Proc Natl Acad Sci USA 93:2224–2228

Reche A, Lázaro JJ, Hermoso R, Chueca A, López Gorgé J (1997) Binding and activation of pea chloroplast fructose-1,6-bisphosphatase by homologous thioredoxins *m* and *f*. Physiol Plant 101:463–470

Reddy AR (1996) Fructose 2,6-bisphosphate-modulated photosynthesis in sorghum leaves grown under low water regimes. Phytochemistry 43:319–322

Reimholz R, Geiger M, Haake V, Deiting U, Krause K-P, Sonnewald U, Stitt M (1997) Potato plants contain multiple forms of sucrose phosphate synthase, which differ in their tissue distributions, their levels during development, and their responses to low temperature. Plant Cell Environ 20:291–305

Roberts A, Griffiths H, Borland AM, Reinert F (1996) Is crassulacean acid metabolism activity in sympatric species of hemi-epiphytic stranglers such as *Clusia* related to carbon cycling as a photoprotective process? Oecologia 106:28–38

Robinson JM (1997) Nitrogen limitation of spinach plants causes a simultaneous rise in foliar levels of orthophosphate sucrose and starch. Int J Plant Sci 158:432–441

Rost S, Frank C, Beck E (1996) The chloroplast envelope is permeable for maltose but not for maltodextrins. Biochim Biophys Acta 1291:221–227

Roughnan PG (1997) Stromal concentrations of coenzyme A and its esters are insufficient to account for rates of chloroplast fatty acid synthesis: evidence for substrate channelling within the chloroplast fatty acid synthase. Biochem J 327:267–273

Roughan PG, Ohlrogge JB (1996) Evidence that isolated chloroplasts contain an integrated lipid-synthesizing assembly that channels acetate into long-chain fatty acids. Plant Physiol 110:1239–1247

Rudi H, Doan DNP, Olsen O-A (1997) A (his)$_6$-tagged recombinant barley (*Hordeum vulgare* L.) endosperm ADP-glucose pyrophosphorylase expressed in the baculovirus-insect cell system is insensitive to allosteric regulation by 3-phosphoglycerate and inorganic phosphate. FEBS Lett 419:124–130

Russin WA, Evert RF, Vanderveer PJ, Sharkey TD, Briggs SP (1996) Modification of a specific class of plasmodesmata and loss of sucrose export ability in the *sucrose export defective1* maize mutant. Plant Cell 8:645–658

Sachs J (1862) Über den Einfluß des Lichtes auf die Bildung des Amylums in den Chlorophyllkörnern. Bot Zeit 20:365–373

Sadimantara GR, Abe T, Suzuki J, Hirano H, Sasahara T (1996) Characterization and partial amino acid sequence of a high molecular weight protein from rice seed endosperm: homology to pyruvate orthophosphate dikinase. J Plant Physiol 149:285–289

Salahas G, Hatzidimitrakis K, Georgiou CD, Angelopoulos K, Gavalas NA (1997) Phosphate and sulfate activate the phosphoenolpyruvate carboxylase from the C_4 plant *Cynodon dactylon* L. Bot Acta 110:309–313

Saliendra NZ, Meinzer FC, Perry M, Thom M (1996) Associations between partitioning of carboxylase activity and bundle sheath leakiness to CO_2, carbon isotope discrimination, photosynthesis, and growth in sugarcane. J Exp Bot 47:907–914

Salvucci ME, Ogren WL (1996) The mechanism of rubisco activase: insights from studies of the properties and structure of the enzyme. Photosynth Res 47:1–11

Sasaki Y, Kozaki A, Hatano M (1997) Link between light and fatty acid synthesis: thioredoxin-linked reductive activation of plastidic acetyl-CoA carboxylase. Proc Natl Acad Sci USA 94:11096–11101

Savitch LV, Gray GR, Huner NPA (1997) Feedback-limited photosynthesis and regulation of sucrose-starch accumulation during cold acclimation and low-temperature stress in a spring and winter wheat. Planta 201:18–26

Schaffer AA, Petreikov M (1997) Inhibition of fructokinase and sucrose synthase by cytosolic levels of fructose in young tomato fruit undergoing transient starch synthesis. Physiol Plant 101:800–806

Schmid R, Dring MJ (1996) Influence of carbon supply on the circadian rhythmicity of photosynthesis and its stimulation by blue light in *Ectocarpus siliculosus*: clues to the mechanism of inorganic carbon acquisition in lower brown algae. Plant Cell Environ 19:373–382

Scholes J, Bundock N, Wilde R, Rolfe S (1996) The impact of reduced vacuolar invertase activity on the photosynthetic and carbohydrate metabolism of tomato leaves. Planta 200:265–272

Schwender J, Seemann M, Lichtenthaler HK, Rohmer M (1996) Biosynthesis of isoprenoids (carotenoids, sterols, prenyl side-chains of chlorophylls and plastoquinone) via a novel pyruvate/glyceraldehyde 3-phosphate non-mevalonate pathway in the green alga *Scenedesmus obliquus*. Biochem J 316:73–80

Shatters RG, West SH (1996) Purification and characterization of non-chloroplastic α-1,4-glucan phosphorylases from leaves of *Digitaria eriantha* Stent. J Plant Physiol 149:501–509

Simon J-P (1996) Molecular forms and kinetic properties of pyruvate P_i dikinase from two populations of barnyard grass (*Echinochloa crus-galli*) from sites of contrasting climates. Aust J Plant Physiol 23:191–199

Skillman JB, Winter K (1997) High photosynthetic capacity in a shade-tolerant crassulacean acid metabolism plant. Plant Physiol 113:441-450

Smeekens S, Rook F (1997) Sugar sensing and sugar-mediated signal transduction in plants. Plant Physiol 115:7–13

Smith LH, Lillo C, Nimmo HG, Wilkins MB (1996) Light regulation of phosphoenolpyruvate carboxylase in barley mesophyll protoplasts is modulated by protein synthesis and calcium, and not necessarily correlated with phosphoenolpyruvate carboxylase kinase activity. Planta 200:174–180

Spencer WE, Wetzel RG, Teeri J (1996) Photosynthetic phentoype plasticity and the role of phosphoenolpyruvate carboxylase in *Hydrilla verticillata*. Plant Sci 118:1–9

Stadler R, Brandner J, Schulz A, Gahrtz M, Sauer N (1995) Phloem loading by the PmSUC2 sucrose carrier from *Plantago major* occurs into companion cells. Plant Cell 7:1545–1554

Steiger S, Pfeifer T, Ratajczak R, Martinoia E, Lüttge U (1997) The vacuolar malate transporter of *Kalanchoe daigremontiana*: A 32 kDa polypeptide? J Plant Physiol 151:137–141

Stieger PA, Feller U (1997) Degradation of stromal proteins in pea (*Pisum sativum* L.) chloroplasts under oxidising conditions. J Plant Physiol 151:556–562

Strand Å, Hurry V, Gustafsson P, Gardeström P (1997) Development of *Arabidopsis thaliana* leaves at low temperatures releases the suppression of photosynthesis and photosynthetic gene expression despite the accumulation of soluble carbohydrates. Plant J 12:605–614

Sweetlove LJ, Burrell MM, ap Rees T (1996) Starch metabolism in tubers of transgenic potato (*Solanum tuberosum*) with increased ADP glucose pyrophosphorylase. Biochem J 320:493–498

Tallman G, Zhu J, Mawson BT, Amodeo G, Nouhi Z, Levy K, Zeiger E (1997) Induction of CAM in *Mesembryanthemum crystallinum* abolishes the stomatal response to blue light and light-dependent zeaxanthin formation in guard cell chloroplasts. Plant Cell Physiol 38:236–242

Tamoi M, Ishikawa T, Takeda T, Shigeoka S (1996) Enzymic and molecular characterization of NADP-dependent glyceraldehyde-3-phosphate dehydrogenase from *Synechococcus* PCC 7942: resistance of the enzyme to hydrogen peroxide. Biochem J 316:685–690

Taniguchi M, Sugiyama T (1997) The expression of 2-oxoglutarate/malate translocator in the bundle-sheath mitochondria of *Panicum miliaceum*, a NAD-malic enzyme-type C_4 plant, is regulated by light and development. Plant Physiol 114:285–293

Thorbjørnsen T, Villand P, Denyer K, Olsen O-A, Smith AM (1996) Distinct isoforms of ADPglucose pyrophosphorylase occur inside and outside the amyloplasts in barley endosperm. Plant J 10:243–250

Tolbert NE, Benker C, Beck E (1995) The oxygen and carbon dioxide compensation points of C_3 plants: possible role in regulating atmospheric oxygen. Proc Natl Acad Sci USA 92:11230–11233

Toroser D, Huber SC (1997) Protein phosphorylation as a mechanism for osmotic-stress activation of sucrose-phosphate synthase in spinach leaves. Plant Physiol 114:947–955

Trevanion SJ, Furbank RT, Ashton AR (1997) NADP-malate dehydrogenase in the C_4 plant *Flaveria bidentis*. Plant Physiol 113:1153–1165

Tripathy BC, Brown CS, Levine HG, Krikorian AD (1996) Growth and photosynthetic responses of wheat plants grown in space. Plant Physiol 110:801–806

Uemura K, Suzuki Y, Shikanai T, Wadano A, Jensen RG, Ohmara W, Yokota A (1996) A rapid sensitive method for determination of relative specificity of rubisco from various species by anion-exchange chromatography. Plant Cell Physiol 37:325–331

Ueno O (1996) Immunocytochemical localization of enzymes involved in the C_3 and C_4 pathways in the photosynthetic cells of an amphibious sedge, *Eleocharis vivipara*. Planta 199:394–403

Usami S, Ohta S, Komari T, Burnell JN (1995) Cold stability of pyruvate, orthophosphate dikinase of *Flaveria brownii*. Plant Mol Biol 27:969–980

Venkaiah B, Kumar A (1996) Multiple forms of starch phosphorylase from *Sorghum* leaves. Phytochemistry 41:713–717

Vidal J, Chollet R (1997) Regulatory phosphorylation of C_4 PEP carboxylase. Trends Plant Sci 2:230–237

Walker DA (1995) Manipulating photosynthetic metabolism to improve crops: an inversion of ends and means. J Exp Bot 46:1253–1259

Walker DA (1997) Tell me where all past years are. Photosynth Res 51:1–26

Walker RP, Leegood RC (1996) Phosphorylation of phosphoenolpyruvate carboxykinase in plants. Biochem J 317:653–658

Wang X, Tang X-y, Anderson LE (1996) Enzyme-enzyme interaction in the chloroplast: physical evidence for association between phosphoglycerate kinase and glyceraldehyde-3-phosphate dehydrogenase in vitro. Plant Sci 117:45–53

Wedel N, Soll J, Paap BK (1997) CP12 provides a new mode of light regulation of Calvin cycle activity in higher plants. Proc Natl Acad Sci USA 94:10479–10484

Wild K, Grafmüller R, Wagner E, Schulz GE (1997) Structure, catalysis and supramolecular assembly of adenylate kinase from maize. Eur J Biochem 250:326–331

Wildner GF, Schlitter J, Müller M (1996) Rubisco, an old challenge with new perspectives. Z Naturforsch 51c:263–276

Wingler A, Lea PJ, Leegood RC (1997) Control of photosynthesis in barley plants with reduced activities of glycine decarboxylase. Planta 202:171–178

Woodrow IE, Kelly ME, Mott KA (1996) Limitation of the rate of ribulosebisphosphate carboxylase activation by carbamylation and ribulosebisphosphate carboxylase activase activity: development and tests of a mechanistic model. Aust J Plant Physiol 23:141–149

Xu H-L, Gauthier L, Desjardins Y, Gosselin A (1997) Photosynthesis in leaves, fruit, stem and petioles of greenhouse-grown tomato plants. Photosynthetica 33:113–123

Yamasaki T, Kudoh T, Kamimura Y, Katoh S (1996) A vertical gradient of the chloroplast abundance among leaves of *Chenopodium album*. Plant Cell Physiol 37:43–48

Zhang X-Q, Chollet R (1997a) Phospho*enol*pyruvate carboxylase protein kinase from soybean root nodules: partial purification, characterization, and up/down-regulation by photosynthate supply from the shoots. Arch Biochem Biophys 343:260–268

Zhang X-Q, Chollet R (1997b) Seryl-phosphorylation of soybean nodule sucrose synthase (nodulin-100) by a Ca^{2+}-dependent protein kinase. FEBS Lett 410:126–130

Zhigang A, Löw R, Rausch T, Lüttge U, Ratajczak R (1996) The 32 kDa tonoplast polypeptide D_i associated with the V-type H^+-ATPase of *Mesembryanthemum crystallinum* L. in the CAM state: A proteolytically processed subunit B? FEBS Lett 389:314–318

Zhu YJ, Komor E, Moore PE (1997) Sucrose accumulation in the sugarcane stem is regulated by the difference between the activities of soluble acid invertase and sucrose phosphate synthase. Plant Physiol 115:609–616

Zotz G, Ziegler H (1997) The occurrence of crassulacean acid metabolism among vascular epiphytes from central Panama. New Phytol 137:223–229

Zrenner R, Krause K-P, Apel P, Sonnewald U (1996) Reduction of the cytosolic fructose-1,6-bisphosphatase in transgenic potato plants limits photosynthetic sucrose biosynthesis with no impact on plant growth and tuber yield. Plant J 9:671–681

Dr. Grahame J. Kelly
Centre for Molecular Biotechnology
School of Life Sciences
Queensland University of Technology
Brisbane, Queensland 4000, Australia
e-mail g.kelly@qut.edu.au

Edited by
U. Lüttge

The Costs and Benefits of Oxygen for Photosynthesizing Plant Cells

By Margarete Baier and Karl-Josef Dietz

1 Introduction

Redox reactions of aerobic metabolism are directly or indirectly linked to atmospheric oxygen, a reactant which is essentially present in unlimited amounts nowadays. The accumulation of O_2 was the consequence of the successful colonization by photoautotrophic organisms of the terrestrial and aqueous habitats during evolution. Concomitantly with the development of oxygenic photosynthesis, oxygen became available as oxidizing reactant in chemical reactions, for instance as terminal electron acceptor of the respiratory electron transport chain, of xanthin oxidase, of lipoxigenase and in photorespiratory oxygenation of ribulose-1,5-bisphosphate. Although still subject to some controversial discussion, photorespiratory energy consumption in the chloroplasts and amino acid synthesis in the peroxisomes and mitochondria may constitute beneficial or even essential metabolic pathways of plants under certain growth conditions.

The benefits of the increased oxygen concentration are not without expenses for the organisms. Molecular oxygen has a redox potential of $+1.299\,V$ ($O_2 + 4\,H^+ + 4\,e^- \rightarrow 2\,H_2O$) and therefore can oxidize most constituents of living cells, at least at elevated temperatures. A frequently experienced example of the destructive power of oxygen is the smell and taste of butter turning rancid during storage in air at room temperature. The oxidizing activity of O_2 is produced by the subsequent uptake of up to 4 electrons leading to the superoxide anion ($O_2\bullet^-$), hydrogen peroxide (H_2O_2), the hydroxyl radical ($OH\bullet$) and finally to water (H_2O) (Fig. 1). The incompletely reduced forms of oxygen are termed *reactive oxygen species* (ROS). As compared to O_2, ROS reveal increased reactivity. Even under the moderate conditions facilitating life, they oxidize a large variety of biomolecules. ROS inactivate enzymes, oxidize membrane lipids and cause nicking of nucleic acids with the danger of mutations. At high concentrations, ROS initiate degradation of proteins, alter membrane permeability and induce programmed or acute cell death. The detrimental effects of ROS are so striking that reagents which promote the generation of superoxide, e.g. paraquat (Szögyi et al. 1989),

Progress in Botany, Vol. 60
© Springer-Verlag Berlin Heidelberg 1999

are used as herbicides. ROS formation is an unavoidable consequence of aerobic life. It is difficult to assess the rates of ROS formation in cells. However, it appears that elaborate detoxification mechanisms enable the plants to cope with the normal rate of ROS production. Only under extremely unfavourable conditions does excessive liberation of ROS inhibit growth (Gómez et al. 1995; Roxas et al. 1997). Hence, the health state of plants in terms of redox homeostasis is determined by the relative rates of ROS production, ROS inactivation and repair of oxidative damage.

From a simple metabolic point of view, plants seem to be more exposed to oxidative stress than animals, bacteria and fungi. They not only consume oxygen but also produce oxygen in photosynthetic tissues upon illumination. Although O_2 easily permeates biomembranes, the O_2 concentration may be elevated above ambient at the site of the chloroplast water-splitting complex. For comparison, the highest rates of light- and CO_2-saturated photosynthetic CO_2 fixation are in the order of 500 µmol mg chlorophyll^{-1} h^{-1}. CO_2 saturation of photosynthesis is accomplished at 0.15% CO_2 in air. Under this condition, the internal CO_2 concentration is close to 0.05%. Even if the oxygen concentration gradient is steeper than that for CO_2, there is no good reason to assume that the oxygen concentration in the chloroplasts considerably exceeds ambient because the CO_2 concentration is already quite high in the exterior. Therefore, the slight increase in local oxygen concentration is unlikely to have a major impact on the oxidation state of the leaves.

In addition to oxygen evolution, photosynthesis rapidly produces ROS in the electron transport system (Asada 1994). Photosynthesis-dependent oxygen reduction indeed poses a severe oxidative risk to photosynthesizing cells. In addition, ROS formation is facilitated by photosensitizing pigments like chlorophylls and carotenoids, which transfer energy to O_2 following excitational activation. Several enzymic and non-enzymic mechanisms are involved in the antioxidative defense. Beside the destructive potential of ROS in plant metabolism and the requirement for defense strategies, mechanisms have evolved to sense the redox state and the level of ROS. The nature of the sensor and the signal transduction pathway are unknown. This chapter focuses on ROS metabolism of photosynthesizing cells, including the reactions of ROS formation and ROS scavenging. Emphasis is also placed on the recently emerged regulatory potential of the redox state to trigger metabolic and genetic changes in plants.

2 Light-Dependent Generation of ROS and Their Reactivity

a) Spin-Pairing as a Physical Activation Mechanism of O_2

The energy diagram depicted in Fig. 1 illustrates that the activation of O_2 is an endergonic process. Dioxygen is stable under environmental conditions, a property which was a prerequisite for atmospheric oxygen accumulation. Spontaneous oxidation of reduced carbon compounds occurs only rarely. The low reactivity is the consequence of the specific electron configuration of dioxygen (Halliwell and Gutteridge 1990). The outer valence shell of molecular oxygen contains two unpaired electrons with parallel spin (Fig. 1). Only few reductands fit to the electron configuration of O_2 (Gille and Sigler 1995). Spin inversion of one electron leads to singulet oxygen (Fig. 1). Spin-pairing facilitates further electron uptake. Stepwise transfer of single electrons to dioxygen yields $O_2\bullet^-$, H_2O_2 and $OH\bullet$ in aqueous solution.

The endergonic activation of O_2 can be accelerated by photodynamic processes. Photosensitizing pigments catalyze the activation of oxygen

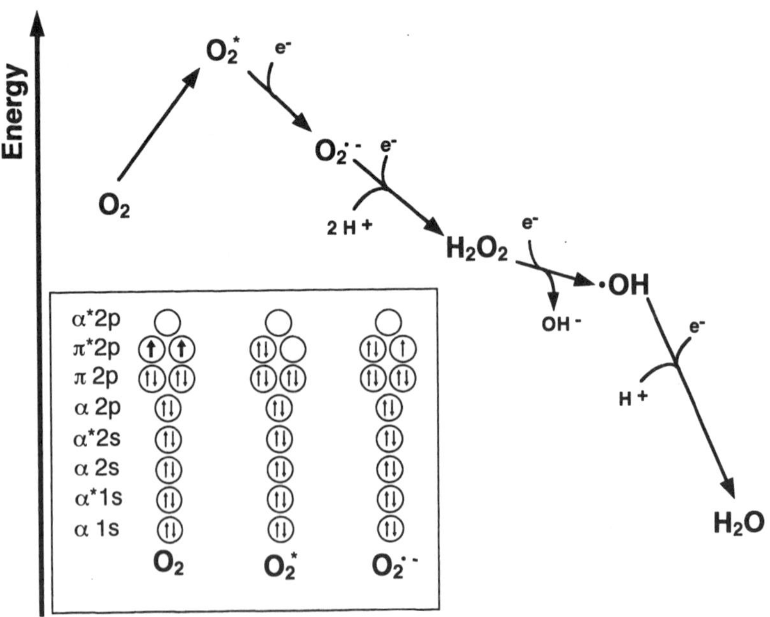

Fig. 1. Formation of reactive oxygen species. Spin-pairing activates dioxygen. The subsequent uptake of electrons is an exergonic process. The figure shows the relative energy contents of the reactive oxygen species. The *inset* illustrates the electron configuration of O_2, singulet oxygen (O_2^*) and superoxide anion (O_2^-). The electron orbitals are assigned on the *left-hand side of the inset.* Hydrogen peroxide (H_2O_2), hydroxyl radicals (•OH) and water (H_2O) are formed by further electron and proton uptake

following excitation by light. The high energy state is transduced to oxygen. The most important activation pathway is the interaction of O_2 with triplet chlorophyll (Anderson et al. 1992). The extent to which chlorophyll-mediated O_2 activation occurs in chloroplasts can be assessed from experiments by Baba et al. (1996). These authors observed a positive relation between photooxidative damage and chlorophyll content. Within 2 h of a high-intensity illumination treatment, photosystem I (PSI) particles lost 50% of their chlorophyll in aerobic atmosphere when their ratio of chlorophyll to P-700 was 180 (PSI-chl$_{180}$ particles). Concomitantly, the electron transfer activity was decreased by 58% as a consequence of oxygen-mediated damage. The PS I-chl$_{180}$ particles were not damaged under anaerobic conditions. Photoinhibition was less or insignificant in PS I particles with decreased chlorophyll content. Damage could not be observed in PS I-chl$_{40}$ particles with a chlorophyll to P-700 ratio of 40. PS I-chl$_{100}$ particles gave intermediate results.

b) ROS Formation by the Photosynthetic Electron Chain

In plants, the photosynthetic electron transport system is the major source of ROS (Asada 1994). In the photosynthetic light reaction, ROS formation takes place at three major sites: (1) ROS may be liberated at the oxygen evolving complex by desactivation of the S_3 *state* during the water splitting cycle (Wydrzynski et al. 1989). This activity can be demonstrated in preparations of PS II particles in vitro, while formation of H_2O_2 at the oxygen-evolving complex cannot be measured under

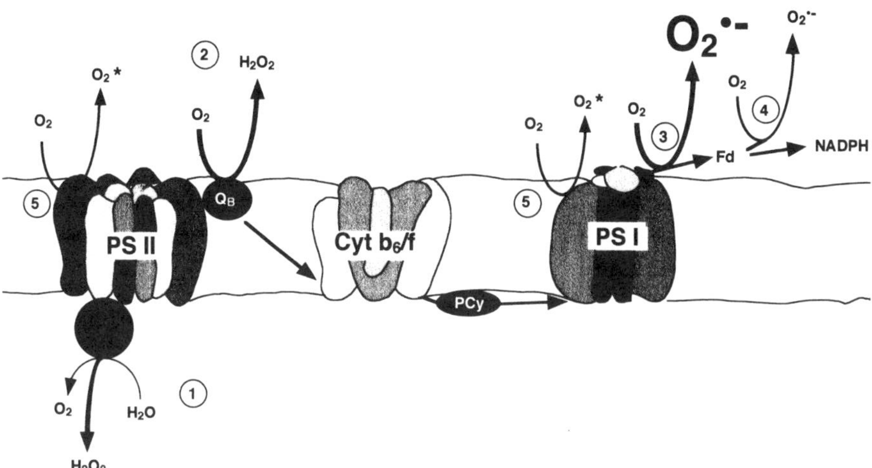

Fig. 2. Sites of ROS formation in the photosynthetic electron transport chain. ROS originates from leakage at the water-splitting complex (*1*), at the acceptor site of photosystem II (*2*) and photosystem I (*3*), from reduced ferredoxin (*4*) and as a consequence of physical activation (*5*). Q_B Plastoquinone bound to PS II; *PCy* plastocyanine

physiological conditions. It only occurs after depriving the complex of chloride and, therefore, for the time being, is an experimental artefact (Wydrzynski et al. 1989). (2) ROS are formed at the reducing site of photosystem II. Electron leakage to O_2 at the acceptor site of PS II (Q_B site) liberates some H_2O_2 at low rate (Elstner and Frommeyer 1979). (3) The major source of ROS is located at the acceptor site of photosystem I. Superoxide is formed either directly (Robinson and Gibbs 1982) or indirectly via reduced ferredoxin (Misra and Fridovich 1971; Fig. 2).

The proportion of photosynthetic electron flow diverted into the reduction of O_2 seems to be very variable. It ranges from 4 to 50% and depends on the plant species, photosynthetic metabolism and environmental conditions (Hodgson and Raison 1991; Osmond and Grace 1995). Hormann et al. (1993) characterized the pH dependency of photosynthetic O_2 reduction in isolated intact chloroplasts. The Mehler reaction displayed a peak activity at pH 5. Such a strong acidification of the thylakoid lumen occurs when the physiological substrates $NADP^+$ and ADP are lacking. The pH optimum suggests that strong luminal acidification activates the oxygen reduction pathway as an alternative electron acceptor leading to energy dissipation. From these data it is concluded that an important physiological function of the Mehler reaction is the regulation of photochemistry by thylakoid energization and prevention of photoinhibition (Neubauer and Yamamoto 1992; see also Polle 1996).

At physiological pH values the superoxide anions spontaneously dismutate into H_2O_2 and O_2 (Asada 1994). Several isoforms of superoxide dismutase (SOD) accelerate this reaction. According to Asada (1994), Mehler reaction combined with SOD action results in about 1 μmol H_2O_2 being produced mg chlorophyll^{-1} h^{-1} at 1% of maximum electron flow. At optimum conditions, the H_2O_2 production increased to 12 μmol H_2O_2 mg chlorophyll^{-1} h^{-1} (Asada 1994, recalculated under the assumption of 40 μl chloroplast volume mg chlorophyll^{-1}). The potential of the Mehler reaction is even higher, and mechanisms are required which decrease the concentration of H_2O_2 in the chloroplast in order to avoid oxidative damage to macromolecules in particular.

c) Destructive Potential of ROS

Beside physical O_2 activation and electron transfer to oxygen in the photosynthetic light reaction, other metabolic reactions also lead to ROS formation. Examples are lipid peroxidation (Wise 1995; Schraudner et al. 1997), oxidase action (Mehdy et al. 1996) and metal-catalyzed Haber-Weiss reactions (Halliwell and Gutteridge 1990; Yruela et al. 1996). It has been well established for years that uncontrolled oxidation inactivates many key enzymes and initiates their degradation, e.g. alcohol dehydro-

Oxidation of fatty acids

Oxidation of amino acids

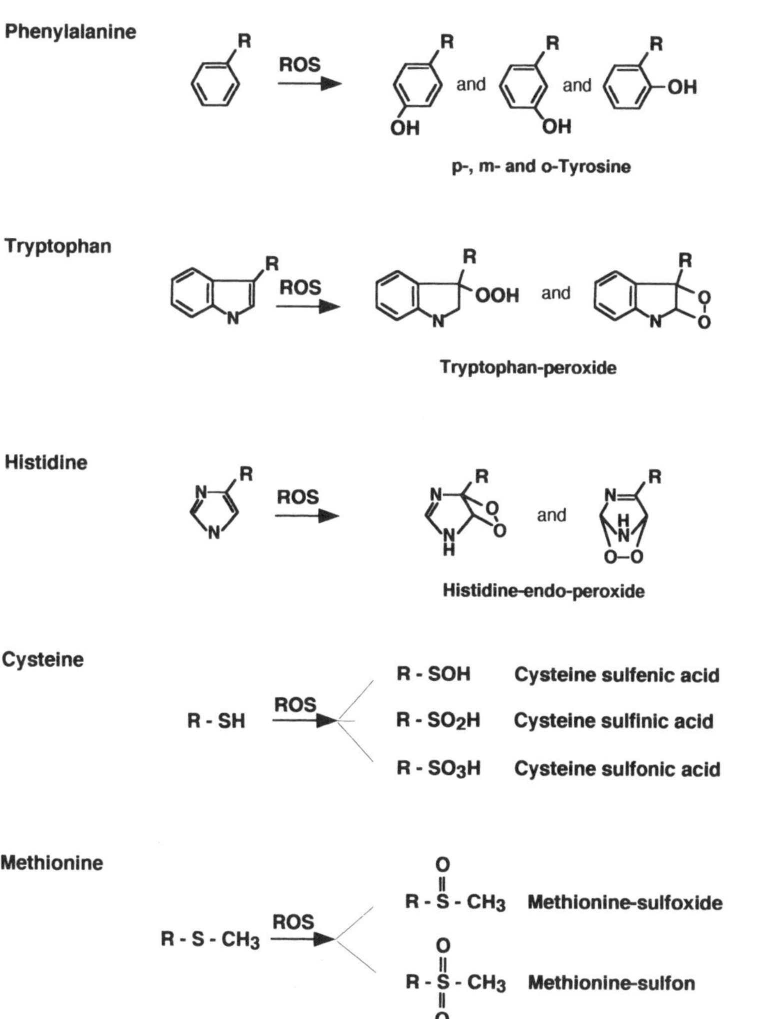

Fig. 3. Oxidative attack on various biomolecules. Unsaturated carbon bonds and sulphur atoms are the main target sites for oxidation

genase, glutamine synthase, glycerate-3-phosphate hydrogenase, super-oxide dismutase and pyruvate kinase (Fucci et al. 1983; Casano and Trippi 1992; Desimone et al. 1996). In addition to enzyme inactivation, ROS facilitate nicking of DNA (Lim et al. 1993) and the non-enzymic breakdown of lipids (Hölzel and Spiteller 1995; Schraudner et al. 1997), as well as the proteolytic degradation of proteins (Fucci et al. 1983; Stieger and Feller 1997; Gómez et al. 1995).

In most cases, ROS attack unsaturated carbon bonds, for instance double bonds in unsaturated fatty acids, in the bases of nucleic acids or in amino acids (Fig. 3). In addition, ROS react with aromatic amino acids (e.g. Shen et al. 1997), also with the sulfur-containing amino acids cysteine and methionine (Stadtman 1992) as well as with histidine, due to its specific conformation (Larson 1988, Elstner 1990). In terms of costs, it is favourable to prevent uncontrolled oxidation of cellular constituents. Enzymic and non-enzymic antioxidants constitute a protective network against ROS.

3 Antioxidants and Their Reactivity

According to the definition by Halliwell (1994), an antioxidant is "a substance that when present at low concentrations compared with those of an oxidizable substrate, significantly delays or prevents oxidation of that substrate". In higher plants, glutathione, ascorbate, tocopherol, flavonoids, a wide range of phenolic and unsaturated aliphatic compounds, various alkaloids and carotenoids (reviewed in Larson 1988) and polyols like mannitol (Ahmad et al. 1979; Bohnert and Jensen 1996; Shen et al. 1997) are important metabolites protecting biomolecules from uncontrolled oxidation. As different as their chemical nature are the reaction properties of these antioxidants.

Compounds containing sulfhydryl groups like reduced glutathione (GSH) readily react with ROS. SH groups are oxidized in a two-electron transfer reaction and disulfide bridges are formed under simultaneous heteromeric or, as in the case of GSH, homomeric dimerization (Alscher 1989). Conversely, one electron is transferred on the oxidant in the case of ascorbate and a monodehydroascorbate radical is formed (Asada 1994; Fig. 4). GSH and ascorbate are highly soluble and accumulate in the stroma of chloroplasts to concentrations as high as 5–25 mM (Kunert and Foyer 1993; Gillham and Dodge 1986; Schöner and Krause 1990).

The group of hydrophilic antioxidants is complemented by a group of highly hydrophobic compounds such as tocopherol, mainly α-tocopherol, and carotenoids which are located in the membranes. The subcellular compartmentation of alkaloids, flavonoids and other phenolic compounds is a function of their physicochemical properties such as hydrophobicity and the presence of specific transport systems in cell

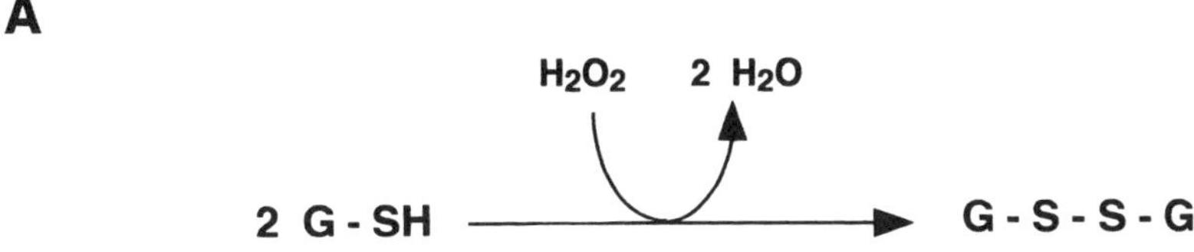

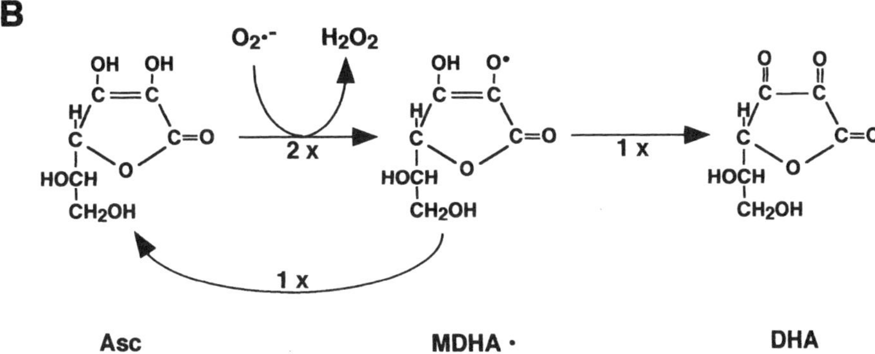

Fig. 4A, B. The reactivity of the low molecular weight antioxidants glutathione and ascorbate. **A** In the presence of O_2 and ROS, respectively, the sulphhydryl groups of glutathione (GSH) are oxidized and a glutatione homodimer is formed. **B** Ascorbate (Asc) is first oxidized to monodehydroascorbate radical (*MDHA•*) which disproportionates into ascorbate and dihydroascorbate (*DHA*)

membranes. Their reactivity is very variable and depends on the type of reactive oxygen species (reviewed in Elstner 1990).

The role of mannitol in the protection of plants against oxidative damage was recently investigated by Shen et al. (1997). These authors targeted mannitol biosynthesis to chloroplasts in transgenic plants. Chloroplast synthesis of mannitol significantly decreased oxidative damage by hydroxyl radicals. Although the reaction mechanism is still unknown, Smirnoff and Cumbes (1989) had demonstrated in vitro that 20 mM mannitol is sufficient to scavenge 60% of the hydroxyl radicals formed in aqueous solutions.

In addition to the low molecular weight antioxidants, enzymic antioxidants provide efficient and ROS-specific protection. Two isoforms of superoxide dismutase (SOD) accelerate the formation of H_2O_2 and O_2 from $O_2•^-$ by about 4 orders of magnitude as compared to the spontaneous dismutation (Badger 1985; Fig. 5A). In a subsequent step, two isoforms of ascorbate peroxidase (APx) reduce the H_2O_2 by ascorbate oxidation. Finally, alkyl hydroperoxide reductases (Ahp) reduce alkyl hydroperoxides to the corresponding alcohols (Fig. 5B).

Each type of antioxidant enzyme is present in a membrane-associated and a soluble form. In the case of APx and SOD, phylogenetically closely related isozymes reside at both locations, in the chloroplast stroma and at the thylakoid membrane. The corresponding genes have been cloned and analyzed. For example, the catalytic domains of the stromal and thylakoid-bound APx-isoforms are highly homologous with 82% primary sequence identity in *Arabidopsis thaliana* (Jespersen et al. 1997). In addition to the common elements, the thylakoid-bound isoform shows a C-terminal extension with a predicted transmembrane segment which is likely to represent the membrane anchor.

In contrast, non-related enzymes have to be assumed to detoxify alkylhydroperoxides (Bleé and Joyard 1996; Baier and Dietz 1997). A membrane-bound alkyl hydroperoxide reductase has been identified in spinach envelope membranes, but a detailed analysis of its molecular character is still missing. The soluble alkyl hydroperoxide reductase BAS1 belongs to the phylogenetically old group of 2-Cys peroxiredoxins. Homologous proteins are present in bacteria, fungi, animals and man (Baier and Dietz 1996a, b). Their unique characteristic is a highly conserved domain composed of four phenylalanine and one cysteine residues, the so-called F motif. A second cysteine is located within another conserved domain. Both cysteine residues undergo a sulfhydryl-disulfide transition during the reduction of alkyl hydroperoxides. Regeneration of active enzyme requires rereduction of the disulfide bridge. On the basis of homology to the yeast enzyme, it may be assumed that thioredoxin directly provides the electrons for disulfide reduction. At this point it seems noteworthy that no enzyme activity towards alkyl hydroperoxides has yet been detected in association with the thylakoid membranes. This is all the more surprising since lipid peroxidation will occur at the thylakoid membrane during photochemistry. Interestingly, Holland et al. (1993) recently identified a third alkyl hydroperoxide reductase denominated phospholipid hydroperoxide glutathione peroxidase, but so far its subcellular localization is still unknown.

Fig. 5A, B. The enzymic antioxidants reduce superoxide and organic and inorganic hydroperoxides in thylakoid-bound and stromal systems (see text for details). **A** Detoxification of superoxide ($O_2^{\cdot-}$) and hydrogen peroxide (H_2O_2). **B** Detoxification of organic hydroperoxides. Abbreviations: *SOD* Superoxide dismutase; *APx* Ascorbate peroxidase; *FTR* Ferredoxin-thioredoxin reductase; *BAS* chloroplast 2-Cys peroxiredoxin; *Asc* Ascorbate; *MDHA*$^{\cdot}$ mono dehydroascorbate radical; *Fd*$_{ox/red}$ oxidized/reduced ferredoxin; *Trx*$_{ox/red}$ oxidized/reduced thioredoxin; *ROOH* organic hydroperoxide; *GSH/GSSG* reduced/oxidized glutathione

Mehler reaction
$2 O_2 + 2 e^- \longrightarrow 2 O_2\bullet^-$
superoxide dismutase
$2 O_2\bullet^- + 2 H^+ \longrightarrow H_2O_2 + O_2$
ascorbate peroxidase
$H_2O_2 + 2 Asc \longrightarrow 2 H_2O + 2 MDHA\bullet$

A

Asc
MDHA•
O_2
$O_2\bullet^-$
SOD
H_2O_2
APx
H_2O
stroma
$O_2\bullet^-$
O_2
SOD
H_2O_2
Asc
MDHA•
APx
H_2O
thylakoid membrane
thylakoid lumen

2-Cys peroxiredoxin BAS1
$ROOH + 2e^- + 2 H^+ \longrightarrow 2 ROH + H_2O$
lipidhydroperoxide reductase
$ROOH + ?_{red} \longrightarrow ROH + ?_{ox}$
glutathione peroxidase
$ROOH + 2 GSH \longrightarrow ROH + GSSG + H_2O$

B

e^-
Fd_{ox}
Fd_{red}
FTR
$O_2\bullet^-$
Trx_{red}
Trx_{ox}
H_2O
OOH
BAS
OH
$O_2\bullet^-$
stroma
???
OOH
thylakoid membrane
thylakoid lumen

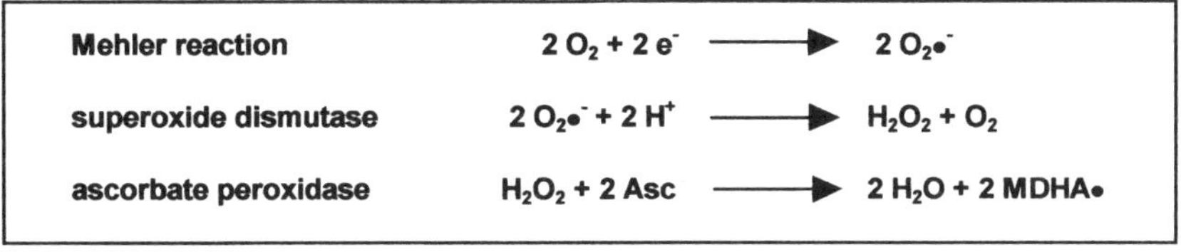

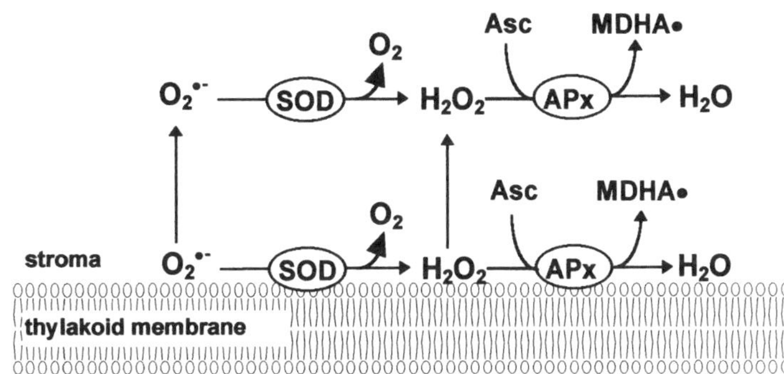

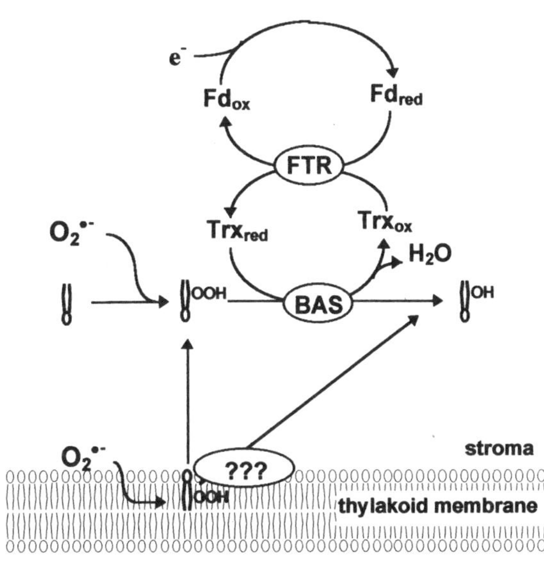

4 The Control of the Redox Homeostasis of the Chloroplast

a) The Antioxidant Network

α) The Interaction of Enzymic and Non-Enzymic Antioxidants

Enzymic and non-enzymic antioxidants are connected in an antioxidant network (Fig. 6). One important link is the coupling of the glutathione pool and the ascorbate pool by glutathione-dependent dehydroascorbate reductase (DHAR; Foyer and Halliwell 1976). As mentioned above, even under favourable growth conditions, the pool size of both antioxidants is large (Schöner and Krause 1990; Kunert and Foyer 1993). Stress conditions such as ozone fumigation, drought stress and herbicide treatment displace the redox system from a highly reduced to a more oxidized state. Frequently, the cells respond with a stimulation of de novo synthesis of small molecular weight antioxidants such as glutathione and

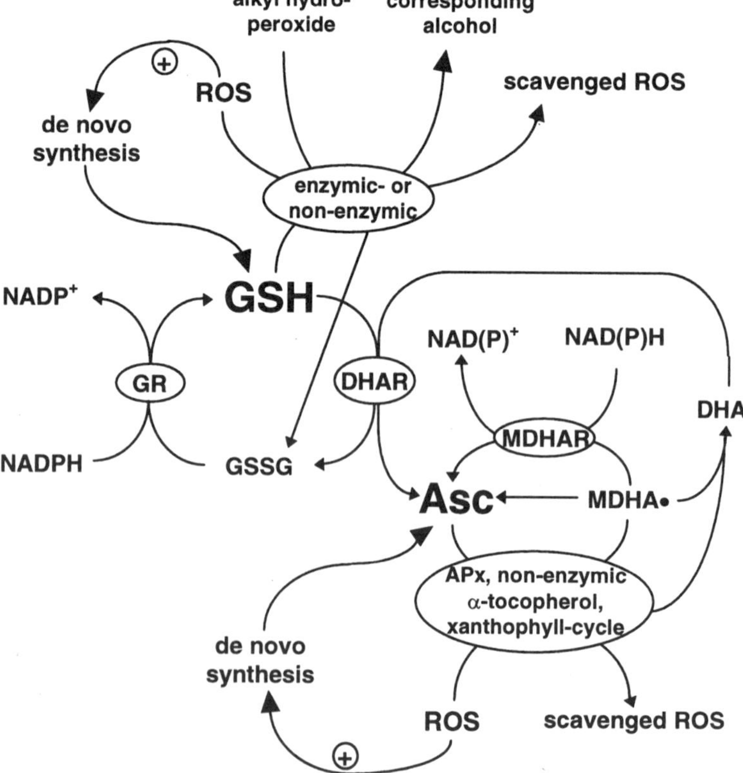

Fig. 6. The antioxidant network. Glutathione (*GSH*) and ascorbate (*Asc*) are central metabolites of the antioxidant network which includes the antioxidant enzymes glutathione reductase (*GR*), dehydroascorbate peroxidase (*DHAR*), monodehydroascorbate reductase (*MDHAR*), ascorbate peroxidase (*APx*), the xanthophyll cycle and α-tocopherol

ascorbate (Alscher 1989; Foyer 1993). Glutathione is synthesized from glutamate, cysteine and glycine by the consecutive action of γ-glutamylcysteine synthase and glutathione synthase (reviewed in Bergmann and Rennenberg 1993). Ascorbate derives from carbohydrates formed in the Calvin cycle (Foyer 1993).

In addition to this role as low molecular weight antioxidant and its participation in regeneration of ascorbate from monodehydroascorbate, GSH also functions as cosubstrate in the reactions catalyzed by glutathione peroxidase (GPx) (Flohé and Günzel 1984) and phospholipid hydroperoxide glutathione peroxidase (PHGPx) Holland et al. 1993). Ascorbate also links parts of antioxidant network. Besides its function as a reductant in H_2O_2 detoxification catalyzed by ascorbate peroxidase (APx; Mehlhorn et al. 1996), it is involved in the regeneration of α-tocopherol (Foyer 1993) and in the deepoxidation of violaxanthin (reviewed in Yamamoto 1985). The ultimate electron donors are NADPH and NADH, which donate electrons to regenerate GSH and ascorbate by glutathione reductase (GR; Foyer and Halliwell 1976) and monodehydroascorbate reductase (MDHAR; Hossain et al. 1984), respectively (Fig. 6). The reductive regeneration is very efficient. A decreasing level of reduction is observed only during acute severe oxidative stress. Activation of GR and MDHAR, on the level of both protein abundance and activity, allows the high reduction state of the redox system to be readjusted on a time scale of minutes to hours, depending on the intensity of the stress (Alscher 1989; Foyer 1993). In this context it is important to note that the redox state of glutathione responds faster and more strongly than the ascorbate system during and after the stress period (Foyer et al. 1994b). Possible causes for the distinct behaviour are the differences in pool size, in metabolic function and in subcellular compartmentation. To some extent, the ascorbate pool may be considered as a slowly reacting backup system to regenerate reduced glutathione. There is one additional mechanism involved in keeping the cytosolic glutatione pool highly reduced. Oxidized glutatione is exported from the cytosol into the vacuole by an ATP-dependent pump in the tonoplast (Dietz et al. 1992; Tommasini et al. 1993). Some data suggest that oxidized glutathione and glutathione conjugates are degraded inside the vacuole in order to recycle the amino acids (Wolf et al. 1996).

β) Developmental Control of the Antioxidative Defence System

The antioxidant network is built up stepwise during early development of the photosynthetic tissue. Most antioxidant enzymes are induced in close relation to the onset of photosynthesis. Conversely, catalase activity appears with a delay of 48 h (Mishra et al. 1995). In all aerobic organisms the antioxidant activity is sufficient to deal with oxidative stress

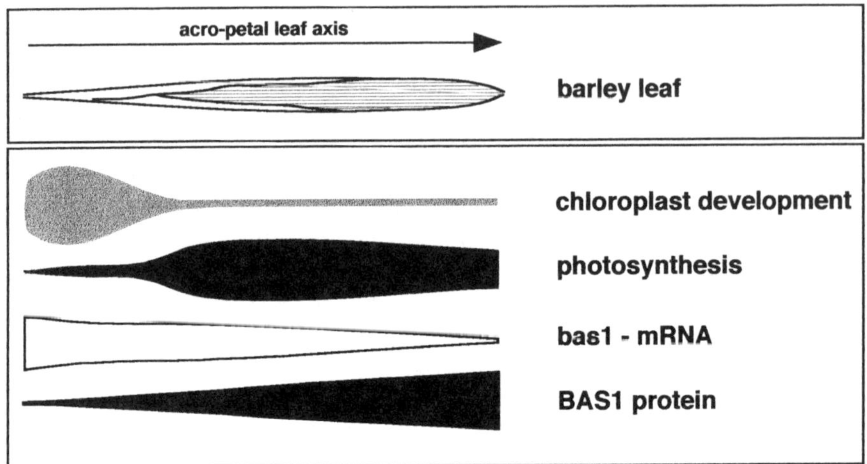

Fig. 7. The 2-Cys peroxiredoxin BAS1 is induced in the cell division and elongation zone of the leaf. Its gene product accumulates in photosynthetically active chloroplasts

characteristics for most growth conditions (Foyer et al. 1994b). Only under severe stress is the antioxidative capacity overburdened and damage develops. The expression of the alkylhydroperoxide reductase BAS1 of the 2-Cys-peroxiredoxin type is a typical example of development control (Fig. 7). BAS1 is restricted to green tissues and is a nuclear-encoded chloroplast protein (Baier and Dietz 1997). The protein amount increases from the base to the tip of the barley leaf, indicating accumulation with age (Baier and Dietz 1996a). Bas1-transcript amounts are maximal in the youngest developing parts of the leaf and decrease towards the tip. This contrasts with the transcript pattern of typical photosynthesis-related genes such as the small and large subunits of ribulose-1,5-bisphosphate carboxylase, chlorophyll a/b-binding protein or D1 protein (Baier et al. 1996). In addition, bas 1 expression is under control of the redox state of the cells, being repressed after external application of the reducing agents β-mercaptoethanol, dithiothreitol and glutathione to the leaves (Baier and Dietz 1997). From this it seems reasonable to suspect that control of expression of antioxidants is partly independent of chloroplast development and that growing parts of plants are subjected to endogenous oxidative stress stimulating expression of antioxidant genes.

γ) Towards a Quantitative Understanding of Antioxidant Activities?

Techniques of molecular biology enable us to specifically modify antioxidative activities of plant metabolism. Yet the data obtained from work with transgenic plants overexpressing or suppressing antioxidant genes are still contradictory at first sight (reviewed by Allen 1995). Expression of Mn-SOD in the chloroplasts reduced sensitivity to paraquat in tobacco (Van Camp et al. 1994), to acifluorofen and freezing in *Medicago sativa* (McKersie et al. 1993) and chilling in cotton (Trolinder and Allen 1994). Conversely, overexpression of chloroplastic Cu/Zn-SOD at a high level did not result in increased stress tolerance in petunia (Tepperman and Dunsmuir 1990), whereas in other experiments a slight induction of Cu/Zn-SOD increased the stress tolerance (Van Camp et al. 1994; McKersie et al. 1993; Trolinder and Allen 1994). Superoxide dismutation has to be followed by hydrogen peroxide reduction in order to avoid uncontrolled oxidation of cell constituents by H_2O_2. Therefore, high expression of SOD may not be advantageous for plants as long as there is no concomitant increase in APx activity in the chloroplasts. A regulation fitting to this line of relationship was described by Sen Gupta et al. (1993), who observed a threefold increase in APx activity and transcript amount in transgenic tobacco plants overexpressing chloroplastic Cu/Zn-SOD. Miyake et al. (1991) showed that some cyanobacteria are devoid of ascorbate peroxidase, suggesting that there is an alternative pathway for the detoxification of H_2O_2 in addition to APx and catalase at least in some species.

The role of plastidic glutathione reductase was investigated by expressing a bacterial gene as chimaeric construct with a plastid targeting address in tobacco (Aono et al. 1993) and poplar (Foyer et al. 1994a). The transgenic plants revealed a slight increase in stress resistance indicating that the rereduction of oxidized glutathione by glutathione reductase may become rate-controlling under stress conditions. In contrast, no major effect on the redox state of the glutathione system was measured upon overexpressing glutathione synthase (Foyer et al. 1996).

Recently, Örvar and Ellis (1997) provided convincing evidence that the cytosolic APx is of central importance in the antioxidant network. Transgenic tobacco mutants with half the APx content of wild-type plants developed severe symptoms of damage at moderate ozone concentrations which had only little effect on wild-type plants. Not surprisingly, the antioxidant network is more or less mirrored in different subcellular compartments such as the chloroplasts, mitochondria, cytosol and even in the apoplast. The subcellular distribution is related to distinct functions. Ozone directly interferes with the apoplastic ascorbate pool. Oxidized ascorbate must be regenerated. Possible mechanisms of regeneration are discussed in a recent review (Dietz 1997). Interestingly, although no or only little glutathione is present in the apoplast, ozone fumigation stimulates oxidation of glutathione. Obviously, the apoplastic ascorbate pool is linked to the cytosolic redox state across the limiting biomembrane (Luwe et al. 1993). Conversely, the transgenic experiment by Örvar and Ellis (1997) described above suggests that the chloroplastic APx cannot take the place of the cytosolic APx during ozone treatment. This conclusion addresses the open question concerning the quality and quantity of subcellular exchange in the antioxidant network. Further, from the data summarized in this section, it has to be concluded that our understanding of antioxidant activities is still very incomplete and far from becoming quantitative.

b) Dynamic Regulation of Photosynthesis

As already mentioned above, the photosynthetic light reaction is linked to ROS formation. ROS formation is high when the incident photon fluence rate exceeds energy consumption in metabolism, for instance in plants exposed to high light intensities at low temperature or at limiting water and CO_2 supply. Effective mechanisms to decrease the rate of ROS formation are the downregulation of photosynthesis and the dissipation of excess energy. The main reactions involved are (1) photorespiration (Heber et al. 1978), (2) Mehler reaction linked to ascorbate oxidation (Osmond and Grace 1995), (3) non-photochemical energy dissipation in dependence of the xanthophyll cycle (Demmig-Adams and Adams 1996), and (4) photoinhibition (Wu et al. 1991). These reactions will be discussed briefly in the following sections.

α) Photorespiration

Oxygenation of ribulose-1,5-bisphosphate is the committed step initiating the metabolic pathway of photorespiration. It is estimated that in unstressed C_3-plants 20–27% of the CO_2 fixed in photosynthesis is released in photorespiration (Canvine 1990). The percentage of CO_2 release considerably increases when the intercellular CO_2 concentration drops as a consequence of stomatal closure. In photorespiration, ATP is consumed during refixation of liberated ammonium and during glycerate kinase reaction. This adds up to the ATP and NADPH consumed in the Calvin cycle during regeneration of RuBP. The released CO_2 may be refixed in normal photosynthesis. Thus, oxygenation of RuBP is a valve diverting metabolites in the energy-dissipating pathway of photorespiration. It allows electron acceptors to be regenerated for the photosynthetic electron transport chain and the transthylakoid proton gradient to be relaxed.

In addition to its function in energy consumption, photorespiration supplies glycine and the cysteine precursor serine and is thereby linked to glutathione synthesis. Recently, it was documented that glycine and cysteine availability limits glutathione synthesis by γ-glutamyl cysteine synthase and glutathione synthase, especially under conditions of photooxidative stress (Strohm et al. 1995; Noctor et al. 1997). On the other hand, Wingler et al. (1997) analyzed barley mutants with decreased glycine decarboxylase activity under photorespiratory conditions. Accumulation of glycine led to an inhibition of photosynthesis. The relevance of this finding for regulating photosynthesis and photorespiration in vivo still needs to be established.

β) Mehler-Ascorbate Peroxidase Reaction

A second mechanism involved in metabolic dissipation of excess energy is the Mehler-ascorbate peroxidase pathway. A total of four light quanta absorbed at PS II and PS I lead to the formation of two molecules of $O_2^{\bullet-}$ and, following dismutation, to one molecule H_2O_2 and one molecule O_2. Four additional quanta are required to produce two molecules of reduced ferredoxin (Fd) (Furbank and Badger 1983) which are indirectly used to reduce H_2O_2 via APx dehydroascorbate reductase and glutathione reductase. In the overall balance, the energy of eight quanta has been dissipated and the thylakoid lumen has been acidified. No redox equivalents have been formed. The simplified reaction sequence can be written as follows:

$$2H_2O + O_2 + 2Fd_{ox} \xrightarrow{\text{8 quanta}} 2O_2^{\bullet-} + 4H^+ + 2Fd_{red} \xrightarrow{\text{SOD}} H_2O_2 + O_2 + 2H^+ + 2Fd_{red}.$$

APx + dehydroascorbate reductase + glutathione reductase

As outlined above, the detoxification of Mehler reaction- and SOD-originated ROS is realized by a twin system of enzymes: one set of APx and SOD is attached to the thylakoid membrane while the second is localized in the stroma. The thylakoid-bound system represents the prime mechanism for ROS inactivation since ROS formation takes place at the thylakoid membrane. The stromal system is suggested to scavenge ROS escaped from the thylakoid-bound protection system (Asada 1994). At maximum rates, the Mehler reaction accepts up to 50% of the electron transport activity (Osmond and Grace 1995); but in contrast to photorespiration, the Mehler reaction does not relax but creates a transthylakoid pH gradient. The pH gradient contributes to membrane energization (Schreiber and Neubauer 1990) and, therefore, to downregulation of the PS II and activation of the xanthophyll cycle (Neubauer and Yamamoto 1992).

γ) The Xanthophyll Cycle

Non-photochemical dissipation of absorbed light energy as heat should be as low as possible in order to perform photosynthesis with high quantum yield. However, under conditions of excess light absorption, it may become favourable to increase dissipation of energy as heat. The xanthophyll cycle provides an effective gear change to adjust the energy flow to the requirements of photochemistry. In the light, violaxanthin is deepoxidized to zeaxanthin via antheraxanthin when the energization state of the thylakoid system is very high. In the dark, epoxidation is stimulated and the ratio of violaxanthin to zeaxanthin increases. The luminal pH of the thylakoids, the supply of carotene and the availability of ascorbate (Hager 1969; Gilmore and Yamamoto 1993; Mohanty and Yamamoto 1995) are major determinants of the state of the xanthophyll cycle. Zeaxanthin and probably antheraxanthin are thought to act as "lightning rod", receiving the energy from excited chlorophyll and dissipating it harmlessly as heat (Demmig-Adams and Adams 1996). ROS generation is avoided, which otherwise could take place during interaction of oxygen with triplet chlorophyll. Violaxanthin, like other carotenoids with less than ten conjugated double bonds, does not quench excited chlorophyll (Demmig-Adams and Adams 1996). The fast redox cycling of the pigments provides a rapid mechanism to adapt plants to momentary changes in the environment.

δ) Photoinhibiton

If the other protection mechanisms do not work properly, photoinhibition is the final consequence of photooxidative stress and the ultimative device to reduce ROS formation. Phenotypically, photoinhibition is defined as strongly reduced quantum yield of photosynthesis. It protects the cells by decreasing the electron flux and the O_2 formation. The extent of damage and the kinetics of regeneration allow dynamic and chronic photoinhibition to be distinguished. Dynamic photoinhibition is short-term and readily reversible. It provides photoprotection by dissipating energy as heat and is tightly linked to the xanthophyll cycle. In contrast, chronic photoinhibition is characterized by long-term loss of photosystem II reaction centre function (Osmond and Grace 1995). It is caused by a rapid degradation of D1 protein (Aro et al. 1993) and chlorophyll a/b binding protein of photosystem II (Lindahl et al. 1995). Regeneration from chronic photoinhibition depends on protein synthesis and structural rearrangement and is therefore slow.

At first sight, all the mechanisms described in this section seem to waste energy and, therefore, resemble flawed pathways of photosynthetic metabolism. However, this view is wrong. A convincing body of evidence has accumulated during the past years that energy-dissipating pathways are essential to protect the photosynthetic apparatus from oxidative destruction and that their activity is under metabolic control.

5 The Regulatory Potential of ROS and the Stromal Redox State

In addition to their effects on metabolism, ROS have gained a prominent function as regulators of cell development and metabolism. A well known example for ROS-mediated signalling is the pathogen-induced oxidative burst which triggers defense gene activation in host cells (reviewed by Mehdy et al. 1996; Low and Merida 1996). Binding of an elicitor to a plasma membrane-bound receptor induces superoxide formation by a membrane-bound NAD(P)H oxygenase via a signal transduction pathway including a G protein, phospholipase C, inositol-1,4,5-trisphosphate, Ca^{2+} and a protein kinase (Fig. 8). The oxidative burst, in turn, is responsible for the induction of defense mechanisms, e.g. upregulation of phenylalanine ammonium lyase, chalcone synthase and isomerase and glutathione-S-transferase, which all belong to the large and variable group of "pathogenesis-related proteins" (PR proteins). The oxidative burst is also involved in the pathogen-dependent induction of apoptosis or hypersensitive response of host cells, a very efficient mechanism to avoid parasitizing by the pathogen (Mehdy et al. 1996).

a) Redox Regulation of Enzyme Activity

Photosynthesis depends on the buildup of the reducing power in the NADP system. Although electron transfer processes lead to the formation of reduced ferredoxin (Fd_{red}) and NADPH as well as of ROS, both types of products differ completely in their reactivity, Fd_{red} and NADPH being strong reductants, while ROS are strong oxidants.

Changes in the cellular reducing power are linked to the redox state of cellular constituents via the thioredoxin system by thiol-/disulfide-exchange (reviewed e.g. in Kunert and Foyer 1993). Thioredoxins constitute a group of small redox proteins of about 12 kDa found in all organisms (reviewed in Holmgren 1989). In bacteria, a deficiency in thiored-

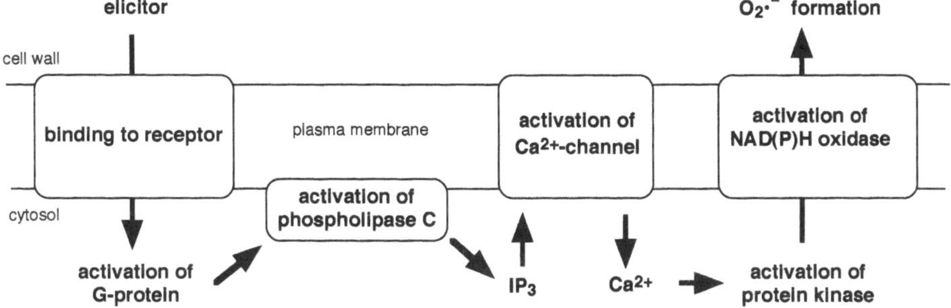

Fig. 8. The oxidative burst in pathogen defense. Elicitor binding stimulates a G-protein which then activates phospholipase C and via inositol-1,4,5-trisphosphate (*IP₃*), subsequently a Ca^{2+}-channel, a protein kinase and a plasma membrane-bound NAD(P)H oxidase which forms superoxide

oxins can be complemented by glutaredoxin (Holmgren 1989). Glutaredoxin was discovered in spinach chloroplasts only recently (Morell et al. 1995). Its function still needs to be established. Regeneration of thioredoxin and glutaredoxin takes place via the ferredoxin-thioredoxin reductase (FTR), ferredoxin and the photosynthetic electron flow. Thioredoxin-dependent activation of enzyme activities in the Calvin cycle and reductive inactivation of the oxidative pentose phosphate cycle guarantee optimum tuning of carbon fluxes in dependence on the metabolic requirements (Scheibe 1996). For example, in the case of phosphoribulokinase (PRK), the cysteine residue at position 55 (Cys-55) interacts with Cys-49 in thioredoxin f (Brandes et al. 1996). In the oxidized form of PRK, Cys-55 is linked to Cys-16 by a disulfide bond.

Thioredoxin-dependent regulation represents a protective feedforward mechanism minimizing ROS production. A high electron pressure in the photosynthetic electron transport chain favours ROS formation, but it also yields a high level of reduction of the ferredoxin and thioredoxin system and, hence, a high state of activation of the enzymes of the Calvin cycle. As a consequence of the enzyme activation, more energy is diverted into the carbon reduction cycle and less in ROS formation by the Mehler reaction.

In addition to thioredoxin and glutaredoxin, other proteineous and non-proteineous sufhydryl compounds act as regulators. Glutathione not only is substrate but may also act as signal molecule (Alscher 1989; Foyer et al. 1994b). In vitro GSH activates and GSSG inactivates hexokinase, glucose-6-phosphate dehydrogenase, several protein kinases and V-type-ATPase (Gilbert 1984; Fig. 9). The physiological significance of these findings remains doubtful, since high concentrations of GSSG were required to promote the inhibitory effects. Such high concentrations of GSSG are unlikely to occur in the cytosol in vivo; but it may well be possible that high concentrations of GSSG and GSH mimic sulfhydryl group modification, which in vivo is realized by more specific and effective redox regulators.

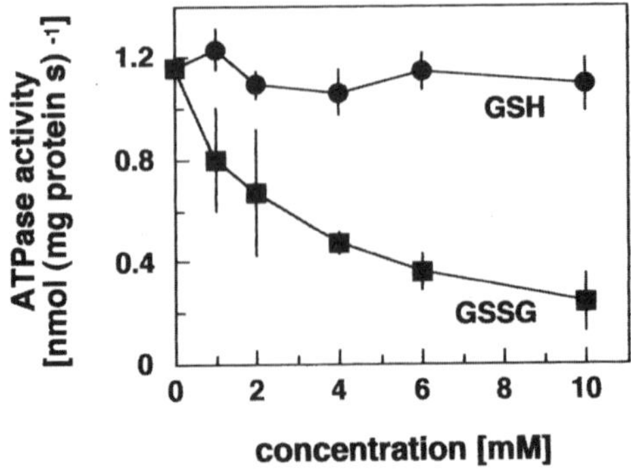

Fig. 9. Effect of oxidized (*GSSG*) and reduced glutathione (*GSH*) on ATP hydrolysis of the V-type-H$^+$-ATPase. (K.-J. Dietz and T. Mimura, unpubl.)

Substrates and cofactors also modify the redox behaviour of target enzymes. For example, increasing amounts of fructose-1,6-bisphosphate reduce the thiol/disulfide redox potential of fructose-1,6-bisphosphatase (Faske et al. 1995). NAD(P)-glyceraldehyde-3-phosphate dehydrogenase is only fully activated in the dark when thiols, 3-phosphoglycerate and ATP are added simultaneously (Baalmann et al. 1995). Obviously, redox regulation is not an independent switch mechanism but is tied into the fine control of metabolism.

There are some indications that redox regulation of target proteins may directly by achieved by ferredoxin-thioredoxin reductase like enzymes. This conclusion is based on the identification of thioredoxin-like domains for instance in the 3'-phosphoadenosine-5'-phosphosulfate reductase of *Arabidopsis thaliana* (Gutierrez-Marcos et al. 1997). Recently, a new perspective was opened on the role of redox modulation in pollen-stigma compatibility in the higher plant *Phalaris coerulescens*. An S-protein exhibited thioredoxin-like activity (Li et al. 1996). Deletions in the C terminus resulted in a decreased thioredoxin-like activity and the development of a phenotype of pollen-stigma incompatibility. From this finding, a working hypothesis was derived that the S protein is involved in redox modulation of a receptor regulating the self-incompatibility reaction.

b) Regulation of Gene Expression by Cellular Redox Homeostasis

A fascinating field of present research is the role of the cellular redox homeostasis in regulation of gene expression. Stressors such as ozone, drought, SO_2, high light and UV-B irradiation increase intracellular ROS formation (Irigoyen et al. 1992; Willekens et al. 1994; Rao et al. 1996; Schraudner et al. 1997) and induce expression of the antioxidant genes encoding SOD, peroxidases, catalase, glutathione reductase and pathogenesis-related proteins (Kangasjärvi et al. 1994; Willekens et al. 1994; Kubo et al. 1995; Schwarz et al. 1996). ROS or ROS-dependent metabolites may be the common link between the various stressors, leading to the similar stress responses.

α) Redox-Sensitive Regulation of Gene Expression
and Identification of Promoter Elements

Lee et al. (1994) and Schubert et al. (1997) introduced deletions in the promoter sites of phenylalanine ammonium lyase and resveratol synthase, respectively, and were able to identify *cis*-acting promoter regions that mediate the response of the genes to oxidative stress. However, the resolution of the analysis was not sufficient to identify the characteristic sequence motifs, let alone the transcription factors involved in wound-, light-, pathogen- and ozone-induced activation of gene expression.

First insight into such regulatory elements has been gained from work with salicylic acid and methyl jasmonate. Both are endogenous stress signals in plants, in the case of methyl jasmonate preferentially after wounding (Farmer and Ryan 1992) and in the case of salicylic acid during pathogen defense (Malamy et al. 1990; Raskin 1992). Salicylic acid is known to stimulate the production of H_2O_2, but H_2O_2 and salicylic acid also act interdependently in establishing systemic acquired resistance. On the one hand, increased levels of H_2O_2 stimulate cross-linking of the cell wall and enhance the enzymes involved in lignin and salicylic acid synthesis. On the other hand, high levels of salicylic acid inhibit catalase and APx activity at least in *Arabidopsis thaliana* (Rao et al. 1997). The inhibition of H_2O_2-degrading enzymes favours accumulation of ROS. Conversely, in *Glycine max*, Tenhaken and Rübel (1997) found no inhibition of catalase and APx following treatment with salicylic acid. They suggest that a loss of membrane control initiates hypersensitive death and that neither lipid peroxidation nor the oxidative burst or the altered accumulation of ROS is an essential element of signal transduction in the hypersensitive response downstream of salicylic acid. Reporter gene analysis indicates that both salicylic acid and methyl jasmonate require the same hexanucleotide sequence (TGACGT) in the nopaline synthase promotor for gene activation, although the reaction kinetics were different (Kim et al. 1993).

β) The Kinetics of the Genetic Responses

The previous sections have shown that plant cells respond to oxidative stress with a multidimensional programme. The various responses differ

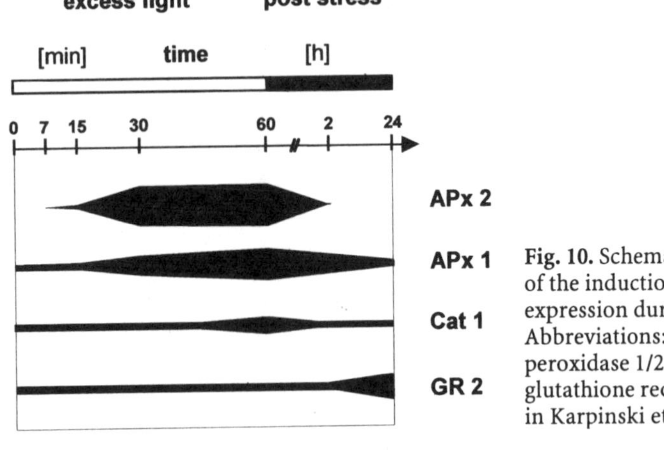

Fig. 10. Schematic respresentation of the induction kinetics of gene expression during oxidative stress. Abbreviations: *APx 1/2* ascorbate peroxidase 1/2; *Cat1* catalase 1; *GR2* glutathione reductase 2. (see Fig. 5 in Karpinski et al. 1997)

in their kinetics (Karpinski et al. 1997) indicating the involvement of distinct signalling pathways (Fig. 10). (1) The fastest genetic response was observed within 7 min and was established after 30 min. Within this period, the transcript of the APx2 gene increased from a level below detection limit to high abundance. (2) A slower and less dramatic increase in gene expression started 15 min after stress exposure. The APx1-mRNA slowly increased from a constitutive background level to about 18-fold in *Arabidopsis thaliana* at excess light. (3) The Cat1 gene which encodes a specific catalase represents the third group of much slower regulatory response. Its transcript amount increased 2.5-fold after 60 min of excess light. (4) The fourth type of induction kinetics is restricted to the poststress period and to chronic stress. For example, the cytosolic glutathione reductase was induced 24 h after stress application. Nothing is known so far about the signalling pathways and the dependence of the late genetic events on the initial responses.

γ) Redox Signalling and Redox-Sensitive Transcription Factors

As pointed out by Lander (1997), redox signalling does not look like a linear "lock and key" system but like a broad chemical recognition system. It has to be assumed that several transcription factors interact with various redox-sensitive binding factors. Escoubas et al. (1995) suggested that the redox state of the plastoquinone pool modulates the expression of the chlorophyll a/b-binding proteins (cab) via protein phosphorylation of an unknown factor. Wingate et al. (1988), Wingsle and Karpinski (1996) and others observed an upregulation of gene expression of phenylalanine ammonium lyase, chalcone synthase, glutathione reductase and Cu/Zn-SOD following oxidative stress. Expression of the cytosolic Cu/Zn-SOD was also stimulated after application of reduced thiols (Hérouart et al. 1993). In contrast, expression of the 2-Cys peroxiredoxin BAS1 was repressed after feeding the reducing compounds GSH, ascorbate or dithiothreitol to excised barley leaves (Baier and Dietz 1997). Again, these results indicate that different mechanisms are involved in redox signalling.

In bacteria and mammalia, the redox state of thiols affects gene expression directly via redox-sensitive transcription factors (Wu et al. 1996). The redox sensitivity is mediated by specific cysteine residues of the transcription factor (Pognonec et al. 1992; Sun and Oberley 1996). Various redox-sensitive regulators have been identified, e.g. OxyR, SoxR, nF-κB and AP-1 (Storz et al. 1990; Meyer et al. 1993; Kullik and Storz 1994; Hidalgo et al. 1997). The most thoroughly studied is OxyR from *Salmonella typhimurium* and *Escherichia coli*. It regulates at least 9 of the 30 genes induced by H_2O_2 (Morgan et al. 1986), for instance catalase and an alkylhydroperoxide reductase of the 2-Cys peroxiredoxin type

(Storz et al. 1989). Following its oxidation, OxyR activates genes transcription. The broad distribution of thiol-mediated transcription regulation in several animals and bacteria (Storz et al. 1990; Berg 1992; Kullik and Storz 1994; Duh et al. 1995; Wu et al. 1996; Sun and Oberley 1996) indicates that similar reaction mechanisms also exist in plants.

ROS may directly interfere with transcription factors which expose redox-sensitive thiol groups. However, long-distance effects of oxidative stress across membranes and between organelles depend on receptors, signals and redox intermediates which transmit ROS activity and redox states. According to Karpinski et al. (1997), possible agents involved in signalling during oxidative stress in plants are salicylic acid, H_2O_2, GSH and GSSG, and photoreceptors, in addition to typical second messengers like calcium, IP_3, G-proteins, protein kinases and the plant hormone methyl jasmonate. The inorganic ions K^+ and Cl^- also seem to be involved. Transmembrane ion fluxes transiently increased in response to stress (Levine et al. 1994; Hahlbrock et al. 1995). This may be due to oxidative activation of K^+ channels which was described in membranes of animal cells (Bolotina et al. 1994). The hyperpolarization of the plasma membrane may then trigger downstream events in the signal transduction pathway.

δ) Intercompartment Redox Signalling

An important aspect in the cellular response to photooxidative stress is the intercompartment signalling. Under conditions of excess light, chloroplasts are the site of ROS formation. However, all antioxidative enzymes are encoded by nuclear genes (Perl-Treves and Galun 1991; Willekens et al. 1994; Baier and Dietz 1997; Jespersen et al. 1997). Signals must be transmitted from the chloroplasts to the nucleus in dependence of the redox state. Different signal transduction pathways may be or have been hypothesized (Fig. 11). (1) Coupled to the cytosolic and stromal malate dehydrogenases the counterexchange of malate and oxalate links the $NADP^+$/NADPH-system of the chloroplast to the NAD^+/NADH-system of the cytosol (malate valve) (Scheibe and Beck 1994). The system allows to transmit redox information between both compartments. (2) Receptors in the inner envelope membrane may perceive the signal and transduce it to the cytosolic site. As assumed for the redox regulation of *cab*-expression, various receptor-dependent transduction cascades could be involved, e.g. G-proteins or kinase-mediated signalling (Wingate et al. 1988; Escoubas et al. 1995). (3) Ion channels may be gated in dependence on the stromal redox state similarly to K^+ channels in the membranes of vascular smooth muscle cells (Bolotina et al. 1994). (4) Efflux of O_2 and H_2O_2 from the chloroplasts may increase ROS formation

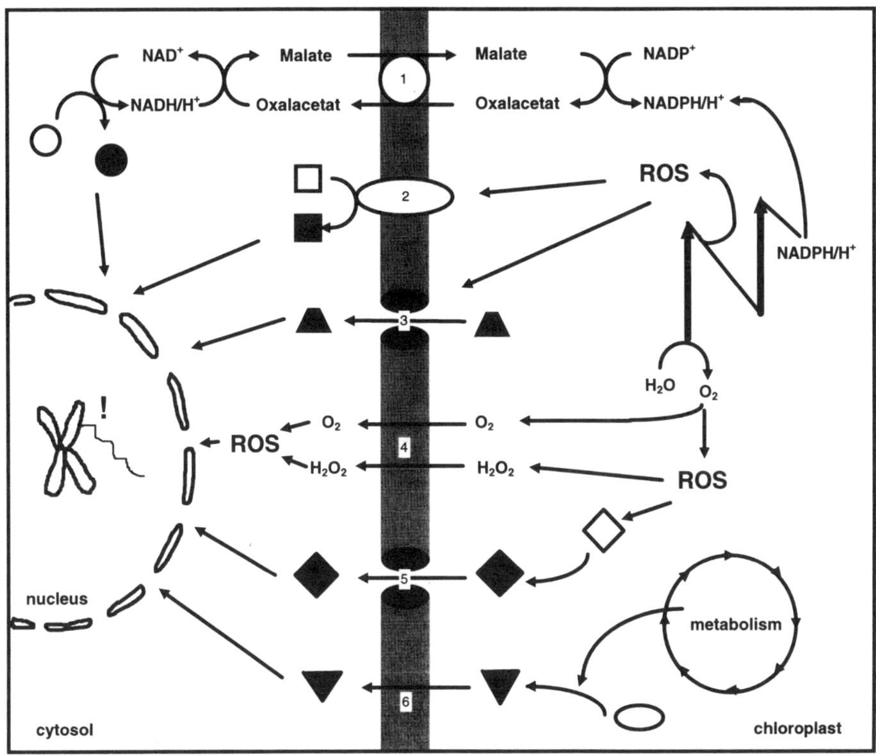

Fig. 11. Hypothetical coupling between the redox systems of the chloroplast and the nucleus. The malate valve (*1*), ROS-stimulated signal transducer (*2*) or translocator (*3*), efflux of O_2 and H_2O_2 (*4*), translocators for second messengers (*5*), and permeation of messengers (*6*)

and displace the redox state in the cytosol. The formed ROS or mediating regulators then influence nuclear gene expression. This regulatory mechanism resembles prokaryotic regulation. The major difference is that in eukaryotes it would have to be composed of two steps of ROS formation, one inside and one outside the chloroplast. (5) Specific carriers reside in the envelope membrane and translocate redox-sensitive metabolites. For example, the ascorbate carrier (Anderson et al. 1983) couples the stromal and cytosolic ascorbate pool. Distinct specificities of the carrier towards either the oxidised or reduced substrate could enhance the signal. Indications for such a mechanism exist for the ascorbate carrier in the plasma membrane, which preferentially accepts the fully oxidized dehydroascorbate as substrate (Horemans et al. 1997). (6) Reaction products of ROS may function as long-distance signals indicating the redox state of the chloroplast stroma. One possible example are aldehydes, which are degradation products of alkyl hydroperoxides (Hölzel and Spiteller 1995). Their rate of generation correlates with oxidative stress (Irigoyen et al. 1992). Due to their lipophilicity, they easily permeate membranes. Jasmonic acid is another metabolite linked to peroxidative degradation of linolenic acid (Vick and Zimmerman 1984). Lipid peroxides also mediate damage of the superoxide dismutase (Lee and Park 1995). Inactivation of SOD could cause an increase in ROS concentration in each compartment and act as feedforward stress signals.

c) Final Remarks on Regulation of Gene Expression
by Cellular Redox Homeostasis

Responses to pathogen attack as well as to photooxidative stress are intermittently linked to the formation of ROS. Detailed expressional analysis indicates that genes of the general stress-response type must be distinguished from genes responding to specific types of stress only (Lee et al. 1994). Obviously, there exist distinct and common signalling pathways. Cross-talk between signalling pathways may be assumed to be similar to what is known from animals and fungi. ROS formation may be a primary or secondary response to stress and represents an essential or a facultative element of the signal transduction pathway. For example, methyl jasmonate signals wounding (Farmer and Ryan 1992) and salicylic acid pathogen attack (Malamy et al. 1990; Raskin 1992). In both cases ROS formation is a secondary answer. Photooxidative stress and ozone treatment involve ROS formation as primary signal. Our understanding of the signalling pathways is only slowly emerging. For instance, little is known about how cells combine and separate the input of information. Naturally, the analysis facing us will primarily aim at the molecular nature of the signalling pathways, the transcription factors and the promotor elements.

6 Conclusion

Reactive oxygen species are formed inevitably during cell metabolism and have a huge deleterious potential. However, plants are well equipped to efficiently inactivate ROS. In fact, as shown above, ROS are employed or even generated as useful signals and as indispensible reactants. Essentially, life is not possible without ROS. Nevertheless, plant growth is inhibited under conditions of excess ROS liberation (Gómez et al. 1995). Conversely, a stimulated accumulation of antioxidants enhances growth in transgenic plants under certain growth conditions (Roxas et al. 1997). H_2O_2 is a cosubstrate of cell wall construction, an inhibitor of peroxidases and of other enzymes, and extracellular signal during oxidative burst. These examples illustrate the two-faced character of oxidative metabolism. Work by Morré et al. (1988) indicated that the balance between ROS production and ROS inactivation is regulated by hormons and antioxidant defence mechanisms. Interestingly, even under non-stress conditions is the glutathione pool not fully reduced (Kunert and Foyer 1993). The GSSG level accounts for up to 10% of the total glutathione pool. This may be taken as evidence for continuous oxidation of reduced antioxidants, or may indicate the requirement for a low level of oxidized antioxidants to maintain a basic level of transcription of genes involved in antioxidant defence. Based on our present knowledge, one can phrase open questions for future research. Reasonably well known are the biochemical and chemical reactions of the enzymic and non-enzymic antioxidant network. However, little is known of the regulatory mechanisms maintaining the redox balance and on the in vivo capacities of the pathways to readjust the redox balance upon displacement following exposure to oxidative stress. Work with transgenic plants has started to provide some first but fragmentary clues on this subject. Only few pieces of the apparently large puzzle have been identified in respect to the receptors, the transduction pathways and the genetic switches involved in signalling redox changes and oxidative stress. It may be envisaged that redox signalling will prove to be similar general importance in signal transduction as changes in proton and free calcium concentrations.

Acknowledgements. We are grateful to Dr. Dortje Golldack for critically reading the manuscript. Parts of the work summarized in this review was performed within the framework of the grant Di 346/6 funded by the Deutsche Forschungsgemeinschaft and the Sonderforschungsbereiche 176 and 549.

References

Ahmad I, Larher F, Stewart GR (1979) Sorbitol, a compatible osmotic solute in *Plantago maritima*. New Phytol 82:671–678

Allen RD (1995) Dissection of oxidative stress tolerance using transgenic plants. Plant Physiol 107:1049–1054

Alscher RG (1989) Biosynthesis and antioxidant function of glutathione in plants. Physiol Plant 77:457–464

Anderson B, Satler A, Virgin I, Styring S (1992) Photodamage to photosystem II – primary and secondary events. J Photochem Photobiol B 15:15–31

Anderson JW, Foyer CH, Walker DA (1983) Light-dependent reduction of dehydroascorbate and uptake of exogenous ascorbate by spinach chloroplasts. Planta 158:442–450

Aono M, Kubo A, Saji H, Tanaka K, Kondo N (1993) Enhanced tolerance to photooxidative stress of transgenic *Nicotiana tabacum* with high chloroplastic glutathione reductase activity. Plant Cell Physiol 34:129–135

Aro E-M, Virgin I, Andersson B (1993) Photoinhibition of photosystem II. Inactivation, protein damage and turnover. Biochim Biophys Acta 1143:113–134

Asada K (1994) Production and action of active oxygen in photosynthetic tissue. In: Foyer CH, Mullineaux PM (eds) Causes of photooxidative stress and amelioration of defense systems in plants. CRC Press, Boca Raton, pp 77–104

Baalmann E, Backhausen JE, Kitzmann C, Scheibe R (1995) Regulation of NADP-dependent glyceraldehyde 3-phosphate dehydrogenase activity in spinach chloroplasts. Bot Acta 107:313–320

Baba K, Itoh S, Hastings G, Hoshina S (1996) Photoinhibition of photosystem I electron transfer activity in isolated photosystem I preparations with different chlorophyll contents. Photosynth Res 47:121–130

Badger M (1985) Photosynthetic oxygen exchange. Annu Rev Plant Physiol 36:27–53

Baier M, Dietz K-J (1996a) Primary structure and expression of a plant homologue of animal and fungal thioredoxin-dependent peroxide reductases and bacterial alkyl hydroperoxide reductases. Plant Mol Biol 31:553–564

Baier M, Dietz K-J (1996b) Thioredoxin-dependent peroxide reductase: a new group of plant peroxidases. In: Obinger C, Burner U, Ebermann R et al. (eds) Plant peroxidases: biochemistry and physiology. IV. International Symposium 1996. University of Geneva, pp 204–209

Baier M, Dietz K-J (1997) The plant 2-Cys peroxiredoxin BAS1 is a nuclear-encoded chloroplast protein: its expressional regulation, phylogenetic origin, and implications for its specific physiological function in plants. Plant J 12:179–190

Baier M, Bilger W, Wolf R, Dietz K-J (1996) Photosynthesis in the basal growing zone of barley leaves. Photosynth Res 49:169–181

Berg JM (1992) Sp1 and the subfamily of zinc finger proteins with guanine-rich binding sites. Proc Natl Acad Sci USA 89:11109–11110

Bergmann L, Rennenberg H (1993) Glutathione metabolism in plants. In: De Kok LJ, Stulen I, Rennenberg H et al. (eds) Sulfur nutrition and assimilation in higher plants. SPB Academic Publishing, The Hague, pp 109–123

Bleé E, Joyard J (1996) Envelope membranes from spinach chloroplasts are a site of metabolism of fatty acid hydroperoxides. Plant Physiol 110:445–454

Bohnert HJ, Jensen RG (1996) Strategies for engineering water-stress tolerance in plants. Trends Biotechnol 14:89–97

Bolotina VM, Najibi S, Placino JJ, Pagano PJ, Cohen RA (1994) Nitric oxide directly activates calcium-dependent potassium channels in vascuolar smooth muscle cells. Nature 368:850–853

Brandes HK, Larimer FW, Hartman FC (1996) The molecular pathway for the regulation of phosphoribulokinase by thioredoxin f. J Biol Chem 271:3333–3335

Canvine DT (1990) Photorespiration and CO_2-concentrating mechanisms. In: Dennis DT, Turpin DH (eds) Plant physiology, biochemistry and molecular biology. Longman, Harlow, pp 253–273

Casano LM, Trippi VS (1992) The effect of oxygen radicals on proteolysis in isolated oat chloroplasts. Plant Cell Physiol 33:329–332

Demming-Adams B, Adams WW III (1996) The role of xantophyll cycle carotenoids in the protection of photosynthesis. Trends Plant Sci 1:21–26

Desimone M, Henke A, Wagner E (1996) Oxidative stress induces partial degradation of the large subunit of ribulose-1,5-bisphosphate carboxylase/oxygenase in isolated chloroplasts of barley. Plant Physiol 111:789–796

Dietz K-J (1997) Functions and responses of the leaf apoplast under stress. Prog Bot 58:221–254

Dietz K-J, Brune A, Pfanz H (1992) *Trans*-tonoplast transport of the sulfur-containing compounds sulfate, methonine, cysteine and glutathione. Phyton 32:37–40

Duh J-L, Zhu H, Shertzer HG, Nebert DW, Puga A (1995) The Y-box motif mediates redox-dependent transcriptional activation in mouse cells. J Biol Chem 270:30499–30507

Elstner EF (1990) Der Sauerstoff: Biochemie, Biologie, Medizin. Wissenschaftsverlag, Mannheim

Elstner EF, Frommeyer D (1979) Analysis of different mechanisms of photosynthetic oxygen reduction. Biochim Biophys Acta 325:182–188

Escoubas J-M, Lomas M, laRoche J, Falkowski PG (1995) Light intensity regulation of cab gene transcription is signalled by the redox state of the pastoquinone pool. Proc Natl Acad Sci USA 92:10237–10241

Farmer EE, Ryan CA (1992) Octadecanoid precursors of jasmonic acid activate the synthesis of wound-inducible proteinase inhibitors. Plant Cell 4:129–134

Faske M, Holtgrefe S, Ocheretina O, Meister M, Backhausen JE, Scheibe R (1995) Redox equilibria between the regulatory thiols of light/dark-modulated enzymes and dithiothreitol: fine-tuning by metabolites. Biochim Biophys Acta 1247:135–142

Flohé L, Günzel WA (1984) Assays for glutathione peroxidase. Methods Enzymol 105:114–121

Foyer CH (1993) Ascorbic acid. In: Alscher RG, Hess JL (eds) Antioxidants in higher plants. CRC Press, Boca Raton, pp 31–58

Foyer CH, Halliwell B (1976) The presence of glutathione and glutathione reductase in chloroplasts: a proposed role in ascorbic acid metabolism. Planta 133:21–25

Foyer CH, Lelandais M, Jouanin L, Kunert KJ (1994a) Overexpression of enzymes of glutathione metabolism in poplar (*Populus tremula* x *P. alba*). Bull Soc Luxemb Biol Clin Special Issue:119–240

Foyer CH, Lelandais M, Kunert KJ (1994b) Photooxidative stress in plants. Physiol Plant 92:696–717

Foyer CH, Souriau N, Perret S, Lelandais M, Kunert KJ, Pruvost C, Jouanin L (1996) Overexpression of glutathione reductase but not glutathione synthetase leads to increase in antioxidant capacity and resistance to photoinhibition in poplar trees. Plant Physiol 109:1047–1057

Fucci L, Oliver CN, Coon MJ, Stadtman ER (1983) Inactivation of key metabolic enzymes by mixed-function oxidation reactions: possible implication in protein turnover and aging. Proc Natl Acad Sci USA 80:1521–1525

Furbank RT, Badger MR (1983) Oxygen exchange associated with electron transport and photophosphorylation in spinach thylakoids. Biochim Biophys Acta 723:400–409

Gilbert HF (1984) Redox control of enzyme activities by thiol/disulfide exchange. Methods Enzymol 107:330–351

Gilbert HF (1995) Thiol/disulfide exchange equilibria and disulfide bond stability. Methods Enzymol 251:8–28

Gille G, Sigler K (1995) Oxidative stress and living cells. Folia Microbiol (Praha) 40:131–152

Gillham DJ, Dodge AD (1986) Hydrogen-scavenging systems within pea chloroplasts: a quantitative study. Planta 167:246–251

Gilmore AM, Yamamoto HY (1993) Linear models relating xanthophylls and lumen acidity to non-photochemical fluorescence quenchin. Evidence that antheraxanthin explains zeaxanthin-independent quenching. Photosynth Res 35:67–78

Gómez LD, Casano LM, Trippi VS (1995) Effect of hydrogen peroxide on degradation of cell wall-associated proteins in growing bean hypocotyls. Plant Cell Physiol 36:1259–1264

Gutierrez-Marcos JF, Roberts MA, Campbell EI, Wray JL (1997) Three members of a novel small gene-familiy from *Arabidopsis thaliana* able to complement functionally an *Escherichia coli* mutant defective in PAPS reductase activity encode proteins with a thioredoxin-like domain and "APS reductase activity". Proc Natl Acad Sci USA 93:13377–13382

Hager A (1969) Lichtbedingte pH-Erniedrigung in einem Chloroplasten-Kompartiment als Ursache der enzymatischen Violaxanthin → Zeaxanthin-Umwandlung; Beziehungen zur Photophosphorylierung. Planta 89:224–243

Hahlbrock K, Scheel D, Logemann E, Nürnberger T, Parniske M, Reinold S, Sacks WR, Schmelzer E (1995) Oligopeptide elicitor-mediated defense gene activation in cultured parsley cells. Proc Natl Acad Sci USA 92:4150–4157

Halliwell B (1994) How to characterize an antioxidant: an update. Biochem Soc Symp 61:73–101

Halliwell B, Gutteridge JMC (1990) Role of free radicals and catalytic metal ions in human disease: an overview. Methods Enzymol 186:1–85

Heber U, Egneus H, Hanch U, Jensen M, Koster S (1978) Regulation of photosynthetic electron transport and photophosphorylation in intact chloroplasts and leaves of *Spinachia oleracea* L. Planta 143:41–53

Hérouart D, Inzé D, Van Montagu M (1993) Redox-activated expression of the cytosolic copper/zinc superoxide dismutase gene in *Nicotiana*. Proc Natl Acad Sci USA 90:3108–3112

Hidalgo E, Ding H, Demple B (1997) Redox signal transduction via iron-sulfur clusters in the SoxR transcription activator. Trends Biochem Sci 22:207–210

Hodgson RA, Raison JK (1991) Superoxide production by thylakoids during chilling and its implication in the susceptibility of plants to chilling-induced photoinhibition. Planta 183:222–228

Holland D, Ben-Hayyim G, Faltin Z, Camoin L, Strosberg AD, Eshdat Y (1993) Molecular characterization of salt-stress associated protein in citrus: protein and cDNA sequence homology to mammalian glutathione peroxidase. Plant Mol Biol 21:923–927

Holmgren A (1989) Thioredoxin and glutaredoxin systems. J Biol Chem 264:13963–13966

Hölzel C, Spiteller G (1995) Zellschädigung als Ursache für die Bildung von hydroperoxiden ungesättigter Fettsäuren. Naturwissenschaften 82:452–460

Horemans N, Asard H, Caubergs RJ (1997) The ascorbate carrier of higher plant plasma membranes preferentially translocates the fully oxidized (dehydroascorbate) molecule. Plant Physiol 114:1247–1253

Hormann H, Neubauer C, Asada K, Schreiber U (1993) Intact chloroplasts display pH 5 optimum of O_2 reduction in the absence of methyl viologen: indirect evidence for a regulatory role of superoxide protonation. Photosynth Res 37:69–89

Hossain MA, Nakano Y, Asada K (1984) Monodehydroascorbate reductase in spinach chloroplasts and its participation in regeneration of ascorbate for scavenging hydrogen peroxide. Plant Cell Physiol 25:385–395

Irigoyen JJ, Emerich DW, Sánchez-Díaz M (1992) Alfalfa leaf senescence induced by drought stress: photosynthesis, hydrogen peroxide metabolism, lipid peroxidation and ethylene evolution. Physiol Plant 84:67–72

Jespersen HM, Kjærsgård IVH, Østergaard L, Welinder G (1997) From sequence analysis of three novel ascorbate peroxidases from *Arabidopsis thaliana* to structure, function and evolution of seven types of ascorbate peroxidase. Biochem J 326:305–310

Kangasjärvi J, Talvinen J, Utriainen M, Karjalainen R (1994) Plant defence systems induced by ozone. Plant Cell Environ 17:783–794

Karpinski S, Escobar C, Karpinska B, Creissen G, Mullineaux PM (1997) Photosynthetic electron transport regulates the expression of cytosolic ascorbate peroxidase genes in *Arabidopsis* during excess light stress. Plant Cell 9:627–640

Kim S-G, Kim Y, An G (1993) Identification of methyl jasmonate and salicylic acid response elements from the nopaline synthase (nos) promotor. Plant Physiol 103:97–103

Kubo A, Saji H, Tanaka K, Kondo N (1995) Expression of *Arabidopsis* cytosolic ascorbate peroxidase gene in response to ozone and sulfur dioxide. Plant Mol Biol 29:479–489

Kullik I, Storz G (1994) Transcriptional regulators of oxidative stress response in prokaryotes and eukaryotes. Redox Rep 1:23–29

Kunert KJ, Foyer C (1993) Thiol/disulfide exchange in plants. In: De Kok LJ et al. (eds) Sulfur nutrition and assimilation in higher plants. SPB Academic Publishing, The Hague, pp 139–151

Lander HM (1997) An essential role of free radicals and derived species in signal transduction. FASEB J 11:118–124

Larson RA (1988) The antioxidants of higher plants. Phytochemistry 27:969–978

Lee MH, Park J-W (1995) Lipid peroxidation products mediate damage of superoxide dismutase. Biochem Mol Biol Int 35:1093–1102

Lee SW, Heinz R, Nazar RN (1994) Differential utilization of alternate initiation sites in a plant defense gene responding to environmental stimuli. Eur J Biochem 226:109–114

Levine A, Tenhaken R, Dixon R, Lamb C (1994) H_2O_2 from the oxidative burst orchestrates the plant hypersensitive disease resistance response. Cell 79:583–593

Li X, Nield J, Hayman D, Langridge P (1996) A self-fertile mutant of *Phalaris* produces an S protein with reduced thioredoxin activity. Plant J 10:505–513

Lim YS, Cha MK, Kim HK, Uhm TB, Park JW, Kim K, Kim IH (1993) Removals of hydroben peroxide and hydroxyl radical by thiol-specific antioxidant protein as a possible role in vivo. Biochem Biophys Res Commun 192:273–280

Lindahl M, Yang D-H, Andersson B (1995) Regulatory proteolysis of the major light-harvesting chlorophyll a/b protein of photosystem II by light-induced membrane-associated enzymic system. Europ J Biochem 231:503–509

Low PS, Merida JR (1996) The oxidative burst in plant defense: function and signal transduction. Physiol Plant 96:533–542

Luwe M, Takahama U, Heber U (1993) Role of ascorbate in detoxifying ozone in the apoplast of spinach (*Spinacia oleracea* L.) leaves. Plant Physiol 101:969–976

Malamy J, Carr JP, Klessing DF, Raskin I (1990) Salicylic acid: a likely endogenous signal in the resistance response of tobacco to viral infection. Science 250:1002–1004

McKersie BD, Chen Y, De Beus M, Bowley SR, Bowler C (1993) Superoxide dismutase enhances tolerance to freezing stress in transgenic alfalfa (*Medicago sativa* L.). Plant Physiol 103:1155–1163

Mehdy MC, Sharma YK, Sathasivan K, Bays NW (1996) The role of activated oxygen species in plant disease resistance. Physiol Plant 98:365–374

Mehlhorn H, Lelandis M, Korth HG, Foyer CH (1996) Ascorbate is the natural substrate for plant peroxidases. FEBS Lett 378:203–206

Meyer M, Schreck R, Baeuerle PA (1993) H_2O_2 and antioxidants have opposite effects on activation of NF-κB and AP-1 in intact cells: AP-1 as secondary antioxidant-responsive factor. EMBO J 12:2005–2015

Mishra NP, Fatama T, Singhal GS (1995) Development of antioxidant defense system of wheat seedlings in response to high light. Physiol Plant 95:77–82

Misra HP, Fridovich I (1971) The generation of superoxide radical during the autoxidation of ferredoxins. J Biol chem 246:6886–6890

Miyake D, Michihata F, Asada K (1991) Scavenging of hydrogen peroxide in prokaryotic and eukaryotic algae: acquisition of ascorbate peroxidase during evolution of cyanobacteria. Plant Cell Physiol 32:33–43

Mohanty N, Yamamoto HY (1995) Mechanism of non-photochemical chlorophyll fluorescence quenching. I. The role of de-epoxidized xanthophylls and sequenstered thylakoid membrane protons as probed by dibucaine. J Plant Physiol 22:231–238

Morell S, Follmann H, Häberlein I (1995) Identification and localization of the first glutaredoxin in leaves of a higher plant. FEBS Lett 369:149–152

Morgan RW, Christman MF, Jacobson FS, Storz G, Ames BN (1986) Hydrogen peroxide-inducible proteins in *Salmonella thyphimurium* overlap with heat shock and other stress proteins. Proc Natl Acad Sci USA 83:8059–8063

Morré DJ, Brightman AO, Wu LY, Barr R, Leak B, Crane FL (1988) Role of plasma membrane redox activities in elongation growth in plants. Physiol Plant 73:187–193

Neubauer C, Yamamoto HY (1992) Ascorbate peroxidase mediates zeaxanthin formation and zeaxanthin-related fluorescence quenching in intact chloroplasts. Plant Physiol 99:1354–1361

Noctor G, Arisi A-CM, Jouanin L, Valadier M-H, Roux Y, Foyer C (1997) The role of glycine in determining the rate of glutathione synthesis in poplar. Possible implications for glutathione production during stress. Physiol Plant 100:255–263

Örvar BL, Ellis BE (1997) Transgenic tobacco plants expressing antisense RNA for cytosolic ascorbate peroxidase show increased susceptibility to ozone injury. Plant J 11:1297–1305

Osmond CB (1994) What is photoinhibition? Some insights from comparison of shade and sun plants. In: Baker NR, Boyer JR (eds) Photoinhibition: molecular mechanisms to the field. Bios, Oxford, pp 1–24

Osmond CB, Grace SC (1995) Perspectives on photoinhibition and photorespiration in the field: quintessential inefficiencies of the light and dark reactions of photosynthesis? J Exp Bot 46:1351–1362

Perl-Treves R, Galun E (1991) The tomato Cu,Zn superoxide dismutase genes are developmentally regulated and respond to light and stress. Plant Mol Biol 17:745–760

Pognonec P, Kato H, Roeder RG (1992) The helix-loop-helix repeat transcription factor USF can be functionally regulated in a redox-dependent manner. J Biol Chem 267:24563–24567

Polle A (1996) Mehler reaction. Friend or foe in photosynthesis? Bot Acta 109:84–89

Rao MV, Paliyath G, Ormrod DP (1996) Ultraviolett-B- and ozone-induced biochemical changes in antioxidant enzymes of *Arabidopsis thaliana*. Plant Physiol 110:125–136

Rao MV, Paliyath G, Ormrod DP, Murr DP, Watkins CB (1997) Influence of salicylic acid on H_2O_2 production, oxidative stress, and H_2O_2-metabolizing enzymes. Plant Physiol 115:137–149

Raskin I (1992) Salicylate, a new plant hormone. Plant Physiol 99:799–803

Robinson JM, Gibbs M (1982) Hydrogen peroxide synthesis in isolated spinach chloroplasts lamellae. Plant Physiol 70:1249–1254

Roxas VP, Smith RK Jr, Allen ER, Allen RD (1997) Overexpression of glutathione S-transferase/glutatjione peroxidase enhances the growth of transgenic tobacco seedlings during stress. Nature [Biotechnol] 15:988–991

Scheibe R (1996) Die Regulation der Photosynthese durch das Licht. Biol Unserer Zeit 26:27–34

Scheibe R, Beck E (1994) The malate valve: flux control at the enzymatic level. In: Schulze ED (ed) Flux control in biological systems. Academic Press, London, pp 3–11

Schöner S, Krause GH (1990) Protective systems against active oxygen species in spinach: response to cold acclimation in excess light. Planta 180:383–389

Schraudner M, Langebartels C, Sandermann H (1997) Changes in the biochemical status of plant cells induced by the environmental pollutant ozone. Physiol Plant 100:274–280

Schreiber U, Neubauer C (1990) O_2-dependent electron flow, membrane energetisation and the mechanism of non-photochemical quenching of chlorophyll fluorescence. Photosynth Res 25:279–293

Schubert R, Fischer R, Hain R, Schreier PH, Bahnweg G, Ernst D, Sandermann H Jr (1997) An ozone-responsive region of grapevine resveratol synthase promotor differs from basal pathogen-responsive sequence. Plant Mol Biol 34:417–426

Schwarz P, Häberle K-H, Polle A (1996) Interactive effects of elevated CO_2, ozone and drought stress on the activities of antioxidant enzymes in needles of norway spruce trees [*Picea abies* (L.) Karsten] grown with luxurious N-supply. J Plant Physiol 148:351–355

Sen Gupta A, Webb RP, Holaday AS, Allen RD (1993) Overexpression of superoxide dismutase protects plants from oxidative stress. Plant Physiol 103:1067–1073

Shen B, Jensen RG, Bohnert HJ (1997) Increased resistance to oxidative stres in transgenic plants by targeting mannitol biosynthesis to chloroplasts. Plant Physiol 113:1177–1183

Smirnoff N, Cumbes QJ (1989) Hydroxyl radical scavenging activity of compatible solutes. Phytochemistry 28:1057–1060

Stadtman ER (1992) Protein oxidation and aging. Science 257:1220–1224

Stieger PA, Feller U (1997) Requirements for the light-stimulated degradation of stromal proteins in isolated pea (*Pisum sativum* L.) chloroplasts. J Exp Bot 48:1639–1645

Storz G, Jacobson FS, Tartaglia LA, Morgan RW, Silveira LA, Ames BN (1989) An alkyl hydroperoxide reductase induced by oxidative stress in *Salmonella typhimurium* and *Escherichia coli*: genetic characterization and cloning of *ahp*. J Bacteriol 171:2049–2055

Storz G, Tartaglia LA, Farr SB, Ames BN (1990) Bacterial defences against oxidative stress. Trends Genet 6:363–368

Strohm M, Jouanin L, Kunert KJ, Pruvost C, Polle A, Foyer CH, Rennenberg H (1995) Regulation of glutathione synthesis in leaves of transgenic poplar (*Populus tremula x P. alba*) overexpressing glutathione synthetase. Plant J 7:141–145

Sun Y, Oberley LW (1996) Redox regulation of transcriptional activators. Free Radic Biol Med 21:335–348

Szögyi M, Cserháti T, Szingeti Z (1989) Action of paraquat and diquat on proteins and phospholipids. Pesticide Biochem Physiol 34:240–245

Tenhaken R, Rübel C (1997) Salicylic acid is needed in hypersensitive cell death in soybean but does not act as a catalase inhibitor. Plant Physiol 115:291–298

Tepperman JM, Dunsmuir P (1990) Transformed plants with elevated levels of chloroplastic SOD are not more resistant to superoxide toxicity. Plant Mol Biol 14:501–511

Tommasini R, Martinoia E, Grill E, Dietz K-J, Amrhein N (1993) Transport of oxidized glutathione into barley vacuoles: evidence for the involvement of the glutathione-S-conjugate ATPase. Z Naturforsch [C]48:867–871

Trolinder NL, Allen RD (1994) Expression of chloroplast localized Mn SOD in transgenic cotton. J Cell Biochem 18A:97

Van Camp W, Willekens H, Bowler C, Van Montague M, Inzé D (1994) Elevated levels of superoxide dismutase protect transgenic plants against ozone damage. Biotechnology 12:165–168

Vick BA, Zimmerman DC (1984) Biosynthesis of jasmonic acid by several plant species. Plant Physiol 75:458–461

Willekens H, Van Camp W, Van Montagu MV, Inzé D, Langebartels C, Sandermann H Jr (1994) Ozone, sulfur dioxide, and ultraviolet B have similar effects on mRNA accumulation of antioxidant genes in *Nicotiana plumbaginifolia* L.. Plant Physiol 106:1007–1014

Wingate VPM, Lawton MA, Lamb CJ (1988) Glutathione causes a massive and selective induction of plant defense genes. Plant Physiol 87:206–210

Wingler A, Lea PJ, Leegood RC (1997) Control of photosynthesis in barley plants with reduced activities of glycine decarboxylase. Planta 202:171–178

Wingsle G, Karpinski S (1996) Differential redox regulation by glutathione of glutathione reductase and CuZn superoxide dismutase genes expression in *Pinus sylvestris* (L.) needles. Planta 198:151–157

Wise RR (1995) Chilling-enhanced photooxidation: the production, action and study of reactive oxygen species produced during chilling in the light. Photosynth Res 45:79–97

Wolf AE, Dietz K-J, Schröder P (1996) Degradation of glutathione-S-conjugates by a carboxypeptidase in the plant vacuole. FEBS Lett 384:31–34

Wu J, Neimanis S, Heber U (1991) Photorespiration is more effective than Mehler reaction in protecting the photosynthetic apparatus against photoinhibition. Bot Acta 104:283–291

Wu X, Bishopric NH, Discher DJ, Murphy BJ, Webster KA (1996) Physical and functional
    sensitivity of zinc finger transcription factors to redox change. Mol Cell Biol 16:1035–
    1046
Wydrzynski T, Ångström J, Vänngård T (1989) H_2O_2 formation by photosystem II. Bio-
    chim Biophys Acta 973:23–28
Yamamoto HY (1985) Xanthophyll cycles. Methods Enzymol 110:303–312
Yruela I, Pueyo JJ, Alonso PJ, Picorel R (1996) Photoinhibition of photosystem II from
    higher plants: effect of copper inhibition. J Biol Chem 271:27408–27415

Dr. Margarete Baier
Julius-von-Sachs-Institut
für Biowissenschaften
mit Botanischem Garten
Lehrstuhl I der Universität
Mittlerer Dallenbergweg 64
D-97082 Würzburg, Germany

Prof. Dr. Karl-Josef Dietz
Lehrstuhl für Stoffwechselphysiologie
und Biochemie der Pflanzen
Fakultät für Biologie
Universität Bielefeld
Universitätsstraße 25
D-33615 Bielefeld, Germany

Edited by
U. Lüttge

Growth: Progress in Auxin Research

By Hartwig Lüthen, Maike Claussen, and Michael Böttger

1 Auxin Physiology – a Never-Ending Story

Almost exactly 70 years ago, Frits Went (1928) characterized the first plant growth substance, auxin. Soon the natural auxin indole-3-acetic acid (IAA) was identified. Seven decades of attempts to clarify its mode of action followed. Although a bulk of data has been accumulated, most fundamental questions are still under intense debate: what protein is the growth-relevant auxin receptor, and where is it localized? What is the architecture of signalling chain leading to growth stimulation? There seems to be general agreement that auxin induces growth via cell wall loosening, but which molecular mechanism accounts for this? The state of auxin research after such a long time may be disappointing, especially if one considers that many plant physiologists – some of whom were very successful in other fields – have worked on the topic. It appears that auxin physiology is a "never-ending story", or at least that the difficulties in solving the problem may have been generally underestimated.

One major difficulty in growth physiology is the diversity of auxin effects. We will focus this chapter on the classical response, the induction of elongation growth and the molecular processes potentially involved in this response.

2 Auxin Perception

a) Membrane-Associated Binding Sites and ABP1

Membrane-associated binding sites for auxin were identified in the 1970s. An activity termed site 1 is localized predominantly at the endoplasmic reticulum (ER), site 2 at the vacuole, and site 3 seems to be identical with the auxin efflux carrier at the plasma membrane (PM) (Dohrmann et al. 1978). Because of the known function of site 3 and the poor substrate specifity of site 2, interest focused on site 1. Site 1 from maize coleoptiles was purified to near homogenity by affinity chroma-

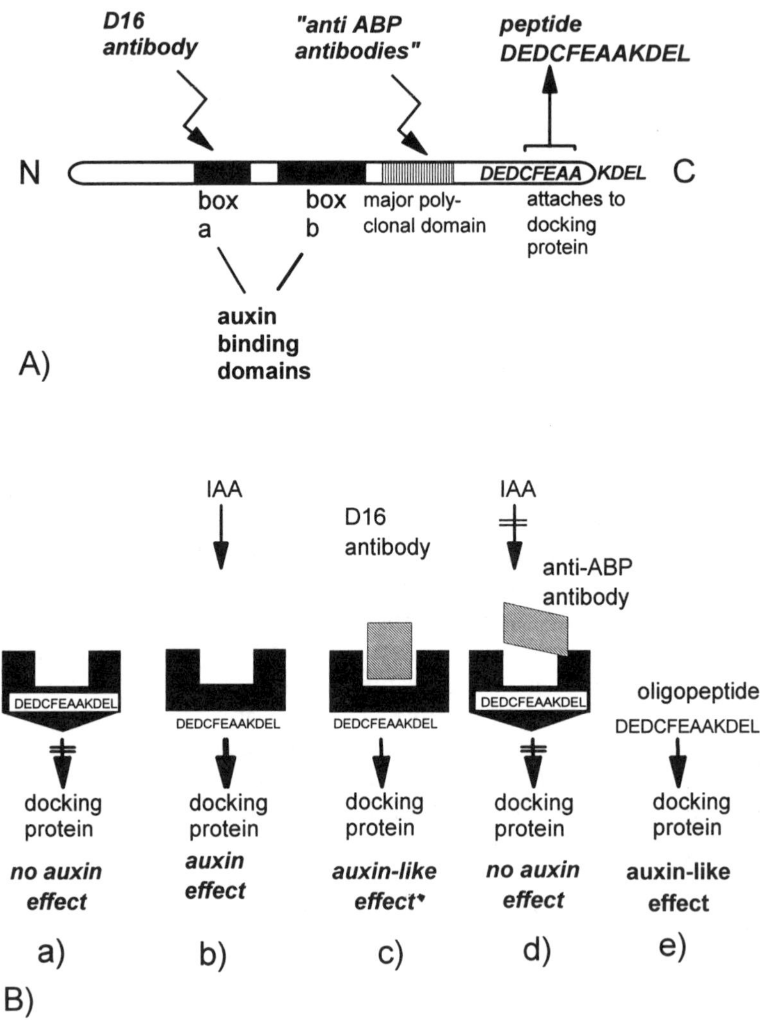

Fig. 1A, B. Molecular structure of auxin-binding protein 1 (ABP1). **A** Origin of bioactive antibodies and peptides. The conservative boxes *a* und *b* are probably the auxin-binding site. Note the C-terminal endoplasmic reticulum retention signal *KDEL*. An antibody raised against box *a* is the *D16 antibody*. Polyclonal antibodies raised against the whole ABP1 molecule (*anti-ABP antibodies*) are often directed against an area around the glycolysation site (*main polyclonal domain*). The C-terminal peptide DEDCFEAAKDEL probably attaches to the docking protein. **B** Possible mode of action of the bioactive antibodies and peptides. **a** Situation in the absence of auxin. Binding of IAA to ABP1 (**b**) causes a conformational change which exposes the C-terminal amino acid sequence DEDCFEAAKDEL. This epitope binds to the docking protein, from which the signalling cascade is started. D16 antibodies with their affinity for the auxin-binding site (**c**) can mimic the action of IAA and therefore have auxin agonist activity. Anti-ABP antibodies may hamper auxin binding (**d**) and/or conformational changes of the ABP1 molecule and inhibit auxin action. Synthetic oligopeptides with the sequence DEDCFEAAKDEL (**e**) can substitute for the IAA-ABP1 complex action by binding to the docking protein and therefore possess auxin-like activity on transmembrane potassium currents of guard cells

tography (Löbler and Klämbt 1985) and termed auxin-binding protein 1, (ABP1). Meanwhile, the maize *ABP1* gene was sequenced and cloned (Hesse et al. 1989; Inohara et al. 1989), as were analogues from *Arabidopsis* (Palme et al. 1992), tobacco, strawberry (see Jones 1994) and radish (Anai et al. 1997). Figure 1A shows an overview of the molecular structure of ABP1: surprisingly, a KDEL sequence occurs at the C terminus. Since KDEL is a signal sequence for retention in the lumen of the ER (Munro and Pelham 1987; Zagouras and Rose 1989; Dennecke et al. 1992), the molecular structure explains the predominant localization of site 1 binding activity of ER. There are two conserved boxes (box A and box B), probably constituing the auxin-binding site. Antibodies have been raised against the whole ABP1 molecule (anti-ABP antibodies; Löbler and Klämbt 1985) or against epitopes of the ABP1 protein sequence (Napier and Venis 1990). D16 antibodies are directed against a synthetic oligopeptide with the sequence of box A (Venis et al. 1992; Jones 1994; Napier 1995).

b) Evidence for ABP1 as the Physiological Auxin Receptor

There is physiological evidence suggesting that ABP1 has receptor function.

1. Anti-ABP1 antibodies (Fig. 1A) block auxin-induced hyperpolarization of tobacco protoplasts by shifting the bell-shaped dose-response curve for auxin towards higher concentrations (Barbier-Brygoo et al. 1989). It seems that binding of the large antibody molecule disturbs binding of IAA or the IAA-induced conformational change (Fig. 1B, d).
2. Adding ABP1 to the protoplast incubation medium has the opposite effect: the curve is shifted towards lower concentrations (Barbier-Brygoo et al. 1989).
3. D16 antibodies (Fig. 1A) – like auxin – induce a membrane hyperpolarization (Venis et al. 1992). Since these antibodies are directed against the auxin-binding domain, their binding changes the conformation of ABP1 in a manner similar to the hormone (Fig. 1B, c).

All these findings were obtained by the tobacco protoplast auxin hyperpolarization assay (Barbier-Brygoo et al. 1989, 1991). However, whatever is measured by this assay, it is not membrane potential: the basal potential of protoplast averages at +5 mV, which is much lower than the potential predicted from the Goldman equation. The auxin-induced hyperpolarization amounts to a mere 4–6 mV. For the question of auxin perception, this objection may, however, be irrelevant: whatever is detected by this assay seems to be auxin-sensitive. Moreover, work from the lab of Hubert Felle (Rück et al. 1993) demonstrates a similar effect of D16 antibodies on an ATP-dependent transmembrane current measured in the whole-cell patch clamp mode, suggesting a stimulation of the proton ATPase by D16 antibodies and auxin. The same authors show that anti-ABP antibodies block the auxin-induced current, but not the fusicoccin (FC)-induced one. This finding confirms that the anti-ABP antibodies act specifically on the auxin response.

c) ABP1 as a Receptor - the Docking Protein Hypothesis

Since antibodies probably do not pass the PM, these data imply that ABP1 acts as a receptor and perceives the auxin signal at the apoplasmic side of the PM. This is indeed puzzling, because ABP1 as a KDEL protein should be strictly localized at the ER. Moreover, ABP1 is not a classical receptor, since it lacks any transmembrane domains. To overcome these difficulties, it was assumed (Klämbt 1990) that a small amount of ABP1 can escape the ER – probably in a regulated manner. It will then show up at the plasma membrane and, after binding auxin, the hormone-ABP1 complex will attach to an as yet unidentified transmembrane docking protein. Here, the signalling chain is started (Fig. 2A). In this sense, the ABP1-IAA complex would be the true hormone, and the docking protein the true receptor.

Is this hypothesis valid? To evaluate this, the answers to three questions are particularly crucial: (1) Is ABP1 secreted to the membrane surface? (2) Is there physiological evidence for the presence of a docking protein? (3) Does ABP1 bind to the plasma membrane and, if so, does the presence of auxin increase this binding? All these questions have been addressed by recent research.

1. ABP1 has been detected at the outer cell surface of protoplasts by immunoepipolarization microscopy (Diekmann et al. 1995), in cell walls by immuno gold labelling (Jones and Herman 1993). This indicates that, in fact, a very low amount of ABP1 can escape to the apoplasm. In a very careful study, Henderson et al. (1997) could not detect any ABP1 exit to the culture medium of coleoptile cells. They estimated the upper limit to be 2% of the total ABP1. Insect cells can be transformed with ABP1 (MacDonald et al. 1994). Although 1% of these cells' total protein content is ABP1, only 1% of the ABP1 is secreted within 2 h (Henderson et al. 1997). This demonstrates the effectiveness of the KDEL rentention system. It has been suggested that auxin stimulates ABP1 exit to the plasma membrane surface by inducing a conformational change shielding the KDEL sequence. However, neither in plant systems nor in transfected insect cells could such an auxin-induced stimulation ever be verified (Henderson et al. 1997). Thus, ABP1 transport to the PM surface may be an unspecific

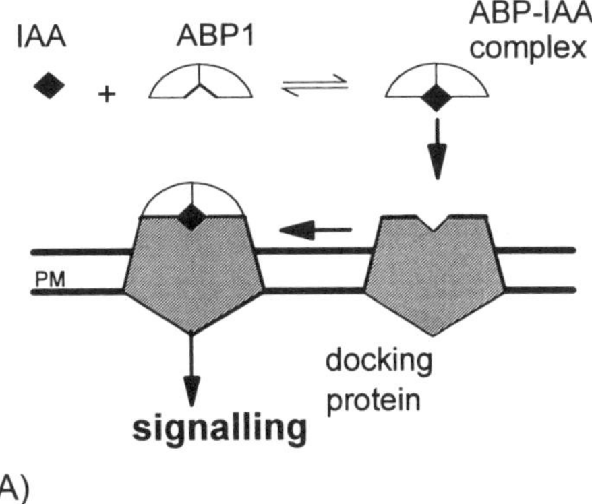

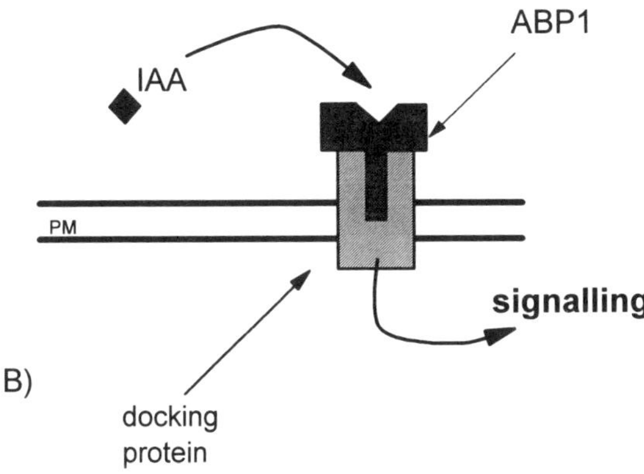

Fig. 2A, B. Two versions of the ABP1 docking protein hypothesis. **A** Original hypothesis proposed by Klämbt (1990), modified. *IAA* binds to the dimer of *ABP1*. The *ABP1-IAA complex* then attaches to a transmembrane docking protein from where the signalling chain is started. Note that the binding of the ABP1-IAA complex is the primary event starting signalling. **B** In this version, docking protein and ABP1 form a functional unit acting as auxin receptor. The docking protein additionally may act as a chaperone that attaches ABP1 to the ER membrane, possibly mediating ABP1 exocytosis, as proposed by MacDonald (1997). Note that the binding of IAA to ABP1 is the event that starts the signalling chain. This model is in line with the finding that the binding of ABP1 to the plasma membrane is not affected by the presence of auxin

escape rather than a regulated transport. It is decisive for the validity of the docking protein hypothesis that it occurs at all.

2. Thiel et al. (1993) treated guard cells with a synthetic oligopeptide with the sequence of the C terminus of ABP1. They observed a modulation of potassium channels similar to the effect of high auxin concentrations in the same system. They postulate that auxin binding to ABP1 results in an exposure of the C terminal sequence, which then attaches to the docking protein (Fig. 1B, a, b). Therefore, appli-

cation of a synthetical peptide with this sequence should induce an auxin-like response (Fig. 1B, e).

3. The Klämbt group (Schiebel et al. 1997) took a closer look at binding ABP1 to isolated plasma membranes. They found – in agreement with the docking protein hypothesis – that ABP1 binds specifically to the plasma membrane. However, their data do not indicate that this binding is enhanced by IAA. This would be clearly predicted by the docking protein hypothesis. In order to save the hypothesis, one can speculate that not the binding of the ABP1-IAA complex triggers the signalling, but that the ABP1-docking-protein complex is a functional unit that forms the receptor (Fig. 2B). A modified model of this type has recently been proposed by MacDonald (1997). She raised the possibility that the docking protein may be a chaperone with attaches ABP1 at the inner surface of ER and PM and possibly facilitates the transport of ABP1 through the membrane flow.

d) Is ABP1 a Red Herring?

Unfortunately, the very clear-cut data in favour of ABP1 as auxin receptor have been, up to now, restricted to an experimental system previously rather uncommon in auxin research. The antibody-sensitive electrophysiological effects occur as early as seconds or 1–2 min after auxin addition, and they have been found up to now only on protoplasts, which are lacking a cell wall and therefore the capacity to display elongation growth. In the same time window, auxin is without any effect on or slightly inhibitory to growth (Tietze-Haß and Dörffling 1977; Lüthen et al. 1990; Fischer et al. 1992), proton secretion (Lüthen et al. 1990; Peters and Felle 1991a, b; Peters et al. 1998), or membrane hyperpolarization (Felle et al. 1991) in intact organs. Stimulatory effects start to build up 10–15 min after auxin addition at the earliest. Thus, the value of the tobacco protoplast hyperpolarization as a model system for the more classical auxin effects is very limited. It still has to be envisaged that only a fraction of the very diverse auxin responses is mediated via ABP1, whereas other responses require auxin binding to other receptors.

Hertel (1995) recently raised serious objection to ABP1 as an auxin receptor, culminating in his claim that ABP1 is a "red herring" rather than a receptor. His arguments are as follows: (1) 4-Cl-IAA appears to bind very poorly to site 1 in comparison to IAA (Hertel 1994). However, in vivo, 4-Cl-IAA is a stronger auxin than IAA (Böttger et al. 1978; Fischer et al. 1992). (2) ABP1 levels vary between pea and maize, whereas auxin sensitivity does not. (3) Overexpression of ABP1 does not change the phenotype significantly. Arguments (2) and (3) indicate that ABP1 levels do not control auxin sensitivity, but, paradoxically, the eletrophysiological effects indicate the contrary. Concerning argument (1), it has to be mentioned that a new study using purified ABP1 instead of site 1 binding resulted in a superior binding of 4-Cl-IAA to ABP1 compared to IAA (Rescher et al. 1996). Arguments (2) and (3) can be overcome by assuming a bottleneck of ABP1 transport to the outer plasma membrane which would uncouple total and active ABP1 levels.

Recent studies in our laboratory indicate that F_{ab} fragments of D16 antibodies and anti-ABP1 antibodies had no auxin-like effect on elongation growth. D16 antibody fragments had instead an inhibitory action on auxin-induced proton secretion (Fig. 3). Whether this response is a homologue to the transient alkalinization evoked by auxin remains unclear. In any case, D16 antibody fragments did not affect auxin-induced acidification or growth of maize coleoptiles (Claussen 1997). There was also no effect whatsoever of ABP1 oligopeptides, which did not induce growth on their own; nor did their presence affect auxin action. Although such negative evidence has to be interpreted with caution, there are by now no hard data to support the validity of the ABP1-docking protein szenario for the triggering of elongation growth.

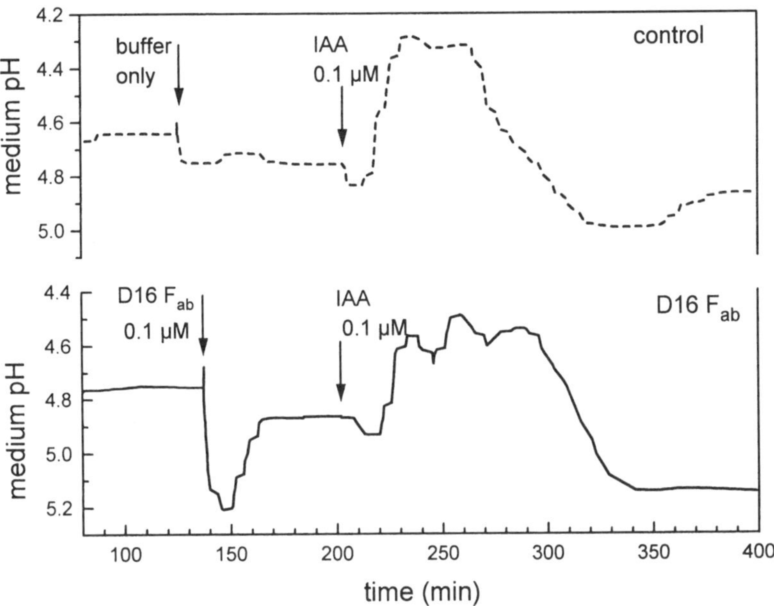

Fig. 3. Effect of D16 F_{ab} antibody fragments and auxin on apoplasmic pH in maize coleoptiles. *Above* IAA causes a pH drop to 4.3 after a short transient alkalinization. *Below* D16 F_{ab} fragments cause a transient alkalinization of the medium, but do not significantly alter the time course or magnitude of a subsequent auxin-induced acidification (Claussen 1997). In contrast to the situation in the tobacco protoplast hyperpolarization assay, D16 antibodies do not mimic the activity of the hormone auxin

e) Auxin Receptor for Growth: Inside or Outside?

There is another threat to the ABP1-docking-protein hypothesis: studies in a number of laboratories indicate that the auxin receptor for the growth response is localized inside the cell. If coleoptiles are allowed to accumulate IAA, and the auxin is then removed from the incubation solution, growth rates will drop rapidly. This is, in fact, compatible with both intracellular and extracellular receptor localization. If, however, the auxin efflux carrier is blocked by the phytotropins naphthylphtalamic acid (NPA; Davies 1974; Vesper and Kuss 1990), pyrenoylbenzoic acid (PBA; Clausen et al. 1996), morphactin (Claussen 1997) or by triiodobenzoic acid (TIBA; Claussen et al. 1996), growth is sustained even when the extracellular auxin is removed from the incubation medium (Fig. 4A). Additionally, we could show that growth did not decline when 2,4-D was removed from the incubation medium even in the absence of any inhibitor. 2,4-D is known to be a very poor substrate for the auxin efflux carrier. NAA, another synthetic auxin, which is rapidly exported, behaves in this test similarly to IAA. These data are readily interpreted by assuming an intracellular binding site for auxin. However, this conclusion is valid for the moment when the auxins are washed off, which was done 1–2 h after treatment, but does not necessarily hold true for the early phases of auxin action. In fact, the auxin efflux inhibitors do not affect the time course of auxin action in the case of short-term auxin pulse applications (Fig. 4B). Thus, it is still possible that extracellular auxin perception (perhaps by ABP1) is crucial in the very early phases of auxin action.

Some mutant studies strengthen the role of intracellular auxin perception. The *aux1* mutant of *Arabidopsis* is resistant to high levels of the auxin herbicide 2,4-D (Maher and Martindale 1980). It could be shown that the auxin-induced root growth inhibition required much higher auxin concentrations in the mutant compared to the wild type (Pickett et al. 1990). The *AUX1* sequence has a high degree of homology to a class of carriers of aromatic amino acids like tryptophan or phenylalanin (Bennett et al. 1996). The idea is that the AUX1 gene product is an auxin uptake carrier. Generally, auxin is taken up by an ion trap mechanism, but in the case of 2,4-D with its very acidic pK_a, carrier-mediated uptake may be favoured. If this uptake mechanism is disrupted in the *aux1* mutant, 2,4-D cannot be taken up efficiently and will not reach the intracellular receptor. It can, of course, also be speculated that AUX1 is, in fact, the auxin receptor that evolved from an amino acid transporter. In either case, the data imply that there must be a pathway leading to auxin responses not starting at extracellular ABP1.

Recently, Boot et al. (1996) treated transgenic tobacco protoplasts harbouring a gene construct consisting of the GUS reporter gene and an

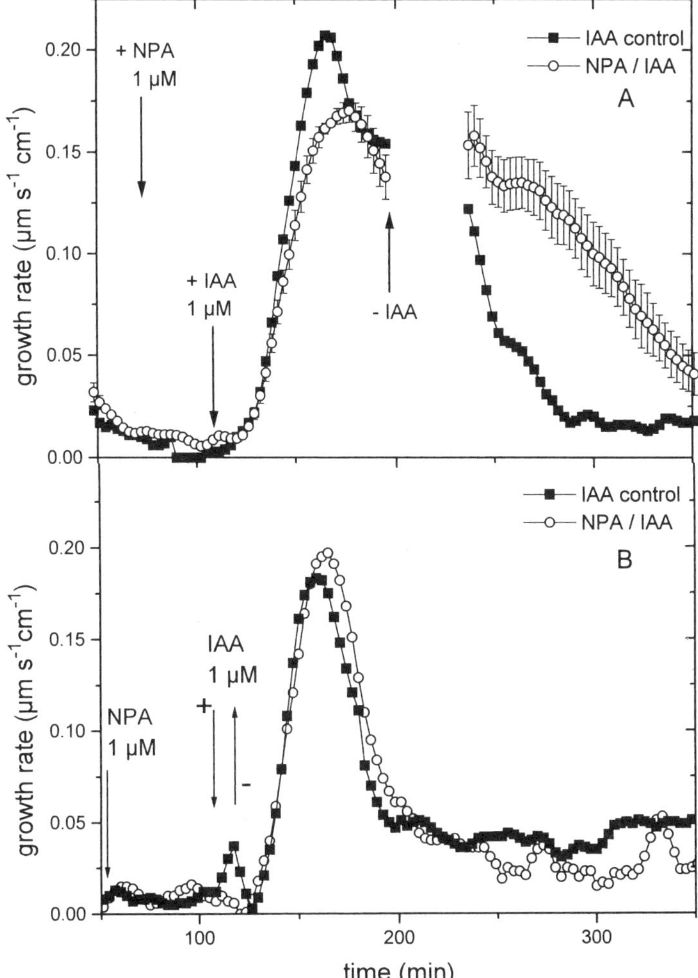

Fig. 4A, B. Experiments to clarify the localizatin of the auxin receptor relevant for growth control. **A** Maize coleoptile segments were treated with the inhibitor of the auxin efflux carrier *NPA* (-O-). (■) are controls not treated with NPA. At the *second arrow*, auxin was applied to both groups. The typical auxin-induced growth stimulation occurs after a short lag phase. After the maximal growth rate was attained, the auxin was removed from the incubation solution. In the untreated controls growth quickly dropped to control values. However, growth of the NPA-treated coleoptiles was sustained in the absence of external auxin. It seems that intracellular auxin, trapped by the blockade of the auxin efflux carrier, is sufficient to control growth. These results indicate the relevance of an intracellular receptor relevant for growth control. **B** Result of an equivalent experiment involving a pulse treatment with auxin of only 8 min. The auxin was washed off during the lag phase. Despite the short duration of the treatment the first peak of auxin action is displayed. Time courses of growth in NPA-treated and untreated coleoptiles do not vary. This result could indicate that in the initial phases of auxin action, an extracellular receptor is involved in the very early phases of auxin action, or that the decay of intracellular auxin is determined by auxin metabolism in such short-term treatments

auxin-induced promoter, with D16 antibodies. D16-F_{ab} significantly inhibited both endogenous and auxin-induced gene activation after 5 h. They also reduced the uptake of 2,4-D to the cells, indicating that D16 antibodies, additionally to their activity on ABP1, interfere with the auxin uptake carrier. Thus, D16 antibodies had no auxin agonist activity in this system. However, the same workers could confirm hyperpolarization in protoplasts induced by D16 antibodies. These results are also in line with the idea that in the very early time window of the electrophysiological measurements, the signal may be perceived by extracellular ABP1, but subsequently, control rapidly shifts to an intracellular binding site. It could also be envisioned that ABP1 and the auxin uptake carrier act as a functional unit. ABP1, by binding to AUX1, could add the auxin selectivity to a rather unspecific uptake carrier for aromatic substances. The D16 antibody, binding to the auxin-binding pocket of ABP1, would disturb this pathway of auxin uptake.

3 Auxin Signal Transduction

Auxin signal transduction is still poorly understood. It has been suggested that a heterometric G protein transmits the signal further downstream (e.g. Scherer 1992). Heterotrimeric G proteins are, in fact, present in higher plants (for a review see Millner and Causier 1996); however, no G protein coupled receptor has been up to now cloned from plants. Ettlinger and Lehle (1988) and Zbell (1988) observed an increase in IP_3 level when carrot cells were treated with auxin. This suggests the involvement of phospholipase C (PLC) in auxin signalling, which would generate diacylglycerol (DAG) and inositol-3-phosphate (IP_3) as second messengers. However, the magnitude of these responses is very low compared to animal systems.

Therefore, it has been suggested that phospholipase A (PLA), not PLC is the target of auxin signalling. Lysolipids would then play a role as second messengers. Inhibitors of PLA abolish auxin-induced proton secretion (Yi et al. 1996) and growth (Scherer and Arnold 1997). Lysolipids like lysophosphatidylcholin (LPC) stimulate in vitro ATPase activity and proton transport (Palmgren and Somarin 1989; Lanfermeijer and Prins 1994; De Michelis et al. 1997). However, in vitro assays of this kind are difficult to interpret, because many detergents also act stimulatory on H^+-ATPase, and LPC has detergent-like activity.

Like most signalling pathways, auxin action seems to involve protein phosphorylation. Kinase inhibitors like staurosporine block auxin-induced acidification, but do not impair proton efflux triggered by the fungal toxin FC (Yi et al. 1996). Fusicoccin activates the ATPase via a very different mechanism involving a 14-3-3 protein as FC receptor (Oecking et al. 1994; de Boer 1997; Jahn et al. 1997) or as signal transducer (Aducci et al. 1995). In vitro phosphorylation of a number of proteins is stimulated in auxin-treated maize mesocotyls (Kato et al. 1996). The kinase activity responds to calmodulin inhibitors, suggesting the involvement of a calmodulin-dependent protein kinase. Mitogen-activated protein (MAP-) kinases have also been suggested to be involved in auxin action (e.g. Mizoguchi et al. 1994). Evidence for this comes from studies using cell cultures and may therefore be relevant for auxin-induced modulation of cell division rather than for the control of elongation growth (for a review on the possible importance of MAP kinase cascades in ethylene and cytokinin signalling and in stress responses refer to Machida et al. 1997).

It is well possible that surprising new evidence will soon come up giving a new direction to the auxin-signalling research. For instance, very recently Ichikawa et al. (1997) isolated an auxin-resistant mutant *axi 141*. A protoplast culture of *Arabidopsis* proliferates in the abscence of the hormone. The *AXI 141* gene has been sequenced and has certain homologies to yeast adenylate cyclase. AXI 141 can complement yeast mutant defective in adenylate cyclase. This evidence tentatively suggests that cAMP may serve as a second messenger in the auxin signalling cascade leading to cell division. This idea, is however, at odds with conventional wisdom suggesting that cAMP in plants, if it occurs at all, is present only in very low, unphysiological concentrations and, therefore, cannot be a potent second messenger. Since *axi 141* plants show a normal phenotype, it may well be that the AXI 141 gene product is not involved in the control of cell elongation.

4 Growth-Relevant Targets of Auxin Signalling

a) Plasma Membrane Proton Pump: ATPase Stimulation or Induction?

It is well established that medium acidification is strongly increased by auxin. Proton pumping is – at least primarily – brought about by the plasma membrane ATPase. Two effects can potentially be responsible for auxin-induced increase of proton pumping: stimulation of ATPase and induction of a rapid de novo synthesis of ATPase. Both possibilities have been experimentally investigated. Rück et al. (1993) could measure by patch clamp methods an ATP-dependent transmembrane current in tobacco protoplasts, which they interpreted as the activity of the plasma membrane proton ATPase. Adding auxin to these protoplasts instantly stimulated this activity. It is well known that the ATPase can be stimulated by LPC, which may link these data to the signalling pathway discussed above. In isolated PM vesicles, stimulation of ATPase by auxin has been detected by several workers. However, this effect may be due to a shift in the K_M of the ATPase, and its physiological significance critically depends on the assumed in vivo ATP concentration.

The alternative model de novo synthesis of ATPase, can easily explain several physiological data obtained with intact coleoptiles and hypocotyls. Auxin-induced acidification requires a lag phase of about 10–15 min, which may be the time needed for synthesis and transport of the ATPase via the membrane flow. It can be rapidly prevented by inhibitors of protein biosynthesis like cycloheximide (CHI; Böttger et al. 1984; Frias et al. 1996). The Hager group could immunologically demonstrate the appearance of de novo-synthesized ATPase at the plasma membranes of maize coleoptiles after about 10 min (Hager et al. 1991). The ATPase level could be strongly suppressed by treatment with CHI, suggesting that the de novo-synthesized ATPase has a very short half-lifetime of about 12 min. This fits nicely with the strong dependence of growth (Cleland 1971; Edelmann and Schopfer 1989) on ongoing protein synthesis. Inhibitor studies suggest that the hypothetical "growth-limiting protein" has a half-lifetime of 12–15 min (Cleland 1971).

Immunological detection of ATPase involves several complications: (1) Auxin may stimulate only a certain isoform of ATPase, whereas the various anti-ATPase antibodies may have a broader range of specificity. (2) Most of the ATPase detected in PM preparations originates from the phloem and from stomata (Villalba et al. 1991), and this ATPase seems not to be induced by auxin. These problems may be the reason why some groups could not confirm auxin-induced synthesis of ATPase (Jahn et al. 1996). It remains unclear, however, why Hager et al. (1991) found a fivefold auxin-induced increase of ATPase protein in intact coleoptiles, whereas Frias et al. (1996) could detect only an 1.7-fold induction of the ATPase stimulation in segments from which the vascular tissues had been removed. In the examples where de novo-synthesized ATPase could be detected, the data leave the question open to what degree the newly produced isoforms contribute to auxin-induced medium acidification and growth. Frias et al. (1996) investigated the auxin-induced expression of a major ATPase isoform gene, *MHA2*, in maize coleoptiles. They found that on both the mRNA and protein level, auxin induced the gene only in non-vascular tissues, whereas the hormone did not affect vascular ATPase expression and synthesis. It is still not clear, however, to what degree auxin-induced ATPase de novo synthesis increases cell wall acidification. Frias et al. (1996) suggest that the auxin-induced acidification response is stronger in the non-vascular tissues, but these data are very difficult to interpret. Future research will show if the rather small increase of ATPase on the protein level by auxin (factor 1.7 in non-vascular tissue) can account for the auxin-induced increase in proton pumping.

b) Channels

Most studies assumed that auxin-induced medium acidification is always an indication of stimulated ATPase. This basic idea is, however, not necessarily correct. Alternative to the conventional Brønsted acid-base theory, pH can be described as a function of the differences in the concentration of strong ions (SID). Any transmembrane transport of strong ions will change SID and, therefore, pH (Stewart 1983). Stimulation of the proton pump only evokes a significant drop in pH if it is balanced by a charge compensating flux of strong ions. K^+ is the major counterion to balance the proton flux by the ATPase. A pH drop may be due to a stimulation of ATPase or, alternatively, may be brought about by an opening of potassium channels. As early as 1973, Haschke and Lüttge could show that auxin-induced medium acidification requires the presence of K^+ ions. However, several workers could detect auxin-induced growth in distilled water or weak NaCl solutions (Terry and Jones 1981; Kutschera and Schopfer 1985). Thus, it was concluded that acidification of the cell wall is an independent side effect and not causal for growth stimulation (e.g. Kutschera 1993). Recently, Claussen et al. (1997) could demonstrate that auxin-induced growth in fact depends very strictly on potassium, if the experiment is carried out in solutions with divalent cations (Fig. 5A). Under these conditions, auxin-induced growth could be reversibly abolished by treatments with potassium channel blockers like tetraethylammonium (TEA; Fig. 5B), Ba^{2+} (Claussen et al. 1997) and Cs^+ (M. Claussen and H. Lüthen, unpubl.). Probably, potassium ions bound to fixed charges in the Donnan free space of the cell wall mask the potassium dependency in solutions of low ionic strength.

Although these data clearly demonstrate that the activity of potassium channels is a necessary condition for auxin-induced growth, they do not really prove that auxin-induced growth is regulated by K^+ channels. Blatt and Thiel (1993) demonstrated in a voltage-clamp study with guard cells that IAA or NAA stimulated the inward-rectifying K^+ current by a factor of 1.5 to 2 in a concentration range between 0.1 and 10 μM. At higher IAA concentrations the inward current was inhibited, and the outward-rectifying current was stimulated. The effects of higher concentration of auxin can be mimicked by C-terminal peptides of ABP1 (Thiel et al. 1993, see above). Unfortunately, patch clamp studies of maize coleoptile protoplasts did not reveal any potassium channel opening induced by auxin (R. Hedrich, pers. comm.).

As with the ATPase, the total ion flux into the cell can be also enhanced by increasing the number of channels in the PM. Ongoing studies in Hedrich's and our laboratories are meanwhile exploring whether the genes of K^+ channels are activated by auxin. Preliminary data

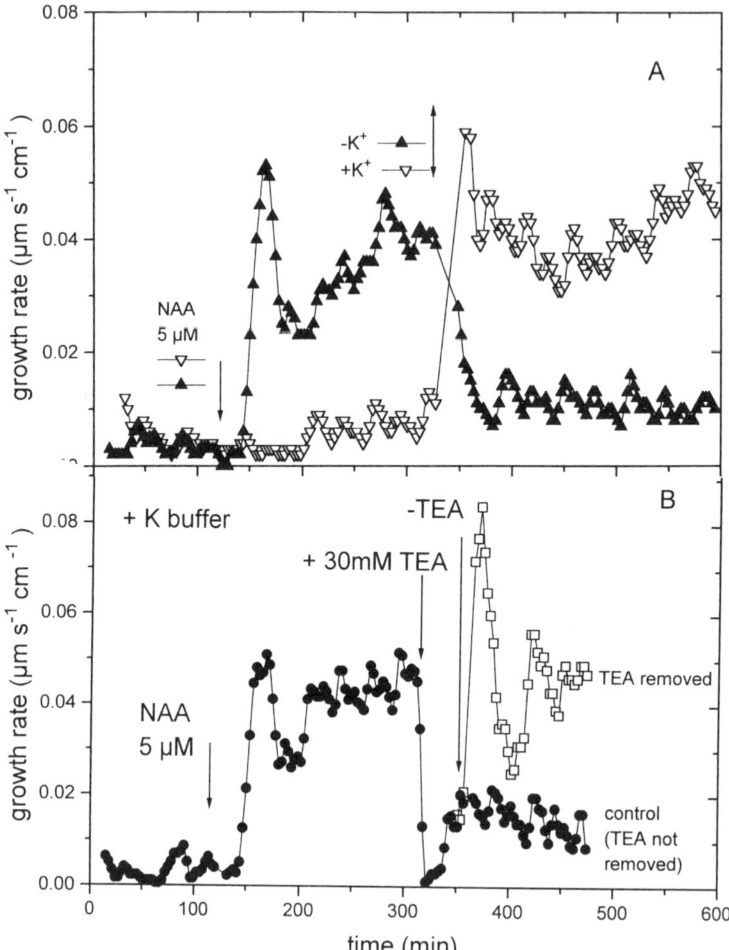

Fig. 5A, B. Demonstration of the dependence of auxin-induced growth on the activity of K⁺ channels. **A** The auxin NAA only induces growth in K⁺-containing buffers (▲) but not in K⁺-free media (∇, -K⁺). It is possible to induce growth without any lag phase under – K-conditions by adding K⁺ (∇) or to stop growth by removing K⁺ und +K conditions (▲). **B** The potassium channel blocker inhibits auxin-induced growth. Removal of TEA (□) causes a rapid recovery, demonstrating reversibility of the action of TEA. (Claussen et al. 1997)

suggest that K⁺ channel gene is induced after 30–45 min of auxin treatment. Although this effect cannot explain the very early phases of growth stimulation, it can easily account for the osmoregulation necessary for sustained auxin-induced growth. Osmoregulation by potassium channels is, for instance, clearly shown in protoplasts, which undergo swelling when treated with auxin. This response is inhibited by potassium channel blockers like TEA (Keller and VanVolkenburgh 1996).

c) Rapid Gene Induction by Auxin

A number of genes are induced by auxin. This topic has been thoroughly reviewed (Abel and Theologis 1996; Sitbon and Perrot-Rechenmann 1997) recently. For the understanding of auxin-induced growth, genes that are very rapidly expressed after auxin addition are most promising. Small auxin upregulated (*SAUR*) mRNAs show up only 3–5 min after treatment (McClure et al. 1989, Yamamoto et al. 1992; Yamamoto 1994). Unfortunately, their function is as yet unknown. The family of *Aux/IAA* genes seems to encode for nuclear proteins which are potential transcription factors. They are also rapidly upregulated. *Aux/IAA* mRNAs have been detected within 5–60 min of the presence of the auxin signal.

A number of rapid genes have known functions. It has long been known that IAA treatment in a variety of tissues triggers ethylene biosynthesis – a beautiful example for hormonal interaction in plants. It was not a big surprise that some auxin-induced genes code for 1-aminocyclopropane-1-carboxylil acid (ACC)-synthase. Another class of auxin-induced genes, the *GH2/4*-like genes, code for glutathione-S-transferase (Hagen et al. 1988; Takahashi and Nagata 1992; Guilfoyle et al. 1993; Bilang and Sturm 1995). The corresponding mRNAs appear after 15–30 min. Gutathione-S-transferase plays a key role in the regulation of the cellular redox state and in detoxification mechanisms. Most thrilling is the fact that this cytosolic enzyme binds auxin and can be photoaffinity-labelled by Azido-IAA (Bilang et al. 1993; Bilang and Sturm 1995).

One problem in the identification of auxin-induced genes is the design of control experiments. It has been shown, for instance, that *GH2/4*-like genes are not induced by abscisic acid (ABA), Cu^{2+}, gibberellic acid (GA) and tryptophan. For *SAUR* genes, it has been checked that their expression is not triggered by ABA, 1-aminocyclopropane-1-carboxylic acid, BA (benzyladenine), GA and tryptophan. More relevant controls for auxin specifity would be the inactivity of 2,3-D or 2-NAA.

The recent demonstration of induction of potassium channels, proton ATPase genes, has been mentioned above.

5 Understanding Wall Loosening

a) Acid-Induced Wall Loosening

The acid growth theory was first proposed by Ulrich Ruge in 1937. His model of auxin-induced growth in *Helianthus* was very progressive. Since ion pumps were still unknown, Ruge explained the auxin-induced acidification by an accumulation of the free acid IAA in the apoplast. The pH drop would then increase cell-wall extensibility, and the turgor pressure then would cause an increase in cell volume, which results in elongation growth. In the subsequent phases, Ruge envisioned other processes like cell-wall synthesis and osmoregulation as more important than acid growth (for more details see Lüthen 1995). The theory was forgotten for decades until it was rediscovered in the 1960s by A. Hager. This work led to the first formulation of this theory in its modern form in 1971 in a pioneering paper (Hager et al. 1971).

Briefly, the theory suggests that auxin-induced cell wall acidification is the cause of auxin-induced wall loosening. It links the membrane activities discussed above to cell-wall loosening and growth control. Like other classical theories in plant physiology (e.g. the Cholodny-Went theory), it survived several challenges (Kutschera and Schopfer 1985; Kutschera 1993), but was never completely disproved.

Doubts have been raised whether the acidification induced by IAA is sufficient for triggering a significant cell-wall loosening. All investigations of the problem up to now measured not the true apoplasmic pH, but the pH of the incubation solution. There were vast differences in the results: Kutschera and Schopfer, in their 1985 paper, reported a pH of 6.5–6 with maize coleoptiles, that dropped to pH 5.5 in the presence of auxin. An acidification of the incubation medium to pH 5.5 did not induce growth in the absence of auxin. Lüthen et al. (1990) found pH of 4.8 without auxin and a pH of 4.1–4.2 with auxin. This pH is able to induce significant growth, about 50% of the value obtained with IAA. Although the result does not prove the exclusive role of protons as cell-wall-loosening factor, it indicates a significant participation of cell-wall acidification in cell-wall loosening. Peters and Felle (1991a, b) could show that in order to obtain equilibrium pH measurements that reflect changes in apoplasmic pH, certain criteria have to be met. Especially a minimal tissue-to-volume ratio has to be used. If this was the case, pH fell to 4.1 in the presence of IAA. The temporal correlation of acidification and growth response was excellent in cases where both parameters were measured simultaneously (Peters et al. 1998). Summarizing the available evidence, it appears that the acid growth theory explains a significant fraction of auxin-induced growth. Many "tests" of the theory assumed that "acid-growth" and "auxin-induced growth" may be additive. Instead, it may be that IAA, besides cell-wall acidification, triggers an additional pH-dependent wall-loosening process, thereby enhancing the sensitivity of the wall to protons. If this is true, many arguments brought forward for or against the theory are missing the point.

In the 1970s, the acid-growth theory was developed as an alternative to the gene-activation hypothesis, since at that time it could not be envisioned that a protein can be synthesized and secreted to the cell wall within 10–20 min. Today, after the finding of de novo synthesis of channels and ATPases in response to auxin and the detection of other "early" auxin-induced genes, the gene-activation and acid-growth theories of auxin action are no longer contradictory.

Progress has been made in the understanding of the cell-wall-loosening process itself. For many years, cell-wall biochemists have failed to exactly pinpoint the cell-wall-loosening (hydrolytic) enzyme that is stimulated by acidic pH.

Cosgrove and coworkers found that boiling cell walls abolishes their ability to undergo acid-induced loosening, probably because of denaturation of a wall-loosening protein. However, a crude cell-wall protein fraction, when added to boiled walls, reconstituted the sensitivity to acid treatment (McQueen-Mason et al. 1992; Li et al. 1993; Cosgrove and Li 1993). The corresponding proteins were isolated and sequenced. It became apparent that these "expansins" were members of a large gene family. Meanwhile, it is obvious that expansins also modulate cell-wall changes in fruit ripening (Rose et al. 1997). Recently, Fleming et al. (1997) could induce the formation of leaf primordia by locally applying expansins. This finding underlines the significance of localized cell-wall-loosening events in the control of morphogenesis and development.

Cell-wall loosening by expansins is not due to a hydrolytic activity, but caused by a removal of hydrogen bridging bonds between cell-wall polymers (McQueen-Mason and Cosgrove 1994). This may solve a puzzle: while dicot and monocot cell walls vastly differ in composition and structure (for a review see Carpita and Gibeaut 1993), cell-wall loosening in response to auxin and pH is very similar in both groups. In fact, dicot expansins can loosen monocot cell walls. McQueen-Mason and Cosgrove (1994) could even demonstrate acid-induced loosening of Whatman filter paper mediated by expansins. Recently, it was found that one group of expansin genes is induced 60 min after auxin treatment (D. J. Cosgrove, pers. comm.).

b)   Other Possible Wall-Loosening Mechanisms

It has to be mentioned that apart from expansin activity, a number of processes have been related to auxin- or acid-induced cell-wall loosening. Masuda and coworkers were able to inhibit auxin-induced elongation and wall loosening with antibodies against various cell-wall polysaccharides like glucans (Hoson et al. 1992) or xyloglucans (Hoson et al. 1991). Recently, a slight inhibition of auxin-induced growth by anti-β-1,3-endoglucanase antibodies was reported (Mutaftschiev et al. 1997). In some cases, auxins caused a long-term increase in cellulase activity (Hayashi et al. 1984; Hayashi 1989; Fry 1989). Some groups consider xyloglucan endotransglycolase (XET) as a key player in auxin-induced wall loosening (Fry et al. 1992; Hetherington and Fry 1993; Hoson et al. 1993). By transglycolysation with an oligosaccharide, XET has the potential to act as a wall-loosening enzyme. Transglycolyzation with a polysaccharide can antagonize this loosening process, which makes the idea of XET as a regulator of cell wall loosening attractive. Since xyloglucans form hydrogen bridges to cellulose fibrils, this concept is not at odds with the idea that expansins modulate wall loosening.

6 Roots: Auxin Research "Down Under"

Most physiological work on auxin signalling has up to now been restricted to coleoptile and hypocotyl segments. These systems have a major advantage: the increase in length is due to cell elongation – cell division is not involved. Roots display very high rates of cell elongation restricted to their very small elongatin zone – in primary roots 3–8 mm from the root apex. Thus, the supply of fresh cells from the meristematic zone may control root elongation, but only on quite long time scales. In the root system, auxin has an inhibitory effect on root growth. These experimental drawbacks may have led to the situation that many basic facts of root auxin physiology are still not established. Thus, it is adequate to begin this section by separating fact and fiction.

There are reports that root growth is stimulated by very low (e.g. 10^{-10} M IAA) auxin concentrations, harking back to Thiman (1937). It is said that the optimum of the bell-shaped dose-response curve is shifted to lower concentrations in roots. Thus, concentrations which are stimulatory in shoots and coleoptiles are overoptimal – and therefore inhibitory – in roots. Since these graphs look so beautiful, they are represented in every botany textbook. However, serious doubts about this pretty picture have to be raised. (1) In all reported cases, stimulation of root growth is a very subtle effect, or even statistically insignificant (Åberg 1957). It is not comparable with the rapid and dramatic growth responses in shoots. There are reports that pretreatment with ethylene biosynthesis blockers is required to see the stimulation (Mulkey et al. 1982), but these could not be confirmed in our laboratory (Lüthen and Böttger 1988). (2) The inhibition of root growth is an instantaneous effect occurring after a short lag phase of some minutes (Evans 1976; Lüthen and Böttger 1993). In contrast, inhibitory high auxin concentrations transiently *stimulate* shoot growth in the first hour after auxin application. After that period, shoots irreversibly cease growing because of cell damage. In 24 h segment biotests, these effects add up. The result is in an apparent inhibition (Cleland 1972), which does not reflect rapid auxin action. Thus, the inhibitory branch of the dose-response curves in roots and coleoptiles are not homologues. (3) The dose-response curves of the root inhibition and stimulation in *abraded* coleoptiles are nearly identical – only the effects are in opposite directions (H. Lüthen and M.

Clausen, unpubl.). Thus, it seems that auxin stimulates shoot growth, but inhibits root cell elongation.

The mechanism of the inhibition of root growth is still unresolved. It has been shown that auxin inhibits proton pumping (Evans and Vesper 1980; Lüthen and Böttger 1988, 1993). This response is precisely to the elongation zone (Lüthen and Böttger 1988). The inhibition of proton pumping may lead to an apoplasmic alkalinization, to cell-wall stiffening, and thus to growth inhibition (e.g. Evans and Vesper 1980; Mulkey et al. 1982). However, again, some facts stand against this type of "upside-down version" of the acid-growth theory: The inhibition cannot be reversed by acidic solutions or by FC. Acid growth – at least in maize roots – is a small and short-lived phenomenon – if it occurs at all. Unpublished data from our laboratory indicate that sensitivity to acid-induced root growth differs considerably between varieties – but that all display the auxin-induced growth inhibition. Thus, it seems that factors other than the apoplasmic pH are causal for an irreversible cell-wall stiffening. Oxidative cross-linking may be a good candidate for such a mechanism. In any case, our group could recently establish that auxin, in fact, decreases cell-wall extensibility in maize root elongation zones (Büntemeyer et al. 1998). This response was also independent of pH.

7 Perspectives in Auxin Research

The "auxin problem", i.e. the question of how auxin causes the dramatic increase in growth rate, is a classical problem in plant physiology. Figure 6 may serve as a summary model that incorporates much of the evidence discussed above. Unfortunately, such a model still has to be very vague and partly speculative. Other auxin physiologists may focus on different aspects. Although auxin research is as old as photosynthesis research, the latter – right from the start – penetrated the molecular mechanisms much more deeply. Thus, one may be disappointed in the current state of auxin research. The question "What is wrong with auxin research?" was first posted by Burström (1974). The basic problem he spotted was the step from in vivo to in vitro research: "Our bad luck is that growth cannot be studied in a test tube, nor can we – as photosynthesis scientists do – fractionate cells and pick out bodies where our process takes place. As soon as a cell is disintegrated we lose control of whether that which we are studying leads to growth or not."

Almost a quarter of a century later, Bruström's problem is still manifest in the state of our knowledge of all stages of the auxin-signalling chain: loosing the growth in the step from in vivo to in vitro begins even in protoplast preparation – as we saw on the ABP1 receptor question.

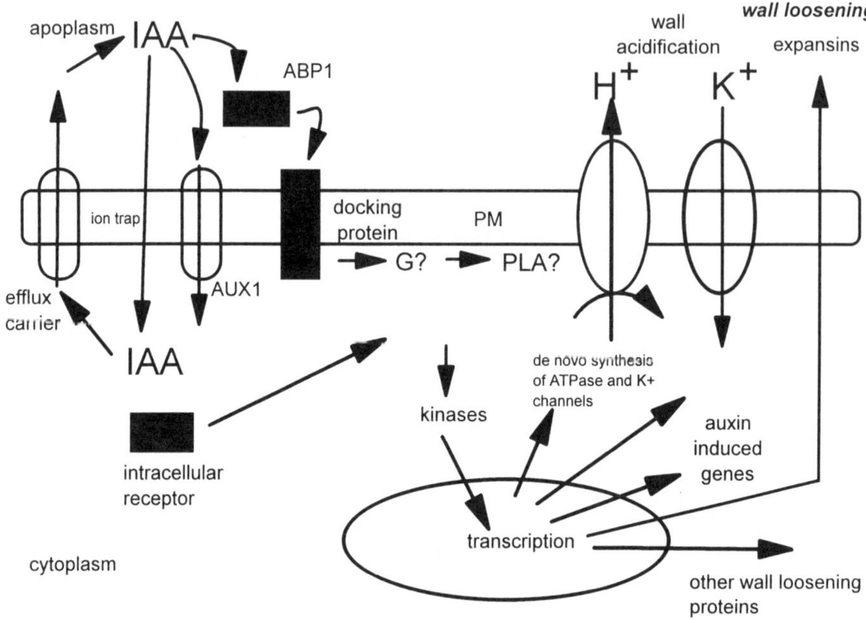

Fig. 6. Outline of a hypothetical model of auxin action on elongation growth. IAA is taken up both via an auxin influx carrier (*AUX1*) and an ion-trap mechanism. It can be exported by an export carrier. Extracellular IAA is sensed by *ABP1*. Additionally, there appears to be an intracellular receptor, which may be ABP1 in the ER or a different intracellular protein. Circumstantial evidence suggest that G proteins and phospholipase A (PLA) are included in a signalling chain, which leads to a stimulation of ATPase. It can be envisioned that a kinase cascade leads to the expression of several genes. The nature of some of them is not very well known, but it seems that both ATPase, K^+ channels, and some expansins are induced by auxin. The concerted activity of K^+ channels and H^+-ATPase leads to a pH drop in the apoplast, which is a critical step in cell-wall loosening. Acid-induced cell-wall loosening depends on the activity of expansins, but other wall-loosening enzymes and cell-wall biosynthesis are certainly necessary for sustained cell elongation

In future, molecular biology can build a bridge between physiology and biochemistry. Mutants of possible key players in the auxin-signalling pathways are becoming more and more available. Auxin physiologists should use imaging techniques to measure growth in *Arabidopsis*, and molecular biologists are already beginning to produce mutants in classical coleoptile plants like maize, and characterizing their growth physiology will allow a deep penetration of the auxin-signalling cascade leading to growth. The number of promising target genes is constantly rising in *Arabidopsis* (for an overview, see Leyser 1997). There is a very interesting auxin-insensitive mutant diageotropica (*dgt*) in tomato, whose molecular biology has not yet been fully elucidated (Zobel 1973; Hicks et al. 1989; Muday et al. 1995; C. Coenen et al., unpubl.). Genes coding for auxin transporters, for putative auxin receptors or for potassium channels have been identified. Is the solution of Burström's problem perhaps next door? Caution is needed: ever since auxin was identified, the hopes for a rapid explanation of its mechanisms of action ran high; but chances are excellent that this time the never-ending story will come to an end.

Note Added in Proof. After submission of the manuscript for the present review, Ichikawa et al. retracted the data presented in their publication. There is no longer any evidence for the involvement of cAMP in auxin singnalling.

Acknowledgements. We thank B. Steffens for critical reading of the manuscript, Winfried Peters and Thomas Jahn for some valuable comments, and K. Philippar, R. Hedrich and C. Coenen for making some of their unpublished data available to us.

References

Abel S, Theologis A (1996) Early genes and auxin action. Plant Physiol 111:9–17

Åberg B (1957) Auxin relations in roots. Annu Rev Plant Physiol 8:153–180

Aducci P, Marra M, Fogliano V, Fullone MR (1995) Fusioccin receptors: perception and transduction of the fusicoccin signal. J Exp Bot 46:1463–1478

Anai T, Takai R, Miyata M, Uchida H, Kosemura S, Yamamura S, Ishizaki R, Hasegawa K (1997) Isolation and characterization of an auxin-binding protein gene from radish, and its expression in insect cells. Physiol Plant 101:606–611

Barbier-Brygoo H, Ephritikhine G, Klämbt D, Ghislain M, Guern J (1989) Functional evidence for an auxin receptor at the plasmalemma of tobacco mesophyll protoplasts. Proc Natl Acad Sci USA 86:891–895

Barbier-Brygoo H, Ephritikhine G, Klämbt D, Maurel C, Palme K, Schell J, Guern J (1991) Perception of the auxin signal at the plasma membrane of tobacco mesophyll proto-plasts. Plant J 1:83–93

Bennett MJ, Marchant A, Green HG, May ST, Ward SP, Millner PA, Walker AR, Schulz B, Feldmann KA (1996) *Arabidopsis AUX1* Gene: a permease-like regulator of root gravitropism. Science 273:948–950

Bilang J, Sturm A (1995) Cloning and characterization of a glutathione S-transferase that can be photolabeled with 5-azido-indole-3-acetic acid. Plant Physiol 109:243–260

Bilang J, MacDonald H, King PJ, Sturm A (1993) A soluble auxin-binding protein from *Hyoscyamus muticus* is a glutathione S-transferase. Plant Physiol 102:29–34

Blatt MR, Thiel G (1993) K^+ channels of stomatal guard cells: bimodal control of the K^+ inward rectifier by auxin. Plant J 5:55–68

Boot CJM, van Duijin B, Mennes AM, Libbenga KR (1996) Regulation of a class of auxin-induced genes in cell suspension cultures from *Nicotiana tabacum*. In: Smith AR, Berry AW, Harpham MV, Moshkov IE, Novikova GV, Kulaeva ON, Hall MA (eds) Plant hormone signal transduction. Kluwer, Dordrecht, pp 41–48

Böttger M, Engvild KC, Soll H (1978) Growth of *Avena* coleoptiles and pH drop of pro-toplast suspension induced by chlorinated indoleacetic acids. Planta 140:89–92

Böttger M, Soll HJ, Bigdon M (1984) Net proton transport in sunflower hypocotyls: com-parative studies of inhibitors. Z Pflanzenphysiol 114:467–475

Büntemeyer K, Lüthen H, Böttger M (1998) Auxin-induced changes in plant cell wall extensibility of maize roots. Planta 204:515–519

Burström HG (1974) What is wrong with auxin research? In: Sumiki Y (ed) Plant growth substances 1973. Hirokawa, Tokyo, pp 789–797

Carpita NC, Gibeaut DM (1993) Structural models of primary cell walls in flowering plants. Consistency of molecular structure with the physical properties of the wall during growth. Plant J 3:1–30

Claussen M (1997) Untersuchungen zur Regulation des auxininduzierten Zellstreckungswachstums und zur Perzeption des Auxinsignals. PhD Thesis, University of Hamburg

Claussen M, Lüthen H, Böttger M (1996) Inside or outside? Localization of the receptor relevant to auxin-induced growth. Physiol Plant 98:861–867

Claussen M, Lüthen H, Blatt M, Böttger M (1997) Auxin-induced growth and its linkage to potassium channels. Planta 201:227–234

Cleland R (1971) Instability of the growth-limiting proteins of the *Avena* coleoptile and their pool size in relation to auxin. Planta 99:1–11

Cleland (1972) The dosage response curve for auxin-induced cell elongation. A reevaluation. Planta 104:1–9

Cosgrove DJ, Li ZC (1993) Role of expansins in cell enlargement of oat coleoptiles: analysis of developmental gradients and photocontrol. Plant Physiol 103:1321–1328

Davies PJ (1974) The uptake and elution of indoleacetic acid by pea stem sections in relation to auxin-induced growth. In: Sumiki Y (ed) Plant growth substances 1973. Hirokawa, Tokyo, pp 767–778

De Boer B (1997) Fusicoccin – a key to multiple 14-3-3-locks? Trends Plant Sci 2:60–66

De Michelis MI, Papini P, Pugliarello MC (1997) Multiple effects of lysophosphatidylcholine on the activity of the plasma membrane H^+-ATPase of radish seedlings. Bot Acta 110:43–48

Dennecke J, De Rycke R, Botterman J (1992) Plant and mammalian sorting systems for protein retention in the endoplasmic reticulum contain a conserved epitope. EMBO J 11:2345–2355

Diekmann W, Venis MR, Robinson DG (1995) Auxins induce clustering of the auxin-binding protein at the surface of maize coleoptile protoplasts. Proc Natl Acad Sci USA 92:3425–3429

Dohrmann U, Hertel R, Kowalik H (1978) Properties of auxin-binding sites in different subcellular fractions from maize coleoptiles. Planta 140:97–106

Edelmann H, Schopfer P (1989) Role of protein and RNA synthesis in the initiation of auxin-mediated growth in coleoptiles of *Zea mays* L. Planta 179:475–489

Ettlinger C, Lehle L (1988) Auxin induces rapid changes in phosphatidylinositol metabolites. Nature 331:176–177

Evans ML (1976) A new sensitive root auxanometer. Plant Physiol 58:599–601

Evans ML, Vesper MJ (1980) An improved method for detecting auxin-induced hydrogen ion efflux from corn coleoptile segments. Plant Physiol 66:561–565

Felle H, Peters W, Palme K (1991) The electrical response of maize to auxin. Biochim Biophys Acta 1064:199–204

Fischer C, Lüthen H, Böttger M, Hertel R (1992) Initial transient growth inhibition in maize coleoptiles following auxin application. J Plant Physiol 141:88–92

Fleming AJ, McQueen-Mason S, Mandel T, Kuhlemeier C (1997) Induction of leaf primordia by cell wall protein expansin. Science 276:1415–1418

Frias I, Caldeira MT, Perez-Castineira MT, Navarro-Avino JP, Culianez-Macia FA, Kuppinger O, Stransky H, Pages M, Hager A, Serrano R (1996) A major isoform of the maize plasma membrane H^+-ATPase: characterization and induction by auxin in coleoptiles. Plant Cell 8:1533–1544

Fry SC (1989) Cellulase, hemicellulase and auxin-stimulated growth: a possible relationship. Physiol Plant 75:532–536

Fry SC, Smith RC, Renwick KF, Martin DJ, Hodge SK, Matthews KJ (1992) Xyloglucan endotransglycolase, a new cell wall loosening enzyme from plants. Biochem J 282:821–828

Guilfoyle TJ, Hagen G, Li Y, Ulmasov T, Liu T, Strabala T, Gee M (1993) Auxin-regulated transcription. Aust J Plant Physiol 20:489–502

Hagen G, Uhrhammer N, Guilfoyle TJ (1988) Regulation of an auxin-induced soybean sequence by cadmium. J Biol Chem 263:6442–6446

Hager A, Menzel H, Krauss A (1971) Versuche und Hypothese zur Primärwirkung des Auxins beim Streckungswachstum. Planta 100:47–52

Hager A, Debus G, Edel HG, Stransky H, Serrano R (1991) Auxin induced exocytosis and a rapid synthesis of a high turnover pool of plasma membrane. ATPase. Planta 185:527–537

Haschke HP, Lüttge U (1973) β-Indolylessigsäure (IES)-abhängiger K⁺-H⁺-Austauschmechanismus und Streckungswachstum bei *Avena*-Koleoptilen. Z Naturforsch [C] 28:555–558

Hayashi T (1989) Xyloglucans in the primary cell wall. Annu Rev Plant Physiol Plant Mol Biol 40:139–168

Hayashi T, Wong YS, MacLachan G (1984) Pea xyloglucan and cellulose. II. Hydrolysis by pea endo-1,4-9-glucanases. Plant Physiol 75:605–610

Henderson J, Bauly JM, Ashford DA, Oliver SC, Hawes CR, Lazarus CM, Venis MA, Napier RM (1997) Retention of maize auxin-binding protein in the endoplasmic reticulum: quantifying escape and the role of auxin. Planta 202:313–323

Hertel R (1994) A critical view on proposed hormone action. The example of auxin. In: Smith CJ, Gallon J, Chiatante D, Zocchi G (eds) Biochemical mechanisms involved in plant growth regulation. Clarendon Press, Oxford, pp 1–15

Hertel R (1995) Auxin-binding protein is a red herring. J Exp Bot 46:461–462

Hesse T, Feldwisch J, Balshüsemann D, Bauw G, Puype M, Vandekerckhove J, Löbler M, Klämbt D, Schell J, Palme K (1989) Molecular cloning and structural analysis of a gene from *Zea mays* L. coding for a putative receptor for the plant hormone auxin. EMBO J 8:2453–2461

Hetherington PR; Fry SC (1993) Xyloglucan endotransglycosylase activity in carrot cell suspension during cell elongation and somatic embryogenesis. Plant Physiol 103:987–992

Hicks GR, Rayle DL, Lomax TL (1989) The diageotropica mutant of tomato lacks high-specific activity auxin binding sites. Science 245:4952–4954

Hoson T, Masuda Y, Sone Y, Misaki A (1991) Xyloglucan antibodies inhibit auxin-induced elongation and cell wall loosening of azuki bean epicotyls but not of oat coleoptiles. Plant Physiol 96:551–557

Hoson T, Masuda Y, Nevins DJ (1992) Comparison of the outer and inner epidermis: inhibition of auxin-induced elongation of maize coleoptiles by glucan antibodies. Plant Physiol 98:1298–1303

Hoson T, Sone Y, Misaki A, Masuda Y (1993) Role of xyloglucan breakdown in epidermal cell walls for auxin-induced elongation of azuki bean epicotyl segments. Physiol Plant 87:142–147

Ichikawa T, Suzuki Y, Czaja I, Schommer C, Lessnick A, Schell J, Walden R (1997) Identification and role of adenyl cyclase in auxin signalling in higher plants. Nature 390:698–701

Inohara N, Shimomura S, Fukui T, Futai M (1989) Auxin binding protein located in the endoplasmic reticulum of maize shoots. Molecular cloning and complete structure. Proc Natl Acad Sci USA 86:3564–3568

Jahn T, Johannson F, Lüthen H, Volkmann D, Larrson C (1996) Reinvestigation of auxin- and fusicoccin stimulation of the plasma membrane ATPase. Planta 199:359–365

Jahn T, Fuglsang AT, Olsson A, Brüntrup IM, Collinge DB, Volkmann D, Sommarin M, Palmgren MG, Larsson C (1997) The 14-3-3 protein interacts directly with the C-terminal region of the plant plasma membrane H⁺-ATPase. Plant Cell 9:1805–1814

Jones AM (1994) Auxin-binding proteins. Annu Rev Plant Physiol Plant Mol Biol 45:393–420

Jones AM, Herman EM (1993) KDEL containing auxin-binding protein is secreted to the plasma membrane and cell wall. Plant Physiol 101:595–606

Kato R, Takatsuna S, Wada T, Narihara Y, Suzuki T (1996) In vitro phosphorylation of proteins in IAA-treated mesocotyls in *Zea mays*. Plant Cell Physiol 37:667–672

Keller CP, Van Volkenburgh E (1996) Osmoregulation by oat coleoptile protoplasts. Plant Physiol 110:1007-1016

Klämbt D (1990) A view about the function of auxin-binding proteins at the plasma membrane. Plant Mol Biol 14:1045–1050

Kutschera U (1993) The current status of the acid growth hypothesis. New Phytol 126:549–569

Kutschera U, Schopfer P (1985) Evidence against the acid growth theory of auxin action. Planta 163:483–493

Lanfermeijer FC, Prins HBA (1994) Modulation of H^+-ATPase by fusicoccin in plasma membrane vesicles from oat (*Avent sativa* L.) roots. A comparison of modulation by fusicocin, trypsin and lysophosphatidylcholine. Plant Physiol 104:1277–1285

Leyser O (1997) Auxin: lessons from a mutant weed. Physiol Plant 100:407–414

Li ZC, Durachko DM, Cosgrove DJ (1993) An oat coleoptile wall protein that induces wall extension in vitro and that is antigenically related to a similar protein from cucumber hypocotyls. Planta 191:349–356

Löbler M, Klämbt D (1985) Auxin binding protein from coleoptile membranes of corn (*Zea mays* L.). I. Purification by immunological methods and characterization. J Biol Chem 260:9848–9853

Lüthen H (1995) Research on auxin action: a never-ending story. Early progress in the 1930s. Mitt Inst Allg Bot Hamburg 25:53–63

Lüthen H, Böttger M (1988) Kinetics of proton secretion and growth in maize roots. Plant Sci Lett 54:37–43

Lüthen H, Böttger M (1993) The role of protons in the auxin-induced root growth inhibition – a critical reexamination. Bot Acta 106:56–63

Lüthen H, Bigdon M, Böttger M (1990) Reexamination of the acid growth theory of auxin action. Plant Physiol 93:931–939

MacDonald H (1997) Auxin perception and signal transduction. Physiol Plant 100:423–433

MacDonald H, Henderson J, Napier RM, Venis MA, Hawes C, Lazarus CM (1994) Authentic processing and targeting of active maize auxin-binding protein in the baculovirus expression system Plant Physiol 105:1049–1057

Machida Y, Nishihama R, Kitakura S (1997) Progress in studies of plant homologs of mitogen-activated protein (MAP) kinase and potential upstream components in kinase cascades. Crit Rev Plant Sci 16:481–496

Maher EP, Martinsdale SJB (1980) Mutants of *Arabidopsis* with altered responses to auxin and gravity. Biochem Genet 18:1041–1053

McClure BA, Hagen G, Brown CS, Gee MA, Guilfoyle TJ (1989) Transcription, organization and sequence of an auxin-regulated gene cluster in soybean. Plant J 1:229–239

McQueen-Mason S, Cosgrove DJ (1994) Disruption of hydrogen bonding between wall polymers by proteins that induce plant wall extension. Proc Natl Acad Sci USA 91:6574–6578

McQueen-Mason S, Durachko DM, Cosgrove DJ (1992) Two endogenous proteins that induce cell wall extension in plants. Plant Cell 4:1425–1433

Millner PA, Causier BE (1996) G-protein coupled receptors in plant cells. J Exp Bot 47:983–992

Mizoguchi T, Gotoh Y, Nishida E, Yamaguchi-Shinozaki K, Hayashida N, Iwasaki T, Kamada H, Shinozaki K (1994) Characterization of two cDNAs that encode for MAP kinase homologues in *Arabidopsis thaliana* and analysis of the possible role of auxin in activating such kinases activities in cultured cells. Plant J 5:111-122

Muday GK, Lomax TL, Rayle DL (1995) Characteristics of the growth and auxin physiology of roots of the tomato mutant diageotropica. Planta 195:548–553

Mulkey TJ, Kuzmanoff K, Evans ML (1982) Promotion of growth and hydrogen ion efflux in roots of maize pretreated with ethylene biosynthesis inhibitors. Plant Physiol 70:186–188

Munro S, Pelham HRB (1987) A C-terminal signal prevents secretion of luminal ER proteins. Cell 48:899–907

Mutaftschiev S, Prat R, Pierron M, Devilliers G, Goldberg R (1997) Relationships between cell wall β-1,3-endoglucanase activity and auxin-induced elongation in mung bean hypocotyl segments. Protoplasma 199:49–56

Napier RM (1995) Towards an understanding of ABP1. J Exp Bot 46:1787–1795

Napier RM, Venis MA (1990) Monoclonal antibodies detect an auxin-induced conformational change in the maize auxin-binding protein. Planta 182:313–318

Oecking C, Eckerson C, Weiler EW (1994) The fusicoccin receptor of plants is a member of the 14-3-3 protein superfamily of eukaryotic regulatory proteins. FEBS Lett 352:163–166

Palme K, Hesse T, Campos N, Garbers C, Yanofsy C, Schell J (1992) Molecular analysis of an auxin binding protein gene located on chromosome 4 of *Arabidopsis*. Plant Cell 4:193–201

Palmgren MG; Somarin M (1989) Lysophosphatidylchloline stimulates ATP-dependent proton accumulation in isolated oat root plasma membrane vesicles. Plant Physiol 90:1009–1014

Peters W, Felle H (1991a) Control of apoplastic pH in corn coleoptile segments. I. The endogenous regulation of cell wall pH. J Plant Physiol 137:655–661

Peters W, Felle H (1991b) Control of apoplastic pH in corn coleoptile segments. II. The effect of various auxins and auxin analogues. J Plant Physiol 137:691–696

Peters WS, Lüthen H, Böttger M, Felle H (1998) The temporal correlation of changes in apoplast pH and growth rate in maize coleoptile segments. Aust J Plant Physiol (in press)

Pickett FB, Wilson AK; Estelle M (1990) The *aux1* mutation of *Arabidopsis* confers both auxin and ethylene resistance. Plant Physiol 94:1462–1466

Rescher U, Walther A, Schiebl C, Klämbt D (1996) In vitro binding affinities of 4-chloro, 2-methyl, and 4-ehtylindoleacetic acid to auxin binding protein 1 (ABP1) correlate with their growth-stimulating activities. Plant Growth Regul 15:1–3

Rose JKC, Lee HH, Bennett AB (1997) Expression of a divergent expansin gene is fruit-specific and ripening-regulated. Proc Natl Acad Sci USA 94:5955–5960

Rück A, Palme K, Venis MA, Napier RM, Felle HH (1993) Patch clamp analysis establishes a role for an auxin-binding protein in the auxin stimulation of plasma membrane current in *Zea mays* protoplasts. Plant J 4:41–46

Ruge U (1973) Untersuchungen über den Einfluß des Heteroauxins auf das Streckungswachstum des Hypokotyls von *Helianthus annuus*. Z Bot 31:1–56

Scherer GFE (1992) Stimulation of growth and phospholipase A2 by the peptides mastoparan and mellitin and by the auxin 2,4-dichlorphenoxyacetic acid. J Plant Growth Regul 11:153–157

Scherer GFE, Arnold B (1997) Inhibitors of animal phospholipase A2 enzymes are selective inhibitors of auxin-dependent growth. Implications for auxin-induced signal transduction. Planta 202:462–469

Schiebl C, Walther A, Rescher U, Klämbt D (1997) Interactions of auxin binding protein 1 with maize coleoptile plasma membranes in vitro. Planta 201:470–476

Sitbon F, Perrot-Rechenmann C (1997) Expression of auxin-regulated genes. Physiol Plant 100:443–455

Stewart PA (1983) Modern quantitative acid-base chemistry. Can J Physiol Pharmacol 61:1444–1461

Takahashi Y, Nagata T (1992) par B: an auxin-regulated gene endocing glutathione S-transferase. Proc Natl Acad Sci USA 86:56–59

Terry ME, Jones RI (1981) Effect of salt on auxin-induced acidification and growth by pea internode sections. Plant Physiol 68:59–64

Thiel G, Blatt MR, Fricker MD, White IR, Millner P (1993) Modulation of K^+ channels in *Vicia* stomatal guard cells by peptide homologues to the auxin-binding protein C-terminus. Proc Natl Acad Sci USA 90:11493–11497

Thiman KV (1937) On the nature of inhibitions caused by auxin. Am J Bot 24:407–412

Tietze-Haß E, Dörffling K (1977) Initial phases of indolylacetic acid induced growth in coleoptile segments of *Avena sativa* L. Planta 135:192–195

Venis MA, Napier RM, Barbier-Brygoo H, Maurel C, Perrot-Rechenmann C, Guern J (1992) Antibodies to a peptide from the maize auxin binding protein have agonist activity. Proc Natl Acad Sci USA 89:7208–7212

Vesper MJ, Kuss CL (1990) Physiological evidence that the primary site of auxin action in maize coleoptiles is an intracellular site. Planta 182:486–491

Villalba JM, Lützelschwab M, Serrano R (1991) Immunocytolocalization of plasma membrane H^+-ATPase in maize coleoptiles and enclosed leaves. Planta 185:485–461

Went FW (1928) Wuchsstoff and Wachstum. Rec Trav Bot Neerl 25:1–116

Yamamoto KT (1994) Further characterization of auxin-regulated mRNA in hypocotyl sections of mung bean [*Vigna radiata* (L.) Wilczek]: sequence homology to genes for fatty acid desaturases and atypical late-embryogenesis-abundant proteins, and the mode of expression of the mRNAs. Planta 192:359–364

Yamamoto KT, Mori H, Imaseki H (1992) cDNA cloning of indole-3-acetic acid-regulated genes: Aux22 and SAUR from mung bean (*Vigna radiata*) hypocotyl tissue. Plant Cell Physiol 33:93–97

Yi H, Park D, Lee Y (1996) In vivo evidence for the involvement of phospholipase a and protein kinase in the signal transduction pathway for auxin-induced corn coleoptile segments. Physiol Plant 96:359–368

Zagouras P, Rose JK (1989) Carboxy-terminal SEKDEL sequences retard but do not retain two secretory proteins in the endoplasmic reticulum. J Cell Biol 109:2633–2640

Zbell B (1988) A modulating role of the inositol exchange on the auxin-mediated phosphoinositide response in plant cell membranes? In: Kutacek M, Bandurski RS, Krekule J (eds) Physiology and biochemistry of auxins in plants. Academia, Praha, pp 187–192

Zobel RW (1973) Some physiological characteristics of the ethylene-requiring tomato mutant diageotropica. Plant Physiol 52:385–389

Dr. Hartwig Lüthen
Dr. Maike Claussen
Professor Dr. Michael Böttger
Universität Hamburg
Institut für Allgemeine Botanik
Ohnhorststraße 18
D-22609 Hamburg, Germany

Edited by
U. Lüttge

Secondary Plant Substances: Sesquiterpenes

By Horst-Robert Schütte

1 Introduction

The sesquiterpenes are the C_{15} representatives of the terpenoid family of the natural products and they diverge from higher isoprenoid biosynthesis at the level of farnesyl pyrophosphate (Ruzicka 1953). Sesquiterpenes comprise one of the largest and most diverse families of natural products, and they are produced by both lower plants such as fungi and other microorganisms and by higher plants, in which they are best known as a component of the essential oils. Most of the nearly 200 skeletal families of sesquiterpenes derived via cyclization of the ubiquitous C_{15} isoprenoid intermediate farnesyl pyrophosphate (Fig. 1; Glasby 1982; Conolly and Hill 1992; Buckingham 1994; Gijsen et al. 1995; Gonzalez and Barrera 1995). The chemistry and biochemistry of sesquiterpenoids are periodically reviewed (Pinder 1977; Cane 1981, 1990; Beale 1990a, b, 1991; Fraga 1991, 1995, 1997; Chappell 1995; Dewick 1995, 1997; McGarvey and Croteau 1995).

The biosynthesis of cyclic sesquiterpenoids is catalyzed by cyclases. Each individual cyclase is capable of converting the universal acyclic precursor farnesyl diphosphate to a distinct sesquiterpene. A unified model for these complex transformations based on a common mechanism involving ionization of the allylic pyrophosphate ester followed by a precise sequence of intramolecular electrophilic addition reactions is established. A major determinant which controls molecular diversity in product structure and stereochemistry is believed to be the precise folding of farnesyl pyrophosphate substrate at the cyclase active site (Hendrickson 1959; Arigoni 1975; Cane 1985, 1989). Little is known about the active site of any cyclase or the manner in which a sesquiterpene synthase imposes a particular conformation on its highly lipophilic substrate, precisely controls the resulting cascade of electrophilic cyclizations and carbon skeletal rearrangements, and ultimatively quenches the positive charge. The sesquiterpenoids show numerous biological activities (Harborne and Tomas-Barberan 1991; Harborne 1993, 1995).

Progress in Botany, Vol. 60
© Springer-Verlag Berlin Heidelberg 1999

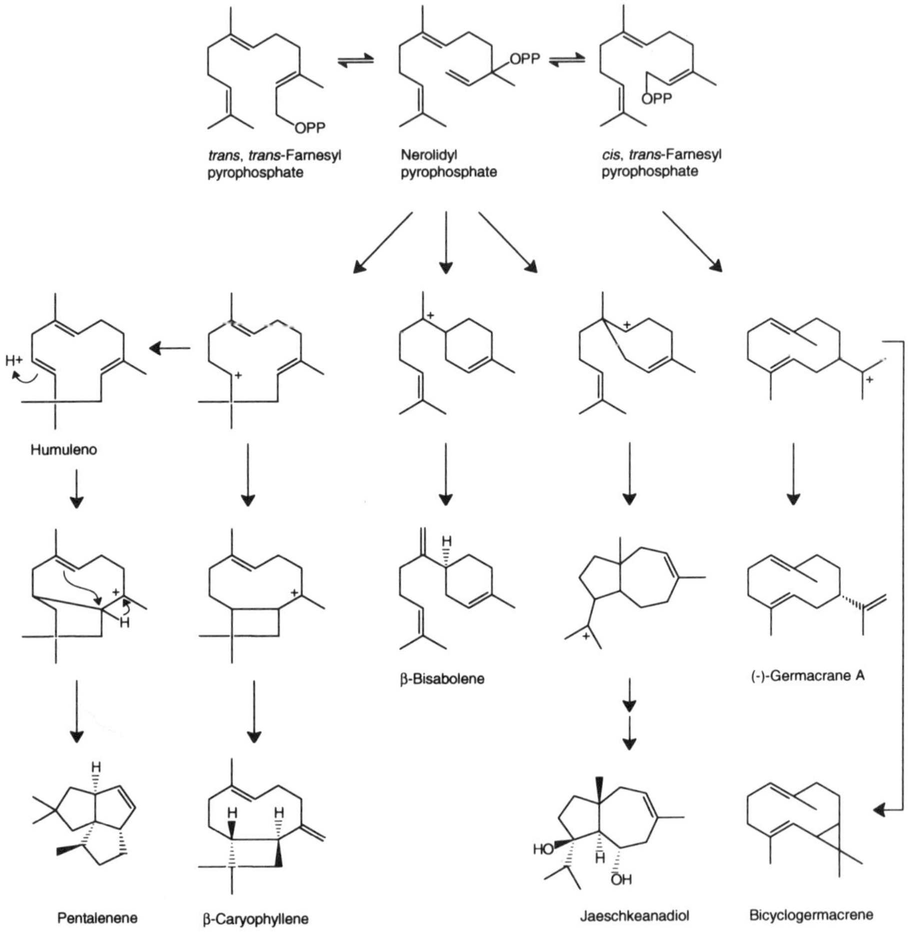

Fig. 1. Different representatives of sesquiterpenoids and their hypothetical formation

Within the sesquiterpenoids, the sesquiterpenoid lactones are a special group (Fig. 2; Fischer et al. 1979; Seaman 1982; Fischer 1991). The majority of these compounds were isolated from the Asteraceae (Compositae) family. The lactones are most commonly located in the glandular hairs (trichomes) on leaves and stems of a plant (Rodriguez et al. 1984; Spring and Bienert 1987). The highly bitter wormwood (*Artemisia absintha* L.) and *Cnicus benedictus* L. have been used for centuries in folk medicine. Both plants are still cultivated for use as bitters in the food industry and in pharmacy. *Cnicus benedictus* with its potent bitter lactone cnicin is the ingredient in bitter liqueurs. The characteristic functionality of the majority of sesquiterpenoid lactones is the α-methylene-γ-lactone group (Fig. 2). Many lactones also contain α,β-unsaturated carbonyls as well as epoxides. These functional groups rep-

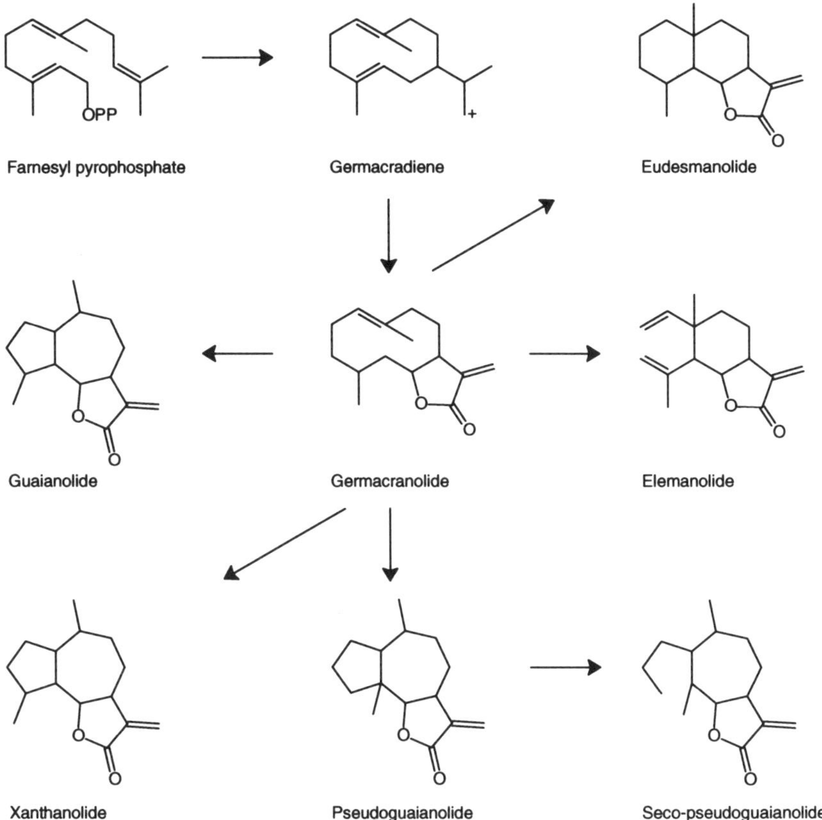

Fig. 2. Different representatives of sesquiterpenoids lactones of the Asteraceae and their biogenetic relationships

resent reactive receptor sites for biological nucleophiles, in particular, thiol and amino groups. Therefore, sesquiterpenoid lactones can cause irreversible alkylations of essential thiol and amino functions of certain enzymes. Consequently, a wide spectrum of biological activities of such lactones has been reported (Rodriguez et al. 1976; Picman 1984, 1986).

2 Farnesyl Pyrophosphate Synthase

Farnesyl pyrophosphate is formed by the sequential head-to-tail condensation of dimethylallyl pyrophosphate with two equivalents of isopentenyl pyrophosphate (Poulter and Rilling 1981; Fig. 3). The reaction has been shown to involve the electrophilic addition of C-1 of dimethylallyl pyrophosphate to C-4 of the cosubstrate, isopentenyl pyrophosphate, thereby generating the C_{10} homologue, geranyl pyrophos-

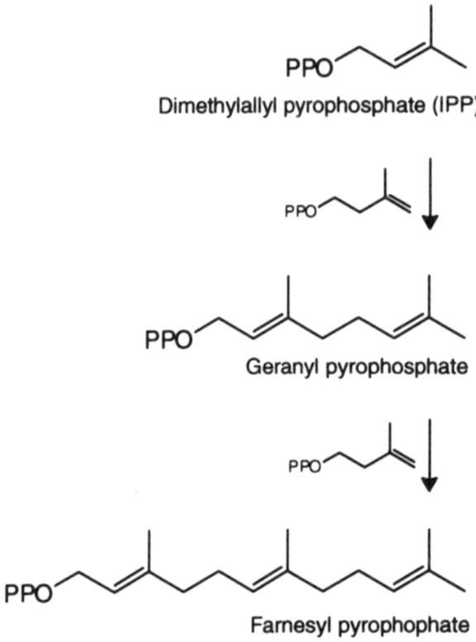

Dimethylallyl pyrophosphate (IPP)

Geranyl pyrophosphate

Farnesyl pyrophophate

Fig. 3. Formation of farnesyl pyrophosphate

phate, which itself undergoes condensation with a second equivalent of isopentenyl pyrosphosphate to yield farnesyl pyrophosphate. These two steps are catalyzed by the same enzyme, farnesyl pyrophosphate synthase, which is a key enzyme in isoprenoid biosynthesis (Dorsay et al. 1966). The enzyme has been isolated from different sources (Cordell 1976). The full stereochemical details of the farnesyl pyrophosphate synthase reaction were established in the course of the classical studies of the biosynthesis of cholesterol.

In contrast to sesquiterpenes synthases, which are confined primarily to plants, fungi, and a limited number of streptomycetes, the farnesyl pyrophosphate synthase appears to occur in all living systems. The enzyme has been purified to homogeneity from several eukaryotic sources, including *Saccharomyces cerevisiae* (Eberhardt and Rilling 1975), avian liver (Reed and Rilling 1975), porcine liver (Yeh and Rilling 1977; Barnard et al. 1978), and human liver (Barnard and Popjak 1980, 1981). In all these organisms, the enzyme is a dimer of 80–84 kDa. The more extensively studied avian liver enzyme has two identical subunits, a motif that is apparently conserved in the enzyme from other eukaryotes. Each subunit has a single catalytic site that catalyzes condensation of isopentenyl pyrophosphate with both dimethylallyl pyrophosphate and geranyl pyrophosphate. The farnesyl pyrophosphate synthase from *Capsicum* fruits was purified (Hugueney and Camara 1990). The enzyme has a molecular mass of 89 ± 5 kDa resulting from the association of two apparently identical subunits having a molecular mass of 43 ± 2 kDa. *Capsicum* farnesyl pyrophosphate synthase is strictly located in the extraplastidial compartment.

The farnesyl pyrophosphate synthase gene has been cloned and sequenced (Ashby and Edwards 1989; Sheares et al. 1989). The structural gene for farnesyl pyrophosphate synthase was isolated on a 4,5-kb EcoRI genomic restriction fragment from the yeast *Saccharomyces cerevisiae* (Anderson et al. 1989). The clone encodes a 40 483-Da polypeptide of 342 amino acids with a high degree of similarity to the protein encoded by a putative rat liver clone of farnesyl pyrophosphate synthase (Clarke et al. 1987) and to an active site protein fragment from avian liver farnesyl pyrophosphate synthase (Brems et al. 1981). When cloned into the yeast shuttle vector YRp17, the 4,5-kb EcoRI fragment directed a two-to threefold overexpression of farnesyl pyrophosphate synthetase activity in transformed yeast cells.

Comparison of the amino acid sequences of a number of prenyltransferases revealed the presence of three conserved regions. Two of these regions have been referred to as domains I and II and are characterized by an aspartate-rich motif which is flanked by other conserved residues. The third region of homology is similar to an active site peptide of avian farnesyl pyrophosphate synthase (Brems and Rilling 1979). It was hypothesized that the conserved regions I and II might be involved in substrate binding. Two of the conserved aspartate residues in domain II of rat farnesyl pyrophosphate synthase were separately changed to glutamate residues (Marrero et al. 1992). The results suggested that the first aspartate in domain II was involved in the catalytic reactions of the enzyme. The conserved aspartates or arginines in domain I and II of rat farnesyl pyrophosphate synthase were individually mutated to glutamate or lysine, respectively (Joly and Edwards 1993; Song and Poulter 1994).

Each mutant enzyme was overexpressed in *E. coli* and purified to apparent homogeneity. The mutagenesis resulted in a decreased V_{max} of appoximately 1000-fold compared to wild type. The results strongly suggest that these amino acids are involved in either the condensation of isopentenyl pyrophosphate and geranyl pyrophosphate to form farnesyl pyrophosphate and/or the release of the farnesyl pyrophosphate product from farnesyl pyrophosphate synthase. A mutant gene was isolated from a *Saccharomyces cerevisiae* strain defective in farnesyl pyrophosphate formation resulting in a Lys-197 $\rightarrow$ Glu substitution (Blanchard and Karst 1993).

The structural gene for the thermostable farnesyl pyrophosphate synthase from *Bacillus stearothermophilus* was cloned, sequenced, and overexpressed in *Escherichia coli* cells (Koyama et al. 1993b, 1994a). A 1260-nucleotide sequence of the cloned fragment was detected. This sequence specifies an open reading frame of 891 nucleotides for farnesyl pyrophosphates synthase. The deduced amino acid sequence shows a 42% similarity with that of *Escherichia coli* farnesyl pyrophosphate synthase. In contrast to thermolabile prenyltransferases, which have four to six cysteine residues, the thermostable farnesyl pyrophosphate synthase carried only two cysteine residues. Neither of the cysteines is essential for catalytic function (Koyama et al. 1994b).

Site-directed mutations into the gene for farnesyl pyrophosphate synthase of *Bacillus stearothermophilus* showed similar catalytic activities of the corresponding enzymes to that of the wild type (Koyama et al. 1993a). Two conserved lys residues in regions I and II contribute to the binding of isopentenyl pyrophosphate and the first and second asp residues in region VI are involved in catalytic function (Koyama et al. 1996). The Phe-gln motif is involved not only in the binding of allylic substrates but also in the catalysis by farnesyl pyrophosphate synthase (Koyama et al. 1995). The crystal structure of avian recombinant farnesyl pyrophosphate synthase has been detected to 2.6-Å resolution (Tarshis et al. 1994). The enzyme exhibits a novel fold composed entirely of α-helixes joined by connecting loops.

3 Aristolochene

Aristolochene (5) is an eremophilane-type sesquiterpenoid that is produced by a variety of organisms. The (–)-enantiomer has been identified from the plant *Aristolochia indica* (Govindachari et al. 1970; Lawrence and Hogg 1973), but also from insect sources (Baker et al. 1981), whereas the (+)-enantiomer has been identified in extracts from the fungus *Aspergillus terreus* (Cane et al. 1987). Aristolochene is also the likely parent compound for a number of sesquiterpenoid toxins produced by filamentous fungi (Sugawara et al. 1985). A family of sesquiterpenoids, which appear to be derived from aristolochene, is produced by the blue cheese mold, *Penicillium roqueforti* (Gorst-Allmann and Steyn 1982).

Aristolochene synthase catalyzes the cyclization of farnesyl pyrophosphate (1) to (+)-aristolochene (5) (Fig. 4). The enzyme is investigated in *Aspergillus terreus* as well as in *Penecillium roqueforti* (Cane et

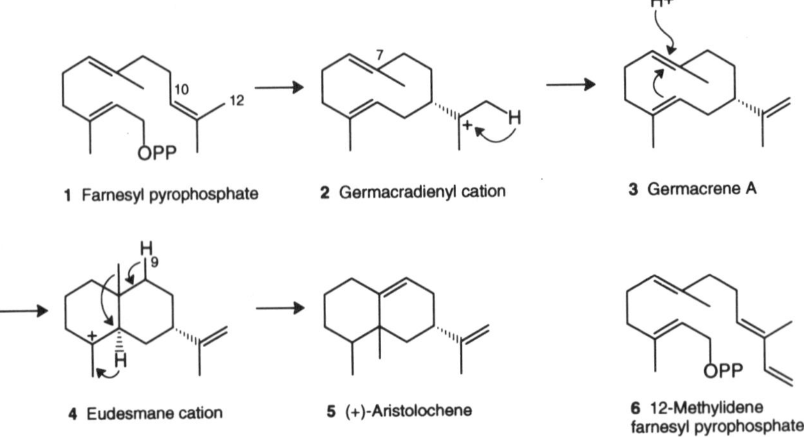

Fig. 4. Biosynthesis of (+)-aristolochene

al. 1989a, 1990c; Hohn and Plattner 1989a). The cyclization of farnesyl pyrophosphate (1) to aristolochene (5) is proceeding with inversion of configuration at C-1 of farnesyl pyrophosphate. Support has been provided for a cyclization mechanism in which ionization of the pyrophosphate moiety is followed by electrophilic attack at C-10 of the distal double bond. Loss of a proton from the C-12 (*cis*) methyl group of the germacradienyl cation (2) is believed to generate the known sesquiterpene hydrocarbon germacrene A (3). The latter intermediate is not released from the active site, but instead undergoes reprotonation at C-7 and further cyclization by attack of the resultant cation on the 3,4-double bond, followed by rearrangement of the transiently generated bicyclic eudesmane cation (4), which itself is converted to aristolochene (5) by successive hydride shift, methyl migration, and deprotonation.

The displacement of pyrophosphate and formation of the new C-C bond takes place with net inversion and the proton which is lost from C-9 of the bicyclic eudesmane cation (4) originates from C-8 of farnesyl pyrophosphate. When the anomalous substrate 6,7-dihydrofarnesyl pyrophosphate is incubated with aristolochene synthase from *Aspergillus terreus*, the resultant dihydrogermacrene cannot be further cyclized and is released from the active site (Cane and Tsantrizos 1996).

Two distinct aristolochene synthases have been isolated from two fungal sources. The *Aspergillus terreus* enzyme has been partially purified some 500-fold and is a monomer with a molecular weight of 58 kDa. By contrast, the enzyme from *Penicillium roqueforti*, which has been purified to homogenetiy, has a subunit molecular weight of 37 kDa and apparently is a monomer in the native state. Antibodies raised to the *P. roqueforti* protein do not cross-react with the crude *A. terreus* enzyme. Partial amino acid sequence information has been used to clone the aristolochene synthase gene (Ari 1) from *P. roqueforti* (Proctor and Hohn 1993). Cloning of the resulting construct into *E. coli* gave transformants which expressed aristolochene synthase at levels up to 40% of soluble protein. Purification of the recombinant protein gave homogenous aristolochene synthase with similar properties when compared to native fungal protein (Cane et al. 1993b). The 12-methylidene analogue (6) of farnesyl pyrophosphate has proved to be an effective mechanism-based inhibitor of aristolochene synthase (Cane and Bryant 1994).

5-epi-Aristolochene (11) is produced in tobacco (*Nicotiana tabacum*) and pepper (*Capsicum annuum*) as part of a sequence to sesquiterpene phytoalexins such as capsidiol (9) and debneyol (14), when the plant tissues are challenged by fungi or other elicitors (Groß 1977; Stoessl 1980; Watson et al. 1985; Threlfall and Whitehead 1988b, 1991; Whitehead and Threlfall 1992). The elicitor-inducible 5-epi-aristolochene cyclase was purified from *Nicotiana tabacum* and resolved as two major polypeptides of 60 and 62 kDa (Vögeli et al. 1990). Capsidiol (9) was first isolated from *Capsicum annuum* and later from both *Datura stramo-*

nium and *Nicotiana tabacum* (Ward et al. 1973; Birnbaum et al. 1974; Bailey et al. 1975; Stoessl et al. 1976a, 1977; Guedes et al. 1982; Watson and Brooks 1984; Brooks et al. 1986). Debneyol (**14**) was isolated from *Nicotiana tabacum* along with the closely related compounds cyclodebneyol (**12**), 7-epi-debneyol, 1-hydroxydebneyol and 8-hydroxydebneyol (**13**) (Burden et al. 1985, 1986; Whitehead et al. 1988).

Tobacco cell suspension cultures produce large amounts of extracellular sesquiterpenoids upon addition of fungal elicitors to the cell cultures (Chappell et al. 1987). The major sesquiterpenoid produced was capsidiol (**9**). The elicitor-induced sesquiterpenoids were de novo synthesized and preceded by a transient activation of 3-hydroxy-3-methylglutaryl coenzyme A reductase, a suspected key enzyme of isoprenoid metabolism (Chappell and Nable 1987; Chappell et al. 1991). The induction of the sesquiterpene cyclase in elicitor-treated cell cultures is primerly regulated by transcriptional control of the cyclase gene (Vögeli and Chappell 1990). Mevinolin, a potent in vitro inhibitor of the tobacco 3-hydroxy-3-methylglutaryl coenzyme A reductase enzyme activity inhibited the elicitor-induced capsidiol accumulation (Vögeli and Chappell 1991).

The rapid synthesis and secretion of large amounts of antibiotic sesquiterpenoids such as capsidiol (**9**) was paralleled by a rapid and large decline in the incorporation rate of radioactivity into sterols. It is suggested that the channelling of isoprenoid intermediates, and especially farnesyl pyrophosphate, into sesquiterpenoids occurred by a coordinated increase in the sesquiterpene cyclase and a decrease in the squalene synthase enzyme activities (Threlfall and Whitehead 1988a; Vögeli and Chappell 1988; Chappell et al. 1989; Zook and Kuc 1991; Hanley et al. 1992).

It is proposed that the hydrocarbon 4-epi-eremophila-9,11-diene (5-epi-aristolochene) (**11**) is a common intermediate in the biosynthesis of all the compounds, and further that 8-hydoxy-debneyol (**13**) is the direct precursor of the phytoalexin cyclodebneyol (**12**) (Fig. 5). 5-epi-Aristolochene (**11**) is a good precursor of capsidiol (**9**) at a rate of 42% in arachidonic acid-treated green pepper seedlings (Yoshizawa et al. 1994). 5-epi-Aristolochene 3-hydroxylase is one of the key enzymes of the pathway leading to capsidiol (**9**) (Hoshino et al. 1995). The enzyme was found in microsomes of elicitor-treated green pepper fruits with a molecular mass of 59 kDa and it is a P450-dependent system, requiring O_2 and NADPH as cofactors.

Elicitation induced the production of P450 isoforms. Ancymidol and ketoconazole, which are P450 inhibitors, have been found to suppress phytoalexin accumulation in green pepper cell cultures (Hoshino et al. 1994). 5-epi-Aristolochene (**11**) is accumulated when the formation of capsidiol (**9**) and debneyol (**14**) from farnesyl pyrophosphate in cell-free preparations of cells from cellulase-elicited cell suspension culture of

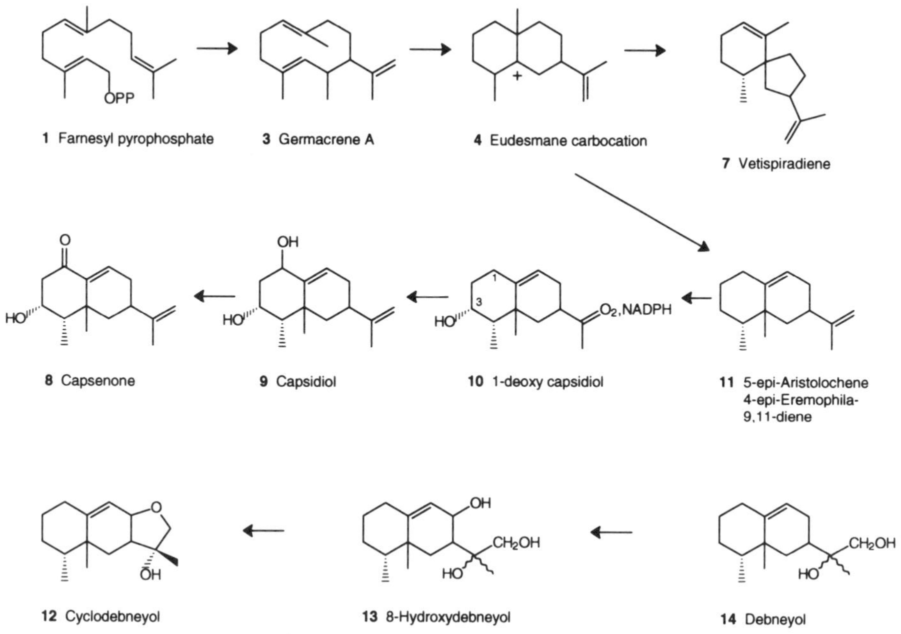

1 Farnesyl pyrophosphate   **3** Germacrene A   **4** Eudesmane carbocation   **7** Vetispiradiene

8 Capsenone   **9** Capsidiol   **10** 1-deoxy capsidiol   **11** 5-epi-Aristolochene 4-epi-Eremophila-9,11-diene

12 Cyclodebneyol   **13** 8-Hydroxydebneyol   **14** Debneyol

Fig. 5. Suggested biosynthetic relationships between eremophilenes from *Nicotiana tabacum*

Nicotiana tabacum is inhibited by the omission of NADPH or the exclusion of molecular oxygen from the incubation mixture (Whitehead et al. 1989). 5-epi-Aristolochene (**11**) has been shown to be a common precursor of the two sesquiterpenoid phytoalexins capsidiol (**9**) and debneyol (**14**) which accumulate in the elicited cultures.

The biosynthesis of capsidiol (**9**) seems to be regulated, in part, by the activity of the first of the two hydroxylases which catalyze its formation from this hydrocarbon and that this 3-hydroxylase is induced during the elicitation process. On the other hand, 5-epi-aristolochene, (**11**) was synthesized from capsidiol (**9**) via 1-deoxycapsidiol (**10**) (Whitehead et al. 1990a). Capsidiol (**9**) is oxydized to the ketone capsenone (**8**) by *Botrytis cinerea* and *Fusarium oxysporum* (Ward and Stoessl 1972; Stoessl et al. 1973). Capsenone (**8**) was less fungitoxic than capsidiol and its formation may be a detoxification mechanism of importance in overcoming host resistance. *cis*-9,10-Dihydrocapsenone could be found as an intermediate on the route by which capsidiol is catabolized in pepper cultures (Whitehead et al. 1987).

In tobacco, the 5-epi-aristolochene synthase appears to be encoded by a complex gene family. A full-length cDNA sequence has been cloned into *E. coli* to allow production of the enzyme at a level of 5–8% of total soluble protein (Facchini and Chappell 1992; Back et al. 1994). Enzymic reaction products from the recombinant protein were identical to those from the native enzyme.

Alternative modifications of the eudesmane cation (4) can lead to vetispiradiene (7) and related compounds. The vetispiran lubimin (18) is one of the most common of the many sesquiterpenoid phytoalexins isolated from members of the Solanaceae (Brindle et al. 1983, 1988). It is accumulated in infected tubers of *Solanum tuberosum* (Katsui et al. 1978; Stoessl and Stothers 1980), in infected fruits of *Solanum melongena* and *Datura stramonium* (Birnbaum et al. 1976, Ward et al. 1976), and in cellulase-treated fruits of *Cyphomandra betaceae*. In infected tubers of *S. tuberosum*, the nor-eudesmane rishitin (16) is also produced, often along with the vetispiranes 3-hydoxylubimin (17), 15-dihydrolubimin (21), isolubimin (22) and solavetivone (15), and the seco-eudesmane phytuberin. In fruit cavities of *D. stramonium* inoculated with *Sclerotinia fructicola*, lubimin (18) accumulates along with 3-hydroxylubimin (17), 2,3-dihydroxygermacrene, and capsidiol (9).

A biosynthetic pathway is proposed from farnesyl pyrophosphate (1), germacrene A (3), and solavetivone (15) to rishitin (16) via 15-hydroxy-solavetivone, 15-dihydrolubimin (21), lubimin (18), and 3-hydroxylubimin (17) involving a 1,2-hydride shift of the eudesmane intermediate shown for capsidiol (9), lubimin (18), 3-hydroxylubimin (17), and rishitin (Fig. 6; Kalan and Osman 1976; Sato et al. 1978; Stoessl et al. 1978; Stoessl and Stothers 1982, 1983). On the other hand, lubimin (18) is metabolized to 15-dihydrolubimin (21) by some fungi and labeled 15-dihydrolubimin yielded labeled isolubimin (22) in healthy potato tissue in absence of fungi (Ward and Stoessl 1977; Stoessl and Stothers 1981). Lubimin (18), 3-hydroxylubimin (17), rishitin (16), and phytuberin are isolated as metabolites of potatoes which had been inoculated with *Monilinia fructicola* (Stoessl et al. 1976b; Coolbear and Threllfall 1985).

The metabolism of capsidiol (9) and rishitin (16) to their 13-hydroxyderivatives could be established by tissue cultures of sweet pepper and potato (Ward et al. 1977). The treatment of root cultures of *Datura stramonium* with copper and cadmium salts has been found to induce the rapid accumulation of high levels of sesquiterpenoid defensive compounds, notably lubimin (18) and 3-hydroxylubimin (17) (Furze et al. 1991). 2-Dehydrolubimin (19) was rapidly and efficiently incorporated into lubimin (18) and may be the direct precursor of lubimin in plants (Whitehead et al. 1990b). 3-Hydroxylubimin (17) was incorporated into rishitin (16). A likely pathway for lubimin detoxification in some strains of the potato pathogener *Gibberella pulicaris (Fusarium sambucinum)* involves cyclization of isolubimin (22) to cyclodehydroisolubimin (23) (Desjardins et al. 1989).

Vetispiradiene synthase from *Hyoscyamus muticus* is encoded by a gene family of some six to eight genes (Back and Chappell 1995). Genomic and cDNA clones have been isolated from cell cultures treated

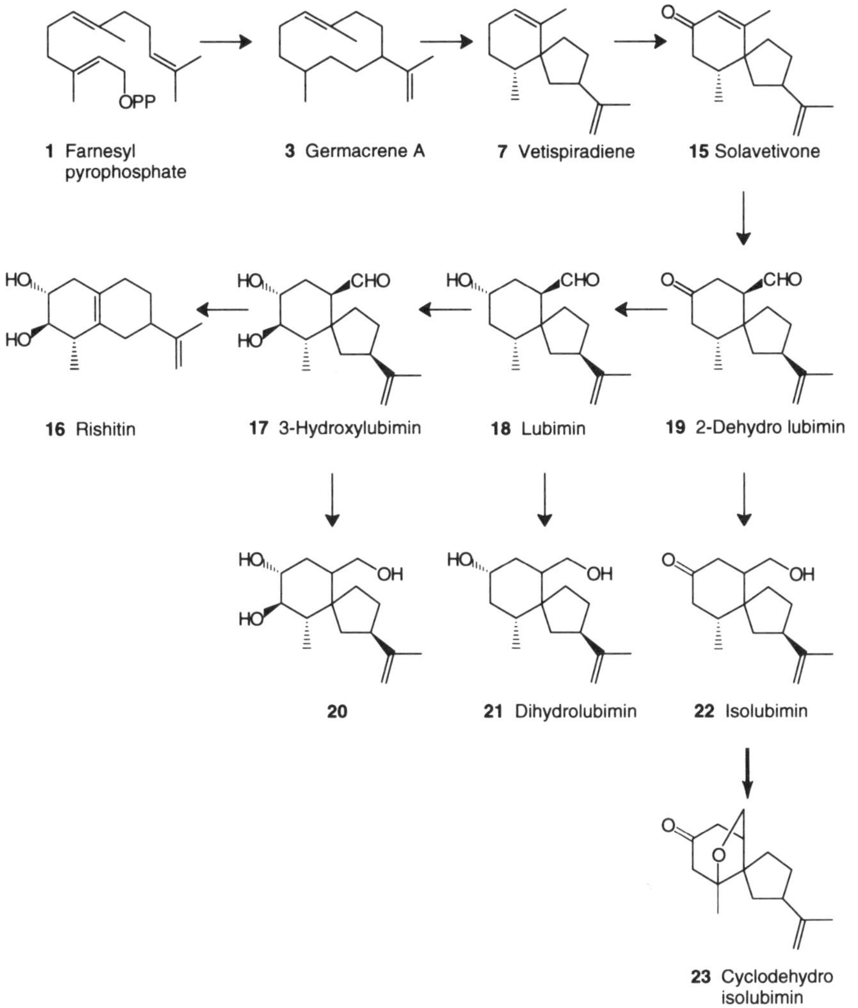

Fig. 6. Biosynthesis and metabolism of lubimin and related compounds

with a fungal elicitor preparation, and bacterial expression of three of these showed that they all coded for enzymes giving the same single reaction product. The deduced amino acid sequence was 77% identical to that of 5-epi-aristolochene synthase from tobacco.

4 Selinen and Ovalicin

A sesquiterpene cyclase enzyme catalyzing the formation of β-selinene (24) from farnesyl pyrophosphate (1) has been partially purified from

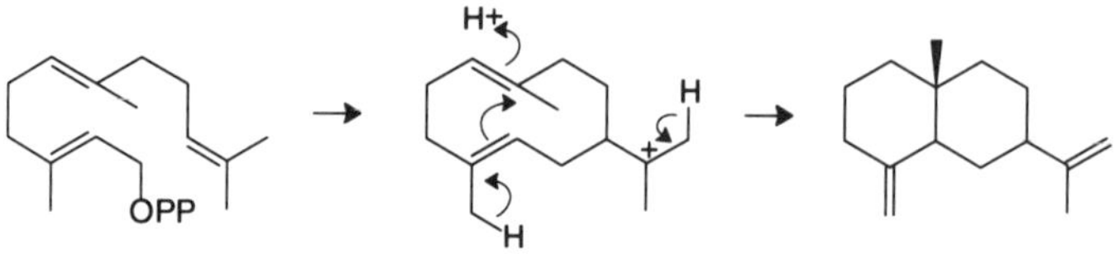

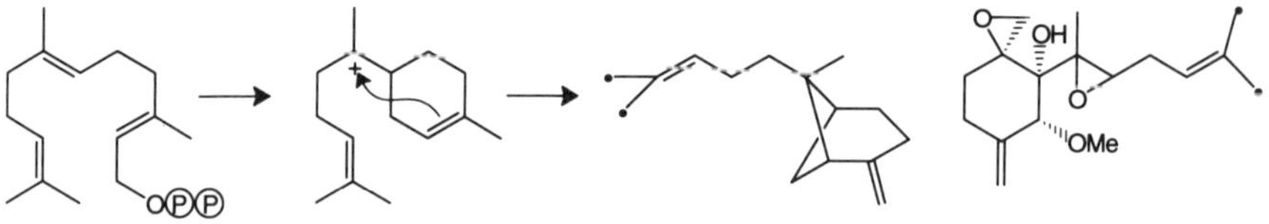

1 Farnesyl
pyrophosphate

24 β-Selinen

Fig. 7. Biosynthesis of selinen

1 Farnesyl   **25a** Bisabolyl cation   **25** β-*trans*-bergamotene   **26** Ovalicin
pyrophosphate

Fig. 8. Biosynthesis of β-*trans*-bergamotene and ovalicin

fruits of calamondin (*Citrofortunella mitis*) (Fig. 7) (Belingheri et al.
1992). A fraction enriched in endoplasmatic reticulum was able to synthesize β–selinene (**24**) and germacrene D, the main C_{15} hydrocarbons
present in the oil of calamondin fruits (Belingheri et al. 1988).

β-*trans*-Bergamotene (**25**) is the parent hydrocarbon in the biosynthesis of ovalicin (**26**) in *Pseudeurotium ovalis* in which farnesyl pyrophosphate is converted to β-bergamotene by cyclization of an intermediate bisabolyl cation (**25a**) (Fig. 8; Cane and King 1976; Cane and
McIlwaine 1987; Cane et al. 1989b, 1990a). The cyclization to the six-
membered ring occurs with net retention of configuration at C-1 of farnesyl pyrophosphate (**1**).

5 Cadinane-Type Sesquiterpenoids

(+)-Epicubenol (**29**) is a sesquiterpene alcohol belonging to the cadinane
family which has been isolated from a variety of soil organisms, including *Streptomyces* sp. (Gerber 1971). The enantiomeric (–)-epicubenol has
also been isolated from cubeb oil as well as from extracts of several other
plants (Tomita and Hirose 1972).

The results of extensive experiments carried out with intact plants
and microorganisms have proposed a stereochemical model for the biosynthesis of a group of biogenetically related cadinane, picrotoxane, and
sativane sesquiterpenes, based on the cyclization of farnesyl diphosphate
(**1**) and the postulated intermediacy of its tertiary allylic isomer nerolidyl diphosphate (**27**). A cell-free extract of *Streptomyces* sp.LL-B7

catalyzed the cyclization of farnesyl diphosphate (1) to (+)-epicubenol (29) (Cane et al. 1993a). The operation of both 1,3- (from C-5 to C-11) and 1,2- (from C-10 to C-9) hydride shifts is documented in the formation of (+)-epicubenol (29) Cane and Tandon 1994, 1995). The migrating hydride from C-1 of farnesyl diphosphate (1) was the 1-pro-S hydrogen, the 1-pro-R hydrogen is residing at C-5 in epicubenol (29). The epicubenol synthase is specific for (3R)-nerolidyl diphosphate (27). The intermediacy of nerolidyl diphosphate (27) is consistent with the demonstrated stereochemistry of the isomerization steps catalyzed by the sesquiterpene cyclase trichodiene synthase (Cane et al. 1985) and the mechanistically related farnesyl pyrophosphate (1) – nerolidyl pyrophosphate (27) isomerase (Cane et al. 1981). Attack of the ionized nerolidyl pyrophosphate (27) by the C-10–C-11 double bond will be on the re face, to generate the *cis*-germacryl cation (28). A second electrophilic cyclization gives the cadinyl cation (30), and finally capture of hydroxy group from water to (+)-epicubenol (29) is in a syn sense with respect to the migrating hydride (Fig. 9).

2,7-Dihydroxycadalene (31) and related compounds are sesquiterpenoid compounds produced in leaves and cotyledons of upland cotton (*Gossypium hirsutum* L.) during the hypersensitive response to incompatible races of the bacterial pathogen *Xanthomonas campestris* (Essenberg et al. 1985, 1990; Pierce and Essenberg 1987). It has been shown a cell-free assay of cyclase activity with farnesyl pyrophosphate (1) as substrate and the identification of the product of the cotton cyclase activity as (+)-δ-cadinene (32) (Davis et al. 1991). (+)-δ-Cadinene (32) is an early enzymatic intermediate in the biosynthesis of the sesquiterpenoid phytoalexins by upland cotton such as 2,7-dihydroxycadalene (31), hemigossypol (33), and gossypol (35) (Fig. 9; Davis and Essenberg 1995). The enzymatic formation of δ-cadinene (32) from farnesyl pyrophosphate (1) is also established in extracts of cotton stems infected with *Verticillium dahliae*, which induces the formation of the sesquiterpenoid phytoalexins desoxyhemigossypol and hemigossypol (33) (Benedict et al. 1995).

The sesquiterpene cyclase whose activity is induced in a glandless, bacterial blight-resistant line of cotton (*Gossypium hirsutum* L.) was purified (Davis et al. 1996). It has a molecular weight of 64–65 kDa. Amino acid sequences of three tryptic peptides from the enzyme are similar to known sequences in other terpene cyclases from plants. Challenging cell suspension cultures of *Gossypium arboreum* with a preparation from *Verticillium dahliae* initiates production of gossypol (35), and this is accompanied by increased mRNA levels. This has led to the isolation of two cDNA clones, and the encoded proteins from these showed a significant degree of sequence identity with known plant terpene cyclases (Chen et al. 1995).

PPO
1 Farnesyl pyrophosphate
PPO
27 Nerolidyl Pyrophosphate
28 cis-Germacryl cation
29 (+)-Epicubenol
30 Cadinyl cation
HO
OH
31 2,7-Dihydroxy-cadalene
32 (+)-δ-Cadinene
CHO
OH
HO
HO
33 Hemigossypol
34 Calamenene
CHO
OH
HO
OH
OH
CHO
HO
35 Gossypol
OH
36 7-Hydroxy-calamenene
OH
37 7-Methoxy-1,2-dihydrocadalene

Cadinane-type sesquiterpenes were also found in the liverwort *Heteroscyphus planus*, such as 7-hydroxycalamenene (**36**) and 7-methoxy-1,2-dihydrocadalene (**37**) (Nabeta et al. 1993). Studies on the biosynthesis of the rare sesquiterpene, 7-methoxy-1,2-dihydrocadalene (**37**), in cell cultures of *Heteroscyphus planus* have shown the presence of a 1,3-hydride shift in the formation of the cadinyl cation, then a 1,2-hydride shift prior to aromatization to calamenene (**34**). The hydroxylation of calamenene must occur without any NIH shift (Fig. 9; Nabeta et al. 1994a, b). The formation of (+)-epicubenol (**29**) and (+)-cubenene in cell-free extracts from *H. planus* has also been reported (Nabeta et al. 1995).

6 Pentalenene

Pentalenene (**43**) isolated from *Streptomyces* UC 5319 is the parent hydrocarbon of the pentalenolactone family of antibiotic metabolites (Cane and Rossi 1979). Pentalenene (**43**) and pentalenolactone (**42**) are biosynthesized via the mevalonoid pathway (Cane et al. 1979, 1981, 1991). Studies on the biosynthesis of pentalenene and pentalenolactone using a cell-free extract from *Streptomyces* UC 5319 have shown that the 9-pro-R proton becomes H-8 of pentalenene (**43**), whilst the 9-pro-S proton undergoes an intramolecular transfer to become H-1 in the final product (Fig. 10; Cane and Tillman 1983; Cane et al. 1990b).

Pentalenene synthase is purified from *Streptomyces* UC 5319. It is a soluble, monomeric enzyme of 42.5 kDa (Cane and Pargellis 1987). A divalent metal cation, either Mg^{2+} or Mn^{2+}, is the only cofactor required. The experiment supports a mechanism in which farnesyl pyrophosphate (**1**) is initially cyclized to humulene (**39**), which undergoes further cyclization initiated by reprotonation at C-10 (Cane et al. 1984, 1988; Harrison et al. 1988; Cane and Weiner 1994).

Consistent with this mechanistic and stereochemical model is the demonstration that cyclization of farnesyl pyrophosphate (**1**) takes place with net inversion of configuration at C-1. Electrophilic attack on the si face of the C-10, 11 double bond is followed by removal of the H-9si proton. The proton that is removed does not undergo exchange with the medium, but is redonated to the re face of the C-9, 10 double bond of the intermediate humulene (**39**), leading to formation of the protoilludyl cation (**40**). Following a 1,2-hydride shift, the resultant cation attacks the remaining double bond, leading after loss of one of the original H-8 protons of farnesyl pyrophosphate (**1**), to formation of pentalenene (**43**). For cyclization of farnesyl pyrophosphate at the distal C-10–C-11 double

Fig. 9. Biosynthesis of epicubenol and cadinyl compounds

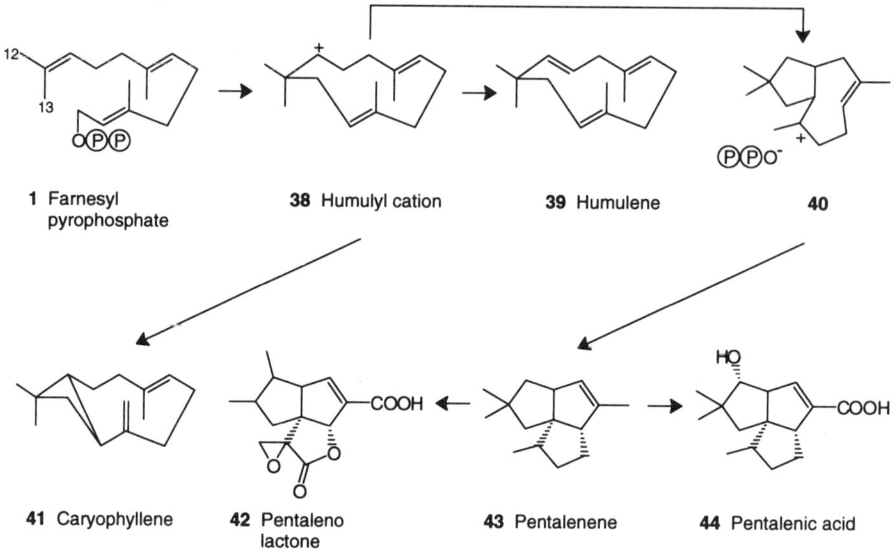

Fig. 10. Biosynthesis of pentalenene and related compounds

bond to form the 11-membered humulyl system (**38**) there is no stereochemical requirement for isomerization to precede cyclization.

Pentalenene synthase from *Streptomyces* has been purified, cloned, and sequenced (Cane et al. 1994). Expression of the gene in *E. coli* has allowed production of a recombinant protein with properties identical to those of the native enzyme. Both enzymes have molecular weights of 41–42 kDa. The recombinant protein has also been crystallized and studied by X-ray diffraction analysis (Lesburg et al. 1995).

The separation and partial purification of cyclases from *Salvia officinalis* are described which catalyze the formation of the olefins humulene (**39**) and caryophyllene (**41**) (Croteau and Gundy 1984; Dehal and Croteau 1988). Experiments with cell-free extracts indicated that these two olefins were formed by different enzymes, which were shown to have similar physical and kinetic parameters.

In the further metabolism of pentalenene (**43**), hydroxylation of C-1 yielding pentalenic acid (**44**) occurred with retention of configuration, but, surprisingly, pentalenic acid is ruled out as an intermediate on the pathway to pentalenolactone (**42**). The rearrangement involving a methyl migration from C-2 to C-1 takes place with stereospecific loss of the 3-pro-R hydrogen, anti to the migrating methy group (Fig. 10).

References

Anderson MS, Yarger JG, Burck CL, Poulter CD (1989) Farnesyl diphosphate synthetase. Molecular cloning, sequence, and expression of an essential gene from *Saccharomyces cerevisiae*. J Biol Chem 264:19176–19184

Arigoni S (1975) Stereochemical aspects of sesquiterpene biosynthesis. Pure Appl Chem 41:219–245

Ashby MN, Edwards PA (1989) Identification and regulation of a rat liver cDNA encoding farnesyl pyrophosphate synthetase. J Biol Chem 264:635–640

Back K, Chappell J (1995) Cloning and bacterial expression of a sesquiterpene cyclase from *Hyoscyamus muticus* and its molecular comparison to related terpene cyclases. J Biol Chem 270:7375–7381

Back K, Yin S, Chappell J (1994) Expression of a plant sesquiterpene cyclase gene in *Escherichia coli*. Arch Biochem Biophys 315:527–532

Bailey JA, Burden RS, Vincent GG (1975) Capsidol: an antifungal compound produced in *Nicotiana tabacum* and *Nicotiana clevelandii* following infection with tobacco necrosis virus. Phytochemistry 14:597

Baker R, Coles HR, Edwards M, Evans DA, Howse PE, Walmsley S (1981) Chemical composition of the frontal gland secretion of Synternes soldiers (Isoptera, Termitidae). J Chem Ecol 7:135–145

Barnard GF, Popjak G (1980) Characterization of liver prenyl transferase and its inactivation by phenylglyoxal. Biochim Biophys Acta 617:169–182

Barnard GF, Popjak G (1981) Human liver prenyltransferase and its characterization. Biochim Biophys Acta 661:87–99

Barnard GF, Langton B, Popjak G (1978) Pseudo-isoenzyme forms of liver prenyl transferase. Biochem Biophys Res Commun 85:1097–1103

Beale MH (1990a) Biosynthesis of C5–C20 terpenoid compounds. Nat Prod Rep 7:25–39

Beale MH (1990b) Biosynthesis of C5–C20 terpenoid compounds. Nat Prod Rep 7:387–407

Beale MH (1991) Biosynthesis of C5–C20 terpenoid compounds. Nat Prod Rep 8:441–454

Belingheri L, Pauly G, Gleizes M, Marpeau A (1988) Isolation by an aqueous two-polymer phase system and identifcation of endomembranes from *Citrofortunella mitis* fruits for sesquiterpenes hydrocarbon synthesis. J Plant Physiol 132:80–85

Belingheri L, Cartayrade A, Pauly G, Gleizes M (1992) Partial purification and properties of the sesquiterpene β-selinene cyclase from *Citrofortunella mitis* fruits. Plant Sci 84:129–136

Benedict CR, Alchanati I, Harvey PJ, Liu J, Stipanovic RD, Bell AA (1995) The enzymatic formation of δ-cadinene from farnesyl diphosphate in extracts of cotton. Phytochemistry 39:327–333

Birnbaum GI, Stoessl A, Grover SH, Stothers JB (1974) Post-infectional inhibitors from plants VIII. carbon-13 NMR studies 40. complete stereostructure of capsidiol. X-ray analysis and carbon-13 nuclear trans-vicinal methyl groups. Can J Chem 52:993–1005

Birnbaum GI, Huber CP, Post ML, Stothers JB, Robinson JR, Stoessl A, Ward EWB (1976) Sequiterpenoid stress compounds of *Datura stramonium*: biosynthesis of the three major metabolites from (1,2-^{13}C) acetate and X-ray structure of 3-hydroxy-lubimin. Chem Commun 300–331

Blanchard L, Karst F (1993) Characterization of a lysine-to-glutamic acid mutation in a conservative sequence of farnesyl diphosphate synthase from *Saccharomyces cerevisiae*. Gene 125:185–189

Brems DN, Rilling HC (1979) Photoaffinity labeling of the catalytic site of prenyl transferase. Biochemistry 18:860–864

Brems DN, Bruenger E, Rilling HC (1981) Isolation and characterization of a photoaffinity-labeled peptide from the catalytic site of prenyltransferase. Biochemistry 20:3711–3718

Brindle PA, Kuhn PJ, Threlfall DR (1983) Accumulation of phytoalexins in potato cell suspension culture. Phytochemistry 22:2719–2721

Brindle PA, Kuhn PJ, Threlfall DR (1988) Biosynthesis and metabolism of sesquiterpenoid phytoalexins and triterpenoids in potato cell suspension cultures. Phytochemistry 27:133–150

Brooks CJW, Watson DG, Freer IM (1986) Elicitation of capsidiol accumulation in suspended callus cultures of *Capsicum annuum*. Phytochemistry 25:1089–1092

Buckingham J (ed) (1994) Dictionary of natural products, vols 1–8. Chapman and Hall, London

Burden RS, Rowell PM, Bailey JA, Loeffler RST, Kemp MS, Brown CA (1985) Debneyol a fungicidal sesquiterpene from TNV-infected *Nicotiana debneyi*. Phytochemistry 24:2191–2194

Burden RS, Loeffler RST, Rowell PM, Bailey JA, Kemp MS (1986) Cyclodebneyol, a fungitoxic sesquiterpene from TNV-infected *Nicotiana debneyi*. Phytochemistry 25:1607–1608

Cane DE (1981) Biosynthesis of sesquiterpenes. In: Porter JW, Spurgeon SL (eds) Biosynthesis of isoprenoid compounds. Wiley, New York, pp 283–374

Cane DE (1985) Isoprenoid biosynthesis. Stereochemistry of the cyclization of allylic pyrophosphate. Acc Chem Res 18:220–226

Cane DE (1989) Stereochemical studies of natural products biosynthesis. Pure Appl Chem 61:493–496

Cane DE (1990) Enzymatic formation of sequiterpenes. Chem Rev 90:1089–1103

Cane DE, Bryant C (1994) Aristolochene synthase. Mechanism-based inhibition of a terpenoid cyclase. J Am Chem Soc 116:12063–12064

Cane DE, Tandon M (1994) Biosynthesis of (+)-epicubenol. Tetrahedron Lett 35:5355–5358

Cane DE, King GGS (1976) Biosynthesis of ovalicin: isolation of β-*trans*-bergamotene. Tetrahedron Lett 4737–4740

Cane DE, McIlwaine DB (1987) The biosynthesis of ovalicin from β-*trans*-bergamotene. Tetrahedron Lett 28:6545–6548

Cane DE, Pargellis C (1987) Partial purification and characterization of pentalenene synthase. Arch Biochem Biophys 254:421–429

Cane DE, Rossi T (1979) The isolation and structural elucidation of pentalenolactone E. Tetrahedron Lett 32:2973–2974

Cane DE, Tandon M (1995) Epicubenol synthase and the stereochemistry of the enzymatic cyclization of farnesyl and nerolidyl diphosphate. J Am Chem Soc 117:5602–5603

Cane DE, Tillman AM (1983) Pentalenene biosynthesis and the emzymatic cyclization of farnesyl pyrophosphate. J Am Chem Soc 105:122–124

Cane DE, Tsantrizos YS (1996) Aristolochene synthase. Elucidation of the cryptic germacrene A synthase activity using the anomalous substrate dihydrofarnesyl diphosphate. J Am Chem Soc 118:10037–10040

Cane DE, Weiner SW (1994) Cyclization of farnesyl diphosphate to pentalenene. Orthogonal stereochemistry in an enzyme-catalyzed SE' reaction. Can J Chem 72:118–127

Cane DE, Rossi T, Pachlatko JP (1979) The biosynthesis of pentalenolactone. Tetrahedron Lett 3639–3642

Cane DE, Rossi T, Tillman AM, Pachlatko JP (1981) Stereochemical studies of isoprenoid biosynthesis. Biosynthesis of pentalenolactone from $(U\text{-}^{12}C_6)$glucose and (6-^2H_2)glucose. J Am Chem Soc 103:1838–1843

Cane DE, Abell C, Tillman AM (1984) Pentalenene biosynthesis and the enzymatic cyclization of farnesyl pyrophosphate : proof that the cyclization is catalyzed by a single enzyme. Bioorg Chem 12:312–328

Cane DE, Ha HJ, Pargellis C, Waldmeier F, Swanson S, Murthy PPN (1985) Trichodiene biosynthesis and the stereochemistry of the enzymatic cyclization of farnesyl pyrophosphate. Bioorg Chem 13:246–265

Cane DE, Rawlings BJ, Yang CC (1987) Isolation of (–)-δ-cadinene and aristolochene from *Aspergillus terreus*. J Antibiot 40:1331–1334

Cane DE, Abell C, Lattman R, Kane CT, Hubbard BR, Harrison PHM (1988) Pentalenene biosynthesis and the enzymatic cyclization of farnesyl pyrophosphate. Anti stereochemistry in a biological SE' reaction. J Am Chem Soc 110:4081–4082

Cane DE, Prabhakaran PC, Salaski EJ, Harrison PHM, Noguchi H, Rawlings BJ (1989a) Aristolochene biosynthesis and enzymatic cyclization of farnesyl pyrophosphate. J Am Chem Soc 111:8914–8916

Cane DE, McIlwaine DB, Harrison PHM (1989b) Bergamotene biosynthesis and the enzymatic cyclization of farnesyl pyrophosphate. J Am Chem Soc 111:1152–1153

Cane DE, McIlwaine DB, Oliver JS (1990a) Absolute configuration of (–)-β-*trans*-bergamotene. J Am Chem Soc 112:1285–1286

Cane DE, Oliver JS, Harrison PHM, Abell C, Hubbard BR, Kane CT, Lattman R (1990b) Biosynthesis of pentalenene and pentalenolactone. J Am Chem Soc 112:4513–4524

Cane DE, Prabhakaran PC, Oliver JS, McIlwaine DB (1990c) Aristolochene biosynthesis. Stereochemistry of the deprotonation steps in the enzymatic cyclization of farnesyl pyrophosphate. J Am Chem Soc 112:3209–3210

Cane DE, Abell C, Harrison PAM, Hubbard BR, Kane CT, Lattmann R, Oliver JS, Weiner SW (1991) Terpenoid biosynthesis and the stereochemistry of enzyme-catalyzed alliylic addition-elimination reactions. Philos Trans R Soc Lond Ser B 332:123–129

Cane DE, Tandon M, Prabhakaran PC (1993a) Epicubenol synthase and the enzymatic cyclization of farnesyl disphosphate. J Am Chem Soc 115:8103–8106

Cane DE, Wu Z, Proctor RH, Hohn TM (1993b) Overexpression in *Escherichia coli* of soluble aristolochene synthase from *Penicillium roqueforti*. Arch Biochem Biophys 304:415–419

Cane DE, Sohng JK, Lamberson CR, Rudnicki SM, Wu Z, Lloyd MD, Oliver JS, Hubbard BR (1994) Pentalenene synthase purification, molecular cloning, sequencing, and high-level expression in Escherichia coli of a terpenoid cyclase from *Streptomyces* UC5319. Biochemistry 33:5846–5857

Chappell J (1995) The biochemistry and molecular biology of isoprenoid metabolism. Plant Physiol 107:1–6

Chappell J, Nable R (1987) Induction of sesquiterpenoid biosynthesis in tobacco cell suspension cultures by fungal elicitor. Plant Physiol 85:469–473

Chappell J, Nable R, Fleming P, Andersen RA, Burton HR (1987) Accumulation of capsidiol in tobacco cell cultures treated with fungal elicitor. Phytochemistry 26:2259–2260

Chappell J, Vonlanken C, Vögeli U, Bhatt P (1989) Sterol and sesquiterpenoid biosynthesis during a growth cycle of tobacco cell suspension cultures. Plant Cell Rep 8:48–52

Chappell J, VonLanken C, Vögeli U (1991) Elicitor-inducible 3-hydroxy-3-methylglutaryl coenzyme A reductase activity is required for sesquiterpene accumulation in tobacco cell suspension cultures. Plant Physiol 97:693–698

Chen XY, Chen Y, Heinstein P, Davisson VJ (1995) Cloning, expression, and characterization of (+)-δ-cadinene synthase: a catalyst of cotton phytoalexin biosynthesis. Arch Biochem Biophys 324:255–266

Clarke CF, Tanaka RD, Svenson K, Wamsley M, Fogelman AM, Edwards PA (1987) Molecular cloning and sequence of a cholesterol-repressible enzyme related to prenyltransferase in the isoprene biosynthetic pathway. Mol Cell Biol 7:3138–3146

Conolly JD, Hill RA (1992) Dictionary of terpenoids, vols 1–3. Chapman and Hall, London

Coolbear T, Threlfall DR (1985) The biosynthesis of lubimin from [1-^{14}C]isopentenyl pyrophosphate by cell-free extracts of potato tuber tissue inoculation with an elicitor preparation from *Phytophthera infestans*. Phytochemistry 24:1963–1971

Cordell GA (1976) Biosynthesis of sesquiterpenes. Chem Rev 76:425–460

Croteau R, Gundy A (1984) Cyclization of farnesyl pyrophosphate to the sesquiterpene olefins humulene and caryophyllene by an enzyme system from sage (*Salvia officinalis*). Arch Biochem Biophys 233:838–841

Davis GD, Essenberg M (1995) (+)-δ-Cadinene is a product of sesquiterpene cyclase activity in cotton. Phytochemistry 39:553–567

Davis GD, Eisenbraun EJ, Essenberg M (1991) Tritium transfer during biosynthesis of cadalene stress compounds in cotton. Phytochemistry 30:197–199

Davis GD, Tsuji J, Davis GD, Pierce ML, Essenberg M (1996) Purification of (+)-δ-cadinene synthase, a sesquiterpene cyclase from bacteria-inoculated cotton foliac tissue. Phytochemistry 41:1047–1055

Dehal SS, Croteau R (1988) Partial purification and characterization of two sesquiterpene cyclases from sage (*Salvia officinalis*) which catalyze the respective conversion of farnesyl pyrophosphate to humulene and caryophyllene. Arch Biochem Biophys 261:346–356

Desjardins AE, Gardner HW, Plattner RD (1989) Detoxification of the potato phytoalexin lubimin by *Gibberella pulicaris*. Phytochemistry 28:431–437

Dewick PM (1995) The biosynthesis of C_5–C_{20} terpenoid compounds. Nat Prod Rep 12:507–534

Dewick PM (1997) The biosynthesis of C_5–C_{20} terpenoid compounds. Nat Prod Rep 14:111–144

Dorsey JK, Dorsey JA, Porter JW (1966) The purification and properties of pig liver gernaly pyrophosphate synthase. J Biol Chem 241:5353–5360

Eberhardt NL, Rilling HC (1975) Prenyltransferase from *Saccharomyces cerevisiae*. Purification to homogeneity and molecular properties. J Biol Chem 250:863–866

Essenberg M, Stoessl A, Stothers JB (1985) Biosynthesis of 2,7-dihydroxycadalene in infected cotton cotyledons: the folding of the farnesyl precursor and possible implications for gossypol biosynthesis. Chem Commun 556–557

Essenberg M, Grover PB, Cover EC (1990) Accumulation of antibacterial sesquiterpeniod in bacterially inoculated *Gossypium* leaves and cotyledons. Phytochemistry 29:3107–3113

Facchini PJ, Chappell J (1992) Gene family for an elicitor-induced sesquiterpene cyclase in tobacco. Proc Natl Acad Sci USA 89:11088–11092

Fischer NH (1991) Sesquiterpenoid lactones. Methods Plant Biochem 7:187–211

Fischer NH, Olivier EJ, Fischer HD (1979) The biogenesis and chemistry of sesquiterpene lactones. Prog Chem Nat Prod 38:47–390

Fraga BM (1991) Sesquiterpenoids. Methods Plant Biochem 7:145–185

Fraga BM (1995) Natural sesquiterpenoids. Nat Prod Rep 12:303–320

Fraga BM (1996) Natural sesquiterpenoids. Nat Prod Rep 13:307–326

Fraga BM (1997) Natural sesquiterpenoids. Nat Prod Rep 14:145–162

Furze JM, Rhodes MCR, Parr AJ, Robins RJ, Whitehead IM, Threlfall DR (1991) Abiotic factors elicit sesquiterpenoid phytoalexin production but not alkaloid production in transformed root cultures of *Datura stamonium*. Plant Cell Rep 10:111–114

Gerber NN (1971) Sesquiterpenoids from actinomycetes. Cadin-4-ene-1-ol. Phytochemistry 10:185–189

Gijsen HJM, Wijnberg JBPA, de Groot A (1995) Structure, occurrence, biosynthesis, biological activity, synthesis, and chemistry of aromadendrane sesquiterpenoids. Prog Chem Org Nat Prod 64:149–193

Glasby JS (1982) Encyclopedia of the terpenoids. John Wiley, Chichester, 2643 pp

Gonzalez AG, Barrera JB (1995) Chemistry and sources of mono- and bicyclic sesquiterpenes from *Ferula* species. Prog Chem Org Nat Prod 64:1–92

Gorst-Allmann CP, Steyn PS (1982) Biosynthesis of PR toxin by *Penicillium roqueforti*. Part 2. Evidence for a hydride shift from 2H N, M, R, spectroscopy. Tetrahedron Lett 23:5359–5362

Govindachari TR, Mohamed PA, Parthasarathi PC (1970) Ishwarane and aristolochene, two new sesquiterpene hydrocarbons from *Aristolochia indica*. Tetrahydron 26:615–619

Groß D (1977) Phytoalexine und verwandte Pflanzenstoffe. Prog Chem Org Nat Prod 34:187–247

Guedes MEM, Kuc J, Hammerschmidt R, Bostock R (1982) Accumulation of six sesquiterpenoid phytoalexins in tobacco leaves infiltrated with *Pseudomonas lacgrymans*. Phytochemistry 21:2987–2988

Hanley KM, Voegeli U, Chappell J (1992) A study of the isoprenoid pathway in elicitor-treated tobacco cell suspension cultures. Environ Sci Res 44:329–336

Harborne JB (1993) Introduction to ecological biochemistry, 4th edn. Academic Press, New York

Harborne JB (1995) Ökologische Biochemie, eine Einführung. Spektrum Akademischer Verlag, Heidelberg

Harborne JB, Tomas-Barberan FA (eds) (1991) Ecological chemistry of plant terpenoids. Clarendon Press, Oxford, 439 pp

Harrison PHM, Oliver JS, Cane DE (1988) Pentalene biosynthesis and the enzymatic cyclization of farnesyl pyrophosphate. Inversion at C-1 during 11-membered-ring formation. J Am Chem Soc 110:5922–5923

Hendrickson JB (1959) Stereochemical implications in sesquiterpene biogenesis. Tetrahedron 7:82–89

Hohn TM, Plattner RD (1989) Purification and characterization of the sesquiterpene cyclase aristolochene synthase from *Penicillium roqueforti*. Arch Biochem Biophys 272:137–143

Hoshino T, Chida M, Yamaura T, Yoshizawa Y, Mizutani J (1994) Phytoalexin induction in green pepper cell cultures treated with arachidonic acid. Phytochemistry 36:1417–1419

Hoshino T, Yamaura T, Imaishi H, Chida M, Yoshizawa Y, Higashi K, Ohkawa H, Mizutani J (1995) 5-epi-Aristolochene 3-hydoxylase from green pepper. Phytochemistry 38:609–613

Hugueney P, Camara B (1990) Purification and characterization of farnesyl pyrophosphate synthase from *Capsicum annuum*. FEBS Lett 273:235–238

Joly A, Edwards PA (1993) Effect of site-directed mutagenesis of conserved aspartate and arginine residues upon farnesyl diphosphate synthase activity. J Biol Chem 268:26983–26989

Kalan EB, Osman SF (1976) Isolubimin: a possible precursor of lubimin in infected potato slices. Phytochemistry 15:775–776

Katsui N, Yagihasi F, Murai A, Masamune T (1978) Structure of oxyglutinosone and epi-oxylubimin, stress metabolites from diseased potato tubers. Chem Lett 1205–1206

Koyama T, Saito K, Ogura K, Obata S, Takeshita A (1993a) Site-directed mutagenesis of farnesyl diphosphate synthase: effect of substitution on the three carboxyl-terminal amino acids. Can J Chem 72:75–79

Koyama T, Obata S, Osabe M, Takeshita A, Yokoyama K, Uchida M, Nishino T, Ogura K (1993b) Thermostable farnesyl diphosphate synthase of *Bacillus stearothermophilus*: molecular cloning, sequence determinating, overproduction and purification. J Biochem (Tokyo) 113:355–363

Koyama T, Obata S, Osabe M, Saito K, Takeshita A, Nishino T, Ogura K (1994a) Thermostable farnesyl diphosphate synthase of *Bacillus stearothermophilus*: crystallization and site-directed mutagenese. Acta Biochim Pol 4:281–291

Koyama T, Obata S, Saito K, Takeshita-Koike A, Ogura K (1994b) Structural and functional roles of the cysteine residues of *Bacillus stearothermophilus* farnesyl diphosphate synthase. Biochemistry 33:12644–12648

Koyama T, Tajima M, Nishino T, Ogura K (1995) Significance of phe-200 and gln-221 in the catalytic mechanism of farnesyl diphosphate synthase of *Bacillus stearothermophilus*. Biochem Biophys Res Commun 212:681–686

Koyama T, Tajima M, Sano H, Doi T, Koike-Takeshita A, Obata S, Nishino T, Ogura K (1996) Identification of significant residues in the substrate-binding site of *Bacillus stearothermophilus* farnesyl diphosphate synthase. Biochemistry 35:9533–9538

Lawrence BM, Hogg JW (1973) Ishwarane in *Bixa orellana* leaf oil. Phytochemistry 12:2995

Lesburg CA, Lloyd MD, Cane DE, Christianson DW (1995) Chrystallization and preliminary X-ray diffraction analysis of recombinant pentalenene synthase. Protein Sci 4:2436–2438

Marrero PF, Poulter CD, Edwards PA (1992) Effects of site-directed mutagenesis of the highly conserved aspartate residues in domain II of farnesyl diphosphate synthase activity. J Biol Chem 267:21873–21878

McGarvey SJ, Croteau R (1995) Terpenoid metabolism. Plant Cell 7:1015–1026

Nabeta K, Katayama K, Nakagawara S, Katoh K (1993) Sesquiterpenes of cadinane type from cultured cells of the liverwort, *Heteroscyphus planus*. Phytochemistry 32:117–122

Nabeta K, Mototani Y, Tazaki H, Okuyama (1994a) Biosynthesis of sesquiterpenes of cadinane type in cultured cells of *Heteroscyphus planus*. Phytochemistry 35:915–920

Nabeta K, Ishikawa T, Kawae T, Okuyama (1994b) Biosynthesis of (1S)-7-methoxy-1,2-dihydrocadalene. Incorporation of ^{2}H- and ^{13}C-labelled mevanolates by cultured cells of *Heteroscyphus planus*. J Chem Soc Perkin Trans 1:3277–3280

Nabeta K, Kigure K, Fujita M, Nagoya T, Ishikawa T, Okuyama H, Takasawa T (1995) Biosynthesis of (+)-cubenene and (+)-epicubenol by cell-free extracts of cultured cells of *Heteroscyphus planus* and cyclization of [^{2}H]farnesyl diphosphate. J Chem Soc Perkin Trans 1:1935–1939

Picman AK (1984) Antifungal activity of the sesquiterpene lactones. Biochem System Ecol 12:13–19

Picman AK (1986) Biological activities of sesquiterpene lactones. Biochem Syst Ecol 14:255–281

Pierce M, Essenberg M (1987) Localization of phytoalexins in fluorescent mesophyll cells isolated from bacterial blight-infected cotton cotyledons and separated from other cells by fluorscence-activated cell sorting. Physiol Mol Plant Pathol 31:273–290

Pinder AR (1977) The chemistry of the eremophilane and related sesquiterpenes. Prog Chem Org Nat Subst 34:81–186

Poulter CD, Rilling HC (1981) Prenyl transferases and isomerases. In: Porter JW, Spurgeon SL (eds) Biosynthesis of isoprenoid compounds, vol 1. Wiley, New York, pp 161–224

Proctor RH, Hohn TM (1993) Aristolochene synthase: isolation, characterization, and bacterial expression of a sesquiterpenoid biosynthetic gene (Ari 1) from *Penicillium roqueforti*. J Biol Chem 268:4543–4548

Reed BO, Rilling HC (1975) Crystallization and partial Characterization of prenyltransferase from avian liver. Biochemistry 14:50–54

Rodriguez E, Towers GAN, Mitchell JC (1976) Biological activities of sesquiterpene lactones. Phytochemistry 15:1573–1580

Rodriguez E, Healey PL, Mehta I (1984) Biology and chemistry of plant trichomes. Plenum, New York

Ruzicka L (1953) The isoprene rule and the biogenesis of terpenic compounds. Experientia 9:357–367

Sato K, Ishiguri Y, Doke N, Tomiyama K, Yagishashi F, Murai A, Katsui N, Masamume T (1978) Biosynthesis of the sesquiterpenoid phytoalexin rishitin from acetate via oxylubimin in potato. Phytochemistry 17:1901–1902

Seaman FC (1982) Sesquiterpene lactones as taxonomic characters in the Asteraceae. Bot Rev 48:121–595

Sheares BT, White SS, Molowa DT, Chan K, Ding VDH; Kroon PA, Bostedor RG, Karkas JD (1989) Cloning, analysis, and bacterial expression of human farnesyl pyrophosphate synthetase and its regulation in Hep G2 cells. Biochemistry 28:8129–8135

Song L, Poulter CD (1994) Yeast farnesyl-diphosphate synthase: site-directed mutagenesis of residues in highly conserved prenyltransferase domains I and II. Proc Natl Acad Sci USA 91:3044–3048

Spring O, Bienert U (1987) Capitate glandular hairs from sunflower leaves: development, distribution and sesquiterpene lactone content. J Plant Physiol 130:441–448

Stoessl A (1980) Phytoalexins – a biogenetic perspective. Phytopathol Z 99:251–272

Stoessl A, Stothers JB (1980) Postinfectional inhibitors from plants. XXXVI. 2-epi and 15-dihydro-2-epi-Lubimin: new stress compounds from the potato. Can J Chem 58:2069–2072

Stoessl A, Stothers JB (1981) Postinfectional inhibitors from plants. XXXVII. A carbon-13 biosynthetic study of stress metabolites from potatoes: the origin of isolubimin. Can J Bot 59:637–639

Stoessl A, Stothers JB (1982) The beta-hop: acetate-2-^{13}C-d$_3$ as a probe for 1,2-hydride shifts in the biosynthesis of natural products. Chem Commun 880–881

Stoessl A, Stothers JB (1983) Carbon-13 magnetic resonance studies. 104. Postinfectional inhibitors from plants. 44. Biosynthesis of antifungal stress metabolites from potato: observation of hydride shifts via beta-hop from incorporation of [2-^{2}H$_3$,2-^{13}C]acetate and carbon-13 magnetic resonance. Can J Chem 61:1766–1770

Stoessl A, Unwin CH, Ward EWB (1973) Postinfectional fungus inhibitors from plants. Fungal oxidation of capsidiol in pepper fruit. Phytopathology 63:1225–1231

Stoessl A, Stothers JB, Ward EWB (1976a) Sesquiterpenoid stress compounds of the Solanaceae. Phytochemistry 15:855–872

Stoessl A, Ward EWB, Stothers JB (1976b) Incorporation of doubly labeled sodium acetate-^{13}C$_2$ into phytuberin and other sesquiterpenes in potatoes: experimental conformation of postulated carbon-carbon cleavage. Tetrahedron Lett 37:3241–3244

Stoessl A, Robinson JR, Rock GL, Ward EWB (1977) Metabolism of capsidiol by sweet pepper tissue: some possible implications of phytoalexin studies. Phytopathology 67:64–66

Stoessl A, Stothers JB, Ward EWB (1978) Carbon-13 NMR studies. Part 76. Postinfectional inhibitors from plants. XXIX. Biosynthetic studies of stress metabolites from potatoes: incorporation of sodium acetate-^{13}C$_2$ into 10 sesquiterpenes. Can J Chem 56:645–653

Sugawara F, Strobel G, Fisher LE, VanDuyne GD, Clardy J (1985) Bipolaroxin, a selective phytotoxin produced by *Bipolaris cynodontis*. Proc Natl Acad Sci USA 82:8291–8294

Tarshis LC, Yan M, Poulter CD, Sacchettini JC (1994) Crystal structure of recombinant farnesyl diphosphate synthase at 2,6-A resolution. Biochemistry 33:10871–10877

Threlfall DR, Whitehead IM (1988a) Co-ordinated inhibition of squalene synthetase and induction of enzymes of sesquiterpenoid phytoalexin biosynthesis in cultures of *Nicotiana tabacum*. Phytochemistry 27:2567–2580

Threlfall DR, Whitehead IM (1988b) The use of biotic and abiotic elicitors to induce the formation of secondary plant products in cell suspension cultures of solanaceous plants. Biochem Soc Trans 16:71–75

Threlfall DR, Whitehead IM (1991) Terpenoid phytoalexins: aspects of biosynthesis, catabolism and regulation. Proc Phytochem Soc Eur 31:159–208

Tomita B, Hirose Y (1972) Identity of cadinenol with (+)-epi-cubenol. Phytochemistry 11:3355–3357

Vögeli U, Chappell J (1988) Induction of sequiterpene cyclase and suppression of squalene synthase activity in plant cell cultures treated with fungal elicitor. Plant Physiol 88:1291–1296

Vögeli U, Chappell L (1990) Regulation of a sesquiterpene cyclase in cellulase-treated tobacco cell suspension cultures. Plant Physiol 94:1860–1866

Vögeli U, Chappell J (1991) Inhibition of a plant sesquiterpene cyclase by mevilonin. Arch Biochem Biophys 288:157–162

Vögeli U, Freeman JW, Chappell J (1990) Purification and characterization of an inducible sesquiterpene cyclase from elicitor-treated tobacco cell suspension cultures. Plant Physiol 93:182–187

Ward EWB, Stoessl A (1972) Postinfectional inhibitors from plants III. Detoxification of capsidiol, an antifungal compound from peppers. Phytopathology 62:1186–1187

Ward EWB, Stoessl A (1977) Phytoalexins from potatoes: evidence for the conversion of lubimin to 15-dihydrolubimin by fungi. Phytopathology 67:468–471

Ward EWB, Unwin CH, Stoessl A (1973) Capsidiol production in pepper fruit infected with bacteria. Phytopathology 63:1537–1538

Ward EWB, Unwin CH, Rock GL, Stoessl A (1976) Post-infectional inhibitors from plants XXIII. Sesquiterpenoid phytoalexins from fruits capsules of *Datura stramonium*. Can J Bot 54:25–29

Ward EWB, Stoessl A, Stothers JB (1977) Metabolism of the sesquiterpenoid phytoalexins capsidiol and rishitin to their 13-hydroxy derivatives by plant cells. Phytochemistry 16:2024–2025

Watson DG, Brooks CJW (1984) Formation of capsidiol in *Capsicum annuum* fruits in response to non-specific elicitors. Physiol Plant Pathol 24:331–337

Watson DG, Rycroft DS, Freer IM, Brooks CJW (1985) Sesquiterpenoid phytoalexins from suspended callus cultures of *Nicotiana tabacum*. Phytochemistry 24:2195–2200

Whitehead IM, Threlfall DR (1992) Production of phytoalexins by plant tissue cultures. J Biotechnol 26:63–81

Whitehead IM, Threlfall DR, Ewing DF (1987) cis-9,10-dihydrocapsenone; a possible catabolite of capsidiol from cell suspension cultures of *Capsicum annuum*. Phytochemistry 26:1367–1369

Whitehead IM, Ewing DF, Threlfall DR (1988) Sesquiterpenoids related to the phytoalexin debneyol from elicited cell suspension cultures of *Nicotiana tabacum*. Phytochemistry 27:1365–1370

Whitehead IM, Threlfall DR, Ewing DF (1989) 5-epi-Aristolochene is a common precursor of the sesquiterpenoid phytoalexins capsidiol and debneyol. Phytochemistry 28:775–779

Whitehead IM, Ewing DF, Threlfall DR, Cane DE, Prabhakaran PC (1990a) Synthesis of (+)-5-epi-aristolochene and (+)-1-deoxycapsidiol from capsidiol. Phytochemistry 29:479–482

Whitehead IM, Atkinson AL, Threlfall DR (1990b) Studies on the biosynthesis and metabolism of the phytoalexin lubimin and related compounds in *Datura stramonium* L. Planta 182:81–88
Yeh LS, Rilling HC (1977) Purification and properties of pig liver prenyltransferase: interconvertible forms of the enzymes. Arch Biochem Biophys 183:718–725
Yoshizawa Y, Yamaura T, Kawaii S, Hoshino T, Mizutani J (1994) Incorporation of 13C-labeled 5-epi-aristolochene into capsidiol in green pepper seedlings. Biosci Biotechnol Biochem 58:305–308
Zook MN, Kuc JA (1991) Induction of sesquiterpene cyclase and suppression of squalene synthetase activity in elicitor-treated or fungal-infected potato tuber tissue. Physiol Mol Plant Pathol 39:377–390

Professor Dr. Horst-Robert Schütte
Institut für Pflanzenbiochemie
Karl-Liebknecht-Straße 15
D-06114 Halle (Saale), Germany

Edited by
U. Lüttge

Systematics and Comparative Morphology

Systematics and Evolution of the Algae: Phylogenetic Relationships of Taxa Within the Different Groups of Algae

By Hans R. Preisig

1 General Aspects

This report deals with progress which has been made in recent years (1990–1996 and part of 1997) in the area of systematics and evolution of algae (excluding cyanobacteria). More than 3000 publications referring to algal systematics appeared in this period, but less than 300 can be cited here. Special reference is given to the phylogenetic relationships within the different groups of algae and the taxonomic implications, whereas results of studies on endocytobioses and evolution of major algal lineages are treated only very briefly. Readers interested in more detailed information on the latter areas are referred to a recent report by M. Melkonian (Prog. Bot. 57:281–311). As in other fields of biology, molecular studies also had a great impact on algal systematics in the past years (for methods used, see Melkonian's report). Together with new information obtained from other approaches (e.g., investigations on ultrastructure, life history studies, cladistic methods of character analysis, etc.), this has resulted in many unexpected new insights into phylogenetic relationships of taxa at all levels. All these tools are powerful in the information each can separately generate in the course of evolutionary investigations, and all are even more powerful when combined. Recognitions of several new taxa of higher rank have been achieved in the past years (e.g., the classes Pelagophyceae and Trebouxiophyceae) and relationships between various groups have been clarified to some degree. However, the new findings cause many taxonomic problems, especially in groups where morphological convergence in not closely related lineages obviously occurs (e.g., in coccoid green algae). Phylogenetic classification should certainly reflect the evolutionary history, i.e., it requires recognition of monophyletic groups. To achieve this will need a considerable effort from a nomenclatural point of view, but in the long term this will be most useful. Before undertaking such a task, however, it is certainly necessary to further explore the complementarity of phylogeny and morphology. The revolution and renaissance in algal systematics are well underway, and the promise for future directions is manifest.

a) Books and General Reviews

Only publications not yet mentioned in Melkonian's report are cited here (for textbooks and reviews on specific groups of algae, see also the following sections on these groups). An excellent new phycology textbook is presented by Van den Hoek et al. (1995; i.e., a translated and extended version of the German edition of 1993). An interesting book on the phylogeny of algae and their plastids, edited by Bhattacharya (1997), outlines the current knowledge and covers all major algal groups (for shorter reviews of phylogeny and evolution of the algae, see also Melkonian et al. 1995; Melkonian 1996). Books on algal ecology include Stevenson et al. (1996), who give special reference to benthic algae in freshwater habitats, and Vymazal (1995), who provides an introduction to algae of wetlands, dealing especially with the requirements and cycling of elements. Terrestrial algae (including algae of lichens) are treated extensively (with keys for their identification) by Ettl and Gärtner (1995). Yamagishi and Akiyama (1990–1997) continued their useful series on *Photomicrographs of the Freshwater Algae*, and Hori (1993) published three volumes with illustrations on algal life histories. Biogeography of freshwater algae is outlined in a book edited by Kristiansen (1996). Review articles of general interest include Norton et al. (1996) on algal biodiversity; and John and Maggs (1997) on species concepts in algae.

Aspects of applied phycology are covered in a book entitled *Algae, Environment and Human Affairs* (Wiessner et al. 1995) and much information on harmful and toxic algae is compiled in a manual which also includes several chapters on taxonomy (Hallegraeff et al. 1995). Harmful and toxic algae are also treated extensively in the *Proceedings of 6th and 7th International Conferences on Toxic Phytoplankton* (Lassus et al. 1995; Yasumoto et al. 1996).

Several books on seaweeds appeared in the past years, e.g., new volumes (nos. 3, 4, 5 and 6) of *Taxonomy of Economic Seaweeds* (Abbott 1992/1994/1995/1997), the *Proceedings of the 15th International Seaweed Symposium* (Lindstrom and Chapman 1996), and a very useful catalog of benthic marine algae of the Indian Ocean (Silva et al. 1996).

2 Euglenophyta

Simpson (1997) places the euglenophytes (euglenoids) with three groups of related organisms, i.e., (1) the kinetoplastids (e.g., *Bodo, Trypanosoma*), (2) the diplonemids (*Diplonema = Isonema, Rhynchopus*), and (3) the new genus *Postgaardi* (see Simpson et al. 1997) in a single taxon, the Euglenozoa (sensu Cavalier-Smith). This is considered to be an ancient group having diverged from the main eukaryotic line shortly after the eukaryotes arose (see, e.g., Cavalier-Smith 1993). From morphological and molecular data, Montegut-Felkner and Triemer (1997) suggest that euglenoids and kinetoplastids arose from a common ancestor and that the phagotrophic euglenoids diverged prior to the photosynthetic forms. They also suggest that the phagotrophic euglenoids with a pellicle composed of longitudinally arranged strips (e.g., *Petalomonas*) diverged prior to those genera with helically arranged pellicular strips (e.g., *Peranema*).

Since a feeding apparatus has been demonstrated in all phagotrophic genera in which ultrastructural studies have been carried out, it appears no longer justified to distinguish the orders Sphenomonadales and Heteronematales (sensu Leedale) depending on the absence or the presence of a special ingestion organelle, and both orders are proposed to be united into one order (Dawson and Walne 1994). There are problems in establishing clear generic boundaries in certain areas because of almost continuous variation among genera such as *Anisonema-Dinema-Entosiphon-Ploeotia* or *Dinema-Heteronema-Metanema* (for a survey of heterotrophic genera, see Larsen and Patterson 1991). This is discussed by Larsen and Patterson (1990), who offer operational critera for the assignment of species, accepting that these may define artificial taxa. *Lentomonas*, a new genus of phagotrophic euglenoids related to *Ploeotia*, is described by Farmer and Triemer (1994). An unusual new genus of a quadriflagellate, phototrophic euglenoid, *Tetreutreptia*, is described from waters of Maritime Canada (McLachlan et al. 1994; Triemer and Lewandoswki 1994). Apart from flagellar number, *Tetreutreptia* has many features in common with other described phototrophic euglenoids (e.g., with species of *Eutreptiella*, cf. Tomas 1993).

Euglenoids are one of several groups of flagellates which are subject to the problems relating to ambiregnal nomenclature, i.e., the naming of taxa may fall under the jurisdiction of both the zoological (ICZN) and botanical (ICBN) codes of nomenclature (Patterson and Larsen 1992). In response to these problems, Larsen and Patterson (1990, 1991) found it appropriate to introduce a double naming system, since there are instances where the correct generic name is different under two codes, e.g., *Dinema* (ICZN) = *Dinematomonas* (ICBN), *Entosiphon* (ICZN) = *Entosiphonomonas* (ICBN), *Peranema* (ICZN) = *Pseudoperanema* (ICBN).

3 Dinophyta (Dinoflagellata)

The relationship of dinoflagellates with ciliates and apicomplexans is now well established and it is also clear that plastids of dinoflagellates are of very diverse origin (for details the reader is referred to Prog. Bot. 52:278–279 and 57:303–304.

a) Books, Monographs, and Reviews

Important books on classification of living and fossil dinoflagellates
(Fensome et al. 1993) and on identification of freshwater and marine
dinoflagellates (Popovsky and Pfiester 1990; Tomas 1996) have been
published since 1990. Comprehensive chapters on Recent and fossil
dinoflagellates, including dinoflagellate cysts, are included in the *Paly-
nonogy* volumes of Jansonius and McGregor (1996). Toxic dinoflagellates, including their taxonomy, are treated extensively in the *Manual on
Harmful Marine Microalgae* (Hallegraeff et al. 1995). Other publications
of general interest deal with the diversity of heterotrophic dinoflagellates
(Larsen and Sournia 1991) and the organization of the flagellar apparatus and cytoskeleton and their use in systematics (Roberts 1991).

b) Taxonomy (Including Information on Life History)

Traditional classification of dinoflagellates is so far based primarily on
the structure of their cell wall (= amphiesma), but new molecular data
are challenging because they bring into question the established models
of dinoflagellate evolution (see below and Saunders et al. 1997c). *Oxyrrhis*, a genus which has been included in the dinoflagellates in the past,
is positioned in all of the recent evolutionary models that consider this
genus as an early divergence, sister to the Dinophyta/Dinoflagellata.
Fensome et al. (1993), for instance, consider *Oxyrrhis* to be a pre-
dinoflagellate, closer to the divergence of the ciliates. In their system of
classification of "true" dinoflagellates, they recognize four classes. The
first class, Syndiniophyceae (which are nonphotosynthetic marine flagellates parasitic in or on protists, invertebrates, and fish eggs), are considered to be an early divergence from the dinoflagellate line and sister to
the "Dinokaryota" (including the three classes Blastodiniphyceae, Nocti-
luciphyceae, and Dinophyceae).

c) Syndiniophyceae

New publications on members of this class are contributed by Fritz and
Nass (1992) on *Amoebophrya ceratii*, a parasite within host dinoflagellates such as *Dinophysis norvegica* and *Scrippsiella* cf. *trochoidea*, and by
Field et al. (1992) on a lobster parasite possibly related to *Hematodinium
perezii*.

d) Blastodiniphyceae

New information on parasitic dinoflagellates related to this group are presented by Landsberg et al. (1994) and Buckland-Nicks and Reimchen (1995).

e) Noctiluciphyceae

Schnepf and Drebes (1993) give evidence that several previously described elements of the sexual life cycle of *Noctiluca* are incorrect, e.g., they consider *Noctiluca* to be anisogamous (not isogamous). They also disagree that *Noctiluca* is a diplont with meiosis occurring during the formation of the swarmers. Höhfeld and Melkonian (1995) make a detailed study of the flagellar apparatus of *N. miliaris* and show that the swarmers are biflagellate (not uniflagellate as previously assumed).

f) Dinophyceae

Steidinger et al. (1996) place dinoflagellates possessing flagellated stages in their life cycle that can transform directly into lobose or filose amoeba-like cells (which form the dominant stage), in the reinstated order Dinamoebales (Fensome et al. unite this order with the Phytodiniales and place it in an uncertain subclass of the Dinophyceae). Steidinger et al. (1996) describe a new family of Dinamoebales, Pfiesteriaceae, including a new genus and species, *Pfiesteria piscicida*. This is a toxic dinoflagellate with a very unusual behavior and complex life cycle including multiple flagellated, amoeboid, and cyst stages. The species has been termed an ambush predator (Burkholder et al. 1995) because it releases a toxin that kills fish and then it phagocytizes the sloughed tissue of these dead or moribund fish. It is structurally a heterotrophic species, but flagellated stages can have cleptochloroplasts in large food vacuoles and can temporarily function as mixotrophs. Species identification is only possible if biflagellated or mastigote armored stages are available. The plate tabulation of the amphiesma is unlike that of any other armored dinoflagellate.

Referring to dinoflagellates of the subclasses Gymnodiniphycidae (orders Gymnodiniales, Suessiales), Perdiniphycidae (orders Gonyaulacales, Peridiniales), Dinophysiphycidae (order Dinophysiales), and Prorocentrophycidae (order Prorocentrales) (sensu Fensome et al. 1993), many important publications dealing with taxonomy and life history have been published since 1990, but only very few can be mentioned here. *Lepidodinium* is a new genus of Gymnodiniaceae, with a green chlorophyll *a*- and *b*-containing endosymbiont probably related to prasinophytes (Watanabe et al. 1990). A similar endosymbiont was found recently in a new species of *Gymnodinium, G. chlorophorum* (Elbrächter and Schnepf 1996; see also Schnepf 1993 for a recent review of dinoflagellate endosymbioses). New data on *Peridinium balticum* and *P. foliaceum (= Glenodinium foliaceum/Kryptoperidinium foliaceum)* suggest that these endosymbionts did originate from a photosynthetic diatom and not, as previously thought, from a chrysophycean alga (Chesnick et al. 1997).

The genus *Alexandrium* (Gonyaulacales), some species of which are producers of most potent neurotoxin, is reviewed in detail by Balech (1995). The recent studies on inter- and intraspecific relationships among representatives of this genus using molecular genetic methods are paradigm examples of the versatility of these methods in providing phylogenetic information down to the level of species and strain (e.g., Costas et al. 1995; Scholin and Anderson 1996).

Molecular data (see Saunders et al. 1997c) support the creation of the order Gonyaulacales, but this order is shown to be a sister to the Gymnodiniales/Peridiniales/Prorocentrales (GPP) complex and most likely should not be considered an order within the Perdiniphycidae. *Amphidinium* (classified in the Gymnodiniales by Fensome et al.) does not combine with other members of this order, but emerges early as a separate lineage. Taxonomy of *Amphidinium* should thus be reevaluated. The GPP complex is probably a relatively recent lineage, which is surprising, since in earlier hypotheses of dinoflagellate evolution the Prorocentrales (and sometimes also the Gymnodiniales) were often considered to be ancestral (cf. Fensome et al. 1993).

Studies on "zooxanthellae" revealed that several dinoflagellate groups have entered a symbiotic mode of existence; i.e., there are symbiotic species among genera of Suessiales (*Symbiodinium*), Gymnodiniales (*Gymnodinium, Amphidinium*), Peridiniales (*Scrippsiella*), Prorocentrales (*Prorocentrum*), and Phytodiniales (*Gloeodinium*) (see Banaszak et al. 1993, McNally et al. 1994; Trench and Thinh 1995). Sequence analyses among zooxanthellae assigned to the genus *Symbiodinium* support their placement in the separate order Suessiales (Saunders et al. 1997c). However, it appears that this order should not be included in the subclass Gymnodiniphycidae, but that it is a major lineage of the Dinophyceae, sister to the GPP complex sensu stricto and worthy of subclass status in the system of Fensome et al. (1993). Molecular data also give evidence for relatively substantial divergence between isolates of *Symbiodinium* reinforcing the conclusion that *Symbiodinium*-like zooxanthellae represent a collection of distinct species (Rowan and Powers 1992; McNally et al. 1994).

4 Prymnesiophyta (Haptophyta)

Molecular data do not support an affiliation of the Prymnesiophyta with the Heterokontophyta or with any other eukaryotic group (see Cavalier-Smith et al. 1996b, Medlin et al. 1997b) and are therefore dealt with as a separate division here. Much information about this group is compiled in a comprehensive book (edited by Green and Leadbeater 1994), and another book (edited by Winter and Siesser 1994) deals specifically with coccolithophorids. Species diversity of both non-coccolithophorids and coccolithophorids is surveyed in a book edited by Tomas (1993), while Moestrup and Thomsen (in Hallegraeff et al. 1995) review the toxic species. A glossary of terms used in studies of prymnesiophytes is presented by Jordan et al. (1995).

In a recent scheme of classification by Jordan and Green (1994), the single class Prymnesiophyceae is subdivided into two subclasses, the Pavlovophycidae (including the single order Pavlovales) and the Prymnesiophycidae (including the single order Prymnesiales). The Pavlovophycidae contains both flagellate organsims plus an undescribed coccoid organism whose taxonomic affinities were only recognized through sequence analysis (Potter et al. 1997b).

Many previous taxonomic schemes have divided the Prymnesiophycidae into several different orders based on morphological characters, but this led to a number of difficulties, such as reconciling the inclusion of some coccolithophorids in typically non-coccolithophorid orders based on the presence of unmineralized scales in the motile stage and the absence of a haptonema, and the problem of how best to classify heterococcolithophorids and holococcolithophorids in view of the fact that some species have both types of coccolith, but at different stages of their life cycle (see, e.g., Faber and Preisig 1994). The creation of the two haptophyte subclasses is supported by molecular data (see Medlin et al. 1997b).

Within the Prymnesiales, (1) a *Phaeocystis* (Phaeocystaceae) group, (2) a *Chrysochromulina/Prymnesium* (Prymnesiaceae) group, and (3) a coccolithophorid group can be distinguished. The taxonomy of *Phaeocystis* is unresolved, but the status of three species (*P. antarctica, P. globosa,* and *P. pouchetii*) was supported by recent sequence analyses (Medlin et al. 1994; for more information on the major bloom-forming genus *Phaeocystis*, see Davidson and Marchant 1992; Vaulot et al. 1994; Lange et al. 1996).

The Prymnesiaceae divide into two different clades, one comprising the *Prymnesium* species and *Chrysochromulina polylepis* (a massive bloom-forming species), and the second comprising most other species of *Chrysochromulina* (which is a paraphyletic genus; see Simon et al. 1997). Edvardsen and Vaulot (1996) describe two motile cell types of

C. polylepis (termed authentic and alternate cells) which differ in size and scale morphology; authentic cells proved to be always haploid, whereas alternate cells can be haploid or diploid (haploid alternate and authentic cells probably function as gametes, whereas diploid alternate cells may be the result of syngamy; see also Green and Leadbeater 1994 for other examples with a heteromorphic haploid-diploid life cycle among prymnesiophyte algae).

The coccolithophorids are considered to form a monophyletic group with the family of Noëlaerhabdaceae (*Emiliania* and *Gephyrocapsa*) sister to the remainder of the lineage (Medlin et al. 1997b). The taxonomy of the coccolithophorid *Emiliania huxleyi*, one of the world's most important species in terms of biomass production, is still unresolved. Recent studies using molecular techniques indicate, however, that there is extensive genetic variety, both on a global scale and within major bloom populations in both space and time (Medlin et al. 1996c; see also special issues of *Sarsia*, vol. 79/4, 1994, and of the *Journal of Marine Systems*, vol. 9/1–2, 1996, for more information on *E. huxleyi*).

Some Prymnesiophycidae which appear to be of significance for phylogenetic considerations, e.g., *Isochrysis and Dicrateria* (Jordan and Green 1994 include them both in a separate family, Isochrysidaceae), have not yet been studied by use of molecular genetic methods. It should be noted that Cavalier-Smith et al. (1996b), who use a scheme of classification different from that of Jordan and Green, place *Isochrysis* and *Dicrateria* in two different orders (Isochrysidales/Dicrateriales) and families (Isochrysidaceae/Dicrateriaceae).

Cavalier-Smith et al. (1996b) also established a new order (Reticulosphaerales) and family (Reticulosphaeraceae) for *Reticulosphaera*, a unique plastid-containing meroplasmodial organism, consisting of a network of cells joined by branching and anastomosing pseudopods. Grell (1990) placed *Reticulosphaera* in the Heterokontophyta, but the new molecular data by Cavalier-Smith et al. indicate that this genus branches well within the Prymnesiophycidae, closest to *Emiliania* and *Phaeocystis*, even though it differs radically from them in external body form. *Reticulosphaera* has apparently lost a haptonema, but it possesses a crescentic structure which is identical in appearance to the haptonematal axoneme base of other prymnesiophytes (Cavalier-Smith et al. 1996b).

5 Heterokontophyta (= Heterokont Chromophytes or Phototrophic Stramenopiles)

There are several groups of colorless flagellates (e.g., bicosoecids) and fungus-like organisms such as Oomycetes, Hyphochytridiomycetes, and Labyrinthulomycetes, which are related to heterokont algae, but only groups with phototrophic representatives are dealt with in this chapter.

Recent molecular analyses show the nonphotosynthetic lineages as early divergencies, whereas the photosynthetic heterokonts emerge later as a monophyletic group (see review by Medlin et al. 1997b). These data also resolve a number of phylogenetic relationships within the heterokontophytes, e.g., between Fucophyceae and Tribophyceae, between the Chrysophyceae and Synurophyceae, between the Dictyochophyceae and the Pelagophyceae (a new class described by Andersen et al. 1993), and between the Sarcinochrysidales sensu stricto and the Pelagophyceae. In some cases, controversial results were produced, e.g., Raphidophyceae form a sister relationship with Tribophyceae and Fucophyceae (according to Medlin et al. 1997b), whereas in the opinion of Cavalier-Smith and Chao (1996b), the Raphidophyceae are a sister taxon to the Eustigmatophyceae and Chrysophyceae/Synurophyceae. All these studies also failed to resolve unequivocally the relationships among deeper-branching heterokonts. For aspects of plastid phylogeny in Heterokontophyta the reader is referred to Prog. Bot. 57:304–305 and to a recent publication by Stoebe et al. (1997). These authors present evidence supporting the endosymbiotic origin of chloroplasts in Heterokontophyta (specifically in diatoms) from a red alga. Almost the entire chloroplast gene complement of the diatom *Odontella* is found within the plastid of the red alga *Porphyra purpurea*, suggesting that chloroplasts of heterokont algae descended from red algal plastids.

a) Eustigmatophyceae

The present knowledge on this class is summarized by Santos (1996). Schnepf et al. (1996) report on a species of *Pseudostaurastrum*, a genus previously assigned to the Tribophyceae. Even though the zoospores of this coccoid organism lack a large extraplastidial eyespot (a feature previously considered to be most characteristic for the Eustigmatophyceae), accommodation in this class is unequivocal due to other characteristics of cell structure and pigment composition. Undoubtedly, many other species which presently are classified as Tribophyceae will prove to belong to the Eustigmatophyceae when detailed studies are made.

b) Dictyochophyceae (Including Pedinellales and Rhizochromulinales)

Moestrup (1995) subdivides the Dictyochophyceae into three orders, Pedinellales, Rhizochromulinales, and Dictyochales (for a different system of classification, see Cavalier-Smith et al. 1995).

Pedinellales. This order currently comprises ten genera, including *Mesopedinella*, a new genus recently described by Daugbjerg (1996a).

Cavalier-Smith et al. (1995) place the phototrophic (mixotrophic) genera in two different families, Pedinellaceae (cells forming a stalk) and Apedinellaceae (cells without a stalk). Daugbjerg (1996b) made a cladistic analysis on ultrastructural data of eight genera of pedinellids, including the colorless *Actinomonas/Pteridomonas* (Actinomonadaceae; cells with a stalk) and *Ciliophrys* (Ciliophryaceae; cells without a stalk), indicating that *Pseudopedinella tricostata* is the most basal pedinellid. The division of the pedinellids in two major lineages based on presence or absence of plastids, as well as the further recognition of subgroups on the basis of presence or absence of stalks, is not supported by molecular data, as presented by Saunders et al. (1997a). Prior to further speculation on the relationships among pedinellid species, it is necessary that more taxa are investigated.

Rhizochromulinales. This order comprises at present only *Rhizochromulina marina*, a pigmented amoeboid species. This order may represent an intermediate group between the Pedinellales and Dictyochales, lacking a siliceous skeleton (as do pedinellids), but possessing bundles of microtubular tentacles similar to those observed in silicoflagellates (Moestrup 1995; O'Kelly and Wujek 1995).

Dictyochales (Silicoflagellates). One of the few extant species, *Dictyocha speculum*, has been studied recently with special reference being given to its autecology, life history, and toxicology (Henriksen et al. 1993). Fossil silicoflagellates, including several new taxa, are described in detail by Desikachary and Prema (1996).

c) Chrysophyceae and Synurophyceae

Detailed data from current research on both classes can be found in the *Proceedings of the International Chrysophyte Symposia* (Sandgren et al. 1995; Kristiansen and Cronberg 1996). Stomatocysts, the characteristic siliceous resting stages of species-specific shape and ornamentation of Chrysophyceae and Synurophyceae, are described and illustrated comprehensively by Duff et al. (1995). These structures, as well as the siliceous scales which are produced by members of these classes, are widely used as valuable indicators for assessing historical lakewater conditions (see, e.g., Smol 1995). Actual knowledge of stomatocyst and scale formation is summarized by Preisig (1994).

α) Chrysophyceae

Preisig (1995) subdivided the Chrysophyceae sensu stricto in three orders, Chromulinales (incl. Ochromonadales), Hibberdiales, and Hydrurales. In Preisig's system of classification, the Chrysophyceae sensu lato also comprise the Chrysomeridales and Sarcinochrysidales (see below).

Chromulinales. Many pigmented and a few colorless genera are included in this order (for a survey of the heterotrophic genera, see Preisig et al. 1991). A species of the colorless genus *Oikomonas* was recently studied by ultrastructural and molecular genetic methods (Cavalier-Smith et al. 1996a) and was clearly shown to be related to the Chrysophyceae sensu stricto (though Cavalier-Smith et al. use a system of classification different from that of Preisig 1995). *Chrysonephele*, a colonial chloroplast-containing genus, has also been clearly shown to be a member of the Chrysophyceae by use of molecular data (Saunders et al. 1997a). This is interesting, since previously this genus was suggested to be a possible phylogenetic link between the Chrysophyceae and Eustigmatophyceae. A new genus of chrysophycean flagellate with organic scales, *Chrysolepidomonas* (representing a new family, Chrysolepidomonadaceae), was described by Peters and Andersen (1993a, b), and includes some species previously assigned to *Sphaleromantis*.

Hibberdiales. Recent studies on *Chromophyton* (see Preisig 1995) and *Lagynion* (O'Kelly and Wujek 1995) revealed a general organization of the flagellar apparatus similar to that of *Hibberdia* (cf. Prog. Bot. 52:285), suggesting that these and possibly a number of other genera (e.g., loricate rhizopodial genera such as *Chrysopyxis* and *Stylococcus*, which are normally classified with *Lagynion* in the family Stylococcaceae), should be classified with the Hibberdiales.

Chrysomeridales. Preisig (1995) included this order in the Chrysophyceae sensu lato, but recent ultrastructural and molecular data appear to support separate class status for this group (see Cavalier-Smith et al. 1995). However, the phylogenetic affinities of this class among the other lineages of heterokonts is not yet clear, though from gene sequence analyses a weak relationship with the Fucophyceae/Tribophyceae clade is suggested (Saunders et al. 1997b).

β) Synurophyceae

Analyses of morphological and molecular genetic data corroborate that the Synurophyceae represent a monophyletic assemblage, sister to the Chrysophyceae (Lavau et al. 1997; Medlin et al. 1997b). *Tessellaria* is weakly resolved as earliest divergence within the Synurophyceae, which is consistent with earlier suggestions based on scale case morphology and development that *Tessellaria* retains some primitive features relative to *Synura* and *Mallomonas* (see Pipes and Leedale 1992; Kristiansen and Vigna 1994). The data presented by Lavau et al. weakly support the studied representatives of the genera *Mallomonas* and *Synura* as monophyletic groupings and upheld several of the sections within these genera that are recognized by current classifications (for a detailed account of *Mallomonas*, see Siver 1991). However, further studies may prove that some changes to the classification and delineation of these genera may be necessary. *Synura lapponica* (section *Lapponicae*), for instance, is very similar to *Tessellaria* based on colony and scale morphology, implying that these taxa may need to be redefined.

d) Diatomophyceae (Bacillariophyceae)

α) Books, Monographs, and Reviews

An important book was published on the biology and morphology of genera of diatoms (Round et al. 1990). Krammer and Lange-Bertalot (1991a, b) finished their treatment of diatoms (vols 2/3 and 2/4) in the book series *Süsswasserflora von Mitteleuropa* (see also Lange-Bertalot 1993). Most valuable are the books of Tomas (1996) for the identification of marine diatoms and the book of Cox (1996) for identification of freshwater diatoms (the latter is the first book which enables the user to work from live specimens). Sims (1996) edited a book entitled *Atlas of British Diatoms* which contains over 6000 illustrations covering almost all known forms of British freshwater, brackish, and littoral marine species. Since the majority of species has been recorded in literature as cosmopolitan, this Atlas is also useful outside Britain. New books (including Proceedings of Diatom Symposia) with many original contributions on taxonomy of diatoms have been edited by Sims (1993), Van Dam (1993), Kociolek (1994), Kociolek and Sullivan (1995), Marino and Montresor (1995), and Jahn et al. (1997). Other useful publications on diatoms appeared in the Journal *Diatom Research* (Biopress, Bristol), and in the series *Bibliotheca Diatomologica* (J. Cramer, Berlin) and *Iconographia Diatomologica: Annotated Diatom Micrographs* (Koeltz Scientific Books, Königstein/Germany). Reviews on subjects with taxonomic implications include Cox (1993) on past and present practice in diatom systematics, Mann (1993) on sexual reproduction in diatoms, Mann and Droop (in Kristiansen 1996) on biodiversity, biogeography, and conservation of diatoms, Mann (1996) on chloroplast morphology, movements and inheritance in diatoms, McQuoid and Hobson (1996), on diatom resting stages, Harwood and Nikolaev (1996) on taxonomy and biostratigraphy of fossil diatoms, and Medlin et al. (1997a) on the origin of diatoms. Gaul et al. (1993) published an important bibliography on the fine structure of diatom frustules, covering more than 4000 diatom names and a period of more than 50 years of electron microscopic studies.

Harmful diatoms are treated by Hallegraeff et al. (1995). The species causing most of the problems are usually *Pseudo-nitzschia multiseries* (syn. *Nitzschia pungens* f. *multiseries*) and *P. australis* (syn. *Nitzschia pseudoseriata*), which may produce massive amnesic shellfish poisoning by means of the toxin domoic acid (for reviews see Hasle et al. 1996 and Fryxell et al. 1997).

β) Taxonomy

Round et al. (1990) classifiy the diatoms in the division Bacillariophyta and recognize three classes, Coscinodiscophyceae (centric diatoms), Fragilariophyceae (araphid pennate diatoms), and Bacillariophyceae (raphid pennate diatoms), and they also establish many new taxa, including subclasses, orders, families, and genera.

Molecular data indicate that the centric and the araphid pennate diatoms represent paraphyletic lineages (Medlin et al. 1996a, b, 1997b). According to these data, the diatoms diverge into two clades (clades 1 and 2). Clade 1 comprises the Thalassiosirales, the pennate diatoms (raphids and araphids) and the bipolar (multipolar) centric diatoms with a central labiate process (Chaetocerotales, Hemiaulales, Cymatosirales, and Lithodesmiales). Clade 2 comprises the radial centric diatoms with peripheral rings of labiate processes (Melosirales, Coscinodiscales, Corethrales, and Rhizosoleniales). The central tube-like structure in fossil diatoms of the Lower Cretaceous may represent an ancestral structrure from which the central strutted process of Thalassiosirales, the central labiate process in the bipolar centric diatoms, and probably the raphe of the pennate diatoms may have evolved. Some ultrastructural features also support the two clades recovered in the molecular tree, e.g., taxa of clade 2 (such as *Coscinodiscus*) have their Golgi bodies associated with a mitochondrion within cisternae of the endoplasmic reticulum, whereas clade 1 taxa (Thalassiosirales and most, if not all, pennates) have their Golgi bodies either as perinuclear shell or in a Plattenband. From the molecular data it is suggested that the traditional features of the morphology of the silica cell wall are only valuable in defining younger branches of the tree.

The enormous number of important publications dealing with traditional aspects of systematics at lower taxonomic level, including descriptions of hundreds of new taxa at generic, specific, or subspecific level, unfortunately cannot be reviewed here due to limitations of space in this report.

e) Raphidophyceae

In a cladistic analysis of combined data sets (nucleotide sequences, ultrastructure, and pigments) the genera of both marine and freshwater raphidophytes were united with high bootstrap values, supporting the hypothesis that they form a monophyletic group (Potter et al. 1997a). Hallegraeff et al. (1995) outline the taxonomy of harmful marine raphidophytes (genera *Chattonella, Fibrocapsa, Heterosigma, Olisthodiscus*), all of which are included in a single order (Chattonellales) and family (Chattonellaceae) by Throndsen (in Tomas 1993). A new marine genus, *Haramonas*, was recently described from Australian mangrove habitats (Horiguchi 1996).

f) Tribophyceae (Xanthophyceae)

Preliminary results from gene sequence analyses on a number of genera indicate that there is litte congruence with the presently used ordinal and familial classification based on morphological features (Potter et al. 1997a), but clearly additional investigations are necessary. Noteworthy studies on Tribophyceae relating to taxonomy have been done on a new coccoid genus, *Pseudogoniochloris* (Krienitz et al. 1993), the filamentous genera *Heterococcus* (Lokhorst 1992) and *Xanthonema* (= *Heterothrix*) (Broady et al. 1997), and the siphonous genera *Vaucheria* (Linne and Kowallik in Kristiansen 1996), and *Pseudodichotomosiphon* (a genus sometimes assigned to the Chlorophyceae in the past; Fukushi-Fujikura et al. 1991).

g) Fucophyceae (Phaeophyceae)

α) Books, Monographs, and Reviews

Much information on ecology and physiology of brown algae (Fucophyceae) can be found in a book of Lobban and Harrison (1994). Other interesting new publications of general interest include Bell (1997) on the evolution of the life cycle, Kawai (1992) on the morphology of chloroplasts and flagellated cells, and Kusel-Fetzmann (1996) on freshwater species of brown algae.

β) Taxonomy

Traditionally, the Fucophyceae are subdivided into several orders which are distinguished mainly on the basis of life cycle type, reproduction, tissue organization, and growth pattern (for a recent system of classification, see Silva and Moe in Lobban and Harrison 1994). However, based on molecular data, the brown algae are grouped in two main clades contrary to traditional views (Druehl et al. 1997). Clade 1 includes Ectocarpales, Scytosiphonales, Chordariales, and Dictyosiphonales, i.e., mostly small filamentous or pseudoparenchymatous forms which were considered by some authors in the past to belong to a single order, Ectocarpales. Clade 2 includes Ralfsiales, Sphacelariales, Syringodermatales, Tilopteridales, Cutleriales, Dictyotales, Desmarestiales, Sporochnales, Laminariales, and Fucales, i.e., all the "seaweeds" with massive thalli, though there are also some small and simple forms in this group. Representatives of clade 1 are characterized by having normal pyrenoids, whereas in representatives of clade 2 pyrenoids are rudimentary or lacking.

So far, not many molecular studies have been made to resolve the phylogenetic relationships between orders and families within the two clades. However, among Sporochnales, Desmarestiales, and Laminariales, close evolutionary relationships appear to exist (Tan and Druehl 1996). Within the Laminariales the molecular data indicate that the "primitive" Pseudochordaceae/Chordaceae/Phyllariaceae families are phylogenetically isolated from the "advanced" Alariaceae/Laminariaceae/Lessoniaceae complex (Druehl et al. 1997). The latter three families, as traditionally conceived, proved to be an unnatural assemblage and should be redefined based on the new data. Molecular phylogenetic analyses on European representatives of Fucales suggest that this order is monophyletic, but among them there are two robust clades, one corresponding to the families Fucaceae and Himanthaliaceae on one side, and a Cystoseiraceae-Sargassaceae group on the other (Rousseau et al. 1997).

Other important publications on brown algal families or major genera include Cheshire et al. (1995) on *Durvillaea* (representing the monogeneric family, Durvillaeacaceae, which was originally assigned to the Fucales but now is usually classified in its own order, Durvillaeales), Clayton (1994) on the Southern Hemisphere family Seirococcaceae (Fucales), and Kilar et al. (in Abbott 1992) on the genus *Sargassum* (Fucales).

h) Pelagophyceae (Including Sarcinochrysidales)

This class was originally described as including *Pelagomonas calceolata*, a new genus of marine ultraplanktonic flagellate (Andersen et al. 1993), but, as presently conceived (Saunders et al. 1997b), it also comprises coccoid members such as *Pelagococcus* and *Aureococcus* (the latter is a common brown tide organism; DeYoe et al. 1995) and genera such as *Sarcinochrysis, Ankylochrysis, Nematochrysopsis,* and *Pulvinaria* (= *Chrysoreinhardia* nom. nud.), which in the past were usually classified in a seperate order (Sarcinochrysidales) of the Chrysophyceae sensu lato (see, e.g., Honda and Inouye 1995; Preisig 1995). Possibly, the new genus *Sulcochrysis* (Honda et al. 1995) should also be classified with the Pelagophyceae. Potter et al. (1997b) revealed that the gene sequences in the flagellate *Pelagomonas calceolata* (type species) may be identical to those in a culture strain which produces only coccoid cells, suggesting a close relationship between the flagellate and coccoid organisms (perhaps they represent two stages in the life history of the same organism).

6 Cryptophyta

From sequence data it is suggested that the cryptophytes form a weakly supported sister group to the glaucocystophytes (cf. Progr. Bot. 57:302; Fraunholz et al. 1997), whereas other phylogenetic trees generated by Cavalier-Smith (1996c) are shown to be consistent with but do not positively support the view that the closest relatives of the cryptophytes are the Chromobiota (heterokontophytes plus haptophytes). The plastids of cryptophytes are considered to be red-algal-like endosymbionts which contain a reduced eukaryotic nucleus (nucleomorph) that is not specifically related to chlorarachniophyte nucleomorphs (Cavalier-Smith et al. 1996c). McFadden et al. (1994) show that a most basal position in the cryptophyte lineage is occupied by the colorless phagotrophic flagellate *Goniomonas* (previously often referred to as *Cyathomonas*; see Larsen and Patterson 1990). However, it is not yet clear whether the cryptophyte common ancestor was primarily colorless or photosynthetic, i.e., *Goniomonas* may have lost its plastid secondarily (Cavalier-Smith et al. 1996c). The colorless, leucoplast-containing flagellate *Chilomonas* is shown to be a sister to all photosynthetic cryptomonads.

The results of Cavalier-Smith et al. (1996c) suggest that genera with nucleomorphs embedded in a chloroplast-envelope invagination into the pyrenoid (*Pyrenomonas/Rhodomonas, Rhinomonas, Storeatula*) represent a clade consistent with the order Pyrenomonadales (see below). Cryptomonads ancestrally having free nucleomorphs are much more diverse. Nucleomorph trees show *Chroomonas* and *Komma* (both with the blue accessory pigment phycocyanin) to form a sister clade to red-pigmented cryptomonads such as *Guillardia* (= *Cryptomonas* sp.Θ *theta*) and *Hanusia* (= *Cryptomonas* sp.Φ *phi*; Deane et al. 1998), but nuclear host sequences support this only weakly. Cavalier-Smith et al. (1996c) suggest that the red and blue cryptomonads diverged early by differential pigment loss.

A new classification system for cryptophytes is proposed by Novarino and Lucas (1993, 1995), recognizing three orders (Goniomonadales/ Goniomonadida, Pyrenomonadales/Pryrenomonadida, and Cryptomonadales/Cryptomonadida). At present, the generic taxonomy of cryptomonads is based mainly on the nature of the furrow-gullet system, the periplast structure (for a review, see Brett et al. 1994), and features of the plastidial complex (for general cell organization, see Kugrens and Lee 1991; for a survey of described genera of cryptophytes, see Throndsen in Tomas 1993). Genera of phagotrophic flagellates such as *Katablepharis/Kathablepharis* and *Leucocryptos* that have previously been placed in the cryptomonads are no longer included in this systematic group (Vørs 1992; Cavalier-Smith 1993).

7 Chlorarachniophyta

Phylogenetic trees indicate that the chlorachniophyte host is closely related to proteomyxids (*Leucodictyon*), plasmodiophorids (*Plasmodiophora*), sarcomonads (e.g., *Cercomonas, Heteromita, Thaumatomonas*) and filose amoebae (e.g., *Fonticula, Vampyrella, Euglypha, Paulinella*), whereas the endosymbiont (nucleomorph-containing plastid) is most closely related to green algae (cf. Prog. Bot. 57:303; Cavalier-Smith 1996; Cavalier-Smith and Chao 1996a; McFadden et al. 1997a). The nucleomorph contains three linear chromosomes with a haploid genome size of 380 kb and is the smallest known eukaryotic genome (Gilson and McFadden 1995). McFadden et al. (1997b) give new information on chlorachniophyte storage products, which are considered to be key characters to define the major groups of algae. It is shown that the principal storage carbohydrate of chlorarchniophytes is a long-chain β-1,3 glucan[1] which is localized within a vacuole in the host cell cytoplasm, suggesting that photosynthate produced by the endosymbiont is stored by the host.

The chlorarachniophytes currently comprise four genera and species, *Chlorarachnion reptans, Cryptochlora perforans, Gymnochlora stellata,* and *Lotharella globosa (= Chlorarachnion globosum)* (see Ishida and Hara 1994; Ishida et al. 1996), but there are several undescribed species which still await full characterization (cf. Hori 1993 and Daugbjerg et al. 1996, who suggest that a species in culture, previously identified as *Pedinomonas minutissima*, is not a prasinophycean flagellate but presumably a chlorachniophyte). Main features used to differentiate the genera include pyrenoid ultrastructure and location of the nucleomorph in the

---

[1] Other groups of algae storing β-1,3 glucans are heterokontophytes (storing laminarans) and euglenophytes (storing paramylon), whereas α-1-4 glucans with occasional α-1-6 side branches (starches) are stored by dinoflagellates, cryptomonads, red algae, and green algae.

periplastidial compartment, whereas vegetative cell morphology and life cycle patterns are considered as features characterizing the species (Ishida et al. 1996). A sexual cycle has been reported from some species (Grell 1990; Beutlich and Schnetter 1993), but no details of meiosis or ploidy levels for various phases have so far been established, so proof of true sexuality is wanting.

8 Rhodopyhta

α) Books, Monographs, and Reviews

A comprehensive book with general information on red algae has been published by Cole and Sheath (1990). Excellent new red algal flora volumes of *Seaweeds of the British Isles* on Ceramiales (Maggs and Hommersand 1993) and Corallinales and Hildenbrandiales (Irvine and Chamberlain 1994) and of the *Marine Benthic Flora of Southern Australia* (Womersley 1994, 1996) appeared recently. Reviews on subjects with implications for taxonomy include Guiry (1992) on species concepts, Ragan and Gutell (1995) on relationships of red algae with other groups of eukaryotes, and Saunders and Bailey (1997) and Saunders and Kraft (1997), among others, on recent advances in red algal systematics that are attributable to molecular studies.

New data on parasitic red algae from molecular studies are contributed by Goff et al. (1996). They compared sequences of hosts and adelphoparasites (i.e., parasites that are closely related to their hosts) in four different red algal orders (Gracilariales, Gigartinales, Plocamiales, Rhodymeniales) and found that each adelphoparasite has evolved either directly from the host on which it is currently found, or it evolved from some other taxon that is closely related to the modern host. Zuccarello and West (1994a, b) describe a new genus of alloparasite (parasite that is not similar to its host), i.e., *Bostrychiocolax* (Choreocolacaceae, Gigartinales), growing on some species of *Bostrychia* and *Stictosiphonia* (Rhodomelaceae, Ceramiales).

β) Taxonomy

The Rhodophyta traditionally consists of one class, Rhodophyceae, and two subclasses, Bangiophycidae and Florideophycidae. Molecular data indicate that the Bangiophycidae (orders Cyanidiales, Porphyridiales, Compsopogonales, Bangiales) are polyphyletic, whereas the Florideophycidae with some 16 orders appears to be monophyletic (Ragan et al. 1994; Saunders and Kraft 1997). For more information on these orders, see Cole and Sheath (1990).

γ) Bangiophycidae

Cyanidiales (sometimes classified as a separate class Cyanidiophyceae beside the Rhodophyceae). Representatives of this group, which all live exclusively in acidic hot springs, are treated extensively in a book edited by Seckbach (1994) who, together with F.D. Ott, proposes to reclassify all the *Cyanidium* and *Galdieria* species in the single genus *Cyanidium* in the subclass Bangiophycidae. The genus *Cyanidioschizon*, which was usually accommodated in the same order in the past, is proposed to be transferred to the Prophyridiales (family Porphyridiaceae), mainly because reproduction is by binary fission (as in *Porphridium* and related genera) and not by formation of endospores as in members of the Cyanidiales.

Prophyridiales. *Glaucosphaera*, a genus previously classified by several authors with the Glaucocystophyta, is shown to be related to unicellular red algae and is now accommodated in the Porphyridiales (Broadwater et al. 1995).

Compsopogonales (including Erythropeltidales). Detailed studies on the freshwater family Compsopogonaceae were made, e.g., by Vis et al. (1992).

Bangiales. From molecular studies on one *Bangia* and several *Porphyra* species, it appears that this order is monophyletic, but there appears to be a remarkable degree of internal divergence (Ragan et al. 1994, Oliveira et al. 1995). Endospore formation, a form of reproduction previously known only from Porphyridiales and Compsopogonales among Bangiophycidae, was observed recently in a species of *Porphyra* (Nelson and Knight 1995). For information on the systematics of *Porphyra* see Lindstrom and Cole (1993).

δ) Florideophycidae

The traditional Kylinian ordinal foundations are severely brought into question by new molecular data, but also additional morphological data (such as ultrastructure of pit connections) and cladistic methods of character analysis provide new insights in the relationships between the major red algal lineages. Saunders and Kraft (1997) review the relevant data and as a result of their considerations recognize four distinct lineages within the Florideophycidae, though the relationships between these lineages are not yet fully resolved.

Lineage 1 comprises a single order (Hildenbrandiales) and family (Hildenbrandiaceae), which appear to have an isolated position relative to the remainder of the Florideophycidae. However, the data are still limited, since so far only one marine species, *Hildenbrandia rubra*, was studied from a molecular standpoint. Freshwater species such as *H. angolensis* closely resemble *H. rubra*, but differ by the absence of tetrasporangia and the presence of gemmae (Sheath et al. 1993a).

Lineage 2 is constituted by six orders: Rhodogorgonales, Corallinales, Batrachospermales, Nemaliales, Acrochaetiales, and Palmariales.

Rhodogorgonales. This order was recently established by Fredericq and Norris (1995) for two bizzare, recently discovered Caribbean genera that combine the lubricous worm-like habit of some Nemaliales, several gonimoblast characters of Batrachospermales, and the deposition of calcite, suggesting a relationship with Corallinales.

Corallinales. Verheij (1993) separates *Sporolithon* from the Corallinaceae and places it in a new family (Sporolithaceae), which is characterized by tetrasporangia that produce cruciately arranged spores and develop within calcified sporangial compartments (Corallinaceae have tetrasporangia that produce zonately arranged spores which do not develop in calcified sporangial compartments; see also Townsend et al. 1994/1995 for a new genus, *Heydrichia*, assigned to the Sporolithaceae). Within the Corallinaceae, two sister clades have been resolved by means of a molecular sequence analysis of 23 species (14 genera and 5 subfamilies; Bailey and Chapman 1996). One clade contains only nongeniculate species of the subfamily Melobesioideae and the second includes geniculate and nongeniculate representatives of the subfamilies Corallinoideae, Amphiroideae, Mastophoroideae, and Metagoniolithoideae. Harvey and Woelkerling (1995) describe two new genera of Corallinaceae (*Austrolithon* and *Boreolithon*), which they place in a new subfamily, Austrolithoideae, based on the presence of multiporate tetrasporangial conceptacles and the lack of cell fusions, secondary pit connections, and genicula.

Batrachospermales. The systematics of this order of freshwater red algae (families Batrachospermaceae, Lemaneaceae, Thoreaceae) is reviewed by Entwisle and Necchi (1992). A new genus, *Rhododraparnaldia*, which appears to be intermediate between Batrachospermales and Acrochaetiales, is described by Sheath et al. (1994). There are many recent publications especially on Batrachospermaceae (e.g., Kumano 1993; Vis and Sheath 1996) but also several on Lemaneaceae (e.g., Sheath et al. 1996b) and Thoreaceae (e.g., Sheath et al. 1993b). The latter familiy is considered to have an uncertain position in the Batrachospermales (Schnepf 1992).

Nemaliales. Preliminary results from molecular studies indicate that this group is a monophyletic sister to the Acrochaetiales/Palmariales complex, although the relationships are not clear (Saunders and Kraft 1997). These authors recognize two families, Galaxauraceae and Liagoraceae.

Acrochaetiales/Palmariales complex. Molecular data indicate that the Acrochaetiales cannot be considered to be ancestral to the remaining Florideophycidae as it was often suggested in the past. The data support an alliance between Acrochaetiales and Palmariales and indicate that the Acrochaetiales is a comparatively recent derivative among the orders of lineage 2 (Saunders et al. 1995; Saunders and Kraft 1997). Saunders and McLachlan (1991) describe a new genus, *Meiodiscus*, for a species previously assigned to *Rhodochorton* (Acrochaetiales) and place it in the family Rhodophysemataceae (Palmariales). *Rhodothamniella*, another genus previously assigned to the Acrochaetiales, is placed by Saunders et al. (1995) in a new family (Rhodothamniellaceae) in the order Palmariales.

Lineage 3 comprises a single order (Ahnfeltiales), family (Ahnfeltiaceae) and genus (*Ahnfeltia*) with three cool- to cold-temperate species. Species of similar habit, assigned to a new genus (*Ahnfeltiopsis*) by Silva and DeCew (1992), are retained in the order Gigartinales (family Phyllophoraceae) due to different life histories, different pit-plugs, internal rather than completely external carposporophytes, and carragenans rather than agar as main nonfibrillar polysaccharides (see below).

Lineage 4 is constituted by Gelidiales, Bonnemaisoniales, Gracilariales, Gigartinales/Plocamiales/Halymeniales, Rhodymeniales, and Ceramiales.

Gelidiales. Acceptance of Gelidiales as a separate order is strongly promoted by every molecular study that has included the group to date (see review by Saunders and Kraft 1997). Freshwater et al. (1995), who made a study of several genera, found ten well-supported major clades representing genera and species complexes. Their data suggest that *Gelidium* and *Pterocladia*, as currently circumscribed, are not monophyletic. For more information on Gelidiales, see Norris in Abbott (1992).

Bonnemaisoniales. Available molecular data for members of this order do not support a close association with any other lineage 4 order (Saunders and Kraft 1997; see also Womersley 1996).

Gracilariales. Molecular data support the ordinal status of this group (Saunders and Kraft 1997), although not in a monophyletic association with the two other agarophyte orders Ahnfeltiales and Gelidiales as previously proposed. Within the order, sequence divergence was found to be relatively large (Bird et al. 1992). For a key to the genera of Gracilariaceae, see Fredericq and Hommersand (1990).

Gigartinales/Plocamiales/Halymeniales. Kraft and coworkers proposed to subsume the large classical order Cryptonemiales into the similarly large Gigartinales, creating a megaorder, Gigartinales, comprising some 40 families (see Womersley 1994; Saunders and Kraft 1997). Recently, Saunders and Kraft (1994) removed the family Plocamiaceae from the Gigartinales to a separate, new order, Plocamiales (to which the families Pseudoanemoniaceae and Sarcodiaceae may also belong). Furthermore, Saunders and Kraft (1996) removed the family Halymeniaceae (originally classified in the Cryptonemiales) to another new order, Halymeniales (to which the families Sebdeniaceae, Corynomorphaceae, Nemastomataceae and Schizymeniaceae may also belong). All these taxonomic decisions are supported by molecular data, now leaving a perhaps monophyletic assemblage of gigartinalean families.

Many papers dealing with the taxonomy of specific taxa in this group have been published since 1990, but because of limited space, reference to only a few of these can be given. Recently, Hommersand et al. (1993, 1994) proposed a revised classification of the family Gigartinaceae in which 69 species are classified into four extant (*Chondrus, Gigartina, Iridaea, Rhodoglossum*) and three reinstated (*Chondracanthus, Sarcothalia, Mazzaella*) genera based on developmental and morphological criteria. In a study on several species of Phyllophoraceae (Gigartinales), the molecular and morphological data presented by Fredericq and Ramirez (1996) challenge the current taxonomic concept that type of life history is a phylogenetically valid criterion for recognition of genera in this family. In a survey of the genera of Nemastomataceae (Halymeniales?, see above), Masuda and Guiry (1995) segregated *Schizymenia, Platoma*, and (with uncertainty) *Titanophora* to their own family, Schizymeniaceae, based on Nematostomataceae tribus Schizymenieae, and transferred *Platoma marginiferum* to a new genus (*Itonoa*) of Nemastomataceae.

Rhodymeniales. The analyses of Saunders and Kraft (1996) provide support for continued recognition of this order as distinct from the Gigartinales sensu stricto, but apparently most closely related to the Halymeniales. Preliminary molecular data indicate that the families of this order are not monophyletic, the Rhodymeniaceae being polyphyletic and the Lomentariaceae paraphyletic (Saunders and Kraft 1996).

Ceramiales. There is molecular evidence that the Ceramiales diverge deep within lineage 4 and the long-standing view that they represent the most "advanced" red algae is not confirmed (Saunders et al. 1996; Saunders and Kraft 1997). This order, the largest of all red algal orders, comprises four families (Ceramiaceae, Delesseriaceae, Dasyaceae, and Rhodomelaceae). These were treated in many recent publications (containing a great number of descriptions of new taxa) but, due to limited space, only very few can be mentioned here. Antithamnioid red algae (subfamily Ceramioideae of Ceramiaceae) are dealt with extensively by Athanasiadis (1996), who describes several new taxa including tribes and genera (see also Cormaci and Furnari 1994, for a new genus, *Halosia*, and tribe, *Halosieae*, of Ceramioideae). Wynne (1996) presents a useful revised key to genera (including several recently described genera) of the family Delesseriaceae. For distribution and systematics of freshwater Ceramiales, see Sheath et al. (1993c).

9 Glaucocystophyta (Glaucophyta)

Seven genera (*Cyanophora, Cyanoptyche, Glaucocystis, Glaucocystopsis, Gloeochaete, Peliaina,* and *Strobilomonas*) are assigned to this division (for a review, see Kies 1992). *Glaucosphaera*, not as previously thought, is shown not to belong to the glaucocystophytes but to be a member of the red algae (Broadwater et al. 1995; Helmchen et al. 1995). Phylogenetic analyses of the nuclear- and plastid-encoded small subunit ribosomal DNA provide evidence for a monophyletic origin of the glaucocystophyte host cell within the eukaryotic crown group radiation (forming a weakly supported sister group to the cryptophytes, cf. Prog. Bot. 57:301–302). The cyanelles of the glaucocystophytes are considered to be true plastids which are of monophyletic origin within this lineage (Bhattacharya et al. 1995; Bhattacharya and Schmidt 1997).

10 Chlorophyta

α) Books, Monographs, and Reviews

No comprehensive books on green algae appeard in the past years (for a book of limited coverage see Vijayaraghavan and Kumari 1995). Phylogeny and evolution of green algae (including charophytes) are reviewed (among others) by Melkonian and Surek (1995), Friedl (1997) and Huss and Kranz (1997). For monographs and reviews of specific groups of green algae see below.

β) Taxonomy

Green plants (including green algae) are monophyletic, consisting of two major lineages; one comprising the Streptophyta (= Klebsormidiophyceae, Zygnematophyceae, Charophyceae and the embryophytes), the second comprising all other green algae (= Chlorophyta sensu stricto, see below). In both lineages several classes and orders are distinguished, which in the following survey are arranged as in the recent scheme of classification by Van den Hoek et al. (1995), except for one class, the Pleurastrophyceae. As shown by Friedl (1996), the type species of *Pleurastrum (P. insigne)* is clearly a member of the Chlorophyceae, and most species/genera included in the Pleurastrophyceae by Van den Hoek et al. (1995) are now accommodated in the Trebouxiophyceae, a new class described by Friedl (1995). New molecular data indicate that several green algal groups as circumscribed by Van den Hoek et al. (1995) should be redefined, since they include species which are phylogenetically not closely related. However, it is certainly necessary that more detailed studies (including a great number of species) are performed before new formal definitions are made.

a) Chlorophyta Sensu Stricto

α) Prasinophyceae (= Micromonadophyceae)

This class comprises a heterogenous assemblage of lineages that arise at the base of the radiation of the Chlorophyta (see, e.g., Steinkötter et al. 1994; Daugbjerg et al. 1995; Marin 1996). For more information on Prasinophyceae see also review by Sym and Pienaar (1993). Some authors consider *Pedinomonas* and related genera (*Resultor* and *Marsupiomonas*) to be phylogenetically isolated and include them in a separate class, Pedinophyceae (Moestrup 1991; Jones et al. 1994; Daugbjerg et al. 1996). However, from recent molecular analyses on *Pedinomonas tuberculata* (Marin 1996) it is suggested that this genus (and related genera) should be classified at ordinal rank (**Pedinomonadales**) in the Prasinophyceae, since it appears to represent a lineage diverging between the **Pseudoscourfieldiales** (*Pseudoscourfieldia, Nephroselmis*) and **Chlorodendrales** (*Tetraselmis, Scherffelia*). Other major lineages within the Prasinophyceae are the **Pyramimonadales** (including *Pyramimonas, Halosphaera, Cymbomonas, Pterosperma,* and *Tasmanites = Pachysphaera*) and the **Mamiellales** (Mamiellaceae, Micromonadaceae) to which the genus *Monomastix* is apparently also related (Marin 1996). The genus *Pycnococcus* (classified in the Mamiellales, family Pycnococcaceae, by Guillard (1991), is considered by Marin (1996) to form a separate evolutionary lineage in the Prasinophyceae, whereas Daugbjerg et al. (1995) consider this genus to be closely related to *Pseudoscourfieldia*. Apart from *Pycnococcus* some other coccoid genera of Prasinophyceae have recently been described (*Bathycoccus, Ostreococcus, Prasinococcus,* and *Prasinoderma*; see, e.g., Hasegawa et al. 1996), but their positions in Mamiellales/Pycnococcales are not yet clearly established. *Ostreococcus* is especially interesting, since it is the smallest eukaryote (diameter ca. 0.8–1.1 x 0.5–0.7 μm), with the smallest DNA content per cell (33.31 fg) known to date (Chrétiennot-Dinet et al. 1995). Another recently described order of the Prasinophyceae is the **Scourfieldiales** (including the single genus *Scourfieldia*; Moestrup 1991), but detailed ultrastructural and molecular data on species of this genus have not yet been published. For *Mesostigma*, see under Streptophyta below.

β) Chlorophyceae

Van den Hoek et al. (1995) subdivide this class into four orders, (1) Volvocales (including Chlamydomonadales, Dunaliellales, Tetrasporales), (2) Chlorococcales (including genera such as *Chlorosarcinopsis, Geminella, Binuclearia, Radiofilum, Cylindrocapsa* and *Sphaeroplea*), (3) Chaetophorales sensu stricto, and (4) Oedogoniales.

Volvocales. Molecular data indicate that this order is not monophyletic (Buchheim et al. 1996, Nakayama et al. 1996b, Friedl 1997). *Chlamydomonas* is a polyphyletic genus within a group consisting of several genera traditionally assigned to Volvocales and Chlorococcales, including *Pleurastrum insigne*, suggesting that the traditional concepts of both Volvocales and Chlorococcales will need revision. Species of the wall-less genera *Dunaliella, Hafniomonas,* and *Polytomella* ally with some species of *Chlamydomonas* (Nakayama et al. 1996b), supporting the contention that multiple losses of glycoprotein cell walls occurred in this group (for taxomony of *Dunaliella*, see Preisig 1992).

The colonial green flagellates apparently had multiple origins from unicells (Buchheim et al. 1994). See Nozaki et al. (1995) for a study on the phylogenetic relationships within the colonial Volvocales; and Nozaki and Ito (1994) for a new family (Tetrabaenaceae) of colonial Volvocales.

A molecular study on *Oltmannsiellopsis*, a genus previously assigned to the Volvocales/Dunaliellales, reveals that this genus is only distantly related to other members of this group (Nakayama et al. 1996b). It is suggested to be an early divergence in the Ulvophyceae/Trebouxiophyceae/Chlorophyceae clade, but so far it has not yet been possible to clarify to which algal class this genus belongs. It has certainly no close relationship to any other known chlorophyte order (family), and therefore the new names Oltmannsiellopsidales (Oltmannsiellopsidaceae) were established.

Chlorococcales. Molecular analyses reveal that coccoid green algae, formerly placed in the single order, Chlorococcales, are distributed over several lineages, including lineages assigned to another green algal class, the Trebouxiophyceae. In fact, many autosporic coccoid green algae are now known which have affinities with the Trebouxiophyceae. Most species of *Chlorella* (including the type species *C. vulgaris*) are members of the Trebouxiophyceae, whereas other species assigned to *Chlorella* (e.g., *C. zofingensis*) belong to the Chlorophyceae. Only those species closely related to *C. vulgaris* may represent the genus *Chlorella*, while those that have their origin in other lineages need to be included in other genera.

Convergence of similar vegetative morphology also occurs among zoospore-forming coccoid green algae. Floyd et al. (1993) show that epiphytic unicellular forms previously assigned to the genus *Characium* belong to three distinct lineages, i.e., Chlorococcales (*Chlamydopodium*), Sphaeropleales (*Characiopodium* gen. nov.) and Trebouxiophyceae (*Fusochloris* gen. nov.). *Chlorococcum* also proves to be a polyphyletic genus with species belonging to different lineages in the Chlorophyceae. To evaluate the taxonomic position of the genus and the delimitation of the order Chlorococcales, sequence data on the type species and on many more taxa of coccoid green algae are needed. Due to limited space, it is not possible to make reference to any other recent publications on coccoid green algae, including descriptions of numerous new taxa.

Sphaeropleales. Deason et al. (1991) present an emended diagnosis of this order, which includes unicellular, filamentous and coenobic non-motile green algae producing motile cells with directly opposed basal bodies. Families assigned to this order are Sphearopleaceae, Hydrodictyaceae, and Neochloridaceae. From molecular data it appears that *Scenedesmus* also belongs to the Sphaeropleales (Wilcox et al. 1992; for taxonomy of *Scenedesmus* see also Kessler 1991 and Trainor 1991). Watanabe and Floyd (1996) list all genera of this order on which ultrastructural data of zoospores are known to date.

Chaetophorales. Only few studies have been done in the last years on members of this order. Molecular analyses on representatives of Chlorophyceae including *Chaetophora incrassata* indicate that Chaetophorales represent a distinct order in this class (Nakayama et al. 1996b).

Chaetopeltidales. This is a new order, established by O'Kelly et al. (1994) to accommodate genera such as *Chaetopeltis, Dicranochaete, Hormotilopsis*, and *Planophila*, i.e., genera that are characterized by having quadriflagellate scaly zoospores with cruciate flagellar apparatuses. Previously, these genera were scattered among the orders Tetrasporales, Chlorococcales, Chlorosarcinales, and Chaetophorales. From ultrastructural and molecular data it appears that this order is ancestral with respect to other Chlorophyceae (O'Kelly et al. 1994; Nakayama et al. 1996b).

Oedogoniales. There is a great need that the position of this order is studied from a molecular viewpoint. Taxonomy of *Oedogonium* and *Bulbochaete* is preliminarily revised by a classical approach (Mrozinska 1991, 1993).

γ) Trebouxiophyceae

This class, described by Friedl (1995), comprises many green algae that completely lack motile stages (autosporic coccoids, e.g., *Chlorella*), but also zoospore-producing forms, including filamentous forms such as the Microthamniales (= Pleurastrales sensu Mattox and Stewart; see Bakker 1995) and probably also multiseriate filamentous forms such as the Prasiolales (see Friedl 1997). The Trebouxiophyceae appear as an array of several independent lineages whose interrelationships are not yet resolved. Most known members of this class live in terrestrial habitats or occur in symbioses with lichen fungi (e.g., *Trebouxia* spp.) or invertebrates (zoochlorellae). The molecular data show that the capacity to exist in lichen associations has multiple independent origins and that many lichen algae are derived from nonsymbiotic terrestrial green algae.

The order Chlorellales, previously considered to belong to the Chlorophyceae, is now classified within the Trebouxiophyceae. Genera such as *Nanochlorum* or the colorless *Prototheca* are also representatives of this order (for taxonomy of *Prototheca* see Wong and Beebee 1994). Some filamentous species previously assigned to the genus *Pleurastrum* (i.e., *P. erumpens* and *P. terrestre*) are now known to be related to trebouxiophycean green algae and Friedl (1996) transferred them to the genus *Leptosira*.

δ) Ulvophyceae

From molecular analyses, it is suggested that the Ulvophyceae/Ulotrichales (Codiolales) occupy a basal position within the Ulvophyceae/Chlorophyceae/Trebouxiophyceae clade (Zechman et al. 1990; Friedl 1997). A new order (Phaeophilales) and family (Phaeophilaceae) of Ulvophyceae is described by Chappell et al. (1990) due to unique zoosporangial structure and developmental sequence, as well as unusual flagellar apparatus features in *Phaeophila*. Friedl (1996) shows that some species previously assigned to the genus *Pleurastrum* (i.e., *P. paucicellulare* and *P. sarcinoideum*) are related to ulotrichalean green algae and these are transferred to the genus *Gloeotilopsis*. A review of the biology of *Enteromorpha* (Ulvales), including aspects of taxonomy, is contributed by Poole and Raven (1997).

ε) Trentepohliophyceae

Molecular data suggest an association of this class with Ulvophyceae (Friedl 1997). A monograph on Trentepohliales, with special reference to the genera *Cephaleuros*, *Phycopeltis*, and *Stomatochroon*, is published by Thompson and Wujek (1997).

ξ) Cladophorophyceae

The hypothesis, based on ultrastructural features, that the Siphonocladales and Cladophorales (S/C complex) are closely related is supported by molecular data suggesting that there is no basis for their independent recognition (Zechman et al. 1990; Bakker et al. 1994). The analyses support two principal lineages, of which one contains predominantly tropical members including almost all siphonocladalean taxa, while the other lineage consists of mostly warm- to cold-temperature species of *Cladophora*. Unclear is the position of *Blastophysa*, a genus which is usually placed in the siphonocladalean family Chaetosiphonaceae based on morphological, cytological and biochemical features (Chappell et al. 1991). The ecologically important genus *Cladophora* (Cladophorales) is reviewed by Dodds and Gudder (1992), with some reference being given to its taxonomy.

η) Bryopsidophyceae

Molecular data indicate that this is a monophyletic group with two lineages; *Bryopsis* and *Codium* (Bryopsidales) comprising one, and *Caulerpa, Halimeda*, and *Udotea* (Halimedales/Caulerpales) comprising the other (Zechman et al. 1990). For information on an extremely harmful tropical species of *Caulerpa* (*C. taxifolia*) introduced into the Mediterranean, see Boudouresque et al. (1994) and Bellan-Santini et al. (1996).

θ) Dasycladophyceae

Morphological, ultrastructural, biochemical, and limited sequence data (Zechman et al. 1990; Bakker et al. 1994) all support monophyly of this class. The phylogenetic relationships based on sequence data of 14 species (representing 8 of the 11 extant genera) of this class were analyzed by Olsen et al. (1994). A comprehensive monograph on this group, including a chapter on systematics, is presented by Berger and Kaever (1992).

b) Streptophyta

This term has been created by Bremer to combine the following groups of green algae and the embryophytes into a single division (see, e.g., Huss and Kranz 1997). Most interestingly, a freshwater flagellate (*Mesostigma viride*), which was previously accommodated in the Prasinophyceae, has recently been shown to belong to the streptophyte lineage (see Melkonian et al. 1995).

α) Klebsormidiophyceae

Wilcox et al. (1993) analyzed gene sequences of three members of this group and found *Chlorokybus* (Chlorokybales) to be the most basal taxon, followed by the branching of *Coleochaete* (Coleochateales) and *Klebsormidium* (Klebsormidiales). There is considerable disagreement among different authors as to whether a member of this group (e.g., *Coleochaete*) or rather a *Chara*-like alga was ancestral to the land plants (for a discussion, see Huss and Kranz 1997). Lokhorst (1996) made detailed comparative taxonomic studies on species of *Klebsormidium*.

β) Zygnematophyceae

Molecular data, as presented, e.g., by Surek et al. (1994), demonstrate an evolutionary relationship between members of this class with other streptophyte groups. The order Zygnematales is shown to be not monophyletic (Bhattacharya et al. 1994; McCourt et al. 1995). Thus, *Zygnemopsis* (filament with twin stellate chloroplasts) appears to be more closely related to *Mesotaenium* (unicell with laminate chloroplasts) than to *Mougeotia* (filament with laminate chloroplasts). The genus *Roya* (usually classified with the Mesotaeniaceae in Zygnematales) appears to be closely related to *Gonatozygon* and *Genicularia* (Gonatozygaceae, Desmidiales; see Park et al. 1996).

Classification and general aspects of biology of desmids are reviewed by Gerrath (1993) and Coesel (in Kristiansen 1996) discusses biogeography. Two useful volumes of a desmid flora of Austria are contributed by Lenzenweger (1996, 1997), and a final volume of the New Zealand desmid flora by Croasdale et al. (1994).

γ) Charophyceae

To date, all molecular data sets support the monophyly of the single extant order Charales (for two additional extinct orders, see Feist and Grambast-Fessard 1991). From analyses of species from each of the six extant genera there is support that the Charales/Characeae comprise two tribes, i.e., a monophyletic (apparently more derived) tribe Chareae (*Chara, Lamprothamnium, Nitellopsis, Lychnothamnus*), whereas the basal topology and relationships of the genera in tribe Nitelleae (*Nitella, Tolypella*) are not so clear (McCourt et al. 1996; Meiers et al. 1997).

Krause (1997) deals extensively with extant Charales in a new volume of the *Süsswasserflora von Mitteleuropa*. Fossil taxa are treated by Feist and Grambast-Fessard (1991), and García (1994) discusses their use in paleolimnology. Information on the oosporangium of Charales is reviewed by Leitch et al. (1990), and Haas (1994) presents a key for the identification of oospores from central Europe. For a current discussion of charophyte evolution and the origin of land plants see Huss and Kranz (1997).

References

Abbott IA (ed) (1992/1994/1995/1997) Taxonomy of economic seaweeds, vols 3, 4, 5, 6. Reports T-CSGCP-023, -031, -035, -040. California Sea Grant College System, LaJolla

Andersen RA, Saunders GW, Paskind MP, Sexton JP (1993) Ultrastructure and 18S rRNA gene sequence for *Pelagomonas calceolata* gen. et sp. nov. and the description of a new algal class, the Pelagophyceae classis nov. J Phycol 29:701–715

Athanasiadis A (1996) Morphology and classification of the Ceramioideae (Rhodophyta) based on phylogenetic principles. Opera Bot 128:1–221

Bailey JC, Chapman RL (1996) Evolutionary relationships among coralline red algae (Corallinaceae, Rhodophyta) inferred from 18S rRNA gene sequence analysis: In: Chaudhary BR, Agrawal SB (eds) Cytology, genetics and molecular biology of algae. SPB Academic Publishing, Amsterdam pp 363–376

Bakker FT, Olsen JL, Stam WT, Van den Hoek C (1994) The *Cladophora* complex (Chlorophyta): new views based on 18S rRNA gene sequences. Mol Phylog Evol 3:365–382

Bakker ME (1995) A deviating pattern of cell division in the green alga *Microthamnion*: ultrastructure of vegetative cell division and zoosporogenesis. Arch Protistenkd 146:117–136

Balech E (1995) The genus *Alexandrium* Halim (Dinoflagellata). Sherkin Island Marine Station, Ireland

Banaszak AT, Iglesias-Prieto R, Trench RK (1993) *Scrippsiella velellae* sp. nov. (Peridiniales) and *Gloeodinium viscum* sp. nov (Phytodiniales), dinoflagellate symbionts of two hydrozoans (Cnidaria). J Phycol 29:517–528

Bell G (1997) The evolution of the life cycle of brown seaweeds. Biol J Linn Soc 60:21–38

Bellan-Santini D, Arnaud PM, Bellan G, Verlaque M (1996) The influence of the introduced tropical alga *Caulerpa taxifolia*, on the biodiversity of the Mediterranean marine biota. J Mar Biol Assoc UK 76:235–238

Berger S, Kaever MJ (1992) Dasycladales: an illustrated monograph of a fascinating algal order. Thieme, Stuttgart

Beutlich A, Schnetter R (1993) The life cycles of *Cryptochlora perforans* (Chlorarachniophyta). Bot Acta 106:441–447

Bhattacharya D (ed) (1997) Origins of algae and their plastids. Plant Syst Evol Suppl 11:1–287

Bhattacharya D, Schmidt HA (1997) Division Glaucocystophyta. Plant Syst Evol Suppl 11:139–148

Bhattacharya D, Surek B, Rüsing M, Damberger S, Melkonian M (1994) Group I introns are inherited through common ancestry in the nuclear-encoded rRNA of Zygnematales (Charophyceae). Proc Natl Acad Sci USA 91:9916–9920

Bhattacharya D, Helmchen T, Bibeau C, Melkonian M (1995) Comparison of nuclear-encoded small-subunit ribosomal RNAs reveal the evolutionary position of the Glaucocystophyta. Mol Biol Evol 12:415–420

Bird CJ, Rice EL, Murphy CA, Ragan MA (1992) Phylogenetic relationships in the Gracilariales (Rhodophyta) as revealed by 18S rDNA sequences. Phycologia 31:510–522

Boudouresque C-F, Meinesz A, Gravez V (eds) (1994) First International Workshop on *Caulerpa taxifolia*. GIS Posidonie, Marseille

Brett SJ, Perasso L, Wetherbee R (1994) Structure and development of the cryptomonad periplast: a review. Protoplasma 181:106–122

Broadwater ST, Scott JL, Goss SPA, Saunders BD (1995) Ultrastructure of vegetative organization and cell division in *Glaucosphaera vacuolata* Korshikov (Porphyridiales, Rhodophyta). Phycologia 34:351–361

Broady PA, Ohtani S, Ingerfeld M (1997) A comparison of strains of *Xanthonema* (= *Heterothrix*, Tribonematales, Xanthophyceae) from Antarctica, Europe and New Zealand. Phycologia 36:164–171

Buchheim MA, McAuley MA, Zimmer EA, Theriot EC, Chapman RL (1994) Multiple origins of colonial green flagellates from unicells: evidence from molecular and organismal characters. Mol Phylog Evol 3:322–343

Buchheim MA, Lemieux C, Otis C, Gutell RR, Chapman RL, Turmel M (1996) Phylogeny of the Chlamydomonadales (Chlorophyceae): a comparison of ribosomal RNA gene sequences from the nucleus and the chloroplast. Mol Phylog Evol 5:391–402

Buckland-Nicks J, Reimchen T (1995) A novel association between an endemic stickleback and a parasitic dinoflagellate: 3. Details of the life cycle. Arch Protistenkd 145:165–175

Burkholder JM, Glasgow HB Jr, Steidinger KA (1995) Stage transformations in the complex life cycle of an ichthyotoxic "ambush predator" dinoflagellate. In: Lassus P, Arzul G, Erard-Le Denn E, Gentien P, Marcaillou-Le Baut C (eds) Harmful marine algal blooms. Lavoisier, Paris, pp 567–572

Cavalier-Smith T (1993) Kingdom Protozoa and its 18 phyla. Microbiol Rev 57:953–994

Cavalier-Smith T (1996) Amoeboflagellates and mitochondrial cristae in eukaryote evolution: megasystematics of the new protozoan subkingdoms Eozoa and Neozoa. Arch Protistenkd 147:237–258

Cavalier-Smith T, Chao EE (1996a) Sarcomonad ribosomal RNA sequences, rhizopod phylogeny, and the origin of euglyphid amoebae. Arch Protistenkd 147:227–236

Cavalier-Smith T, Chao EE (1996b) 18S RNA sequence of *Heterosigma carterae* (Raphidophyceae) and the phylogeny of heterokont algae (Ochrista). Phycologia 35:500–510

Cavalier-Smith T, Chao EE, Allsopp MTEP (1995) Ribosomal RNA evidence for chloroplast loss within Heterokonta: pedinellid relationships and a revised classification of ochristan algae. Arch Protistenkd 145:209–220

Cavalier-Smith T, Chao EE, Thompson CE, Hourihane SL (1996a) *Oikomonas*, a distinctive zooflagellate related to chrysomonads. Arch Protistenkd 146:273–279

Cavalier-Smith T, Allsopp MTEP, Häuber MM, Gothe G, Chao EE, Couch JA, Maier U-G (1996b) Chromobiote phylogeny: the enigmatic alga *Reticulosphaera japonensis* is an aberrant haptophyte, not a heterokont. Eur J Phycol 31:255–263

Cavalier-Smith T, Couch JA, Thorsteinsen KE, Gilson P, Deane JA, Hill DRA, McFadden GI (1996c) Cryptomonad nuclear and nucleomorph 18S rRNA phylogeny. Eur J Phycol 31:315–328

Chappell DF, O'Kelly CJ, Wilcox LW, Floyd GL (1990) Zoospore flagellar apparatus architecture and the taxonomic position of *Phaeophila dendroides* (Ulvophyceae, Chlorophyta). Phycologia 29:515–523

Chappell DF, O'Kelly CJ, Floyd GL (1991) Flagellar apparatus of the biflagellate zoospores of the enigmatic marine green alga *Blastophysa rhizopus*. J Phycol 27:423–428

Cheshire AC, Conran JG, Hallam ND (1995) A cladistic analysis of the evolution and biogeography of *Durvillaea* (Phaeophyta). J Phycol 31:644–655

Chesnick JM, Kooistra WHCF, Wellbrock U, Medlin LK (1997) Ribosomal RNA analysis indicates a benthic pennate diatom ancestry for the endosymbionts of the dinoflagellates *Peridinium foliaceum* and *Peridinium balticum* (Pyrrhophyta). J Euk Microbiol 44:314–320

Chrétiennot-Dinet M-J, Courties C, Vaquer A, Neveux J, Claustre H, Lautier J, Machado MC (1995) A new marine picoeucaryote: *Ostreococcus tauri* gen. et sp. nov. (Chlorophyta, Prasinophyceae). Phycologia 34:285–292

Clayton MN (1994) Circumscription and phylogenetic relationships of the Southern Hemisphere family Seirococcaceae (Phaeophyceae). Bot Mar 37:213–220

Cole KM, Sheath RG (eds) (1990) Biology of the red algae. Cambridge University Press, Cambridge

Cormaci M, Furnari G (1994) *Halosia elisae* gen. et sp. nov. (Ceramiaceae, Rhodophyta) from the Mediterranean Sea and *Halosieae* trib. nov. Phycologia 33:19–23

Costas E, Zardoya R, Bautista J, Garrido A, Rojo C, Lopez-Rodas V (1995) Morphospecies vs. genospecies in toxic marine dinoflagellates: an analysis of *Gymnodinium catenatum/Gyrodinium impudicum* and *Alexandrium minutum/A. lusitanicum* using antibodies, lectins, and gene sequences. J Phycol 31:801–807

Cox EJ (1993) Diatom systematics: a review of past and present practice and a personal vision for future development. Nova Hedwigia Beih 106:1–20

Cox EJ (1996) Identification of freshwater diatoms from live material. Chapman and Hall, London

Croasdale H, Flint EA, Racine MM (1994) Flora of New Zealand. Desmids, vol 3. Manaaki Whenua Press, Lincoln, New Zealand

Daugbjerg N (1996a) *Mesopedinella arctica* gen. et. sp. nov. (Pedinellales, Dictyochophyceae). I. Fine structure of a new marine phytoflagellate from Artic Canada. Phycologia 35:435–445

Daugbjerg N (1996b) *Mesopedinella artica* (Pedinellales). II. Phylogeny of *Mesopedinella*, including a cladistic analysis of Dictyochophyceae. Phycologia 35:563–568

Daugbjerg N, Moestrup Ø, Arctander P (1995) Phylogeny of genera of Prasinophyceae and Pedinophyceae (Chlorophyta) deduced from molecular analysis of the *rbcL* gene. Phycol Res 43:203–213

Davidson AT, Marchant HJ (1992) The biology and ecology of *Phaeocystis* (Prymnesiophyceae). Prog Phycol Res 8:1–46

Dawson NS, Walne PL (1994) Evolutionary trends in euglenoids. Arch Protistenkd 144:221–225

Deane JA, Hill DRA, Brett SJ, McFadden GI (1998) *Hanusia phi* gen. et sp. nov. (Cryptophyceae): characterization of "*Cryptomonas* sp. Φ". Eur J Phycol 33:149–154

Deason TR, Silva PC, Watanabe S, Floyd GL (1991) Taxonomic status of the species of the green algal genus *Neochloris*. Plant Syst Evol 177:213–219

Desikachary TV, Prema P (1996) Silicoflagellates (Dictyochophyceae). Bibl Phycol 100:1–298

DeYoe HR, Chan AM, Suttle CA (1995) Phylogeny of *Aureococcus anophagefferens* and a morphologically similar bloom-forming alga from Texas as determined by 18S ribosomal RNA sequence analysis. J Phycol 31:413–418

Dodds WK, Gudder DA (1992) The ecology of *Cladophora*. J Phycol 28:415–427

Druehl LD, Mayes C, Tan IH, Saunders GW (1997) Molecular and morphological phylogenies of kelp and associated brown algae. Plant Syst Evol Suppl 11:221–235

Duff KE, Zeeb BA, Smol JP (1995) Atlas of chrysophycean cysts. Dev Hydrobiol 99:1–189

Edvardsen B, Vaulot D (1996) Ploidy analysis of the two motile forms of *Chrysochromulina polylepis* (Prymnesiophyceae). J Phycol 32:94–102

Elbrächter M, Schnepf E (1996) *Gymnodinium chlorophorum*, a new green bloom forming dinoflagellate (Gymnodiniales, Dinophyceae) with a vestigial prasinophyte endosymbiont. Phycologia 35:381–393

Entwisle TJ, Necchi O Jr (1992) Phylogenetic systematics of the freshwater red algal order Batrachospermales. Jpn J Phycol 40:1–12

Ettl H, Gärtner G (1995) Syllabus der Boden-, Luft- und Flechtenalgen. G. Fischer, Stuttgart

Faber WW, Preisig HR (1994) Calcified structures and calcification in protists. Protoplasma 181:78–105

Farmer MA, Triemer RE (1994) An ultrastructural study of *Lentomonas applanatum* (Preisig) n. g. (Euglenida). J Euk Microbiol 41:112–119

Feist M, Grambast-Fessard N (1991) The genus concept in Charophyta: evidence from Paleozoic to Recent. In: Riding R (ed) Calcareous algae and stromatolites. Springer, Berlin Heidelberg New York, pp 189–203

Fensome RA, Taylor FJR, Norris G, Sarjeant WAS, Wharton DI, Williams GL (1993) A classification of living and fossil dinoflagellates. Micropaleontology Spec Publ 7:1–351

Field RH, Chapman CJ, Taylor AC, Neil DM, Vickerman K (1992) Infection of the Norway lobster *Nephrops norvegicus* by a *Hematodinium*-like species of dinoflagellate on the west coast of Scotland. Dis Aquat Organisms 13:1–15

Floyd GL, Watanabe S, Deason TR (1993) Comparative ultrastructure of the zoospores of eight species of *Characium* (Chlorophyte). Arch Protistenkd 143:63–73

Fraunholz MJ, Wastl J, Zauner S, Rensing SA, Scherzinger MM, Maier U-G (1997) The evolution of crytophytes. Plant Syst Evol Suppl 11:163–174

Fredericq S, Hommersand MH (1990) Diagnoses and key to the genera of the Gracilariaceae (Gracilariales, Rhodophyta). Hydrobiologia 204/205:173–178

Fredericq S, Norris JN (1995) A new order (Rhodogorgonales) and family (Rhodogorgonaceae) of red algae composed of two tropical calciferous genera, *Renouxia* gen. nov. and *Rhodogorgon*. Cryptog Bot 5:316–331

Fredericq S, Ramirez ME (1996) Systematic studies of the antarctic species of the Phyllophoraceae (Gigartinales, Rhodophyta) based on *rbcL* sequence analysis. Hydrobiologia 326–327:137–143

Freshwater DW, Fredericq S, Hommersand HM (1995) A molecular phylogeny of the Gelidiales (Rhodophyta) based on analysis of plastid *rbcL* nucleotide sequences. J Phycol 31:616–632

Friedl T (1995) Inferring taxonomic positions and testing genus level assignments in coccoid green lichen algae: a phylogenetic analysis of 18S ribosomal RNA sequences from *Dictyochloropsis reticulata* and from members of the genus *Myrmecia* (Chlorophyta, Trebouxiophyceae cl. nov.). J Phycol 31:632–639

Friedl T (1996) Evolution of the polyphyletic genus *Pleurastrum* (Chlorophyta): inferences from nuclear-encoded ribosomal DNA sequences and motile cell ultrastructure. Phycologia 35:456–469

Friedl T (1997) The evolution of the green algae. Plant Syst Evol Suppl 11:87–111

Fritz L, Nass M (1992) Development of the endoparasitic dinoflagellate *Amoebophrya ceratii* within host dinoflagellate species. J Phycol 28:312–320

Fryxell GA, Villac MC, Shapiro LP (1997) The occurrence of the toxic diatom genus *Pseudo-nitzschia* (Bacillariophyceae) on the West Coast of the USA, 1920–1996: a review. Phycologia 36:419–437

Fukushi-Fujikura Y, Harada N, Othuru O, Maeda M (1991) Cell wall polysaccharides of *Pseudodichotomosiphon constrictus* with special reference to its systematics. Phytochemistry 30:201–203

García A (1994) Charophyta: their use in paleolimnology. J Paleolimnol 10:43–52

Gaul U, Geissler U, Henderson M, Mahoney R, Reimer CW (1993) Bibliography on the fine-structure of diatom frustules (Bacillariophyceae). Proc Acad Nat Sci Phila 144:69–238

Gerrath JF (1993) The biology of desmids: a decade of progress. Prog Phycol Res 9:79–192

Gilson P, McFadden GI (1995) The chlorarachniophyte: a cell with two different nuclei and two different telomeres. Chromosoma 103:635–641

Goff LJ, Moon DA, Nyvall P, Stache B, Mangin K, Zuccarello G (1996) The evolution of parasitism in the red algae: molecular comparison of adelphoparasites and their hosts. J Phycol 32:297–312

Green JC, Leadbeater BSC (eds) (1994) The haptophyte algae. Syst Assoc Spec Vol 51. Clarendon Press, Oxford pp 1–446

Grell K (1990) Indications of sexual reproduction in the plasmodial protist *Chlorarachnion reptans* Geitler. Z Naturforsch 45c:112–114

Guillard RRL, Keller MD, O'Kelly CJ, Floyd GL (1991) *Pycnococcus provasolii* gen. et sp. nov., a coccoid prasinoxanthin-containing phytoplankter from the western North Atlantic and Gulf of Mexico. J Phycol 27:39–47

Guiry M (1992) Species concepts in marine red algae. Prog Phycol Res 8:251–278

Haas JN (1994) First identification key for charophyte oospores from central Europe. Eur J Phycol 29:227–235

Hallegraeff GM, Anderson DM, Cembella AD (eds) (1995) Manual on harmful marine microalgae. IOC Manuals and Guides 33:1–551. Intergovernmental Oceanographic Commission (of UNESCO), Paris

Harvey AS, Woelkerling WJ (1995) An account of *Austrolithon intumescens* gen. et. sp. nov. and *Boreolithon van-heurckii* (Heydrich) gen. et comb. nov. (Austrolithoideae subfam. nov., Corallinaceae, Rhodophyta). Phycologia 34:362–382

Harwood DM, Nikolaev VA (1996) Cretaceous diatoms: morphology, taxonomy, biostratigraphy. Short courses in paleontology 8:81–106. Paleontological Society, Knoxville, Tennessee

Hasegawa T, Miyashita H, Kawachi M, Ikemoto H, Kurano N, Miyachi S, Chihara M (1996) *Prasinoderma coloniale* gen. et sp. nov., a new pelagic coccoid prasinophyte from the western Pacific Ocean. Phycologia 35:170–176

Hasle GR, Lange CB, Syvertsen EE (1996) A review of *Pseudo-nitzschia*, with special reference to the Skagerrak, North Atlantic, and adjacent waters. Helgoländer Meeresunters 50:131–175

Helmchen TA, Bhattacharya D, Melkonian M (1995) Analyses of ribosomal RNA sequences from glaucocystophyte cyanelles provide new insights into the evolutionary relationships of plastids. J Mol Evol 41:203–210

Henriksen P, Knipschildt F, Moestrup Ø, Thomsen HA (1993) Autecology, life history and toxicology of the silicoflagellate *Dictyocha speculum* (Silicoflagellata, Dictyochophyceae). Phycologia 32:29–39

Höhfeld I, Melkonian M (1995) Ultrastructure of the flagellar apparatus of *Noctiluca miliaris* Suriray swarmers (Dinophyceae). Phycologia 34:508–513

Hommersand MH, Guiry MD, Fredericq S, Leister GL (1993) New perspectives in the taxonomy of the Gigartinaceae (Gigartinales, Rhodophyta). Hydrobiologia 260/261:105–120

Hommersand MH, Fredericq S, Freshwater DW (1994) Phylogenetic systematics and biogeography of the Gigartinaceae (Gigartinales, Rhodophyta) based on sequence analysis of *rbc*L. Bot Mar 37:193–203

Honda D, Inouye I (1995) Ultrastructure and reconstruction of the flagellar apparatus architecture in *Ankylochrysis lutea* (Chrysophyceae, Sarcinochrysidales). Phycologia 34:215–227

Honda D, Kawaichi M, Inouye I (1995) *Sulcochrysis biplastida* gen. et sp. nov. (Chrysophyta): cell structure and absolute configuration of the flagellar apparatus. Phycol Res 43:1–16

Hori T (ed) (1993) An illustrated atlas of the life histories of algae, vols 1–3. Uchida Rokakuho, Tokyo

Horiguchi T (1996) *Haramonas dimorpha* gen. et sp. nov. (Raphidophyceae), a new marine raphidophyte from Australian mangrove. Phycol Res 44:143–150

Huss VAR, Kranz HD (1997) Charophyte evolution and the origin of land plants. Plant Syst Evol Suppl 11:103–114

Irvine LM, Chamberlain YM (1994) Seaweeds of the British Isles, vol 1: Rhodophyta, part 2B, Corallinales, Hildenbrandiales. HMSO, London

Ishida K, Hara Y (1994) Taxonomic studies on the Chlorarachniophyta. I. *Chlorarachnion globosum* sp. nov. Phycologia 33:351–358

Ishida K, Nakayama T, Hara Y (1996) Taxonomic studies on the Chlorarachniophyta. II. Generic delimitation of *Gymnochlora stellata* gen. et sp. nov. and *Lotharella* gen. nov. Phycol Res 44:37–45

Jahn R, Meyer B, Preisig HR (eds) (1997) Microalgae: aspects of diversity and systematics. Nova Hedwigia 65:1–452

Jansonius J, McGregor DC (eds) (1996) Palynology: principles and applications. 3 volumes. American Association of Stratigraphic Palynologists Foundation, Dallas, Texas, 1330 pp

John DM, Maggs CM (1997) Species problems in eukaryotic algae: a modern perspective. In: Claridge MF, Dawah HA, Wilson MR (eds) Species: the units of biodiversity. Chapman and Hall, London, pp 83–107

Jones HLJ, Leadbeater BSC, Green JC (1994) An ultrastructural study of *Marsupiomonas pelliculata* gen. et sp. nov., a new member of the Pedinophyceae. Eur J Phycol 29:171–181

Jordan RW, Green JC (1994) A check-list of the extant Haptophyta of the world. J Mar Biol Assoc UK 74:149–174

Jordan RW, Kleijne A, Heimdal BR, Green JC (1995) A glossary of the extant Haptophyta of the world. J Mar Biol Assoc UK 75:769–814

Kawai H (1992) A summary of the morphology of chloroplasts and flagellated cells in the Phaeophyceae. Korean J Phycol 7:33–43

Kessler E (1991) *Scenedesmus*: Problems of a highly variable genus of green algae. Bot Acta 104:169–171

Kies L (1992) Glaucocystophyceae and other protists harbouring procaryotic endosymbionts. In: Reisser W (ed) Algae and symbioses: plants, animals, fungi, viruses, interactions explored. Biopress, Bristol, pp 353–377

Kociolek JP (ed) (1994) Proceedings of the 11th International Diatom Symposium. Mem Calif Acad Sci 17:1–670

Kociolek JP, Sullivan MJ (eds) (1995) A century of diatom research in North America. Koeltz Scientific Books, Champaign, Illinois

Krammer K, Lange-Bertalot H (1991a) Bacillariophyceae, 3. Teil. In: Ettl H, Gerloff J, Heynig H, Mollenhauer D (eds) Süsswasserflora von Mitteleuropa 2(3):1–576. G Fischer, Stuttgart

Krammer K, Lange-Bertalot H (1991b) Bacillariophyceae, 4. Teil. In: : Ettl H, Gärtner G, Gerloff J, Heynig H, Mollenhauer D (eds) Süsswasserflora von Mitteleuropa 2(4):1–437. G Fischer, Stuttgart

Krause W (1997) Charales (Charophyceae). In: : Ettl H, Gärtner G, Heynig H, Mollenhauer D (eds) Süsswasserflora von Mitteleuropa 18:1–202. G Fischer, Stuttgart

Krienitz L, Hegewald E, Reymond OL, Peschke T (1993) Variability of LM, TEM and SEM characteristics of *Pseudogoniochloris tripus* gen. et comb. nov. (Xanthophyceae). Arch Hydrobiol Suppl 97:67–82

Kristiansen J (ed) (1996) Biogeography of freshwater algae. Hydrobiologia 336:1–161

Kristiansen J, Cronberg G (eds) (1996) Chrysophytes: progress and new horizons. Nova Hedwigia Beih 114:1–266

Kristiansen J, Vigna MS (1994) Tubular scales in *Synura* and the possible origin of bristles in *Mallomonas*: Phycologia 33:67–70

Kugrens P, Lee RE (1991) Organization of cryptomonads. In: Patterson DJ, Larsen J (eds) The biology of free-living heterotrophic flagellates. Syst Assoc Spec Vol 45. Clarendon Press, Oxford, pp 219–233

Kumano S (1993) Taxonomy of the family Batrachospermaceae (Batrachospermales, Rhodophyta). Jpn J Phycol 41:253–274

Kusel-Fetzmann EL (1996) New records of freshwater Phaeophyceae from lower Austria. Nova Hedwigia 62:79–89

Landsberg JH, Steidinger KA, Blakesley BA, Zondervan RL (1994) Scanning electron microscope study of dinospores of *Amyloodinium* cf. *ocellatum*, a pathogenic dinoflagellate parasite of marine fish, and comments on its relationship to the Peridiniales. Dis Aquat Organisms 20:23–32

Lange M, Gouillou L, Vaulot D, Simon N, Amann RI, Ludwig W, Medlin LK (1996) Identification of the class Prymnesiophyceae and the genus *Phaeocystis* with ribosomal RNA-targeted nucleic acid probes detected by flow cytometry. J Phycol 32:858–868

Lange-Bertalot H (1993) 85 neue Taxa und über 100 weitere neu definierte Taxa ergänzend zur Süsswasserflora von Mitteleuropa Band 2/1–4. Bibl Diatomol 27:1–454. Cramer, Berlin

Larsen J, Patterson DJ (1990) Some flagellates (Protista) from tropical marine sediments. J Nat Hist 24:801–937

Larsen J, Patterson DJ (1991) The diversity of heterotrophic euglenids. In: Patterson DJ, Larsen J (eds) The biology of free-living heterotrophic flagellates. Syst Assoc Spec Vol 45. Clarendon Press, Oxford, pp 205–217

Larsen J, Sournia A (1991) The diversity of heterotrophic dinoflagellates. In: Patterson DJ, Larsen J (eds) The biology of free-living heterotrophic flagellates. Syst Assoc Spec Vol 45. Clarendon Press, Oxford, pp 313–332

Lassus P, Arzul G, Erard-Le Denn E, Gentien P, Marcaillou-Le Baut C (eds) (1995) Harmful marine algal blooms. Proc 6th Int Conf on Toxic Marine Phytoplankton. Lavoisier, Paris

Lavau S, Saunders GW, Wetherbee R (1997) A phylogenetic analysis of the Synurophyceae using molecular data and scale case morphology. J Phycol 33:135–151

Leitch AR, John DM, Moore JA (1990) The oosporangium of the Characeae (Chlorophyta, Charales). Prog Phycol Res 7:213–268

Lenzenweger R (1996) Desmidiaceenflora von Österreich. Teil 1. Bibl Phycol 101:1–162

Lenzenweger R (1997) Desmidiaceenflora von Österreich. Teil 2. Bibl Phycol 102:1–216

Lindstrom SC, Chapman DJ (eds) (1996) Proceedings of the 15th International Seaweed Symposium. Hydrobiologia 326/327:1–534

Lindstrom SC, Cole KM (1993) The systematics of *Porphyra*: character evolution in closely related species. Hydrobiologia 260/261:151–157

Lindstrom SC, Olsen JL, Stam WT (1996) Recent radiation of the Palmariaceae (Rhodophyta). J Phycol 32:457–468

Lobban CS, Harrison PJ (1994) Seaweed ecology and physiology. Cambridge University Press, Cambridge

Lokhorst GM (1992) Taxonomic studies in the genus *Heterococcus* (Tribophyceae, Tribonematales, Heteropediaceae): a combined cultural and electron microscopy study: Cryptogamic Studies 3:1–246. G Fischer, Stuttgart

Lokhorst GM (1996) Comparative taxonomic studies on the genus *Klebsormidium* (Charophyceae) in Europe. Cryptogamic Studies 5:1–132. G Fischer, Stuttgart

Maggs CA, Hommersand MH (1993) Seaweeds of the British Isles, vol 1: Rhodophyta, part 3A, Ceramiales. HMSO, London

Mann DG (1993) Patterns of sexual reproduction in diatoms. Hydrobiologia 269/270:11–20

Mann DG (1996) Chloroplast morphology, movements and inheritance in diatoms. In: Chaudhary BR, Agrawal SB (eds) Cytology, genetic and molecular biology of algae. SPB Academic Publishers, Amsterdam, pp 249–274

Marin B (1996) Die frühe Evolution der Grünalgen. PhD Thesis, University of Cologne

Marino D, Montresor M (eds) (1995) Proc 13th Int Diatom Symposium. Biopress, Bristol

Masuda M, Guiry MD (1995) Reproductive morphology of *Itonoa marginifera* (J. Agardh) gen. et comb. nov. (Nemastomataceae, Rhodophyta). Eur J Phycol 30:57–67

McCourt RM, Karol KG, Kaplan S, Hoshaw RW (1995) Using *rbc*L sequences to test hypotheses of chloroplast and thallus evolution in conjugating green algae (Zygnematales, Charophyceae). J Phycol 31:989–995

McCourt RM, Karol KG, Guerlesquin M, Feist M (1996) Phylogeny of extant genera in the family Characeae (Charales, Charophyceae) based on *rbc*L sequences and morphology. Am J Bot 83:125–131

McFadden GI, Gilson PR, Hill DRA (1994) *Goniomonas*: rRNA sequences indicate that this phagotrophic flagellate is a close relative of the host component of cryptomonads. Eur J Phycol 29:29–32

McFadden GI, Gilson PR, Hofmann CJB (1997a) Division Chlorarachniophyta. In: Bhattacharya D (ed) Origins of the algae and their plastids. Plant Syst Evol Suppl 11:175–185

McFadden GI, Gilson PR, Sims IM (1997b) Preliminary characterization of carbohydrate stores from chlorarachniophytes (Division Chlorarachniophyta). Phycol Res 45:145–151

McLachlan JL, Seguel MR, Fritz L (1994) *Tetreutreptia pomquetensis* gen. et sp. nov. (Euglenophyceae): a quadriflagellate, phototrophic marine euglenoid. J Phycol 30:538–544

McNally KL, Govind NS, Thomé PE, Trench RK (1994) Small-subunit ribosomal DNA sequence analyses and a reconstruction of the inferred phylogeny among symbiotic dinoflagellates (Pyrrophyta). J Phycol 30:316–329

McQuoid MR, Hobson LA (1996) Diatom resting stages. J Phycol 32:889–902

Medlin LK, Lange M, Baumann MEM (1994) Genetic differentiation among three colony-forming species of *Phaeocystis*: further evidence for the phylogeny of the Prymnesiophyta. Phycologia 33:199–212

Medlin LK, Gersonde R, Kooistra WHCF, Wellbrock U (1996a) Evolution of the diatoms (Bacillariophyta): II. Nuclear-encoded small-subunit rRNA sequence comparisons confirm a paraphyletic origin for the centric diatoms. Mol Biol Evol 13:67–75

Medlin LK, Kooistra WHCF, Gersonde R, Wellbrock U (1996b) Evolution of the diatoms (Bacillariophyta): III. The age of the Thalassiosirales. Nova Hedwigia Beih 112:221–234

Medlin LK, Barker GLA, Campbell L, Green JC, Hayes PK, Marie D, Wrieden S, Vaulot D (1996c) Genetic characterization of *Emiliania huxleyi* (Haptophyta). J Mar Syst 9:13–31

Medlin LK, Kooistra WHCF, Gersonde R, Sims PA, Wellbroek U (1997a) Is the origin of diatoms related to the end-Permian mass extinction? In: Jahn R, Meyer B, Preisig HR (eds) Microalgae: aspects of diversity and systematics. Nova Hedwigia 65:1–11

Medlin LK, Kooistra WHCF, Potter D, Saunders GW, Andersen RA (1997b) Phylogenetic relationships of the "golden algae" (haptophytes, heterokont chromophytes) and their plastids. Plant Syst Evol Suppl 11:187–219

Meiers ST, Rootes WL, Proctor VW, Chapman RL (1997) Phylogeny of the Characeae (Charophyta) inferred from organismal and molecular characters. Arch Protistenkd 148:308–317

Melkonian M (1996) Phylogeny of photosynthetic protists and their plastids. Verh Dtsch Zool Ges 89:71–96

Melkonian M, Surek B (1995) Phylogeny of the Chlorophyta: congruence between ultrastructural and molecular evidence. Bull Soc Zool Fr 120:191–208

Melkonian M, Marin B, Surek B (1995) Phylogeny and evolution of the algae. In: Arai R, Kato M, Doi Y (eds) Biodiversity and evolution. National Science Museum Foundation, Tokyo, pp 153–176

Moestrup Ø (1991) Further studies of presumedly primitive green algae, including the description of Pedinophyceae class. nov. and *Resultor* gen. nov. J Phycol 27:119–133

Moestrup Ø (1995) Current status of chrysophyte "splinter groups": synurophytes, pedinellids, silicoflagellates. In: Sandgren CD, Smol JP, Kristiansen J (eds) Chrysophyte algae: ecology, phylogeny and development. Cambridge University Press, Cambridge, pp 75–91

Montegut-Felkner AE, Triemer RE (1997) Phylogenetic relationships of selected euglenoid genera based on morphological and molecular data. J Phycol 33:512–519

Mrozinska T (1991) A preliminiary investigation of the taxonomical classification of the genus *Oedogonium* Link (Oedogoniales) based on the phylogenetic relationship. Arch Protistenkd 139:85–101

Mrozinska T (1993) A preliminary investigation of the taxonomical classification of the genus *Bulbochaete* Agardh (Oedogoniales, Chlorophyta) based on the phylogenetic relationship. Arch Protistenkd 143:113–123

Nakayama T, Watanabe S, Mitsui K, Uchida H, Inouye I (1996a) The phylogenetic relationship between the Chlamydomonadales and Chlorococcales inferred from 18S rDNA sequence data. Phycol Res 44:47–55

Nakayama T, Watanabe S, Inouye I (1996b) Phylogeny of wall-less green flagellates inferred from 18S rDNA sequence data. Phycol Res 44:151–161

Nelson WA, Knight GA (1995) Endosporangia: a new form of reproduction in the genus *Porphyra* (Bangiales, Rhodophyta). Bot Mar 38:17–20

Norton TA, Melkonian M, Andersen RA (1996) Algal biodiversity. Phycologia 35:308–326

Novarino G, Lucas IAN (1993) Some proposals for a new classification system of the Cryptophyceae. Bot J Linn Soc 111:3–21

Novarino G, Lucas IAN (1995) A zoological classification system of cryptomonads. Acta Protozool 34:173–180

Nozaki H, Itoh M (1994) Phylogenetic relationships within the colonial Volvocales (Chlorophyta) inferred from cladistic analysis based on morphological data. J Phycol 30:353–365

Nozaki H, Itoh M, Sano R, Uchida H, Watanabe MM, Kuroiwa T (1995) Phylogenetic relationships within the colonial Volvocales (Chlorophyta) inferred from *rbc*L gene sequence data. J Phycol 31:970–979

O'Kelly JC, Wujek DE (1995) Status of the Chrysamoebales (Chrysophyceae): Observations on *Chrysamoeba pyrenoidifera*, *Rhizochromulina marina* and *Lagynion delicatulum*. In: Sandgren CD, Smol JP, Kristiansen J (eds) Chrysophyte algae: ecology, phylogeny and development. Cambridge University Press, Cambridge, pp 361–372

O'Kelly CJ, Watanabe S, Floyd GL (1994) Ultrastructure and phylogenetic relationships of Chaetopeltidales ord. nov. (Chlorophyta, Chlorophyceae). J Phycol 30:118–128

Oliveira MC, Kurniawan J, Bird CJ, Rice EL, Murphy CA, Singh RK, Gutell RR, Ragan MA (1995) A preliminary investigation of the order Bangiales (Bangiophycidae, Rhodophyta) based on the sequences of nuclear small-subunit ribosomal RNA genes. Phycol Res 43:71–79

Olsen JL, Stam WT, Berger S, Menzel D (1994) 18S rDNA and evolution in the Dasycladales (Chlorophyta): modern living fossils. J Phycol 30:729–744

Park NE, Karol KG, Hoshaw RW, McCourt RM (1996) Phylogeny of *Gonatozygon* and *Genicularia* (Gonatozygaceae, Desmidiales) based on *rbc*L sequences. Eur J Phycol 31:309–313

Patterson DJ, Larsen J (1992) A perspective on protistan nomenclature. J Protozool 39:125–131

Peters MC, Andersen RA (1993a) The fine structure and scale formation of *Chysolepidomonas dendrolepidota* gen. et sp. nov. (Chrysolepidomonadaceae fam. nov., Chrysophyceae). J Phycol 29:469–475

Peters MC, Andersen RA (1993b) The flagellar apparatus of *Chrysolepidomonas dendrolepidota* (Chrysophyceae), a single-cell monad covered with organic scales. J Phycol 29:476–485

Pipes LD, Leedale GF (1992) Scale formation in *Tessellaria volvocina* (Synurophyceae). Br Phycol J 27:11–19

Poole LJ, Raven JA (1997) The biology of *Enteromorpha*. Prog Phycol Res 12:1–148

Popovsky J, Pfiester LA (1990) Dinophyceae (Dinoflagellida). In: Ettl H, Gerloff J, Heynig H, Mollenhauer D (eds) Süsswasserflora von Mitteleuropa 6:1–272. G Fischer, Jena

Potter D, Saunders GW, Andersen RA (1997a) Phylogenetic relationships of the Raphidophyceae and Xanthophyceae as inferred from nucleotide sequences of the 18S ribosomal RNA gene. Am J Bot 84:966–972

Potter D, LaJeunesse TC, Saunders GW, Andersen RA (1997b) Convergent evolution masks extensive biodiversity among marine coccoid picoplankton. Biodivers Conserv 6:99–107

Preisig HR (1992) *Dunaliella*: morphology and taxonomy. In: Avron M, Ben-Amotz A (eds) *Dunaliella*: physiology, biochemistry and biotechnology. CRC Press, Boca Raton, pp 1–15

Preisig HR (1994) Siliceous structures and silicification in flagellated protists. Protoplasma 181:29–42

Preisig HR (1995) A modern concept of chrysophyte classification. In: Sandgren CD, Smol JP, Kristiansen J (eds) Chrysophyte algae: ecology, phylogeny and development. Cambridge University Press, Cambridge, pp 46–74

Preisig HR, Vørs N, Hällfors G (1991) Diversity of heterotrophic heterokont flagellates. In: Patterson DJ, Larsen J (eds) The biology of free-living heterotrophic flagellates. Syst Assoc Spec Vol 45. Clarendon Press, Oxford, pp 361–399

Ragan MA, Gutell RR (1995) Are red algae plants? Bot J Linn Soc 118:81–105

Ragan MA, Bird CJ, Rice EL, Gutell RR, Murphy CA, Singh RK (1994) A molecular phylogeny of the marine red algae (Rhodophyta) based on the nuclear small-subunit rRNA gene. Proc Natl Acad Sci USA 91:7276–7280

Roberts K (1991) The flagellar apparatus and cytoskeleton of dinoflagellates: organization and use in systematics. In: Patterson DJ, Larsen J (eds) The biology of free-living heterotrophic flagellates. Syst Assoc Spec Vol 45. Clarendon Press, Oxford, pp 285–302

Round FE, Crawford RM, Mann DG (1990) The diatoms: biology and morphology of the genera. Cambridge University Press, Cambridge

Rousseau F, Leclerc M-C, De Reviers B (1997) Molecular phylogeny of European Fucales (Phaeophyceae) based on partial large-subunit rDNA sequence comparisons. Phycologia 36:438–446

Rowan R, Powers DA (1992) Ribosomal RNA sequences and the diversity of symbiotic dinoflagellates (zooxanthellae). Proc Natl Acad Sci USA 89:3639–3643

Sandgren CD, Smol JP, Kristiansen J (eds) (1995) Chrysophyte algae: ecology, phylogeny and development. Cambridge University Press, Cambridge

Santos LMA (1996) The Eustigmatophyceae: actual knowledge and research perspectives. Nova Hedwigia Beih 112:509–512

Saunders GW, Bailey JC (1997) Phylogenesis of pit-plug associated features in the Rhodophyta: inferences from molecular systematic data. Can J Bot 75:1436–1447

Saunders GW, Kraft GT (1994) Small-subunit rRNA gene sequences from representatives of selected families of the Gigartinales and Rhodymeniales (Rhodophyta). 1. Evidence for the Plocamiales. Can J Bot 72:1250–1263

Saunders GW, Kraft GT (1996) Small-subunit rRNA gene sequences from representatives of selected families of the Gigartinales and Rhodymeniales (Rhodophyta). 2. Recognition of the Halymeniales. Can J Bot 74:694–707

Saunders GW, Kraft GT (1997) A molecular perspective on red algal evolution: focus on the Florideophycidae. Plant Syst Evol Suppl 11:115–138

Saunders GW, McLachlan JL (1991) Morphology and reproduction of *Meiodiscus spetsbergensis* (Kjellmann) gen. et comb. nov., a new genus of Rhodophysemataceae (Rhodophyta). Phycologia 30:272–286

Saunders GW, Bird CJ, Ragan MA, Rice EL (1995) Phylogenetic relationships of species of uncertain taxonomic position within the Acrochaetiales-Palmariales complex (Rhodophyta): Inferences from phenotypic and 18S rDNA sequence data. J Phycol 31:601–611

Saunders GW, Strachan IM, West JA, Kraft GT (1996) Nuclear small-subunit ribosomal RNA gene sequences from representative Ceramiaceae (Ceramiales, Rhodophyta). Eur J Phycol 31:23–29

Saunders GW, Hill DRA, Tyler PA (1997a) Phylogenetic affinities of *Chrysonephele palustris* (Chrysophyceae) based on inferred nuclear small-subunit ribosomal RNA sequence. J Phycol 33:132–134

Saunders GW, Potter D, Andersen RA (1997b) Phylogenetic affinities of the Sarcinochrysidales and Chrysomeridales (Heterokonta) based on analyses of molecular and combined data. J Phycol 33:310–318

Saunders GW, Hill DRA, Sexton JP, Andersen RA (1997c) Small-subunit ribosomal RNA sequences from selected dinoflagellates: testing classical evolutionary hypotheses with molecular systematic methods. Plant Syst Evol Suppl 11:237–259

Schnepf E (1992) Electron microscopical studies of *Thorea ramosissima* (Thoreaceae, Rhodophyta). Taxonomic implications of *Thorea* pit plug ultrastructure. Plant Syst Evol 181:233–244

Schnepf E (1993) From prey via endosymbiont to plastid: comparative studies in dinoflagellates. In: Lewin RA (ed) Origins of plastids. Chapman and Hall, New York, pp 53–76

Schnepf E, Drebes G (1993) Anisogamy in the dinoflagellate *Noctiluca*? Helgoländer Meeresunters 47:265–273

Schnepf E, Niemann A, Wilhelm C (1996) *Pseudostaurastrum limneticum*, a eustigmatophycean alga with astigmatic zoospores: morphogenesis, fine structure, pigment composition and taxonomy. Arch Protistenkd 146:237–249

Scholin CA, Anderson DM (1996) LSU rDNA-based RFLP assays for discriminating species and strains of *Alexandrium* (Dinophyceae). J Phycol 32:1022–1035

Seckbach J (ed) (1994) Evolutionary pathways and enigmatic algae: *Cyanidium caldarium* (Rhodophyta) and related cells. Development Hydrobiology 91. Kluwer, Dordrecht, 349 pp

Sheath RG, Kaczmarczyk D, Cole KM (1993a) Distribution and systematics of freshwater *Hildenbrandia* (Rhodophyta, Hildenbrandiales) in North America. Eur J Phycol 28:115–121

Sheath RG, Vis ML, Cole KM (1993b) Distribution and systematics of the freshwater red algal family Thoreaceae in North America. Eur J Phycol 28:231–241

Sheath RG, Vis ML, Cole KM (1993c) Distribution and systematics of freshwater Ceramiales (Rhodophyta) in North America. J Phycol 29:108–117

Sheath RG, Whittick A, Cole KM (1994) *Rhododraparnaldia oregonica*, a new freshwater red algal genus and species intermediate between the Acrochaetiales and the Batrachospermales. Phycologia 33:1–7

Sheath RG, Müller KM, Whittick A, Entwisle TJ (1996a) A re-examination of the morphology and reproduction of *Nothocladus lindaueri* (Batrachospermales, Rhodophyta). Phycol Res 44:1–10

Sheath RG, Müller KM, Vis ML, Entwisle TJ (1996b) A re-examination of the morphology, ultrastructure and classification of genera in the Lemaneaceae (Batrachospermales, Rhodophyta). Phycol Res 44:233–246

Silva PC, Basson PW, Moe R (1996) Catalogue of the benthic marine algae of the Indian Ocean. Univ Calif Publ Bot 79:1–1259

Simon N, Brenner J, Edvardsen B, Medlin LK (1997) The identifcation of *Chrysochromulina* and *Prymnesium* species (Haptophyta, Prymnesiophyceae) using fluorescent or chemiluminescent oligonucleotide probes: a means for improving studies on toxic algae. Eur J Phycol 32:393–401

Simpson AGB (1997) The identity and composition of the Euglenozoa. Arch Protistenkd 148:318–328

Simpson AGB, Van den Hoff J, Bernard C, Burton HR, Patterson DJ (1997) The ultrastructure and systematic position of the euglenozoon *Postgaardi mariagerensis*, Fenchel et al. Arch Prostistenkd 147:213–225

Sims PA (ed) (1993) Progress in diatom studies: contributions to taxonomy, ecology and nomenclature. Nova Hedwigia Beih 106:1–377

Sims PA (ed) (1996) An atlas of diatoms of Great Britain. Biopress, Bristol

Siver PA (1991) The biology of *Mallomonas*: Morphology, taxonomy and ecology. Dev Hydrobiol 63:1–230

Smol JP (1995) Application of chrysophytes to problems in paleoecology. In: Sandgren CD, Smol JP, Kristiansen J (eds) Chrysophyte algae: ecology, phylogeny and development. Cambridge University Press, Cambridge, pp 303–329

Steidinger KA, Burkholder JM, Glasgow HB Jr, Hobbs CW, Garrett JK, Truby EW, Noga EJ, Smith SA (1996) *Pfiesteria piscicida* gen. et sp. nov. (Pfiesteriaceae fam. nov.), a new toxic dinoflagellate with a complex life cycle and behavior. J Phycol 32:157–164

Steinkötter J, Bhattacharya D, Semmelroth I, Bibeau C, Melkonian M (1994) Prasinophytes form independent lineages within the Chlorophyta: evidence from ribosomal RNA sequence comparisons. J Phycol 30:340–345

Stevenson RJ, Bothwell ML, Lowe RL (1996) Algal ecology: freshwater benthic ecosystems. Academic Press, San Diego

Stoebe B, Schaffran I, Von der Lippe C, Laatsch T, Behn W, Nitsch T, Freier U, Kowallik KV (1997) Molecular evolution of green and non-green chloroplasts. Phycologia 36 (Suppl):109

Surek B, Beemelmanns U, Melkonian M, Bhattacharya D (1994) Ribosomal RNA sequence comparisons demonstrate an evolutionary relationship between Zygnematales and Charophytes. Plant Syst Evol 191:171–181

Sym SD, Pienaar RN (1993) The class Prasinophyceae. Prog Phycol Res 9:281–376

Tan IH, Druehl LD (1996) A ribosomal DNA phylogeny supports the close evolutionary relationships among the Sporochnales, Desmarestiales, and Laminariales. J Phycol 32:112–118

Thompson RH, Wujek DE (1997) Trentepohliales: *Cephaleuros, Phycopeltis*, and *Stomatochroon*: morphology, taxonomy, and ecology. Science Publishers, Enfield, New Hampshire

Tomas CR (ed) (1993) Marine phytoplankton: a guide to naked flagellates and coccolithophorids. Academic Press, San Diego

Tomas CR (ed) (1996) Marine phytoplankton. Identifying marine diatoms and dinoflagellates. Academic Press, San Diego

Townsend RA, Chamberlain YM, Keats DW (1994) *Heydrichia woelkerlingii* gen. et sp. nov., a newly discovered non-geniculate red alga (Corallinales, Rhodophyta) from Cape Province, South Africa. Phycologia 33:177–186

Townsend RA, Woelkerling WJ, Harvey AS, Borowitzka M (1995) An account of the red algal genus *Sporolithon* (Sporolithaceae, Corallinales) in Southern Australia. Aust Syst Bot 8:85–121

Trainor FR (1991) The format for a *Scenedesmus* monograph. Arch Hydrobiol Suppl 88:47–54

Trench RK, Thinh L-V (1995) *Gymnodinium linucheae* sp. nov.: the dinoflagellate symbiont of the jellyfish *Linuche unguiculata*. Eur J Phycol 30:149–154

Triemer RE, Lewandowski CL (1994) Ultrastructure of the basal apparatus and putative vestigial feeding apparatuses in a quadriflagellate euglenoid (Euglenophyta). J Phycol 30:28–38

Van Dam H (ed) (1993) Proceedings of the 12th International Diatom Symposium. Hydrobiologia 269/270:1–540

Van den Hoek C, Mann DG, Jahns HM (1995) Algae: an introduction to phycology. Cambridge University Press, Cambridge

Vaulot D, Birrien J-L, Marie D, Casotti R, Veldhuis MJW, Kraay GW, Chrétiennot-Dinet M-J (1994) Morphology, ploidy, pigment composition and genome size of cultured strains of *Phaeocystis* (Prymnesiophyceae). J Phycol 30:1022–1035

Verheij E (1993) The genus *Sporolithon* (Sporolithaceae fam. nov., Corallinales, Rhodophyta) from the Spermonde Archipelago, Indonesia. Phycologia 32:184–196

Vijayaraghavan MR, Kumari S (1995) Chlorophyta: structure, ultrastructure and reproduction. Studies in cryptogamic botany 3:1–324. Bishen Singh Mahendra Pal Singh, Dehra Dun, India

Vis ML, Sheath RG (1996) Distribution and systematics of *Batrachospermum* (Batrachospermales, Rhodophyta) in North America. 9. Section *Batrachospermum*: description of five new species. Phycologia 35:124–134

Vis ML, Sheath RG, Cole KM (1992) Systematics of the freshwater red algal family Compsopogonaceae in North America. Phycologia 31:564–575

Vørs N (1992) Ultrastructure and autecology of the marine, heterotrophic flagellate *Leucocryptos marina* (Braarud) Butcher 1967 (Katablepharidaceae-Kathablepharidae), with a discussion of the genera *Leucocryptos* and *Kathablepharis/Katablepharis*). Eur J Protistol 28:369–389

Vymazal J (1995) Algae and element cycling in wetlands. Lewis Publishers, Boca Raton

Watanabe MM, Suda S, Inouye I, Sawaguchi T, Chihara M (1990) *Lepidodinium viride* gen. et sp. nov. (Gymnodiniales, Dinophyta), a green dinoflagellate with a chlorophyll *a*- and *b*-containing endosymbiont. J Phycol 26:741–751

Watanabe S, Floyd GL (1996) Considerations on the systematics of coccoid green algae and related organisms based on the ultrastructure of swarmers. In: Chaudhary BR, Agrawal SB (eds) Cytology, genetics and molecular biology of algae. SPB Academic Publishing, Amsterdam, pp 1–19

Wiessner W, Schnepf E, Starr RC (eds) (1995) Algae, environment and human affairs. Biopress, Bristol

Wilcox LW, Lewis LA, Fuerst PA, Floyd GL (1992) Assessing the relationships of autosporic and zoosporic chlorococcalean green algae with 18S rDNA sequence data. J Phycol 28:381–386

Wilcox LW, Fuerst PA, Floyd GL (1993) Phylogenetic relationships of four charophycean green algae inferred from complete nuclear-encoded small subunit rRNA gene sequences. Am J Bot 80:1028–1033

Winter A, Siesser WG (eds) (1994) Coccolithophores. Cambridge University Press, Cambridge

Womersley HBS (1994) The marine benthic flora of Southern Australia. IIIA. Rhodophyta. Bangiophyceae and Florideophyceae (Acrochaetiales, Nemaliales, Gelidiales, Hildenbrandiales and Gigartinales senso lato). Australian Biological Resources Study, Canberra

Womersley HBS (1996) The marine benthic flora of Southern Australia. Rhodophyta – IIIB. Gracilariales, Rhodymeniales, Corallinales and Bonnemaisoniales. Australian Biological Resources Study, Canberra

Wong A, Beebee T (1994) Identification of a unicellular, non-pigmented alga that mediates growth inhibition in anuran tadpoles: a new species of the genus *Prototheca* (Chlorophyceae: Chlorococcales). Hydrobiologia 277:85–96

Wynne M (1996) A revised key to genera of the red algal family Delesseriaceae. Nova Hedwigia Beih 112:171–190

Yamagishi T, Akiyama M (eds) (1990–1997) Photomicrographs of the freshwater algae, vols 10–18. Uchida Rokakuho, Tokyo

Yasumoto T, Oshima Y, Fukuyo Y (eds) (1996) Harmful and toxic algal blooms. Proc 7th Int Conf on Toxic Phytoplankton. Intergovernmental Oceanographic Commission, UNESCO, Paris

Zechman FW, Theriot EC, Zimmer EA, Chapman RL (1990) Phylogeny of the Ulvophyceae (Chlorophyta): cladistic analysis of nuclear encoded rRNA sequence data. J Phycol 26:700–710

Zuccarello GC, West JA (1994a) Comparative development of the red algal parasites *Bostrychiocolax australis* gen. et sp. nov. and *Dawsoniocolax bostrychiae* (Choreocolacaceae, Rhodophyta). J Phycol 30:137–146
Zuccarello GC, West JA (1994b) Host specificity in the red algal parasites *Bostrychiocolax australis* and *Dawsoniocolax bostrychiae* (Choreocolacaceae, Rhodophyta). J Phycol 30:462–473

Hans R. Preisig
Botanischer Garten
und Institut für systematische Botanik
der Universität Zürich
Zollikerstraße 107
CH-8008 Zürich, Switzerland

Edited by
J. W. Kadereit

Systematics of the Pteridophytes

By Stefan Schneckenburger

Several important conferences were devoted to various fields of pteridology. Important papers of a biogeography symposium of 1990 were published in 1993 (see Sect. 4). A symposium entitled Use of Molecular Data in Evolutionary Studies of Pteridophytes was held in Knoxville in 1994. The proceedings were collected in a special issue of the American Fern Journal in 1995. The centenary of Eric Holttum's birthday was commemorated with the Holttum Memorial Pteridophyte Symposium held at Kew in 1995. Proceedings were published in 1996 under the title Pteridology in Perspective (Camus et al. 1996). Of nearly equal importance are the papers collected in the Holttum Memorial Volume (Johns 1997a). A fine book concerning all fields of pteridology (morphology, classification, geography, ecology, biosystematics, and chemotaxonomy) was presented by Kramer et al. (1995). It is a dad duty to mention the deaths of Karl. U. Kramer (1928–1994), who prepared these reports during the past decade, of Lenette Rogers Atkinson (1899–1996), a pioneer in the study of fern gametophytes, and of Tadeus Reichstein (1897–1996), especially known for his studies on the cytotaxonomy of *Asplenium*.

1 Systematics

A useful overview of species concepts and speciation was given by Haufler (1996). He states that "probably all pteridophyte species conform at least to the morphological species concept and most are also good biological species". A highly recommended paper on the recent state of non-molecular phylogenetic hypotheses for ferns was published by Smith (1995). An overview of characters and recent phylogenies poses problems. There is a consensus about the isolated positions of the eusporangiate ferns and about the ancient lineages and general agreement as to the major evolutionary groups (such as dryopteroid, thelypteroid, blechnoid, asplenioid, polypodioid, dennstaedtioid, and pteridoid/cheilanthoid ferns). On the other hand, there are different opinions about their origins and interrelationships. This holds especially for the position of such highly modified groups as the filmy ferns or the heterosporous ferns. Stevenson and Laconte (1996) presented an overall cladistic analysis of familial and ordinal relationships of pteridophyte genera and the position of the larger groups of cormophytes. The pteridophytes are demonstrated as paraphyletic to seed plants with the Ophioglossales considered as fern allies instead of ferns.

Wolf (1996) discussed the use of different genes for phylogenetic analysis of pteridophytes. He emphasized the necessity of not relying on data from single genes in analyzing such large groups at various taxonomic levels. Whereas the chloroplast *rbc*L gene comprises the largest molecular data set for pteridophytes, it seems to have its best phylogenetic resolution within and among closely related families. The nuclear ribosomal 18S rRNA gene is evolving more slowly and seems to be suitable for the deeper nodes of vascular plant families. Another slowly evolving region, the chloroplast 16S ribosomal RNA gene, appears to provide a good phylogenetic signal for inferring relationships among the major pteridophyte groups. On the other hand, the chloroplast gene *atp*B is evolving slightly faster than *rbc*L and seems to have potential for work at the family and generic level. Similar regions are the ITS-1 region of the nuclear ribosomal repeat and another ITS region between mitochondrial ribosome genes. They seem to be valuable within fern genera and even species.

Some of the problems posed by Smith (1995) were brought nearer to an answer by analysis of the nucleotide variation in *rbc*L from 99 genera, representing 31 of the 33 extant families (Hasebe et al. 1995). Neighbour joining, maximum parsimony, and maximum likelihood methods resulted in optimal trees that were similar. The main results are: (1). Osmundaceae are the most basal lineage of the leptosporangiate ferns. (2) Polypodiaceae, Grammitidaceae and *Pleurosoriopsis*, a genus which was placed in various families, form a monophyletic group which is most derived among indusiate ferns. Together with the Davalliaceae, Oleandraceae, Nephrolepidaceae, Lomariopsidaceae, Dryopteridaceae, Thelypteridaceae, Blechnaceae and Aspleniaceae, they form a monophyletic group. The group consisting of the above-mentioned clade and Dennstaedtiaceae, Monachosoraceae, Pteridaceae and Vittariaceae is monophyletic. (3) Dryopteridaceae, and Dennstaedtiaceae are polyphyletic. (4) Tree ferns in the Cyatheaceae, Metaxyaceae, and Dicksoniaceae form a monophyletic group that emerged early in the evolution of the leptosporangiate ferns. Furthermore, Plagiogyriaceae and Loxomataceae emerge with the tree ferns. (5) Prior to the tree ferns the heterosporous ferns diverged, which form a monophyletic group. (6) Schizaeaceae, Cheiropleuriaceae, Dipteridaceae, Gleicheniaceae, Matoniaceae, and Hymenophyllaceae are basal to the aquatic heterosporous ferns. (7) *Psi-*

lotum and *Tmesipteris* form a monophyletic group and have no close relationship with any leptosporangiate ferns.

Cladistic analysis of extant ferns based solely on morphological and molecular data and on a combination of both sets was presented by Pryer et al. (1995). This highly valuable study presents hypotheses also on the character evolution in ferns. First of all, it clearly showed the value of combining molecular and non-molecular data sets. In all analyses, monophyly was supported for the following clades: leptosporangiate ferns (with *Osmunda* as the most basal genus), heterosporous ferns, *Cheiropleuria-Dipteris, Diplopterygium-Stromatopteris*, tree ferns, schizaeoid ferns, and a large clade consisting of a group of derived leptosporangiate ferns that excludes dennstaedtioids and pteridoids. The dennstaedtioid ferns proved to be paraphyletic.

Rothwell and Stokey (1994) reported fossil evidence for a new heterosporous fern genus, *Hydropteris*, from the Late Cretaceous. Cladistic analysis demonstrated that heterosporous ferns form a monophyletic group, which led them to adopt the order Hydropteridales, including Marsileaceae, Salviniaceae, Azollaceae, and Hydropteridaceae fam. nov. Hasebe et al. (1994, 1995) demonstrated the monophyly of the heterosporous ferns based on *rbcL* sequences. They seem to be more closely related to the most derived leptosporangiate ferns than they are to the Schizaeaceae, as was considered in the past. The results of Pryer et al. (1995) also strongly corroborate a single origin of heterospory in leptosporangiate ferns.

A comprehensive overview of the fossil history of pteridophytes was presented by Collinson (1996). Rothwell (1996) discussed the evolution and phylogenetic relationships of ferns. After origin of the filicales near the base of the Lower Carboniferous and their significant diversification during the Carboniferous, all of the well-documented families of this time became extinct during the Permian. They were replaced by relatively primitive families with living representatives from the Permian and Jurassic. The majority of modern groups originated during the Cretaceous and, at family and even generic level, by the beginning of the Tertiary.

Valuable contributions to the systematics and floristics of *Ophioglossum* (**Ophioglossaceae**) in southern Africa were made by Burrows (1992, 1996). With 16 species, this is the region with the highest species diversity of this genus in the world. A complete revision of southern African *Ophioglossum* was presented by Burrows (1992). The different species tend to grow in close proximity to each other. The role of these "genus communities" in south-central Africa was pointed out by Burrows (1996). This knowledge allows accurate morphological comparisons to be made under identical climatic and edaphic conditions. The chromosome numbers and the ploidal levels of North American species of *Botrychium* were documented by Wagner (1993). A list of sterile hybrids, together with the hypothetical parents for polyploidal botrychiums, is added.

Our poor knowledge of the marattialean ferns was improved by a revision of the genus *Christensenia* (**Marattiaceae**), which proved to consist of two species (Rolleri 1993). Spore morphology correlates completely with characters of gross morphology (Rolleri et al. 1996).

A comprehensive paper on the taxonomy of *Asplenium* (**Aspleniaceae**) was published by Viane (1993). The correlations of the most diverse characters in the family – with full justification treated as monogeneric, with a single isolated genus – are outlined from the most diverse angles, with, of course, the perispore as the trait d'union. Altogether, 70% of the species were tested by multivariate analysis. Correlations between "microcharacters" of the spore and many other characters of the species are generally very satisfactory, but the usefulness of the characters is very uneven in different species groups, evolutionary change having evidently been quite slow in some, fairly fast in others. Moreover, a great deal of parallel evolution must have taken place in a large number of species groups. So, "genera" with a single deviating character like rudimentary or lacking indusium, reticulate veins, etc. should be included.

A series of studies of section *Hymenasplenium* of *Asplenium* was carried out by Murakami and collaborators. Described from the Old World first with less than 10 species, it was shown that there are about 50–60 species, including 10 closely related neotropical ones. The neotropical species were revised by Murakami and Moran (1993). It was shown by Murakami and Schaal (1994) by molecular methods that there are two major clades within neotropical species of *Hymenasplenium*. Molecular studies proved the group to be the most basally diverged monophyletic group distantly related to any of the remaining species of Aspleniaceae (Murakami 1995). The treatment of *Hymenasplenium* as a separate genus, defined by its peculiar dorsiventral creeping rhizome and the basic chromosome number $n = 39$, was proposed. A summary of the current knowledge concerning the biosystematics and evolution of the *Asplenium trichomanes* complex in Europe and the Macaronesian Islands was given by Bennert and Fischer (1993).

The chromosome number of the Hawaiian endemic *Sadleria* (**Blechnaceae**) was determined by F. S. Wagner (1995) as $n = 33$ and therefore similar to *Blechnum*, its closest relative.

In a series of papers, Conant and collaborators presented their investigations of the relationships within the **Cyatheaceae** (Conant et al. 1994, 1996a, b; Stein et al. 1997), which mark important progress in their study. Starting point was the analysis of restriction site data from chloroplast genome of New World tree ferns. Three major evolutionary lineages were identified: the *Alsophila*, *Cyathea* and *Sphaeropteris* clade. Each of them proposed to be recognized at generic level. It remains to be seen whether there are additional, well-defined monophyletic groups among the Old World Cyatheaceae. A combined analysis of molecular and morphological data resulted in trees that are similar to those derived from cpDNA data alone, and enabled the definition of the derived states for some morphological characters which had been controversially discussed in the past.

The **Davalliaceae** were completely revised by Nooteboom (1992, 1994). Besides *Davallia*, only the genera *Davallodes* with 7, *Leucostegia* with 2, and *Gymnogrammitis* with 1 species are accepted. The smaller ones were revised by Nooteboom (1992). By far the largest is *Davallia*, which includes the formerly accepted *Araiostegia, Humata,* and *Pachypleuria* (Nooteboom 1994). The formerly estimated number of ca. 90 species was reduced to 34. Rödl-Linder and Nooteboom (1997) studied the spore morphology of the members of Davalliaceae and its taxonomical value. Cytological studies support the isolated position of *Gymnogrammitis* with a basic chromosome number of $n = 36$, while all other Davalliaceae known up to now show $n = 40$ (Kato et al. 1992).

The systematic position of the **Dennstaedtiaceae** was investigated by Wolf et al. (1994) by analysis of *rbcL*. They demonstrated that they emerge in a basal clade within the leptosporangiate ferns. Insight into the systematics of this complex family was gained by molecular analysis also. Wolf (1995) pointed out that this family, as treated formerly, is not monophyletic. The genera are grouped in four main clades; even the genus *Dennstaedtia* in the classical sense seems to be paraphyletic. *Monachosorum*, a genus which was thought to constitute a monogeneric family in the past, occurs in a clade within Dennstaedtiaceae s.str. This was also supported by Hasebe et al. (1995). It remains unsolved whether Dennstaedtiaceae s.l. are paraphyletic (Wolf et al. 1994) or polyphyletic (Hasebe et al. 1994, 1995).

Palmer (1994) gives a monographic treatment of the Hawaiian species of *Cibotium* (**Dicksoniaceae**), one of the highly polymorphic genera of the archipelago.

The discussion about generic limits within **Dryopteridaceae** was opened again by the discovery of the sterile generic hybrid x *Dryostichum singulare* by Wagner et al. (1992). This taxon, found in several localities in Ontario, is a natural hybrid between *Dryopteris goldiana* and *Polystichum lonchitis*. Yatskievych (1996) presented a revision of the little-known genus *Phanerophlebia*. He considered *Phanerophlebia* as a monophyletic unit which deserves generic rank within the Dryopteridaceae, and accepted eight species. As a regional contribution, the West Indian species of *Polystichum* were revised by Mickel (1997a). They are floristically tied more narrowly to eastern Asia than to continental America. A monograph of the highly variable Hawaiian species of *Dryopteris* was presented by Fraser-Jenkins (1994). They are ideal objects for further studies of speciation processes.

Four new genera of **Grammitidaceae** were proposed recently. The authors argue that all segregates deserve generic rank as undoubted monophyletic units within *Grammitis* sensu latissimo. These are *Enterosora* (Bishop and Smith 1992), *Micropolypodium* (Smith 1992), *Melpomene* (Smith and Moran 1992), and *Terpsichore* (Smith 1993a). Smith (1993a) provides a key to the neotropical genera of Grammitidaceae. It is highly interesting, and suggestive of antiquity for this family, that most of the genera of neotropical Grammitidaceae are also represented by one or few species in Africa and islands of the Indian Ocean. Most of these genera are absent from southeast Asia and the Pacific. The existence of receptacular paraphyses was reported by Parris (1997) for Grammitidaceae from Asia, Australia, and the Pacific Islands. The taxonomic significance of their occurrence and their types will help to define natural groups within this complex and difficult family.

Evolutionary relationships within the genus *Trichomanes* s.l. (**Hymenophyllaceae**) were studied by Dubuisson (1996), using morphological and anatomical characters as well as *rbcL* nucleotide sequences. Some groups, traditionally considered as natural, seemed to be polyphyletic. A technique for the investigation of genetic variation in filamentous gametophytes of *Trichomanes* was presented by Ji et al. (1994). Windisch (1992) presented a monograph of the difficult, mainly neotropical *Trichomanes crispum* group of *Achomanes*, the largest and most taxonomically complex subgenus of *Trichomanes*. The representation of the genus *Hymenophyllum* s.l. in the Greater Antilles was investigated by Sánchez and Caluff (1996). Twenty six taxa are named, 23% of which are endemic. The endemism rate of the hymenophyllaceous ferns is not as marked as in other fern genera, where one finds rates of 50%. Analysis of the genetic variation of the endangered filmly fern *Trichomanes speciosum* in Europe led to the identification of two forms (Rumsey et al. 1996). From all sites investigated, only one form was present; only at the northwestern extremes of the species range do both occur in close proximity.

After more than 80 years, the insufficiently known fern *Thysanosoria pteridiformis* (**Lomariopsidaceae**) was collected again in New Guinea. Detailed analysis of habitat, morphology, and ecology of this climbing fern could be made for the first time (Johns 1996a).

The relic genus *Matonia* (**Matoniaceae**) was revised by Kato (1993a), who accepted two truly distinct species.

One of the main problems in the systematics of filicales are the **Polypodiaceae**, a highly puzzling group, which has seen more divergent treatments than any other fern family. Exemplified with the highly puzzling *Selligua* group, Hovenkamp (1996) demonstrated the inevitable instability of generic circumscriptions dependent on cladistic analyses with data sets containing parallelisms. Especially within some polypodiaceous groups, many characters formerly considered as important for generic delimitation are now recognized as showing large numbers of parallelisms.

Morphology, development, and evolution of the spore wall were studied by van Uffelen. The study of sporogenesis of several taxa (van Uffelen 1992, 1993) was followed by an analysis of the value of spore characters in phylogenetic analysis (van Uffelen 1993, 1997). As four main types of spores are known, and as exospore formation has been studied in only a small number of species, the succession of wall surface patterns during exospore formation cannot often be used to elucidate relationships between species or genera at the moment. Of seven perispore types which were distinguished earlier, none elucidates relationships in the family, as the number of species of a certain type is either very limited or very large and occurring in distantly related groups. Studies of sporogenesis, especially of the exospore, are urgently needed in related groups.

The *Polypodium vulgare* complex was studied by several authors. In a broad circumscription, it contains 17 species, which are connected by reticulate speciation (Haufler et al. 1995b). Neuroth (1996) emphasized the study of the relationships between some diploid taxa: Macaronesien *P. macaronesicum* s.l. and *P. cambricum*, widespread around the Mediterranean area. They are closely related, but should be separated at an infraspecific level. Chloroplast DNA restriction site data suggest the monophyly of this complex. Two distinct diploid species groups were found and multiple, and in three cases, reciprocal origins of tetraploids were revealed (Haufler et al. 1995a). Two little-known small genera of Malesian ferns were revised: *Thylacopteris* (Rödl-Linder 1994a) and *Polypodiopteris* (Rödl-Linder 1994b). Microsoroid ferns were monographed by Nooteboom (1997). He recognized *Leptochilus* (9 species and 1 hybrid), the monotypic *Podosorus*, and *Microsorum* (49 species, including *Phymatosorus* and *Neocheiropteris*). Zink (1993) revised the African and Madagascan species of *Lepisorus*, maintaining 9 of them. A nomenclatural and bibliographic list of the Asiatic species – the majority – is added. The Old World genus *Belvisia (8 species)*, characterized by sterile/fertile hemidimorphism and reaching from tropical Africa to China, Polynesia, and Australia, was revised by Hovenkamp and Franken (1993).

A series of very interesting papers (Gay 1993b, Gay and Hensen 1992; Gay et al. 1994) deals with the taxonomy, morphology, ecology, and biogeography of the epiphytic Malesian ant fern *Lecanopteris*. The genus comprises 13 species with expanded rhizomes which shelter ants in a mutualistic association. After detailed investigations of the morphology, rhizome architecture, and evolution (Gay 1993b), a taxonomic revision is given by Gay et al. (1994). The loose taxon specifity between ants and hosts, together with the behaviour of the insects in respect to the different morphology of the host plants is discussed by Gay and Hensen

(1992). *rbcL* studies revealed the monophyly of the genus (Grammer and Haufler 1997).

A very puzzling group are the cheilanthoid ferns (**Pteridaceae** s.l.). *rbcL*-based studies of 25 from 200 species allowed new and promising insights in the phylogeny and the circumscription of genera (Gastony and Rollo 1995). Some genera proved to be polyphyletic (e.g. *Pellaea, Cheilanthes*). Some recent removals from *Notholaena* to *Cheilanthes* and the separation of the genus *Argyrochosma* were supported. Unfortunately, the newly descibed segregate genus *Astrolepis* (Benham and Windham 1992) was not taken into account. The monotypic *Llavea* proved to be no member of the cheilanthoids s. str. *Paragymnopteris*, another segregate genus of *Hemionitis* s.l., was proposed by Shing (1993).

The genus *Mohria* (**Schizaeaceae**) was revised by Roux (1995). It was thought to consist of only three species in the past, but intense research on anatomy, morphology, and karyology led to the conclusion that seven species have to be distinguished.

The **Vittariaceae**, a family with a distinctive circumscription but some controversities in generic and subgeneric taxonomy was analyzed by Crane et al. (1995). *rbcL* analysis confirmed the limits of the family and led to the recognition of more strictly circumscribed monophyletic genera, partly formerly recognized at a subgeneric level of *Vittaria* and *Antrophytum*, respectively.

Wagner and Beitel (1992) discussed the generic classification of North American **Lycopodiaceae**, together with a convincing treatment of genus recognition in pteridophytes overall, differentiating characters, and possible character trends. Seven genera in three subfamilies are recognized. Wagner (1992) presented a list of the published chromosome numbers of *Lycopodium* s. l. together with a discussion of cytological problems within this family. Evidence in support of a base number of 11 is presented. Furthermore, the author evaluates the role of allohomoploid nothospeciation, rarely found in true ferns, but relatively common in Lycopodiaceae. An attempt to survey the diversity and variation of the nearly 185 species in the genus *Lycopodium* s. l. in the neotropics was given by Øllgaard (1992).

Habitat, evolution, and speciation in *Isoetes* (**Isoetaceae**) were discussed by Taylor and Hickey (1992). Diversity has evolved by a series of habitat adaptations that have resulted in morphological simplicity, homoplasy, and reticulate evolution of this genus, which primitively was aquatic.

2 Bibliography, Collections, Nomenclature

Supplements 6 and 7 of *Index Filicum*, covering the period from 1976 to 1995 (Johns 1996b, 1997b) were published by the Royal Botanic Gardens, Kew, which will be responsible for the continuation of the series.

Pichi Sermolli (1996) compiled the authors of scientific names in pteridophytes. He includes information on the citation and spelling of authors' names, particularly Chinese names, together with a selected bibliography. Many bibliographic corrections for the citations of family names for pteridophytes are given by Pichi Sermolli (1993). Standardization of the classification for the families of pteridophyta for the users of taxonomic information was proposed as a Scientific Consensus Classification by Hennipman (1996). The author combined the taxonomic work between 1947 and 1990 with the cladograms of Hasebe et al. (1995), and proposed a comprehensible and practical classification which seems to ensure a minimum of stability of names for the future. Biographies and bibliographies of some important pteridologists were published. To be mentioned are R. E. Holttum (Stearn 1996; Price 1996; Edwards et al. 1997), K. U. Kramer (Zink 1995), T. Reichstein (Schneller 1997), L. R. Atkinson (Mickel 1997b), and R. E. G. Pichi Sermolli (Bizzarri 1992).

A Group of Pteridologists (GEP) was found in 1994 (Viane 1996). The main goal of this group, not residing under the umbrella of IAP, would be to promote cooperation between specialists and amateurs.

3 Floristics

a) Asia, Australia, Pacific

Parris et al. (1992) published an annotated checklist of ferns and fern allies of Mount Kinabalu, one of the most interesting floristic regions of Malesia. They list about 610 species, belonging to 28 families and 145 genera. About 50 species seem to be endemic to Mt. Kinabablu. Various fern families of Ambon and Seram were treated by Kato (1992, 1994a, b, 1996), and Kato and Parris (1992). Iwatsuki et al. (1995) published the first volume of the *Flora of Japan*, which covers pteridophytes and gymnosperms. Within the *Flora of Australia* series, the pteridophytes of the offshore islands (Norfolk, Lord Howe, Christmas, Macquarie, and others) were treated by DuPuy and Orchard (1993) and Green and Tindale (1994). A "picture book" with 130 coloured photographs of the 90 native pteridophyte species (8 endemic) of the Mariana Islands was produced by Raulerson and Rinehart (1992). Within the framework of a pteridophyte flora of Hawaii, which includes approximately 225 native and naturalized taxa, W. H. Wagner (1995) published an interesting overview of evolution and conservation aspects. Valier (1995) presents a selection of 60 Hawaiian native ferns and fern allies in a nicely illustrated booklet.

b) Africa, Macaronesia

An annotated checklist of Malawi's pteridophytes, including 219 species, was presented by Burrows and Burrows (1993). Since the last floristic treatment in the *Flora Zambesiaca* (1970), an increase of 24% in the total number of taxa is reported. A checklist of vascular plants of Macaronesia (Azores, Madeira archipelago, Salvage Islands, Canary Islands, and Cape Verde Islands) was presented by Hansen and Sunding (1993). The pteridophytes of Canary Islands and Madeira are covered by Hohenester and Welss (1993) and Gibby and Paul (1994), respectively.

c) Europe

Volume I of the 2nd edition of the *Flora Europaea* was published by Tutin et al. (1993) and covers the pteridophytes of continental Europe. Fischer and Lobin (1995) presented an illustrated treatment of the European pteridophytes. Additional lists provide the species found in Turkey, on the Azores, Madeira, the Canary Islands, and Cape Verde Islands. Distribution maps of all pteridophytes occurring in Eastern Germany are included in an atlas of vascular plants of the former GDR area (Benkert et al. 1996). After first records in Great Britain, persisting colonies of gametophytes of the filmy fern *Trichomanes speciosum* were found at various places in Europe (NE France, adjacent Germany, and Luxembourg: Rasbach et al. 1993, 1995; NW Germany: Bennert et al. 1994; E-Germany, Czech Republic: Vogel et al. 1993). Sporophytes of this species in Europe are of an extreme oceanic distribution. The highly specialized gametophyte communities can be regarded as relicts of times with a warmer and moister climate. The habitats were studied in detail by Rasbach et al. (1995).

d) America

Beginning in the north, the excellent pteridophyte part of the multivolume *Flora of North America – North of Mexico* was published in 1993 by the Flora of North America Editorial Committee. It contains keys, descriptions, distribution maps, and beautiful figures. Volume 1 of the *Flora Mesoamericana* (Davidse and Sousa 1995; in Spanish) deals with Central America (geographically from the Mexican Chiapas and Tabasco provinces to southern Panama). One hopes that such regional floras, which are understandable for many local workers in floristics and conservation, will help these countries towards a better understanding of their natural resources. A treatment of the pteridophytes of Nueva Galicia (Western Mexico) was presented by Mickel (1992). The pteridophytes represent a mixture of species of the wetter regions in southern Mexico and the drier regions in northern Mexico. The Hymenophyllaceae of Mexico were completely revised by Pacheco (1994). A detailed phytogeographical analysis of the pteridophytes of the state of Veracruz (Mexico) was presented by Palacios-Rios and Goméz-Pompa (1997). All ferns and fern allies of Honduras were listed by Nelson Sutherland et al. (1995). There are 651 species in 109 genera; 5 species are considered as endemic.

Within the *Flora of the Venezuelan Guayana* series, Volume 2 was published in 1995 (Steyermark et al. 1995).The pteridophyte section was coordinated by A. R. Smith, who also wrote most of the treatments with assistance from seven specialists. The beautifully illustrated Flora contains 671 species in 92 genera with 93 species (14%) endemic to the Flora region and 144 species (22%), including the family Hymenophyllopsidaceae and nearly all of the genus *Pterozonium* are endemic to the Guiana Shield. Five families are treated in two new fascicles of the *Flora of the Guianas* (Görts-van Rijn 1993, 1994). A complete checklist of the pteridophytes of the Guianas was published by Boggan et al. (1997). The first part of a well-illustrated guide to the vascular plants of central French Guiana was presented by Mori et al. (1997). It provides keys and descriptions for nearly 200 pteridophyte species.

The Ecuadorean members of the tribe Physematieae (Dryopteridaceae) were revised by Stolze et al. (1994). The *Flora of Peru* series on pteridophytes was completed by Tryon et al. (1992, 1993, 1994). The last-mentioned fascicle contains a comprehensive index to names, a list of species to be added to the foregoing parts, and considerations of pteridophyte diversity in respect to ecology and geography of Peru. Endemism and diversity of Peruvian pteridophytes were analysed by León and Young (1996) in detail. Both concentrate at higher elevations, especially in humid montane forests. With the Atlantic Ocean draining basin containing 71% of Peru's surface area and 97% of the pteridophyte species, Peru shows a strong assymetry in distribution patterns. The main part of Vol. 1 of *Flora de Chile* (Marticorena and Rodríguez 1995, in Spanish) is dedicated to ferns and fern allies. In total, 167 taxa are treated in this well-illustrated book.

4 Geography, Ecology, and Biodiversity

A series of papers dealing with changing concepts in the biogeography of pteridophytes resulting from an earlier symposium were published in 1993. The use of phytogeographic principles in explaining fern relationships was elucidated by Smith (1993b), together with a discussion of the role of special aspects of fern biology (e.g. long-distance dispersal by spores). Kato (1993b) reviewed the recent literature on dispersal and vicariance as historical explanations for fern distribution. Barrington (1993) analysed the power of explanations for disjunctions and endemic centres in ferns, with emphasis on isolating mechanisms. Similar to angiosperms, the pteridophyte flora of tropical Africa is depauperate in comparison with South America and Southeast Asia. The importance of shifting of climatic zones due to continental drift during the Tertiary and climatic oscillations during the Quaternary was discussed by Kornas (1993). Kramer (1993) compared the distribution patterns of pteridophytes and spermatophytes. Whereas one finds similar patterns at the level of principal families, the pteridophyte genera are, in general, more widespread. Primitive genera tend to occupy special habitats and/or show special growth forms in ferns, rather than concentrating in geographic relict areas as angiosperms do. Aspects of endemism, diversity centres, threats, and conservation are discussed by Given (1993) in consideration of increasing threats to strongly habitat-specific ferns.

The variability of small isolated populations of *Asplenium septentrionale* was studied by Holderegger and Schneller (1994). All three populations investigated by isozyme electrophoresis could be distinguished genetically from each other. Only one showed internal variation. Herbarium studies proved that these strongly isolated populations of plants of only a few individuals with low genetic variability are capable of surviving for long periods with a very low extinction rate, a fact of importance in conservation biology. These studies were followed with the investigation of colonisation effects and genetic variability within populations of *Asplenium ruta-muraria* (Schneller and Holderegger 1996a). Populations on recently built walls showed almost no genetic variability. It increased on older walls and reached a maximum in natural habitats. General methodological discussions are found in Schneller and Holderegger (1996b).

Poulsen and Nielsen (1995) documented the contribution of pteridophytes to the flora of 1 ha of Amazonian primary rainforests in Ecuador (elevation ca. 250 m). They found a total of 50 species, comprising 16 terrestrial species, 3 scandent species, 6 species of climbers and 25 epiphytic ones. In total, they represent less than 5% of the total number of the vascular plant species in this area. Similar studies with comparable results were carried out in Brunei and in northwest Borneo (Poulsen 1996a, b).

The role of soil spore banks of temperate ferns was analysed by Dyer and Lindsay (1992). At least some temperate fern species form a persistent soil spore bank. Spores thus achieve dispersal of genotypes in time as well as space. The widespread existence of spore banks reinforces the interpretation of gametophytes as non-competitive weedy opportunists colonising temporary open habitats, even when the sporophyte phase of the same life cycle is a long-lived perennial of late-successional vegetation. Their role as a resource for conservation measurements was discussed by Dyer and Lindsay (1996). Studies of three sympatric species of *Botrychium* in Alberta (Canada) by Lesica and Ahlenslager (1996) revealed a short lifespan and association with disturbed habitats which led to the suggestion of the necessity of natural disturbance regimes for long-term persistence.

Aspects of spore dispersal in *Selaginella* in respect to heterospory and the effect of wind were discussed by Filippini-De Giorgi et al. (1997). A conference held in the New York Botanical Garden in 1993 delt with biodiversity and conservation of neotropical montane forests. Øllgaard (1995) showed that 83 species of *Huperzia* are known from this area, whereas the Andes include more than a half, almost 50% of them endemic. Eighteen of 24 species of the Brazilian montane region are endemic, many of them taxonomically isolated. A detailed analysis of montane pteridophytes of Costa Rica was presented by Mehltreter (1995), who adds a complete list of species. Of 1099 peridophytes recorded for Costa Rica, only 282 occur in the higher montane area (< 2500 m), but only 56 are restricted to these regions, with 22 being endemic. The importance of mountains to neotropical pteridophytes by impeding migration and promoting species richness and endemism was pointed out by Moran (1995). The Andes with 2000 species, and the montane regions of SE Brazil with 600 are in sharp contrast to the Amazonian lowland with only 300 species, making it the most species-poor region for pteridophytes in the Neotropics. Comparable numbers are given by Parris et al. (1992) for SE Asia: the species and genus richness in the Mount Kinabalu region is highest around 1500 m. Pteridophyte diversity in Malesia and New Zealand was measured by counting taxa in several 0.5-ha plots by Parris (1996). The highest numbers of taxa were found in lower montane forest in Sabah and Seram (ca. 95 taxa/plot), the lowest in coastal lowland mixed dipterocarp forest in Sabah (18 species/plot) and in *Nothofagus* forest (18 species/plot) and subalpine scrub in New Zealand (4 species/plot).

The ecology of the Malesian ant-fern *Lecanopteris* (Polypodiaceae) was investigated by Gay and others (see Sec. 1). Barthlott et al. (1994) were able to show that young nymphs of the neotropical, semiaquatic grasshopper *Paulinia acuminata* are well camouflaged by colour and surface structure which cannot be distinguished from the leaves of their food plant, *Salvinia auriculata*. Young nymphs as well as the leaves are extremely water-repellent because of epicuticular waxes with similar ultrastructure based on functional analogy. The salt tolerance of the mangrove fern *Acrostichum danaefolium* was studied under laboratory conditions as well as in natural sites (Sánchez Peña 1994). The sporophytes proved to have a high salt tolerance, which is caused by the ability to retain Na and Cl within the roots. Growth is promoted by low salinity.

5 Conservation

The Pteridophyte Specialist Group of the IUCN elaborated an action plan for future conservation aims (Given 1992). Thorough investigations of the pteridophyte records from several sources from Trinidad and Tobago, combined with a risk index rating, show that 52.8% of the 307 species are at risk to varying degrees (Baksh-Comeau 1996). An annotated list of threatened ferns and fern allies of Cuba was published by Villaverde and Caluff (1997). They mention 82 of 600 species (14%) under threat, an astonishingly low number. The pteridophytes of Italy, classified as being threatened in Italy by the Bern Convention, are presented by Cellinese et al. (1996) with detailed distribution maps, brief plant profiles, and figures. A red data book on Cape Verde Flora and Fauna includes a detailed red list of the native pteridophytes (Lobin and Ormonde 1996). Threatened and local plants are listed regularly for New Zealand, latterly Cameron et al. (1995). A new red data book of German plants was published by Korneck et al. (1996). From 77 pteridophyte species recorded from Germany, 49% are under threat, a higher percentage than in spermatophytes (31%). Ecology and conservation principles of endangered ferns were discussed by Bennert et al. (1995). In germination experiments with 31 species of pteridophytes at risk in Germany, Bennert and Danzebrink (1996) observed a germination rate close to 100%, but were successful in raising young sporophytes only in those 20 species which require light for germination. Raine and Sheffield (1997) described a method for aseptic culture of gametophytes of the filmy fern *Trichomanes speciosum* from gemmae. Douglas and Sheffield (1992) tested existing and novel artificial growth systems for the production of fern gametophytes. They showed that agar-based plate culture does not yield maximum growth results; the greatest biomass developed in air-lift fermenter cultures.

Conservation status and distribution of two serpentine restricted *Asplenium* species in Central Europe were discussed by Vogel (1996). Ecology and conservation of the filmy fern *Trichomanes speciosum* in Britain and Ireland were treated by Ratcliffe et al. (1993). There are approximately only 50–100 plants in 43 colonies. Sporophyte establishment appears to be most critical under a climate that is marginal for its temperature requirements.

6 Morphology and Anatomy

A very important contribution to our knowledge of pteridophyte anatomy was presented by Schneider (1996a, b) with a comparative study of root anatomy. The great number of 608 species from 170 genera of all families was investigated, mostly for the first time. While rhizodermis and central cylinder showed only slight variations, the most variable structure was the cortex. According to the degree of complexity of the cortex tissues, altogether ten root types were distinguished. It was demonstrated that root anatomy is very valuable in fern systematics, especially at family level. In most fern families (e.g. Polypodiaceae) all species have the same root type independently of their growth forms such as epiphytes, rheophytes, etc. Stützel and Gailing (1995) showed that the irregular phyllotaxis of ferns with horizontal rhizomes with the fronds arranged only in its dorsal part is not due to secondary shifts, but to the fact that ramenta and leaves are part of the same, regular ontogenetic pattern. They demonstrated that paleae show the same ontogenetic mode and pattern as leaves. Gailing (1995) investigated ferns with erect rhizomes. The fronds are not replaced by ramenta in one ontogenetic spiral. The ramenta are always inserted in the area between the fronds in the same way as is described for their emergence in angiosperms. *Lomagramma guianensis* (Dryopteridaceae) shows a unique vegetative complexity with two completely distinct rhizome forms. Architectural field studies (Gay 1993a) and detailed analysis of morphogenesis (Hébant-Mauri and Gay 1993) showed that *Lomagramma* employs a single type of rhizome, similar in morphogenesis and structure but varying in morphological expression and behaviour. Similar control and complexity of the growth cycle can be found in some angiosperms such as the Araceae. Andersen and Øllgaard (1996) proposed a recommendable standardization of the morphological terms of the unique leaf in the Gleicheniaceae. A clearer distinction between terms used for shoots and shoot branchings on the one had and for leaves and their divisions on the other should be aimed at in future. A recommendable terminology for the classification of ornamentation of pteridophyte spores was proposed by Lellinger and Taylor (1997).

Fertilization in *Athyrium filix-femina* was studied in situ by using video microscopy (Fasciati et al. 1994a). Plasmogamy and karyogamy in the same species were investigated by light and electron microscopy (Fasciati at al. 1994b). Fascinating photographs and detailed analysis of the processes with many new findings are presented.

The occurrence of vascular tissue in older female gametophytes of *Phegopteris polypodioides* (Thelypteridaceae) was reported by Bao et al. (1997). This phenomenon was not seen until now in gametophytes of higher ferns. Frey et al. (1994) analysed the sporophyte-gametophyte junction in *Tmesipteris* (Psilotaceae). They detected gametophytic transfer cells and intruding sporophytic haustorial cells. Similar structures are known from Anthocerotae, and a closer relationship between the Psilotatae and the hornworts seems to be probable. Whittier and Braggins (1992) were successful in growing young gametophytes of the poorly known *Phylloglossum drummondii* (Lycopodiaceae) from spores in axenic culture. The results suggest that *Phylloglossum* is not as similar to the subgenus *Lepidotis* of *Lycopodium* as once thought and remains of enigmatic systematic position. Whittier (1996) proved the possibility of extending the viability of the short-lived spores of *Equisetum hyemale* for more than a year by freezing them at –70 °C.

To complete this section, the comprehensive review of White and Turner (1995) on the anatomy and development of the fern sporophyte (apical meristems, development of branches, and vascular tissues) should be mentioned. Especially the role of the shoot apical cell as a single apical initial of the meristem was reconfirmed during the past years.

7 Chemosystematics

A modern overview on chemistry and chemotaxonomy of pteridophytes was given by Wollenweber (in Kramer et al. 1995). A review of the occurrence of diterpenes and triterpenes in leaf exudates of angiosperms and pteridophytes was published by Wollenweber (1996). Phloroglucinol derivatives in *Dryopteris* sect. *Fibrillosae* and related taxa were studied by Widén et al. (1996) on a worldwide basis. They are remarkably constant in most taxa, a reflection of their common origin from a few diploid apomictic taxa by hybridisation. The highly specific leaf flavonoids of some *Dicranopteris* taxa were studied by Yusuf (1995). Major flavonoids are flavonols and flavones with glycosidic combinations. A series of papers is dedicated to the study of the terpenes of cheilanthoid ferns. An overview of their structures is found in Arriaga-Giner et al. (1997). Flavonoid aglycones and a novel dihydrostilbene from the frond exudate of *Notholaena nivea*, together with the review of notholaenic and isonotholaenic acids in further species of this genus, were reported by Wollenweber et al. (1993). In *Macrothelypteris torresiana* drimane sesquiterpenes were found in ferns for the first time (Siems et al. 1996).

Smith and Seawright (1995) reviewed the carcinogenicity of bracken fern (*Pteridium* spp.) for humans. It is well known as causing cancer naturally in sheep and cattle. The major carcinogenic substance is ptaquiloside, a water-soluble norsequiterpenoid glycoside, which can be transferred through milk (Alonso-Amelot et al. 1996). There is epidemiological evidence that the bracken causes cancer in man especially by indirect consumption of milk of local cows. Ptaquiloside at rates which are dangerous for cattle has also been reported from the Australian *Cheilanthes sieberi* (Smith et al. 1992). Fifteen other fern species from Dennstaedtiaceae, Dicksoniaceae, and Pteridaceae containing ptaquilisodes or ptaquiloside-like mutagenic compounds are listed by Smith (1997).

8 Ethnobotany, Uses

Christensen (1997) studied the knowledge and importance of ferns within two small indigenous rural communities in Sarawak (Borneo). Their most important uses are as vegetables, less intensively as fibres, and for medical purposes. They are recognised as a separate life-form, and a well-developed taxonomy indicates their long continual use. Wagner (1997) reviewed the biology and utilisation of *Azolla*. After a treatment of its taxonomy, distribution, morphology, physiology, reproduction, and development, a comprehensive review of its manifold utilisation is given. Besides its use as biofertilizer, *Azolla* can be used as an animal feed. (*A. caroliniana* in carp diet, Mohanty and Dash 1995), a medicine, and a water purifier.

Mickel (1994) and Denkewitz (1995) produced two attractive and lavishly illustrated books on ferns in horticulture which should be mentioned at the end of this section. Whereas Mickel gives a comprehensive account on species, varieties, etc. (more than 500 kinds of hardy ferns are mentioned), Denkewitz puts emphasis on garden and landscape architecture.

References

Alonso-Amelot ME, Castillo M, Smith BL, Lauren D (1996) Bracken ptaquiloside in milk. Nature 382:587

Andersen EØ, Øllgaard B (1996) A note on some morpholigical terms of the leaf in the Gleicheniaeae. Am Fern J 86:52–57

Arriaga-Giner FJ, Rumbero A, Wollenweber E (1997) *Notholaena* terpenoids: two new epidermic diterpenes from frond exudate of the fern, *Notholaena rigida*. Z Naturforsch [c] 52:292–294

Baksh-Comeau YS (1996) Index rating of threatened ferns in Trinidad and Tobago. In: Camus JM, Gribby M, Johns RJ (eds) Pteridology in perspective. Royal Botanic Gardens, Kew, pp 139–151

Bao WM, Aur CW, Wang QX (1997) Studies in the development of the gametophytes of Thelypteridaceae of North-eastern China. In: Johns RJ (ed) Holttum memorial volume. Royal Botanic Gardens, Kew, pp 91–94

Barrington DS (1993) Ecological and historical factors of fern biogeography. J Biogeogr 20:275–280

Barthlott W, Riede K, Wolter M (1994) Mimicry and ultrastructural analogy between the semi-aquatic grasshopper *Paulinia acuminata* (Orthoptera: Pauliniidae) and its foodplant, the water-fern *Salvinia auriculata* (Filicatae: Salviniaceae). Amazoniana 13:47–58

Benham DM, Windham MD (1992) Generic affinities of the star-scaled cloak ferns. Am Fern J 82:47–58

Benkert D, Fukarek F, Korsch H (1996) Verbreitungsatlas der Farn- und Blütenpflanzen Ostdeutschlands. Fischer, Jena

Bennert HW, Danzebrink B (1996) Spore germination of pteridophytes at risk in Germany. Mem Accad Lunigian Sci 66:37–50

Bennert HW, Fischer G (1993) Biosystematics and evolution of the *Asplenium trichomanes* complex. Webbia 48:743–760

Bennert HW, Jäger W, Leonhards W, Rasbach H, Rasbach K (1994) Prothallien des Hautfarns *Trichomanes speciosum* (Hymenophyllaceae) auch in Nordrhein-Westfalen. Flor Rundbr (Bochum) 28:80

Bennert HW, Danzebrink B, Heiser T, Paeger J, Schiemionek A, Stoor AM (1995) Ökologie und Schutz gefährdeter Farnpflanzen. Farnblätter 26/27:111–131

Bishop LE, Smith AR (1992) The fern genus *Enterosora* (Grammitididaceae) in the New World. Syst Bot 17:345–362

Bizzarri MP (1992) L'attività scientifica del prof. Rodolfo E. G. Pichi Sermolli. Webbia 48:701–733

Boggan J, Funk V, Kelloff C, Hoff M, Cremers G, Feuillet C (1997) Checklist of the plants of the Guianas. Biological diversity of the Guianas program. Smithsonian Institution, Washington, DC

Burrows JE (1992) The taxonomy of the genus *Ophioglossum* L. (Ophioglossaceae) in southern Africa. MSc Thesis. University of Natal, Pietermaritzburg

Burrows JE (1996) The genus *Ophioglossum* L. in south-central Africa. In: Camus JM, Gibby M, Johns RJ (eds) Pteridology in perspective. Royal Botanic Gardens, Kew, pp 329–336

Burrows JE, Burrows SM (1993) An annotated checklist of the pteridophytes of Malawi. Kirkia 14:78–99

Cameron E, de Lange P, Given DR, Johnson P, Ogle C (1995) New Zealand Botanical Society threatened and local plant list. N Z Bot Soc Newslett 39:15–28

Camus JM, Gibby M, Johns RJ (eds) Pteridology in perspective. Royal Botanic Gardens, Kew

Cellinese N, Jarvis CE, Pichi Sermolli REG, Press JR, Short MJ, Viciani D (1996) Threatened plants of Italy – Pteridophyta. Mem Accad Lunigian Sci 66:117–145

Christensen H (1997) Uses of ferns in two indigenous communities in Sarawak, Malaysia. In: Johns RJ (ed) Holttum memorial volume. Royal Botanic Gardens, Kew, pp 177–192

Collinson ME (1996) "What use are fossil ferns?" – 20 years on: with a review of the fossil history of extant pteridophyte families and genera. In: Camus JM, Gibby M, Johns RJ (eds) Pteridology in perspective. Royal Botanic Gardens, Kew, pp 349–394

Conant DS, Stein DB, Valinski AEC, Sudarsanam P (1994) Phylogenetic implications of chloroplast DNA variation in the Cyatheceae. I. Syst Bot 19:60–72

Conant DS, Raubeson LA, Attwood DK, Stein DB (1996a) The relationships of Papuasian Cyatheaceae to New World tree ferns. Am Fern J 85:328–340

Conant DS, Raubeson LA, Attwood DK, Perera S, Zimmer EA, Sweere JA, Stein DB (1996b) Phylogenetic and evolutionary implications of combined analysis of DNA and morphology in the Cyatheaceae. In: Camus JM, Gibby M, Johns RJ (eds) Pteridology in perspective. Royal Botanic Gardens, Kew, pp 231–248

Crane EH, Farrar DR, Wedel JF (1995) Phylogeny of the Vittariaceae: convergent simplification leads to a polyphyletic *Vittaria*. Am Fern J 85:283–305

Davidse G, Sousa MS (eds) (1995) Flora Mesoameriana, vol 1. Psilotaceae a Salviniaceae. Universidad Nacional Autónoma de México, México (in Spanish)

Denkewitz L (1995) Farngärten. Ulmer, Stuttgart

Douglas GE, Sheffield E (1992) The investigation of existing and novel artificial growth systems for the production of fern gametophytes. Fern horticulture: past and present and future perspectives. Intercept, Andover, pp 183–187

Dubuisson J-Y (1996) Evolutionary relationships within the genus *Trichomanes* sensu lato (Hymenophyllaceae) based on anatomical and morphological characters and a comparison with *rbcL* nucleotide sequences: preliminary results. In: Camus JM, Gibby M, Johns RJ (eds) Pteridology in perspective. Royal Botanic Gardens, Kew, pp 285–287

DuPuy DF, Orchard AE (1993) Pteridophyta. Flora Aust 50:530–570

Dyer AF, Lindsay S (1992) Soil spore banks of temperate ferns. Am Fern J 82:89–123

Dyer AF, Lindsay S (1996) Soil spore banks – a new resource for conservation. In: Camus JM, Gibby M, Johns RJ (eds) Pteridology in perspective. Royal Botanic Gardens, Kew, pp 153–160

Edwards PJ, FitzGerald SMD, Johns RJ, Lynas L, Newman M, Price MG, Wortley PJ (1997) R. E. Holttum publications, 1919–1996: a bibliography. In: Johns RJ (ed) Holttum memorial volume. Royal Botanic Gardens, Kew, pp 7–32

Fasciati R, Scheller J, Roos U-P (1994a) Fertilization in the fern *Athyrium filix-femina* (Pterophyta). I. Live observations. Crypt Bot 4:329–335

Fasciati R, Schneller J, Jenni V, Roos U-P (1994b) Fertilization in the fern *Athyrium filix-femina*. II. Ultrastructure. Crypt Bot 4:356–367

Filippini-De Giorgi A, Holderegger R, Schneller JJ (1997) Aspects of spore dispersal in *Selaginella*. Am Fern J 87:93–103

Fischer E, Lobin W (1995) Farnpflanzen. In: Frey W, Frahm JP, Fischer E, Lobin W Die Moos- und Farnpflanzen Europas, 6th edn. Kleine Kryphogamenflora, vol 4. Fischer, Stuttgart, pp 319–391

Flora of North America Editorial Committee (1993) Flora of North America north of Mexico. vol 2. Pteridophytes and gymnosperms. Oxford University Press, New York

Fraser-Jenkins CR (1994) *Dryopteris* (Pteridophyta) of Hawai'i – a monographic study. Thaiszia 4:15–47

Frey W, Campbell EO, Hilger HH (1994) The sporophyte-gametophyte junction in *Tmesipteris* (Psilotaceae, Psilotopsida). Beitr Biol Pflanz 68:105–111

Gailing O (1995) Untersuchungen der Scheitelmorphologie bei *Dryopteris*-Arten und *Athyrium filix-femina* L. sowie der Verzweigung bei *Athyrium filix-femina* L. Beitr Biol Pflanz 69:31–41

Gastony GJ, Rollo DR (1995) Phylogeny and generic circumscription of cheilanthoid ferns. Am Fern J 85:341–360

Gay H (1993a) The architecture of dimorphic clonal fern, *Lomagramma guianensis* (Aublet) Ching (Dryopteridaceae). Bot J Linn Soc 111:343–358

Gay H (1993b) Rhizome structure and evolution in the ant-associated epiphytic fern *Lecanopteris* Reinw. (Polypodiaceae). Bot J Linn Soc 113:135–160

Gay H, Hensen R (1992) Ant specifity and behaviour mutualism with epiphytes: the case of *Lecanopteris* (Polypodiaceae). Biol J Linn Soc 47:261–284

Gay H, Hennipman E, Huxley CR, Parrot FJE (1994) The taxonomy, distribution, and ecology of the epiphytic Malesian ant-fern *Lecanopteris* Reinw. (Polypodiaceae). Gard Bull (Singapore) 45:293–335

Gibby M, Paul AM (1994) Pteridophyta. In: Press JR, Short MJ (eds) Flora of Madeira. Natural History Museum, London, pp 25–53

Given DR (1992) The Pteridophyte Specialist Group action plan. Species 19:66

Given DR (1993) Changing aspects of endemism and endangerment in pteridophyta. J Biogeogr 20:293–302

Görts-van Rijn ARA (ed) (1993) Flora of the Guianas Ser B, 6: Dryopteridaceae (Cremers G, Kramer KU, Moran RC, Smith AR), Nephrolepidaceae, Oleandraceae (Cremers G, Kramer KU), Thelypteridaceae (Smith AR). Koeltz Scientific Books, Königstein

Görts-van Rijn ARA (ed) (1994) Flora of the Guianas Ser B, 3 Hymenophyllaceae (Lellinger DB). Koeltz Scientific Books, Königstein

Grammer WA, Haufler CH (1997) Applying chloroplastic and nuclear DNA sequences to test phylogenetic hypotheses: a case study using the ant-fern genus *Lecanopteris*. Am J Bot 84 Suppl:164

Green PS, Tindale MD (1994) Pteridophyta. Flora Aust 49:546–615

Hansen A, Sunding P (1993) Flora of Macaronesia. Checklist of vascular plants, 4th edn. Sommerfeltia 17:1–295

Hasebe M, Omori T, Nakazawa M, Sano T, Kato M, Iwatsuki K (1994) *RbcL* gene sequences provide evidence for the evolutionary lineages of leptosporangiate ferns. Proc Natl Acad Sci USA 91:5730–5734

Hasebe M, Wolf PG, Pryer M, Ueda K, Ito M, Sano R, Gastony GJ, Yokoyama J, Manhart JR, Murakami N, Crane EH, Haufler CH, Hauk WG (1995) Fern phylogeny based on *rbc*L nucleotide sequences. Am Fern J 85:134–181

Haufler CH (1996) Species concepts and speciation in pteridophytes. In: Camus JM, Gibby M, Johns RJ (eds) Pteridology in perspective. Royal Botanic Gardens, Kew, pp 291–305

Haufler CH, Soltis DE, Soltis PS (1995a) Phylogeny of the *Polypodium vulgare* complex: insights from chloroplast DNA restriction site data. Syst Bot 20:110–119

Haufler CH, Windham MD, Rabe EW (1995b) Reticulate evolution in the *Polypodium vulgare* complex. Syst Bot 20:89–109

Hébant-Mauri R, Gay H (1993) Morphogenesis and its relation to architecture in the dimorphic clonal fern *Lomagramma guianensis* (Aublet) Ching (Dryopteridaceae). Bot J Linn Soc 112:257–276

Hennipman E (1996) Scientific consensus classification of pteridophyta. In: Camus JM, Gibby M, Johns RJ (eds) Pteridology in perspective. Royal Botanic Gardens, Kew, pp 191–202

Hohenester A, Welss W (1993) Exkursionsflora für die Kanarischen Inseln mit Ausblicken auf ganz Makaronesien. Ulmer, Stuttgart

Holderegger R, Schneller JJ (1994) Are small isolated populations of *Asplenium septentrionale* variable? Biol J Linn Soc 51:377–385

Hovenkamp PH (1996) The inevitable instability of generic circumscriptions in Old World Polypodiaceae. In: Camus JM, Gibby M, Johns RJ (eds) Pteridology in perspective. Royal Botanic Gardens, Kew, pp 249–260

Hovenkamp PH, Franken NAP (1993) An account of the fern genus *Belvisia* Mirbel (Polypodiaceae). Blumea 37:511–527

Iwatsuki K, Yamazaki T, Boufford DE, Ohba H (eds) (1995) Flora of Japan, vol 1. Pteridophyta and Gymnospermae. Kodansha, Tokyo

Ji J, Russell SJ, Gibby M, Vogel JC, Sheffield E, Peng H-Z (1994) A technique for the investigation of genetic variation in filamentous gametophytes of *Trichomanes* (Hymenophyllaceae). Am Fern J 84:126–130

Johns (1996a) A recollection of *Thysanosoria* A. Gepp in Gibbs (Lomariopsidaceae). In: Camus JM, Gibby M, Johns RJ (eds) Pteridology in perspective. Royal Botanic Gardens, Kew, pp 175–177

Johns RJ (1996b) Index Filicum – supplementum sextum pro annis 1976–1990. Royal Botanic Gardens, Kew

John RJ (ed) (1997a) Holttum memorial volume. Royal Botanic Gardens, Kew

Johns RJ (1997b) Index Filicum – supplementum septimum pro annis 1991–1995. Royal Botanic Gardens, Kew

Kato M (1992) Taxonomic studies of pteridophytes of Ambon and Seram (Moluccas) collected by Indonesian-Japanese botanical expeditions. VIII. Aspleniaceae, Blechnaceae, and Lomariopsidaceae. J Fac Sci Univ Tokyo Sect III Bot 15:135–153

Kato M (1993a) A taxonomic study of the genus *Matonia* (Matoniaceae). Blumea 38:167–172

Kato M (1993b) Biogeography of ferns: dispersal and vicarance. J Biogeogr 20:265–274

Kato M (1994) Taxonomic studies of pteridophytes of Ambon and Seram (Moluccas) collected by Indonesian-Japanese botanical expeditions. IX. Woodsiaceae, Lindsaeaceae, and Adiantaceae. J Fac Sci Univ Tokyo Sect III Bot 15:315–347

Kato M (1996) Taxonomic studies of pteridophytes of Ambon and Seram (Moluccas) collected by the Indonesian-Japanese botanical expeditions X. Dennstaedtiaceae, Vittariaceae, and Tectarioideae (Dryopteridaceae). Acta Phytotax Geobot 47:61–90

Kato M, Parris BS (1992) Taxonomic studies of pteridophytes of Ambon and Seram (Moluccas) collected by Indonesian-Japanese botanical expeditions. VII. Grammitidaceae. J Fac Sci Univ Tokyo Sect III Bot 15:111–133

Kornas J (1993) The significance of historical factors and ecological perference in the distribution of African pteridophytes. J Biogeogr 20:281–286

Korneck D, Schnittler M, Vollmer I (1996) Rote Liste der Farn- und Blütenpflanzen (Pteridophyta et Spermatophyta) Deutschlands. Schriftenr Vegetationskd 28:21–187

Kramer KU (1993) Distribution patterns in major pteridophyte taxa relative to those of angiosperms. J Biogeogr 20:287–291

Kramer KU, Schneller JJ, Wollenweber E (1995) Farne und Farnverwandte. Thieme, Stuttgart

Lellinger DB, Taylor WC (1997) A classification of spore ornamentation in the pteridophyta. In: Johns RJ (ed) Holttum memorial volume. Royal Botanic Gardens, Kew, pp 32–42

León B, Young KR (1996) Distribution of pteridophyte diversity and endemism in Peru. In: Camus JM, Gibby M, Johns RJ (eds) Pteridology in perspective. Royal Botanic Gardens, Kew, pp 77–91

Lesica P, Ahlenslager K (1996) Demography and life history of three sympatric species of *Botrychium* subg. *Botrychium* in Waterton Lakes National Park, Alberta. Can J Bot 74:538–543

Lobin W, Ormonde J (1996) Lista vermelha para los pteridóphitos (Pteridophyta). Cour Forschungsinst Senckenb 193:38–42

Marticorena C, Rodríguez R (eds) (1995) Flora de Chile, vol 1. Pteridophyta. Gymnospermae. Universidad de Concepcíon, Concepcíon (in Spanish)

Mehltreter K (1995) Species richness and geographical distribution of montane pteridophytes of Costa Rica, Central America. Feddes Repert 106:563–584

Mickel JT (1992) Pteridophytes. In: McVaugh R (ed) Flora Galiciana, vol 17. Gymnosperms and pteridophytes. University of Michigan Herbarium, Ann Arbor

Mickel JT (1994) Ferns for American gardens. Macmillan, New York

Mickel JT (1997a) A review of the West Indian species of *Polystichum*. In: Johns RJ (ed) Holttum memorial volume. Royal Botanic Gardens, Kew, pp 119–143

Mickel JT (1997b) Bibliography of Lenette Rogers Atkinson (1899–1996). Am Fern J 87:41–42

Mohanty SN, Dash SP (1995) Evaluation of *Azolla caroliniana* for inclusion in carp diet. J Aqua Trop 10:343–353

Moran RC (1995) The importance of mountains to pteridophytes, with emphasis on Neotropical montane forests. In: Churchill SP, Balslev H, Forero E, Luteyn JL (eds) Biodiversity and conservation of neotropical montane forests. New York Botanical Garden, New York, pp 359–363

Mori SA, Cremers G, Gracie C, de Granville JJ, Hoff M, Mitchell JD (1997) Guide to the vascular plants of Central French Guiana. I. Pteridophytes, Gymnosperms, and Monocotyledons. Mem NY Bot Gard 76:1–442

Murakami N (1995) Systematics and evolutionary biology of the fern genus *Hymenasplenium* (Aspleniaceae). J Plant Res 108:257–268

Murakami N, Moran RC (1993) Monograph of the neotropical species of *Asplenium* sect. *Hymenasplenium* (Aspleniaceae). Ann Mo Bot Gard 80:1–38

Murakami N, Schaal BA (1994) Chloroplast DNA variation and the phylogeny of *Asplenium* sect. *Hymenasplenium* (Aspleniaceae) in the new world tropics. J Plant Res 107:245–251

Nelson Sutherland C, Gamarra Gamarra R, Fernández Casas J (1995) Hondurensis plantarum vascularium catalogus. Pteridophyta. Fontqueria 43:1–139

Neuroth R (1996) Biosystematik und Evolution des *Polypodium vulgare*-Komplexes (Polypodiaceae; Pteridophyta). Diss Bot 256:1–209

Nooteboom HP (1992) Notes on Davalliaceae. I. The genera *Araiostegia, Davallodes, Leucostegia,* and *Gymnogrammitis*. Blumea 37:165–187

Nooteboom HP (1994) Notes on Davalliaceae. II. A revision of the genus *Davallia*. Blumea 39:151–214

Nooteboom HP (1997) The microsoroid ferns. Blumea 42:261–395

Øllgaard B (1992) Neotropical Lycopodiaceae – an overview. Ann Mo Bot Gard 79:687–717

Øllgaard B (1995) Diversity of *Huperzia* (Lycopodiaceae) in neotropical montane forests. In: Churchill SP, Balslev H, Forero E, Luteyn JL (eds) Biodiversity and conservation of neotropical montane forests. New York Botanical Garden, New York, pp 349–358

Pacheco L (1994) Flora de México, vol 6 no 2. Pteridofitas: familia Hymenophyllaceae. Consejo Nacional de la Flora de México, México, DF, 56 pp

Palacios-Rios M, Gómez-Pompa A (1997) Phytogeographical analysis of the pteridophytes of Veracruz, México. In: Johns RJ (ed) Holttum memorial volume. Royal Botanic Gardens, Kew, pp 217–234

Palmer DD (1994) The Hawaiian species of *Cibotium*. Am Fern J 84:73–85

Parris BS (1996) Measurements of pteridophyte species diversity in Malesia and New Zealand. In: Camus JM, Gibby M, Johns RJ (eds) Pteridology in perspective. Royal Botanic Gardens, Kew, pp 43–51

Parris BS (1997) Receptacular paraphyses in Asian, Australian and Pacific Islands taxa private to Grammitidaceae. In: Johns RJ (ed) Holttum memorial volume. Royal Botanic Gardens, Kew, pp 81–90

Parris BS, Beaman RS, Beaman JH (1992) The plants of Mount Kinabalu. I. Ferns and fern allies. Royal Botanic Gardens, Kew

Pichi Sermolli REG (1993) New studies on some family names of Pteridophyta. Webbia 47:121–143

Pichi Sermolli REG (1996) Authors of scientific names in Pteridophyta. Royal Botanic Gardens, Kew

Poulsen AD (1996a) The herbaceous ground flora of the Batu Apoi Forest Reserve, Brunei Darussalam. In: Edwards DS, Booth WE, Choy SC (eds) Tropical rainforest research – current issues. Kluwer, Dordrecht, pp 43–47

Poulsen AD (1996b) Species richness and density of ground herbs within a plot of lowland rainforest in north-west Borneo. J Trop Ecol 12:177–190

Poulsen AD, Nielsen IH (1995) How many ferns are there in one hectare of tropical rain forest? Am Fern J 85:29–35

Price M (1996) Holttum and ferns. In: Camus JM, Gibby M, Johns RJ (eds) Pteridology in perspective. Royal Botanic Gardens, Kew, pp 13–26

Pryer KM, Smith AR, Skog JE (1995) Phylogenetic relationships of extant ferns based on evidence from morphology and *rbc*L sequences. Am Fern J 85:205–282

Raine CA, Sheffield E (1997) Establishment and maintenance of aseptic culture of *Trichomanes speciosum* gametophytes from gemmae. Am Fern J 87:87–92

Rasbach H, Rasbach K, Jérôme C (1993) Über das Vorkommen des Hautfarns *Trichomanes speciosum* (Hymenophyllaceae) in den Vogesen (Frankreich) und dem benachbarten Deutschland. Carolinea 51:51–52

Rasbach H, Rasbach K, Jérôme C (1995) Weitere Beobachtungen über das Vorkommen des Hautfarns *Trichomanes speciosum* Willd. in den Vogesen und dem benachbarten Deutschland. Carolinea 53:21–32

Ratcliffe DA, Birks HJB, Birks HH (1993) The ecology and conservation of the Killarney fern *Trichomanes speciosum* Willd. in Britain and Ireland. Biol Conserv. 66:231–257

Raulerson L, Rinehart AF (1992) Ferns and orchids of the Mariana Islands. Agana, Guam

Rödl-Linder G (1994a) A monograph of the fern genus *Thylacopteris* (Polypodiaceae). Blumea 39:351–364

Rödl-Linder G (1994b) Revision of the fern genus *Polypodiopteris* (Polypodiaceae). Blumea 39:365–371

Rödl-Linder G, Nooteboom HP (1997) Notes on Davalliaceae III. In: Johns RJ (ed) Holttum memorial volume. Royal Botanic Gardens, Kew, pp 67–79

Rolleri C (1993) Revision of the genus *Christensenia*. Am Fern J 83:3–19

Rolleri C, Lavalle MC, Mengascini A, Rodríguez M (1996) Spore morphology and systematics of the genus *Christensenia*. Am Fern J 86:80–88

Rothwell GW (1996) Phylogenetic relationships of ferns: a palaeobotanical perspective. In: Camus JM, Gibby M, Johns RJ (eds) Pteridology in perspective. Royal Botanic Gardens, Kew, pp 395–404

Rothwell GW, Stokey RA (1994) The role of *Hydropteris pinnata* gen. et sp. nov. in reconstructing the cladistics of heterosporous ferns. Am J Bot 8:479–492

Roux JP (1995) Systematic studies in the genus *Mohria* (Pteridophyta: Anemiaceae). VI. Taxonomic review. Bothalia 25:1–12

Rumsey FJ, Russel SR; Ji J, Barret JA, Gibby M (1996) Genetic variation in the endangered filmy fern *Trichomanes speciosum*. Willd. In: Camus JM, Gibby M, Johns RJ (eds) Pteridology in perspective. Royal Botanic Gardens, Kew, pp 161–165

Sánchez C, Caluff MG (1996) The genus *Hymenophyllum* Smith in the Greater Antilles. In: Camus JM, Gibby M, Johns RJ (eds) Pteridology in perspective. Royal Botanic Gardens, Kew, pp 61–73

Sánchez Peña RO (1994) Zur Wirkung hoher NaCl-Konzentrationen auf das ökophysiologische Verhalten des Mangrovenfarms *Acrostichum danaefolium* Langsd. & Fisch. Diss Bot 216:1–174

Schneider H (1996a) Vergleichende Wurzelanatomie der Farne. PhD Thesis, University of Zurich, Schaker, Aachen

Schneider H (1996b) The root anatomy of ferns: a comparative study. In: Camus JM, Gibby M, Johns RJ (eds) Pteridology in perspective. Royal Botanic Gardens, Kew, pp 271–283

Schneller JJ (1997) Tadeus Reichstein (1897–1996). Am Fern J 87:33–40

Schneller JJ, Holderegger R (1996a) Colonization events and genetic variability within populations of *Asplenium ruta-muraria* L. In: Camus JM, Gibby M, Johns RJ (eds) Pteridology in perspective. Royal Botanic Gardens, Kew, pp 571–580

Schneller JJ, Holderegger R (1996b) Genetic variations in small, isolated fern populations. J Veg Sci 7:113–120

Shing KH (1993) A new genus *Paragymnopteris* Shing separated from *Gymnopteris* Bernh. Indian Fern J 10:226–231

Siems K, Weigt F, Wollenweber E (1996) Drimanes from the epicuticular wax of the fern *Macrothelypteris torresiana*. Phytochemistry 41:1119–1121

Smith AR (1992) A review of the fern genus *Micropolypodium* (Grammitidaceae). Novon 2:419–425

Smith AR (1993a) *Terpsichore*, a new genus of Grammitidaceae (Pteridophyta). Novon 3:478–489

Smith AR (1993b) Phytogeographic principles and their use in understanding fern relationships. J Biogeogr 20:255–264

Smith AR (1995) Non-molecular phylogenetic hypotheses for ferns. Am Fern J 85:104–122

Smith AR, Moran RC (1992) *Melpomene*, a new genus of Grammitidaceae (Pteridophyta). Novon 2:426–432

Smith BL (1997) The toxicity of bracken fern (genus *Pteridium*) to animals and its relevance to man. In: D'Mello JPF (ed) Handbook of plant and fungal toxicants. CRC Press, Boca Raton, pp 63–76

Smith BL, Seawright AA (1995) Bracken fern (*Pteridium* spp.) carcinogenicity and human health – a brief review. Nat Toxins 3:1–5

Smith BL, Embling PP, Lauren DR, Agnew MP (1992) Carcinogenicity of *Pteridium esculentum* and *Cheilanthes sieberi* in Australia and New Zealand. In: James LF, Keeler RF, Bailey EM, Cheeke PR, Hegarty MP (eds) Poisonous plants. Proceedings of the 3rd International Symposium. Iowa State University Press, Iowa City, pp 448–452

Stearn WT (1996) A short account of the life of Richard Eric Holttum (1895–1990) tropical botanist and religious thinker. In: Camus JM, Gibby M, Johns RJ (eds) Pteridology in perspective. Royal Botanic Gardens, Kew, pp 3–11

Stein DB, Conant DS, Valinski AEC (1997) The implications of chloroplast DNA restriction site variation on the classification and phylogeny of the Cyatheaceae. In: Johns RJ (ed) Holttum memorial volume. Royal Botanic Gardens, Kew, pp 235–254

Stevenson DW, Laconte H (1996) Ordinal and familial relationships of pteridophyte genera. In: Camus JM, Gibby M, Johns RJ (eds) Pteridology in perspective. Royal Botanic Gardens, Kew, pp 435–468

Steyermark J, Berry PA, Holst BK (eds) (1995) Flora of the Venezuelan Guayana, vol 2. Pteridophytes and spermatophytes (Acanthaceae-Araceae). Timber Press, Portland, Oregon

Stolze RG, Pacheco L, Øllgaard B (1994) Polypodiaceae-Dryopteridoideae-Physematieae. In: Harling G, Andersson L (eds) Flora of Ecuador 49. Council for Nordic Publications in Botany, Copenhagen

Stützel T, Gailing O (1995) Blätter und Spreuschuppen bei Farnen. Beitr Biol Pflanz 69:17–29

Taylor WC, Hickey RJ (1992) Habitat, evolution, and speciation in *Isoetes*. Ann Mo Bot Gard 79:613–622

Tryon RM, Stolze RG, Smith AR (1992) Pteridophyta of Peru, part 3: 16. Thelypteridaceae. Fieldiana Bot 29

Tryon RM, Stolze RG, Leon B (1993) Pteridophyta of Peru, part 5: 18. Aspleniaceae – 21. Polypodiaceae. Fieldiana Bot 32

Tyron RM, Stolze RG, Hickey RJ, Øllgaard B (1994) Pteridophyta of Peru, part 6: 22. Marsileaceae – 28. Isoetaceae. Fieldiana Bot

Tutin TG, Burges NA, Chater AO, Edmondson JR, Heywood VH, Moore DM, Valentine DH, Walters SM, Webb DA (eds) (1993) Flora Europaea, vol 1. Psilotaceae to Platanaceae. Cambridge University Press, Cambridge

Valier K (1995) Ferns of Hawai'i. University of Hawai'i Press, Honolulu

Villaverde CS, Caluff MG (1997) The threatened ferns and allied plants from Cuba. In: Johns RJ (ed) Holttum memorial volume. Royal Botanic Gardens, Kew, pp 203–215

Van Uffelen GA (1992) Sporogenesis in Polypodiaceae (Filicales). II. The genera *Microgramma* Presl and *Belvisia* Mirbel. Blumea 36:515–540

Van Uffelen GA (1993) Sporongenesis in Polypodiaceae (Filicales). III. Species of several genera. Spore characters and their value in phylogenetic analysis. Blumea 37:529–561

Van Uffelen GA (1997) The spore wall in Polypodiaceae: development and evolution. In: Johns RJ (ed) Holttum memorial volume. Royal Botanic Gardens, Kew, pp 95–117

Viane RLL (1993) A multivariate morphological-anatomical analysis of the perispore in Aspleniaceae (Pteridophyta). PhD Thesis, University of Gent

Viane RLL (1996) On the origin and foundation of the Group of European Pteridologists (GEP). Mem Accad Lunigian Sci 66:159–160

Vogel JC (1996) Conservation status and distribution of two serpentine restricted *Asplenium* species in Central Europe. In: Camus JM, Gibby M, Johns RJ (eds) Pteridology in perspective. Royal Botanic Gardens, Kew, pp 187–188

Vogel JC, Jeßen S, Gibby M, Jermy AC, Ellis L (1993) Gametophytes of *Trichomanes speciosum* (Hymenophyllaceae: Pteridophyta) in Central Europe. Fern Gaz 14:227–232

Wagner FS (1992) Cytological problems in *Lycopodium* sens. lat. Ann Mo Bot Gard 79:718–729

Wagner FS (1993) Chromosomes of North American grapeferns and moonworts (Ophioglossaceae: *Botrychium*). Contrib Univ Mich Herb 19:83–92

Wagner FS (1995) The chromosomes of *Sadleria* (Blechnaceae). Contrib Univ Mich Herb 20:239–240

Wagner GM (1997) *Azolla*: a review of its biology and utilization. Bot Rev 63:1–27

Wagner WH Jr (1995) Evolution of Hawaiian ferns and fern allies in relation to their conservation status. Pac Sci 49:31–41

Wagner WH Jr, Beitel JM (1992) Generic classification of modern North American Lycopodiaceae. Ann Mo Bot Gard 79:676–686

Wagner WH Jr, Wagner FS, Reznicek AA, Werth CR (1992) x *Dryostichum singulare* (Dryopteridaceae), a new fern nothogenus from Ontario. Can J Bot 70:245–253

White RA, Turner MD (1995) Anatomy and development of the fern sporophyte. Bot Rev 61:281–305

Whittier DP (1996) Extending the viability of *Equisetum hyemale* spores. Am Fern J 86:114–118

Whittier DP, Braggins JE (1992) The young gametophyte of *Phylloglossum* (Lycopodiaceae). Ann Mo Bot Gard 79:730–736

Widén C-J, Fraser-Jenkins CR, Reichstein T, Gibby M, Sarvela J (1996) Phloroglucinol derivatives in *Dryopteris* sect. *Fibrillosae* and related taxa (Pteridophyta, Dryopteridaceae). Ann Bot Fenn 33:69–100

Windisch PG (1992) *Trichomanes crispum* L. (Pteridophyta, Hymenophyllaceae) and allied species. Bradea 6:78–117

Wolf PG (1995) Phylogenetic analysis of *rbcL* and nuclear ribosomal RNA gene sequences in Dennstaedtiaceae. Am Fern J 85:306–327

Wolf PG (1996) Pteridophyte phylogenies based on analysis of DNA sequences: a multiple gene approach. In: Camus JM, Gibby M, Johns RJ (eds) Pteridology in perspective. Royal Botanic Gardens, Kew, pp 203–215

Wolf PG, Soltis PS, Soltis DE (1994) Phylogenetic relationships of dennstaedtioid ferns: evidence from *rbcL* sequences. Mol Phylog Evol 3:383–392

Wollenweber E (1996) Flavonoids, phenolics and terpenoids in leaf exudates of angiosperms and pteridophytes. In: Antus S, Gábor M, Vetschera K (eds) Flavonoids and bioflavonoids. Akadémiai Kiadó, Budapest, pp 211–220

Wollenweber E, Doerr M, Waton H, Favre-Bonvin J (1993) Flavonoid aglycones and a dihydrostilbene from the frond exudate of *Notholaena nivea*. Phytochemistry 33:611–612

Yatskievych G (1996) A revision of the fern genus *Phanerophlebia* (Dryopteridaceae). Ann Mo Bot Gard 83:168–199

Yusuf UK (1995) The taxonomic significance of leaf flavonoids in West Malesian *Dicranopteris* taxa (Gleicheniaceae). Blumea 40:211–215

Zink MJ (1993) Systematics of the fern genus *Lepisorus* (J. Smith) Ching (Polypodiaceae-Lepisoreae) with special reference to Africa. PhD Thesis, Univ of Zurich

Zink MJ (1995) In memoriam Prof. Dr. Karl U. Kramer. Farnblätter 26/27:1–28

Dr. Stefan Schneckenburger
TU Darmstadt
Botanischer Garten
Institut für Botanik (FB 10)
Schnittspahnstraße 3–5
D-64287 Darmstadt, Germany

Edited by
J. W. Kadereit

Lichenized and Lichenicolous Fungi 1995–96

By Harrie J. M. Sipman

1 Introduction

The exploration of lichen biodiversity continued at a high rate in 1995–1996. The number of newly described species again exceeded the figures for the preceding 2-year period (see Fig. 1). At least 22 genera of lichenized fungi and 11 of lichenicolous fungi were described. At species level these figures are 418 and 144. The discovery of so many new species is certainly strongly supported by recently opened character complexes like secondary product chemistry and ascus tip structure, but perhaps even more by the availability of ample new herbarium collections and an increasing attention to crustose lichens. Particularly spectacular is the result of increased attention to lichenicolous fungi (Fig. 1).

Once more, over one third of the new taxa was found in the northern extratropics. This may be unexpected, because it is the world's best-

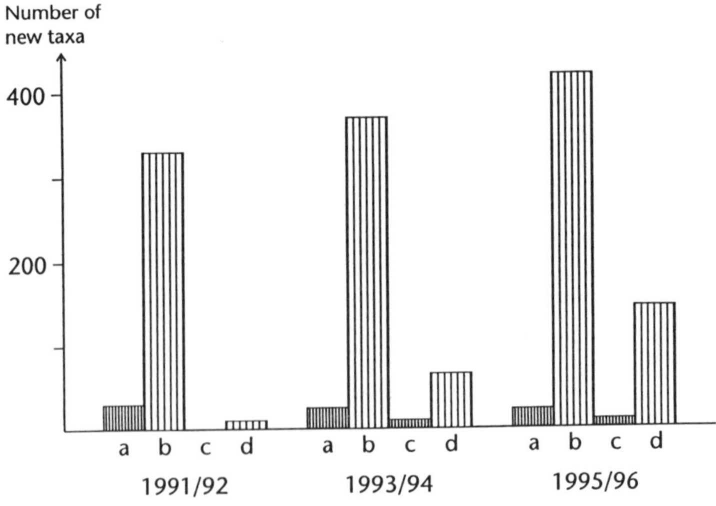

Fig. 1a-d. Comparison of numbers of newly described genera (**a**) and species (**b**) of lichenized fungi and genera (**c**) and species (**d**) of lichenicolous fungi over the periods 1991–1992, 1993–1994 and 1995–1996. Figures based on the literature evaluated for the lichenology contributions in Progress in Botany, without claim to completeness

investigated area. It shows that even this area is very incompletely known, in spite of its concentration of taxonomic infrastructure, experts, libraries and herbaria. Within the northern extratropics the exploration of North America was particularly productive. The figures for the southern extratropics are distinctly lower this time, about one fifth of the total, while the number of taxa described from the tropics equals the figures from the northern extratropics. This suggests that the study of the tropical lichen flora is receiving more attention than before. Provisionally excluded from the figures of Fig. 1 for reasons presented below is the investigation of *Lecanora* s.l. by Motyka (1995/1996), which claims another 7 new genera and 200 new species from Europe.

Recognition of taxa from descriptions is often more difficult in lichens than in other groups of organisms. This is caused by the plastic morphology of lichens, which is difficult to catch in words. Consequently, experts in the past gave the urgent advice never to trust a result obtained with an identification key alone, but always to compare with reliably identified herbarium material, which means the restriction of serious lichen study to the few places in the world with an adequate lichen herbarium. Illustrations so far have been able to overcome this only to a very limited extent, e.g. for some anatomical details. Now Wirth (1995) has produced a flora for southwestern Germany with many colored pictures of such high quality that the dependence on a comparison in the herbarium is much reduced. Indeed, this work has a very stimulating effect on lichen research in Germany and beyond.

Three Festschrifts were published, for Christian Leuckert on the occasion of his 65th birthday and retirement (Bibliotheca Lichenologica 57), Antonin Vezda on the occasion of his 70th birthday (Bibliotheca Lichenologica 58) and Gerhard Follmann on the occasion of his 65th birthday and retirement (Daniëls et al. 1995). They constitute important collections of literature on a wide range of subjects. Another important group is formed by the Proceedings of the 2nd IAL meeting in Lund (Cryptogamic Botany 5). A sad event was the death on 3 June 1995 of Josef Poelt, the father of the postwar revival of lichenology in central Europe.

A new textbook became available, Lichen Biology (Nash 1996), with chapters on photobionts, morphology, anatomy and morphogenesis of thallus, physiology, biochemistry, population biology, geography, systematics, and air pollution, written by various specialists.

2 Character Investigation

a) Morphology and Anatomy of the Mycobiont

Perhaps the most surprising contribution was made by Gilbert (1996a), who reports the occurrence of albino ascomata in lichens. This seems quite reasonable, since in secondary compound studies zero content races are regularly found. He found albinos in Lecanorales (*Bacidia, Catillaria, Lecidea, Micarea, Porpidia, Protoblastenia*), Graphidales (*Phaeographis*) and in pyrenocarps (*Acrocordia gemmata, Polyblastia, Verrucaria*). They have been misunderstood in the past, and a new genus was even based on them, *Leucocarpopsis*.

Some interesting new or little-known character complexes used: Rambold (1995) demonstrated taxonomically significant differences in the septa of medullary hyphae, seemingly of value at higher systematic levels. Hammer (1996) uses meristeme development as distinguishing character between *Cladonia floridana* and *C. atlantica*. A neglected character is the presence of mucilaginous attachments of ascospores. Apart from the well-known gelatinous layer ("halo"), other forms may occur, as shown by Roux and Sérusiaux (1995). They observed mucoid appendages of ascospores in *Raciborskiella*, which are similar to those of conidia in *Strigula* and *Raciborskiella* (Fig. 2). A frequently used surface structure in lichens, which causes problems in particular to beginners, is the so-called pruina, with a supposed similarity to hoarfrost. It is usually

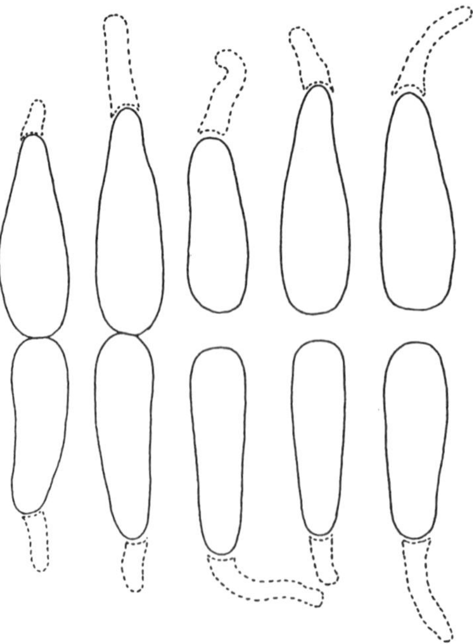

Fig. 2. Ascospores with characteristic mucilaginous appendages of *Raciborskiella janeirensis* (after Sérusiaux 1995); spores much reduced, total length is about 30 µm.

composed of oxalate crystals, sometimes anthraquinone crystals. Heid-marsson (1996) investigated the cortex of *Dermatocarpon* by TEM and found the "pruina" to reflect a different cortex type with noncollapsed, broken cells.

b) Chemotaxonomy

The taxonomic significance of chemical differences is controversially discussed. Some pigments seem highly specific at genus level, e.g., the epithecium pigments of *Aspicilia* or *Lecidea*. Accordingly, Hafellner and Kalb (1995) propose a new generic division of Trichotheliaceae using ascocarp wall pigments (see below).

In the Teloschistaceae many species are covered by anthraquinones giving them a yellow to red color. Slight differences in color are often helpful to recognize species, which suggest that they are based on chemical differences. However, Arnold and Poelt (1995) analyzed an-thraquinone content of *Xanthoria* species and found that mostly the same substances were present. The color differences are probably de-pendent on concentration differences.

An improved method to recognize gyrophoric acid and its satellites in Umbilicariaceae was proposed by Narui et al. (1996), who used an im-proved gradient high-performance liquid chromatographic method. Tabacchi et al. (1995) used tandem mass spectrometry (MS-MS) to identify triterpenes in *Evernia prunastri*. Seaward et al. (1995) use FT-Raman microscopic studies to investigate the contact zone of *Haema-tomma ochroleucum* var. *porphyrium* with its substrate. The method allows recognition of calcium oxalates and phenolic lichen compounds.

c) Photobiont

After the discovery that different *Trebouxia* species can be associated with the same mycobiont species, evidence is now presented that the association can also be very specific. Tschermak-Woess (1995a) found the phycobiont of *Phlyctis argena*, *Dictyochloropsis splendida*, to be constant in 150 samples from 50 localities throughout Europe. Nakano and Ihda (1996) found *Trentepohlia lagenifera* as the only photobiont in nine specimens of *Pyrenula japonica*.

Reduced size of the photobiont of *Sticta canariensis*, as compared with the same species in other lichens or in free state, was reported by Tschermak-Woess (1995b).

d) Molecular Biology

Various investigations were made using macromolecules for taxonomic purposes. The redelimitation of the genera *Ionapsis* and *Hymenelia* (Lutzoni and Brodo 1995) was supported by an enzyme electrophoretic study. For immunological studies Pelkonen et al. (1995) offer an improved method of ultracryomicrotomy.

Preparatory studies for the application of DNA and RNA included a method for purification of DNA from lichens (Armaleo and Clerc 1995) and a quantitative investigation of RNA production in *Cladonia cristatella* (Brink et al. 1995). DePriest (1995) investigated ribosomal DNA in the *Cladonia chlorophaea* complex and found extensive length and restriction site variation resulting from variable sequence insertions (introns). She presents a discussion of introns, and concludes that the variation has relevance at species level.

Beard and DePriest (1996) studied rDNA in *Cladina subtenuis* and found polymorphism in SSU rDNA between mats but not within mats, suggesting that the latter are genetic individuals, and that macromolecular evolution has occurred in the species.

An analysis of 18S rDNA sequences was used by Eriksson and Strand (1995) to investigate the relationships of *Nephroma, Peltigera,* and *Solorina. Peltigera* and *Solorina* appeared to be closest relatives, and all three together are closer to Lecanorales than to Leotiales or Pezizales.

The phylogenetic relations of *Omphalina* were investigated by Lutzoni and Vilgalys (1995a) by means of a ribosomal DNA analysis. They found that the lichenized species form a monophyletic group, derived from saprophytic ancestors. They tend to be more frequently in a uninucleate stage. A thallus composed of globules is the more primitive stage.

e) Biology, Culture

Lichens are often believed to be variable organisms, but little is known of the genetic background of such variation. Schipperges et al. (1995) found differences in morphology and physiology between populations of *Cetraria nivalis* from different areas in Europe, They supposed that this variation depends on genetic differences, caused by isolation in different refugia during the Pleistocene. Genetic polymorphism in *Umbilicaria cylindrica* was demonstrated by Fahselt et al. (1995) using enzyme banding studies. A comparison of electrophoretic isoenzyme banding patterns of sexually and non-sexually reproducing species (Fahselt 1995) showed that in the former more differences between populations occur. Enzyme polymorphism was also observed by Swanson et al. (1996) in

Umbilicaria americana from North America. It was found on sites with higher UV radiation, and negatively correlated with phenolic content.

Results of probably the largest program for cultivating lichen mycobionts were presented by Crittenden et al. (1995). Of 1183 species, (2238 specimens), 42% were successfully isolated from spores or thallus fragments. Lichenicolous fungi and lichens with cyanobacteria cultured less readily. Isolated mycobiont cultures of *Cladonia grayi* were made by Armaleo (1995) to investigate factors influencing the production of grayanic acid. Hamada (1996) and Hamada et al. (1996) found that lichen substance production in cultures of lichenized and nonlichenized lichens was stimulated by added sucrose or ethylene glycol in the medium. Hamada and Miyagawa (1995) found the same for salazinic acid production in *Ramalina siliquosa* and gyrophoric acid in *Lobaria discolor*. A promising new technique may be the one described by Vicente et al. (1995), who use polyacrylamide gel to immobilize lichen cells and enzymes for bioproduction of lichen metabolites.

f) Phyologeny

Cladistic analyses are being used increasingly in taxonomic practice, not least for the interpretation of macromolecular results. Therefore it is important that weaknesses of the method are explored. One problem was met by Tehler (1995) when he compared phylogeny reconstructed from morphological data with data from 18S ribosomal RNA. He found three different phylogenies. Lutzoni and Vilgalys (1995b) discuss the integration of morphological and molecular data sets in estimating phylogenies. They suggest using Rodrigo's homogeneity test to determine if the data sets reflect the same phylogenetic history. An interesting experiment was a test of the current superspecific classification in Cladoniaceae by a cladistic analysis, as performed by Hyvönen et al. (1995) for 37 species of *Cladina* and *Cladonia* sect. *Unciales*. The results did not support traditional taxonomy and suggested that *Unciales* are polyphyletic. Their conclusion: "the chemical and morphological information available even in fairly well-known groups of lichens ... is at present inadequate ... of little help in resolving the phylogeny of the group".

3 Classification

a) Lichenized Ascomycetes

α) Discocarpous Orders, Lecanorales

In Lecanoraceae, Lumbsch and collaborators continue a revision of the genus *Lecanora* with treatments of usnic acid-containing species (Lumbsch et al. 1995a), species with a dark hypothecium (Lumbsch et al. 1996), and the *L. subcarnea* group (Dickhäuser et al. 1995). Much discussion was raised by the publication of a study on the Lecanoraceae of Europe, which was made by Motyka in 1961–1978 and published posthumously almost 20 years later (Motyka 1995/1996). Having started at a time when lichen chemotaxonomy was hardly developed and ascus characters not usually observed, the author used only the traditional techniques and came to taxonomic concepts very different from those developed in the past 20 years using new techniques. It may be questioned whether the author would have published the work in this form had he been still alive and able to incorporate the new developments. Now a heavy burden has been put on the scientific community. All these taxa, presented in Polish and Latin, have to be restudied using modern methods, in particular the 7 new genera and over 200 new species, to find out what should be accepted as replacement for current taxonomy and what can be disregarded.

Many new taxa in the Ramalinaceae were described by Spjut (1996), who studied the luxuriant lichen vegetation along the Pacific coast of California and Baja California. By carefully sampling the diversity of the area, he unraveled what were thus far thought to be a few polymorphic species. These were more or less turned into genera, including the new *Vermilacinia* (Spjut 1995), and the number of species was raised to almost tenfold. In New Zealand Blanchon et al. (1996) revised the genus *Ramalina*, recognizing 18 taxa. Lumbsch et al. (1995b) recognized in Australia a crustose representative, the first known in this family, for which they proposed the genus *Ramalinora*.

In the family Physciaceae, Giralt, Mayrhofer and collaborators continued the revision of the southern European representatives of *Rinodina* (Giralt and Barbero 1995; Giralt et al. 1995; Giralt and Mayrhofer 1995; Ropin and Mayrhofer 1995). Further treatments deal with corticolous, vegetatively reproducing species in North America (Sheard 1995) and rock-inhabiting species of South Africa (Matzer and Mayrhofer 1996). Moberg (1995) presents a treatment of *Phaeophyscia* for China and the Russian Far East.

The genus *Caloplaca* in the Teloschistaceae, with several hundred species, is by now among the largest species conglomerates in lichens for which no convincing infrageneric natural classification is available.

Some more or less artificially defined groups were revised: the species on coastal rock in western North America (20 species, Arup 1995a, b); the *Caloplaca sideritis* group in North and Central America (Wetmore 1996); the *Caloplace lactea* group in Europe (Navarro-Rosinés and Hladun 1996).

Important new revisions in the Parmeliaceae include: *Relicina* worldwide (Elix 1996); *Anzia* in New Guinea (Yoshimura et al. 1995) and in Central and South America (Yoshimura 1995). Nash et al. (1995) revised the genus *Xanthoparmelia* in South America, raising the number of species to 77 and demonstrating that the genus has a much higher diversity in this area than assumed before. The secondary product chemistry shows less variation than on other continents, but endemism is high, with a majority of the species restricted to South America. In the framework of the Flora of Australia project, Elix (1994) with collaborators produced a treatment of Parmeliaceae covering 31 genera and 395 species.

In the affinitites of *Parmelia* s.l. only one more genus was proposed: the monotypic *Bulbotricella* from Venezuela (Marcano et al. 1996b). In the fruticose affinities Common and Brodo (1995) propose a new genus *Nodobryoria* for *Bryoria oregana* and a few related species. Thell et al. (1995a) review the ascus types throughout the family and Henssen (1995b) confirms the close affinity of the crustose *Protoparmelia badia* with Parmeliaceae.

The rearrangement of the cetrarioid lichens made further progress. One more new genus was proposed, *Kaernefeltia*, for *Cetraria californica* and *C. merrillii* by Thell and Goward (1996). Interesting is the rearrangement of species across the border of the cetrarioid lichens proper. Kämefelt and Thell (1996) transferred two species from *Dactylina* to *Allocetraria*, and Thell (1995) transferred the *Cetraria commixta* group to *Melanelia*, based on ascoma and conidioma characters. A new delimitation for *Nephromopsis, Cetrariopsis,* and *Cetreliopsis* is given by Randlane et al. (1995). A survey of *Allocetraria* is presented by Thell et al. (1995b). Saag and Randlane (1995) made a cladistic analysis of 83 cetrarioid lichens based on 42 morphological and chemical characters. The recently segregated genera appeared as monophyletic, except for *Tuckermannopsis*. A comparison was made between cladograms based on different character interpretations; substances as such or in biochemical groupings. The latter method produced results closer to the accepted system.

Two new families in Lecanorales were proposed: Aphanopsidaceae for the small genera *Aphanopsis* and *Steinia*, based on ascus type and morphological and anatomical similarities (Printzen and Rambold 1995), and Gloeoheppiaceae for *Gloeoheppia, Pseudopeltula* and *Gudelia* based on apothecial development (Henssen 1995a).

Several further new genera of Lecanorales were published: *Labyrintha* (Porpidiaceae) from New Zealand (Malcolm et al. 1995); *Fellhaneropsis* (Pilocarpaceae) for *Bacidia myrtillicola* and *B. vezdae* (Sérusiaux 1996; *Jarmania* (Bacidiaceae), with byssoid thallus, from Tasmania (Kantvilas 1996); *Loxosporopsis*, of uncertain affinity, from N. America (Brodo and Henssen 1995); and *Podotara* (incertae sedis), foliicolous from New Zealand, with remarkable, stalked apothecia (Malcolm and Vezda 1996).

Among the more important generic treatments are: a worldwide revision of the genus *Physcidia* with 7 species (Kalb and Elix 1995); a worldwide revision of *Haematomma*, accepting 35 species (Staiger and Kalb 1995); a redefinition and revision of *Biatora* in Europe with 17 species (Printzen (1995); a key to 55 species of Lecideaceae in Europe (Hertel 1995); a treatment of *Leptogium* for southern Chile, with 16 species (Galloway and Jørgensen 1995). Hafellner (1995b) continued his work on the redefinition of Acarosporaceae based on ascus structure and reintroduces *Piccolia* for the *Biatorella conspersa* group. Lutzoni and Brodo (1995) redelimit the genera *Ionaspis* and *Hymenelia*, based on morphological/anatomical and on enzyme electrophoretic study. They discuss the significance of the ascus type for the taxonomy.

β) Other Discocarpous Orders

Two major contributions in the order Caliciales were published. Tibell (1996) prepared a Flora Neotropica treatment with 17 genera and 51 species. Wedin (1995) revised the Sphaerophoraceae of the temperate areas of the Southern Hemisphere, 23 species in 3 genera, based on morphology and TLC.

In the Arthoniales, three new, monotypic genera were described: *Tania* from Sabah (Egea et al. 1995); *Feigeana* from Socotra (Roccellaceae) with stromatic, labyrinthoid ascocarps and crustose (subfruticose) thallus (Mies et al. 1995); *Follmanniella* from N. Chile (Peine and Werner 1995). Larger revisions treated the genus *Arthothelium* in India (Makhija and Patwardhan 1995) and lichenicolous *Arthoniae* with reddish, K+ pigments (Grube et al. 1995). Lücking and Matzer (1996) discussed the systematic position of *Mazosia*.

In the Graphidales Guderley and Lumbsch (1996) produced a revision for the genus *Diploschistes* in South Africa, and Purvis et al. (1995) for *Thelotrema* (s.l.) in Europe (incl. Azores).

γ) Pyrenocarpous Orders

A new survey with determination key to all genera of pyrenocarpous lichens was presented by Harris (1995). He also presents a new generic arrangement for the lichenized Pleosporales, reintroducing the genus *Naetrocymbe* for part of the species previously assigned to *Arthopyrenia* and establishing a family Naetrocymbaceae for it, characterized by pseudoparaphyses with refractive globules near the septa.

The largest order, Verrucariales, is among the least-known lichen groups. Many representatives are crustose with a thin thallus and small, blackish perithecia, and only careful microscopic examination can reveal their affinities. Moreover, they tend to grow on rock, which hampers adequate sampling. Consequently, knowledge of the group is very incomplete and largely restricted to Europe. After long neglect the group is receiving increased attention. McCarthy (1995a) prepared a treatment for the ten aquatic *Verrucaria* species known from Australia. *Dermatocarpon* sect. *Polyrhizion* was revised by Breuss (1995), who also revised the genus *Placidiopsis* (Breuss 1996a). The genus *Catapyrenium* was divided into eight genera: *Involucropyrenium, Anthracocarpon, Heteroplacidium, Clavascidium, Catapyrenium* s.str., *Placidium, Neocatapyrenium* and *Scleropyrenium* (Breuss 1996b).

The family Trichotheliaceae, thus far considered as of uncertain affinity, was decided to be different at the order level, and attributed to an order of its own, Trichotheliales (Hafellner and Kalb 1995). These authors also propose a division of *Porina* based on ascocarp wall pigment, introducing the genus *Pseudosagedia* with subgenera *Pseudosagedia* and *Limosagedia*. *Trichothelium* is restricted to species with dark pigment. This divison was criticized later by, e.g., Harris (1995), who proposed a division into *Trichothelium* (s.l.), *Segestria, Porina* and *Clathroporina* (s.l.). McCarthy (1995b) revised the genus *Clathroporina* and considered it insufficiently distinct from *Porina*. A new genus, *Myeloconis*, was described to accommodate species with pigmented medulla (McCarthy and Elix 1996). Another new genus, *Polycornum*, was described from New Zealand (Malcolm and Vezda 1995).

A survey of the corticolous representatives of *Julella* is presented by Aptroot and van den Boom (1995), and Harada describes from Japan two new genera with some resemblance to Verrucariales, but considered to be of uncertain affinities: *Cyanopyrenia* (Harada 1995) and *Hyalopyrenia* (Harada 1996).

b) Incertae Sedis

The redefinition of species in *Lepraria* s.l. has reached a stage of consolidation. A survey is presented by Leuckert et al. (1995) and for Britain by Orange (1995).

c) Lichenicolous Fungi

The increased attention to lichenicolous fungi has led to the definition of several new genera: *Helicobolomyces*, coelomycetous anamorph of *Arthonia* (Grube et al. 1995); *Diederimyces* (Dothideales, teleomorph of *Phaeosporobolus alpinus*) from *Fuscidea* in the Pyrenees (Etayo 1995); *Perigrapha* (Arthoniales), lichenicolous on *Parmelia* s.str. (Hafellner 1996b); *Zevadia*, a hyphomycete on *Usnea* in W Ireland (David and Hawksworth 1995b); *Coniambigua*, *Lichenobactridium* and *Pycnopsammina*, three genera of lichenicolous deuteromycetes (Etayo and Diederich 1995); *Lanatosphaera* and *Paradoxomyces*, foliicolous Dothideales viz. Arthoniales (Matzer 1996); *Stygiomyces* on *Pseudocyphellaria* from Tasmania and *Pseudonitschkia* from South America (Coppins and Kondratyuk 1995); *Reichlingia* from sterile lichens with *Trentepohlia* in Central Europe (Diederich and Scheidegger 1996); *Wernerella* (Dothidales) on *Rinodina* in SW Europe and Morocco (Navarro-Rosinés et al. 1996); and *Clauzadella* (Verrucariaceae) from *Verrucaria* in France (Navarro-Rosinés and Roux 1996).

Further important revisions include: a conspectus of *Phacopsis* by Triebel et al. (1995); a treatment of lichenicolous *Arthoniae* with reddish, K+ pigments (Grube et al. 1995); and of selected groups of lichenicolous ascomycetes with fissitunicate asci on foliicolous lichens (Matzer 1996). Diederich (1996) revised the lichenicolous heterobasidomycetes, a group which frequently makes galls on lichen thalli. Identification is hampered because the basidia degenerate quickly, so that in many samples the principal characters are not observable. The basidiocarps may sometimes be mistaken for degenerated ascocarps. Consequently, the group was poorly known, and 41 out of 54 accepted species were new.

The lichen genus with the largest known lichenicolous flora nowadays is probably *Pseudocyphellaria*. Kondratyuk and Galloway (1995) report 51 species, found on 53 *Pseudocyphellaria* species; the area with the richest lichenicolous flora, as far as is known, is the Hohe Tauern National Park in Austria, from where Hafellner and Türk (1995) report over 140 lichenicolous fungi and lichens.

4 Floristics and Phytogeography

a) The Northern Extratropics

α) *Europe, Mediterranean Area and Atlantic Islands*

Publications with significant contributions are far too numerous to mention and only a very small selection can be treated here.

The activities towards a checklist of the Mediterranean lichens lead to a bundle with lists from Morocco (1100 taxa, Egea 1996), Tunisia (415 taxa, Seaward 1996) Israel (234 taxa, Galun and Mukhtar 1996) and Mediterranean Turkey (459 taxa, John 1996). Puntillo (1996) presents a thorough survey of lichens of Calabria (Italy), based on literature and personal observations, with notes on distribution and ecology. An important aid for the study of the Atlantic islands is produced by Hafellner (1995c): a checklist and bibliography of lichens and lichenicolous fungi of insular Laurimacaronesia containing ca. 2000 taxa. It was followed by further reports from this author on noteworthy lichens and lichenicolous fungi from Macaronesia (Hafellner 1995a, 1996a). From the Ukraine a checklist was published containing 1222 species (Kondratyuk et al. 1996).

β) North America

An updated checklist for continental US and Canada, including the lichenicolous and allied fungi, was produced by Esslinger and Egan (1995).

Much attention was paid to the Pacific coast of the northern USA and Canada. Arup (1995b) presented a key to 20 coastal saxicolous species of *Calopaca*. The Cladoniaceae were treated for the NW USA (Hammer 1995) and the Queen Charlotte Islands (Brodo and Ahti 1996). Goward et al. (1995) present a treatment of *Peltigera* for British Columbia. Of wider coverage is the treatment of the corticolous and lignicolous species of *Bacidia* (27) and *Bacidina* (12) by Ekman (1996).

γ) Other Areas

A flora was published for Nagaland (Singh and Sinha 1994), which treats 346 species. For Mongolia a checklist containing 912 species was published by Cogt (1995), and Andreev et al. (1996) present a checklist of lichens and lichenicolous fungi of the Russian Arctic, including 1078 species. An example of useful exchange of knowledge by visiting scientists is given by a visit of Thor to Japan, where he was able to recognize various inconspicuous species familiar to him from Europe (Kashiwadani and Thor 1995). Zhurbenko (1996) reports 660 lichens and 61 lichenicolous fungi from the northern Krasnoyarsk Territory, Central Siberia, including many new to Russia.

b) The Tropics

The poor knowledge of tropical lichen taxonomy is still an impediment
to floristic work, and consequently publications are scarce. For the Neotropics a significant contribution is a checklist of the Venezuelan Andes
with 745 taxa (Marcano et al. 1996b). It is based mainly on the work of
López Figueiras, probably the most productive lichenologist residing in
the tropics so far. Lücking and Lücking (1995) report 98 foliicolous lichens from Cocos Island, Costa Rica. Sipman (1995) brings a preliminary survey of the lichen flora of the montane forests in Colombia, in
particular the taxonomic composition and geographical affinities.

   For Papua New Guinea, Archer et al. (1995) produced a first survey of
the genus *Pertusaria*, recognizing 29 species. Egea et al. (1996) treat the
genus *Lecanactis* et al., with 14 taxa. Aptroot et al. (1995) report 63 species from Laing Island, a tiny island off the north coast measuring less
than 100 x 30 m, but with a largely original coastal forest.

   Tropical Africa remains probably the least known of the three tropical
land masses, and only few taxa are reported. David and Hawksworth
(1995a) present 29 species from Zanzibar, and Becker and Lücking
(1995) describe foliicolous lichens from the Ivory Coast.

c) The Southern Extratropics

The most important publication from this area is no doubt the volume of
the *Flora of Australia* treating the Parmeliaceae, 31 genera and 395 species (Elix 1994). For Australia also a new checklist of lichens and fungi
became available (Filson 1996). For South Africa a revision of *Diploschistes* was published (Guderley and Lumbsch 1996).

   Among the results in southern South America should be mentioned: a
treatment of Stictaceae (25 species, *Pseudocyphellaria* and *Sticta* only)
from Tierra del Fuego (Galloway et al. 1995); a bibliography 1986–1995
for Argentina (Osorio 1996); a treatment of Cladoniaceae for Chile,
treating 4 genera and 58 species (Stenroos 1995). Larger floristic reports
are those by Osorio (1995) from Southern Rocha, Uruguay (106 species),
and by Fleig (1995) from Rio Grande do Sul, Brazil (163 species). Ferraro
(1995) presents characteristics of the lichen flora in the phytogeographical regions of Corrientes, Argentina.

5 Ecology and Physiology

a) Analysis of Vegetation Structure

Connected with the recognition of forest as an important lichen habitat, various ecological studies were performed. Sillet (1995) compared epiphyte assemblages of interior and clear-cut edges of a 700-year-old forest in western Oregon and found that the canopy assemblages are lower at the edge. In an analysis of environmental factors connected with the distribution patterns of *Lobaria oregana* and *L. pulmonaria*, Shirazi et al. (1996) found no difference in heat tolerance. Sympatric populations show differences in drought resistance but infraspecific variation is just as wide. In an ecological investigation of *Nephroma occultum*, Goward (1995) found three types of different ecological behavior, each with a different range, considered primary, secondary, and tertiary. In the (humid) primary range, the species is common in the canopy, while in the much larger tertiary range it is found incidentally on humid tree bases. Pfefferkorn (1996) presents a phytosociological survey of the epiphytic vegetation in Vorarlberg (Austria), in particular forest. Sorediate crustose lichens were found to form 24% of the epiphytic lichen flora in a forest in Switzerland (Dietrich and Scheidegger 1996); on individual trees this percentage was even higher.

Lichens are known to be not normally restricted to a single tree species, but the relations between phorophyte and lichen flora are still poorly known. The correlations between *Lepraria* and *Leproloma* species and phorophyte species in Baden-Württemberg (Germany) are described by Wirth and Heklau (1995). Kuusinen (1996a) investigated the lichen flora on the basal trunks of six tree species in old boreal forest in Finland, and found *Populus* to have the most distinct and richest flora; some lichens were common on all or most tree species, while others were more selective; the average species number per tree ranged from 18 to 31. Jarman and Kantvilas (1995) report on a *Lagarostrobos franklinii* tree with 75 lichen species, with a vertical zonation.

Investigation of the tropical forest lichen flora has hardly started. Gradstein et al. (1996) develop methods to sample epiphytic diversity in tropical rainforest, incl. corticolous and foliicolous lichens. A zonation of foliicolous lichens on Cecropiaceae was observed by Lücking (1995a) in Costa Rica. In an investigation of the lichen flora in mangroves near São Paulo, Marcelli (1995) recognized six groups with different substrate preferences, depending on substrate age and microclimate.

Many other interesting topics are dealt with. A survey of the lichen flora on chalk and limestone in streams was made by Gilbert (1996b). He found around 50 species. Trampling by cattle in the headwaters, silt load, and eutrophication are detrimental to the vegetation. Metal-enriched environments often offer habitats for unusual lichen species.

Purvis and Halls (1996) present a review which shows that in particular Fe and Cu are important for the species composition. Mine spoil heaps, considered toxic, may be of considerable importance lichenologically. The function of different spore forms is little understood. Smith (1995) compared spore features in the lichen flora of the Hawaiian Islands, the Galapagos Islands, Britain and North America. The first two appear to have more large- and dark-spored taxa, while there is no difference in thick-walled taxa. A classification of lignicolous communities was presented by Sarrión and Burgaz (1995) for central Spain, including a new association, *Buellietum cedricolae*. The use of thallus fragments for dispersal is often hypothesized, but evidence is still restricted. Eldridge (1996) studied thallus fragments present in water runoff on soil in semi-arid woodland in Australia, and recognized eight species, which are apparently dispersed in this way. A subfossil lichen flora, about 1300–1400 years old, is reported from Greenland by Alstrup (1995). It had been buried under the ice. Forty five lichens and six lichenicolous fungi were recognized. The flora is similar to the recent flora of similar conditions locally. Also its chemistry agrees with the present situation, as found by Huneck et al. (1995) for *Umbilicaria cylindrica*.

Lichen colonization on rock 4–10 years after forest fire was observed by Garty (1995) in Israel. Sipman and Raus (1995) studied lichen colonization on lava in Santorini and found microclimate to have more influence on the species number than time. Two species are restricted to young lava, *Stereocaulon vesuvianum* and *Lecanora conferta*.

b) External Relations

The wide presence and high diversity of lichens, though in contrast with their low biomass and productivity, make it likely that they have effects on their environment. Observations are still scarce, however. An example of negative interaction was found by Solhaug et al. (1995) when they measured reduced photosynthesis in stems of *Populus tremula* covered by lichens. Light transmission was reduced by 10–55%, and photosynthetic O_2-evolution by 50%. The reduction is probably due to adaptation of bark cells to shade.

Much influence can be expected in cold environments, where the performance of higher plants is much reduced while that of lichens is not. The significance for biomass production was dealt with by several authors. Hahn et al. (1996) investigated the role of macrolichens as carbon and nutrient pools in arctic vegetation. Gremmen et al. (1995) investigated relations between in situ nutrient availability and standing crop of lichens in Antarctica. Kappen et al. (1995) discuss carbon acquisition of lichens in polar regions. Nash and Olafsen (1995) predict increasing lichen productivity in arctic ecosystems due to global warm-

                                                    453

ing: photosynthesis is primarily light-limited while N-fixation is temperature-limited; N frequently limits productivity.

A clear effect of lichens on the ecosystem in a warmer environment was demonstrated by Knops et al. (1996). They investigated the effect of *Ramalina menziesii* in oak forest in California. Removal of lichens affected the interception of rainfall by the canopy and the deposition of water and nutrients. Hesbacher et al. (1995) surveyed the presence of sequestered lichen compounds in 103 wild-caught imagines of Arctiidae, and found lichen compounds in 24, belonging to 11 species. Parietin and atranorin were among the most frequently detected.

c) Biodeterioration

A number of studies dealt with the effect of lichens on rocky substrate and its biodegradation. Ascaso and Wierzchos (1995a, b) present a survey of investigations and used SEM equipped with a backscattered electron detector. Also Wilson (1995) reviews interactions between rock and lichen, and processes contributing to rock weathering.

Effects on antique stonework received particular attention. Seaward and Edwards (1995) used Raman microscopic analysis to study fresco deterioration by *Dirina massiliensis* fo. *sorediata*. They found a damaging effect of calcium oxalate deposition. Ariño and Saiz-Jimenez (1996) investigated the effect of lichens on resin applied to archeological monuments for protection against weathering. Some colonized the resin surface, some were able to penetrate it. A protective effect was reported by Ariño et al. (1995) on Roman pavement. As an explanation they presume reduction of weathering by the lichen cover.

d) Physiology

The large number of physiological investigations published in 1995/96 is only very fragmentarily treated here. Various studies deal with the conditions under which lichens undergo photosynthesis. It turns out that a high water content of the thallus may have a negative effect. This was found in xeric habitats by Lange et al. (1995) while studying the relation between net photosynthesis and thallus water content in soil lichens. It is explained by reduced CO_2 diffusion caused by the increased water content. It also occurs in humid habitats: Green et al. (1995) measured net photosynthesis in a temperate rainforest in New Zealand and found that maximal net photosynthesis was seldom reached. Low light availability and high water content were the main limiting factors. Lange and Green (1995) demonstrate the same phenomenon for *Xanthoria calcicola* and *Lecanora muralis* from central Europe, and Lange et al. (1996a) for *Peltigera* species in the same area. The latter used fluorescence techniques to show that there is a real fall in photosynthetic rate and not an increased recycling of carbon dioxide: for a survey see Lange and Meyer (1996). Thomas et al. (1996) measured that *Coenogonium* sp. did not show such a photosynthetic depression at high water contents, probably because of the filamentous thallus structure, which minimizes CO_2 diffusion resistance.

Evidence that lichens have photosynthesis at low temperature is accumulating. Kappen et al. (1996a) measured photosynthetic performance of *Lasallia pustulata* in the field and found activity between –2 and +5 °C and a water content between 50 and 500% of dry weight. The coldest day was almost as efficient as the warmest day. Light was the main limiting factor, as could be demonstrated by laboratory experiments. In another paper (Kappen et al. 1996b), metabolic activity below zero was reported. Optimum for net photosynthesis of polar lichens is between 0 and 15 °C. Lange et al. (1996b) measured influence of light, water content, and temperature on CO_2 uptake in seven arctic lichen species, in controlled laboratory conditions and in the field. They found large differences between species, and uptake at –10 °C. Leisner et al. (1996) observed chlorophyll fluorescence of lichens under field conditions and found activity stopped only by drought or temperatures under –5 °C. Photoinhibitory damage was observed only after such occasions and was quickly repaired after even brief periods of normal metabolic activity.

Though lichens are generally slow growers, their performance is sometimes surprising. Renhorn and Esseen (1995) measured annual growth in transplanted alectorioid lichens and found growth rates between 6 and 40%. Fragmentation of thalli increased biomass production. McCune et al. (1996) measured growth rates in lichen transplants: pendants of lichen fragments. With methods for adjusting moisture content to a standard, it was possible to measure growth rates. These were usually around 10–40% annually, highest for *Evernia prunastri*, lowest for *Lobaria* spp.

6 Applied Aspects

a) Pollution Monitoring

Various studies using lichen vegetation to map air pollution were published. Since SO_2 concentrations have decreased greatly in many places in the western world, attention is now directed more to other contaminants. Loppi (1996) used an IAP study to monitor air pollution by geothermal power plants in Italy. Pollutants are mainly mercury and boron. Søchting (1995) developed a method to monitor nitrogen deposition by *Cladonia portentosa* and *Hypogymnia physodes*, and mapped deposition in Denmark and NW Europe. Other elements were taken into account by Glenn et al. (1995), who found a decline in lichens and elevated levels of Pb, Zn and Cu in lichens near roads in NE Spain. Brown et al. (1995) present effects of agricultural chemicals on lichens (*Peltigera* spp., *Parmelia sulcata*).

While most monitoring studies use epiphytic lichens, Mezger (1996) used also epilithic lichens for pollution monitoring in Berlin. He measured presence and frequency of species, and analyzed their correlation with pollution and other environmental parameters connected with large cities. The effect of reduced SO_2 is documented by, e.g., van Dobben (1996). He observed a partial recovery of the epiphytic lichen vegetation around Den Bosch (The Netherlands). Also Kirschbaum et al. (1996) noted a decrease in pollution load as compared to 25 years ago around Giessen and Wetzlar. A neutralizing effect of bark wound runoff was documented by Gauslaa (1995). He found a correlation with Lobarion occurrence in SE Norway, supposedly caused by acidification of the remaining bark through acid rain. Vonarburg (1993) measured microclimate, lichen growth and air pollution in Switzerland. His results explain the poor correlation between ozone concentration and lichen damage on free-standing trees: lichens are dry and inactive during periods with high ozone concentration. In urban areas high levels of acid deposition coincide with physiological activity of lichens and cause damage.

Only a small section of papers has been treated here. Much more literature can be found in the series *Lichens and Air Pollution*, published in the journal *The Lichenologist*.

b) Conservation

The steady decline in the lichen flora over many years is leading to more and more concern among lichenologists, and a rapidly increasing number of studies deals with conservation aspects.

The difficulties encountered are perhaps well illustrated by the observation by Follmann (1995), who revisited "lichen oases" in Northern Chile after 30 years and noted a strong decline, due to unknown reasons. The area is rarely visited by lichen experts and for conservation measures investigations would have to start from almost zero. A very different situation exists in the case of the endangered lichen *Erioderma pedicellatum* in Europe. Holien et al. (1995) assessed the present state and ecology, and discuss the potential for protection. In the case of *Teloschistes flavicans* in Great Britain, Gilbert and Purvis (1996) investigated its present distribution and ecology and showed that in the last century it was found from central England (Yorkshire, Lancashire, Leicestershire) southward, while it is now restricted to SW England, where few stations with more than a few plants remain.

Various studies present arguments for conservation need and discuss opportunities. To give a small selection: Gilbert (1995) treats chalk grassland lichens in England, their significance, and causes of decline. Guerra et al. (1995) list the lichens of gypsiferous outcrops in SE Spain and treat their conservation significance. From The Netherlands, van den Boom et al. (1996) present a list of species on megalithic monuments and discuss the influence of human activities and conservation needs. Schlechter (1995) treats 18 macrolichens in the Eifel Mts. (Germany) and their conservation status. Fryday (1996) reports ca. 250 species from N Wales mountains and assesses their conservation value; Wolseley et al. (1996) report 248 species from Skomer Island (GB), an inventory for conservation purposes.

Lists of threatened species (Red Lists) are now widely accepted as an important instrument for the protection of lichens, and are becoming available for a growing number of regions. Kuusinen (1995a) presents a list of threatened lichens from Finland; Kashiwadani and Kurokawa (1995) a list of threatened lichens of Japan; Wirth et al. (1996) a Red List for Germany. Thor (1995) gives a survey of all available Red Lists and explains their potential.

The significance of (old) forest as habitat for lichen conservation has become an important research theme, and information from various parts of the world is becoming available. Kuusinen (1995b, 1996b) identified the value of old-growth *Salix caprea* and *Populus tremula* and of spruce swamp forest for the conservation of lichen diversity in the boreal forests of Finland. Holien (1996) investigated the distribution of Caliciales in differently managed forests in Scandinavia and found that old forest is preferred. Some threatened species prefer forest on rich soil. Heiman (1996) analyzed the changes in a forest reserve in the Appalachians and presents a survey of the macrolichens as a baseline study for conservation projects. Hauck (1995) presents information about lichens of old-growth forest in Niedersachsen, with a list of extinct and threatened species and causes of decline, and stresses the importance of conservation. Hébrard et al. (1995) investigated the effect of forest clearing on lichen vegetation in SE France. They noticed a significant increase in bryophytes and suggest using the changes for a disturbance index. Elimination of undergrowth had a similar effect, probably because it suppresses the lateral protection against drying action of wind. Rosentreter (1995) presents activities to develop a management scheme to conserve high lichen diversity in managed forest of NW USA. A survey of lichens of special interest and recommendations for management are given.

From tropical forest little information is available so far. Lücking (1995b), while demonstrating patterns of biodiversity of foliicolous lichens in Costa Rica, proposes *Badimia* as indicative of rich sites of interest for conservation.

In many densely populated regions disturbance of the forest has reached a stage where the remaining populations of sensitive lichens appear unable to colonize new substrate. Therefore Scheidegger et al. (1995) developed transplantation methods for foliose lichens by means of propagules and thallus fragments. A successful transplantation of *Lobaria pulmonaria* was reported (Scheidegger 1995).

This survey makes no claim to completeness, and for additional titles the reader is referred in particular to two sources, the series *Recent Literature on Lichens* and the journal *The Bryologist* (alphabetically arranged by the author; also accessible as a database on internet; address: <http://www.toyen.uio.no/botanisk/bot-mus/lav/sok_rll.htm>) and *Eléments de bibliographie lichénologique récente* in the *Bulletin d'informations de l'association française de lichénologie* (arranged according to subjects, in French).

References

Alstrup V (1995) In situ cryo-subfossil vegetation in Northwest Greenland. Cryptog Bot 5:172–176

Andreev M, Kotlov Y, Makarova I (1996) Checklist of lichens and lichenicolous fungi of the Russian arctic. Bryologist 66:137–169

Aptroot A, van den Boom PPG (1995) Stigula lateralis spec. nov. with notes on the genus Julella (Ascomycetes). Mycotaxon 56:1–8

Aptroot A, Diederich P, Sérusiaux E, Sipman HJM (1995) Lichens and lichenicolous fungi of Laing Island (Papua New Guinea). Bibl Lichenol 57:19–48

Archer AW, Elix JA, Streimann H (1995) The lichen genus Pertusaria (Lichenised Ascomycotina) in Papua New Guinea. Mycotaxon 56:387–401

Ariño X, Saiz-Jimenez C (1996) Lichen deterioration of consolidants used in the conservation of stone monuments. Lichenologist 28:391–394

Ariño X, Ortega-Calvo JJ; Gomez-Bolea A, Saiz-Jimenez C (1995) Lichen colonization of the Roman pavement at Baelo Claudia (Cadiz, Spain): biodeterioration vs. bioprotection. Sci Total Environ 167:353–363

Armaleo D (1995) Factors affecting depside and depsidone biosynthesis in a cultured lichen fungus. Cryptog Bot 5:14–21

Armaleo D, Clerc P (1995) A rapid and inexpensive method for the purification of DNA from lichens and their symbionts. Lichenologist 27:207–213

Arnold N, Poelt J (1995) Über Anthrachinon-Pigmente bei einigen Arten der Flechtengattung *Xanthoria*, insbesondere aus der Verwandschaft von *Xanthoria elegans* (Teloschistaceae). Bibl Lichenol 57:49–58

Arup A (1995a) Eight species of *Caloplaca* in coastal western North America. Bryologist 98:92–111

Arup A (1995b) Littoral species of *Caloplaca* in North America: a summary and a key. Bryologist 98:129–140

Ascaso C, Wierzchos J (1995a) Study of the biodeterioration zone between the lichen thallus and the substrate. Cryptog Bot 5:270–281

Ascaso C, Wierzchos J (1995b) Estudio de la interfase talo liquenico-sustrato litico con microscopia electronica de barrido en modo de electrones retrodispersados. In: Daniëls FJA, Schulz M, Peine J (eds) Flechten Follmann. Geobotanical and Phytotaxonomical Study Group, Botanical Institute, University of Cologne, Cologne, Germany, pp 43–54

Beard KH, DePriest PT (1996) Genetic variation within and among mats of the reindeer lichen, *Cladina subtenuis*. Lichenologist 28:171–182

Becker U, Lücking R (1995) Foliikole Flechten aus dem Taï-Nationalpark, Elfenbeinküste (Tropisches Afrika), I. Neue Arten. In: Daniëls FJA, Schulz M, Peine J (eds) Flechten Follmann. Geobotanical and Phytotaxonomical Study Group, Botanical Institute, University of Cologne, Cologne, Germany, pp 161–173

Blanchon DJ, Braggins JE, Stewart A (1996) The lichen genus *Ramalina* in New Zealand. J Hattori Bot Lab 79:43–98

Breuss O (1995) Bemerkungen zur Sektion *Polyrhizion* der Flechtengattung *Dermatocarpon (Verrucariaceae)*. Osterr Z Pilzkd 4:137–145

Breuss O (1996a) Revision der Flechtengattung *Placidiopsis (Verrucariaceae)*. Osterr Z Pilzkd 5:65–94

Breuss O (1996b) Ein verfeinertes Gliederungskonzept für *Catapyrenium* (lichenisierte Ascomyceten, Verrucariaceae). Ann Naturhist Mus Wien 89B Suppl:35–50

Brink JJ, Ahmadjian V, Fröberg L, Goldsmith S (1995) Time course of RNA accumulation in cultures of lichen fungi. Crytog Bot 5:55–59

Brodo IM, Ahti T (1996) Lichens and lichenicolous fungi of the Queen Charlotte Islands, British Columbia, Canada. 2. The Cladoniaceae. Can J Bot 74:1147–1180

Brodo IM, Henssen A (1995) A new isidiate crustose lichen in northwestern North America. Bibl Lichenol 58:27–41

Brown DH, Standell CJ, Miller JE (1995) Effects of agricultural chemicals on lichens. Cryptog Bot 5:220–223

Cogt U (1995) Die Flechten der Mongolei. Willdenowia 25:289–397

Common RS, Brodo IM (1995) *Bryoria* sect. *Subdivergentes* recognized as the new genus *Nodobryoria* (Lichenized Ascomycotina). Bryologist 98:189–206

Coppins BJ, Kondratyuk SY (1995) *Stygiomyces* and *Pseudonitschkia*: two new genera of lichenicolous fungi. Edinb J Bot 52:229–236

Crittenden PD, David JC, Hawksworth DL, Campbell FS (1995) Attempted isolation and success in the culturing of a broad spectrum of lichen-forming and lichenicolous fungi. New Phytol 130:267–297

Daniëls FJA, Schulz M, Peine J (eds) (1995) Flechten Follmann. Geobotanical and Phytotaxonomical Study Group, Botanical Institute, University of Cologne, Cologne, Germany

David JC, Hawksworth DL (1995a) Lichens of Mauritius. I. Some new species and records. Bibl Lichenol 57:93–111

David JC, Hawksworth DL (1995b) *Zevadia*: a new lichenicolous hyphomycete from western Ireland. Bibl Lichenol 58:63–71

DePriest PT (1995) Phylogenetic analyses of the variable ribosomal DNA of the *Cladonia chlorophaea* complex. Cryptog Bot 5:60–70

Dickhäuser A, Lumbsch HT, Feige GB (1995) A synopsis of the *Lecanora subcarnea* group. Mycotaxon 56:303–323

Diederich P (1996) The lichenicolous heterobasidiomycetes. Bibl Lichenol 61:1–198

Diederich P, Scheidegger C (1996) *Reichlingia leopoldii* gen. et sp. nov., a new lichenicolous hyphomycete from Central Europa. Bull Soc Nat Luxemb 97:3–8

Dietrich M, Scheidegger C (1996) The importance of sorediate crustose lichens in the epiphytic lichen flora of the Swiss Plateau and the Pre-Alps. Lichenologist 28:245–256

Egea JM (1996) Catalogue of lichenized and lichenicolous fungi of Morocco. Bocconea 6:19–114

Egea JM, Tehler A, Torrente P, Sipman H (1995) *Tania*, a new genus with byssoid thallus in the order Arthoniales and new data on *Sagenidiopsis*. Lichenologist 27:351–359

Egea JM, Sérusiaux E, Torrente P (1996) The lichen genus *Lecanactis* and allied genera in Papua New Guinea. Mycotaxon 59:47–59

Ekman S (1996) The corticolous and lignicolous species of *Bacidia* and *Bacidina* in North America. Opera Bot 127, 148 pp

Eldridge DJ (1996) Dispersal of microphytes by water erosion in an Australian semi-arid woodland. Lichenologist 28:97–100

Elix JA (1994) Flora of Australia 55, Lichens – Lecanorales 2. Australian Biological Resources Study, Canberra, xviii, 360 pp

Elix JA (1996) A revision of the lichen genus *Relicina*. Bibl Lichenol 62:1–150

Eriksson OE, Strand Å (1995) Relationships of the genera *Nephroma, Peltigera* and *Solorina (Peltigerales, Ascomycotina)* interferred from 18S rDNA sequences. Syst Ascomycetum 14:33–39

Esslinger TL, Egan RS (1995) A sixth checklist of the lichen-forming, lichenicolous, and allied fungi of the continental United States and Canada. Bryologist 98:467–549

Etayo J (1995) Two new species of lichenicolous fungi from the Pyrenees. Nova Hedwigia 61:189–197

Etayo J, Diederich P (1995) Lichenicolous fungi from the Western Pyrenees, France and Spain. I. New species of Deuteromycetes. In: Daniëls FJA, Schulz M, Peine J (eds) Flechten Follmann. Geobotanical and Phytotaxonomical Study Group, Botanical Institute, University of Cologne, Cologne, Germany, pp 205–221

Fahselt D (1995) Lichen sexuality from the perspective of multiple enzyme forms. Cryptog Bot 5:137–143

Fahselt D, Alstrup V, Tavares S (1995) Enzyme polymorphism in *Umbilicaria cylindrica* in northwestern Greenland. Bryologist 98:118–122

Ferraro LI (1995) Comentarios sobre la distribucion de los liquenes en las diferentes regiones fitogeograficas de la provincia de Corrientes, Nordeste de Argentina, America del Sur. In: Daniëls FJA, Schulz M, Peine J (eds) Flechten Follmann. Geobotanical and Phytotaxonomical Study Group, Botanical Institute, University of Cologne, Cologne, Germany, pp 403–413

Filson RB (1996) Checklist of Australian lichens and allied fungi. Flora Aust Suppl Ser 7:204 pp

Fleig M (1995) Lichens from "Casa de Pedra" and surroundings, Bagé, Rio Grande do Sul, Brazil. In: Daniëls FJA, Schulz M, Peine J (eds) Flechten Follmann. Geobotanical and Phytotaxonomical Study Group, Botanical Institute, University of Cologne, Cologne, Germany, pp 415–426

Follmann G (1995) On the impoverishment of the lichen flora and the retrogression of the lichen vegetation in coastal central and northern Chile during the last decades. Cryptog Bot 5:224–231

Fryday AM (1996) The lichen vegetation of some previously overlooked high-level habitats in North Wales. Lichenologist 28:521–541

Galloway DJ, Jørgensen PM (1995) The lichen genus *Leptogium* (Collemataceae) in southern Chile, South America. In: Daniëls FJA, Schulz M, Peine J (eds) Flechten Follmann. Geobotanical and Phytotaxonomical Study Group, Botanical Institute, University of Cologne, Cologne, Germany, pp 227–247

Galloway DJ, Stenroos S, Ferraro LI (1995) Lobariaceae y Stictaceae. In: Guarrera SA, Gamundi de Amos I, Matteri CM (eds) Flora criptogámica de Tierra del Fuego XIII(6). Buenos Aires, Argentina, pp 1–78

Galun M, Mukhtar A (1996) Checklist of the lichens of Israel. Bocconea 6:149–171

Garty J (1995) Establishment of pioneer lithobiontic cyanobacteria, algae, microfungi, and lichens subsequent to a severe forest fire in Israel: In: Daniëls FJA, Schulz M, Peine J (eds) Flechten Follmann. Geobotanical and Phytotaxonomical Study Group, Botanical Institute, University of Cologne, Cologne, Germany, pp 111–121

Gauslaa Y (1995) The *Lobarion*, an epiphytic community of ancient forests threatened by acid rain. Lichenologist 27:59–76

Gilbert OL (1995) The conservation of chalk grassland lichens. Cryptog Bot 5:232–238

Gilbert OL (1996a) The occurrence of lichens with albino fruit bodies (ascomata) and their taxonomic significance. Lichenologist 28:94–97

Gilbert OL (1996b) The lichen vegetation of chalk and limestone streams in Britain. Lichenologist 28:145–159

Gilbert OL, Purvis OW (1996) *Teloschistes flavicans* in Great Britain: distribution and ecology. Lichenologist 28:493–506

Giralt M, Barbero M (1995) The saxicolous species of the genus *Rinodina* in the Iberian Peninsula containing atranorin, pannarin or gyrophoric acid. Mycotaxon 56:45–80

Giralt M, Mayrhofer H (1995) Some corticolous and lignicolous species of the genus *Rinodina* (lichenized Ascomycetes, Physciaceae) lacking secondary compounds and vegetative propagules in Southern Europe and adjacent regions. Bibl Lichenol 57:127–160

Giralt M, Mayrhofer H, Sheard JW (1995) The corticolous and lignicolous sorediate, blastidiate and isidiate species of the genus *Rinodina* in southern Europe. Lichenologist 27:3–24

Glenn MG, Gomez-Bolea A, Lobello R (1995) Metal content and community structure of cryptogam bioindicators in relation to vehicular traffic in Montseny Biosphere Reserve (Catalonia, Spain). Lichenologist 27:291–304

Goward T (1995) *Nephroma occultum* and the maintenance of lichen diversity in British Columbia. Mitt Eidgenoss Forschungsanst Wald Schnee Landsch 70(1):93–101

Goward T, Goffinet B, Vitikainen O (1995) Synopsis of the genus *Peltigera* (lichenized Ascomycetes) in British Columbia, with a key to the North American species. Can J Bot 73:91–111

Gradstein SR, Hietz P, Lücking R, Lücking A, Sipman HJM, Vester HFM, Wolf JHD, Gardette E (1996) How to sample the epiphytic diversity of tropical rain forests. Ecotropica 2:59–72

Green TGA, Meyer A, Büdel B, Zellner H, Lange OL (1995) Diel patterns of CO_2-exchange for six lichens from a temperate rain forest in New Zealand. Symbiosis 18:251–273

Gremmen NJM, Huiskes AHL, Francke JW (1995) Standing crop of the coastal macrolichen *Mastodia tesselata*, and its relationship to nutrient concentration on Petermann Island, Antarctica. Lichenologist 27:387–394

Grube M, Matzer M, Hafellner J (1995) A preliminary account of the lichenicolous *Arthonia* species with reddish, K+ reactive pigments. Lichenologist 27:25–42

Guderley R, Lumbsch HT (1996) The lichen genus *Diploschistes* in South Africa (Thelotremataceae). Mycotaxon 58:269–292

Guerra J, Ros RM, Cano MJ, Casares M (1995) Gypsiferous outcrops in SE Spain, refuges of rare, vulnerable and endangered bryophytes and lichens. Cryptog Bryol Lichenol 16:125–135

Hafellner J (1995a) Bemerkenswerte Funde von Flechten und lichenicolen Pilzen auf makaronesischen Inseln. II. Einige bisher übersehene arthoniale Arten. Herzogia 11:133–142

Hafellner J (1995b) Über *Piccolia*, eine lichenisierte Pilzgattung der Tropen (Ascomycotina, Lecanorales). Bibl Lichenol 58:107–122

Hafellner J (1995c) A new checklist of lichens and lichenicolous fungi of insular Laurimacaronesia including a lichenological bibliography for the area. Fritschia 5:1–132

Hafellner J (1996a) Bemerkenswerte Funde von Flechten und lichenicolen Pilzen auf makaronesischen Inseln. V. Über einige Neufunde und zwei neue Arten. Herzogia 12:133–145

Hafellner J (1996b) Studien an lichenicolen Pilzen und Flechten VIII. *Perigrapha*, eine neue Ascomycetengattung für "*Melanotheca*" *superveniens* Nyl. (Arthoniales). Nova Hedwigia 63:173–181

Hafellner J, Kalb K (1995) Studies in Trichotheliales ordo novus. Bibl Lichenol 57:161–186

Hafellner J, Türk R (1995) Über Funde lichenicoler Pilze und Flechten im Nationalpark Hohe Tauern (Kärnter Anteil, Österreich). Carinthia II 185(105):599–635

Hahn SC, Oberbauer SF, Gebauer R, Grulke NE, Lange OL, Tenhunen JD (1996) Vegetation structure and aboveground carbon and nutrient pools in the Imnavait Creek Watershed. Ecol Stud 120:109–128

Hamada N (1996) Introduction of the production of lichen substances by non-metabolites. Bryologist 99:68–70

Hamada N, Miyagawa H (1995) Secondary metabolites from isolated lichen mycobionts cultured under different osmotic conditions. Lichenologist 27:201–295

Hamada N, Miyagawa H, Miyawaki H, Inoue M (1996) Lichen substances in mycobionts of crustose lichens cultured on media with extra sucrose. Bryologist 99:71–74

Hammer S (1995) A synopsis of the genus *Cladonia* in the northwestern United States. Bryologist 98:1–28

Hammer S (1996) Meristeme initials: distinguishing characters in two *Cladonia* species. Bryologist 99:397–400

Harada H (1995) *Cyanopyrenia japonica* gen. et sp. nov., a peculiar pyrenocarpous cyanolichen from Japan. Lichenologist 27:249–254

Harada H (1996) *Hyalopyrenia japonica*, a peculiar new pyrenocarpous lichen genus from Japan. Lichenologist 28:415–419

Harris RC (1995) More Florida lichens including the 10 ₵ tour of the pyrenolichens. New York

Hauck M (1995) Epiphytische Flechtenflora ausgewählter buchen- und eichenreicher Laubalthölzer in Niedersachsen. Informationsdienst Naturschutz Niedersachsen 15(4):55–70

Hébrard JP, Loisel R, Roux C, Gomila H, Bonin G (1995) Incidence of clearing on phanerogamic and cryptogamic vegetation in South-Eastern France: disturbance indices. In: Bellan D, Bonin G, Emig C (eds) Functioning and dynamics of natural and perturbed ecosystems. Technique et Documentation, Lavoisier, Intercept Ltd, pp 747–758

Heidmarsson S (1996) Pruina as a taxonomic character in the lichen genus *Dermatocarpon*. Bryologist 99:315–320

Heiman K (1996) Macrolichens of the Blue Ride Parkway in North Carolina. Evansia 13:47–57

Henssen A (1995a) The new lichen family *Gloeoheppiaceae* and its genera *Gloeoheppia*, *Pseudopeltula* and *Gudelia* (Lichinales). Lichenologist 27:261–290

Henssen A (1995b) Apothecial structure and development in *Protoparmelia badia* (Parmeliaceae s. lat). In: Daniëls FJA, Schulz M, Peine J (eds) Flechten Follmann. Geobotanical and Phytotaxonomical Study Group, Botanical Institute, University of Cologne, Germany, Cologne, pp 55–62

Hertel H (1995) Schlüssel für die Arten der Flechtenfamilie Lecideaceae in Europa. Bibl Lichenol 58:137–180

Hesbacher S, Giez I, Embacher G, Fiedler K, Max W, Trawöger A, Türk R, Lange OL, Proksch P (1995) Sequestration of lichen compounds by lichen-feeding members of the Arctiidae (Lepidoptera). J Chem Ecol 21(12):2079–2089

Holien H (1996) Influence of site and stand factors on the distribution of crustose lichens of the Caliciales in a suboceanic spruce forest area in Central Norway. Lichenologist 28:315–330

Holien H, Gaarder G, Håpnes A (1995) *Erioderma pedicellatum* still present, but highly endangered in Europe. Graphis Scripta 7:79–84

Huneck S, Schmidt J, Alstrup V (1995) Lichen substances from subfossil and recent *Umbilicaria cylindrica*. Bibl Lichenol 57:231–239

Hyvönen J, Ahti T, Stenroos S, Gowan SP (1995) Genus *Cladina* and the section *Unciales* of the Genus *Cladonia* (Cladoniaceae, lichenized Ascomycotina), a preliminary phylogenetic analysis. J Hattori Bot Lab 78:243–253

Jarman SJ, Kantvilas G (1995) Epiphytes on an old Huon pine tree (*Lagarostrobos franklinii*) in Tasmanian rainforest. N Z J Bot 33:65–78

John V (1996) Preliminary catalogue of lichenized and lichenicolous fungi of Mediterranean Turkey. Bocconea 6:173–216

Kalb K, Elix JA (1995) The lichen genus *Physcidia*. Bibl Lichenol 57:265–296

Kantvilas G (1996) A new byssoid lichen genus from Tasmania. Lichenologist 28:229–237

Kappen L, Sommerkorn M, Schroeter B (1995) Carbon acquisition and water relations of lichens in polar regions – potentials and limitations. Lichenologist 27:531–545

Kappen L, Schroeter B, Hestmark G, Winkler JB (1996a) Field measurements of photosynthesis of umbilicarious lichens in winter. Bot Acta 109:292–298

Kappen L, Schroeter B, Scheidegger C, Sommerkorn M, Hestmark G (1996b) Cold resistance and metabolic activity of lichens below 0 °C. Adv Space Res 18(12):119–128

Kärnefelt I, Thell A (1996) A new classification for the *Dactylina/Dufourea* complex. Nova Hedwigia 62:487–511

Kashiwadani H, Kurokawa S (1995) Threatened lichens in Japan. Mitt Eidgenoss Forschungsanst Wald Schnee Landsch 70, 1:141–146

Kashiwandi H, Thor G (1995) Northern circumpolar crustose lichens new to Japan. J Jpn Bot 70:303–321

Kashiwadani H, Moon KH, Inoue M (1996) Lichens of Mt. Nishi-Azuma, Tohoku, Japan. Mem Nat Mus Tokyo 29:71–92

Kirschbaum U, Marx A, Schick JE (1996) Beurteilung der lufthygienischen Situation Gießens und Wetzlars mittels epiphytischer Flechten (1995). Angew Bot 70:78–96

Knops JMH, Nash TH III, Schlesinger WH (1996) The influence of epiphytic lichens on the nutrient cycling of an oak woodland. Ecol Monogr 66(2):159–179

Kondratyuk SY, Galloway DJ (1995) Lichenicolous fungi and chemical patterns in *Pseudocyphellaria*. Bibl Lichenol 57:327–345

Kondratyuk SY, Navrotskaya I, Khodosovtsev A, Solonina A (1996) Checklist of Ukrainian lichens. Bocconea 6:217–294

Kurokawa S (1996) Checklist of lichens of Toyama with notes on floristic features. Bull Bot Gard Toyama 1:1–21

Kuusinen M (1995a) Threatened lichens in Finland. Cryptog Bot 5:247–251

Kuusinen M (1995b) Epiphytic lichen diversity on *Salix caprea* and *Populus tremula* in old-growth forests of Finland. Mitt Eidgenoss Forschungsanst Wald Schnee Landsch 70(1):125–132

Kuusinen M (1996a) Epiphyte flora and diversity on basal trunks of six old-growth forest tree species in southern and middle boreal Finland. Lichenologist 28:443–463

Kuusinen M (1996b) Importance of spruce swamp-forests for epiphyte diversity and flora on *Picea abies* in southern and middle boreal Finland. Ecography 19:41–51

Lange OL, Green TGA (1995) High thallus water content severely limits photosynthetic carbon gain of central European epilithic lichens under natural conditions. Oecologia 108:13–20

Lange OL, Meyer A (1996) Photosynthetische Primärproduktion der Flechten: ökologische Bedeutung von hohem Thallus-Wassergehalt und CO_2-Konzentrierungsmechanismus. Verh Ges Okol 25:153–166

Lange OL, Reichenberger H, Meyer A (1995) High thallus water content and photosynthetic CO_2 exchange of lichens. Laboratory experiments with soil crust species from local xerothermic steppe formations in Franconia, Germany. In: Daniëls FJA, Schulz M, Peine J (eds) Flechten Follmann. Geobotanical and Phytotaxonomical Study Group, Botanical Institut, University of Cologne, Cologne, Germany, pp 139–153

Lange OL, Green TGA, Reichenberger H, Meyer A (1996a) Photosynthetic depression at high thallus water content in lichens: concurrent use of gas exchange and fluorescence techniques with a cyanobacterial and a green algae *Peltigera* species. Bot Acta 109:43–50

Lange OL, Hahn SC, Müller G, Meyer A, Tenhunen JD (1996b) Upland tundra in the foothills of the Brooks Range, Alaska: influence of light, water content and temperature in CO_2 exchange of characteristic lichen species. Flora 191:67–83

Leisner JMR, Bilger W, Lange OL (1996) Chlorophyll fluorescence characteristics of the cyanobacterial lichen *Peltigera rufescens* under field conditions. I. Seasonal patterns of photochemical activity and the occurrence of photosytem II inhibition. Flora 191:261–273

Leuckert C, Kümmerling H, Wirth V (1995) Chemotaxonomy of *Lepraria* Ach. and *Leproloma* Nyl. ex Crombie, with particular reference to Central Europe. Bibl Lichenol 58:245–259

Loppi S (1996) Lichens as bioindicators of geothermal air pollution in central Italy. Bryologist 99:41–48

Lücking R (1995a) Foliikole Flechten auf Cecropiaceen im Kronendach eines tropischen Regenwaldes. Bibl Lichenol 58:261–274

Lücking R (1995b) Biodiversity and conservation of foliicolous lichens in Costa Rica. Mitt Eidgenoss Forschungsanst Wald Schnee Landsch 70(1):63–92

Lücking R, Lücking A (1995) Foliicolous lichens and bryophytes from Cocos Island, Costa Rica. A taxonomical and ecogeographical study I. Lichens. Herzogia 11:143–174

Lücking R, Matzer M (1996) Ergänzungen und Verbesserungen zur Kenntnis der foliikolen Flechtenflora Costa Ricas. Die Familie Opegraphaceae (einschließlich der Gattung *Mazosia*). Nova Hedwigia 63:109–144

Lumbsch HT, Feige GB, Elix JA (1995a) A revision of the usnic acid containing taxa belonging to *Lecanora sensu stricto* (Lecanorales: Lichenized Ascomycotina). Bryologist 98:561–577

Lumbsch HT, Rambold G, Elix JA (1995b) *Ramalinora* (Ramalinaceae) – a new lichen genus from Australia. Aust Syst Bot 8:521–530

Lumbsch HT, Guderley R, Elix JA (1996) A revision of some species in *Lecanora sensu stricto* with a dark hypothecium (Lecanorales, Ascomycotina). Bryologist 99:269–291

Lutzoni F, Brodo IM (1995) A generic redelimitation of the *Ionaspis-Hymenelia* complex (lichenized Ascomycotina). Syst Bot 20(3):224–258

Lutzoni F, Vilgalys R (1995a) *Omphalina* (Basidiomycota, Agaricales) as a model system for the study of coevolution in lichens. Cryptog Bot 5:71–81

Lutzoni F, Vilgalys R (1995b) Integration of morphological and molecular data sets in estimating fungal phylogenies. Can J Bot 73 Suppl 1:S649–S659

Makhija U, Patwardhan PG (1995) The lichen genus *Arthothelium* (family Arthoniaceae) in India. J Hattori Bot Lab 78:189–235

Malcolm WM, Vezda A (1995) New foliicolous lichens from New Zealand 1. Folia Geobot Phytotaxon 30:91–96

Malcolm WM, Vezda A (1996) New foliicolous lichens from New Zealand 3. Folia Geobot Phytotaxon 31:263–268

Malcolm WM, Elix JA, Owe-Larsson B (1995) *Labyrintha implexa* (Porpidiaceae), a new genus and species from New Zealand. Lichenologist 27:241–248

Marcano V, Mohali S, Palacios-Prü E, Morales Méndez A (1996a) The lichen genus *Bulbotricella*, a new segregate in the *Parmeliaceae* from Venezuela. Lichenologist 28:421–430

Marcano V, Morales Méndez A, Caldaron L (1996b) A first checklist of the lichen-forming fungi of the Venezuelan Andes. Trop Bryol 12:193–235

Marcelli MP (1995) Habitat selection of epiphytic lichens on *Rhizophora mangle* in the mangroves of the Itanhaem river, São Paulo, Brazil. In: Daniëls FJA, Schulz M, Peine J (eds) Flechten Follmann. Geobotanical and Phytotaxonomical Study Group, Botanical Institute, University of Cologne, Cologne, Germany, pp 533–541

Matzer M (1996) Lichenicolous ascomycetes with fissitunicate asci on foliicolous lichens. Mycol Pap 17:1–202

Matzer M, Mayrhofer H (1996) Saxicolous species of the genus *Rinodina* (lichenized Ascomycetes, Physciaceae) in Southern Africa. Bothalia 26, 1:11–30

McCarthy PM (1995a) Aquatic species of *Verrucaria* in eastern Australia. Lichenologist 27:105–126

McCarthy PM (1995b) A reappraisal of *Clathroporina* Müll. Arg. (Trichotheliaceae). Lichenologist 27:321–350

McCarthy PM, Elix JA (1996) *Myeloconis*, a new genus of pyrenocarpous lichens from the Tropics. Lichenologist 28:401–414

McCune B, Derr CC, Muir PS, Shirazi A, Sillett SC, Daly WJ (1996) Lichen pendants for transplant and growth experiments. Lichenologist 28:161–169

Mezger U (1996) Biomonitoring mit epilithischen und epiphytischen Flechten in einem Belastungsgebiet (Berlin). Ein Verfahrensvergleich. Bibl Lichenol 63:1–164

Mies B, Lumbsch HT, Tehler A (1995) *Feigeana socotrana*, a new genus and species from Socotra, Yemen (Roccellaceae; Euascomycctidae). Mycotaxon 54:155–162

Moberg R (1995) The lichen genus Phaeophyscia in China and Russian Far East. Nord J Bot 15:319–335

Motyka J (1995, 1996) Porosty (Lichenes), vols 1–4. Rodzina Lecanoraceae. Lublin

Nakano T, Ihda T-A (1996) The identity of photobionts from the lichen *Pyrenula japonica*. Lichenologist 28:437–442

Narui T, Culberson CF, Culberson WM, Johnson A, Shibata S (1996) A contribution to the chemistry of the lichen family Umbilicariaceae (Ascomycotina). Bryologist 99:199–211

Nash TH III (ed) (1996) Lichen biology. Cambridge University Press, Cambridge

Nash TH III, Olafsen AG (1995) Climate change and the ecophysiological response of arctic lichens. Lichenologist 27:559–565

Nash TH III, Gries C, Elix JA (1995) A revision of the lichen genus *Xanthoparmelia* in South America. Bibl Lichenol 56:1–157

Navarro-Rosinés P, Hladun N (1996) Las especies saxícolo-calcícolas del grupo de *Caloplaca lactea* (Teloschistaceae, líquenes), en las regiones mediterránea y medioeuropea. Bull Soc Linn Provence 47:139–166

Navarro-Rosinés P, Roux C (1996) Le *Clauzadella gordensis* gen. et sp. nov., ascomycète lichénicole non lichénisé (Verrucariales, Verucariaceae). Can J Bot 74:1533–1538

Navarro-Rosinés P, Roux C, Giralt M (1996) *Wernerella* gen. nov (Dothideales, Ascomycetes) un género para incluir *Leptosphaeria maheui* Werner. Bull Soc Linn Provence 47:167–177

Orange A (1995) The British species of *Lepraria* and *Leproloma*: chemistry and identification. Br Lichen Soc Bull 76:1–9

Osorio HS (1995) Contribution to the lichen flora of Uruguay. XXVIII. Lichens from southern Rocha. Comun Bot Mus Hist Nat Montevideo 103(5):1–12

Osorio HS (1996) Contribution to the lichen flora of Argentina. XIX. Bibliography covering 1986–1995. Comun Bot Mus Hist Nat Montevideo 107(6):1–18

Peine J, Werner B (1995) *Follmanniella scutellata* gen. et sp. nov., a new genus and species of Roccellaceae (Opegraphales) from the Atacama desert, North Chile, South America. In: Daniëls FJA, Schulz M, Peine J (eds) Flechten Follmann. Geobotanical and Phytotaxonomical Study Group, Botanical Institute, University of Cologne, Cologne, Germany, pp 287–299

Pelkonen V-P, Hyvärinen M, Tarhanen S (1995) Ultracryomicrotomy in immunological studies on lichens. Br Lichen Soc Bull 77:29–33

Pfefferkorn V (1996) Epiphytische Flechtenvereine in Vorarlberg (Österreich) unter besonderer Berücksichtigung der Hemerobie von Waldökosystemen. Vorarlberger Natursch 1:9–152

Printzen C (1995) Die Flechtengattung *Biatora* in Europa. Bibl Lichenol 60:1–275

Printzen C, Rambold G (1995) Aphanopsidaceae – a new family of lichenized ascomycetes. Lichenologist 27:99–103

Puntillo D (1996) I licheni di Calabria. Museo Region Sci Nat (Torino) Monogr 22, 229 pp

Purvis OW, Halls C (1996) A review of lichens in metal-enriched environments. Lichenologist 28:571–601

Purvis OW, Jørgensen PM, James PW (1995) The lichen genus *Thelotrema* Ach in Europe. Bibl Lichenol 58:335–360

Rambold H (1995) Observations on hyphal, ascus and ascospore wall characters in Lecanorales *s.l.* Cryptog Bot 5:111–119

Randlane T, Thell A, Saag A (1995) New data about the genera *Cetrariopsis*, *Cetreliopsis* and *Nephromopsis* (fam. Parmeliaceae, lichenized Ascomycotina). Cryptog Bryol Lichenol 16:35–60

Renhorn K-E, Esseen P-A (1995) Biomass growth in five alectorioid lichen epiphytes. Mitt Eidgenoss Forschungsanst Wald Schnee Landsch 70(1):133–140

Ropin K, Mayrhofer H (1995) Über corticole Arten der Gattung *Rinodina* (Physciaceae) mit grauem Epihymenium. Bibl Lichenol 58:361–382

Rosentreter R (1995) Lichen diversity in managed forests of the Pacific Northwest, USA. Mitt Eidgenoss Forschungsanst Wald Schnee Landsch 70(1):103–124

Roux C, Sérusiaux E (1995) Présence d'appendices mucoïdes sur les ascospores de *Raciborskiella janeirensis* (Müll. Arg.) R. Sant. Bull Soc Linn Provence 46:91–94

Saag A, Randlane T (1995) Phylogenetic affinities of cetrarioid lichens. Cryptog Bot 5:128–136

Sarrión FJ, Burgaz AR (1995) Comunidades lignícolas del sector central de Sierra Morena (SW de España). Cryptog Bryol Lichenol 16:137–144

Scheidegger C (1995) Early development of transplanted isidioid soredia of *Lobaria pulmonaria* in an endangered population. Lichenologist 27:361–374

Scheidegger C, Frey B, Zoller S (1995) Transplantation of symbiotic propagules and thallus fragments: methods for the conservation of threatened epiphytic lichen populations. Mitt Eidgenoss Forschungsanst Wald Schnee Landsch 70(1):41–62

Schipperges B, Kappen L, Sonesson M (1995) Intraspecific variations of morphology and physiology of temperate to Arctic populations of *Cetraria nivalis*. Lichenologist 27:517–529

Schlechter E (1995) Auswertung eines Verbreitungsatlas der Makrolichenen des Eifelberglandes (Westdeutschland). I. Neufunde und Wiederfunde. In: Daniëls FJA, Schulz M, Peine J (eds) Flechten Follmann. Geobotanical and Phytotaxonomical Study Group, Botanical Institute, University of Cologne, Cologne, Germany, pp 461–474

Seaward MRD (1996) Checklist of Tunisian lichens. Bocconea 6:115–148

Seaward MRD, Edwards HGM (1995) Lichen-substratum interface studies, with particular reference to Raman microscopic analysis. 1. Deterioration of works of art by *Dirina massiliensis* forma *sorediata*. Cryptog Bot 5:282–287

Seaward MRD, Edwards HGM, Farwell D (1995) FT-Raman microscopic studies of *Haematomma ochroleucum* var. *porphyrium*. Bibl Lichenol 57:395–407

Sérusiaux E (1996) Foliicolous lichens from Madeira, with the description of a new genus and two new species and a worldwide key of foliicolous *Fellhanera*. Lichenologist 28:197–227

Sheard JW (1995) Disjunct distributions of some North American, corticolous, vegetatively reproducing *Rinodina* species (Physciaceae, lichenized Ascomycetes). Herzogia 11:115–132

Shirazi AM, Muir PS, McCune B (1996) Environmental factors influencing the distribution of the lichens *Lobaria oregana* and *L. pulonaria*. Bryologist 99:12–18

Sillet SC (1995) Branch-epiphyte assemblages in the forest interior and on the clearcut edge of a 700-year-old Douglas Fir canopy in western Oregon. Bryologist 98:301–312

Singh KP, Sinha GP (1994) Lichen flora of Nagaland. Bishen Sing Mahendra Pal Singh, Dehra Dun

Sipman HJM (1995) Preliminary review of the lichen biodiversity of the Colombian montane forests: In: Churchill SP, Balslev H, Forero E, Luteyn JL (eds) Biodiversity and conservation of neotropical montane forests. The New York Botanical Garden, Bronx, New York, pp 313–320

Sipman HJM, Raus T (1995) Lichen observations from Santorini (Greece). Bibl Lichenol 57:409–428

Smith CW (1995) Notes on long-distance dispersal in Hawaiian lichens: ascospore characters. Cryptog Bot 5:209–213

Søchting U (1995) Lichens as monitors of nitrogen deposition. Cryptog Bot 5:264–269

Solhaug KA, Gauslaa Y, Haugen J (1995) Adverse effect of epiphytic crustose lichens upon stem photosynthesis and chlorophyll of *Populus tremula* L. Bot Acta 108:233–239

Spjut RW (1995) *Vermilacinia* (Ramalinaceae, Lecanorales), a new genus of lichens. In: Daniëls FJA, Schulz M, Peine J (eds) Flechten Follmann. Geobotanical and Phytotaxonomical Study Group, Botanical Institute, University of Cologne, Cologne, Germany, pp 337–351

Spjut RW (1996) Niebla and Vermilacinia (Ramalinaceae) from California and Baja California. Sida Bot Miscell 14, 209 pp

Staiger B, Kalb K (1995) Die Flechtengattung *Haematomma*. Bibl Lichenol 59:3–198

Stenroos S (1995) Cladoniaceae (Lecanorales, lichenized Ascomycotina) in the flora of Chile. Gayana Bot 52(2):89–101

Swanson A, Fahselt D, Smith D (1996) Phenolic levels in *Umbilicaria americana* in relation to enzyme polymorphisms, altitude and sampling date. Lichenologist 28:331–339

Tabacchi R, Tsoupras G, Allemand P (1995) Identification of triterpenes from lichens by tandem mass spectrometry (MS-MS). Bibl Lichenol 57:429–442

Tehler A (1995) Arthoniales phylogeny as indicated by morphological and rDNA sequence data. Cryptog Bot 5:82–97

Thell A (1995) A new position of the *Cetraria commixta* group in *Melanelia* (Ascomycotina, Parmeliaceae). Nova Hedwigia 60:407–422

Thell A, Goward T (1996) The new cetrarioid genus *Kaernefeltia* and related groups in the Parmeliaceae (Lichenized Ascomycotina). Bryologist 99:125–136

Thell A, Mattson J-E, Kärnefelt I (1995a) Lecanoralean ascus types in the lichenized families Alectoriaceae and Parmeliaceae. Cryptog Bot 5:120–127

Thell A, Randlane T, Kärnefelt I, Gao X-Q, Saag A (1995b) The lichen genus *Allocetraria* (Ascomycotina, Parmeliaceae). In: Daniëls FJA, Schulz M, Peine J (eds) Flechten Follmann. Geobotanical and Phytotaxonomical Study Group, Botanical Institute, University of Cologne, Cologne, Germany, pp 353–370

Thomas MA, Nash TH III, Tucker SC (1996) *Coenogonium*: a green algal lichen without photosynthetic depression at high water contents. Lichenologist 28:341–345

Thor G (1995) Red lists – aspects of their compilation and use in lichen conservation. Mitt Eidgenoss Forschungsanst Wald Schnee Landsch 70(1):29–39

Tibell L (1996) Calicales. Flora Neotropica Monogr 69:1–78

Triebel D, Rambold G, Elix JA (1995) A conspectus of the genus *Phacopsis* (Lecanorales). Bryologist 98:71–83

Tschermak-Woess E (1995a) *Dictyochloropsis splendida (Chlorophyta)*, the correct phycobiont of *Phlyctis argena* and the high degree of selectivity or specificity involved. Lichenologist 27:169–187

Tschermak-Woess E (1995b) The taxonomic position of the green phycobiont of *Sticta canariensis* (Ach.) Bory ex Delise and its extraordinary modification in the lichenized state. Bibl Lichenol 58:433–438

van Dobbben HF (1996) Decline and recovery of epiphytic lichens in an agricultural area in The Netherlands (1900–1988). Nova Hedwigia 62:477–485

van den Boom PPG, Aptroot A, van Herk C (1996) The lichen flora of megalithic monuments in the Netherlands. Nova Hedwigia 62:91–104

Vicente C, Pereyra MT, Pedrosa MM, Solas MT, Pereira EC (1995) Immobilization of lichen cells and enzymes for bioproduction of lichen metabolites. Technical requirements and optimization of product recovering. In: Daniëls FJA, Schulz M, Peine J (eds) Flechten Follmann. Geobotanical and Phytotaxonomical Study Group, Botanical Institute, University of Cologne, Cologne, Germany, pp 97–110

Vonarburg C (1993) Das Mikroklima an Standorten epiphytischer Flechten. Veroff Nat Mus Luzern 5:1–122

Wedin M (1995) The lichen family Sphaerophoraceae (Caliciales, Ascomycotina) in temperate areas of the Southern Hemisphere. Symb Bot Ups 31(1):1–102

Wetmore CM (1996) The *Caloplacae sideritis* group in North and Central America. Bryologist 99:292–314

Wilson MJ (1995) Interactions between lichens and rocks; a review. Cryptog Bot 5:299–305

Wirth V (1995) Die Flechten Baden-Württembergs, part 1, 2. Ulmer, Stuttgart

Wirth V, Heklau M (1995) Die epiphytische Arten der Flechtengattung *Lepraria* und *Leproloma* in Baden-Württemberg. Bibl Lichenol 57:443–457

Wirth V, Schöller H, Scholz P, Ernst G, Feuerer T, Gnüchtel A, Hauck M, Jacobsen P, John V, Litterski B (1996) Rote Liste der Flechten (*Lichenes*) der Bundesrepublik Deutschland. Schriftenr Vegetationskd 28:307–368

Wolseley PA, James PW, Coppins BJ, Purvis OW (1996) Lichens of Skomer Island, West Wales. Lichenologist 28:543–570

Yoshimura I (1995) The lichen genus *Anzia* (Parmeliaceae, Lecanorales) in Central and South America. In: Daniëls FJA, Schulz M, Peine J (eds) Flechten Follmann. Geobotanical and Phytotaxonomical Study Group, Botanical Institute, University of Cologne, Cologne, Germany, pp 377–287

Yoshimura I, Sipman HJM, Aptroot A (1995) The lichen genus *Anzia* in New Guinea. Bibl Lichenol 58:439–469

Zhurbenko M (1996) Lichens and lichenicolous fungi of the northern Krasnoyarsk Territory, Central Siberia. Mycotaxon 58:185–232

Dr. Harrie J. M. Sipman
Botanischer Garten
und Botanisches Museum
Freie Universität Berlin
Königin-Luise-Straße 6–8
D-14191 Berlin, Germany

Edited by
H. J. Kadereit

Ecology and Vegetation Science

Mycorrhizae:
Ectotrophic and Ectendotrophic Mycorrhizae

By Reinhard Agerer

1 Ectomycorrhiza

a) Symbiotic Organisms and Morphology/Anatomy
of the Symbiotic Organs

Ectomycorrhizae forming Ascomycetes are compiled by Maia et al. (1996). Lepage et al. (1997) describe fossil *Rhizopogon*- or *Suillus*-like ectomycorrhizae on pine from Eocene time. In a review regarding trends in root-microbe symbiosis, Fitter and Moyersoen (1996) concluded that ectomycorrhizal symbiosis has probably evolved at least twice, once or more in the branching to the Pinaceae and Gnetales, and once in the root of the clade that includes all the rosid and asterid lineages of Chase et al. (1993), and possibly the Hamamelids as well. A historical overview on characterization of ectomycorrhizae has been provided by Agerer (1996a). A delta-based system for characterization and DEtermination of EctoMYcorrhizae (DEEMY) has been published on CD-ROM (Agerer and Rambold 1996; Rambold and Agerer 1997).

α) Comprehensive Descriptions of Selected Ectomycorrhizae

Albatrellus ovinus + *Picea abies* (Agerer et al. 1996a), *Bankera fuligineo-alba* + *Pinus* (Agerer and Otto 1997), *Descolea antarctica* + *Nothofagus* (Palfner 1997), *Entoloma sinuatum* + *Salix* (Agerer 1997a), *Hydnum rufescens* + *Picea abies* (Agerer et al. 1996b), *Hysterangium crassirhachis* + *Pseudotsuga* (Müller and Agerer 1996a), *Inocybe appendiculata* + *Picea* (Beenken et al. 1996a), *I. fuscomarginata* + *Salix/Populus* (Beenken et al. 1996b), *I. lacera* + *Populus* (Cripps 1997a), *I. obscurobadia* (Beenken et al. 1996c), *I. terrigena* + *Pinus* (Beenken et al. 1996d), *Lactarius chrysorrheus* + *Quercus* (Palfner and Agerer 1996b), *L. lilacinus* + *Alnus* (Pritsch et al. 1997), *L. obscuratus* + *Alnus* (Pritsch et al. 1997), *L. omphaliformis* + *Alnus* (Pritsch et al. 1997), *L. salmonicolor* + *Abies* (Pillukat 1996), *L. serifluus* + *Quercus* (Palfner and Agerer 1996b), *Naucoria escharoides* + *Alnus* (Pritsch et al. 1997), *N. subconspersa* + *Alnus*

(Pritsch et al. 1997), *Paxillus involutus* + *Pinus* (Mleczko 1997a), *Piloderma croceum* + *Pseudotsuga* (Goodman and Trofymow 1996), *Pisolithus aurantioscabrosus* + *Shorea parviflora* (Watling et al. 1995), *Ramaria aurea* + *Fagus* (Agerer 1996c), *R. largentii* + *Picea* (Agerer 1996d), *R. spinulosa* + *Fagus* (Agerer 1996e), *R. subbotrytis* + *Quercus* (Agerer 1996f), *Rhizopogon subcaerulescens* + *Tsuga* (Agerer et al. 1996c), *Russula fuegiana* + *Nothofagus* (Palfner and Godoy 1996b), *R. pumila* + *Alnus* (Pritsch et al. 1997), *Thelephora terrestris* + *Eucalyptus* (Ingleby and Mason 1996), *Thaxterogaster albocanus* + *Nothofagus* (Palfner 1996), *Tomentella albomarginata* + *Pinus* (Agerer 1996a), *T. ferruginea* + *Fagus* (Raidl and Müller 1996), *Tricholoma scalpturatum* + *Populus* (Cripps 1997b), *Truncocolmella citrina* + *Pseudotsuga* (Eberhart and Luoma 1996), *Tuber aestivum* + *Corylus* (Müller et al. 1996a), *T. borchii* + *Corylus* (Rauscher et al. 1996), *T. himalayense* + *Quercus* (Comandini and Pacioni 1997), *T. indicum* + *Pinus* (Zambonelli et al. 1997), *T. indicum* + *Quercus* (Comandini and Pacioni 1997; Zambonelli et al. 1997), *T. melanosporum* + *Corylus* (Rauscher et al. 1995), *T. mesentericum* + *Corylus* (Rauscher et al. 1995), *T. rufum* + *Corylus* (Rauscher et al. 1995), *T. uncinatum* + *Corylus* (Müller et al. 1996b).

β) Unidentified Ectomycorrhizae Named Binomially

An index of names published in the period 1993–1994 is given by Agerer (1995). The following descriptions are provided: "*Alnirhiza atroverrucosa*" (Pritsch et al. 1997), "*A. cana*" (Pritsch et al. 1997), "*A. cremicolor*" (Pritsch et al. 1997), "*A. cystidiobrunnea*" (Pritsch et al. 1997), "*A. suffusa*" (Pritsch et al. 1997), "*A. texta*" (Pritsch et al. 1997), "*A. violacea*" (Pritsch et al. 1997), "*Nothofagirhiza vinicolor*" (Palfner and Godoy 1996a), "*Piceirhiza cornuta*" (Montecchio and Agerer 1997), "*P. stagonopleres*" (Beenken and Agerer 1996), "*Pinirhiza arachnorosea*" (Wöllecke et al. 1997a), "*P. granulosa*" (Golldack et al. 1996a), "*P. lactogelatinosa*" (Golldack et al. 1997a), "*P. lactariosimilis*" (Golldack et al. 1997b), "*Pinirhiza lutea*" (Mleczko 1996), "*P. luteoalba*" (Wöllecke et al. 1997b), "*P. stellannulata*" (Golldack et al. 1996b), "*Populirhiza pustulosa*" (Mleczko 1997b), "*Pseudotsugaerhiza baculifera*" (Müller and Agerer 1996b), "*Quercirhiza argenteobrunneola*" (Fischer and Agerer 1996), "*Q. fibulocystidiata*" (Jakucs et al. 1997), "*Q. squamosa*" (Palfner and Agerer 1996a), "*Tetraberliniaerhiza bicolor*" (Moyersoen 1996a), "*T. cerviformis*" (Moyersoen 1996b), "*T. heterocystidiae*" (Moyersoen 1996c), "*Tsugaerhiza luteoannulata*" (Agerer and Molina 1997), "*T. tripigmentata*" (Agerer and Trappe 1997).

γ) Verification of Ectomycorrhizal Nature of Fungi, Including Short Descriptions

Diagnostic ectomycorrhizal features of eight *Tuber* species of economic importance are given (Granetti 1995) to facilitate determination work by technicians. Possibly misidentified ectomycorrhizae of *Boletu edulis* in New Zealand have been depicted; a *Descolea* ectomycorrhiza has been described instead (Wang et al. 1995). Parlade et al. (1996) proved *Lyophyllum decastes* as being ectomycorrhizal with some conifers. Ectomycorrhizae of *Tricholoma caligatum* + *Cedrus atlantica* and some unidentified morphotypes are briefly described by Abourouh and Najim (1995). The imperfect fungus *Leptodontidium orchidicola* formed a Hartig net and a patchy mantle on roots of *Salix glauca* seedlings (Fernando and Currah 1996). Graf and Brunner (1996) describe briefly the *Salix herbacea* ectomycorrhizae of *Cenococcum geophilum*, *Cortinarius favrei*, and *Entoloma alpicola*. Marchetti and Varese (1996) synthesized and characterized briefly *Laccaria laccta* + *Picea abies* and *Hebeloma crustuliniforme* + *Picea abies* ectomycorrhizae. McGee (1996) found and synthesized, respectively, ectomycorrhizae of the zygomycetous fungi *Densospora tubaeformis* and *D. solicarpa* with *Melaleuca uncinata*.

δ) Strain Variability of Fungi Regarding Ectomycorrhiza Formation

Cairney and Chambers (1997) concluded in a review on considerable intraspecific variation between *Pisolithus tinctorius* isolates in host specificity, growth form of extramatrical mycelia and organic nitrogen utilization, but to what extent differential host compatibility reflects taxonomic variations is not clear yet. Age and maintenance conditions of pure cultures of *Laccaria bicolor* strains exhibited strong differences in mycelial growth of pure cultures and fructification ability, as expressed in size and features of fruitbodies when ectomycorrhizal with *Pseudotsuga menziesii* (di Battista et al. 1996). Kropp (1997) found evidence in ectomycorrhization experiments on *Pinus strobus* with different mono- and dikaryotic strains of *Laccaria bicolor* that the percentage of converting short roots to ectomycorrhizae is polygenically controlled.

ε) Population Studies

Dikaryon-monokaryon crossings of *Laccaria bicolor* strains and their mycorrhiza formation were found in the root system of *Pinus banksiana* (de la Bastide et al. 1995). Gryta et al. (1997) studied fruitbody populations of *Hebeloma cylindrosporum* in a coastal sand dune forest ecosystem by DNA comparison at an interval of 3 years, and after 3 years were not able to detect the genets previously found. Due to application of PCR/RFLPs in the rDNA-ITS regions Erland (1995) was able to study the abundance of *Tylospora* + *Picea abies* ectomycorrhizae; it could be concluded that this ectomycorrhiza was one of the main mycorrhizal fungi in the south Sweden spruce forest studied. Gardes and Bruns (1996) compared with molecular tools the occurrence of fruitbodies and the corresponding ectomycorrhizae, and came to the conclusion that there was in several cases no correlation between above- and below-ground abundances. Overrepresentation and underrepresentation of ectomycorrhizae occurred. Cloned *Corylus avellana* were used to investigate their receptivity to *Tuber melanosporum* (Mamoun and Olivier 1996). The authors could show that the percentage of roots colonized by *T. melanosporum* reached a higher level on a special clone than on uncloned seedlings. The relation between *Tuber* and other symbionts appeared to depend on the morphology of the root system. *Tuber melanosporum* spread more easily on medium-sized root systems, whereas other symbionts developed in parallel to the root volume (Mamoun and Olivier 1996). The survival of two ectomycorrhizal fungi (*Hebeloma westraliense, Setchelliogaster* sp.) on roots of *Eucalyptus globulus* after outplanting were studied by Thomson et al. (1996b); depending upon the soil conditions, different amounts of new roots were colonized by the inoculated fungi, but after 12 months ectomycorrhizae of "resident" fungi increased in number. Reddy et al. (1997) found a higher total number of ectomycorrhizae on seedlings of *Pinus patula* when coinoculated with *Thelephora terrestris* and *Laccaria laccata* in comparison to a separate inoculation with either fungus. Simard et al. (1997) studied the influence of *Pseudotsunga menziesi* and *Betula papyrifera* seedlings in single and dual culture in natural soil upon morphotype composition. They obtained evidence that the abundance and frequency of some morphotypes on either plant species were affected by neighbouring seedlings.

b) Ontogeny and Ultrastructure

In several naturally grown *Picea abies* ectomycorrhizae cysteine-rich proteins were found in mantle and rhizomorph cells, in intrahyphal matrix and in proteinaceous inclusions of hyphal cells (Franz and Acker 1995). Balestrini et al. (1996) detected cellulose and xyloglucane in the plant cell walls of *Tuber magnatum* + *Corylus avellana* ectomycorrhizae, unesterified pectins in the junction between three root cells; methylesterified pectins were not found, and arabinosylated-β-(1,6)-glucans were only scarcely present in tissue sections. None of these substances was found in the matrix between the mantle and Hartig net hyphae. In fungal cell walls the presence of chitin, high-mannose side chains of glycoproteins, and β-1,3-glucans was proven. Wallander et al. (1997) found significant variations in protein and chitin concentrations and in the ergosterol-to-chitin ratios, both between seasons and mycorrhizal morphotypes of *Pinus sylvestris*; chitin concentrations peaked in both early summer and winter for all morphotypes, but remained low during midsummer and fall; ergosterol-to-chitin ratios increased till October. Kottke et al. (1995) were able to obtain relative quantification of Al, P and N. Using cerium as tracer element, their analysis revealed apoplastic flow through the fungal cell walls of the hyphal mantle and the Hartig net; no apoplastic barrier was found. A high capacity for sequestration of aluminium in polyphosphate bodies was also found, and Al and P increased in bodies in active mycelia with exposure time, but were partly replaced by nitrogen afterwards. In this study, calcium was found in phenolic material incorporated in the mantle at different levels depending on ectomycorrhizal type and soil treatment. Kottke (1997) describes adhesion pad formation and characterizes the penetration of root cuticle of early stages of *Laccaria amethystina* + *Picea abies* ectomycorrhizae. Pargney and Prevost (1996) found at the mantle-soil interface of *Lactarius blennius* + *Fagus sylvatica* ectomycorrhizae a mucilagenous material rich in chitin; an ascomycete infection of *Lactarius subdulcis* + *F. sylvatica* ectomycorrhizae revealed a decrease in chitin content in the walls between the hyphae present in the mantle and intracellularly. Prevost and Pargney (1997) found in ectomycorrhizal rhizomorphs of *Lactarius blennius* and *L. subdulcis* dead small hyphae with very thick electron-dense walls and live hyphae showing relatively large diameter and walls less thick than the others.

c) Physiology

Cairney and Burke (1996) review the physiological heterogeneity within fungal mycelia and stress this as an important concept for a functional understanding of the ectomycorrhizal symbiosis.

α) Substances Assumed as Important for Formation
of Ectomycorrhizae

Hormones. Gay et al. (1995) provide genetical, biochemical and ultrastructural evidence that fungal auxin is involved in ectomycorrhiza formation of IAA overproducer mutants of *Hebeloma cylindrosporum*. Mycelial cultures of *Suillus bovinus* showed decreased production of IAA after addition of even small amounts (10 mg/l) of aluminium (Kieliszewska-Rokicka et al. 1995).

Phenolics. Coumarin-like compounds were found in *Tuber* spp. + *Quercus pubescens* ectomycorrhizae, which were not found in either non-mycorrhizal roots or fruitbodies; *T. magnatum* and *T. borchii* caused different patterns of coumarin-like compounds (Tirillini and Granetti 1995). Weiss et al. (1997) found in *Larix decidua* ectomycorrhizae of *Suillus tridentinus* and *Boletinus cavipes* tissue-specific and development-dependent accumulation of diverse phenylpropanoids.

Proteins (Polypeptides). Martin et al. (1995) found during ectomycorrhiza formation several abundant transcripts which showed a significant amino acid sequence similar to a family of secreted morphogenetic fungal proteins, the so-called hydrophobins; such transcripts were high in aerial hyphae of *Pisolithus tinctorius* and during mantle formation. Hydrophobin-encoding cDNAs were cloned and characterized from *Pisolithus tinctorius* aerial mycelium and during early stages of ectomycorrhiza formation on *Eucalyptus globulus*; they were barely detectable in mycelium grown in liquid cultures (Tagu et al. 1996). Beguiristain and Lapeyrie (1997) found that *Eucalyptus globulus* stimulates hypaphorine accumulation in *Pisolithus tinctorius* hyphae during ectomycorrhizal colonization while excreted fungal hypaphorine controls root hair development. During ectomycorrhizal development, Burgess and Dell (1996) found a decrease in all plant root proteins and differential accumulation of fungal proteins. This domination of the fungal partner in the protein biosynthesis of developing ectomycorrhizae is probably a consequence of stimulated fungal growth and the corresponding decrease in plant meristematic activity. Niini et al. (1996) detected in *Pinus sylvestris* roots during ectomycorrhiza formation with *Suillus bovinus* two new α-tubulins in the acidic α-tubulin cluster, but no change occurred in the fungus; a high level of plant actin at an early stage of ectomycorrhiza formation suggests a significant role of this protein in the interaction between plant cells and fungal hyphae. E.C. Diaz et al. (1996) found differences in α-tubulin production in *Eucalyptus globulus* roots during formation of ectomycorrhizae with an aggressive and non-aggressive strain of *Pisolithus tinctorius*; the more aggressive strain enhanced α-tubulin expression, and this coincided with the increase in lateral root formation. The authors suggest that the changes found in α-tubulin expression support a role for cytoskeleton components in ectomycorrhizal development. Data obtained from *Pinus contorta* + *Suillus variegatus* ectomycorrhizae suggest that plant and fungal α-tubulin are growth-related proteins, subject to changes, while the amount of actin reflects the general metabolic activity of the mycorrhizae (Timonen et al. 1996).

Other Substances. The first contact of *Picea abies* roots with hyphae of *Pisolithus tinctorius* was significantly accelerated upon treatment with 0.5 µM jasmonic acid (Regvar and Gogala 1996).

β) Recognition, Elicitors

As rapid reactions of ectomycorrhizal elicitors (*Hebeloma crustuliniforme*) on suspension-cultured spruce cells, Salzer et al. (1996) detected reactions of Cl, K, Ca-fluxes, phosphorylization and dephosphorylization of two different proteins, respectively, alkalinization of the medium, and a transient synthesis of H_2O_2. The authors suggest that the efficacy of elicitors released from fungal cell walls is controlled by apoplastic enzymes of the host; the plant itself is able to reduce the activity of fungal elicitors on their way through the plant cell walls; but those elicitors which finally reach the plasma membrane of the host cells induce reactions that are similar to the early defense reactions in plant-pathogene interactions. Salzer et al. (1997) studied the effect of purified spruce chitinases and β-1,3-glucanases on the activity of elicitors from ectomycorrhizal fungi, and suggested that apoplastic chitinases in the root cortex destroy elicitors from the ectomycorrhizal fungi without damaging the fungus, and by this mechanism the host plant could attenuate the elicitor signal and adjust its own defense reactions to a level allowing symbiotic interaction.

γ) Enzymes (cf. also c.α)

Mycorrhization of *Picea abies* with *Amanita muscaria* increased sucrose phosphate synthase in the seedlings (Hampp et al. 1995) and the concentration of fructose-2,6-bisphosphate, a potent inhibitor of fructose-1,6-bisphosphatase, was decreased in needles; the authors also obtained evidence that in fungal extracts phosphofructokinase was stimulated by fructose-2,6-bisphosphate, while in host tissue it was phosphotransferase. Schaeffer et al. (1996) found evidence for an upregulation of the host and a downregulation of fungal phosphofructokinase activity in ectomycorrhizae of *Picea abies* + *Amanita muscaria*. Roots of *Eucalyptus pilularis* and *Pinus sylvestris* ectomycorrhizal with *Pisolithus tinctorius* had no increased N-acetyl-glucosamine activity, in contrast to the pathogenic fungus *Heterobasidion annosum* (Hodge et al. 1995). Hodge et al. (1996) were able to detect in *Lactarius* sp. + *Fagus sylvatica* ectomycorrhizae exochitinase activity in the mantle and in the Hartig net; emanating hyphae showed substantial β-N-acetylglucosamidase activity. Leprince and Quiquampoix (1996) studied the adsorption of two extracellular acid phosphatases of *Hebeloma cylindrosporum* on montmorillonite and found evidence for a pH-dependent modification of the enzyme conformation due to mainly electrostatic interactions with the clay surface. At low pH, the two positively charged enzymes unfold on the negatively charged montmorillonite surface; at high pH, both the enzymes and the clay are negatively charged, and adsorption decreases;

adsorption and modifications are largely irreversible. In contrast to the ericoid mycorhizal fungus, *Hymenoscyphus ericae*, cultures of ectomycorrhizal fungi tested had no extracellular phenoloxidases and therefore no solubilization capacity of tanninic acid (Bending and Read 1996b). Münzenberger et al. (1997) studied the laccase and peroxidase activity of *Larix decidua* and *Picea abies* + *Laccaria amethystina* ectomycorrhizae and found, in comparison to the already high contents of laccase in the mycelium, the highest laccase activities in ectomycorrhizae of both tree species; ectomycorrhizae contained the lowest proxidase activities. The results showed that four acidic peroxidases of the host with specific iP values were almost completely suppressed due to ectomycorrhiza formation.

δ) Carbon Nutrition of Ectomycorrhizae

Differently aged zones of *Amanita muscaria* + *Picea abies* ectomycorrhizae showed a decrease of sucrose in the zones exhibiting the highest fungal proportion and these zones of interaction were characterized by high levels of fructose-2,6-bisphosphate, which exceeded those resulting from a mere addition of the levels contained in the single partners (Hampp et al. 1995). Eltrop and Marschner (1996a, b) concluded that a possible growth depression of *Picea abies* seedlings ectomycorrhizal with *Pisolithus tinctorius* as compared to non-mycorrhizal is caused by increased root respiration, whereas the production of extramatrical mycelium of mycorrhizal plants is of minor importance. Colpaert et al. (1996) found a consistently high demand of the fungi in a semihydroponic system of *Pinus sylvestris* ectomycorrhizal with *Thelephora terrstris*, *Suillus bovinus* and *Scleroderma citrinum* for carbohydrates due to their large biomass of external mycelia and the increased below-ground respiration of mycorrhizal plants. This could explain the lower rate of shoot growth of ectomycorrhizal seedlings.

ε) Phosphate Nutrition

Marschner and Godbold (1995) concluded that element content (P, Fe, Ca) of the cortex cell walls of ectomycorrhizal roots of *Paxillus involutus* + *Picea abies* is strongly influenced by age of the mycorrhiza, while the distance from the tip seems to be of minor importance. Influences of different phosphorus forms on number and length of mycorrhizae were found by McElhinney and Mitchell (1995). The extramatrical mycelium mass of the ectomycorrhizal system *Paxillus involutus* + *Pinus sylvestris* peaked (ergosterol method) when P was low and other nutrients were high (N, K; Ca, Mg and S had no effect). This investment resulted in a

660% higher biomass in mycorrhizal compared to non-mycorrhizal seedlings (Ekblad et al. 1995). In the same experiment the root/shoot ratio in both mycorrhizal and non-mycorrhizal *P. sylvestris* seedlings decreased to increased N contents of the substrate, whereas phosphorus had no effect. The same design applied to *Paxillus involutus* + *Alnus incana* + *Frankia* resulted in no reaction of the root/shoot relation to increased N but resulted in a decrease reaction to P (Ekblad et al. 1995). The latter results have to be interpreted cautiously, as *P. involutus* is – at least in nature – not a specific fungus of *Alnus*. Cumming (1996) highlights the role of mycorrhizal fungi in altering the allocation of P between roots and shoots, as in P_i-limiting soils P was retained in roots at the expense of translocation to the foliage of *Pinus rigida* ectomycorrhizal with any of the species *Paxillus involutus, Laccaria bicolor* or *Pisolithus tinctorius*. Moreover, the activities of ATPase systems of the mycorrhizal species tested do not support the hypothesis that this enzyme plays an important role in P_i acquisition under P-limiting conditions. Colpaert et al. (1997) found evidence that acid phosphatase activity of *Thelephora terrestris* and *Suillus luteus* + *Pinus sylvestris* ectomycorrhizae were positively correlated with the ergosterol content in the substrate and decreased with increasing P nutrition. *Thelephora terrestris* grew in contrast to *S. luteus* best in high P_i nutrition. The experiment did not support that phytate is a useful P source for ectomycorrhizae of these fungi.

ζ) Nitrogen Nutrition (cf. also d.δ, γ)

Observations by Turnbull et al. (1995) indicated that mycorrhizal associations confer on species of *Eucalyptus* the ability to broaden resource base substantially with respect to nitrogen. This ability to use organic nitrogen was not directly related to that of the fungal symbiont in isolation, since seedlings mycorrhizal with *Pisolithus* sp. were able to assimilate sources of nitrogen (in particular histidine and protein) on which the fungus in pure culture appeared to grow weakly. Michelsen et al. (1996), using $\delta^{15}N$ of subarctic plants, found evidence that ericoid, ectomycorrhizal and non- and VA-mycorrhizal species access different sources of soil nitrogen. $\delta^{15}N$ in ectomycorrhizae was 2‰ higher than in non-mycorrhizal roots, in ectomycorrhizal mantles it was weven 2.4–6.4‰ higher (Högberg et al. 1996). Nitrogen source regulates the biosynthesis of NADP-glutamate dehydrogenase in the ectomycorrhizal basidiomycete *Laccaria bicolor* (Lorillou et al. 1996). The amount of extramatrical mycelium of *Picea abies* seedlings ectomycorrhizal with *Pisolithus tinctorius* was significantly lower in nitrate- than in ammonium-supplied plants (Eltrop and Marschner 1996a). N uptake rates were increased in *Picea abies* seedlings ectomycorrhizal with *Pisolithus tincto-*

rius only when they were supplied with ammonium but not with nitrate (Eltrop and Marschner 1996b). Bending and Read (1996a) could show that cultures of ectomycorrhizal fungi did not have access to N contained in protein complexed by tanninic acid, while ericoid mycorrhizal fungi did.

η) Other Macronutrients

Studies on in vitro weathering of phlogopite by two strains of *Pisolithus tinctorius* and *Paxillus involutus* suggest that the bioweathering mechanisms could be related to the release of fungal organic acids or other complex-forming molecules (Paris et al. 1995). The uptake of ^{45}Ca in an unlimed peat system by *Picea abies* and *Betula pendula* ectomycorrhizal with *Paxillus involutus*, was increased by mycorrhizal colonization; in limed substrates the uptake was as high or even higher than with non-mycorrhizal seedlings (Andersson et al. 1996). Paris et al. (1996) suggest due to studies with *Pisolithus tinctorius* and *Paxillus involutus* mycelia and phlogopite a key role for fungal oxalic acid during mineral weathering in response to nutrient deficiency. Frey et al. (1997a) detected in P-rich granules of *Picea abies* + *Hebeloma crustuliniforme* ectomycorrhizae several elements including Ca, K, Cl, S, Cs and Sr, with highest concentrations for S.

ι) Micronutrients (see also ε)

Hauer and Dawson (1996) found evidence that ectomycorrhizae of *Quercus palustris* influence iron sequestering under iron-limiting conditions.

κ) Water

Increased drought resistance of *Pinus taeda* seedlings ectomycorrhizal with *Pisolithus tinctorius* could, in part, be attributed to drought-induced colonization by mycorrhizae and the ability of mycorrhizal plants to maintain high transpiration rates as a result of greater lateral root formation and lower shoot mass; the osmotic adjustment of droughted plants was not affected (Davies et al. 1996).

d) Ecology

α) Ecological Laboratory Research

Different degrees of ectomycorrhiza development of *Laccaria bicolor,* *Rhizopogon luteolus* and *Suillus bovinus* on *Pinus sylvestris* were related to potassium concentration, organic matter content and the pH of the soils, suggesting that composition of soils affects ectomycorrhizal development (Baar and Elferink 1996). Colpaert and van Laere (1996) studied the activity of cellulase, β-xylosidase, β-glucosidase, phenoloxidase, phosphomonoesterase, and protease in beech leaf litter incubated with the saprophytic fungus *Lepista nuda* or with *Pinus sylvestris* seedlings ectomycorrhizal with either *Thelephora terrestris* or *Suillus bovinus.* They found that phosphomonoesterase was higher in *L. nuda* and *S. bovinus* treatments, and was intermediate with *T. terrestris.* For all other enzymes, the activity in litter inoculated with the white-rot fungus *L. nuda* was considerably larger than with the ectomycorrhizal fungi; phenoloxidase activity was only clearly increased in the *S. bovinus* treatment. Colpaert and van Tichelen (1996a) incubated fresh beech leaf litter with *L. nuda* or *Pinus sylvestris* seedlings ectomycorrhizal with *T. terrestris, S. bovinus* or *Paxillus involutus* and pointed out that the mycorrhizal fungi caused only a low decomposition of the litter as compared to *L. nuda,* and nitrogen was released only by *L. nuda.* It therefore seemed likely that the studied mycorrhizal fungi did not have the ability to decompose efficiently the lignocellulose matrix. In a competition experiment of *Picea abies* and *Betula pendula* with the common ectomycorrhizal fungus *Scleroderma citrinum,* Ek et al. (1996) found that the mycorrhizal mycelium received carbohydrates mainly from the birch plant and the nitrogen transfer by the fungus to the plants was largely directed towards the birch; there was no conclusive evidence of a net transfer of carbon between the two seedlings. Thomson et al. (1996a) found evidence that ectomycorrhizal fungi which colonized roots most extensively increased plant growth to the greatest extent, but growth was less related to hyphal development in soil. In a combined experiment with *Alnus glutinosa* and *Pinus sylvestris* and the connecting fungus *Paxillus involutus* (a species usually not forming ectomycorrhizae with *Alnus*) and *Frankia,* Ekblad and Huss-Danell (1995) pointed out that the proportion of *Frankia*-fixed nitrogen in pine, transferred from alder, was greatest (9%) when the pine was nitrogen starved and mycorrhizal and the alder was fixing maximally (low N and high P); however, the amount of fixed N transferred to pine was not statistically different from zero.

β)   Researches in Natural Habitats (see also a.ε and c.ζ)

In *Eucalyptus* forest soil, mycelial systems of vesicular-arbuscular my-corrhizal and ectomycorrhizal fungi are apparently localized in different domains, and there were also zones where non-mycorrhizal roots (most-ly cluster roots produced by members of the Proteaceae) predominated (Brundrett and Abbott 1995). Studies on spatial patterns of mycorrhizal infectiveness of soils along a successional chronosequence emphasize the need to explicitly evaluate spatial heterogeneity in mycorrhizal infectivity in studies of the role of mycorrhiza in succession (Boerner et al. 1996). Helm et al. (1996) studied the chronosequence near a glacier in Alaska and found that morphotype diversity increased from early successional stages to later stages mostly due to an increase in evenness rather than in richness. See and Alexander (1996), studying the ectomycorrhizal community of *Shorea leprosula* seedlings, showed that overall level of colonization and total number of morphotypes per seedling were higher at a site which appeared to have experienced the least disturbances and at which there were the greatest number of adult *S. leprosula* trees. Morphologically different mats of mat-forming fungi did not overlap, but there was a tendency to clustering, and mats can persist for at least 2 years after their host tree has been cut (Griffiths et al. 1996). Rao et al. (1997) found in a *Pinus kesiya* stand a positive correlation between the number of ectomycorrhizae and mycorrhizal infection with soil moisture, soil pH, total nitrogen, available phosphorus, exchangeable potassium, and organic matter of the soil. In an enclosure experiment with *Salix* and *Populus*, Rossow et al. (1997) found evidence that in an early successional taiga ecosystem browsing by mammals induced reduction of ectomycorrhizal infection and this reduction is suggested as playing a central role in the shift of this salicaceous species to a vegetation with more *Alnus* and spruce trees.

γ)   Coexistence with Other Organisms

Saprophytic and Parasitic Fungi. Seedlings of *Pseudotsuga menziesii* ectomycorrhizal with certain strains of *Hebeloma crustuliniforme*, *Laccaria laccata* and *Thelephora terrestris* showed fewer dead roots in comparison with uninoculated seedlings when confronted with the pathogen *Rhizinia undulata* (Zak and Ho 1994). The number of colony-forming units of *Fusarium moniliforme* was significantly reduced when seedlings of *Pinus banksiana* were inoculated with *Paxillus involutus* and *Bacillus subtilis* alone or in combination; *Suillus tomentosus*, on the other hand, had no effect (Hwang et al. 1995). Ursic et al. (1997) found in nursery-grown *Pinus strobus* seedlings that *Thelephora terrestris* ectomycorrhizae were highly significantly negatively correlated with root rot. *Rhizo-*

pogon luteolus and *Pisolithus tinctorius* reduced damping off of *Eucalyptus tereticornis* by *Phytium aphanidermatum* due to antagoistic effects (Bhat et al. 1997). Marchetti and Varese (1996) pointed out that *Verticillim bulbillosum* significantly inhibited formation of ectomycorrhizae by *Laccaria laccata* + *Picea abies* most likely by means of toxic metabolites and mycoparsitism; on the contrary, it had no deleterious effects on formation of ectomycorrhizae by *Hebeloma crustuliniforme* on *Picea abies*.

Bacteria. Extramatrical mycelium of *Paxillus involutus, Thelephora terrestris, Laccaria proxima, Suillus variegatus* and *Hebeloma crustuliniforme* ectomycorrhizae of *Pinus contorta* seedlings reduced soil bacterial activity as measured by thymidine incorporation (Olsson et al. 1996). Results obtained by Shishido et al. (1996) suggest that some strains of fluorescent pseudomonads enhanced spruce seedling growth through mechanisms unrelated to increased mycorrhizal colonization, but growth promotion of pine was facilitated by an interaction with ectomycorrhizae. Results of Frey-Klett et al. (1997) raise questions concerning the mycorrhization helper bacteria concentrations in the soil which is effective for promotion of mycorrhizal establishment and the timing of the bacterial effect, as in an experiment the promoting effect became apparent only after the concentration of the strain of MHBs tested dropped below a certain concentration. Dual cultures between *Suillus grevillei* and bacteria isolated from fruitbodies or ectomycorrhizae with *Larix decidua* showed that Gram-positive bacteria seldom stimulated fungal growth; among Gram-negative bacteria a *Pseudomonas fluorescens* and a *P. putida* strain showed the greatest enhancement of growth, whereas *Streptomyces* always caused a significant inhibition of the fungus (Varese et al. 1996). Frey et al. (1997b) concluded that *Laccaria bicolor* exerts a trehalose-mediated selection on the fluorescent pseudomonads present in the vincinity of the mycorrhizae.

Animals. Pastor et al. (1996) could detect spores of 22 ectomycorrhizal fungal genera in faeces of two rodents. Ruess and Dighton (1996) found that the population of a fungal feeder nematode developed faster and to a greater extent with mycorrhizal fungi as food source than it did with saprobic fungal species. In a microcosm experiment with ectomycorrhizal *Pinus sylvestris* seedlings with or without reinoculation with soil fauna, and N-poor or N-rich soil, Setälä et al. (1997) found that ca. ten times more ectomycorrhizal fungal biomass was found on pine roots growing in N-poor than in N-rich soils. In addition, the amount of ectomycorrhizal fungi was significantly reduced by the complex and abundant faunal community, particular in N-poor soils, where the amount of ectomycorrhizal fungi was less than 16% of that found in system with reduced grazing pressure. Gehring et al. (1997) reported on a removal experiment of pinyon pine (*Pinus edulis*) scales (*Matsucoccus acalyptus*) that the scales negatively affect ectomycorrhiza. Secondly, they found no ectomycorrhizal effect on mortality of scales, when levels of ectomycorrhiza were enhanced, and thirdly, high-scale densities suppressed ectomycorrhizal colonization, but only on trees susceptible to scales.

Plants. Sylvia and Jarstfer (1997) found in a *Pinus elliotti* plantation that weedy plant competition significantly affected the number of ectomycorrhizae and the distribution of morphotypes. Taylor and Bruns (1997) obtained evidence by RFLPs and sequence analysis that the mycorrhizal fungus of the non-photosynthetic orchid *Cephalanthera austinae* also formed ectomycorrhizae with trees; this fungus apparently was a member of Thelephoraceae. Mamoun and Olivier (1997) found that *Festuca ovina* reduced colonization of young *Corylus avellana* by *Tuber melanosporum*; coculture with *F. ovina* resulted in necrosis of hazel roots.

δ) Influence by Man

Fertilization with Lime or Nitrogen (cf. also e.ε). In a fertilization experiment of *Pinus patula* stands with a mixture of limestone, ammonium nitrate, potassium cloride, gypsum and monoammonium phosphate, Carlson (1994) found a change in abundance of four roughly characterized morphotypes of ectomycorrhizae, while the total number of mycorrhizal tips was not affected. Container-grown *Laccaria bicolor* + *Pseudotsuga menizesii* seedlings fertilized with four different levels of nitrogen had taken up less N (concentrations and total amounts) for the production of the same biomass than the non-mycorrhizal ones (Gagnon et al. 1995). The authors conclude that a total seedling N concentration of 1.6% and a substrate fertility of 52 ppm N are appropriate to optimize both the ectomycorrhizal development and the growth of the seedlings. Majdi and Nylund (1996) suggest that water and nitrogen input, due to liquid fertilization, lower longevity of mycorrhizal roots and promote fine root production at deeper soil layers. Andersson et al. (1997) concluded that limiting of N-fertilized peat in which *Pinus sylvestris* + *Paxillus involutus* was growing induced chemical or microbial immobilization of the added N; this is suggested to be the main reason for the decreased uptake of N in lime treatments. Frey et al. (1997a) found in *Picea abies* + *Hebeloma crustuliniforme* ectomycorrhizae no Hartig net due to application of high NH_4^+ contents in the nutrient solution. In an N-free fertilization experiment with (P, K, Ca, Mg, S) Kåren and Nylund (1996) suggested that moderate levels of N-free fertilization are not likely to drastically affect the community structure of the dominating ectomycorrhizal morphotypes of *Picea abies*. Ek (1997) supplied ammonium or nitrate exclusively to the extramatrical mycelium of

Paxillus involutus + *Betula pendula* ectomycorrhizae and found that respiration in the fungal compartment represents 11–29% of total fungal and root respiration, and the mycelium in the fungal compartment received 20–29% of shoot net assimilation, and 43–64% of the carbon allocated to the mcyelium was respired.

Pesticides. Zambonelli et al. (1995) obtained evidence that the application of triadimefon in the open field against the powdery mildew on oaks infected with *Tuber borchii* does not enhance the growth of *Hebeloma sinapizans*, a highly competitive fungus. Pedersen and Sylvia (1997) found that benomyl can successfully inhibit development of vesicular-arbuscular mycorrhizal fungi under controlled conditions in the greenhouse with no inhibitory effects on the ectomycorrhizal fungus *Pisolithus tinctorius*.

Pollution – Acidification. Fruitbodies originating from an acid rain-polluted area had smaller amounts of IAA than had those of control stands (Kieliszewska-Rokicka et al. 1995). In substrates of neutro-basic pH, short-term exposures of *Pinus halepensis* seedlings to acid rain (pH 4.5, 3) positively affected ectomycorrhizal fungi, in particular *Suillus* species (Honrubia and Diaz 1996).

Pollution – Effects of Ozone (and SO$_2$). Studying *Hebeloma crustuliniforme* + *Pinus ponderosa* systems, Anderson and Rygiewicz (1995) found that shoot-applied ozone significancly reduced ^{14}C activity in the fungus of mycorrhizal plants; the results suggest a substantial impact of ozone on the carbon balance of the mycorrhiza; but mycorrhizal and non-mycorrhizal seedlings did not appear to react differently. Ectomycorrhizal *Pinus halepensis* seedlings treated with gaseous SO$_2$ (40 ppb) showed only a slight reduction in percentage of mycorrhizal colonization. However, in combination with 50 ppb ozone, mycorrhizal colonization was significantly reduced. Morphological alterations of mycorrhizae were also observed, with a reduction in the coraloid structures in favour of simple ones. Moreover, a change in species composition was noted, the ectomycorrhizae probably formed by *Suillus* species being replaced by ectendomycorrhizae (G. Diaz et al. 1996).

Pollution – Effects of Nitrogen. In the vicinity of a pulp mill, Holopainen et al. (1996) found ultrastructural changes of several ectomycorrhizal types, the clearest of which were increased tannin depositions in cortical cells, intracellular growth of hyphae in cortical cells and the appearance of electron-dense accumulations in the vacuoles of the fungal cells; nitrogen deposition is suspected to be the primary cause of root decline.

Pollution – Greenhouse Effect. Hodge (1996) reviewed the impact of elevated CO_2 on mycorrhizal associations and implications for plant growth. The total uptake of sulphur by *Quercus robur* was enhanced by elevated CO_2 and further enhanced by elevated CO_2 and mycorrhization with *Laccaria laccata*, and more reduced sulphur was transported from the leaves to the roots in mycorrhizal plants at elevated CO_2 concentrations as compared to non-inoculated seedlings (Seegmüller et al. 1996). The biomass allocation between the plant and the fungi of naturally ectomycorrhizal *Pinus sylvestris* seedlings was not affected by a doubled ambient CO_2 concentration, and the proportion of fungi in the roots remained constant (Markkola 1996). Walker et al. (1995a) found in an experiment with enriched atmospheric CO_2 concentrations (700 µl 1^{-1}, 525 µl 1^{-1}, and ambient) that the highest mycorrhizal percent infections within each P treatment (68, 43, 18 µg g^{-1} soil) of *Pinus ponderosa* + *Pisolithus tinctorius* varied temporally. Mycorrhizal colonization was not affected by the atmospheric treatments after 4 months, while seedlings grown in ambient CO_2 exhibited the highest percent infections within each P treatment after 8 months; those grown in 700 µl 1^{-1} CO_2 had the highest percentages after 1 year. Lewis and Strain (1996) found no interactions between CO_2 (35.5 Pa, 71.0 Pa), phosphorus supply (high, low) and mycorrhizal status on dry mass of 60- to 120-day-old seedlings of *Pinus taeda* ectomycorrhizal with *Pisolithus tinctorius*. They found, however, that, due to reductions in specific leaf area, elevated CO_2 reduced the relative cost of the symbiosis. Kottke (1997) pointed out that no structural changes in mycorrhization of *Picea abies* + *Laccaria amethystina* related to elevated CO_2 occurred, but fine roots and mycorrhizae developed faster. Walker et al. (1995b) found in an CO_2 enrichment experiment (700 µl 1^{-1}, 525 µl 1^{-1}, and ambient) on *Pinus ponderosa* ectomycorrhizal with *Pisolithus tinctorius* that the response of juvenile plants to CO_2 enrichment is ephemeral, with the effects on roots more pronounced and persistent overall than those on shoots, and that the response is dependent on nitrogen availability. Lewis et al. (1994) obtained results that suggest that although elevated CO_2 may significantly increase root carbohydrate levels of *Pinus taeda* seedlings, the increases may not affect the percent of fine roots that were ectomycorrhizal with *Pisolithus tinctorius*. Godbold and Berntson (1997) found, due to elevated CO_2 concentrations, significant changes in the composition of ectomycorrhizal morphotypes on *Betula papyrifera* toward morphotypes with a higher incidence of emanating hyphae and rhizomorphs. Markkola et al. (1996) showed in an CO_2 enrichment (700 ppm) experiment that during a winter acclimation period the fungal biomass in the soil and total fungal biomass both in the roots and in the soil increased, while the ratio of needle biomass:fungal biomass decreased; and N concentration in previous-year needles was lower in the 700-ppm CO_2 environment.

Pollution – Effects of Toxic Metals. Ectomycorrhizal colonization of *Pinus strobus* with *Pisolithus tinctorius* reduced Al toxicity on the seedlings, possibly due to enhanced uptake of nutrients, especially phosphorus (Schier and Macquattie 1995). The ectomycorrhizae *Rhizopogon roseolus* + *Pinus sylvestris* were shown to accumulate Cd and Al in the fungal mantle and to decrease these elements gradually along the Hartig net towards the inside of the root, suggesting a filtering effect (Turnau et al. 1996). Comparing Al-adapted and non-Al-adapted *Suillus bovinus* mycelia, Gerlitz (1996) came to the conclusion that in the Al-adapted fungus a detoxification of freely mobile Al-ions into a stable and insoluble complex is considered to be due to a capture of intracellular Al by mobile polyphosphate of shorter chain length. Lead treatment of *Picea abies* seedlings (in part ectomycorrhizal with *Paxillus involutus, Pisolithus tinctorius, Laccaria laccata*) suggested that these ectomycorrhizal fungi differ in their effect on Pb accumulation in the roots; the binding capacity of the extramatrical mycelium seems to be an important factor; *P. involutus* being the most efficient (Marschner et al. 1996). At a higher ratio than 1 between ^{134}Cs/K, *Picea abies* seedlings ectomycorrhizal with *Hebeloma crustuliniforme* contained a lower concentration of ^{134}Cs than non-inoculated seedlings, and in ectomycorrhizal seedlings the amount of ^{134}Cs was significantly lower in needles and roots. Highest X-ray net counts ere found in the fungal mantle, suggesting an accumulation and slow translocation to other plant parts (Brunner et al. 1996). Cadmium added to cultures of *Rhizopogon roseolus* in the concentrations of 0.3–90 mM reduced superoxiddismutase activity and changed the isoform pattern of the enzyme and a new isoform appeared (Miszalski et al. 1996). *Picea abies* seedlings ectomycorrhizal with *Hebeloma crustuliniforme* reduced the uptake of ^{134}Cs and ^{85}Sr. The degree of ectomycorrhizal colonization was of crucial importance and seemed to be governed by the period during which ectomycorrhizae were allowed to develop and by the ammonium concentration in the nutrient solution (Riesen and Brunner 1996).

Pollution – Damage of Trees. *Fagus sylvatica* trees, both healthy and unhealthy growing in close vicininy, were found to grow in significantly different soil chemistry, with healthy trees generally growing in soils containing higher concentrations of calcium, magnesium and potassium, and lower aluminium/calcium ratios. Contemporarily, healthy trees were found to have significantly higher proportions of ectomycorrhizal roots than their unhealthy neighbours (Power and Ashmore 1996). Causin et al. (1996) demonstrated a significant decrease in the proportion of ectomycorrhizae between healthy and declining *Quercus robur* trees, but the probability of infection and of finding vital ectomycorrhizae was alike. The authors found a variable association of some morphotypes with declining intensity.

Other Toxic Substances. Meharg et al. (1997a) found a mineralization capacity of *Suillus variegatus* and *Paxillus involutus* mycelium for 2,4-dichlorophenol. The mineralization was even greater when ectomycorrhizae on *Pinus sylvestris* were involved. Meharg et al. (1997b) found that four ectomycorrhizal fungi tested were able to biotransform 2,3,6-trinitrotoluene in pure culture.

e) Application of Ectomycorrhizae

Bencivenga et al. (1994) found in a field study that *Populus alba* and *Quercus pubescens* were the best tree species out of seven for mycorrhization with *Tuber magnatum*. Mature oak trees at the margin of a truffle yard were found to influence negatively the number of *Tuber melanosporum* ectomycorrhizae which were previously synthesized on *Quercus pubescens* seedlings (Bencivenga et al. 1995b). Laurence et al. (1996) concluded from their experiments that ectomycorrhizal fungi could be a suitable tool for improving rooting in vitro and survival at acclimation of micropropagated conifer cuttings.

f) Methods

Tagu and Martin (1995) used random sequencing of cDNA clones to generate expressed sequence tags and found them useful to analyze changes in gene expression during ectomycorrhiza formation. Timonen (1995) avoided the problems of autofluorescence of ectomycorrhizae by using time-resolved fluorescence microscopy and Europium labeling. A sampling and counting technique was selected to evaluate objectively the percentage of mycorrhizae in truffle plants with acceptable errors (Bencivenga et al. 1995a). Rootlet formation of somatic embryo-derived *Larix* plantlets was improved by ectomycorrhizal fungi (Piola et al. 1995). An improved HPLC method was developed to determine the chitin content of ectomycorrhizal roots based on analysis of glucosamine content (Ekblad and Näsholm 1996). Piombo et al. (1996) developed a method to measure the amounts of oxalate around ectomycorrhizal roots. A method for synchronization of lateral root tip emergence of seedlings and ectomycorrhiza formation is described by Burgess et al. (1996). Wiemken and Ineichen (1996) developed a method to access the belowground part of a model spruce ecosystem. Hampp et al. (1996) found no effect of a transgenic hybrid aspen expressing T-DNA indoleacetic-biosynthesis genes of *Agrobacterium tumefaciens* in the roots regarding ectomycorrhiza formation with *Amanita muscaria* as compared to wild type of *Populus tremula x P. tremuloides*. Schelkle et al. (1996) successfully used laser scanning confocal microscopy for characterization of ectomycorrhizae and for localization of associated bacteria. Sweeney et al. (1996) selected and cloned a repetitive 319-bp DNA probe specific to *Laccaria amethystina, L. proxima, L. bicolor, L. laccata* (not to *L. tortilis*). Carnero et al. (1997) used ITS-5.8S as a specific probe to estimate fungal or plant rRNA in the symbiotic tissue of *Eucalyptus globulus* + *Pisolithus tinctorius* ectomycorrhizae. It was also possible to determine whether an mRNA was down- or upregulated. Kreuzinger et al. (1996a, b) were able to identify *Lactarius deterrimus* ectomycorrhizae by PCR amplification of the glycerinaldehyde-3-phosphate dehydrogenase encoding genes of fruitbody and ectomycorrhizae. Frey et al. (1997a) used energy-dispersive X-ray microanalysis on ultrathin cryosections of high pressure-frozen ectomycorrhizae. Gandeboeuf et al. (1997) applied sequence-characterized amplified region (SCAR) primers to distinguish several *Tuber* species, including *T. melanosporum* and *T. indicum*. Using esterase isozyme profiles, Timonen et al. (1997) could show that all analyzed ten morphotypes of a microcosm system containing *Pinus sylvestris* and unsterilie, sieved humus could be distinguished.

g) Additional Reviews

Finlay (1995) discusses interactions between soil acidification, plant growth and nutrient uptake in ectomycorrhizal associations of forest trees, focuses on methodological problems, and works out research priorities for future investigations. Two of the major problems are seen in the lack of possibilities to identify the ectomycorrhizae involved in tree nutrition and in the distinction of mycorrhizal biomass and activity from that of the other soil microorganisms. Entry et al. (1996) proposed ectomycorrhizal plants as a viable and cost-effective method to remove redionuclides from soils that have been contaminated by nuclear testing and nuclear reactor accidents. Tagu and Martin (1996) focus on molecular analysis of cell-wall proteins expressed during the early stage of ectomycorrhiza development. Siderophore production of mycorrhizal fungi was reviewed by Haselwandter (1995). Plant-microbe mutualisms dependent community structures are reviewed by Read (1995). Colpaert and van Tichelen (1996b) compile the roles of environmental stress (excess nitrogen, excess CO_2, heavy metals) on ectomycorrhizae.

The long-desired, completely revised second edition of *Mycorrhizal Symbiosis* has now been published (Smith and Read 1997).

2 Ectendomycorrhiza

Wilcoxina mikolae and *W. rehmii* produced in culture the siderophore ferricrocin (Prabhu et al. 1996).

3 Monotropoid Mycorrhiza

Using molecular identifications methods, Cullings et al. (1996) could show that some Monotropaceae are highly specific in their fungal associations, and at least one species, *Pterospora andromedea*, is specialized on a single species group within the genus *Rhizopogon*; phylogenetic analysis of the Monotropaceae showed that specialization had been derived from narrowing of fungal associations within the lineage containing *P. andromedea*. Kasuya et al. (1995) compared the mycorrhizae of *Monotropastrum globosum* and the intermingling ectomycorrhizae of *Fagus crenata,* and concluded due to anatomical similarities that the typical monotropoid and the ectomycorrhizae are formed by the same fungus.

4 Some Highlights of This Report Period

The fossil record of typical *Rhizopogon-* or *Suillus*-like pine ectomycorrhizae from Eocone time suggests that ectomycorrhizae have not changed much during the past 50 million years. The beginning of ectomycorrhizal evolution must date back considerably longer. The new advice for determination of ectomycorrhizae, called DEEMY, should help, particularly when a more updated version is available, to determine ectomycorrhizae on an anatomical basis. An increasing amount of DNA studies will possibly result in future in more identifications and species-specific determinations of ectomycorrhizae. The discovery of a new zygomycetous genus (*Densospora*) which is capable of forming ectomycorrhizae can be a motive to search for more such species of this fungal relationship.

The application of species-specific DNA primers is obviously a very good possibility for population studies on ectomycorrhizal fungi in the field. Some success has already been published not only with respect to persistence of fungal clones but also regarding relations between aboveground and below-ground components.

The studies of hydrophobins and their role during ectomycorrhiza formation are now coming into focus. Considerable more research has to be done to understand the purpose of these compounds, as is true of studies regarding the cytoskeleton.

Studies on the impact of elevated CO_2 on mycorrhizae and carbon allocation have increased during the past period; but still considerably more investigations have to be performed to obtain a more consistent picture. Too diverse are the plant and fungal species and the soil conditions which have to be considered.

References

Abourouh M, Najim L (1995) Les different types d'ectomycorhizes naturelles de *Cedrus atlantica* Manetti au Maroc. Crypt Bot 5:332–340

Agerer R (1995) Index of unidentified ectomycorrhizae. IV. Names and identifications published 1993–1994. Mycorrhiza 5(6):449–450

Agerer R (1996a) Ectomycorrhizae of *Tomentella albomarginata* (Thelephoraceae) on Scots pine. Mycorrhiza 6:1–7

Agerer R (1996b) Characterization of ectomycorrhizae: a historical overview. Descr Ectomyc 1:1–22

Agerer R (1996c) *Ramaria aurea* (Schaeff.: Fr.) Quél. + *Fagus sylvatica*. L. Descr Ectomyc 1:107–112

Agerer R (1996d) *Ramaria largentii* Marr & D. E. Stuntz + *Picea abies* (L.) Karst. Descr Ectomyc 1:113–118

Agerer R (1996e) *Ramaria spinulosa* (Fr.) Quél. + *Fagus sylvatica* L. Descr Ectomy 1:119–124

Agerer R (1996f) *Ramaria subbotrytis* (Coker) Corner + *Quercus robur* L. Descr Ectomyc 1:125–130

Agerer R (1997a) *Entoloma sinuatum* (Bull.: Fr.) Kummer + *Salix* spec. Descr Ectomyc 2:13–18

Agerer R, Molina R (1997) "*Tsugaerhiza luteoannulata*" + *Tsuga heterophylla* (Rafin.) Sarg. Descr Ectomyc 2:79–84

Agerer R, Otto P (1997) *Bankera fuligineo-alba* (J. C. Schmidt: Fr.) Pouzar + *Pinus sylvestris* L. Descr Ectomyc 2:1–6

Agerer R, Rambold G (1996) DEEMY a DELTA-based information system for characterization and DEtermination of EctoMYcorrhizae. Section Mycology, Institute for Systematic Botany, University of München

Agerer R, Trappe JM (1997) "*Tsugaerhiza tripigmentata* + *Tsuga heterophylla* (Rafin.) Sarg. Descr Ectomyc 2:85–89

Agerer R, Klostermeyer D, Steglich W, Franz F, Acker G (1996a) Ectomycorrhizae of *Albatrellus ovinus* (Scutigeraceae) on Norway spruce with some remarks on the systematic position of the family. Mycotaxon 59:289–307

Agerer R, Kraigher H, Javornik B (1996b) Identification of ectomycorrhizae of *Hydnum rufescens* on Norway spruce and the variability of the ITS region of *H. rufescens* and *H. repandum* (Basidiomycetes). Nova Hedwigia 63:183–194

Agerer R, Müller WR, Bahnweg G (1996c) Extomycorhizae of *Rhizopogon subcaerulescens* on *Tsuga heterophylla*. Nova Hedwigia 63(3–4):397–415

Anderson CP, Rygiewicz P (1995) Allocation of carbon in mycorrhizal *Pinus ponderosa* seedlings exposed to ozone. New Phytol 13(4):471–480

Andersson S, Jensen P, Söderström B (1996) Effects of mycorrhizal colonization by *Paxillus involutus* on uptake of Ca and P by *Picea abies* and *Betula pendula* grown in unlimed and limed peat. New Phytol 133(4):695–704

Andersson S, Ek H, Söderström B (1997) Effects of liming on the uptake of organic and inorganic nitrogen by mycorrhizal (*Paxillus involutus*) and non-mycorrhizal *Pinus sylvestris* plants. New Phytol 135(4):763–771

Baar J, Elferink MO (1996) Ectomycorrhizal development on Scots pine (*Pinus sylvestris* L.) seedlings in different soils. Plant Soil 179(2):287–292

Balestrini R, Hahn MG, Bonfante P (1996) Location of cell-wall components in ectomycorrhizae of *Corylus avellana* and *Tuber magnatum*. Protoplasma 191(1–2):55–69

Beenken L, Agerer R (1996) "*Piceirhiza stagonopleres*" + *Picea abies* (L.) Karst. Descr Ectomyc 1:71–76

Beenken L, Agerer R, Bahnweg G (1996a) *Inocybe appendiculata* Kühn. + *Picea abies* (L.) Karst. Descr Ectomyc 1:35–40

Beenken L, Agerer R, Bahnweg G (1996b) *Inocybe fuscomarginata* Kühn. + *Salix* spec./*Populus nigra* L. Descr Ectomyc 1:41–46

Beenken L, Agerer R, Bahnweg G (1996c) *Inocybe obscurobadia* (J. Favre) Grund & D.E. Stuntz + *Picea abies* (L.) Karst. Descr Ectoymc 1:47–52

Beenken L, Agerer R, Bahnweg G (1996d) *Inocybe terrigena* (Fr.) Kuyper + *Pinus sylvestris* L. Descr Ectomyc 1:53–58

Beguiristain T, Lapeyrie F (1997) Host plant stimulates hypaphorine accumulation in *Pisolithus tinctorius* hyphae during ectomymcorrhizal infection while excreted fungal hypaphoric controls root hair development. New Phytol 136(3):525–532

Bencivenga M, Donnini D, Cenci CA, Hasko A (1994) The development of symbiont plants and the analysis of mycorrhizas in a cultivated truffle bed of *Tuber magnatum* Pico. Ann Fac Agric Univ Perugia 45:321–338

Bencivenga M, Donnini D, Tanfulli M, Guiducci M (1995a) Sampling technique of the roots and the root tips to evaluate truffle plants. Micol Ital 24(2):35–47

Bencivenga M, Di Massimo G, Donnini D, Tanfullu M (1995b) Analysis of mycorrhization and development of truffles set out in 1982 at varying distances from mature oak trees. Micol Ital 24(3):110–116.

Bending GD, Read DJ (1996a) Nitrogen mobilization from protein-polyphenol complex by ericoid and ectomycorrhizal fungi. Soil Biol Biochem 28(12):1603–1612

Bending GD, Read DJ (1996b) Effects of the soluble polyphenol tanninic acid on the activites of ericoid and ectomycorrhizal fungi. Soil Biol Biochem 28(12):1595–1602

Bhat M, Jeyarajan R, Ramarj B (1997) Biocontrol of damping off of *Eucalyptus tereticornis* Sm. using ectomycorrhizae. Indian For 123(4):307–312

Boerner REJ, DeMars BG, Leicht PN (1996) Spatial patterns of mycorrhizal infectiveness of soils along a successional chronosequence. Mycorrhiza 6(2):79–90

Brundrett MC, Abbott LK (1995) Mycorrhizal fungus propagules in the jarrah forest. II. Spatial variability in inoculum levels. New Phytol 131(4):461–469

Brunner I, Frey B, Riesen TK (1996) Influence of ectomycorrhization and cesium/potassium ratio on uptake and localization of cesium in Norway spruce seedlings. Tree Physiol 16(8):705–711

Burgess T, Dell B (1996) Changes in protein biosynthesis during the differentiation of *Pisolithus-Eucalyptus grandis* ectomycorrhiza. Can J Bot 74(4):553–560

Burgess T, Dell B, Malajczuk N (1996) In vitro synthesis of *Pisolithus-Eucalyptus* ectomycorrhizae: synchronization of lateral root tip emergence and ectomycorrhizal development. Mycorrhiza 6(3):189–196

Cairney JWG, Burke RM (1996) Physiological heterogeneity within fungal mycelia: an important concept for a functional understanding of the ectomycorrhizal symbiosis. New Phytol 134:685–695

Cairney JWG, Chambers SM (1997) Interactions between *Pisolithus tinctorius* and its hosts: a review of current knowledge. Mycorrhiza 7(3):117–131

Carlson CA (1994) The influence of fertilization on ectomycorrhizal colonization of *Pinus patula* roots. S Afr Bosboutydskr 171:1–6

Carnero D, Carnero E, Tagu D, Martin F (1997) Ribosomal DNA internal transcribed spacers to estimate the proportion of *Pisolithus tinctorius* and *Eucalyptus globulus* RNAs in ectomycorrhiza. Appl Environ Microbiol 63(3):840–843

Causin R, Montecchio L, Mutto Accordi S (1996) Probability of ectomycorrhizal infection in a declining stand of common oak. Ann Sci For 53:743–752

Chase MW, 41 others (1993) Phylogenetics of seed plants: an analysis of nucleotide sequences from the plastid gene rbcL. Ann Mo Bot Gard 80:528–580 (cited in Fitter and Moyersoen 1996)

Colpaert JV, van Laere A (1996) A comparison of the extracellular enzyme activities of two ectomycorrhizal and leaf-saprotrophic basidiomycetes colonizing beach leaf litter. New Phytol 134(1):133–141

Colpaert JV, van Tichelen KK (1996a) Decomposition, nitrogen and phosphorus mineralization from beech leaf litter colonized by ectomycorrhizal or litter-decomposing basidiomycetes. New Phytol 134(1):123–132

Colpaert JV, van Tichelen KK (1996b) Mycorrhizas and environmental stress. In: Frankland JC, Magan N, Gadd GM (eds) Fungi and environmental stress. Cambridge, Univ Press, Cambridge, pp 109–128

Colpaert JV, Van Laere A, Van Assche JA (1996) Carbon and nitrogen allocation in ectomycorrhizal and non-mycorrhizal *Pinus sylvestris* L. seedlings. Tree Physiol 16(9):787–793

Colpaert JV, van Laere A, van Tichelen KK, van Assche A (1997) The use of inositol hexaphosphate as a phosphorus source by mycorrhizal and non-mycorrhizal Scots pine (*Pinus sylvestris*). Functional Ecol 11(4):407–415

Comandini O, Pacioni G (1997) Mycorrhizae of Asian black truffles, *Tuber himalayense* and *T. indicum*. Mycotaxon 63:77–86

Cripps CL (1997a) *Inocybe lacera* (Fr.: Fr.) Kumm. + *Populus tremuloides* Michx. Descr Ectomyc 2:19–23

Cripps CL (1997b) *Tricholoma scalpturatum* (Fr.) Quél. + *Populus tremuloides* Michx. Descr Ectomyc 2:73–78

Cullings KW, Szaro TM, Bruns TD (1996) Evolution of extreme specialization within a lineage of ectomycorrhizal epiparasites. Nature (Lond) 379(6560):63–66

Cumming JR (1996) Phosphate-limitation physiology in ectomycorrhizal pitch pine (*Pinus rigida*) seedlings. Tree Physiol 16(11–12):977–983

Davies FT Jr, Svenson SE, Cole JC, Phavaphutanon L, Duray SA, Olalde-Portugal V, Meier CE, Bo SH (1996) Non-nutritional stress acclimation of mycorrhizal woody plants exposed to drought. Tree Physiol 16(11–12):985–993

De la Bastide PY, Kropp BR, Piche Y (1995) Vegetative interactions among mycelia of *Laccaria bicolor* in pure culture and symbiosis with *Pinus banksiana*. Can J Bot 73(11):1768–1779

Di Battista C, Selosse M-A, Bouchard D, Stenström E, le Tacon F (1996) Variation in symbiotic efficiency, phenotypic characters and ploidy level among different isolates of the ectomycorrhizal basidiomycete *Laccaria bicolor* strain S 238. Mycol Res 100(11):1315–1324

Diaz EC, Martin F, Tagu D (1996) Eucalypt α-tubulin: cDNA cloning and increased level of transcripts in ectomycorrhizal root system. Plant Mol Biol 31(4):905–910

Diaz G, Barrantes O, Honrubia M, Garcia C (1996) Effect of ozone and sulfur dioxide on mycorrhizae of *Pinus halepensis* Miller. Ann Sci For (Paris) 53(4):849–856

Eberhart J, Luoma D (1996) *Truncocolumella citrina* Zeller + *Pseudotsuga menziesii* (Mirb.) Franco. In: Goodman DM, Durall DM, Trofymow JA, Berch SM (eds) Concise descriptions of some North American ectomycorrhizae. Canada-BC Forest Resource Development Agreement, Canadian Forest Service, Victoria, BC, pp CDE9.1–CDE9.4

Ek H (1997) The influence of nitrogen fertilization on the carbon economy of *Paxillus involutus* in ectomycorrhizal association with *Betula pendula*. New Phytol 135:133–142

Ek H, Andersson S, Söderström B (1996) Carbon and nitrogen flow in silver birch and Norway spruce connected by a common mycorrhizal mycelium. Mycorrhiza 6(6):465–467

Ekblad A, Huss-Danell K (1995) Nitrogen fixation by *Alnus incana* and nitrogen transfer from *A. incana* to *Pinus sylvestris* influenced by macronutrients and ectomycorrhiza. New Phytol 131(4):453–459

Ekblad A, Näsholm T (1996) Determination of chitin in fungi and mycorrhizal roots by an improved HPLC analysis of glucosamine. Plant Soil 178(1):29–35

Ekblad A, Wallander H, Carlsson R, Huss-Danell K (1995) Fungal biomass in roots and extramatrical mycelium in relation to macronutrients and plant biomass of ectomycorrhizal *Pinus sylvestris* and *Alnus incana*. New Phytol 131(4):443–451

Eltrop L, Marschner H (1996a) Growth and mineral nutrition of non-mycorrhizal and mycorrhizal Norway spruce (*Picea abies*) seedlings grown in semi-hydroponic sand culture. II. Carbon partitioning in plants supplied with ammonium or nitrate. New Phytol 133(3):479–486

Eltrop L, Marschner H (1996b) Growth and mineral nutrition of non-mycorrhizal and mycorrhizal Norway spruce (*Picea abies*) seedlings grown in semi-hydroponic sand cultures. I. Growth and mineral nutrient uptake in plants supplied with different forms of nitrogen. New Phytol 133(3):469–478

Entry JA, Vance NH, Hamilton MA, Zabowski D, Watrud LS, Adriano DC (1996) Phytoremediation of soil contaminated with low concentrations of radionuclides. Water Air Soil Pollut 88(1–2):167–176

Erland S (1995) Abundance of *Tylospora fibrillosa* ectomycorrhizas in a south Swedish spruce forest measured by RFLP analysis of the PCR-amplified rDNA ITS region. Mycol Res 99(12):1425–1428

Fernando AA, Currah RS (1996) A comparative study of the effects of the root endophytes *Leptodontidium orchidicola* and *Phialocephala fortinii* (Fungi imperfecti) on the growth of some subalpine plants in culture. Can J Bot 74(7):1071–1078

Finlay RD (1995) Interactions between soil acidification, plant growth and nutrient uptake in ectomycorrhizal associations in forest trees. Ecol Bull 44:197–214

Fischer CR, Agerer R (1996) "*Quercirhiza argenteobrunneola*" + *Quercus ilex* L. Descr Ectomyc 1:101–105

Fitter AH, Moyersoen B (1996) Evolutionary trends in root-microbe symbioses. Philos Trans R Soc Lond B 351:1367–1375

Franz F, Acker G (1995) Ultrastructural location of proteins in rhizomorphs and mantles of natural spruce ectomycorrhizae. Nova Hedwigia 61(3–4):367–376

Frey B, Brunner I, Walther P, Scheidegger C, Zierold K (1997a) Element localization in ultrathin cryosections of high-pressure frozen ectomycorrhizal spruce roots. Plant Cell Environ 20(7):929–937

Frey P, Frey-Klett P, Garbaye J, Berge O, Heulin T (1997b) Metabolic and genotypic fingerprinting of fluorescent pseudomonads associated with the Douglas-fir *Laccaria bicolor* mycorrhizoshere. Appl Environ Microbiol 63(5):1852–1860

Frey-Klett P, Pierrat JC, Garbaye J (1997) Location and survical of mycorrhiza helper *Pseudomonas fluorescens* during establishment of mycorrhizal symbiosis between *Laccaria bicolor* and Douglas fir. App Environ Microbiol 63(1):139–144

Gagnon J, Langlois CG, Bouchard D, Le Tacon F (1995) Growth and ectomycorrhizal formation of container-grown Douglas-fir seedlings inoculated with *Laccaria bicolor* under four levels of nitrogen fertilization. Can J Forest Res 25(12):1953–1961

Gardes M, Bruns TD (1996) Community structure of ectomycorrhizal fungi in a *Pinus muricata* forest: above- and below-ground views. Can J Bot 74(10):1572–1583

Gandeboeuf D, Depré C, Roeckel-Drevét P, Nicolas P, Chevalier G (1997) Typing *Tuber* ectomycorrhizae by polymerase chain amplification of the internal transcribed spacer of rDNA and the sequence characterized amplified region markers. Can J Microbiol 43:723–728

Gay G, Sotta B, Tranvan H, Gea L, Vian B (1995) Fungal auxin is involved in ectomycorrhiza formation: genetical, biochemical and ultrastructural studies with IAA overproducer mutants of *Hebeloma cylindrosporum*. In: Sandermann H Jr, Bonnet-Masimert M (eds) Contribution to forest tree physiology. INRA, Paris, pp 215–231

Gehring CA, Cobb NS, Whitham TG (1997) Three-way interactions among ectomycorrhizal mutualists, scale insects, and resistant and susceptible pinyon pines. Am Nat 149(5):824–841

Gerlitz TGM (1996) Effects of aluminium on polyphosphate mobilization of the ectomycorrhizal fungus *Suillus bovinus*. Plant Soil 178(1):133–140

Godbold DL, Berntson GM (1997) Elevated atmospheric CO_2 concentration changes ectomycorrhizal morphotype assemblages on *Betula papyrifera*. Tree Physiol 17:347–350

Golldack J, Münzenberger B, Agerer R, Hüttl RF (1996a) "*Pinirhiza granulosa*" + *Pinus sylvestris* L. Descr Ectomyc 1:77–81

Golldack J, Münzenberger B, Agerer R, Hüttl RF (1996b) "*Pinirhiza stellannulata*" + *Pinus sylvestris* L. Descr Ectomyc 1:89–93

Golldack J, Münzenberger B, Hüttl RF (1997a) "*Pinirhiza lactogelatinosa*" + *Pinus sylvestris* L. Descr Ectomyc 2:49–53

Golldack J, Münzenberger B, Hüttl RF (1997b) "*Pinirhiza lactariosimilis*" + *Pinus sylvestris* L. Descr Ectomyc 2:43–48

Goodman DM, Trofymow JA (1996) *Piloderma fallax* (Libert) Stalpers + *Pseudotsuga menzesii* (Mirb.) Franco. In: Goodman DM, Durall DM, Trofymow JA, Berch SM (eds) Concise descriptions of some North American ectomycorrhizae. Canada-BC Forest Resource Development Agreement, Canadian Forest Service, Victoria, BC, pp CDE1.1–CDE1.4

Graf F, Brunner I (1996) Natural and synthesized ectomycorrhizas of the alpine dwarf willow *Salix herbacea*. Mycorrhiza 6:227–235

Granetti B (1995) Caratteristiche morfologiche, biometriche e strutturali delle micorrize di *Tuber* di interesse economico. Mic Ital 24(2):101–117

Griffiths RP, Bradshaw GA, Marks B, Lienkkaemper GW (1996) Spatial distribution of ectomycorrhizal mats in coniferous forests of the Pacific Northwest. Plant Soil 180(1):147–158

Gryta H, Debaud JC, Effosse A, Gay G, Marmeisse R (1997) Fine-scale structure of populations of the ectomycorrhizal fungus *Hebeloma cylindrosporum* in coastal sand dune forest ecosystems. Mol Ecol 6(4):353–364

Hampp R, Schaeffer C, Wallenda T. Stulten C, Johann P, Einig W (1995) Changes in carbon partitioning or allocation due to ectomycorrhiza formation: biochemichal evidence. Can J Bot 73(Suppl 1):S548–S556

Hampp R, Ecke M, Schaeffer C, Wallenda T, Wingler A, Kottke I, Sundberg B (1996) Axenic mycorrhization of wild type and transgenic hybrid aspen expressing T-DNA indoleacetic acid-biosynthesis genes. Trees (Berl) 11(1):59–64

Haselwandter K (1995) Mycorrhizal fungi: sidephore production. Crit Rev Biotechnol 15(3/4):287–291

Hauer RJ, Dawson JO (1996) Growth and iron sequestering of pin oak (*Quercus palustris*) seedlings inoculated with soil containing ectomycorrhizal fungi. J Arboricult 22(3):122–130

Helm DJ, Allen EB, Trappe JM (1996) Mycorrhizal chronosequence near Exit Glacier, Alaska. Can J Bot 74(9):1496–1506

Hodge A (1996) Impact of elevated CO_2 on mycorrhizal associations and implications for plant growth. Biol Fertil Soils 23:388–398

Hodge A, Alexander IJ, Gooday GW (1995) Chitinolytic activities of *Eucalyptus pilularis* and *Pinus sylvestris* root systems challenged with mycorrhizal and pathogenic fungi. New Phytol 131(2):255–261

Hodge A, Alexander IJ, Gooday GW, Williamson FA (1996) Localization of chitinolytic activities in *Fagus sylvatica* mycorrhizas. Mycorrhiza 6(3):181–187

Högberg P, Hogbom L, Schinkel H, Högberg M, Johannisson C, Wallmarki H (1996) ^{15}N abundance of surface soils, roots and mycorrhizae in profiles of European forest soils. Oecologia (Berl) 108(2):207–214

Holopainen T, Heinonen-Tanski H, Halonen A (1996) Injuries to Scots pine mycorrhizas and chemical gradients in forest soil in the environment of a pulp mill in central Finland. Water Air Soil Pollut 87(1–4):111–130

Honrubia M, Diaz G (1996) Effect of simulated acid rain on mycorrhiza of Aleppo pine (*Pinus halepensis* Miller) in calcareous soil. Ann Sci For (Paris) 53(5):947–954

Hwang S-H, Chakravarty P, Chang K-F (1995) The effect of two ectomycorrhizal fungi, *Paxillus involutus* and *Suillus tomentosus,* and of *Bacillus subtilis* on *Fusarium* damping-off in Jack pine seedlings. Phytoprotection 76(2):57–66

Ingleby K, Mason PA (1996) Ectomycorrhizas of *Thelephora terrestris* formed with *Eucalyptus globulus.* Mycologia 88(4):548–553

Jakucs E, Agerer R, Bratek Z (1997) "*Quercirhiza fibulocystidiata*" + *Quercus* spec. Descr Ectomyc 2:67–71

Kåren O, Nylund J-E (1996) Effects of N-free fertilization on ectomycorrhiza community structure in Norway spruce stands in southern Sweden. Plant Soil 181:295–305

Kasuya MCM, Masaka K, Igarashi T (1995) Mycorrhizae of *Monotropastrum globosum* growing in a *Fagus crenata* forest. Mycoscience 36(4):461–464

Kieliszewska-Rokicka B, Rudawska M, Leski T (1995) Effects of acid rain and aluminium on ectomycorrhiza symbiosis: alteration of IAA-synthesizing activity in ectomycorrhizal fungi. Bulgar J Plant Physiol 21(2–3):111–119

Kottke I (1997) Fungal adhesion pad formation and penetration of root cuticle in early stage mycorrhizas of *Picea abies* and *Laccaria amethystea.* Protoplasma 196:55–64

Kottke I, Pargney JC, Qian XM, Le Disquet I (1995) Passage and deposition of solutes in the hyphal sheath of ectomycorrhizas – the soil-root interface. In: Sandermann J Jr, Bonnet-Masimbert M (eds) Contribution to tree physiology. INRA Paris, pp 255–271

Kreuzinger N, Podeu R, Göbl F, Gruber F, Kubicek CP (1996a, b) Identification of some ectomycorrhizal basidiomycetes by PCR-amplification of their gpd (glycerinaldehyde-3-phosphate dehydrogenase encoding) genes. Phytol (Horn) 36(4):109–118; Applied Environ Microbiol 62(9):3432–3438

Kropp BR (1997) Inheritance of the ability for mycorrhizal colonization of *Pinus strobus* by *Laccaria bicolor.* Mycologia 89(4):578–585

Laurence N, Bartschi H, Debaud J-C, Gay G (1996) Rooting and acclimatization of micropropagated cuttings of *Pinus pinaster* and *Pinus sylvestris* are enhanced by the ectomycorrhizal fungus *Hebeloma cylindrosporum.* Physiol Plant 98(4):759–766

Lepage BA, Currah RS, Stockey RA, Rothwell GW (1997) Fossil ectomycorrhizae from the middle Eocene. Am J Bot 84(3):410–412

Leprince F, Quiquampoix H (1996) Extracellular enzyme activity in soil: effect of pH and ionic strength on the interaction with montmorillonite of two acid phosphatases secreted by the ectomycorrhizal fungus *Hebeloma cylindrosporum.* Eur J Soil Sci 47(4):511–522

Lewis JD, Strain BR (1996) The role of mycorrhizas in response of *Pinus taeda* seedlings to elevated CO_2. New Phytol 133:431–443

Lewis JD, Thomas RB, Strain BR (1994) Effect of elevated CO_2 on mycorrhizal colonization of loblolly pine (*Pinus taeda* L.) seedlings. Plant Soil 165:81–88

Lorillou S, Botton B, Martin F (1996) Nitrogen source regulates the biosynthesis of NADP-glutamate dehydrogenase in the ectomycorrhizal basidiomycete *Laccaria bicolor.* New Phytol 132(2):289–296

Maia LC, Yano AM, Kimbrough JW (1996) Species of Ascomycota forming ectomycorrhizae. Mycotaxon 57:371–390

Majdi H, Nylund J-E (1996) Does liquid fertilization affect fine root dynamics and lifespan of mycorrhizal short roots? Plants Soil 185(2):305–309

Mamoun M, Olivier J-M (1996) Receptivity of cloned hazels to artificial ectomycorrhizal infection by *Tuber melanosporum* and symbiotic competitors. Mycorrhiza 6(1):15–19

Mamoun M, Olivier JM (1997) Mycorrhizal inoculation of cloned hazels by *Tuber melanosporum*: effect of soil disinfectation and co-culture with *Festuca ovina*. Plant Soil 188(2):221–226

Marchetti M, Varese GC (1996) Influence of *Verticillium bulbillosum* on in vitro formation of mycorrhizae by *Laccaria laccata* and *Hebeloma crustuliniforme* with *Picea abies*. Allionia 34:45–54

Markkola AM (1996) Resource allocation in ectomycorrhizal symbiosis in Scots pine affected by environmental changes. Acta Univ Ouluensis Ser A Sci Rer Nat 278:1–43

Markkola AM, Ohtonen A, Ahonen-Jonnarth U, Ohtonen R (1996) Scots pine responses to CO_2 enrichment. I. Ectomycorrhizal fungi and soil fauna. Environ Pollut 94(3):309–316

Marschner P, Godbold DL (1995) Mycorrhizal infection and ageing affect element localization in short roots of Norway spruce (*Picea abies* (L.) Karst.). Mycorrhiza 5(6):417–422

Marschner P, Godbold DL, Jentschke G (1996) Dynamics of lead accumulation in mycorrhizal and non-mycorrhizal Norway spruce (*Picea abies* (L.) Karst). Plant Soil 178(2):239–245

Martin F, Laurent P, de Carvalho D, Burgess T, Murphy P, Nehls U, Tagu D (1995) Fungal gene expression during ectomycorrhiza formation. Can J Bot 73(Suppl 1):S541–S547

McElhinney C, Mitchell DT (1995) Influence of mycorrhizal fungi on the response of Sitka spruce and Japanese larch to forms of phosphorus. Mycorrhiza 5(6):409–415

McGee PA (1996) The Australian zygomycetous mycorrhizal fungi: the genus *Densospora* gen. nov. Aust Syst Bot 9:329–336

Meharg AA, Cairney WG, Maguire N (1997a) Mineralization of 2,4-dichlorophenol by ectomycorrhizal fungi in axenic culture and in symbiosis with pine. Chemosphere 34(12):2495–2504

Meharg AA, Dennis GR, Cairney JWG (1997b) Biotransformation of 2,4,6-trinitrotoluene (TNT) by ectomycorrhizal basidiomycetes. Chemosphere 35(3):513–521

Michelsen A, Schmidt IK, Jonasson S, Quarmby C, Sleep D (1996) Leaf ^{15}N abundance of subarctic plants provides field evidence that ericoid, ectomycorrhizal, and non- and arbuscular mycorrhizal species access different sources of soil nitrogen. Oecologia (Berl) 105(1):53–63

Miszalski Z, Botton B, Turnau K (1996) New SOD isoform in *Rhizopogon roseolus* (Corda in Sturm) in the presence of cadmium. Acta Physiol Plant 18(2):129–134

Mleczko P (1996) "*Pinirhiza lutea*" + *Pinus sylvestris* L. Descr Ectomyc 1:83–88

Mleczko P (1997a) *Paxillus involutus* (Batsch) Fr. + *Pinus sylvestris* L. Descr Ectomyc 2:25–30

Mleczko P (1997b) "*Populirhiza pustulosa*" + *Populus tremula* L. Descr Ectomyc 2:61–66

Montecchio L, Agerer R (1997) "*Piceirhiza cornuta*" + *Picea abies* (L.) Karst. Descr Ectomyc 2:31–35

Moyersoen B (1996a) "*Tetraberliniaerhiza bicolor*" + *Tetraberlinia bifoliolata* (Harms) Haumann. Descr Ectomyc 1:137–141

Moyersoen B (1996b) "*Tetraberliniaerhiza cerviformis*" + *Tetraberlinia bifoliata* (Harms) Haumann. Descr Ectomyc 1:143–147

Moyersoen B (1996c) "*Tetraberliniaerhiza hetercystidiae*" + *Tetraberlinia bifoliolata* (Harms) Haumann. Descr Ectomyc 1:149–153

Müller WR, Agerer R (1996a) *Hyterangium crassirhachis* Zeller & Dodge + *Pseudotsuga menziesii* (Mirb.) Franco. Descr Ectomyc 1:29–34

Müller WR, Agerer R (1996b) "*Pseudotsugaerhiza baculifera*" + *Pseudotsuga menziesii* (Mirb.) Franco. Descr Ectomyc 1:95–100

Müller WR, Rauscher T, Agerer R, Chevalier G (1996a) *Tuber aestivum* Vitt. + *Corylus avellana* L. Descr Ectomyc 1:167–172

Müller WR, Rauscher T, Agerer R, Chevalier G (1996b) *Tuber unicatum* Chat.. + *Corylus avellana* L. Descr Ectomyc 1:173–178

Münzenberger B, Otter T, Wüstrich D, Polle A (1997) Peroxidase and laccase activities in mycorrhizal and non-mycorrhizal fine roots of Norway spruce (*Picea abies*) and larch (*Larix decidua*). Can J Bot 75(6):932–938

Niini SS, Tarkka MT, Raudaskoski M (1996) Tubulin and actin protein patterns in Scots pine (*Pinus sylvestris*) roots and developing ectomycorrhiza with *Suillus bovinus*. Physiol Plant 96(2):186–192

Olsson PA, Chalot M, Baath E, Finlay RD, Söderström B (1996) Ectomycorrhizal mycelia reduce bacterial activity in a sandy soil. FEMS Microbiol Ecol 21(2):77–86

Palfner G (1996) *Thaxterogaster albocanus* Horak & Moser + *Nothofagus pumilio* (Poepp. et Endl.) Kramer. Descr Ectomyc 1:155–160

Palfner G (1997) *Descolea antarctica* Singer + *Nothofagus alpina* (Poepp. et Endl.) Oerst. Descr Ectomyc 2:7–12

Palfner G, Agerer R (1996a) "*Quercirhiza squamosa*": an unidentified ectomycorrhiza on *Quercus robur*. Sendtnera 3:137–145

Palfner G, Agerer R (1996b) Ectomycorrhizae of *Lactarius chrysorrheus* and *Lactarius serifluus* on *Quercus robur*. Sendtnera 3:119–136

Palfner G, Godoy R (1996a) "*Nothofagirhiza vinicolor*" + *Nothofagus pumilio* (Poepp. et Endl.) Kramer. Descr Ectomyc 1:65–70

Palfner G, Godoy R (1996b) *Russula fuegiana* Singer + *Nothofagus pumilio* (Poepp. et Endl.) Kramer. Descr Ectomyc 1:131–136

Pargney JC, Prevost A (1996) Comparison of natural ectomycorrhizae between beech (*Fagus sylvatica*) and two fungi (*Lactarius blennius* var. *viridis* and *Lactarius subdulcis*). II. Cytochemical characterization of interfaces. Ann Sci For 53:991–1003

Paris F, Bonnaud P, Ranger J, Lapeyrie F (1995) In vitro weathering of phlogopite by ectomycorrhizal fungi: I. Effect of K^+ and Mg^{2+} deficiency on phyllosilicate evolution. Plant Soil 177(2):191–201

Paris F, Botton B, Lapeyrie F (1996) In vitro weathering of phlogopite by ectomycorrhizal fungi. II. Effect of K^+ and Mg^{2+} deficiency and N sources on accumulation of oxalate and H^+. Plant Soil 179(4):141–150

Parlade J, Alvarez IF, Pera J (1996) Ability of native ectomycorrhizal fungi from northern Spain to colonize Douglas-fir and other introduced conifers. Mycorrhiza 6(1):51–55

Pastor J, Dewey B, Christian DP (1996) Carbon and nutrient mineralization and fungal spore composition of fecal pellets from voles in Minnesota. Ecography 19(1):52–61

Pedersen CT, Sylvia DM (1997) Limitations to using benomyl in evaluating mycorrhizal functioning. Biol Fertil Soils 25(2):163–168

Pillukat A (1996) *Lactarius salmonicolor* R. Heim & Leclair + *Abies alba* Mill. Descr Ectomyc 1:59–64

Piola F, Rohr R, von Aderkas P (1995) Controlled mycorrhizal initiation as a means to improve root development in somatic embryo plantlets of hybrid larch (*Larix x eurolepis*). Physiol Plant 95(4):575–580

Piombo G, Babre D, Fallavier P, Cazevieille P, Arvieu JC, Callot G (1996) Oxalate extraction and determination by ionic chromatography in calcareous soils and in mycorrhized roots environment. Commun Soil Sci Plant Anal 27(5–8):1663–1667

Power SA, Ashmore MR (1996) Nutrient relations and root mycorrhizal status of healthy and declining beech (*Fagus sylvatica* L.) in southern Britain. Water Air Soil Pollut 86(1–4):317–333

Prabhu V, Biolchini PF, Boyer GL (1996) Detection and identification of ferricrocin produced by ectendomycorrhizal fungi in the genus *Wilcoxina*. Biometals 9(3):229–234

Prevost A, Pargney JC (1997) Comparison of natural ectomycorrhizae between beech (*Fagus sylvatica*) and two fungi (*Lactarius blennius* var. *viridis* and *Lactarius subdulcis*). III. Rhizomorphs. Ann Sci For (Paris) 54(1):117–124

Pritsch K, Munch JC, Buscot F (1997) Morphological and anatomical characterization of black alder *Alnus glutinosa* (L.) Gaertn. ectomycorrhizas. Mycorrhiza 7(4):201–216

Raidl S, Müller WR (1996) *Tomentella ferruginea* + *Fagus sylvatica* L. Descr Ectomyc 1:161–166

Rambold G, Agerer R (1997) DEEMY – the concept of a characterization and determination system for ectomycorrhizae. Mycorrhiza 7(2):113–116

Rao CS, Sharma GD, Shukla AK (1997) Distribution of ectomycorrhizal fungi in pure stands of different age groups of *Pinus kesiya*. Can J Microbiol 43(1):85–91

Rauscher T, Agerer R, Chevalier G (1995) Ektomykorrhizen von *Tuber melanosporum*, *T. mesentericum* und *T. rufum* an *Corylus avellana*. Nova Hedwigia 61:281–322

Rauscher T, Müller WR, Agerer R, Chevalier G (1996) *Tuber borchii* Vitt. + *Corylus avellana* L. Descr Ectomyc 1:173–178

Read DJ (1995) Plant-microbe mutualisms and community structure. In: Schulze E-D, Mooney HA (eds) Biodiversity and ecosystem function. Springer, Berlin Heidelberg New York, pp 181–209

Reddy M, Sudhakara M, Natarajan K (1997) Coinoculation efficacy of ectomycorrhizal fungi on *Pinus patula* seedlings in a nursery. Mycorrhiza 7(3):133–138

Regvar M, Gogala N (1996) Changes in root growth patterns of (*Picea abies*) spruce roots by inoculation with an ectomycorrhizal fungus *Pisolithus tinctorius* and jasmonic acid treatment. Trees (Berl) 10(6):410–414

Riesen TK, Brunner I (1996) Effect of ectomycorrhizae and ammonium on ^{134}Cs and ^{85}Sr uptake into *Picea abies* seedlings. Environ Pollut 93(1):1–8

Rossow LJ, Bryant JP, Kielland K (1997) Effects of aboveground browsing by mammals on mycorrhizal infection in an early successional taiga ecosystem. Oecologia (Berl) 110(1):94–98

Ruess L, Dighton J (1996) Cultural studies on soil nematodes and their fungal hosts. Nematologia 42(3):330–346

Salzer P, Hebe G, Reith A, Zitterell-Haid B, Stransky H, Gaschler K, Hager A (1996) Rapid reactions of spruce cells to elicitors released from the ectomycorrhizal fungus *Hebeloma crustuliniforme*, and inactivation of these elicitors by extracellular spruce cell enzymes. Planta 198(1):118–126

Salzer P, Hübner B, Sirrenberg A, Hager A (1997) Differential effect of purified spruce chitinases and β-1,3-glucanases on the activity of elicitors from ectomycorrhizal fungi. Plant Physiol 114(3):957–968

Schaeffer C, Johann P, Nehls U, Hampp R (1996) Evidence for an up-regulation of the host and a down-regulation of the fungal phosphofructokinase activity in ectomycorrhizas of Norway spruce and fly agaric. New Phytol 124(4):697–702

Schelkle M, Ursic M, Farquhar M, Peterson RL (1996) The use of laser scanning confocal microscopy to characterize mycorrhizas of *Pinus strobus* L. and to localize associated bacteria. Mycorrhiza 6(5):431–440

Schier GA, Mcquattie CJ (1995) Effect of aluminium on the growth, anatomy, and nutrient content of ectomycorrhizal and nonmycorrhizal eastern white pine seedlings. Can J For Res 25(8):1252–1262

See LS, Alexander IJ (1996) The dynamics of ectomycorrhizal infection of *Shorea leprosula* seedlings in Malaysian rain forests. New Phytol 132(2):297–305

Seegmüller S, Schulte M, Herschbach C, Rennenberg H (1996) Interactive effects of mycorrhization and elevated atmospheric CO_2 on sulphur nutrition of young pedunculate oak (*Quercus robur* L.) trees. Plant Cell Environ 19(4):418–426

Setälä H, Rissanen J, Markkola AM (1997) Conditional outcomes in the relationship between pine and ectomycorrhizal fungi in relation to biotic and abiotic environment. Oikos 80:112–122

Simard SW, Molina R, Smith JE, Perry DA, Jones MD (1997) Shared compatibility of ectomycorrhizae on *Pseudotsuga menziesii* and *Betula papyrifera* seedlings grown in mixture in soils from southern British Colombia. Can J For Res 27:331–342

Shishido M, Petersen DJ, Massicotte HB, Chanway CP (1996) Pine and spruce seedling growth and mycorrhizal infection after inoculation with plant growth promoting *Pseudomonas* strains. FEMS Microbiol Ecol 21(2):109–119

Smith SE, Read DJ (1997) Mycorrhizal symbiosis, 2nd edn. Academic Press, San Diego

Sweeney M, Harmey MA, Mitchell DT (1996) Detection and identification of *Laccaria* species using a repeated DNA sequence from *Laccaria proxima*. Mycol Res 100(12):1515–1521

Sylvia DM, Jarstfer AG (1997) Distribution of mycorrhiza on competing pines and weeds in a southern pine plantation. Soil Sci Am J 61(1):139–144

Tagu D, Martin F (1995) Expressed sequence tags of randomly selected cDNA clones from *Eucalyptus globulus-Pisolithus tinctorius* ectomycorrhiza. Mol Plant Microbe Interact 8(5):781–783

Tagu D, Martin F (1996) Molecular analysis of cell wall proteins expressed during the early stage of ectomycorrhiza development. New Phytol 133(1):73–85

Tagu D, Nasse B, Martin F (1996) Cloning and characterization of hydrophobins-encoding cDNAs from the ectomycorrhiza basidiomycete *Pisolithus tinctorius*. Gene 168(1):93–97

Taylor DL, Bruns TD (1997) Independent, specialized invasions of ectomycorrhizal mutualism by two nonphotosynthetic orchids. Proc Natl Acad Sci USA 94(9):4510–4515

Thomson BD, Grove TS, Malajczuk N, Hardy GES (1996a) The effect of soil pH on the ability of ectomycorrhizal fungi to increase the growth of *Eucalyptus globulus* Labill. Plant Soil 178:209–214

Thomson BD, Hardy GES, Malajczuk N, Grove TS (1996b) The survival and development of inoculant ectomycorrhizal fungi on roots of outplanted *Eucalyptus globulus* Labill. Plant Soil 178(2):247–253

Timonen S (1995) Avoiding autofluorescence problems: time-resolved fluorescence microscopy with plant and fungal cells in ectomycorrhiza. Mycorrhiza 5(6):455–458

Timonen S, Söderström B, Raudaskoski M (1996) Dynamics of cytoskeletal proteins in developing pine ectomycorrhiza. Mycorrhiza 6(5):423–429

Timonen S, Tammi H, Sen R (1997) Characterization of the host genotype and fungal diversity in Scots pine ectomycorrhiza from natural humus microcosms using isozyme and PCR-RFLP analyses. New Phytol 135:313–323

Trillini B, Granetti B (1995) Coumarin-like compounds in mycorrhizal infection of *Quercus pubescens* Willd. with *Tuber magnatum* Pico and *T. borchii* Vitt. Micol Ital 24(2):179–184

Turnau K, Kottke I, Dexheimer J (1996) Toxic element filtering in *Rhizopogon roseolus/Pinus sylvestris* mycorrhizas collected from calamine dumps. Mycol Res 100(1):16–22

Turnbull MH, Goodall R, Stewart GR (1995) The impact of mycorrhizal colonization upon nitrogen source utilization and metabolism in seedlings of *Eucalyptus grandis* Hill. ex Maiden and *Eucalyptus maculata* Hook. Plant Cell Environ 18(12):1386–1394

Ursic M, Peterson RL, Husband B (1997) Relative abundance of mycorrhizal fungi and frequency of root rot on *Pinus strobus* seedlings in a southern Ontario nursery. Can J For Res 27(1):54–62

Varese GC, Portinaro S, Trotta A, Scannerini S, Luppi-Mosca AM, Martinotti MG (1996) Bacteria associated with *Suillus grevillei* sporocarps and ectomycorrhizae and their effects on in vitro growth of the mycobiont. Symbiosis 21(2):129–147

Walker RF, Geisinger DR, Johnson DW, Ball JT (1995a) Enriched atmospheric CO_2 and soil P effects on growth and ectomycorrhizal colonization of juvenile ponderose pine. For Ecol Manage 78:207–215

Walker RF, Geisinger DR, Johnson DW, Ball JT (1995b) Interactive effects of atmospheric CO_2 enrichment and soil N upon growth and ectomycorrhizal colonization of ponderosa pine seedlings. For Sci 41(3):491–500

Wallander H, Massicotte H, Nylund J-E (1997) Seasonal variation in protein, ergosterol and chitin in five morphotypes of *Pinus sylvestris* L. ectomycorrhizae in a mature Swedish forest. Soil Biol Biochem 29(1):45–53

Wang Y, Sinclair L, Hall IR, Cole ALJ (1995) *Boletus edulis* sensu lato: a new record for New Zealand. N Z J Corp Horticult Sci 23(2):227–231

Watling R, Taylor A, Lee Su See, Sims K, Alexander I (1995) A rainforest *Pisolithus*: its taxonomy and ecology. Nova Hedwigia 61(3–4):417–429

Weiss M, Mikoljewski S, Peipp H, Schmitt U, Schmidt J, Wray V, Strack D (1997) Tissue-specific and development-dependent accumulation of phenylpropanoids in larch mycorrhizas. Plant Physiol 114(1):15–27

Wiemken V, Ineichen K (1996) A method to access the below-ground part of a model spruce ecosystem. Funct Ecol 10(3):417–420

Wöllecke J, Münzenberger B, Hüttl RF (1997a) "*Pinirhiza arachnorosea*" + *Pinus sylvestris* L. Descr Ectomyc 2:37–42

Wöllecke J, Münzenberger B, Hüttl RF (1997b) "*Pinirhiza luteoalba*" + *Pinus sylvestris* L. Descr Ectomyc 2:55–60

Zak B, Ho I (1994) Resistance of ectomycorrhizal fungi to rhizina root rot. Ind J Mycol Plant Pathol 24(3):192–195

Zambonelli A, Penjor D, Pisi A (1995) Effects of tradimefon on *Tuber borchii* Vitt. and *Hebeloma sinapizans* (Paulet) Gill. infected seedlings. Micol Ital 24(3):65–73

Zambonelli A, Tibiletti E, Pisi A (1997) Anatomical-morphological characterization of *Tuber indicum* Cooke & Massee on *Pinus pinea* L. and *Quercus cerris* L. Micol Ital 28(1):29–36

Prof. Dr. Reinhard Agerer
Section Mykologie
Institut für systematische
Botanik der Universität München
Menzinger Straße 67
D-80638 München, Germany

Edited by
M. Runge

Plant Population Ecology

By Cornelia Lehmann, Franz Rebele, and Uwe Starfinger

1 Introduction

Population ecology is an ever-increasing field in plant ecology. The large number of papers published in the field since the last review in this series (Starfinger and Stöcklin 1996) makes it impossible to review them all. Our selection reflects our interest in how processes at the population level influence the distribution of plant populations and, consequently, their role in communities. Clonality, which is of important impact on the dynamics of many plant populations, further attracted our attention. For this reason, we have chosen three main subjects: (1) new insights in the dispersal of diaspores as essential prerequisites for the spatial and temporal organization of plant populations and community composition, (2) clonality and implications of clonal growth in the light of the considerable progress concerning the work on clonal plants, and (3) interactions between plants with special reference to ongoing debates on the importance of competitive mechanisms and positive interactions as well.

2 Seed Dispersal

The distribution of seed plants is initially determined by seed dispersal, and only later by a number of biotic and abiotic factors. General questions relating to quantitative aspects include: How is dispersal realized? What distances can dispersal cover, and where are the seed deposited? Of central importance in recent papers on the quality of dispersal are the terms efficiency and effectiveness as proposed by Reid (1989). Efficiency is the probability that a seed is transported to a safe site and germinates, effectiveness is the proportion of plants in a population dispersed by a particular vector (but note different definitions, e.g., Schupp 1993). Because effectiveness is difficult to assess, Bustamante and Canals (1995) propose a model for the indirect estimation using the efficiency of frugivores.

a) Wind Dispersal

Little quantitative theory exists on wind dispersal of trees. Greene and Johnson (1995) developed two simple models for the relationship between source strength (number of seeds produced per unit area), dispersal distance, and deposition. They field-tested them and found them acceptable at the scale of about 1 km. Results indicated that rare trees in forest fragments are isolated rather than forming a metapopulation. Limited dispersal was also found as one factor that limits the spread of pines in a large old field (Pinder et al. 1995). Secondary dispersal on snow can lead to larger dispersal distances; it was found to be of consequence only in species that release a high proportion of seeds (i.e, > 15%) in winter, e.g. *Betula*. Its importance is greater in small-statured plants (Greene and Johnson 1997). Even seed without adaptation to wind dispersal can be transported over considerable distances by rarely occurring high wind speeds. Van Dorp et al. (1996) used a wind tunnel experiment to predict maximum dispersal distances of 10 to 30 m for selected grassland species with wind speeds that occur about once a year.

In samaras, seed mass is a main determinant for dispersal capacity. In pines, an isometric relation between seed mass and samara shape exists in smaller-seeded species, in large-seeded species samara length decreases with seed mass. In consequence, large-seeded pines have a tendency toward animal-dispersal modes (Benkman 1995). Aerodynamic behavior of *Acer saccharinum* samaras was also influenced strongly by factors like mass distribution and wing shape, which varied greatly between samaras from different parens (Sipe and Linneroth 1995).

b) Hydrochory

Hydrochory plays a major role in structuring riparian floras: Johansson et al. (1996) found a correlation between the frequency of species in river corridors and floating capacity of their seeds. They stress the importance of continuous river corridors for regional biodiversity. The temporal variation in water levels in floodplains has consequences for the effect of dispersal timing (van Splunder et al. 1995): seedlings of different *Salix* species occur in narrow belts along the river, those of the early-dispersing species were found at higher elevations than those of later-dispersing species. This was in accordance with the water level during the respective dissemination period. Rivers ususally transport seeds in one direction only. The rare case of legitimate seed dispersal by a fish described by Horn (1997) is an interesting example of how upstream movement of seeds can be maintained.

c) Zoochory

Endozoochorous dispersal of fleshy fruits is often seen as a bird- or mammal-dispersal syndrome reflecting coevolution of plant and animal traits. Typically, bird-dispersed seeds are thought to be small and colorful, mammal-dispersed seeds larger, dull-colored, with distinct odors. Several recent papers have highlighted different problems in the understanding of these complex relationships. Tamboia et al. (1996) studied fruit characters, and fruit preference by birds and mammals in several *Solanum* species. They found that fruit character suites were not good predictors for preference by birds and mammals as in the general model. Jordano (1995) analyzed fleshy fruit characteristics of 910 angiosperm species and found strong phylogenetic autocorrelation for 11 of 16 fruit traits examined. He concludes that dispersal syndromes are not entirely interpretable as current adaptations to seed dispersers. Fukui (1996) points out the conflicts between fruits and frugivores that lie in those aspects where the benefits are not parallel for the plant and the disperser. Gut retention time, which is usually below 60 min and often below 10 min, is a key issue: plants should be favored by long gut retention because it increases dispersal distance and reproductive success. Birds, however, prefer food that is processed faster, and select fruits accordingly, i.e., fruits with smaller seeds and more pulp. This again leads to the deposition of fecal pellets with numerous seeds with the consequence of sibling competition. More attention should be on these conflicts to understand the exact role of seed dispersers in the reproductive success of fleshy-fruited plants.

Efficiency of dispersal has quantitative and qualitative ascpects, including removal of fruits and their deposition at sites suitable for germination and establishment. In general, a plant can enhance efficiency by offering large numers of fruits to generalist dispersers or by producing fruits to specialist dispersers which should offer better quality dispersal. Often disperser assemblies include both types. Larson (1996) found the specialist disperser of a desert mistletoe to deliver the highest-quality dispersal. Still, in this case the plant produced large numbers of fruit, which were partly eaten by generalists and not dispersed to potential growing sites. She interpreted this finding in the light of the very low recruitment rates found for the plant: attracting the specialist with a display of many fruits was vital for this plant and exerted a stronger selective force than the cost of producing more fruits. She calls this dispersal system a compromise between the framework proposed for specialists and that proposed for generalists. Germination success of seeds may also be enhanced by gut passage, a factor that needs to be considered in order to assess the efficiency of a disperser. It is studied in detail relatively rarely. Traveset and Willson (1997) have compared germination success of several Alaskan rainforest plants with and without passage through bird and bear gut. Though gut passage enhanced germination in two species, it did not in others, and no differences were found between bear- and bird-treated seeds. These results suggest that the actual movement of the seeds away from the parent is more important than the treatment in the gut.

A methodological problem is raised by Okamoto (1996). The ratio of fruit production to fruit removal, as an important index of the quantitative aspect of dispersal, is usually assessed with fruit traps placed under fruiting trees. These, however, capture also seeds from conspecific trees. Okamoto (1996) suggests an "apparent rate of escaped seed (AES)" instead of the normally calculated removal rate as an estimate of effectively escaped seed and demonstrates its use in the comparison of efficiencies in two dispersal traits: extended vs. synchronous ripening.

Most studies of endozoochory have focused on dispersal syndromes concerning fleshy-fruited plants. Another plant group for which endozoochory is important are pasture species. Grazing animals induce disturbance to pastures with their dung pats, but these are important microsites for the seeds carried in the dung. Malo and Suárez (1995) found that this combination of disturbance and seed dispersal results in similarity between, but variation within, pasture communities.

Similarly, epizoochory of seeds of pasture species on grazing animals is of great importance in maintaining species richness. Fischer et al. (1996) used a "sheep dummy" and tame sheep to study retention time of diaspores and sheep locomotion, and found surprisingly high numbers of individual diaspores, and species, which could stay on the fleece for as long as 7 months, resulting in potential dispersal over the entire roaming area of the sheep. Kiviniemi (1996) tested the ability of seeds of *Agrimonia eupatoria, Geum rivale*, and *Triglochin palustre* to remain attached to the fur of fallow deer and cattle. She found dispersal distances of tens of meters to a kilometer. Both papers stress the importance of epizoochorous dispersal of pasture species for the maintenance of species richness in a fragmented landscape. A similar function can be carried out by mowing machinery. Strykstra et al. (1996) demonstrated that *Rhinantus angustifolius*, a hemiparasitic annual, was successfully established in plots after dispersal by mowers.

Dispersal is a key factor in succession, and limited dispersal may influence the speed of succession as well as the community composition. Tree establishment in grasslands may be severely limited. Deposition rates of seeds by birds and bats in an abandoned pasture in Amazonia were much lower than in a forest or its gaps (Nepstad et al. 1996). Under treelets in the pasture they were much higher. In a central European abandoned grassland, fleshy-fruited and wind-dispersed woody species were also very rare; here a multistaged dispersal by jays and mice led to *Quercus petraea* and *Corylus avellana* being the most abundant woody species (Kollmann and Schill 1996).

d) Seed Predation and Dispersal

In many plant species, seed predators can damage a considerable proportion of seeds. Predation can be the main factor influencing seed production as in a rainforest palm studied by Cunningham (1997). As a consequence, seed predators may have a strong impact on the demography of the plants whose seed they consume (Hulme 1997). Where predators consume seeds of different species selectively, they can even influence community composition. Evidence for this was found in an early successional North American forest (Meiners and Stiles 1997) as well as in a Mediterranean scrubland (Hulme 1997). Consequently, selection may favor traits that help avoid or reduce seed predation. This can relate to seed color where the predator finds seeds visually. Nystrand and Grandström (1997) found the variation in seed color of *Pinus sylvestris* consequential for predator escape: ground-feeding finches found more pale than dark seeds on burned ground and more dark seeds on mineral soil. They conclude that with visually searching predators, a single cryptic seed color will be selected for where a plant regenerates on uniform-colored substrates. Another route to seed predator avoidance may be the timing of flowering and seed release. *Actaea spicata* has a bimodal flowering phenology, with early flowers sufferig seed loss to a moth predator but not later flowers, suggesting a predator avoidance hypothesis of flowering asynchrony (Eriksson 1995). Myrmechory is often a complex of seed predation and dispersal. Several studies favor the predator avoidance hypothesis of ant dispersal, e.g., Ruhren and Dudash (1996) focusing on a North American forest lily, or Espadaler and Gómez (1996, 1997) on a Mediterranean spurge.

3 Clonality and Implications of Clonal Growth

In the plant kingdom, clonality is a significant phenomenon which has attracted a lot of research work. Recently, new insight has been developed about its ecological implications. A clone is defined as the sum of genetically identical individuals. As it represents all the tissue originating from a single zygote, it is also called the genet. In plants, clones may be produced in two ways: by agamospermy, which is asexual seed production without meiosis and the fusion of gametes, and by clonal growth, which is the vegetative reproduction of new genetically identical units, the ramets. Both modes of clonal propagation implicate some peculiarities for the spatial distribution of genetic variation in populations. In the case of clonal growth, an individual genet occupies space and increaseses in size, producing more or less aggregated patches. Generally, distribution of ramets belonging to a unique genet is restricted to a single population with the exception of occasional transport of clonal fragments over long distances and the ramets of floating aquatic clones, which can be found in several populations. In the case of agamospermy, populations are mosaics comprising groups of genetically distinct individuals. As agamospermous seeds can be dispersed deliberately over large geographical distances, individuals belonging to an agamospermous clone are not restricted to a single population.

a) Sexual vs. Asexual Reproduction

Recently, it has become a consensus that clonal plant populations are usually genetically variable, more or less similar to nonclonal perennials, although monoclonal populations do occur (Eriksson 1997). Almost all clonal plants also reproduce sexually, at least in some populations (McLellan et al. 1997). Agamospermous species occasionally reproduce sexually. For example, sexual diploid and apomictic triploid plants of *Taraxacum* section *ruderalia* are less genetically isolated than has previously been supposed. Menken et al. (1995) found evidence for a high level of gene flow between the diploid and triploid components of a population, e.g., due to facultative apomixis.

In plants mulitplying by both sexual reproduction and clonal growth, there is a notion of a tradeoff between seed production and clonal propagation (Eriksson 1997). Schmid et al. (1995) demonstrated a size threshold for sexual reproduction in two clonal perennials, but no size dependency for clonal growth (rhizome formation). The absence of a clear size threshold for clonal growth emphasizes the similarity between clonal growth (vegetative reproduction) by rhizomes and growth of other vegetative parts, as opposed to sexual reproductive allocation. Young seed-derived plants did not allocate resources to propagation. This was interpreted as a bet-hedging strategy using resources as insurance against hazards such as herbivory. According to this hypothesis, a clone shifts away from the conservative bet-hedging strategy towards more reproduction and clonal growth as it spreads.

b) Clonal Diversity

Molecular markers can be used to measure genetic variation and further allow the identification and determination of the spatial arrangements of ramets belonging to a genet. McLellan et al. (1997) distinguish two measures of molecular variation: genotypic variation (= number and frequency of individual genets) and genetic variation (number and frequency of alleles, expected heterozygosity), and point out that calculations of genetic variation in clonal plants should be carried out at the genet level, i.e., each identified genotype should be represented only once in the data set. They review the peculiarities of applying the methods of population genetics on clonal plant populations.

Recently, several studies used molecular markers like enzyme electrophoresis (Cronberg 1995; Cole and Voskuil 1996, Hossaert-McKey et al. 1996; Jonsson et al. 1996; Lehmann 1997), RAPD-PCR (Fernando and Cass 1996; Schaal and Leverich 1996; Steinger et al. 1996) as well as DNA fingerprinting (Piquot et al. 1996) to obtain information about the spatial distribution of genets.

In some species, electrophoretic studies are restricted by their limited variability and DNA-based methods may be more useful. For example, protein electrophoresis failed to detect any difference in banding pattern in populations of the colonizing clonal aquatic species *Butomus umbelllatus*, but with help of RAPD, three genets out of 150 ramets from two populations could be distinguished (Fernando and Cass 1996). Piquot et al. (1996) revealed genotypic diversity in ten populations of *Sparganium erectum* by allozymes and oligonucleotide DNA fingerprinting. The number of distinct genotypes detected with DNA probes was nearly twice the number of multilocus isozyme genotypes. However, despite showing less variability, allozymes provided a fairly reliable estimation of the clonal diversity occurring within and among populations.

Cole and Voskuil (1996) studied *Lemna minor* (duckweed) a floating, aquatic plant whose vegetative reproduction allows it to colonize freshwater habitats rapidly around the world. They report a low number of genotypes in a population and a high degree of differentiation at allozyme loci between populations. They found a low level of gene flow and apparent low frequency of sexual reproduction that has produced substantial levels of genetic divergence among populations, despite an absence of morphological differentiation.

In *Calamagrostis epigejos*, clonal diversity varied with respect to heavy metal pollution of the habitats. On unpolluted habitats, five and ten clones, respectively, were distinguished on areas of 100 m^2 each. On the other hand, high levels of clonal diversity were determined on two polluted sites; 59 clones were found on an abandoned sewage farm and 92 clones near a copper smelter (Lehmann 1997).

Patterns of genetic microdifferentiation within a natural population of *Lathyrus sylvestris* indicate that genetic differentiation in time was much less marked than differentiation in space (Hossaert-McKey et al. 1996).

Schaal and Leverich (1996) identified individual genotypes of the prairie plant *Asclepias meadii* and determined the location and number of ramets for each genotype within a population. Further, they studied the effects of two different management regimes on genotypic composition. Mown populations were composed of a few (usually less than seven), but quite large, genotypes. On the other hand, populations managed by burning had many (< 30) small genotypes. These different population structures were due to differences in mode of reproduction. As mowing occurred usually before seed set, these populations could not reproduce sexually, while the timetable of the burning regime did not curtail sexual reproduction.

c) Long-Term Dynamics of Clonal Growth

Long-term studies of clones revealed phases of growth and senescence. Falinska (1995) described the patterns of growth and senescence of individual genets of *Filipendula ulmaria* in a long-term field study of 17 years. She distinguished five phases in the life history of *F. ulmaria*. After genet establishment, growth and integration of the genet leads to exclusive occupation of its area. In the absence of competition by other species, genets can attain 0.5–2 m^2, which is a critical size for disintegration. These phases are followed by weakening genet integrity, senescence, and disintegration of the genets. Subsequently, free space within genet area is colonized by other species.

Brodie et al. (1995) recorded the spatial dispersion patterns of a northern clone of *Populus balsamifera* and reconstructed the clonal development. Their data suggest three phases of clone development each of about 15–20 years: a first phase of postfire colonization, a second of consolidation, and a more recent one of directional expansion.

Petersen and Squires (1995) examined the change in spatial pattern due to mortality over a period of 10 years in mature *Populus tremuloides* clones. In contrast to theoretical predictions, the distribution of *Populus* living in 1989 tended toward greater clumping than that expected from random mortality of *Populus* living in 1979. The pattern was interpreted to be evidence of senescence of the *Populus* clones, with mortality occurring in smaller trees after they were no longer connected to the remainder of the clone.

Clonal plants may regulate their ramet production in relation to stand density and avoid shoot self-thinning. The possible mechanism for this intraclonal growth regulation are reviewed by Suzuki and Hutchings (1997). However, in some clonal species genet mortality due to self-thinning has been reported, for example in *Sasa kurilensis* (Makita 1996).

Long-term dynamics of clonal growth may include the production of a ramet bank in clonal trees. In *Ailanthus altissima*, a light-demanding species, "clonal oskars" persist many years in the shade awaiting an opening in the tree layer (Kowarik 1995).

d) Implications of Clonal Growth

Individual ramets of a given clone may be located in microhabitats of differing quality and thus perceive environmental heterogeneity on the whole clone level. Stuefer (1996) drew attention to the fact that heterogeneity may refer to a number of fundamentally different aspects of environmental variability (e.g., scale, contrast, predictability, temporal vs. spatial heterogeneity), and that each of these aspects may severely constrain the viability of potentially adaptive traits. Clonal growth may serve as a means for buffering environmental heterogeneity and local stress in various ways: by physiological integration, semiautonomous clonal fragments, foraging, and division of labor.

α) Physiological Integration Within Clones

In many clonal species, ramets may stay connected by stolons or rhizomes, which may be physiologically integrated. The scale of such integration determines whether the ramets respond to the environment as independent individuals or as functionally integrated parts of a clone. Physiological integration offers advantages like extended support of new ramets, recycling and sharing of resources, and regulation of intraclonal competition among ramets through developmental control. Jónsdóttir and Watson (1997) evaluated the ecological and evolutionary significance of physiological integration of clonal plants and conclude from available data a relationship between average environmental resource availability and environmental stability and the extent of integration. Plants with small clonal fragments (group of physically interconnected ramets of a clone) and restrictive integration seem to be more common in relatively productive environments of high to intermediate disturbance. In contrast, full integration in large clonal fragments is of great significance in more stable and resource-poor environments.

Several studies provided experimental evidence how physiological integration facilitates resource sharing. ^{14}C labeling of photoassimilate revealed patterns of resource distribution within clones of *Glechoma hederacea* (Price et al. 1996). Alpert (1996) examined nutrient sharing in natural clonal fragments of *Fragaria chilonensis* by tracing the movement of ^{15}N. Within fragments, nitrogen was shared between all ramets along a stolon, but large net transfer took place only from older to younger ramets. A significant increase in the total biomass of younger ramets was the result, a possible decrease in the biomass of some older ramets, and an increase in allocation to new stolons in ramets that imported nutrients.

De Kroon et al. (1996) studied water translocations between interconnected mother and daughter ramets in two *Carex* species using a newly developed quantitative method based on deuterium tracing. When a ramet pair was exposed to a heterogenous water supply, water translocation became unidirectional and strongly increased to a level at which 30–60% of the water acquired by the wet ramet was exported towards the dry ramet. The wet ramet increased water uptake in response to the dry conditions experienced by the entire interconnected ramet.

β) Sectoriality

Physiological integration is not necessarily a function on the whole clone level. Despite persistent physical connections between ramets, physiological integration may break down at various structural levels. For example, clones may become assemblages of semiautonomous integrated physiological units (IPU), where the IPU has been defined as that level of morphological organization within which the assimilation, distribution, and utilization of carbon is regulated. IPUs physiologically subdivide physical coherent plant structures resulting in sectoriality, which may restrict the influence of lethal, localized environmental factors, within the affected IPU, as could be shown for zinc transport in *Glechoma hederacea* (Price et al. 1996).

An experiment in which competition was either present or absent throughout the space occupied by the clone or in patchy distribution showed that *Glechoma hederacea* did not respond at the whole clone level. Instead, connected stolons (IPUs) responded independently to local competition (Price and Hutchings 1996). Clover clones responded to neighboring clones in increasing the likelihood of growing away from competitors. The changes provoked by conspecific neighbors were located within the parts of the clones encountering the neighbors, rather than spread throughout the whole clone (Hutchings et al. 1996).

γ) Foraging in Clonal Plants

The ability of a clone to exploit favorable patches or avoid unfavorable patches via selective (i.e., nonrandom) placement of ramets has been interpreted as "foraging" behavior in plants analogous to patch sampling by a foraging animal. De Kroon and Hutchings (1995) reconsidered the foraging concept and proposed a widening to a more general framework into which processes of resource acquisition in clonal and nonclonal plants can be placed with equal validity. They consider two reasons to reformulate the foraging concept.

First, foraging is not only achieved by selective habitat choice via morphological plasticity of spacers (stolons and rhizome internodes). Localized morphological responses of roots and shoots are also important expressions of foraging. Consequently, in a new concept developed for all plants, leaves and root tips as well as ramets in clonal plants would be regarded as resource-acquiring structures that are projected into a heterogenous habitat by spacers such as orthotropic stems and root branches, and plagiotropic stolons and rhizomes in clonal plants.

Second, in response to different environmental conditions, morphological plasticity can generate different patterns of placement of resource-acquiring structures in some species, while other species may be relatively unresponsive, analogous to a variety of animal behaviors from continuous movement around the habitat in search of food to more passive "sit-and-wait" foraging. Each of these behaviors may be adaptive in terms of resource acquisiton in the habitats in which they have evolved. The whole range of possible behaviors, from highly plastic to hardly plastic, must therefore be regarded as expression of foraging. Accordingly, de Kroon and Hutchings (1995) define foraging as "the process whereby an organism searches or ramifies within its habitat, which enhances its acquisition of essential resources" (p. 150).

In the view of Oborny and Cain (1977) such a broad definition runs the risk of interpreting whatever a plant does as foraging and does not help to clarify which plants really are "foragers". Consequently, they propose a more specific definition: "Plant foraging occurs when a plant exhibits any morphological plasticity that is selectively advantageous for resource acquisition at a particular spatio-temporal distribution in the resource" (p. 165). These authors review models of spatial spread and foraging in clonal plant species and compare theoretical predictions with empirical data.

Hutchings et al. (1996) studied the responses of different genotypes of *Trifolium repens* at two soil-nutrient levels. Their results were consistent with the hypothesis that clones consolidate occupation at favorable habitat patches and increase their probability of escape from unfavorable conditions. Evans and Cain (1995) tested explicitly the foraging behavior of the clonal plant *Hydrocotyle bonariensis* and obtained direct evidence of an effective foraging response. They tested directly whether plants can preferentially locate "good" patches or avoid "bad" patches when grown in a heterogeneous environment. In the heterogeneous treatment, *Hydrocotyle* rhizomes exhibited a previously undocumented behavior: they appeared to veer away from "bad" patches.

Foraging behavior of plants is achieved by morphological plasticity in spacer length and branching intensity. Clonal grasses produce two types of spacers: stolons and rhizomes, each serving unique functional roles. Comparing three clonal grasses, *Agrostis stolonifera* (a stoloniferous grass), *Holcus mollis* (a rhizomatous grass), and *Cynodon dactylon* (a grass forming both stolons and rhizomes), Dong and Pierdominici (1995) found that stolons serve primarily as foraging organs for light, whereas the main function of rhizomes is storage of meristem carbohydrates.

The efficiency with which plants can forage, and therefore the growth they can achieve, depends on the type of heterogeneity encountered and on species-specific properties. The capacity to forage for nutrients in *Glechoma hederacea* was clearly dependent on the spatial scale of heterogeneity. When patches were larger and spatial predictability higher, plants could select better-quality sites for root growth, accumulating in greater clone biomass. If the patches were small in comparison with the distance between ramets *G. hederacea* could not adjust its morphology rapidly enough to respond to less predictable environments where changes in patch quality were more frequent. Thus, *G. hederacea* responded to environments with small-scale patchiness as if they were homogeneously poor (Wijeshinge and Hutchings 1997).

Species from contrasting habitats show different foraging behavior for light. *Hydrocotyle vulgaris* from open fernlands seems to be able to forage for light by plasticity in petiole length, whereas in *Lamiastrum galeobdolon* from forest understory, physiological integration evened out local responses of ramets (Dong 1995).

δ) Division of Labor in Clonal Plants

If the patterns of abundance of different resources do not coincide, parts of clones located in patches of different quality should specialize to preferentially acquire the most abundant resource in the patch and transport it to parts of the plants sited where this resource is scarce. In parallel to the economic concept of division of labor, with the two essential elements specialization and cooperation, Alpert and Stuefer (1997) define division of labor in clonal plants as the coordinated specialization of individual ramets for the acquisition of different resources, combined with the reciprocal exchange of these resources between ramets. The authors distinguish two types of divsion of labor. The first type is the programmed specialization, coupled with cooperation, among connected ramets in rhizomatous sedges and grasses from arctic and alpine tundra. The second type of division of labor is specialization induced by environmental heterogeneity, which has been observed in at least eight clonal plant species.

To explicitly test the hypothesis that division of labor between ramets can enhance clone growth under heterogeneous conditions, Stuefer et al. (1994, 1996) conducted experiments on *Potentilla anserina, P. reptans,* and *Trifolium repens,* placing clonal fragments in an artificial environment in which the availabilities of two resources (light and water) were patchy and negatively correlated. Thus, the sites of abundance of one resource were always correlated with shortage of the other resource. Ramets of clones perceiving such "complementary resource patchiness" achieved higher yields than in treatments providing all ramets of a clone with high levels of one resource and low levels of the other. The enhanced yield appeared to result from within-clone variation in root:shoot ratio. Shaded ramets had high root:shoot ratios, and unshaded had low ratios. The authors inferred that the high light level had increased evapotranspiration and decreased water availability, that the resulting complementary patchiness of light and water had induced "specialization for abundance" in between ramets, i.e., shaded ramets allocated more ramet biomass to roots where water was abundant and lit ramets allocated more ramet biomass to shoots where light was abundant. Reciprocal resource transport combined with increased overall efficiency of resource uptake enabled clones to grow best in the patchy habitat. On the other hand, the uniform root:shoot ratio of clones in homogeneous habitats failed to optimize the acquisition of any essential resource. Specialization for abundance is a property based on clonal integration. When ramets are severed from one another, each specializes to acquire scarce resources, just like a nonclonal plant.

4 Interactions Between Plants

a) Positive Interactions Among Plants (Facilitation)

Whereas the role of competitive mechanisms in plant interactions may have been overemphasized in the past, positive interactions have been largely overlooked as important factors in community structure. Recent research reveals that facilitation and competition may operate simultaneously and that the overall effects of one species on another may vary among different habitats as mechanisms shift in relative importance (Bertness and Callaway 1994; Greenlee and Callaway 1996; Rebele 1996; Callaway and Walker 1997; for review see Callaway 1995).

Bertness and Callaway (1994) hypothesized that the importance of facilitation in plant communities increases, whereas the importance of competition decreases along a gradient of increasing abiotic stress. Greenlee and Callaway (1996) conducted field experiments on xeric, rocky slopes of the Garnet Range in western Montana to test this hypothesis. Spatial pattern analysis of the short-lived rosette- and taproot-forming endemic perennial herb *Lesquerella carinata* var. *languida* and associated bunchgrass species (*Pseudoregneria spicata, Koeleria cristata, Festuca idahoensis*), bitterbrush (*Purshia tridentata*), and *Pinus ponderosa* showed a positive spatial association of *Lesquerella* with bunchgrasses. Extreme climatological differences between 2 years of investigation corresponded with a shift in relative importance of facilitation and competition. In the wet, cool year 1993 (low drought stress), bunchgrasses competed with *Lesquerella*, in the dry, hot year 1994 (high drought stress), bunchgrasses facilitated *Lesquerella*.

Other recent studies which showed cooccurrence of competitive and facilitative effects investigated interactions of *Artemisia tridentata* and montane pines (*Pinus ponderosa* and *P. monophylla*) in western Nevada (Callaway et al. 1996) and interactions of wetland plants (*Typha latifolia, Salix exigua*, and *Myosotis laxa*) in western Montana (Callaway and King 1996).

Callaway and Walker (1997) give a synthesis of species interactions involving a complex balance of competition and facilitation. They discuss the role of these two processes on the background of gradients of abiotic stress and consumer pressure with respect to different life stages, sizes, and densities of the interacting species. In 1997 a *Special Feature* was published by *Ecology* reexamining the role of positive interactions in communities, where Kareiva and Bertness in their editorial note point out that positive interactions are pervasive forces in communities and that incorporating them into our understanding of natural systems may resolve many long-standing conceptual problems in ecology.

b) The Relative Importance of Root and Shoot Competition Along Productivity Gradients

An ongoing debate in population and community ecology is whether competition is important in unproductive environments or more or less restricted to productive habitats. This question is closely related to the relationship of shoot competition for light in productive and root competition for nutrients in unproductive habitats, and is also related to the question whether there are inherent traits in determining competitive ability.

Grime (1973a, 1979) and Keddy (1989) have suggested that the intensity of interspecific competition increases along productivity gradients. Competition may be most intense in productive habitats because such habitats support high growth rates and large amounts of biomass that result in preemption of space and light. Unproductive habitats support lower growth rates and less above-ground biomass, have less shading, and may have lower intensities of competition. Species with high competitive ability might dominate productive habitats, while species with low competitive ability may be displaced to less productive habitats where competition is less intense. This perspective suggests a quantitative change in the intensity of competition along productivity gradients.

In contrast, Newman (1973, 1983) and Tilman (1988) suggest that unproductive habitats should be characterized by intense competition for soil resources. Tilman's theory of resource competition predicts that there may be no quantitative change in the intensity of competition along a productivity gradient, but that there may be an important qualitative change, with plants mainly competing for soil resources in unproductive habitats and mainly competing for light in more productive areas. This theory suggests that each species is specialized for a particular ratio of soil resources and light and, consequently, the species that characterize a particular habitat are also the superior competitors for the particular resource ratio of that habitat. Therefore, competition may be important at all points along a productivity gradient, but its quality may vary.

In the past years, several investigations were carried out to test the relationship of above- and below-ground competition along productivity (and disturbance) gradients. In most cases, competitive response was measured with individual target plants, where transplants are grown with no neighbors present, with shoots of neighbors present, with roots of neighbors present, or with both shoots and roots of neighbors present.

Wilson and Tilman (1995) studied competitive responses of eight old-field plant species in four environments where fertility and disturbance were varied. The competitive effect of neighboring vegetation shifted from roots to shoots as nitrogen increased, but decreased in disturbed plots. Gerry and Wilson (1995) tested the influence of initial size on the competitive responses of six plant species. They found no evidence that initially smaller plants were weaker competitors than larger plants under conditions where competition was primarily belowground.

Grubb et al. (1997) examined the question whether root or shoot competition is more important for perennial plants in chalk grassland. They concluded that root competition is more important than shoot competition in enabling some species to be regularly more abundant than others in turf 5–10 cm tall, but that shoot competition is more important in turf ca. 20 cm tall, and paradoxically also more important locally in very short turf where wide-leaved species form flat rosettes very close to the ground.

Belcher et al. (1995) investigated root and shoot competition along a soil-depth gradient in an alvar vegetation system, where plants are growing in thin soil over limestone rock. Above-ground biomass was strongly correlated to soil depth, indicating that soil resources increased and light decreased along the gradient. Analysis based on measures of competition intensity (for discussion of competition intensity measures see also Grace 1995) using the annual *Trichostema brachiatum* as a phytometer species suggests that competition in this system was primarily below ground, but competition intensity did not vary significantly along the soil-depth gradient. According to these and other field studies, the authors suggest a graphical model which hypothesized changes in the intensity of mutualism/facilitation and competition along a gradient of decreasing stress. Mutualism and facilitation occur in low biomass sites, competition occurs in higher biomass sites. The model further suggests that a study along a short gradient within a site of modest productivity would detect major changes in the relative importance of root and shoot competition. In contrast, a study along a (longer) natural gradient would find the intensity of root and shoot competition positively correlated, as both would go from being negligible to being important. Thus depending upon scale, one could either emphasize changes in resource ratios, or increase in competition intensity above and below ground.

Goldberg and Novoplansky (1997) reviewed studies on the relative importance of competition in unproductive environments, coming to the conclusion that empirical results bearing on this question are quite variable and that a consistent answer has not yet emerged. They propose a new general hypothesis that includes the contradictory predictions as special cases which relate to different types of resource dynamics and different types of interactions between the growth and survival components of fitness. They call this the two-phase resource dynamics hypothesis of plant interactions along productivity gradients based on the assumption that soil resources are usually supplied in pulses rather than continuously. When soil resource supply is temporally variable, individual plants will experience two distinct phases of resource availability: pulse periods when resources are high and most growth and resource accumulation occurs, and interpulse periods when resources are too low for most plants to take up and most mortality due to resource deficits takes place. According to the two-phase resource dynamics hypothesis they postulate Grime's hypothesis that competition is unimportant at low productivity will hold when soil resource availability between pulses in unproductive environments is controlled by abiotic factors and when survival during interpulse intervals is independent of or even negatively correlated with growth during pulse periods. In contrast, Newman's and Tilman's hypothesis that competition is equally important along productivity gradients should apply when either of these conditions is not true. Goldberg and Novoplansky predict that the conditions for Grime's hypothesis to apply are more likely for productivity gradients driven by water than by mineral nutrients and when response to competition is measured for community structure or individual survival rather than for individual growth.

In their two-phase resource dynamics hypothesis Goldberg and Novoplansky do not relate to the facilitation-competition debate mentioned above, where a shift in relative importance of competition and facilitation was noticed in unproductive habitats.

c) Competitive Exclusion and Coexistence

Observations from natural gradients of plant productivity, eutrophication of aquatic and terrestrial ecosystems, as well as fertilization experiments, mostly reveal decreasing diversity of plant communities with increasing productivity (for references see Huston 1994). The mechanism generally presumed to cause this decrease in diversity is an increased intensity of competition at higher levels of productivity. Since Grime (1973b, 1979) and Al-Mufti et al. (1977) pointed out that species diversity of herbaceous vegetation in Britain shows a "humped" pattern with increasing total above-ground productivity this unimodal productivity-diversity relationship has been accepted as a general dogma by many ecologists (Real 1995). Unimodal "hump-shaped" curves have been attributed to increased competitive exclusion as the result of decreased heterogeneity in limiting resources at high productivities, especially when light supply at the soil surface decreases.

Abrams (1995) questions the validity of the overall pattern of this productivity-biodiversity relationship and gives other possible explanations for reduced diversity at high productivity that are unrelated to competition and competitive exclusion. He also suggests that even when light supply decreases at nutrient-rich highly productive habitats, species diversity can still increase monotonically with productivity.

In his conclusions, Abrams demands additional theory to determine how productivity affects competitive exclusion under the full range of possible mechanisms for coexistence, more attention to definitions of productivity and issues of scale when gathering data, and observations that will provide data on the various non-competition-based explanations for diversity gradients.

Whereas Abrams' hypothesis that coexistence and thus high diversity are even possible in highly productive environments is provocative and demands further validation, e.g., for other vegetation types in different climatic zones, it is broadly accepted that species coexist in less productive habitats like calcareous or serpentine grassland despite competition. Temporal and spatial heterogeneity and variation in species' competitive abilities among microsites is supposed to be an important mechanism of species coexistence (see Reynolds et al. 1997 for serpentine and Gigon and Leutert 1996 for calcareous grassland).

Long-term experimental field studies using the substitutive replacement design (de Wit 1960) showed that species of tall herbaceous communities can coexist for several years with a reversal in dominance of the species pairs investigated. Rebele (1996) tested the relative competitive abilities of the clonal perennials *Tanacetum vulgare, Solidago canadensis,* and *Calamagrostis epigejos* along a nutrient gradient for a period of 5 years. The experiment confirmed that there is no innate quality of competitive power as a property of species. Competitive abilities depend on site factors (nutrient regimes, climatic fluctuations) and on the particular species to compete against. During the 5-year period, there was only one case of competitive exclusion: *C. epigejos* excluded *S. canadensis* on the most fertile substrate in the fifth growing season. In most cases, the species coexisted with positive effects on biomass and nutrient levels even upon the inferior competitor. Each of the three species was able to dominate in at least one combination of substrate type and mixture.

Mal et al. (1997) conducted a 4-year field replacement experiment with *Lythrum salicaria* and *Typha angustifolia* having four initial densities and four relative proportions of each species. The overall rate of ramet production differed significantly, between density treatments and between years. They found evidence for significant intraspecific competition in both species and for interspecific competition as well. Overall ramet production in *Thypha* was greater in the 1st year, but from the 2nd year onward the situation reversed and *Lythrum* became dominant until the 4th year.

d) Symmetric and Asymmetric Competition

The term asymmetric competition has different meanings in the ecological literature. Keddy and Shipley (1989) and Shipley and Keddy (1994) defined asymmetric competition phenomenologically and independently of the hypothesized mechanics. The ways of competitive interactions between species are classified as symmetric when both species experience either more or less intense interspecific than intraspecific interactions. Asymmetric competition occurs when the dominant species experiences less intense interspecific than intraspecific interactions, while the subordinate species experiences more intense interspecific than intraspecific interactions. This concept of asymmetric competition is part of the theory of competitive hierarchies and transitive networks (Shipley and Keddy 1994) in plant communities. The asymmetric type of two-species interaction permits the competitive exclusion of the subordinate species. This type is necessary but not sufficient for the existence of multispecies hierarchies of competitive ability in which all but the competitive dominant can be competitively excluded, and therefore precludes the possibility of a stable multispecies equilibrium (Shipley and Keddy 1994). In an additive competition pot experiment, Keddy et al. (1997) found that interspecific competitive asymmetry increases with soil productivity.

Harper (1977) and Weiner (1990) use the term asymmetric competition in a quite different sense. Symmetric and asymmetric competition is related to the different ways in which competing plants gain resources. Plants may acquire resources in proportion to biomass, or larger plants may capture more resources per unit biomass, because they have the ability to preempt resources from smaller neighbors. The first of this type is called symmetric or two-sided competition, the second is called asymmetric or one-sided competition (Schwinning and Fox 1995). Competition for soil resources is thought to be symmetric, whereas competition for light is often asymmetric (Wilson 1988; Weiner 1990). Conolly and Wayne (1996) proposed an "index of interspecific competitive asymmetry" to quantify the magnitude of asymmetric competition for interactions between members of different species.

Weiner et al. (19979 studied below-ground competition between individuals of *Kochia scoparia* of different size in pot experiments with dividers above ground. There was no evidence that larger individuals had a disproportionate effect on smaller individuals. The effect of a small neighbor on the growth rate of a plant was similar for large and small plants, as was the effect of a large neighbor. Thus, initial size differences were not exacerbated by (symmetric) competition.

Schwinning and Fox (1995) developed a new neighborhood model, which reflects seedling size effect with respect to the type of competitive symmetry. The model was implemented in a population growth model for two species, one at low density (the invader), and one at high density (the resident). In this model, the species differ only in their seedling biomass distributions. Under these conditions, they found that asymmetric competition always favors invasion by the species with larger average seedling size, but impairs invasion by other species. Based on this invasibility criterion, they concluded that asymmetric competition always favors competitive exclusion in their model. However, by modifying some of the model assumptions, they suggest scenarios in which asymmetric competition may promote coexistence.

In a field experiment with *Pennisetum americanum*, competition for light was symmetric at low density but asymmetric at high density (Schwinning 1996). However, size variation at low density decreased during growth, because small plants had greater relative growth rates than larger plants. Size variations stayed constant at high density, since plants of all sizes had equal average relative growth rates. Based on these results and a general discussion, Schwinning proposed that the type of resource limitation does not determine the mode of competition. Competition for light can be symmetric, and foraging for heterogeneously distributed soil resources can produce asymmetric competition below ground. Furthermore, the mode of competition alone does not determine size structure dynamics. Size-dependence of resource conversion efficiency and allocation can modify the effects of resource uptake on growth.

In recent ecological literature on competition, the term asymmetry has been used in a further context. Many plants and animals have morphological structures that are bilaterally symmetrical. Deviations from bilateral symmetry may represent developmental perturbations. Random deviations from symmetry, known as fluctuating asymmetry, are used to detect such perturbations and monitor ecological stresses within and between populations (Rettig et al. 1997). In a field experiment with an even-aged poplar (*Populus euamericana* cv. Eugenei) clone Rettig et al. studied intraspecific (intraclonal) and interspecific competition (against "weeds") of that clone. They found that increases in intra- and interspecific competition (increasing density treatments) increased fluctuating asymmetry in the leaves of the poplar ramets.

e) Effects of Herbivory on Competition

Herbivores can affect competitive interactions in a specific way. Similar to the debate on the importance of competition along a productivity gradient, the role of herbivory effecting competition intensity is under contest. It is presumed that the importance of competition declines with increasing levels of frequencies of disturbance such as those caused by herbivores. Another theory, however, has presumed that while disturbance may alter the outcome of competition, the importance of competition will be unchanged. Some authors have suggested that applying such generalizations about disturbances to grazing may be oversimplistic (for reference see Taylor et al. 1997).

Taylor et al. studied the effects of herbivory (primarily by nutria, *Myocastor coypus*) on neighbor interactions between three dominant grasses (*Panicum virgatum, Spartina patens* and *S. alterniflora*). The grasses studied are dominant species in the fresh, oligohaline, and mesohaline marshes, respectively. Additive mixtures and monocultures of transplants were used in conjunction with exclosure fences to determine the impact of herbivory on neighbor interactions in the different marsh types. Herbivory had a strong effect on all three species and was important in all three marshes. In the absence of herbivores, the impact of neighbors was significant for two of the species (*Panicum virgatum* and *Spartina patens*) and varied considerably between environments, with competition intensifying for *Panicum virgatum* and decreasing for *Spartina patens* with increasing salinity. Indications of positive neighbor effects were observed for both of these species, though in contrasting habitats and to differing degrees. In the presence of herbivores, competitive and positive effects were eliminated. It was observed that in this case, intense herbivory was able to override other biotic interactions such as competition and mutualism, which were not detectable in the presence of herbivores.

Bonser and Reader (1995) investigated whether effects of competition and herbivory on plant growth depend on the above-ground biomass of vegetation. Transplants of *Poa compressa*, a perennial grass, were planted in plots within six old fields and two herbaceous plant communities near water with a range of mean above-ground biomass from 64 to 776 g/m^2. The experimental design included plots which were caged to prevent herbivory and plots where neighbors were removed in caged and uncaged plots. Transplant shoot mass was significantly greater where herbivores were excluded and neighbors were removed, especially at sites with high biomass. Shoot mass due to neighbor removal was significantly greater at sites with higher biomass. The authors conclude that competition and herbivory (primarily by small mammals and mollusks) each had a greater effect of plant growth at sites with higher biomass and that herbivory has less effect than competition on plant growth at sites with relatively low biomass.

References

Abrams PA (1995) Monotonic or unimodal diversity-productivity gradients: what does competition theory predict? Ecology 76:2019–2027

Al-Mufti MM, Sydes CL, Furness SB, Grime JP, Brand SR (1977) A quantitative analysis of shoot phenology and dominance in herbaceous vegetation. J Ecol 65:759–791

Alpert P (1996) Nutrient sharing in natural clonal fragments of *Fragaria chiloensis*. J Ecol 84:395–406

Alpert P, Stuefer JF (1997) Division of labour in clonal plants. In: de Kroon H, van Groenendael J (eds) The ecology and evolution of clonal plants. Backhuys, Leiden, pp 137–154

Belcher JW, Keddy PA, Twolan-Strutt L (1995) Root and shoot competition intensity along a soild depth gradient. J Ecol 83:673–682

Benkman CW (1995) Wind dispersal capacity of pine seeds and the evolution of different seed dispersal modes in pines. Oikos 73:221–224

Bertness MD, Callaway R (1994) Positive interactions in communities. Tree 9:191–193

Bonser SP, Reader RJ (1995) Plant competition and herbivory in relation to vegetation biomass. Ecology 76:2176–2183

Brodie C, Houle G, Fortin MJ (1995) Development of a *Populus balsamifera* clone in subarctic Québec reconstructed from spatial analyses. J Ecol 83:309–320

Bustamente RO, Canals M (1995) Dispersal quality in plants: how to measure efficiency and effectiveness of a seed disperser. Oikos 73:133–136

Callaway RM (1995) Positive interactions among plants. Bot Rev 61:306–349

Callaway RM, King L (1996) Temperature-driven variation in substrate oxygenation and the balance of competition and facilitation. Ecology 77:1189–1195

Callaway RM, Walker LW (1997) Competition and facilitation: a synthetic approach to interactions in plant communities. Ecology 78:1958–1965

Callaway RM, DeLucia EH, Moore D, Nowak R, Schlesinger WH (1996) Competition and facilitation: contrasting effects of *Artemisia tridentata* on desert vs. Montane pines. Ecology 77:2130–2141

Cole CT, Voskuil MI (1996) Population genetic structure in duckweed (*Lemna minor*, Lemnaceae). Can J Bot 74:22–230

Conolly J, Wayne P (1996) Asymmetric competition between plant species. Oecologia 108:311–320

Cronberg N (1995) Clonal structure and fertility in a sympatric population of the peat mosses *Sphagnum rubellum* and *Sphagnum capillifolium*. Can J Bot 74:1375–1385

Cunningham SA (1997) Predator control of seed production by a rain forest understory palm. Oikos 79:282–290

De Kroon H, Hutschings MJ (1995) Morphological plasticity in clonal plants: the foraging concept reconsidered. J Ecol 83:143–152

De Kroon H, Fransen B, van Rheenen JWA, van Dijk A, Kreulen R (1996) High levels of interrramet water translocation in two rhizomatous *Carex* species, as quantified by deuterium labelling. Oecologia 106:73–84

De Wit CT (1960) On competition. Versl Landbouwkd Onderz 687:3–30

Dong M (1995) Morphological responses to local light conditions in clonal herbs from contrasting habitats, and their modification due to physiological integration. Oecologia 101:282–288

Dong M, Pierdomionici MG (1995) Morphology and growth of stolons and rhizomes in three clonal grasses, as affected by different light supply. Vegetatio 116:25–32

Eriksson O (1995) Asynchronous flowering reduces seed predation in the perennial forest herb *Actaea spicata*. Acta Oecol 16:195–203

Eriksson O (1997) Clonal life histories and the evolution of seed recruitment. In: de Kroon H, van Groenendael J (eds) The ecology and evolution of clonal plants. Backhuys, Leiden, pp 211–226

Espadaler X, Gómey C (1996) Soil surface searching and transport of *Euphorbia characias* seeds by ants. Acta Oecol 18:39–46

Espadaler X, Gómez C (1997) Seed production, predation and dispersal in the Mediterranean myrmechore of *Euphorbia characias* (Euphorbiaceae). Ecography 19:7–15

Evans JP, Cain ML (1995) A spatially explicit test for foraging behaviour in a clonal plant. Ecology 76:1147–1155

Falinska K (1995) Genet disintegration in *Filipendula ulmaria*: consequences for population dynamics and vegetation succession. J Ecol 83:9–21

Fernando DD, Cass DD (1996) Genotypic differentiation in *Butomus umbellatus* (Butomaceae) using isozymes and random amplified polymorphic DNAs. Can J Bot 74:647–652

Fischer SF, Poschlod P, Beinlich B (1996) Experimental studies on the dispersal of plants and animals on sheep in calcareous grasslands. J Appl Ecol 33:1206–1222

Fukui A (1996) Retention time of seeds in bird guts: costs and benefits of fruiting plants and frugivorous birds. Plant Species Biol 11:141–147

Gerry AK, Wilson SD (1995) The influence of initial size on the competitive responses of six plant species. Ecology 76:272–279

Gigon A, Leutert A (1996) The dynamic keyhole-key model of coexistence to explain diversity of plants in limestone and other grasslands. J Veg Sci 7:29–40

Goldberg D, Novoplansky A (1997) On the relative importance of competition in unproductive environments. J Ecol 85:409–418

Grace JB (1995) On the measurement of plant competition intensity. Ecology 76:305–308

Greene DF, Johnson EA (1995) Long-distance dispersal of tree seeds. Can J Bot 73:1036–1045

Greene DF, Johnson EA (1997) Secondary dispersal of tree seeds on snow. J Ecol 85:329–340

Greenlee JT, Callaway RM (1996) Abiotic stress and the relative importance of interference and facilitation in montane bunchgrass communities in Western Montana. Am Nat 148:386–396

Grime JP (1973a) Competitive exclusion in herbaceous vegetation. Nature 242:344–347

Grime JP (1973b) Control of species density in herbaceous vegetation. J Environ Man 1:151–167

Grime JP (1979) Plant strategies and vegetation processes. Wiley, Chichester

Grubb PJ, Ford MA, Rochefort L (1997) The control of relative abundance of perennials in chalk grassland: is root competition or shoot competition more important? Phytocoenologia 27:289–309

Harper JL (1977) Population biology of plants. Academic Press, London

Horn MH (1997) Evidence for dispersal of fig seeds by the fruit-eating characid fish *Brycon guatemalensis* Regan in a Costa Rican tropical rain forest. Oecologia 109:259–264

Hossaert-McKey M, Valero M, Magda D, Jarry M, Cuguen J, Vernet P (1996) The evolving genetic history of a population of of *Lathyrus sylvestris*: evidence from temporal and spatioal genetic structure. Evolution 50:1808–1821

Hulme PE (1997) Post-dispersal seed predation and the establishment of vertebrate dispersed plants in Mediterranean scrublands. Oecologia 111:91–98

Huston MA (1994) Biological diversity. Cambridge University Press, Cambridge

Hutchings MJ, Turkington R, Carey P, Klein E (1996) Morphological plasticity in *Trifolium repens* L.: the effects of clone genotype, soil nutrient level, and the genotype of conspecific neighbours. Can J Bot 75:1382–1393

Johansson ME, Nilsson C, Nilsson E (1996) Do rivers function as corridors for plant dispersal? J Veg Sci 7:593–598

Jónsdóttir IS, Watson MA (1997) Extensive physiological integration: an adaptive trait in resource-poor environments? In: de Kroon H, van Groenendael J (eds) The ecology and evolution of clonal plants. Backhuys, Leiden, pp 109–136

Jonsson BO, Jónsdóttir IS, Cronberg N (1996) Clonal diversity and allozyme variation in populations of the arctic sedge *Carex bigelowii* (Cyperaceae). J Ecol 84:449–459

Jordano P (1995) Angiosperm fleshy fruits and seed dispersers: a comparative analysis of adaptation and constraints in plant-animal interactions. Am Nat 145:163–191

Kareiva PM, Bertness MD (1997) Re-examining the role of positive interactions in communities. Ecology 78:1945

Keddy PA (1989) Competition. Chapman and Hall, London

Keddy A, Shipley B (1989) Competitive hierarchies in herbaceous plant communities. Oikos 54:234–241

Keddy PA, Twolan-Strutt L, Shipley B (1997) Experimental evidence that interspecific competitive asymmetry increases with soil productivity. Oikos 80:253–256

Kiviniemi K (1996) A study of adhesive seed dispersal of three species under natural conditions. Acta Bot Neerl 45:73–83

Kollmann J, Schill H-P (1996) Spatial patterns of dispersal, seed predation and germination during colonization of abandoned grassland by *Quercus petraea* and *Corylus avellana*. Vegetatio 125:193–205

Kowarik I (1995) Clonal growth in *Ailanthus altissima* on a natural site in West Virginia. J Veg Sci 6:853–856

Larson D (1996) Seed dispersal by specialist versus generalist foragers: the plant's perspective. Oikos 76:113–120

Lehmann C (1997) Clonal diversity of *Calamagrostis epigejos* in relation to environmental stress and habitat heterogeneity. Ecography 20:483–490

Makita A (1996) Density regulation during the regeneration of two monocarpic bamboos: self-thinning or intraclonal regulation? J Veg Sci 7:281–288

Mal TK, Lovett-Doust J, Lovett-Doust L (1997) Time-dependent competitive displacement of *Typha angustifolia* by *Lythrum salicaria*. Oikos 79:26–33

Malo JE, Suárez F (1995) Establishment of pasture species on cattle dung: the role of endozoochorous seeds. J Veg Sci 6:169–174

McLellan AJ, Prati D, Kaltz O, Schmid B (1997) Structure and analysis of phenotypic and genetic variation in clonal plants. In: de Kroon H, van Groenendael J (eds) The ecology and evolution of clonal plants. Backhuys, Leiden, pp 185–210

Meiners SJ, Stiles EW (1997) Selective predation on the seeds of woody plants. J Torrey Bot Soc 124:67–70

Menken SBJ, Smit E, Den Nijs HCM (1995) Genetic population structure in plants: gene flow between diploid sexual and triploid asexual dandelions (*Taraxacum* section *ruderalia*). Evolution 49:1108–1118

Nepstad DC, Uhl C, Pereira CA, da Silva JMC (1996) A comparative study of tree establishment in abandoned pasture and mature forest of eastern Amazonia. Oikos 76:25–39

Newman EI (1973) Competition and diversity in herbaceous vegetation. Nature 244:310

Newman EI (1983) Interactions between plants. In: Lange OL, Nobel PS, Osmond CB, Ziegler H (eds) Physiological plant ecology III. Springer, Berlin Heidelberg New York, pp 679–710

Nystrand O, Grandström A (1997) Post-dispersal seed predation on *Pinus sylvestris* seeds by *Fringilla* spp: ground substrate affects selection for seed color. Oecologia 110:353–359

Oborny B, Cain ML (1997) Models of spatial spread and foraging in clonal plants. In: de Kroon H, van Groenendael J (eds) The ecology and evolution of clonal plants. Backhuys, Leiden, pp 155–183

Okamoto M (1996) What can we learn about seed dispersal from seed trap experiments at fruiting trees? Efficiency of dispersal traits. Plant Species Biol 11:149–155

Petersen CJ, Squiers ER (1995) An unexpected change in spatial pattern across 10 years in an aspen-white-pine forest. J Ecol 83:847–855

Pinder JE III, Golley FB, Lide RF (1995) Factors affecting limited reproduction by loblolly pine in a large oldfield. Bull Torrey Bot Club 122:306–311

Piquot Y, Saumitou-Laprade P, Petit D, Vernet P, Epplen JT (1996) Genotypic diversity revealed by allozymes and oligonucleotide DNA fingerprinting in French populations of the aquatic macrophyte *Sparganium erectum*. Mol Ecol 5:251–258

Price EA, Hutchings MJ (1996) The effects of competition on growth and form in *Glenchoma hederacea*. Oikos 75:279–290

Price EA, Hutchings MJ, Marshall C (1996) Causes a consequences of sectoriality in the clonal herb *Glechoma hederacea*. Vegetatio 127:41–54

Real LA (1995) Emphasizing new ideas to stimulate research in ecology (Editor's note). Ecology 76:2019

Rebele F (1996) Konkurrenz and Koexistenz bei ausdauernden Ruderalpflanzen. Kovac, Hamburg

Reid N (1989) Dispersal of mistletoes by honey eaters and flowerpeckers: components of seed dispersal quality. Ecology 70:137–145

Rettig JE, Fuller RC, Corbett AL, Getty T (1997) Fluctuating asymmetry indicates levels of competition in an even-aged poplar clone. Oikos 80:123–127

Reynolds HL, Hungate BA, Chapin FS III, D'Antonio CM (1997) Soil heterogeneity and plant competition in an annual grassland. Ecology 78:2076–2090

Ruhren S, Dudash MR (1996) Consequences of the timing of seed release of *Erythronium americanum* (Liliaceae), a deciduous forest myrmechore. Am J bot 83:633–640

Schaal BA, Leverich WJ (1996) Molecular Variation in isolated plant populations. Plant Species Biol 11:33–40

Schmid B, Bazzaz FA, Weiner J (1995) Size dependency of sexual reproduction and of clonal growth in two perennial plants. Can J Bot 73:1831–1837

Schupp EW (1993) Quantity, quality, and the effectiveness of seed dispersal by animals. Vegetatio 107/108:15–29

Schwinning S (1996) Decomposition analysis of competitive symmetry and size structure dynamics. Ann Bot 77:47–57

Schwinning S, Fox GA (1995) Population dynamic consequences of competitive symmetry in annual plants. Oikos 72:422–432

Shipley B, Keddy PA (1994) Evaluating the evidence for competitive hierarchies in plant communities. Oikos 69:340–345

Sipe TW, Linneroth AR (1995) Intraspecific variation in samara morphology and flight behavior in *Acer saccharinum* (Aceraceae). Am J Bot 82:1412–1419

Starfinger U, Stöcklin J (1996) Seed, pollen, and clonal dispersal and their role in structuring plant populations. Prog Bot 57:336–355

Steinger T, Körner C, Schmid B (1996) Long-term persistence in a changing climate: DNA analysis suggests very old ages of clones of alpine *Carex curvula*. Oecologia 105:94–99

Strykstra RJ, Bekker RM, Verweij GL (1996) Establishment of *Rhinanthus angustifolius* in a successional hayfield after seed dispersal by mowing machinery. Acta Bot Neerl 45:557–562

Stuefer JF (1996) Potential and limitations of current concepts regarding the response of clonal plants to environmental heterogeneity. Vegetatio 127:55–70

Stuefer JF, During HJ, de Kroon H (1994) High benefits of clonal integration in two stoloniferous species in response to heterogenous light environments. J Ecol 82:511–518

Stuefer JF, de Kroon H, During HJ (1996) Exploitation of environmental heterogeneity by spatial division of labour in a clonal plant. Funct Ecol 10:328–334

Suzuki JI, Hutchings MJ (1997) Interactions between shoots in clonal plants and the effects of stored resources on the structure of shoot populations. In: de Kroon H, van Groenendael J (eds) The ecology and evolution of clonal plants. Backhuys, Leiden, pp 311–329

Tamboia T, Cipollini ML, Levey DJ (1996) An evaluation of vertebrate seed dispersal syndromes in four species of black nightshade (*Solanum* sect. *Solanum*). Oecologia 107:522–532

Taylor KL, Grace JB, Marx BD (1997) The effects of herbivory on neighbor interactions along a coastal marsh gradient. Am J Bot 84:709–715

Tilman D (1988) Plant strategies and the dynamics and structure of plant communities. Princeton University Press, Princeton, NJ

Traveset A, Willson MF (1997) Effects of birds and bears on seed germination of fleshy-fruited plants in temperate rain forests of southeast Alaska. Oikos 80:89–95

Van Dorp D, van den Hoek WPM, Daleboudt C (1996) Seed dispersal capacity of six perennial grassland species measured in a wind tunnel at varying wind speed and height. Can J Bot 74:1956–1963

Van Splunder I, Coops H, Voeseneck LACJ, Blom CWPM (1995) Establishment of alluvial forest species in floodplains: the role of dispersal timing, germination characteristics and water level fluctuations. Acta Bot Neerl 44:269–278

Weiner J (1990) Asymmetric competition in plant populations. Tree 5:360–364

Weiner J, Wright DB, Castro S (1997) Symmetry of below-ground competition between *Kochia scoparia* individuals. Oikos 79:85–91

Wijeshinge DK, Hutchings MJ (1997) The effects of spatial scale of environmental heterogeneity on the growth of a clonal plant: an experimental study with *Glechoma hederacea*. J Ecol 85:17–28

Wilson JB (1988) The effect of initial advantage on the course of plant competition. Oikos 51:19–24

Wilson SD, Tilman D (1995) Competitive responses of eight old-field plant species in four environments. Ecology 76:1169–1180

Dr. Cornelia Lehmann
Priv.-Doz. Dr. Franz Rebele
Dr. Uwe Starfinger
Institut für Ökologie und Biologie
Technische Universität Berlin
Rothenburgstraße 12
D-12165 Berlin, Germany

Edited by
M. Runge

Subject Index

Printing: Mercedesdruck, Berlin
Binding: Buchbinderei Lüderitz & Bauer, Berlin